Morse Index of Minimal Submanifolds

Mathematical Surveys and Monographs

Volume 292

Morse Index of Minimal Submanifolds

Francisco Urbano

American Mathematical Society
Real Sociedad Matemática Española

2020 *Mathematics Subject Classification.* Primary 53A10, 53C42, 49Q05, 58E12, 35J15.

For additional information and updates on this book, visit
www.ams.org/bookpages/surv-292

Library of Congress Cataloging-in-Publication Data

Cataloging-in-Publication Data has been applied for by the AMS.
See http://www.loc.gov/publish/cip/.
DOI: https://doi.org/10.1090/surv/292

∞ The paper used in this book is acid-free and falls within the guidelines
established to ensure permanence and durability.
Visit the AMS home page at `https://www.ams.org/`

10 9 8 7 6 5 4 3 2 1 30 29 28 27 26 25

This book is dedicated to Ana María and my daughters María and Ana.
Without their help and encouragement it would never have seen the light of day.

Contents

Foreword

The minimal submanifolds of a Riemannian manifold are those that have first variation formula of volume equal to zero and appear naturally in several central questions of geometric analysis. The critical points of the volume functional that have second variation formula greater than or equal to zero are the stable minimal submanifolds.

Among several problems in this area, which have played a fundamental role in the development of mathematics and which are treated in this text, we highlight the study of the complete stable minimal hypersurfaces of the Euclidean space $\mathbb{R}^{n+1}$ (see Sections 4.1 and 4.2 for a presentation of the classical case of surfaces in $\mathbb{R}^3$ and Section 4.3 for the relevant differences between the cases $n < 7$ and $n \geq 7$). Recent remarkable results are also presented for the dimension $n = 3$ (see Section 4.3).

As a consequence of the above theorems, if we consider submanifolds that minimize the volume within different natural families (most of which verify different topological constraints), these results allow us to control the regularity of volume minimizing submanifolds.

In a broader context, in addition to stable minimal submanifolds, minimal submanifolds of finite index are also of great interest. These objects appear when constructing critical submanifolds using variational techniques of the min-max type. In the applications of this method, it is fundamental to control the topology and geometry of the minimal submanifolds of given index. This book presents, in a systematic and exhaustive way, the known results about this condition. Just to give a few examples, results are collected on the index of compact minimal surfaces in positively curved 3-manifolds, minimal hypersurfaces of the n-sphere, minimal Lagrangian submanifolds of complex projective space, and other important families in various areas of geometric analysis.

The book contains a wealth of preparatory material and covers the various techniques used in the field, the strategies, the known results and many of the open problems.

Although the Morse index has many applications, this is practically the only text that takes this approach. It is recommended for both graduate students and researchers in geometric analysis.

Antonio Ros, University of Granada

Preface

Minimal surfaces, or in general minimal submanifolds, are a very important topic in geometric analysis, both for their importance in themselves and for their relation to other fields: mathematical physics, partial differential equations, complex analysis, among others.

From the different points of view that they can be studied, in this text we will focus on one, namely that minimal submanifolds are solutions to a problem of the calculus of variations. Minimal submanifolds are the critical points of the volume functional, and therefore it is important to study the second variation of the volume, which will give us, at least to order one, information about the minima of the volume functional.

Until the 1980s, only the index and the nullity of compact minimal submanifolds had been studied, as well as their stability without any compactness hypothesis. The works of D. Fischer-Colbrie and R. Schoen ([**FC85**], [**FCS80**]) made possible to define the index of two-sided minimal hypersurfaces that are not necessarily compact, making an exhaustive study of the index of Schrödinger operators on Riemannian manifolds. Subsequently, this index concept was extended to one-sided minimal hypersurfaces that are not necessarily compact. Apart from these cases, we have not seen the concept of index defined for non-compact submanifolds of codimension at least two, nor have we found any results in this respect. Therefore, we do not include it in the text.

From this point of view, the main objective of the book is to present, in a reasoned and systematic way, the main results on the index of minimal submanifolds in which the Jacobi operator of the second variation of the volume is used in a relevant way. It is also significant the large number of examples shown as well as the calculation of their indices. The book is therefore not a text on minimal submanifolds, but is aimed at researchers who want to know the main results and techniques used on the second variation of the volume.

Writing such a text involves the use of many different tools, which sometimes means that certain results are not proved and the index of certain examples is not calculated. Most of the prerequisites for reading the text are covered in Chapter 1 of this manual.

Since the proof of the Willmore conjecture by F.C. Marques and A. Neves, [**MN14**], where the min-max theory was fundamental for its solution, this technique has been widely used to obtain examples of minimal hypersurfaces with a strong control over the behavior of their indices, which is becoming very relevant in this theory (see [**MN16, MN21**] among others). One motivation for the study of the index comes from min-max constructions. Despite their importance, these results will not be included in the text, since the techniques of this theory are very different from those used in the book and, frankly, the author finds serious difficulties to make a worthy exposition of them.

Chapter 1 introduces the notation, concepts, and most interesting results on Riemannian manifolds that will be used throughout the text. Only in some special cases the results are proved. The basic notions of Levi-Civita connection, curvature, isometry group, integration, and the classical operators: Laplacian, divergence, Hessian, as well as the relation of the above concepts for two conformal metrics are introduced. A small excursion into Sobolev spaces and Schrödinger operators will also be made.

A short introduction to differential forms, cohomology groups and Hodge theory will be the content of Section 1.3. In Section 1.5, a brief introduction to Riemann surfaces is given, with special attention to the Riemann-Roch theorem, which is used several times in the book. Finally, the basic notions of Kähler manifolds are treated in Section 1.6.

In Chapter 2, we begin with a brief introduction to the theory of submanifolds and prove the first variation formula of the volume, which allows us to characterize the critical points of the volume functional as the minimal submanifolds, i.e., those with zero mean curvature vector. Next, we obtain the second variation formula of the volume for a minimal submanifold, which will provide the definition of the Jacobi operator $\mathcal{L}$. This is a second order operator acting on sections of the normal bundle of the submanifold, whose associated quadratic form $\mathcal{Q}$, acting on compactly supported sections of the normal bundle, will be called the index form of the minimal submanifold. This will allow us to define the Morse index and the nullity of a compact minimal submanifold, as well as the concept of stability of any minimal submanifold. In the special case of two-sided hypersurfaces, the Jacobi operator $\mathcal{L}$ reduces to a Schrödinger operator acting on functions defined on the hypersurface, and thanks to the work of D. Fischer-Colbrie and R. Schoen mentioned above, the index is defined as a generalization of the compact case. Note that the index and nullity in the compact case are non-negative integers, while the index in the case of non-compact hypersurfaces can be infinite. For one-sided hypersurfaces, we also define their indices as a generalization of the compact case. Finally, in the last two sections, we prove some general results on the index and nullity, and provide simple and important examples of minimal submanifolds, whose indices and nullities are computed.

The content of Chapter 3 has two very different parts. In the first one we study the index and stability of hypersurfaces of non-negatively curved Riemannian manifolds, emphasizing the case of surfaces. In this case, and in the first two sections, we study the index of Schrödinger operators on complete surfaces reflecting the Jacobi operators of minimal surfaces of a 3-manifold with non-negative scalar curvature, as well as the Jacobi operator of minimal surfaces of Euclidean space $\mathbb{R}^3$. This abstract study will be applied in Section 3.3 to obtain relevant results on complete minimal surfaces of a 3-manifold with non-negative scalar curvature (or non-negative Ricci curvature) that are either stable or have finite index. As a culmination of this study it is worth noting the result that states that in a 3-manifold with positive Ricci curvature, there are no two-sided complete minimal surfaces that are stable.

Chapter 4 is entirely devoted to the study of stability and the index of minimal submanifolds of Euclidean space, with special emphasis on the case of hypersurfaces. Only in the last section of the chapter, we will study surfaces in $\mathbb{R}^4$.

The first two sections are devoted to the case of minimal surfaces of $\mathbb{R}^3$, both orientable and nonorientable. As a direct application of some results of Chapter 3 it is proved that the planes of $\mathbb{R}^3$ are the only minimal complete orientable surfaces that are stable. The result of D. Fischer-Colbrie, [**FC85**], will reduce the study of the index of the minimal complete orientable surfaces of $\mathbb{R}^3$ to that of the family of surfaces of finite total curvature, since the rest have infinite Morse index. There are many results obtained on the index in the above family, perhaps the star result being that of O. Chodosh and D. Maximo, [**CM23**], in which they obtain a lower bound on the index of a minimal orientable surface of $\mathbb{R}^3$, involving the genus, number of ends, and multiplicity of ends of such a surface of finite total curvature. A similar bound is also obtained in the nonorientable case, with the most important consequence being the non-existence of stable or index one surfaces.

In Section 4.3 we also consider with the case of hypersurfaces of Euclidean space of dimension at least three. In almost all sections we study the stable hypersurfaces, always keeping in mind that for dimension at least seven there are examples of minimal and stable complete orientable hypersurfaces which are not hyperplanes. We solve Berstein's problem for dimension less than or equal five and culminate with the extension to dimension three of the classical result that hyperplanes are the only complete orientable minimal stable hypersurfaces in Euclidean space.

Chapter 5 studies the index and nullity of **compact** minimal submanifolds of the sphere, the real projective space, the product of two spheres, and of elliptic Berger spheres. The first two sections deal with the case where the ambient manifold is a sphere, first proving the classical results of J. Simons, [**Sim68**], which ensure the non-existence of stable minimal submanifolds and classify those of lower index and nullity. The special case of hypersurfaces is also studied, obtaining a lower bound on the index in terms of the first Betti number of the hypersurface. In the special case of surfaces, the Clifford torus on $\mathbb{S}^3$ is characterized by its index.

In Section 5.3 the study of cones of Euclidean space on minimal submanifolds of the sphere by J. Simons is collected. In it, a stability criterion is obtained, which allows us to prove that the cones on minimal hypersurfaces of the sphere of dimension at most 6 are unstable, giving stable examples when the dimension of the cones is 7.

In Sections 5.4 and 5.5 the minimal stable submanifolds of the real projective space and of the product of two spheres are classified, as well as the minimal two-sided hypersurfaces of index one. In both cases, a lower bound of the index in terms of the first Betti number of the hypersurface is obtained. In the last section, the case where the ambient manifold is an elliptic Berger sphere is studied, showing that in some cases there are minimal stable submanifolds and classifying them.

Finally, Chapter 6 deals with the case where the ambient manifold is a Kähler manifold, considering mainly the cases where the manifold is the complex Euclidean space or the complex projective space and the minimal submanifold is Lagrangian. However, we begin by proving that any complex submanifold of a Kähler manifold is stable and even calibrated for the calibration given by some power of the Kähler form. In the case of minimal Lagrangian submanifolds, it is worth noting the result of Y. Oh, [**Oh90**], in which he proves that if the Ricci curvature of the ambient manifold is non-positive, they are all stable. In the special case where the ambient manifold is the complex Euclidean space, or more generally a Calabi-Yau manifold, such submanifolds are even calibrated to the special Lagrangian calibration. In Section 6.3, and following the work of J. Harvey and H.B. Lawson, [**HL82**], a foray is made into the study of such calibrations and their calibrated submanifolds, giving interesting examples.

When the ambient manifold is the complex projective space, a classical result of H.B. Lawson and J. Simons is proved, [**LS73**], stating that its minimal stable compact submanifolds are the complex ones. This result is a stimulus for the study of the index of the minimal Lagrangian submanifolds of the complex projective space. Y. Oh, defined the notion of Hamiltonian stability of such submanifolds, as those submanifolds which are stable under Hamiltonian variations of the complex projective space. His study is closely related to that of the first nonzero eigenvalue of the Laplacian of such submanifolds. The rest of the chapter is mainly devoted to the study of the Hamiltonian stability and the index of the compact minimal Lagrangian surfaces and not only of the complex projective plane but also of the complex quadric $\mathbb{S}^2 \times \mathbb{S}^2$.

I would like to thank my colleague from the Department of Geometry and Topology of the University of Granada, Francisco Torralbo, for his patience and help in composing the text using LaTeX software. I am also indebted to my colleagues Francisco López, Joaquín Pérez, and Antonio Ros for their help in preparing the fourth chapter of the book. Finally, I would like to thank Joaquín Pérez, director of IMAG, for allowing me to use the facilities of the institute during the past year.

The author would like to thank the referees for their valuable comments and remarks about the book.

Francisco Urbano

CHAPTER 1

Foundations

1.1. Riemannian manifolds

Let (M^n, g) be a Riemannian manifold of dimension n, where g denotes the Riemannian metric on M. Along the book, the Riemannian metric is also denoted by $\langle , \rangle$ and, until mentioned, manifolds are considered connected, and compact manifolds will always be considered without boundary.

Given two Riemannian manifolds (M^n, g) and $(\widehat{M}^n, \widehat{g})$, an **isometry** (or **local isometry**) from (M^n, g) to $(\widehat{M}^n, \widehat{g})$ is a diffeomorphism (or local diffeomorphism) $\Phi : M \to \widehat{M}$ such that $\Phi^*(\widehat{g}) = g$, where $\Phi^*(\widehat{g})$ denotes the pull-back of the metric $\widehat{g}$. We will denote by $\mathrm{Iso}(M, g)$ the group of isometries of (M, g).

If $\gamma : [a, b] \to (M, g)$ is a piecewise smooth curve on M, then the **length** of α is defined by

$$L(\gamma) = \int_a^b \|\gamma'(t)\| \, \mathrm{d}t.$$

Now, if p, q are points on M, we define the **distance** from p to q as

$$d(p, q) = \inf \{L(\gamma) \mid \gamma \text{ is a piecewise smooth curve connecting p and q}\}.$$

Then (M, d) is a metric space and the topology of M is the topology associated with the distance d. The Riemannian manifold M is said to be **complete** if (M, d) is a complete metric space. Thus, any compact Riemannian manifold is complete.

In the next result we summarize some well-known properties about covering metrics which will be used often in the book.

PROPOSITION 1.1. *Let (M, g) be a Riemannian manifold and $G \subset \mathrm{Iso}(M, g)$ a subgroup acting properly discontinuous on M. If $\Pi : M \to M/G$ is the covering map, then g projects into a metric $\widehat{g}$ on M/G which becomes $\Pi : (M, g) \to (M/G, \widehat{g})$ in a local isometry.*

Conversely, if $\Pi : \widetilde{M} \to (M, g)$ is a covering map, then the covering metric $\widetilde{g} = \Pi^*(g)$ on $\widetilde{M}$ is complete if g is complete and the automorphism group of the covering Π satisfies $\mathrm{Aut}(\widetilde{M}, \Pi) \subset \mathrm{Iso}(\widetilde{M}, \widetilde{g})$.

If (M, g) is a nonorientable Riemannian manifold, then M has a two-fold covering $\Pi : \widetilde{M} \to M$, where $\widetilde{M}$ is an orientable manifold. If the automorphism group of Π is $\mathrm{Aut}(\widetilde{M}, \Pi) = \{Id, \tau\}$, then τ is an isometry of $(\widetilde{M}, \widetilde{g})$ without fixed points. $(\widetilde{M}, \Pi)$ is called the orientable double covering of M.

If $\mathfrak{X}(M)$ is the Lie algebra of smooth vector fields on M, we will denote by

$$\nabla : \mathfrak{X}(M) \times \mathfrak{X}(M) \to \mathfrak{X}(M)$$
$$(X, Y) \mapsto \nabla_X Y$$

the **Levi-Civita connection** of the Riemannian metric g. It is characterized by the relation

$$
\begin{aligned}
g(\nabla_X Y, Z) = \frac{1}{2}\{&X(g(Y,Z)) + Y(g(X,Z)) - Z(g(X,Y)) \\
&+ g(Z,[X,Y]) + g(Y,[Z,X]) - g(X,[Y,Z])\}
\end{aligned}
\tag{1.1}
$$

for all vector fields X, Y, Z, where $[,]$ is the Lie bracket on $\mathfrak{X}(M)$. In fact, ∇ is the only connection on M that is torsion-free and compatible with the metric g, i.e.,

$$
\nabla_X Y - \nabla_Y X = [X,Y],
$$
$$
X(g(Y,Z)) = g(\nabla_X Y, Z) + g(Y, \nabla_X Z).
$$

The tensorial character of ∇ on the first variable allows to define, for each $X \in \mathfrak{X}(M)$ and $p \in M$, an endomorphism

$$
\begin{aligned}
\nabla X : T_pM &\to T_pM \\
v &\mapsto \nabla_v X.
\end{aligned}
$$

Also, if $\gamma : (a,b) \to M$ is a smooth curve on M, we denote by $\frac{DX(t)}{dt}$ the covariant derivative of a vector field $X(t)$ along γ. The curve γ is called **geodesic** if $\frac{D\gamma'(t)}{dt} = 0$, $\forall t \in (a,b)$. Given any point $p \in M$ and any vector $v \in T_pM$, there exists a unique geodesic γ satisfying $\gamma(0) = p$ and $\gamma'(0) = v$. The **exponential map** $\exp : TM \to M$ can be defined in a neighborhood of the zero section of the tangent bundle by $\exp(p,v) = \gamma(1)$, where γ is the geodesic with initial conditions $\{p, v\}$. So $\gamma(t) = \exp(p, tv)$ and the problem is when the geodesic can be defined for any $t \in \mathbb{R}$. This problem is solved by the following classical theorem.

THEOREM 1.2 (Hopf-Rinow's theorem [**dC81, Spi79**]). *Let (M,g) be a Riemannian manifold. Then the following properties are equivalent:*

(1) *(M,g) is complete.*
(2) *The exponential map is defined on the whole TM.*
(3) *$\{A \subset M \mid A \text{ is compact}\} = \{A \subset M \mid A \text{ is closed and bounded}\}$.*
(4) *The length of any divergent curve $\alpha : [0,\infty) \to M$, (i.e., for any compact $K \subset M$ there exists $t \in [0,\infty)$ such that $\alpha(t) \notin K$), is infinite.*

Furthermore, if (M,g) is complete, then:

a) *For any $p \in M$ and any $R>0$, $B_p(R) := \exp_p(B_0(R)) = \{q \in M | d(p,q) < R\}$, where $B_0(R) = \{v \in T_pM \mid \|v\| < R\}$. $B_p(R)$ will be called the **geodesic ball** of center p and radius R.*
b) *For any $p, q \in M$, there exists a geodesic α connecting p and q with $L(\alpha) = d(p,q)$.*

A geodesic **ray** in a Riemannian manifold M is a geodesic $\alpha : [0,\infty) \to M$ such that α minimizes the distance between any two points of it. It is well-known that if M is complete and non-compact, then for any point $p \in M$ there exists a ray $\alpha : [0,\infty) \to M$ with $\alpha(0) = p$. It is interesting to note the relation between a ray and the topological notion of end. To do so, we recall the definition of end.

DEFINITION 1.3. Let M be a manifold and consider the set, partially ordered by the inclusion, $\{K \subset M \mid K \text{ is compact}\}$. If $\pi_0(M \setminus K)$ denotes the set of connected components of $M \setminus K$, then $\{\pi_0(M \setminus K) \mid K \text{ is compact}\}$ is an inverse system of sets. The set of ends of M is the inverse limit of this inverse system and an **end** of M is an element of that inverse limit.

It is clear that if M is compact then M has no ends. Also, if $\{K_n \mid n \in \mathbb{N}\}$ is a sequence of compact subsets of M with $K_n \subset K_{n+1}$ and $\cup_{n=1}^{\infty}\mathrm{Int}(K_n) = M$, then $\{\pi_0(M \setminus K_n) \mid n \in \mathbb{N}\}$ is also an inverse system and the set of ends of M is also given by the inverse limit of $\{\pi_0(M \setminus K_n) \mid n \in \mathbb{N}\}$. In such a case, an end of M is a sequence $\{E_n \mid n \in \mathbb{N}\}$ where $E_{n+1} \subset E_n$ and being E_n a connected component of $M \setminus K_n$.

If M is complete and not compact, then there are at least as many rays as ends of M. In fact, if $\{E_n \mid n \in \mathbb{N}\}$ is an end of M, let p be a point of M and for each n let $p_n \in E_n$. If α_n is the minimizing geodesic in M connecting p and p_n with $\|\alpha_n'\| = 1$, then $\{\alpha_n'(0)\}$ converges, after extracting a subsequence, to a unit vector $v \in T_pM$ and the geodesic α with $\alpha(0) = p$ and $\alpha'(0) = v$ is a ray.

The **curvature operator** R of the Riemannian manifold (M, g) is defined by

$$R : \mathfrak{X}(M) \times \mathfrak{X}(M) \times \mathfrak{X}(M) \to \mathfrak{X}(M)$$
$$R(X,Y)Z = \nabla_X\nabla_Y Z - \nabla_Y\nabla_X Z - \nabla_{[X,Y]}Z,$$

for arbitrary vector fields $X, Y, Z \in \mathfrak{X}(M)$. We will write

$$R(X, Y, Z, W) = g(R(X, Y)Z, W).$$

The tensorial character of R allows us to define R at any point of M, and so, if Π is a plane of T_pM and $\{v, w\}$ is a basis of Π, then the **sectional curvature** of the plane Π is defined by

$$K(\Pi) = \frac{R(v, w, w, v)}{\|v \wedge w\|^2}.$$

The Riemannian manifold (M, g) has **constant curvature** $c \in \mathbb{R}$, if for any point $p \in M$ and any plane Π of T_pM, the sectional curvature $K(\Pi) = c$. The curvature operator R of such a Riemannian manifold is given by

$$R(X, Y)Z = c\{g(Y, Z)X - g(X, Z)Y\}.$$

The **Euclidean space** $(\mathbb{R}^n, \langle , \rangle)$ with the Euclidean metric

$$\langle v, w\rangle = \sum_{i=1}^{n} v_i w_i, \quad v = (v_1, \dots, v_n),\ w = (w_1, \dots, w_n),$$

is a complete Riemannian manifold with constant curvature 0. Also, the **sphere** of radius r

$$\mathbb{S}^n(r) = \{p \in \mathbb{R}^{n+1} \mid \langle p, p\rangle = r^2\},$$

endowed with the Riemannian metric induced from the Euclidean metric $\langle , \rangle$, is a complete (compact) Riemannian manifold of constant curvature $1/r^2$. We will denote $\mathbb{S}^n = \mathbb{S}^n(1)$.

Finally, if $(\mathbb{R}_1^{n+1}, \langle\langle , \rangle\rangle)$ is the Lorentz space with the Lorentz product

$$\langle\langle v, w\rangle\rangle = \sum_{i=1}^{n} v_i w_i - v_{n+1}w_{n+1}, \quad v = (v_1, \dots, v_{n+1}),\ w = (w_1, \dots, w_{n+1}),$$

then the **hyperbolic space**

$$\mathbb{H}^n(-r) = \{p \in \mathbb{R}_1^{n+1} \mid \langle\langle p, p\rangle\rangle = -r^2\},$$

endowed with the Riemannian metric induced by $\langle\langle , \rangle\rangle$, is a complete Riemannian manifold of constant curvature $-1/r^2$. We will denote $\mathbb{H}^n = \mathbb{H}^n(-1)$.

A well-known result says that every simply connected complete Riemannian manifold of constant curvature c is isometric to the Euclidean space if $c = 0$, to the sphere $\mathbb{S}^n(r)$ if $c = 1/r^2$, or to the hyperbolic space $\mathbb{H}^n(-r)$ if $c = -1/r^2$. Thus these models are the universal coverings of the complete Riemannian manifolds with constant curvature. A classification of them can be found in [**Wol11**].

Among the compact Riemannian manifolds with constant positive curvature, it is interesting to note the **real projective space** $\mathbb{RP}^n(r)$, defined as $\mathbb{RP}^n(r) = \mathbb{S}^n(r)/\mathbb{Z}_2$, where $\mathbb{Z}_2$ is the subgroup of the isometry group of $\mathbb{S}^n(r)$ generated by the antipodal map. $\mathbb{RP}^n(r)$ is an orientable manifold if and only if n is odd. So, if n is even, $\mathbb{S}^n(r)$ is the orientable double covering of $\mathbb{RP}^n(r)$. As before, we write $\mathbb{RP}^n = \mathbb{RP}^n(1)$.

The **Ricci tensor** S, in a Riemannian manifold (M^n, g), is defined as the bilinear symmetric map $S : T_pM \times T_pM \to \mathbb{R}$ given by

$$S(v, w) = \sum_{i=1}^{n} R(v, e_i, e_i, w)$$

for $v, w \in T_pM$, $p \in M$, where $\{e_1, \dots, e_n\}$ is an orthonormal basis of T_pM. The **Ricci curvature** of a vector $v \in T_pM$ will be $\operatorname{Ricc}(v) = S(v, v)$. The manifold M is called **Einstein** if $S = \lambda g$ for some $\lambda \in \mathbb{R}$. The function ρ on M defined by $\rho = \sum_{i=1}^n \operatorname{Ricc}(e_i)$ is called the **scalar curvature** of (M, g).

On a Riemannian manifold (M^n, g) we remind the classical operators:

(1) The **gradient** $\nabla : C^\infty(M) \to \mathfrak{X}(M)$ defined by

$$g(\nabla f, X) = X(f), \quad f \in C^\infty(M), \quad X \in \mathfrak{X}(M).$$

(2) The **divergence** $\operatorname{div} : \mathfrak{X}(M) \to C^\infty(M)$, defined by

$$(\operatorname{div} X)(p) = \sum_{i=1}^{n} g(\nabla_{e_i} X, e_i),$$

where $\{e_1, \dots, e_n\}$ is an orthonormal basis of T_pM.

(3) The **hessian** of a function $f \in C^\infty(M)$, $\nabla^2 f : \mathfrak{X}(M) \times \mathfrak{X}(M) \to C^\infty(M)$ defined by

$$(\nabla^2 f)(X, Y) = g(\nabla_X \nabla f, Y) = X(Y(f)) - g(\nabla f, \nabla_X Y), \quad X, Y \in \mathfrak{X}(M).$$

(4) The **laplacian** $\Delta : C^\infty(M) \to C^\infty(M)$, defined by

$$\Delta f = \operatorname{div} \nabla f = \sum_{i=1}^{n} (\nabla^2 f)(e_i, e_i), \quad f \in C^\infty(M),$$

where $\{e_1, \dots, e_n\}$ is an orthonormal local reference on M.

Among the properties of these operators we note the Leibniz rule

$$\operatorname{div}(fX) = g(\nabla f, X) + f \operatorname{div} X. \tag{1.2}$$

Finally, a function $f \in C^\infty(M)$ is said to be **harmonic** if $\Delta f = 0$. In various parts of this book we will use harmonic functions and their properties. Although there are many monographs on harmonic functions, we will mainly use the one by P. Li, [**Li04**].

The next result is known as **Bochner's formula**.

PROPOSITION 1.4. *Let (M^n, g) be a Riemannian manifold and $f \in C^\infty(M)$. Then*

$$\frac{1}{2}\Delta\|\nabla f\|^2 = S(\nabla f, \nabla f) + g(\nabla\Delta f, \nabla f) + \|\nabla^2 f\|^2.$$

PROOF. Let p be an arbitrary point of M and $\{e_1, \dots, e_n\}$ be a local orthonormal reference around p such that $(\nabla_{e_i} e_j)_p = 0$ for all $1 \le i, j \le n$. Then

$$e_i(\|\nabla f\|^2) = 2g(\nabla_{e_i}\nabla f, \nabla f) = 2g(\nabla_{\nabla f}\nabla f, e_i),$$

and thus

$$\begin{aligned}
\frac{1}{2}\Delta\|\nabla f\|^2(p) &= \sum_{i=1}^n g(\nabla_{e_i}\nabla_{\nabla f}\nabla f, e_i)(p)\\
&= S(\nabla f, \nabla f)(p) + \sum_{i=1}^n \Big\{g(\nabla_{\nabla f}\nabla_{e_i}\nabla f, e_i) + g(\nabla_{[e_i,\nabla f]}\nabla f, e_i)\Big\}(p)\\
&= S(\nabla f, \nabla f)(p) + \nabla f(\Delta f)(p) + \sum_{i=1}^n g(\nabla_{e_i}\nabla f, [e_i, \nabla f])(p)\\
&= S(\nabla f, \nabla f)(p) + g(\nabla\Delta f, \nabla f)(p) + \sum_{i=1}^n \|\nabla_{e_i}\nabla f\|^2(p),
\end{aligned}$$

and the result follows. □

Two Riemannian metrics g and g^* on a manifold M are **conformal** if $g^* = e^{2u}g$, where $u \in C^\infty(M)$. In the next result, we relate some geometric objects corresponding to both metrics.

PROPOSITION 1.5. *Let (M^n, g) be a Riemannian manifold and $g^* = e^{2u}g$ a conformal metric to g. Then*

(1) *If ∇^* and ∇ are the Levi-Civita connections of g^* and g, respectively, then*

$$\nabla^*_X Y = \nabla_X Y + g(X, \nabla u)Y + g(Y, \nabla u)X - g(X, Y)\nabla u.$$

(2) *If $f \in C^\infty(M)$ and $X \in \mathfrak{X}(M)$, then*

$$\begin{aligned}
\nabla^* f = e^{-2u}\nabla f, \quad \operatorname{div}^*(X) &= \operatorname{div}(X) + ng(X, \nabla u),\\
(\nabla^*)^2 f &= \nabla^2 f - \mathrm{d}f \otimes \mathrm{d}u - \mathrm{d}u \otimes \mathrm{d}f + g(\nabla f, \nabla u)g,\\
\Delta^* f &= e^{-2u}\{\Delta f + (n-2)g(\nabla u, \nabla f)\}.
\end{aligned}$$

(3) *If S^* and S are the Ricci tensors of g^* and g, respectively, then*

$$S^* = S - \big(\Delta u + (n-2)\|\nabla u\|^2\big)g - (n-2)\nabla^2 u + (n-2)\mathrm{d}u \otimes \mathrm{d}u.$$

(4) *If ρ^* and ρ are the scalar curvatures of g^* and g, respectively, then*

$$\rho^* = e^{-2u}\big(\rho - 2(n-1)\Delta u - (n-1)(n-2)\|\nabla u\|^2\big).$$

In particular, if M is a surface ($n = 2$), then

$$K^* = e^{-2u}\big(K - \Delta u\big),$$

where K^* and K are the Gauss curvatures of g^* and g, respectively.

It is known, [**MS39**], that the isometry group $\mathrm{Iso}(M,g)$ of a Riemannian manifold is a Lie group with respect to the compact-open topology. Its Lie algebra is identified with the algebra $\mathcal{K}(M)$ of **Killing** vector fields on M:

$$\mathcal{K}(M) = \{X \in \mathfrak{X}(M) \mid L_X g = 0\},$$

where L is the Lie derivative on M. With repect to the Levi-Civita connection ∇, $X \in \mathfrak{X}(M)$ is a Killing field if and only if $g(\nabla_V X, V) = 0$, $\forall V \in \mathfrak{X}(M)$.

The isometry group $\mathrm{E}(n)$ of the Euclidean space $\mathbb{R}^n$, the isometry group $\mathrm{O}(n+1)$ of the sphere $\mathbb{S}^n(r)$ and the isometry group $\mathrm{O}^1(n+1)$ of the hyperbolic space $\mathbb{H}^n(-r)$ have dimension $n(n+1)/2$. Also, the isometry group $\mathrm{PO}(n+1) = \mathrm{O}(n+1)/\{\pm I\}$ of the real projective space $\mathbb{RP}^n(r)$ has dimension $n(n+1)/2$. The next result characterizes these examples in terms of their isometry groups.

THEOREM 1.6 ([**Kob72**]). *Let (M^n, g) be a Riemannian manifold. Then*

$$\dim \mathrm{Iso}(M,g) \leq n(n+1)/2,$$

and the equality holds if and only if M is isometric to one of the following spaces of constant curvature:

(1) *An n-dimensional Euclidean space $\mathbb{R}^n$.*
(2) *An n-dimensional sphere $\mathbb{S}^n(r)$.*
(3) *An n-dimensional hyperbolic space $\mathbb{H}^n(-r)$.*
(4) *An n-dimensional real projective space space $\mathbb{RP}^n(r)$.*

Given two Riemannian manifolds (M^n, g) and $(\widehat{M}^n, \widehat{g})$, a **conformal transformation** from (M^n, g) to $(\widehat{M}^n, \widehat{g})$ is a diffeomorphism $\Phi : M \to \widehat{M}$ such that $\Phi^*(\widehat{g})$ and g are conformal metrics. If we denote by $\mathrm{Con}(M,g)$ the group of conformal transformations of (M,g), it is clear that $\mathrm{Iso}(M,g) \subset \mathrm{Con}(M,g)$. In the book we will use conformal transformations of the sphere, and with respect to them it is interesting to note the following fact. If $B(1) = \{a \in \mathbb{R}^{n+1} \mid \|a\| < 1\}$ is the unit disk, then for any $a \in B(1)$ the map $\psi_a : \mathbb{S}^n \to \mathbb{S}^n$ defined by

$$\psi_a(p) = a + \frac{1 - \|a\|^2}{\|p+a\|^2}(p+a) \tag{1.3}$$

is a conformal transformation of the sphere with $(\psi_a)^* g = \frac{(1-\|a\|^2)^2}{\|p+a\|^4} g$, where g is the standard metric on the sphere $\mathbb{S}^n$. Moreover, any conformal transformation of the sphere is given by $F \circ \psi_a$ with $a \in B(1)$ and F an isometry of $(\mathbb{S}^n, g)$.

1.2. Integration

If (M^n, g) is a Riemannian manifold, the **measure** on M associated to the metric g is denoted by $\mathrm{d}v$ and the integral of a measurable function $f : M \to \mathbb{R}$ is denoted by $\int_M f \,\mathrm{d}v$.

If $1 \leq p < \infty$, we denote by $L^p(M,g)$, or simply $L^p(M)$, the space of measurable functions f on M such that $|f|^p$ is integrable, i.e., $\int_M |f|^p \,\mathrm{d}v < \infty$. This linear space is a Banach space with norm

$$\|f\|_p = \left(\int_M |f|^p \,\mathrm{d}v\right)^{1/p}.$$

$L^2(M,g)$ is also a Hilbert space with the product $(f,g)_2 = \int_M fg \,\mathrm{d}v$.

We will denote by $L^\infty(M)$ the space of essentially bounded measurable functions on M. This space is also a Banach space with the norm

$$\|f\|_\infty = \inf\{c > 0 \mid c \text{ is an essential bound of } f\}.$$

In the next theorem, we recall some classical results in integration theory that will be used throughout the book.

THEOREM 1.7. *Let (M^n, g) be a Riemannian manifold.*

(1) (Monotone convergence theorem) *If $\{f_m : M \to \mathbb{R},\ m \in \mathbb{N}\}$ is a monotone increasing sequence of measurable non-negative functions on M with pointwise limit f, then*

$$\int_M f\,\mathrm{d}v = \lim_{m\mapsto\infty} \int_M f_m\,\mathrm{d}v.$$

(2) (Dominated convergence theorem) *Let $\{f_m : M \to \mathbb{R},\ m \in \mathbb{N}\}$ be a sequence of measurable functions with pointwise limit f. If there exists a function $h \in L^1(M)$ such that $|f_m| \le h$, $\forall m \in \mathbb{N}$, then $f \in L^1(M)$ and*

$$\int_M f\,\mathrm{d}v = \lim_{m\mapsto\infty} \int_M f_m\,\mathrm{d}v.$$

(3) (Change of variable formula) *Let $F : (M, g) \to (\widehat{M}, \widehat{g})$ be a diffeomorphism between Riemannian manifolds. If f is an integrable function on $\widehat{M}$, then $(f \circ F)|\operatorname{Jac} F|$ is integrable on M and*

$$\int_M (f \circ F)|\operatorname{Jac} F|\,\mathrm{d}v = \int_{\widehat{M}} f\,\mathrm{d}\widehat{v},$$

where $|\operatorname{Jac} F|$ is the absolute value of the Jacobian of F. In particular, if F is an isometry, then

$$\int_M (f \circ F)\,\mathrm{d}v = \int_{\widehat{M}} f\,\mathrm{d}\widehat{v}.$$

*Thus the **volume**, $\operatorname{vol}(M, g) = \int_M \mathrm{d}v$, is invariant under isometries.*

(4) (Leibniz integral rule) *Let $f : (a, b) \times M \to \mathbb{R}$ be a map satisfying*

a) For every $t \in (a, b)$, $f_t \in L^1(M)$.

b) For every $p \in M$, $t \in (a, b) \mapsto \frac{\partial f}{\partial t}(t, p)$ is continuous in (a, b) and there exists $h \in L^1(M)$ such that for every $t \in (a, b)$, $|\frac{\partial f}{\partial t}(t, p)| \le h(p)$.

Then, $t \mapsto \int_M f_t\,\mathrm{d}v$ is of class one and

$$\frac{\mathrm{d}}{\mathrm{d}t} \int_M f_t\,\mathrm{d}v = \int_M \frac{\partial f}{\partial t}(t, p)\,\mathrm{d}v.$$

(5) (Divergence theorem) *If $X \in \mathfrak{X}_0(M)$, then*

$$\int_M \operatorname{div}(X)\,\mathrm{d}v = 0,$$

where $\mathfrak{X}_0(M) = \{X \in \mathfrak{X}(M) \mid X \text{ has compact support}\}$.

(6) (Conformal invariance of the Dirichlet integral) *If $g^* = e^{2u} g$ is a metric conformal to g, then the measures associated with both metrics are related by $\mathrm{d}v^* = e^{nu}\mathrm{d}v$ and*

$$\int_M \|\nabla^* f\|_*^n\,\mathrm{d}v^* = \int_M \|\nabla f\|^n\,\mathrm{d}v, \tag{1.4}$$

for any $f \in C_0^\infty(M) = \{f \in C^\infty(M) \mid f$ *has compact support*$\}$.

(7) (Fubini theorem) *Let* (M_i, g_i), $i = 1, 2$, *be Riemannian manifolds and* $(M_1 \times M_2, g = g_1 \times g_2)$ *their Riemannian product. If* f *is an integrable function on* $M_1 \times M_2$, *then*

$$\int_{M_1 \times M_2} f\,\mathrm{d}v = \int_{M_1} \left(\int_{M_2} f \mathrm{d}v_2 \right) \mathrm{d}v_1 = \int_{M_2} \left(\int_{M_1} f \mathrm{d}v_1 \right) \mathrm{d}v_2,$$

where $\mathrm{d}v, \mathrm{d}v_1$ *and* $\mathrm{d}v_2$ *are the measures associated to* g, g_1 *and* g_2.

Before introducing the Sobolev spaces of a Riemannian manifold, we will prove a technical result that we will use sometimes along the book.

Proposition 1.8 ([**LY82**]). *Let* (M^n, g) *be a Riemannian manifold with finite volume and* $F : M \to \mathbb{S}^m$, $m \geq 2$, *a measurable map such that* $F^{-1}(q)$ *has measure zero in* M *for any point* $q \in \mathbb{S}^m$.

(1) *Then there exists* $a \in B(1) = \{a \in \mathbb{R}^{m+1} \mid \|a\| < 1\}$ *such that*

$$\int_M (\psi_a \circ F)\,\mathrm{d}v = 0,$$

where ψ_a *is the conformal transformation of* $\mathbb{S}^m$ *given in* (1.3).

(2) *If* u *is a smooth and positive function satisfying* $u^{n/2} \in L^1(M)$, *then there exists* $a \in B(1) \subset \mathbb{R}^{m+1}$ *such that*

$$\int_M u^{n/2}(\psi_a \circ F)\,\mathrm{d}v = 0.$$

Proof. **Proof of 1**. Let $H : B(1) \to \mathbb{R}^{m+1}$ be the map given by

$$H(a) = \frac{1}{\mathrm{vol}(M, g)} \int_M (\psi_a \circ F)\,\mathrm{d}v.$$

Then, using the Schwartz inequality we have

$$\begin{aligned} \|H(a)\|^2 &= \frac{1}{\mathrm{vol}(M, g)^2} \sum_{i=1}^{m+1} \left(\int_M (\psi_a \circ F)_i\,\mathrm{d}v \right)^2 \\ &\leq \frac{1}{\mathrm{vol}(M, g)} \sum_{i=1}^{m+1} \int_M (\psi_a \circ F)_i^2\,\mathrm{d}v = 1, \end{aligned}$$

and so $H : B(1) \to \overline{B(1)}$. It is also clear that H is a continuous function.

Let $a \in \mathbb{S}^m = \partial B(1)$ be any point and $\{a_n\}$ be a sequence of points of $B(1)$ converging to a. Since $F^{-1}(-a)$ has measure zero in M, we have

$$H(a_n) = \frac{1}{\mathrm{vol}(M, g)} \int_{M \setminus F^{-1}(-a)} (\psi_{a_n} \circ F)\,\mathrm{d}v.$$

Then the sequence $\{\psi_{a_n} \circ F : M \setminus F^{-1}(-a) \to \mathbb{R}\}$ has pointwise limit a and $\|\psi_{a_n} \circ F\| \leq 1 \in L^1(M \setminus F^{-1}(-a))$ for any n. Then Theorem 1.7,2, says that $\lim_{n\to\infty} H(a_n) = a$ and so we have extended H to a continuous function $G : \overline{B(1)} \to \overline{B(1)}$ with $G_{|\mathbb{S}^m} = Id$. Under these conditions, if $G(a) \neq 0$ for any $a \in \overline{B(1)}$, then the map $a \mapsto \frac{G(a)}{\|G(a)\|}$ is a retraction from $\overline{B(1)}$ to $\mathbb{S}^m$, which is a contradiction. So there exists a point $a \in \overline{B(1)}$ such that $G(a) = 0$. Since $a \notin \mathbb{S}^m$, we prove 1.

Proof of 2. Consider the metric $g^* = ug$. Then, as $u^{n/2} \in L^1(M)$, we have that

$$\operatorname{vol}(M, g^*) = \int_M dv^* = \int_M u^{n/2}\, dv < \infty.$$

So, applying part 1 to the map $F : (M, g^*) \to S^m$, we get a point $a \in B(1)$ such that

$$0 = \int_M (\psi_a \circ F)\, dv^* = \int_M u^{n/2}(\psi_a \circ F)\, dv,$$

which proves 2. □

Now we will introduce the Sobolev spaces in a Riemannian manifold. If (M, g) is a Riemannian manifold, we denote by $L^2(M, TM)$ the space of measurable vector fields X on M such that $\|X\|^2$ is integrable, i.e., $\int_M \|X\|^2\, dv < \infty$. This space is also a Hilbert space with the product

$$(X, Y)_2 = \int_M g(X, Y)\, dv.$$

Now, if $f \in C^\infty(M)$ and $Y \in \mathfrak{X}_0(M)$, from (1.2) and Theorem 1.7,3, it follows that

$$0 = \int_M \{g(\nabla f, Y) + f \operatorname{div}(Y)\}\, dv.$$

This expression justifies the next definition.

DEFINITION 1.9. Given a function $f \in L^2(M)$, a vector field $X \in L^2(M, TM)$ is called a **weak derivative** of f if X satisfies

$$0 = \int_M \{g(X, Y) + f \operatorname{div}(Y)\}\, dv, \quad \forall Y \in \mathfrak{X}_0(M).$$

If X exists, it is unique and is denoted by ∇f.

The **Sobolev space** $H^{1,2}(M, g)$, or simply $H^{1,2}(M)$, is defined by

$$H^{1,2}(M) = \{f \in L^2(M) \mid f \text{ has a weak derivative}\}.$$

$H^{1,2}(M)$ is a Hilbert space with the product

$$(f, g)_{1,2} = (f, g)_2 + (\nabla f, \nabla g)_2.$$

We will denote by $H_0^{1,2}(M)$ the closure of $C_0^\infty(M)$ in the space $\big(H^{1,2}(M), (,)_{1,2}\big)$. It is clear that $H_0^{1,2}(M)$ is a Hilbert subspace of $H^{1,2}(M)$.

A function $u : M \to \mathbb{R}$ is said to be **Lipschitz** if there exists a constant $C \geq 0$ such that $|u(p) - u(q)| \leq C d(p, q)$, $\forall p, q \in M$, where d is the distance associated to g. The smaller constant C that satisfies the above relation is called the Lipschitz constant of u. It is interesting to note that the function u defined by $u(p) = d(p, p_0)$, where p_0 is a point of M, is a Lipschitz function with Lipschitz constant one. We also recall that every function $u \in C_0^1(M)$ is Lipschitz.

In the next result we collect some important results about Sobolev spaces which will be used in the book.

THEOREM 1.10 ([**Aub82**]). *Let (M, g) be a Riemannian manifold. Then*

(1) *If $u : M \to \mathbb{R}$ is a Lipschitz function with compact support, then $u \in H_0^{1,2}(M)$.*
(2) *If (M, g) is complete, then $H^{1,2}(M) = H_0^{1,2}(M)$.*

(3) (Rellich-Kondrakov) *If Ω is a relatively compact open subset of M, then the inclusion $i : H_0^{1,2}(\Omega) \to L^2(\Omega)$ is a compact operator, i.e., if $\{u_k\} \subset H_0^{1,2}(\Omega)$ is a $(\,,\,)_{1,2}$-bounded sequence, then there exists a subsequence $\{u'_k\}$ that converges in $L^2(\Omega)$.*

1.3. Hodge theory

Given a manifold M^n, we will denote by $\Omega^p(M)$, $0 \le p \le n$, the space of differential p-forms on M, where $\Omega^0(M) = C^\infty(M)$. If $\wedge : \Omega^p(M) \times \Omega^q(M) \to \Omega^{p+q}(M)$ is the exterior product, then $\Omega(M) = \oplus_{p=0}^n \Omega^p(M)$ is the graduated algebra of differential forms on M.

The **exterior differential** $\mathrm{d} : \Omega^p(M) \to \Omega^{p+1}(M)$ is the linear map defined by

$$\begin{aligned}(\mathrm{d}\omega)(X_0, \dots, X_p) &= \sum_{i=0}^{p} (-1)^i X_i\big(\omega(X_0, \dots, \widehat{X_i}, \dots, X_p)\big) \\ &\quad + \sum_{0 \le i < j \le p} (-1)^{i+j} \omega\big([X_i, X_j], X_0, \dots, \widehat{X_i}, \dots, \widehat{X_j}, \dots, X_p\big),\end{aligned}$$

for any $\omega \in \Omega^p(M)$ and $X_0, \dots, X_p \in \mathfrak{X}(M)$, where $\widehat{X}$ means to delete X. It is interesting to note that $\mathrm{d} \circ \mathrm{d} = 0$ and that

$$\mathrm{d}(\omega_1 \wedge \omega_2) = (\mathrm{d}\omega_1) \wedge \omega_2 + (-1)^p \omega_1 \wedge (\mathrm{d}\omega_2),$$

for arbitrary forms ω_1 and ω_2 with $\omega_1 \in \Omega^p(M)$.

A differentiable singular p-chain c in M is $c = \sum_{i=1}^k a_i \sigma_i$, where $a_i \in \mathbb{R}$ and σ_i are differentiable singular p-simplices in M. The space of differentiable singular p-chains in M is denoted by $C_p(M, \mathbb{R})$. We denote by $\partial : C_p(M, \mathbb{R}) \to C_{p-1}(M, \mathbb{R})$ the boundary operator satisfying $\partial \circ \partial = 0$.

Now, if $\omega \in \Omega^p(M)$ and $c \in C_p(M, \mathbb{R})$ we represent by $\int_c \omega$ the integral of ω on the chain c (see [**War83**]).

THEOREM 1.11 (Stokes theorem [**War83**]). *Let c be a differentiable singular p-chain ($p \ge 1$) in a manifold M and $\omega \in \Omega^{p-1}(M)$. Then*

$$\int_c \mathrm{d}\omega = \int_{\partial c} \omega.$$

Under these conditions, we introduce the homology and cohomology groups of M. In fact, we consider the spaces of closed p-forms and exact p-forms on M, respectively, defined by

$$\begin{aligned} Z^p(M) &= \{\omega \in \Omega^p(M) \mid \mathrm{d}\omega = 0\} \\ B^p(M) &= \{\mathrm{d}\omega \in \Omega^p(M) \mid \omega \in \Omega^{p-1}(M)\}. \end{aligned}$$

The above property of d ($\mathrm{d}^2 = 0$) means that $B^p(M) \subset Z^p(M)$ and so, we can define the p-th **de Rham cohomology space** of M by

$$H^p_{dR}(M) = \frac{Z^p(M)}{B^p(M)} = \{[\omega] \mid \mathrm{d}\omega = 0\}.$$

It is important to note that if M^n is an orientable compact manifold, then the linear map

$$H^n_{dR}(M^n) \to \mathbb{R}, \quad [\omega] \mapsto \int_M [\omega] := \int_M \omega,$$

defines an isomorphism ([**Spi79**]). The correct definition of this isomorphism is a consequence of Theorem 1.11.

Similarly, if $Z_p(M,\mathbb{R}) = \{c \in C_p(M,\mathbb{R}) \mid \partial c = 0\}$ and $B_p(M,\mathbb{R}) = \{\partial c \mid c \in C_{p+1}(M,\mathbb{R})\}$, the above property of ∂ ($\partial^2 = 0$) means that $B_p(M,\mathbb{R}) \subset Z_p(M,\mathbb{R})$ and so, we can define the **p-th differentiable singular homology group of M with coefficients in $\mathbb{R}$** by

$$_\infty H_p(M,\mathbb{R}) = \frac{Z_p(M,\mathbb{R})}{B_p(M,\mathbb{R})} = \{[c] \mid \partial c = 0\}.$$

THEOREM 1.12 (de Rham theoren [**War83**]). *Given a manifold M and $p \geq 0$, if $_\infty H_p(M,\mathbb{R})^*$ represents the dual space of $_\infty H_p(M,\mathbb{R})$, then the map*

$$F : H^p_{dR}(M) \to {_\infty H_p(M,\mathbb{R})^*}$$
$$F([\omega])([c]) = \int_{[c]} [\omega] := \int_c \omega$$

is a well-defined isomorphism. The correct definition of F is a consequence of the Stokes theorem (Theorem 1.11).

Suppose (M^n, g) is a Riemannian manifold. If $\nabla\omega$ denotes the covariant derivative of a p-form ω, i.e.,

$$(1.5)\quad (\nabla_X\omega)(X_1,\dots,X_p) = X(\omega(X_1,\dots,X_p)) - \sum_{i=1}^p \omega(X_1,\dots,\nabla_X X_i,\dots,X_p),$$

then we define the **exterior codifferential** $\delta : \Omega^{p+1}(M) \to \Omega^p(M)$ as

$$(\delta\omega)(X_1,\dots,X_p) = \sum_{i=1}^n (\nabla_{e_i}\omega)(e_i, X_1,\dots,X_p),$$

where $\{e_1,\dots,e_n\}$ is a local orthonormal reference on M and $X_1,\dots,X_p \in \mathfrak{X}(M)$.

The operator $\Delta := \mathrm{d}\delta + \delta\mathrm{d} : \Omega^p(M) \to \Omega^p(M)$ is called the **Hodge laplacian** and a p-form α satisfying $\Delta\alpha = 0$ is called **harmonic**. We denote by

$$\mathcal{H}^p(M) = \{\alpha \in \Omega^p(M) \mid \Delta\alpha = 0\}.$$

Note that when $p = 0$, the Hodge laplacian is the laplacian of functions defined in Section 1.1.

We will denote by $L^2(M, \Lambda^p T^*M)$ the space of measurable p-forms ω on M such that $\|\omega\|^2$ is integrable. This space is also a Hilbert space with the product

$$(\omega_1,\omega_2)_2 = \int_M g(\omega_1,\omega_2)\,\mathrm{d}v.$$

If the Riemannian manifold M is compact, then the operators d and δ are adjoint with respect to $(,)_2$, i.e. $(\alpha,\delta\beta)_2 = (d\alpha,\beta)_2$ and thus every p-form α satisfies

$$\int_M g(\Delta\alpha,\alpha)\,\mathrm{d}v = \int_M \|\mathrm{d}\alpha\|^2\,\mathrm{d}v + \int_M \|\delta\alpha\|^2\,\mathrm{d}v.$$

So a p-form α is harmonic if and only if $\mathrm{d}\alpha = \delta\alpha = 0$.

THEOREM 1.13 (Hodge decomposition theorem). *If (M^n, g) is a compact Riemannian manifold, then*

(1) $\Omega^p(M) = \mathcal{H}^p(M) \oplus \mathrm{d}(\Omega^{p-1}(M)) \oplus \delta(\Omega^{p+1}(M))$, $0 \leq p \leq n$.

(2) *$\mathcal{H}^p(M)$ is finite dimensional and isomorphic to $H^p_{dR}(M)$. Hence*

$$\dim \mathcal{H}^p(M) = \beta_p(M),$$

where $\beta_p(M)$ is the p-th Betti number of M.

Some properties of the Hodge theory for compact manifolds remain valid for complete Riemannian manifolds.

THEOREM 1.14 ([**Kod49**]). *Let (M^n, g) be a complete Riemannian manifold. Then*

(1) *If $\alpha \in \mathcal{H}^p(M) \cap L^2(M, \Lambda^p T^*M)$, then $\mathrm{d}\alpha = \delta\alpha = 0$.*
(2) *$\Omega^p(M) \cap L^2(M, \Lambda^p T^*M) = \big(\mathcal{H}^p(M) \cap L^2(M, \Lambda^p T^*M)\big) \oplus \overline{\mathrm{d}\Omega_0^{p-1}(M)} \oplus \overline{\delta\Omega_0^{p+1}(M)}$, where the clausures are taken in the L^2-topology and*

$$\Omega_0^k(M) = \{\alpha \in \Omega^k(M) \mid \alpha \textit{ has compact support}\}.$$

When $p = 1$, there exists a natural isomorphism $X \in \mathfrak{X}(M) \mapsto X^\flat \in \Omega^1(M)$ defined by $X^\flat(Y) = g(X, Y)$, for any $Y \in \mathfrak{X}(M)$. The inverse of this isomorphism is $\omega \in \Omega^1(M) \mapsto \omega^\sharp \in \mathfrak{X}(M)$ given by $g(\omega^\sharp, Y) = \omega(Y)$, for any $Y \in \mathfrak{X}(M)$. It is easy to check that

$$\delta X^\flat = \operatorname{div} X, \qquad \operatorname{div} \omega^\sharp = \delta\omega.$$

So we can rewrite the divergence theorem as follows

THEOREM 1.15. *If (M, g) is a Riemannian manifold and $\omega \in \Omega_0^1(M)$, then*

$$\int_M \delta\omega \, \mathrm{d}v = 0.$$

Assuming that M is complete, we say that a vector field $X \in \mathfrak{X}(M)$ is **harmonic** if the 1-form $X^\flat$ is harmonic, i.e., $(\delta\mathrm{d} + \mathrm{d}\delta)(X^\flat) = 0$. Using the above definitions, it is not difficult to see that

$$X \in \mathfrak{X}(M) \text{ is harmonic if and only if } \quad \Delta X = \operatorname{Ricc}(X), \tag{1.6}$$

where $\Delta : \mathfrak{X}(M) \to \mathfrak{X}(M)$ is the **rough Laplacian** defined by

$$\Delta = \sum_{i=1}^n \{\nabla_{e_i}\nabla_{e_i} - \nabla_{\nabla_{e_i} e_i}\},$$

where $\{e_1, \ldots, e_n\}$ is an orthonormal local reference on M. The harmonicity of X is also equivalent to $\mathrm{d}X^\flat = 0$ and $\delta X^\flat = 0$, which means that

$$\operatorname{div}(X) = 0, \qquad g(\nabla_v X, w) = g(\nabla_w X, v), \quad \forall\, v, w \in TM.$$

We will denote by $\mathcal{H}(M)$ the space

$$\mathcal{H}(M) = \{X \in \mathfrak{X}(M) \mid X \text{ is harmonic}\}.$$

Hence $\mathcal{H}(M)$ is isomorphic to $\mathcal{H}^1(M)$ and so, if M is compact, its dimension is the first Betti number of M.

Now we will introduce the star operator.

Let (M^n, g) be an oriented Riemannian manifold. We denote by $\eta \in \Omega^n(M)$ the **volume form** on M, i.e., $\eta_p(e_1, \ldots, e_n) = 1$ for any positively oriented orthonormal basis $\{e_1, \ldots, e_n\}$ of T_pM and any $p \in M$. We define the **star operator** as the operator

$$\star : \Omega^p(M) \to \Omega^{n-p}(M)$$
$$(\star\alpha)(X_1, \ldots, X_{n-p})\eta = \alpha \wedge X_1^\flat \wedge \cdots \wedge X_{n-p}^\flat,$$

for any $0 \leq p \leq n$ and any $X_1, \ldots, X_{n-p} \in \mathfrak{X}(M)$, where $\wedge$ is the exterior product of forms.

The homomorphism $\star$ is an isomorphism with $\star^2 = (-1)^{p(n-p)} Id$. Furthermore the exterior codifferential $\delta : \Omega^{p+1}(M) \to \Omega^p(M)$ can be given by

$$\delta = (-1)^{p+1} \star^{-1} \mathrm{d} \star .$$

To conclude this section, we will prove a Kato's inequality for harmonic vector fields. This result will be used in the book.

LEMMA 1.16. *Let X be a harmonic vector field on a Riemannian manifold (M^n, g). Then*

(1) $\frac{1}{2}\Delta\|X\|^2 = S(X,X) + \|\nabla X\|^2$,
(2) $\|X\|^2\|\nabla X\|^2 \geq \frac{n}{4(n-1)}\|\nabla\|X\|^2\|^2$ (Kato's inequality),

where S is the Ricci tensor of M.

PROOF. **Proof of 1**. Let p be any point of M and let $\{e_1, \ldots, e_n\}$ be a local orthonormal reference around p such that $(\nabla_{e_i} e_j)_p = 0$ for any $1 \leq i, j \leq n$. Then

$$e_i(\|X\|^2) = 2g(\nabla_{e_i} X, X) = 2g(\nabla_X X, e_i),$$

and thus

$$\begin{aligned}
\frac{1}{2}\Delta\|X\|^2(p) &= \sum_{i=1}^{n} g(\nabla_{e_i}\nabla_X X, e_i)(p) \\
&= S(X,X)(p) + \sum_{i=1}^{n}\{g(\nabla_X\nabla_{e_i} X, e_i) + g(\nabla_{[e_i,X]}X, e_i)\}(p) \\
&= S(X,X)(p) + \sum_{i=1}^{n} g(\nabla_{e_i} X, [e_i, X])(p) \\
&= S(X,X)(p) + \sum_{i=1}^{n} \|\nabla_{e_i} X\|^2(p),
\end{aligned}$$

and 1 follows.

Proof of 2. It is clear that we can work on the open subset $\{p \in M \mid X_p \neq 0\}$. On such an open subset we can choose an orthonormal reference $\{e_1, \ldots, e_n\}$ such that $X = \|X\| e_1$. So, using that X is harmonic, we have

$$\begin{aligned}
\|\nabla X\|^2 &= \sum_{i,j=1}^{n} g(\nabla_{e_i} X, e_j)^2 \geq \sum_{j=1}^{n} g(\nabla_{e_1} X, e_j)^2 + \sum_{i=2}^{n}\{g(\nabla_{e_i} X, e_1)^2 + g(\nabla_{e_i} X, e_i)^2\} \\
&\geq \sum_{j=1}^{n} g(\nabla_{e_1} X, e_j)^2 + \sum_{i=2}^{n} g(\nabla_{e_i} X, e_1)^2 + \frac{\left(\sum_{i=2}^{n} g(\nabla_{e_i} X, e_i)\right)^2}{n-1} \\
&= \frac{n}{n-1} g(\nabla_{e_1} X, e_1)^2 + 2\sum_{j=2}^{n} g(\nabla_{e_1} X, e_j)^2 \geq \frac{n}{n-1}\sum_{j=1}^{n} g(\nabla_{e_1} X, e_j)^2.
\end{aligned}$$

Also, since $X = \|X\|e_1$, we have

$$\|X\|^2 \sum_{j=1}^{n} g(\nabla_{e_1} X, e_j)^2 = \sum_{j=1}^{n} g(\nabla_X X, e_j)^2 = \sum_{j=1}^{n} g(\nabla_{e_j} X, X)^2 = \frac{1}{4}\|\nabla \|X\|^2\|^2,$$

and 2 follows. □

1.4. Schrödinger operators

Let (M^n, g) be a Riemannian manifold and $\Delta : C^\infty(M) \to C^\infty(M)$ the Laplacian operator. A **Schrödinger operator** on M is an operator $L = \Delta + q$, where $q \in C^\infty(M) \cap L^\infty(M)$. If Ω is a relatively open subset of M, the eigenvalue problem for L with Dirichlet condition for Ω is the problem given by

$$Lu + \lambda u = 0, \quad u \in C^\infty(\Omega) \cap H_0^{1,2}(\Omega), \quad u \neq 0, \quad \lambda \in \mathbb{R}.$$

If (u, λ) is a solution to the above problem, u is called an **eigenfunction** of L with **eigenvalue** λ. The eigenspace associated to the eigenvalue λ is denoted by

$$V_\lambda = \{u \in C^\infty(\Omega) \cap H_0^{1,2}(\Omega) \mid Lu + \lambda u = 0\}.$$

The eigenspace V_λ is finite dimensional and its dimension is called the **multiplicity** of the eigenvalue λ. Furthermore, eigenspaces corresponding to different eigenvalues are orthogonal in $L^2(\Omega)$.

We define the symmetric bilinear form $Q : H_0^{1,2}(\Omega) \times H_0^{1,2}(\Omega) \to \mathbb{R}$ by

$$Q(u, w) = \int_\Omega \{g(\nabla u, \nabla w) - quw\}\, dv.$$

Q will also denote the corresponding quadratic form $Q : H_0^{1,2}(\Omega) \to \mathbb{R}$ given by $Q(u) = Q(u, u)$. It is interesting to note that the quadratic form Q is continuous with respect to $(,)_{1,2}$.

THEOREM 1.17. *Let Ω be a relatively compact open subset of a Riemannian manifold (M, g) and $L = \Delta + q$ a Schrödinger operator on M. Then, there exists a sequence $\{(\lambda_n, \varphi_n) \mid n \in \mathbb{N}\}$ where $\{\lambda_n \mid n \in \mathbb{N}\}$ are all the eigenvalues of L and $\varphi_n \in V_{\lambda_n}$. This sequence satisfies the following properties:*

(1) *$\lambda_1 \leq \lambda_2 \leq \dots$ and $\{\lambda_n \mid n \in \mathbb{N}\}$ diverges to $+\infty$.*
(2) *$\{\varphi_n \mid n \in \mathbb{N}\}$ is a Hilbert basis of $L^2(\Omega)$.*
(3) *$\lambda_n = \inf\left\{ \frac{Q(u)}{\|u\|_2^2} \middle| u \in H_0^{1,2}(\Omega), (u, \varphi_k)_2 = 0, 1 \leq k \leq n-1 \right\}$. Furthermore, if $u \in H_0^{1,2}(\Omega)$ satisfies $(u, \varphi_k)_2 = 0$, $1 \leq k \leq n-1$ and $Q(u) = \lambda_n \|u\|_2^2$, then $u \in V_{\lambda_n}$.*

*The set $\{\lambda_n \mid n \in \mathbb{N}\}$ is called the **spectrum** of L in Ω.*

As an illustration of the above theorem, we recall the eigenvalues of the Laplacian operator for two important examples used in the book.

THEOREM 1.18 ([**BGM71**]).

(1) *The spectrum of the Laplacian operator of the sphere $\mathbb{S}^n(r)$ is given by*

$$\{\lambda_k = \frac{1}{r^2}(k-1)(k+n-2), \quad k \geq 1\}.$$

The multiplicity of the eigenvalue λ_k *is* 1 *for* $k = 1$, $n+1$ *for* $k = 2$ *and* $\binom{n+k-1}{k-1} - \binom{n+k-3}{k-3}$ *for* $k \geq 3$. *Furthermore, the eigenspace* V_{λ_k} *corresponding to the eigenvalue* λ_k *is*

$$V_{\lambda_k} = \{P_{|\mathbb{S}^n(r)} \,|\, P \text{ is a harmonic homogeneous polynomial of degree } k-1 \text{ on } \mathbb{R}^{n+1}\}.$$

(2) *Let* (M_i, g_i), $i = 1, 2$ *be compact Riemannian manifolds. Then*

$$\mathrm{Spec}(M_1 \times M_2, g_1 \times g_2) = \{\lambda + \mu \mid \lambda \in \mathrm{Spec}(M_1, g_1), \mu \in \mathrm{Spec}(M_2, g_2)\}.$$

Furthermore, the eigenspace V_ρ *corresponding to the eigenvalue* ρ *is*

$$V_\rho = \sum_{\lambda+\mu=\rho} \mathrm{Span}\{f(p,q) = f(p)g(q) \mid f \in V_\lambda^{M_1},\ g \in V_\mu^{M_2}\}.$$

In the following result we collect some important and known results about Schrödinger operators, which will be useful throughout the book.

THEOREM 1.19 ([**Aub82, GT01**]). *Let* (M, g) *be a Riemannian manifold and* $L = \Delta + q$ *a Schrödinger operator on* M.

(1) (Maximum principle) *Let* Ω *be a relatively compact open subset of* M. *Suppose that* $q \leq 0$ *on* Ω *and* $u \in C^\infty(\Omega)$ *satisfies* $Lu \geq 0$ *on* Ω. *If* u *attains its maximum, then either* u *is always negative or* u *is constant.*

(2) (Harnack inequality) *Let* Ω *be an open subset of* M *and* u *a non-negative solution to the equation* $Lu = 0$ *on* Ω. *Then for any relatively compact open subset* Ω' *of* Ω *we have*

$$\sup\{u(x) \mid x \in \Omega'\} \leq C \inf\{u(x) \mid x \in \Omega'\},$$

where C *is a constant depending on* Ω *and* Ω'.

(3) (Fredholm alternative) *Let* Ω *be a relatively compact open subset of* M. *If* $u = 0$ *is the unique solution to the problem* $Lu = 0$ *on* Ω *and* $u = 0$ *on* $\partial\Omega$, *then for any* $f \in C^\infty(\Omega)$ *there exists a unique solution to the problem* $Lu = f$ *in* Ω *and* $u = 0$ *on* $\partial\Omega$.

As an application of the maximum principle we prove a result that will be used later.

PROPOSITION 1.20. *Let* Ω *be a relatively compact open subset of a Riemannian manifold* (M, g) *and let* $L = \Delta + q$ *be a Schrödinger operator on* M. *If* u *is a non-negative smooth function on* Ω *with* $Lu = 0$, *then either* $u = 0$ *or* u *has no zeros on* Ω.

PROOF. Suppose that there exists a point $p \in \Omega$ such that $u(p) = 0$. Then $v = -u$ is a non-positive smooth function which has a maximum at p. Let $a = \inf\{q(x) \mid x \in \Omega\}$ and $b = \min\{a, 0\}$. If $L^* = \Delta + b$, since $bv \geq qv$ because $v \leq 0$, then $L^*v \geq Lv = 0$. Now the maximum principle (Theorem 1.19) for the operator L^* says that v is constant and since $v(p) = 0$, we get that $u = -v = 0$. □

Finally, in the next result we study an important property of the first eigenvalue of a Schrödinger operator.

THEOREM 1.21. *Let* Ω *be a relatively compact connected open subset of a Riemannian manifold* (M, g) *and* $L = \Delta + q$ *a Schrödinger operator on* Ω. *Then:*

(1) *If* $u \in V_{\lambda_k}$ *for some* $k \in \mathbb{N}$ *and* u *is positive on* Ω, *then* $\lambda_k = \lambda_1$.

(2) V_{λ_1} *has dimension one and there exists a positive function in* V_{λ_1}.

PROOF. **Proof of 1.** Given a function $f \in C_0^\infty(\Omega)$, then f can be written as $f = hu$ for some function h on Ω. Then, using that $\Delta u + qu + \lambda_k u = 0$ and Theorem 1.7,5, we get

$$\begin{aligned} Q(f) &= \int_\Omega \{\|\nabla f\|^2 - qf^2\}\, dv \\ &= \int_\Omega \{u^2\|\nabla h\|^2 + h^2\|\nabla u\|^2 + 2hug(\nabla u, \nabla h) - qu^2h^2\}\, dv \\ &= \int_\Omega \{u^2\|\nabla h\|^2 + \operatorname{div}(uh^2\nabla u) + \lambda_k h^2u^2\}\, dv \geq \lambda_k \|f\|_2^2. \end{aligned}$$

Hence $\lambda_k \leq \frac{Q(f)}{\|f\|_2^2}$ for any $f \in C_0^\infty(\Omega)$. Since $H_0^{1,2}(\Omega) = \overline{C_0^\infty(\Omega)}$ and Q is continuous, we get that $\lambda_k \leq \frac{Q(f)}{\|f\|_2^2}$ for any $f \in H_0^{1,2}(\Omega)$. Theorem 1.17 says that $\lambda_k = \lambda_1$, which proves 1.

Proof of 2. Now, let $u \in C^\infty(\Omega) \cap H_0^{1,2}(\Omega)$ be a nonzero eigenfunction of λ_1. Since $u \in H_0^{1,2}(\Omega)$, then $|u| \in H_0^{1,2}(\Omega)$ too (see, for example [**Aub82**]). Furthermore

$$Q(|u|) = \int_\Omega \{\|\nabla |u|\|^2 - qu^2\}\, dv = Q(u) - \int_{u^{-1}(\{0\})} \|\nabla u\|^2\, dv.$$

Since $\{p \in \Omega \mid u(p) = 0 \text{ and } (\nabla u)_p \neq 0\}$ is a set of measure zero, we get that $\int_{u^{-1}(\{0\})} \|\nabla u\|^2\, dv = 0$ and thus $Q(|u|) = Q(u) = \lambda_1\|u\|_2^2 = \lambda_1\||u|\|_2^2$. From Theorem 1.17 it follows that $|u| \in V_{\lambda_1}$ and thus $|u| \in C^\infty(\Omega)$.

If $L^* = L + \lambda_1$, we have that $L^*|u| = 0$ and $|u| \geq 0$, and so from Proposition 1.20 and using that $u \neq 0$, it follows $|u| > 0$. So any nonzero eigenfuction of λ_1 has no zeroes on Ω.

Finally, if $\dim V_{\lambda_1} \geq 2$, then we can find orthogonal nonzero functions u and $\widetilde{u}$ in V_{λ_1}. Since u and $\widetilde{u}$ have no zeroes, then $\int_\Omega u\widetilde{u}\, dv \neq 0$, which contradicts their orthogonality. So, $\dim V_{\lambda_1} = 1$. It is also clear that there is a positive function on V_{λ_1}. □

1.5. Riemann surfaces

A **Riemann surface** is a complex manifold M of complex dimension one. The maximal atlas in M with charts $z_\alpha : U_\alpha \to \mathbb{C}$ satisfying that

$$z_\beta \circ z_\alpha^{-1} : z_\alpha(U_\alpha \cap U_\beta) \to z_\beta(U_\alpha \cap U_\beta)$$

are holomorphic maps, is called the **conformal structure** of M. It is clear that the conformal structure of M defines a differentiable structure in M which becomes M in an orientable surface.

A Riemannian metric g on a Riemann surface M is called **conformal** if, for any complex local coordinate $z : U \to \mathbb{C}$, the metrics g and $z^*\langle,\rangle$ are conformal on U, where $\langle,\rangle$ is the Euclidean metric on $\mathbb{C}$. It is well known that every Riemann surface admits a conformal Riemannian metric. Also, if $g, \widehat{g}$ are two conformal Riemannian metrics on a Riemann surface M, then they are conformal, i.e., $\widehat{g} = e^{2u}g$, for some smooth function u on M.

On the other hand, if (M, g) is an orientable Riemannian surface, then around any point $p \in M$ there exist isothermal coordinates $\phi : U \to \mathbb{R}^2$, i.e., g and $\phi^*\langle,\rangle$ are conformal metrics on U, where $\langle,\rangle$ is the Euclidean metric on $\mathbb{R}^2$. Under these conditions, the set of isothermal neighborhoods defines a conformal structure on

M, so M is a Riemann surface and g is a conformal metric on M. Moreover, if $\widehat{g}$ is another Riemannian metric on M conformal to g, i.e., $\widehat{g} = e^{2u}g$, then $\widehat{g}$ defines the same conformal structure on M as g. These facts establish a one-to-one correspondence between the set of conformal structures of an orientable surface and the set of conformal classes of metrics on M.

A continuous map $F : M \to \widehat{M}$ between Riemann surfaces is called **holomorphic** if, in local coordinates $\{(U_\alpha, z_\alpha)\}$ on M and $\{(\hat{U}_\beta, w_\beta)\}$ in $\widehat{M}$, all the maps $w_\beta \circ F \circ z_\alpha^{-1}$ are holomorphic whenever they are defined. A **biholomorphism** $F : M \to \widehat{M}$ between Riemann surfaces is a homeomorphism such that F and F^{-1} are holomorphic maps. In such a case, we will say that M and $\widehat{M}$ are **conformally equivalent**.

A function $f : M \to \mathbb{C}$ on a Riemann surface is holomorphic if and only if $\frac{\partial f}{\partial \bar{z}} = 0$, where $\frac{\partial}{\partial z}$ and $\frac{\partial}{\partial \bar{z}}$ are the Cauchy-Riemann operators

$$\frac{\partial}{\partial z} = \frac{1}{2}\left(\frac{\partial}{\partial x} - i\frac{\partial}{\partial y}\right), \qquad \frac{\partial}{\partial \bar{z}} = \frac{1}{2}\left(\frac{\partial}{\partial x} + i\frac{\partial}{\partial y}\right),$$

where $z = x + iy$ any complex coordinate neighborhood on M.

The **complex line** $\mathbb{C}$ and its open subsets are Riemann surfaces. A conformal metric on $\mathbb{C}$ is the Euclidean metric. The **sphere** $\mathbb{S}^2$ is also a Riemann surface, and the standard metric on it is a conformal metric. A conformal metric on the Riemann surface $\mathbb{D} = \{z \in \mathbb{C} \mid \|z\| < 1\}$ (open subset of $\mathbb{C}$) is

$$g = \frac{4}{(1 - \|z\|^2)^2}\langle , \rangle,$$

where $\langle , \rangle$ is the Euclidean metric, and $(\mathbb{D}, g)$ is isometric to the hyperbolic space $\mathbb{H}^2$.

A crucial result in the theory of Riemann surfaces is the following:

THEOREM 1.22 (Uniformization theorem). *Let M be a simply connected Riemann surface. Then M is conformally equivalent to $\mathbb{S}^2$, $\mathbb{C}$, or the unit disk $\mathbb{D}$. Furthermore, every Riemann surface M is conformally equivalent to $\widehat{M}/G$, where $\widehat{M}$ is $\mathbb{S}^2$, $\mathbb{C}$, or $\mathbb{D}$, and G is a subgroup of the group of biholomorphisms of $\widehat{M}$ which acts properly discontinuous on $\widehat{M}$.*

Now, let $f : M \to \widehat{M}$ be a non-constant holomorphic map. For any point $p \in M$, we can find local coordinates around p and $f(p)$ such that the map f is given by

$$f(z) = z^n,$$

where n is a positive integer. The number n depends only on f and p. We say that p is a **ramification point** of f if $n > 1$. We call $n - 1$ the **order of ramification** of f at p, and it is denoted by $b_f(p)$.

Suppose $f : M \to \widehat{M}$ is a non-constant holomorphic map between compact Riemann surfaces M and $\widehat{M}$. Then, for any point $q \in \widehat{M}$, the integer number

$$d = \sum_{p \in f^{-1}(q)} (b_f(p) + 1)$$

is independent of q and it is called the **degree** of f. This means that f takes each value of $\widehat{M}$ exactly d times, where we have taken the multiplicities into account.

THEOREM 1.23 (Riemann-Hurwitz formula [**FK92**, **Jos06**]). *Let M and $\widehat{M}$ be compact Riemann surfaces of genera h and $\widehat{h}$, respectively. If $f : M \to \widehat{M}$ is a non-constant holomorphic map, then*

$$h - 1 = d(\widehat{h} - 1) + \frac{b(f)}{2},$$

where $b(f) = \sum_{p \in M} b_f(p)$ is the total number of ramification of f.

Let $\Pi : E \to M$ be a **holomorphic line bundle** over a Riemann surface M. If $h_\alpha : \Pi^{-1}(U_\alpha) \to U_\alpha \times \mathbb{C}$ and $h_\beta : \Pi^{-1}(U_\beta) \to U_\beta \times \mathbb{C}$ are local trivializations of the bundle with $U_\alpha \cap U_\beta \neq \emptyset$, then the corresponding transition functions $g_{\alpha,\beta}(p) =: U_\alpha \cap U_\beta \to \mathbb{C}^*$ defined by $(p, g_{\alpha,\beta}(p)) = (h_\alpha \circ h_\beta^{-1})(p, 1)$ are holomorphic maps satisfying $g_{\alpha\alpha} = 1$ and $g_{\alpha\beta} g_{\beta\gamma} = g_{\alpha\gamma}$. Up to isomorphisms of holomorphic line bundles, the transition functions determine the holomorphic line bundle. This means that if M is a Riemann surface, $\{U_\alpha\}$ is an open covering of M and $\{g_{\alpha,\beta} : U_\alpha \cap U_\beta \to \mathbb{C}^*\}$ a family of holomorphic maps satisfying $g_{\alpha\alpha} = 1$ and $g_{\alpha\beta} g_{\beta\gamma} = g_{\alpha\gamma}$, then, up to bundle isomorphisms there exists a unique holomorphic line bundle over M whose transition functions are $g_{\alpha,\beta}$.

Using local trivializations of the bundle, a holomorphic section σ of a holomorphic line bundle over M can be identified with a family $\{f_\alpha : U_\alpha \to \mathbb{C}\}$ of holomorphic functions such that on $U_\alpha \cap U_\beta$ we have $f_\alpha = g_{\alpha\beta} f_\beta$. In this context we define a **meromorphic section** of E as a family $\{f_\alpha : U_\alpha \to \overline{\mathbb{C}}\}$ of meromorphic functions such that on $U_\alpha \cap U_\beta$ we have $f_\alpha = g_{\alpha\beta} f_\beta$.

The **Picard group** of M, denoted by $\mathrm{Pic}(M)$, is the set of isomorphic classes of holomorphic line bundles over M. The product is given by

$$\begin{gathered} \mathrm{Pic}(M) \times \mathrm{Pic}(M) \to \mathrm{Pic}(M) \\ ([E], [F]) \mapsto [E \otimes F], \end{gathered}$$

where $E \otimes F$ is the tensorial product of E and F. The isomorphy class of the trivial bundle is the unit of the group and the inverse of $[E]$ is given by $[E^*]$, where E^* stands for the dual bundle of E.

Among the holomorphic line bundles of a Riemann surface M it is interesting to note the holomorphic tangent bundle TM and its dual TM^*. This bundle is called the **canonical bundle** of M and it is denoted by $K_M = TM^*$. The sections of the canonical bundle are the holomorphic 1-forms of M.

On the other hand, a **divisor** D of a Riemann surface is a map $D : M \to \mathbb{Z}$ such that for any compact subset $K \subset M$ the set $\{p \in K \mid D(p) \neq 0\}$ is finite. We write $D = \sum_{p \in M} D(p)p$. If M is compact, then $\deg(D) := \sum_{p \in M} D(p)$ is an integer called the degree of D.

If $\mathrm{Div}(M)$ denotes the set of the divisors of M, then $\mathrm{Div}(M)$ is a group defining the sum of two divisors by $(D_1 + D_2)(p) = D_1(p) + D_2(p)$.

Now, if $f : M \to \mathbb{S}^2 \equiv \overline{\mathbb{C}}$ is a meromorphic map, we define a divisor $(f) \in \mathrm{Div}(M)$ by $(f)(p) = 0$ if $f(p) \in \mathbb{C}^*$, $(f)(p) = k$ if p is a zero of f of order k, and $(f)(p) = -k$ if p is a pole of f of order k. We note that if M is compact, then $deg(f) = 0$ for any meromorphic map f on M. Similarly, if σ is a meromorphic section of a holomorphic line bundle E, then we can define a divisor (σ) of M.

If M is a compact Riemann surface, we will associate to each divisor D of M a holomorphic line bundle E_D in the following way. Let D be a divisor of M and $\{p_1, \ldots, p_n\}$ be the set of points where D is not zero. Let $\{(U_\alpha, z_\alpha)\}$ be a covering of M by complex neighborhoods such that p_i belongs only to a unique U_{α_i} with $z_{\alpha_i}(p_i) = 0$. If $U_\alpha \cap U_\beta \neq \emptyset$, we define $g_{\alpha\beta} : U_\alpha \cap U_\beta \to \mathbb{C}^*$ by

$$g_{\alpha\beta} = \frac{f_\alpha}{f_\beta},$$

where $f_{\alpha_i} = z_{\alpha_i}^{D(p_i)}$ for $1 \leq i \leq n$ and $f_\alpha = 1$ otherwise. Then there exists a holomorphic line bundle E_D with transition functions $\{g_{\alpha\beta}\}$. This defines a map

$$F : \mathrm{Div}(M) \to \mathrm{Pic}(M)$$
$$D \mapsto [E_D]$$

satisfying $F(D + D') = [E_D] \otimes [E_{D'}]$. Note that the family $\{f_\alpha\}$ defines a meromorphic section σ_D on E_D such that $(\sigma_D) = D$. Also, if E is a holomorphic line bundle over M, it is possible to choose a meromorphic section σ on E such that $[E_{(\sigma)}] = [E]$. So F is an epimorphism. Finally,

$$\ker F = \mathrm{Div}_0(M) := \{(f) \mid f \text{ is a meromorphic map on } M\}.$$

So F induces an isomorphism

$$\frac{\mathrm{Div}(M)}{\mathrm{Div}_0(M)} \equiv \mathrm{Pic}(M).$$

THEOREM 1.24 (Riemann-Roch theorem [**Jos06**]). *Let M be a compact Riemann surface of genus h and let D be a divisor of M. Then*

$$\dim_{\mathbb{C}} H^0(M, \mathcal{O}(E_D)) - \dim_{\mathbb{C}} H^0(M, \mathcal{O}(K \otimes (E_D)^*)) = \deg(D) + 1 - h,$$

where $H^0(M, \mathcal{O}(E))$ is the complex linear space of holomorphic sections of the holomorphic line bundle E and K is the canonical line bundle of M.

To conclude this section, we will define the first Chern class of a holomorphic line bundle on a compact Riemann surface. This will allow us to reformulate the Riemann-Roch theorem.

DEFINITION 1.25 ([**Jos06**]). Let E be a holomorphic line bundle over a compact Riemann surface M, with transition functions $\{g_{\alpha\beta}, U_\alpha\}$.

(1) A hermitian metric λ^2 on E is a collection of positive smooth real valued functions $\{\lambda_\alpha^2 : U_\alpha \to \mathbb{R}\}$ such that on $U_\alpha \cap U_\beta$ we have $\lambda_\beta^2 = \lambda_\alpha^2 g_{\alpha\beta} \overline{g_{\alpha\beta}}$.

(2) The first Chern form $c_1(E, \lambda^2)$ of a hermitian bundle E is defined as

$$c_1(E, \lambda^2)_{|U_\alpha} = \frac{1}{2\pi i} \frac{\partial^2}{\partial z \partial \overline{z}} (\log \lambda_\alpha^2)\, dz \wedge d\overline{z},$$

where z is a complex coordinate on U_α.

(3) The cohomology class defined by $c_1(E, \lambda^2)$ is independent of the choice of the hermitian metric on E. It is denoted by

$$c_1(E) \in H_{dR}^2(M)$$

and is called the first Chern class of E.

It is known, [**GH78**], that the first Chern class $c_1(E)$ is an integer cohomology class, i.e., $\int_M c_1(E) \in \mathbb{Z}$. We call this number the degree of E, $\deg E = \int_M c_1(E)$. If $E = E_D$ for a divisor D on M, then $\deg E_D = \deg D$.

Now we can reformulate the Riemann-Roch theorem as follows.

THEOREM 1.26 (Riemann-Roch theorem). *Let M be a compact Riemann surface of genus h and let E be a holomorphic line bundle over M. Then*

$$\dim_{\mathbb{C}} H^0(M, \mathcal{O}(E)) - \dim_{\mathbb{C}} H^0(M, \mathcal{O}(K \otimes E^*)) = \deg(E) + 1 - h.$$

1.6. Kähler manifolds

DEFINITION 1.27. A **Kähler manifold** is a Riemannian manifold of even dimension (M^{2n}, g) with a $(1,1)$-tensor J satisfying

$$J^2 = -Id, \quad g(Jv, Jw) = g(v, w), \quad \nabla J = 0,$$

for any $v, w \in TM$.

In a Kähler manifold M^{2n},

$$\omega(X, Y) := g(JX, Y), \quad X, Y \in \mathfrak{X}(M),$$

is a closed 2-form on M, called the **Kähler form**. Then, $\omega^n = \omega \wedge \cdots \wedge \omega$ is a $2n$-form on M without zeros. Thus M is orientable and ω^n is called the canonical orientation on M.

Also, the Nijenhuis tensor of M vanishes, and so by Newlander-Nirenberg's theorem, [**KN96**], M is a complex manifold of complex dimension n.

The curvature R and the Ricci tensor S of a Kähler manifold M have the following properties

$$R(X, Y)JZ = JR(X, Y)Z, \quad R(JX, JY)Z = R(X, Y)Z,$$
$$S(JX, JY) = S(X, Y),$$

for all vector fields $X, Y, Z \in \mathfrak{X}(M)$. Using these properties,

$$\eta(X, Y) = S(JX, Y), \quad X, Y \in \mathfrak{X}(M),$$

defines a 2-form on M, which is closed, and is called de **Ricci form** of M.

Let $K_M = \Lambda^n T^*M$ be the highest exterior power of the holomorphic cotangent bundle T^*M of M. Then K_M is a holomorphic line bundle over M, called the **canonical bundle** of M. Its sections are the holomorphic n-forms of M.

The **first Chern class** of the Kähler manifold M, $c_1(M) \in H^2_{dR}(M)$, is the first Chern class of the canonical bundle K_M (the generalization to higher dimension of Definition 1.25). It is well know, see [**GH78**] for details, that

$$[\eta] = 2\pi c_1(M),$$

where η is the Ricci form of M.

If $\Pi \subset T_pM$ is a plane invariant by the complex structure J, then the sectional curvature of Π is called the **holomorphic sectional curvature** of the J-invariant vectorial plane Π. It is clear that if $0 \neq v \in \Pi$ then

$$K(\Pi) = \frac{R(v, Jv, Jv, v)}{\|v\|^4}.$$

The Kähler manifold M has **constant holomorphic sectional curvature** $c \in \mathbb{R}$, if for any point $p \in M$ and any J-invariant vectorial plane Π of T_pM, the holomorphic sectional curvature $K(\Pi) = c$. The curvature operator R of such a Kähler manifold is given by

(1.7)
$$R(X,Y)Z = \frac{c}{4}\{g(Y,Z)X - g(X,Z)Y + g(JY,Z)JX - g(JX,Z)JY - 2g(JX,Y)JZ\}.$$

If a Kähler manifold M^{2n} has constant holomorphic sectional curvature c, then its Ricci tensor is given by

$$S = \frac{(n+1)c}{2} g,$$

and thus, M is an Einstein manifold.

The **complex Euclidean space** $\mathbb{C}^n \equiv \mathbb{R}^{2n}$ endowed with the Euclidean metric and the complex structure $Jv = iv$, for any $v \in \mathbb{C}^n$, is a flat Kähler manifold. In this case, the Kähler 2-form ω of $\mathbb{C}^n$ is exact and $\omega = \mathrm{d}\Lambda/2$, where Λ is the **Liouville 1-form** on $\mathbb{C}^n$ given by $\Lambda(v) = \langle v, Jp\rangle$, for all $v \in T_p\mathbb{C}^n$. The restriction of Λ to the sphere of radius r, $\mathbb{S}^{2n+1}(r) \subset \mathbb{C}^{n+1}$, is the contact 1-form of the canonical **Sasakian structure** on $\mathbb{S}^{2n+1}(r)$.

Let $\mathbb{S}^1 \times \mathbb{S}^{2n+1}(r) \to \mathbb{S}^{2n+1}(r)$ be the action of $\mathbb{S}^1$ on $\mathbb{S}^{2n+1}(r)$ given by

$$e^{i\theta}(z_1, \dots, z_{n+1}) = (e^{i\theta} z_1, \dots, e^{i\theta} z_{n+1}),$$

and $\Pi : \mathbb{S}^{2n+1}(r) \to \mathbb{CP}^n(4/r^2) := \mathbb{S}^{2n+1}(r)/\mathbb{S}^1$ the Hopf fibration of $\mathbb{S}^{2n+1}(r)$ on the **complex projective space** $\mathbb{CP}^n(4/r^2)$. Since Π is a submersion, we define the Fubini-Study metric g on $\mathbb{CP}^n(4/r^2)$ by

$$g(v, w) = \langle v^*, w^*\rangle, \quad \forall\, v, w \in T\mathbb{CP}^n(4/r^2),$$

where v^* and w^* are the horizontal lifts of v and w to $T\mathbb{S}^{2n+1}(r)$, respectively. The complex structure J on $\mathbb{CP}^n(4/r^2)$ is defined by

$$Jv = \mathrm{d}\Pi(iv^*), \quad \forall\, v \in T\mathbb{CP}^n(4/r^2).$$

Then $(\mathbb{CP}^n(4/r^2), g, J)$ is a complete (compact) Kähler manifold of constant holomorphic sectional curvature $4/r^2$.

Similarly, let $(,)$ be the hermitian form on $\mathbb{C}^{n+1}$ given by

$$(z, w) = \sum_{i=1}^{n} z_i \overline{w}_i - z_{n+1}\overline{w}_{n+1},$$

where $\overline{z}$ is the conjugate of z. If

$$\mathbb{H}_1^{2n+1} = \{z \in \mathbb{C}^{n+1} \mid (z, z) = -1/r^2\}$$

is the anti-de Sitter space, $\Re(,)$ induces on $\mathbb{H}_1^{2n+1}$ a Lorentzian metric of constant curvature $-1/r^2$, where $\Re$ denotes the real part. Let $\mathbb{S}^1 \times \mathbb{H}_1^{2n+1} \to \mathbb{H}_1^{2n+1}$ be the action of $\mathbb{S}^1$ on $\mathbb{H}_1^{2n+1}$ given by

$$e^{i\theta}(z_1, \dots, z_{n+1}) = (e^{i\theta} z_1, \dots, e^{i\theta} z_{n+1}),$$

and $\Pi : \mathbb{H}_1^{2n+1} \to \mathbb{CH}^n(-4/r^2) := \mathbb{H}_1^{2n+1}/\mathbb{S}^1$ the Hopf fibration of $\mathbb{H}_1^{2n+1}$ on the **complex hyperbolic space** $\mathbb{CH}^n(-4/r^2)$. We define a metric g on $\mathbb{CH}^n(-4/r^2)$ by

$$g(v, w) = \Re(v^*, w^*), \quad \forall\, v, w \in T\mathbb{CH}^n(-4/r^2),$$

where v^* and w^* are the horizontal lifts of v and w to $T\mathbb{H}_1^{2n+1}$, respectively. The complex structure J on $\mathbb{CH}^n(-4/r^2)$ is defined by

$$Jv = \mathrm{d}\Pi(iv^*), \quad \forall\, v \in T\mathbb{CH}^n(-4/r^2).$$

Then $(\mathbb{CH}^n(-4/r^2), g, J)$ is a complete Kähler manifold with constant holomorphic sectional curvature $-4/r^2$.

A result of J.-I. Igusa, [**Igu54**], says that any simply connected complete Kähler manifold of constant holomorphic sectional curvature c is holomorphically isometric to complex Euclidean space if $c = 0$, to complex projective space $\mathbb{CP}^n(4/r^2)$

if $c = 4/r^2$, or to complex hyperbolic space $\mathbb{CH}^n(-4/r^2)$ if $c = -4/r^2$. From a well-known result of J.L. Synge, a complete Kähler manifold of positive constant holomorphic sectional curvature is necessarily simply connected and thus holomorphically isometric to the complex projective space.

Kähler manifolds can also be characterized among the Riemannian manifolds by the behavior of their holonomy groups. In fact, the following result is known:

PROPOSITION 1.28. *Let (M^{2n}, g) be a Riemannian manifold of even dimension $2n$ and $\mathrm{Hol}(g) \subset \mathrm{O}(2n)$ its holonomy group. Then M is a Kähler manifold if and only if $\mathrm{Hol}\,(g) \subset \mathrm{U}(n)$, where $\mathrm{U}(n)$ is the unitary group.*

Among Kähler manifolds, it is worth mentioning an important family of them, the Calabi-Yau manifolds, since they admit parametric families of calibrations (Definition 2.16). In Chapter 6 we will study their calibrated submanifolds.

DEFINITION 1.29. A **Calabi-Yau manifold** is a Kähler manifold (M^n, g, J) of complex dimension n whose holonomy group $\mathrm{Hol}(g) \subset \mathrm{SU}(n)$, where $\mathrm{SU}(n)$ is the special unitary group.

The following characterization of the Calabi-Yau manifolds is important.

PROPOSITION 1.30. *Let (M^n, g, J) be a Kähler manifold of complex dimension n. Then M is Calabi-Yau if and only if M admits a non-trivial parallel holomorphic n-form. In particular, every Calabi-Yau manifold is Ricci-flat.*

CHAPTER 2

Variation Formulas. Stability. Index and Nullity

2.1. Introduction

In this chapter we will first obtain the Euler-Lagrange equation of the volume functional on immersions of a manifold into an arbitrary Riemannian manifold. The critical points of such a functional will be the minimal immersions. For minimal immersions we will obtain the second variation of the volume functional, which gives us the Jacobi operator of a minimal immersion and its corresponding index form. This allows us to define the stability of minimal submanifolds.

The next goal of the chapter will be to define the index and the nullity of any compact minimal submanifold as well as the index of any minimal hypersurface. In the first case the Jacobi operator acts on sections of the normal bundle of the minimal immersion, while in the second case we can reduce the problem to study a Jacobi operator acting on functions. This will be done not only when the normal bundle (which is of rank one) was trivial (two-sided hypersurfaces), but also when the normal bundle is not trivial (one-sided hypersurfaces).

In the last two sections we present some general properties of the index and nullity of minimal submanifolds. We will also compute the index and nullity of important examples of minimal submanifolds which will be characterized by the behavior of their index and nullity.

2.2. Submanifolds

Let $(\overline{M}, g)$ be an m-dimensional Riemannian manifold and $\Phi : M \to \overline{M}$ an immersion of an n-dimensional manifold M into $\overline{M}$. The induced metric $\Phi^*(g)$ on M is again denoted by g (unless it might lead to confusion). If $T^\perp M$ is the **normal bundle** of Φ, then the pullback by Φ of the tangent bundle to $\overline{M}$ decomposes orthogonally into

$$\Phi^* T\overline{M} = TM \oplus T^\perp M.$$

The metric g endowes a metric on $T^\perp M$, which will also be denoted by g. We will denote by $\mathfrak{X}^\perp(M)$ the linear space of normal vector fields along Φ.

Let $\overline{\nabla}$ be the connection on $\Phi^* T\overline{M}$ induced by the Levi-Civita connection of $(\overline{M}, g)$ and let $\overline{\nabla} = \nabla + \nabla^\perp$ be the corresponding decomposition, where ∇ is the Levi-Civita connection of the induced metric on M and $\nabla^\perp$ is the **normal connection** on $T^\perp M$. Then the **Gauss and Weingarten formulas** are given by

$$\begin{aligned} \overline{\nabla}_X Y &= \Phi_*(\nabla_X Y) + \sigma(X, Y) \\ \overline{\nabla}_X \xi &= -\Phi_*(A_\xi X) + \nabla^\perp_X \xi, \end{aligned} \tag{2.1}$$

where $X, Y \in \mathfrak{X}(M)$, $\xi \in \mathfrak{X}^\perp(M)$ and σ (respectively A_ξ) denotes the **second fundamental form** of the immersion Φ (respectively the **Weingarten endomorphism** associated to a normal vector field ξ). It is interesting to note the relationship between σ and A_ξ, as well as that $\nabla^\perp$ is a connection on $T^\perp M$ compatible with the metric induced on $T^\perp M$,

$$g(A_\xi X, Y) = g(\sigma(X,Y),\xi), \quad X(g(\xi,\eta)) = g(\nabla^\perp_X\xi,\eta) + g(\xi,\nabla^\perp_X\eta).$$

The immersion Φ is called **totally geodesic** if $\sigma = 0$. The **mean curvature vector** of the immersion Φ is the normal vector field $\vec{H}$ defined by

$$\vec{H} = \frac{1}{n}\sum_{i=1}^{n}\sigma(e_i,e_i),$$

where $\{e_1,\ldots,e_n\}$ is an orthonormal reference on TM.

Let $R^\perp$ be the **normal curvature operator**, defined by

$$R^\perp(X,Y)\xi = \nabla^\perp_X\nabla^\perp_Y\xi - \nabla^\perp_Y\nabla^\perp_X\xi - \nabla^\perp_{[X,Y]}\xi,$$

for all $X, Y \in \mathfrak{X}(M)$ and $\xi \in \mathfrak{X}^\perp(M)$. If $\overline{R}$ (respectively R) is the curvature operator of $\overline{M}$ (respectively M), then the **Gauss, Weingarten and Ricci equations** of the immersion Φ are respectively given by

$$\begin{aligned}
&\overline{R}(V,W,X,Y) = R(V,W,X,Y) - g(\sigma(V,Y),\sigma(W,X)) + g(\sigma(V,X),\sigma(W,Y)),\\
&\overline{R}(V,W,X,\xi) = g((\nabla\sigma)(V,W,X),\xi) - g((\nabla\sigma)(W,V,X),\xi),\\
&\overline{R}(V,W,\xi,\eta) = R^\perp(V,W,\xi,\eta) - g([A_\xi,A_\eta]V,W),
\end{aligned}\tag{2.2}$$

for all $V, W, X, Y \in \mathfrak{X}(M)$ and $\xi,\eta \in \mathfrak{X}^\perp(M)$, where $[A_\xi, A_\eta]$ is the commutator and $\nabla\sigma$ is the covariant derivative of the second fundamental form defined by

$$(\nabla\sigma)(V,W,X) = \nabla^\perp_V\sigma(W,X) - \sigma(\nabla_V W,X) - \sigma(W,\nabla_V X).$$

An important special case is when M is a **hypersurface** of $\overline{M}$, i.e., when Φ is an immersion with $m = n+1$. In this case, the normal bundle is a rank-one vector bundle over M, and this fact suggests the following definition.

DEFINITION 2.1. A hypersurface $\Phi : M^n \to (\overline{M}^{n+1}, g)$ is called **two-sided** if the normal bundle $T^\perp M$ is trivial. Since the normal bundle has rank one, this means that there exists a smooth unit normal vector field to M, which will be denoted by N. In the same way, we will say that Φ is an **one-sided** immersion if the normal bundle is not trivial.

The following result is easy to prove :

LEMMA 2.2. *Let $(\overline{M}, g)$ be an $(n+1)$-dimensional orientable Riemannian manifold and $\Phi : M^n \to \overline{M}$ be a hypersurface. Then Φ is two-sided if and only if M is an orientable manifold.*

PROOF. Suppose that Φ is two-sided and let N be a unit normal vector field on M. We define an n-form ω on M by

$$\omega_p(v_1,\ldots,v_n) = \eta_{\phi(p)}(N_p, \Phi_*(v_1),\ldots,\Phi_*(v_n)), \quad \forall p \in M,$$

where $\eta \in \Omega^{n+1}(\overline{M})$ is an orientation on $\overline{M}$. It is clear that ω have no zeros on M, so M is orientable.

Conversely, if $\eta \in \Omega^{n+1}(\overline{M})$ and $\omega \in \Omega^n(M)$ are the volume forms on the oriented Riemannian manifolds $\overline{M}$ and M respectively, then for each point $p \in M$ we define N_p as the unique unit normal vector at p such that

$$\eta_{\phi(p)}(N_p, \Phi_*(e_1),\ldots,\Phi_*(e_n)) = 1,$$

where $\{e_1, \dots, e_n\}$ is a positively oriented orthonormal basis of T_pM, i.e., the volume form $\omega_p(e_1, \dots, e_n) = 1$. The unit normal vector field N trivializes the normal bundle, so Φ is a two-sided hypersurface. □

In the two-sided case, the unit normal vector field N is parallel, $\nabla^\perp N = 0$. Also, the mean curvature vector $\vec{H}$ defines a function H on M by $\vec{H} = HN$, which is called the **mean curvature** of Φ.

When $(\overline{M}, g)$ is nonorientable the situation is quite different, as the following examples show.

EXAMPLE 2.3.

(1) *Let* $\Phi : \mathbb{S}^{2n+1} \to (\mathbb{S}^{2n+2}, g)$ *be the totally geodesic embedding given by* $\Phi(x) = (x, 0)$ *for any* $x \in \mathbb{S}^{2n+1}$. *Then* Φ *is two-sided and a unit normal vector field to* Φ *is given by* $N_p = (0, \dots, 0, 1)$, $\forall p \in \mathbb{S}^{2n+1}$. *It is clear that* Φ *is invariant under the actions of the antipodal maps on* $\mathbb{S}^{2n+1}$ *and* $\mathbb{S}^{2n+2}$, *and so* Φ *induces a totally geodesic embedding* $\Psi : \mathbb{RP}^{2n+1} \to \mathbb{RP}^{2n+2}$. *Then* Ψ *is a one-sided hypersurface of the orientable manifold* $\mathbb{RP}^{2n+1}$ *into the nonorientable manifold* $\mathbb{RP}^{2n+2}$. *We claim that* Φ *is one-sided.*

In fact, if Ψ *is two-sided, then its normal bundle is trivial and so there exists a normal vector field* V *to* Ψ *with* $V_x \neq 0$, $\forall x \in \mathbb{RP}^{2n+1}$. *This vector field can be lifted to a smooth normal vector field* $\widetilde{V}$ *to* $\mathbb{S}^{2n+1}$ *by*

$$\widetilde{V}_p = \left(d\Pi_{\Phi(p)}\right)^{-1}(V_{\widetilde{\Pi}(p)}), \quad \forall p \in \mathbb{S}^{2n+1},$$

where $\widetilde{\Pi} : \mathbb{S}^{2n+1} \to \mathbb{RP}^{2n+1}$ *and* $\Pi : \mathbb{S}^{2n+2} \to \mathbb{RP}^{2n+2}$ *are the canonical projections. So* $\widetilde{V} = fN$ *with* f *a smooth function without zeros. But this is impossible because the vector field* $\widetilde{V}$ *satisfies* $\widetilde{V}_{-p} = -\widetilde{V}_p$, $\forall p \in \mathbb{S}^{2n+1}$, *and so* $f(-p) = -f(p)$, $\forall p \in \mathbb{S}^{2n+1}$, *which is a contradiction.*

(2) *Let* $\Psi \times Id : \mathbb{RP}^{2n+1} \times \mathbb{RP}^2 \to \mathbb{RP}^{2n+2} \times \mathbb{RP}^2$ *be the product of the above embedding and the identity map of* $\mathbb{RP}^2$. *Then, reasoning as in 1,* $\Psi \times Id$ *is a one-sided embedding of the nonorientable manifold* $\mathbb{RP}^{2n+1} \times \mathbb{RP}^2$ *into the nonorientable manifold* $\mathbb{RP}^{2n+2} \times \mathbb{RP}^2$.

(3) *Let* $\Psi : \mathbb{RP}^{2n} \to (\mathbb{RP}^{2n} \times \mathbb{R}, g \times dt^2)$ *be the embedding given by* $\Psi(x) = (x, t_0)$ *with* $t_0 \in \mathbb{R}$. *The unit vector field* $N(x) = (0, \frac{\partial}{\partial t})$, $\forall x \in \mathbb{RP}^{2n}$ *is a trivialization of the normal bundle of* Ψ. *So* Ψ *is a two-sided embedding of the nonorientable manifold* $\mathbb{RP}^{2n}$ *into the nonorientable manifold* $\mathbb{RP}^{2n} \times \mathbb{R}$.

(4) *If* $\Pi : \mathbb{S}^{2n} \to \mathbb{RP}^{2n}$ *is the canonical projection and* Ψ *the embedding given in 3, then* $\Psi \circ \Pi : \mathbb{S}^{2n} \to \mathbb{RP}^{2n} \times \mathbb{R}$ *is a two-sided hypersurface of the orientable manifold* $\mathbb{S}^{2n}$ *into the nonorientable manifold* $\mathbb{RP}^{2n} \times \mathbb{R}$.

2.3. First variation of the volume. Minimal submanifolds

DEFINITION 2.4. A **compactly supported variation** of an immersion $\Phi : M \to \overline{M}$ is a smooth map

$$F : (-\varepsilon, \varepsilon) \times M \to \overline{M}$$

which satisfies the following conditions:

(1) Each map $\Phi_t = F(t, -) : M \to \overline{M}$ is an immersion.
(2) $\Phi_0 = \Phi$.

(3) The variational vector field $X = \frac{\partial F}{\partial t}\big|_{t=0}$, which is a section of $\Phi^*T\overline{M}$, has compact support on M.

LEMMA 2.5. *Let $\Phi : M \to (\overline{M}, g)$ be an immersion and X be a compactly supported section of $\Phi^*T\overline{M}$. Then there exists a compactly supported variation F of the immersion Φ whose variational vector field is X.*

PROOF. If $C \subset M$ is the support of X, we define $F : (-\epsilon, \epsilon) \times M \to \overline{M}$ by

$$F(t,p) = \begin{cases} \exp(\Phi(p), tX_p) & \text{on } C \\ \Phi(p) & \text{on } M \setminus C \end{cases}$$

where exp is the exponential map in $(\overline{M}, g)$. Note that since C is compact, it is possible to choose ϵ such that for all $p \in C$ the set $\{\exp(\Phi(p), tX_p) \mid t \in (-\epsilon, \epsilon)\}$ is contained in a geodesic ball of $\overline{M}$ around $\Phi(p)$.

Now it is easy to check that F satisfies the properties of the lemma. □

Under these conditions, if F is a variation of the immersion $\Phi : M \to \overline{M}$, we define the volume function of the variation $V_F : (-\varepsilon, \varepsilon) \to \mathbb{R}$ by

$$V_F(t) = \operatorname{vol}(\Omega, g_t) := \int_\Omega \mathrm{d}v_t \tag{2.3}$$

where g_t is the metric induced on M by the immersion Φ_t, $\mathrm{d}v_t$ is the Riemannian measure associated to g_t, and Ω is a relatively compact domain of M containing the support of X. Although V_F also depends of Ω, its derivative with respect to t, at $t = 0$, does not. In fact:

THEOREM 2.6 (First variation formula). *If $\Phi : M^n \to (\overline{M}^m, g)$ is an immersion and F is a compactly supported variation of Φ with variational vector field X, then we have that*

$$\left.\frac{\mathrm{d}V_F(t)}{\mathrm{d}t}\right|_{t=0} = -n \int_M g(X, \vec{H})\,\mathrm{d}v,$$

where $\vec{H}$ is the mean curvature vector of Φ and V_F the volume function defined in (2.3).

PROOF. Following the notation used above, we consider the identity map $I_t : (M, g) \to (M, g_t)$. From the change of variable formula, Theorem 1.7,3, we get that

$$V_F(t) = \int_\Omega \mathrm{d}v_t = \int_\Omega |\operatorname{Jac} I_t|\,\mathrm{d}v = \int_\Omega \sqrt{\det g_t}\,\mathrm{d}v,$$

where $(g_t)_{ij}(p) = g_t(e_i, e_j)$, $1 \le i, j \le n$ and $\{e_1, \ldots, e_n\}$ is an orthonormal basis of (T_pM, g). Using Theorem 1.7,4 again, we have

$$V_F'(t) = \int_\Omega (\sqrt{\det g_t})'\,\mathrm{d}v = \frac{1}{2}\int_\Omega \sqrt{\det g_t}\,\operatorname{trace}(g_t'(g_t)^{-1})\,\mathrm{d}v. \tag{2.4}$$

Now we will calculate g_t'. Given any point $p \in \Omega$, let $\{E_1, \ldots, E_n\}$ be a local orthonormal reference around p such that $(E_i)_p = e_i$ and $(\nabla_{E_i}E_j)_p = 0$. We denote $\widetilde{E}_i = (\Phi_t)_* E_i$. Then, since $[\frac{\partial}{\partial t}, E_i] = 0$ in $(-\epsilon, \epsilon) \times M$, we have

$$\begin{aligned}(g_t)'_{ij}(t,p) &= \frac{\mathrm{d}}{\mathrm{d}t} g(\widetilde{E}_i, \widetilde{E}_j)(t,p) = g\left(\overline{\nabla}_{\frac{\partial F}{\partial t}}\widetilde{E}_i, \widetilde{E}_j\right)(t,p) + g\left(\overline{\nabla}_{\frac{\partial F}{\partial t}}\widetilde{E}_j, \widetilde{E}_i\right)(t,p) \\ &= g\left(\overline{\nabla}_{\widetilde{E}_i}\frac{\partial F}{\partial t}, \widetilde{E}_j\right)(t,p) + g\left(\overline{\nabla}_{\widetilde{E}_j}\frac{\partial F}{\partial t}, \widetilde{E}_i\right)(t,p).\end{aligned}$$

Therefore

$$(2.5)\quad (\operatorname{trace} g_t')(t,p) = 2\sum_{i=1}^{n} g\left(\nabla_{\widetilde{E}_i}\frac{\partial F}{\partial t}, \widetilde{E}_i\right)(t,p)$$
$$= 2\sum_{i=1}^{n} E_i\left(g\left(\frac{\partial F}{\partial t}, \widetilde{E}_i\right)\right)(t,p) - 2\sum_{i=1}^{n} g\left(\frac{\partial F}{\partial t}, \nabla_{\widetilde{E}_i}\widetilde{E}_i\right)(t,p).$$

Now, if V^t is the vector field on M defined by $V^t(p) = (\frac{\partial F}{\partial t})^\top(t,p)$, where $\top$ stands for the tangent component, we have that

$$g\left(\frac{\partial F}{\partial t}, \widetilde{E}_i\right)(t,p) = g((\Phi_t)_*(V^t), (\Phi_t)_*E_i))(t,p) = g_t(V^t, E_i)(p).$$

So, if α_t is the 1-form on M defined by $\alpha_t(Y) = g_t(V^t, Y)$, we have that the codifferential of α with respect to the metric g is

$$\delta\alpha_t(p) = \sum_{i=1}^{n}\{E_i(g_t(V^t, E_i)) - g_t(V^t, \nabla_{E_i}E_i)\}(p) = \sum_{i=1}^{n} E_i(g_t(V^t, E_i))(p).$$

Now we get from (2.5) that

$$(2.6)\qquad (\operatorname{trace} g_t')(t,p) = 2\delta\alpha_t(p) - 2\sum_{i=1}^{n} g\left(\frac{\partial F}{\partial t}, \nabla_{\widetilde{E}_i}\widetilde{E}_i\right)(t,p),$$

and so, since $g_0(p) = Id$, from (2.4) and Theorem 1.15 it follows that

$$V_F'(0) = \frac{1}{2}\int_\Omega \operatorname{trace} g_t'(0)\, \mathrm{d}v = \int_\Omega (\delta\alpha_0 - ng(X, \vec{H}))\, \mathrm{d}v$$
$$= -n\int_\Omega g(X, \vec{H})\, dv = -n\int_M g(X, \vec{H})\, \mathrm{d}v. \qquad \square$$

DEFINITION 2.7. An immersion $\Phi : M^n \to (\overline{M}^m, g)$ is called **minimal** if it is a critical point of the volume functional, i.e., if $V_F'(0) = 0$ for any compactly supported variation F of Φ. From Theorem 2.6, it follows that Φ is a minimal immersion if and only if its mean curvature vector vanishes $\vec{H} = 0$.

2.4. Second variation of the volume. Jacobi operator. Stability

We begin this section by studying the second derivative of the volume functional at a critical point, that is, for a minimal immersion.

THEOREM 2.8 (Second variation formula). *Let $\Phi : M^n \to (\overline{M}^m, g)$ be a minimal immersion and $F : (-\epsilon, \epsilon) \times M \to \overline{M}$ be a compactly supported variation of Φ with variational vector field X. Then*

$$\left.\frac{\mathrm{d}^2 V_F(t)}{\mathrm{d}t^2}\right|_{t=0} = \int_M \left\{\|\nabla^\perp\xi\|^2 - \|A_\xi\|^2 - \sum_{i=1}^{n}\overline{R}(\xi, E_i, E_i, \xi)\right\}\mathrm{d}v,$$

where ξ is the normal component of X and $\{E_1, \dots, E_n\}$ is a local orthonormal reference on M.

PROOF. From (2.4) and using again Theorem 1.7,4 we have that

$$V_F''(0) = \frac{1}{2}\int_\Omega (\sqrt{\det g_t}\ \operatorname{trace}(g_t'(g_t)^{-1}))'(0)\, \mathrm{d}v.$$

Now, using that $g_0(p) = Id$, we get

$$\left(\sqrt{\det g_t}\ \mathrm{trace}(g_t'(g_t)^{-1})\right)'(0) = \frac{1}{2}(\mathrm{trace}\, g_t'(0))^2 + \mathrm{trace}(g_t''(0) - (g_t'(0))^2).$$

If V is the tangential component of the vector field X, then by the proof of Theorem 2.6 we have that $\mathrm{trace}\, g_t'(0) = 2\delta\alpha_0 - 2ng(\vec{H}, X) = 2\delta\alpha_0$. So we get that

$$V_F''(0) = \frac{1}{2}\int_\Omega \{2(\delta\alpha_0)^2 + \mathrm{trace}(g_t''(0) - (g_t'(0))^2)\}\, dv. \tag{2.7}$$

We will calculate the integrands of (2.7). From (2.6) we get that

$$\begin{aligned}
\mathrm{trace}\, g_t''(0) &= 2\frac{\mathrm{d}}{\mathrm{d}t}_{|t=0}\delta\alpha_t - 2\sum_{i=1}^n g(\nabla_{\frac{\partial F}{\partial t}}\frac{\partial F}{\partial t}, \nabla_{\widetilde{E}_i}\widetilde{E}_i)(0) - 2\sum_{i=1}^n g(\frac{\partial F}{\partial t}, \nabla_{\frac{\partial F}{\partial t}}\nabla_{\widetilde{E}_i}\widetilde{E}_i)(0)\\
&= 2\frac{\mathrm{d}}{\mathrm{d}t}_{|t=0}\delta\alpha_t - 2\sum_{i=1}^n \overline{R}(X, E_i, E_i, X) - 2\sum_{i=1}^n g(\frac{\partial F}{\partial t}, \overline{\nabla}_{\tilde{E}_i}\overline{\nabla}_{\frac{\partial F}{\partial t}}\tilde{E}_i)(0)\\
&= 2\frac{\mathrm{d}}{\mathrm{d}t}_{|t=0}\delta\alpha_t - 2\sum_{i=1}^n \overline{R}(X, E_i, E_i, X) - 2\sum_{i=1}^n E_i(g(X, \overline{\nabla}_{E_i}X) + 2\sum_{i=1}^n g(\overline{\nabla}_{E_i}X, \overline{\nabla}_{E_i}X)\\
&= 2\frac{\mathrm{d}}{\mathrm{d}t}_{|t=0}\delta\alpha_t - 2\sum_{i=1}^n \overline{R}(X, E_i, E_i, X) - \Delta\|X\|^2 + 2\sum_{i=1}^n g(\overline{\nabla}_{E_i}X, \overline{\nabla}_{E_i}X).
\end{aligned}$$

If $X = V + \xi$ is the decomposition of X into tangential and normal components, then using (2.1)

$$\overline{\nabla}_{E_i}X = \nabla_{E_i}V - A_\xi E_i + \sigma(V, E_i) + \nabla^\perp_{E_i}\xi,$$

and so

$$\begin{aligned}
\sum_{i=1}^n g(\overline{\nabla}_{E_i}X, \overline{\nabla}_{E_i}X) &= \|\nabla V\|^2 + \|A_\xi\|^2 + \|\nabla^\perp\xi\|^2\\
&\quad + \sum_{1=1}^n \{\|\sigma(V, E_i)\|^2 - 2g(\nabla_{E_i}V, A_\xi E_i) + 2g(\sigma(V, E_i), \nabla^\perp_{E_i}\xi)\}.
\end{aligned} \tag{2.8}$$

Since Φ is a minimal immersion, we have from the Gauss and Codazzi equations (2.2) that

(2.9)

$$\begin{aligned}
\sum_{i=1}^n \overline{R}(X, E_i, E_i, X) &= \sum_{i=1}^n \{\overline{R}(V, E_i, E_i, V) + \overline{R}(\xi, E_i, E_i, \xi) + 2\overline{R}(V, E_i, E_i, \xi)\}\\
&= \mathrm{Ricc}(V) + \sum_{i=1}^n \{\|\sigma(V, E_i)\|^2 + \overline{R}(\xi, E_i, E_i, \xi) - 2g(\nabla\sigma)(E_i, V, E_i), \xi)\}\\
&= \mathrm{Ricc}(V) + \sum_{i=1}^n \{\|\sigma(V, E_i)\|^2 + \overline{R}(\xi, E_i, E_i, \xi)\} - 2\,\mathrm{div}(A_\xi V)\\
&\quad + 2\sum_{i=1}^n \{g(\nabla_{E_i}V, A_\xi E_i) + g(\sigma(V, E_i), \nabla^\perp_{E_i}\xi)\}
\end{aligned}$$

and so, taking into account (2.8) and (2.9), we get that

$$\begin{aligned}\operatorname{trace} g_t''(0) = {} & 2\frac{\mathrm{d}}{\mathrm{d}t}_{|t=0}\delta\alpha_t - 2\operatorname{Ricc}(V) - \Delta\|X\|^2 \\ & + 2\|\nabla V\|^2 + 2\|A_\xi\|^2 + 2\|\nabla^\perp\xi\|^2 + 4\operatorname{div}(A_\xi V) \\ & - 2\sum_{1=1}^{n}\{4g(\nabla_{E_i}V, A_\xi E_i) + \overline{R}(\xi, E_i, E_i, \xi)\}.\end{aligned} \tag{2.10}$$

On the other hand, considering that $X = V + \xi$ and (2.1), we have

$$\begin{aligned}\operatorname{trace}(g_t'(0))^2 &= \sum_{i,j=1}^{n}(g_t'(0))_{ij}(g_t'(0))_{ji} = \sum_{i,j=1}^{n}\left(g(\overline{\nabla}_{\tilde{E}_i}\frac{\partial F}{\partial t}, \tilde{E}_j) + g(\overline{\nabla}_{\tilde{E}_j}\frac{\partial F}{\partial t}, \tilde{E}_i)\right)^2(0) \\ &= \sum_{i,j=1}^{n}\left(g(\overline{\nabla}_{E_i}X, E_j) + g(\overline{\nabla}_{E_j}X, E_i)\right)^2 \\ &= \sum_{i,j=1}^{n}\left(g(\nabla_{E_i}V, E_j) + g(\nabla_{E_j}V, E_i) - 2g(A_\xi E_i, E_j)\right)^2 \\ &= 2\|\nabla V\|^2 + 4\|A_\xi\|^2 - 8\sum_{i=1}^{n} g(\nabla_{E_i}V, A_\xi E_i) + 2\sum_{i=1}^{n} g(\nabla_{\nabla_{E_i}V}V, E_i).\end{aligned}$$

From (2.10) it follows that

$$\begin{aligned}\operatorname{trace}(g_t''(0) - (g_t'(0))^2) = {} & 2\frac{\mathrm{d}}{\mathrm{d}t}_{|t=0}\delta\alpha_t - 2\operatorname{Ricc}(V) - \Delta\|X\|^2 + 2\|\nabla^\perp\xi\|^2 - 2\|A_\xi\|^2 \\ & + 4\operatorname{div}(A_\xi V) - 2\sum_{i=1}^{n}\{g(\nabla_{\nabla_{E_i}V}V, E_i) + \overline{R}(\xi, E_i, E_i, \xi)\}.\end{aligned}$$

On the other hand, the Bochner formula tells us that

$$\operatorname{div}\left(\nabla_V V - (\operatorname{div} V)V\right) = S(V, V) - (\operatorname{div} V)^2 + \sum_{i=1}^{n} g(\nabla_{\nabla_{E_i}V}V, E_i),$$

and so we get that

(2.11)

$$\begin{aligned}\operatorname{trace}(g_t''(0) - (g_t'(0))^2) = {} & 2\frac{\mathrm{d}}{\mathrm{d}t}_{|t=0}\delta\alpha_t - \Delta\|X\|^2 + 2\|\nabla^\perp\xi\|^2 \\ & - 2\|A_\xi\|^2 - 2\sum_{i=1}^{n}\overline{R}(\xi, E_i, E_i, \xi) \\ & - 2(\operatorname{div} V)^2 - 2\operatorname{div}\left(\nabla_V V - (\operatorname{div}(V)V - 2A_\xi V\right).\end{aligned}$$

Now, since $\alpha_0 = V^\flat$, then $\delta\alpha_0 = \operatorname{div}(V)$, and so, from (2.7) and (2.11) it follows

$$\begin{aligned}V_F''(0) = {} & \int_\Omega \left\{\|\nabla^\perp\xi\|^2 - \|A_\xi\|^2 - \sum_{i=1}^{n}\overline{R}(\xi, E_i, E_i, \xi)\right\}\mathrm{d}v \\ & + \int_\Omega \left\{\frac{\mathrm{d}}{\mathrm{d}t}_{|t=0}\delta\alpha_t - \frac{\Delta\|X\|^2}{2} - \operatorname{div}\left(\nabla_V V - (\operatorname{div} V)V - 2A_\xi V\right)\right\}\mathrm{d}v.\end{aligned}$$

Since V is a vector field with compact support, the result follows from Theorems 1.7,4,5 and 1.15. □

COROLLARY 2.9 (Second variation formula). *Let $\Phi : M^n \to (\overline{M}^m, g)$ be a minimal immersion and $F : (-\epsilon, \epsilon) \times M \to \overline{M}$ be a compactly supported variation of Φ with variational vector field X. Then*

$$\left.\frac{\mathrm{d}^2 V_F(t)}{\mathrm{d}t^2}\right|_{t=0} = -\int_M g(\Delta^\perp \xi + \mathcal{A}\xi + \mathcal{R}\xi, \xi)\, \mathrm{d}v,$$

where $\Delta^\perp : \mathfrak{X}^\perp(M) \to \mathfrak{X}^\perp(M)$ is the ***normal Laplacian*** *defined by*

$$\Delta^\perp = \sum_{i=1}^n \{\nabla^\perp_{E_i}\nabla^\perp_{E_i} - \nabla^\perp_{\nabla_{E_i}E_i}\},$$

$\mathcal{A}$ and $\mathcal{R}$ are endomorphisms of $\mathfrak{X}^\perp(M)$ defined by

$$\mathcal{A}\xi = \sum_{i=1}^n \sigma(A_\xi E_i, E_i) \quad , \quad \mathcal{R}\xi = \sum_{i=1}^n (\overline{R}(\xi, E_i), E_i)^\perp$$

($\perp$ means normal component), where $\{E_1, \ldots, E_n\}$ is a local orthonormal reference on M, and ξ is the normal component of X.

PROOF. Since ξ also has compact support, i.e., $\xi \in \mathfrak{X}^\perp_0(M)$, we define a 1-form α on M by

$$\alpha(v) = g(\nabla^\perp_v \xi, \xi), \quad v \in TM.$$

Then the codifferential of α is

$$\delta\alpha = E_i(\alpha(E_i)) - \alpha(\nabla_{E_i}E_i) = \|\nabla^\perp \xi\|^2 + g(\Delta^\perp \xi, \xi),$$

where $\{E_1, \ldots, E_n\}$ is an orthonormal reference on M.

Since ξ also has compact support, the divergence theorem (Theorem 1.15) implies that

$$V_F''(0) = -\int_M g(\Delta^\perp \xi + \mathcal{A}\xi + \mathcal{R}\xi, \xi)\, \mathrm{d}v,$$

which proves the corollary. □

This new expression of the second derivative of the volume functional of a minimal immersion suggests the following definition.

DEFINITION 2.10. Let $\Phi : M^n \to (\overline{M}^m, g)$ be a minimal immersion. The **Jacobi operator** associated with Φ is the operator $\mathcal{L} : \mathfrak{X}^\perp(M) \to \mathfrak{X}^\perp(M)$ defined by

$$\mathcal{L}\xi = \Delta^\perp \xi + \mathcal{A}\xi + \mathcal{R}\xi.$$

If $\xi, \eta \in \mathfrak{X}^\perp_0(M)$, then we define $\alpha \in \Omega^1_0(M)$ by

$$\alpha(v) = g(\nabla^\perp_v \xi, \eta) - g(\xi, \nabla^\perp_v \eta), \quad \forall v \in TM.$$

Then we have $\delta\alpha = g(\Delta^\perp \xi, \eta) - g(\xi, \Delta^\perp \eta)$ and so

$$\int_M g(\Delta^\perp \xi, \eta)\, \mathrm{d}v = \int_M g(\xi, \Delta^\perp \eta)\, \mathrm{d}v.$$

This implies that the Jacobi operator is self-adjoint and that the corresponding symmetric bilinear form $\mathcal{Q} : \mathfrak{X}^\perp_0(M) \times \mathfrak{X}^\perp_0(M) \to \mathbb{R}$ is given by

$$\begin{aligned}\mathcal{Q}(\xi, \eta) &= -\int_M g(\mathcal{L}\xi, \eta)\, \mathrm{d}v \\ &= \int_M \Big\{ g(\nabla^\perp \xi, \nabla^\perp \eta) - g(A_\xi, A_\eta) - \sum_{i=1}^n \overline{R}(\xi, E_i, E_i, \eta) \Big\}\, \mathrm{d}v.\end{aligned} \tag{2.12}$$

$\mathcal{Q}$ will also denote the quadratic form $\mathcal{Q} : \mathfrak{X}_0^\perp(M) \to \mathbb{R}$ defined by $\mathcal{Q}(\xi) = \mathcal{Q}(\xi, \xi)$. This quadratic form will be called the **index form** of Φ.

If Φ is a hypersurface, i.e., $m = n + 1$, the index form of Φ is given by

$$\mathcal{Q}(\xi) = \int_M \{\|\nabla^\perp \xi\|^2 - \|A_\xi\|^2 - \overline{\mathrm{Ricc}}(\xi)\}\, dv. \tag{2.13}$$

DEFINITION 2.11. If $\Phi : M \to (\overline{M}, g)$ is a minimal immersion, then a **Jacobi vector field** on M is a normal vector field $\xi \in \mathfrak{X}^\perp(M)$ such that $\mathcal{L}\xi = 0$. The space of Jacobi fields on M, i.e., the kernel of $\mathcal{L}$ is denoted by

$$\mathcal{J}(M) = \{\xi \in \mathfrak{X}^\perp(M) \mid \mathcal{L}\xi = 0\}.$$

PROPOSITION 2.12. *Let $\Phi : M \to (\overline{M}, g)$ be a minimal immersion and X be a Killing vector field on $(\overline{M}, g)$. Then the normal component of the restriction of X to M is a Jacobi vector field on M.*

REMARK 2.13. Since the Jacobi equation is the derivative of the mean curvature vector along the diffeomorphisms generated by X, and X is a Killing vector field whose diffeomorphisms are isometries, the proof of Proposition 2.12 is geometrically obvious. However, we include an analytic demonstration of it.

PROOF. Denoting by $\overline{\nabla}$ the Levi-Civita connection of $(\overline{M}, g)$, since X is a Killing vector field, we have that

$$g(\overline{\nabla}_Z X, W) + g(\overline{\nabla}_W X, Z) = 0, \tag{2.14}$$

for arbitrary vector fields Z, W on $\overline{M}$. Now, we will prove that for any vector field Z on $\overline{M}$ we have that

$$\overline{\nabla}_Z \overline{\nabla}_Z X - \overline{\nabla}_{\overline{\nabla}_Z Z} X + \overline{R}(X, Z)Z = 0. \tag{2.15}$$

In fact, for any vector field W on $\overline{M}$ and using (2.14) we have that

$$\begin{aligned} g(\overline{\nabla}_Z \overline{\nabla}_Z X, W) &= Z(g(\overline{\nabla}_Z X, W)) - g(\overline{\nabla}_Z X, \overline{\nabla}_Z W) \\ &= -Z(g(\overline{\nabla}_W X, Z)) - g(\overline{\nabla}_Z X, \overline{\nabla}_Z W) \\ &= -g(\overline{\nabla}_Z \overline{\nabla}_W X, Z) - g(\overline{\nabla}_W X, \overline{\nabla}_Z Z) - g(\overline{\nabla}_Z X, \overline{\nabla}_Z W). \end{aligned}$$

Again using (2.14) we have

$$\begin{aligned} \overline{R}(X, Z, Z, W) &= \overline{R}(Z, W, X, Z) \\ &= g(\overline{\nabla}_Z \overline{\nabla}_W X, Z) - g(\overline{\nabla}_W \overline{\nabla}_Z X, Z) + g(\overline{\nabla}_Z X, [Z, W]) \\ &= g(\overline{\nabla}_Z \overline{\nabla}_W X, Z) - W(g(\overline{\nabla}_Z X, Z)) + g(\overline{\nabla}_Z X, \overline{\nabla}_Z W) \\ &= g(\overline{\nabla}_Z \overline{\nabla}_W X, Z) + g(\overline{\nabla}_Z X, \overline{\nabla}_Z W). \end{aligned}$$

Adding the last two equations we get (2.15).

We will also denote by X the restriction of X to M and by V and ξ the tangential and normal components of X to Φ. If $\{E_i \mid 1 \leq i \leq n\}$ is an orthonormal reference on M, we have that $\sum_{i=1}^n \overline{\nabla}_{E_i} E_i = \sum_{i=1}^n \nabla_{E_i} E_i$ because Φ is a minimal immersion. Now, if η is any normal vector field to Φ, taking $Z = E_i$ in (2.15) we get that

$$0 = \sum_{i=1}^{n} \{g(\overline{\nabla}_{E_i}\overline{\nabla}_{E_i}(V+\xi),\eta) - g(\sigma(\nabla_{E_i}E_i, V),\eta)\}$$
$$+ \sum_{i=1}^{n} \{g(\nabla^{\perp}_{\nabla_{E_i}E_i}\xi,\eta)) - \overline{R}(V+\xi, E_i, E_i, \eta)\}.$$

With (2.1) and (2.2), the last equation becomes

$$0 = g(\Delta^{\perp}\xi,\eta) + g(\mathcal{R}\xi,\eta) - g(\mathcal{A}\xi,\eta) + 2\sum_{i=1}^{n} g(\sigma(E_i, \nabla_{E_i}V),\eta). \tag{2.16}$$

Now, for any point $p \in M$, we can assume, without loss of generality, that $e_i = E_i(p)$ diagonalizes to $A_{\eta(p)}$, i.e., $A_{\eta(p)}e_i = \lambda_i e_i$. So, from (2.14) it follows that

$$g(\overline{\nabla}_{E_i}X, A_\eta E_i)(p) = -g(\overline{\nabla}_{A_\eta E_i}X, E_i)(p) = -g(\overline{\nabla}_{e_i}X, \lambda_i e_i) = -g(\overline{\nabla}_{E_i}X, A_\eta E_i)(p).$$

Therefore

$$0 = \sum_{i=1}^{n} g(\overline{\nabla}_{E_i}X, A_\eta E_i) = \sum_{i=1}^{n} \{g(\nabla_{E_i}V, A_\eta E_i) - g(A_\xi E_i, A_\eta E_i)\}$$
$$= \sum_{i=1}^{n} g(\sigma(E_i, \nabla_{E_i}V),\eta) - g(\mathcal{A}\xi,\eta),$$

and (2.16) becomes in $0 = g(\mathcal{L}\xi,\eta)$. Since η is an arbitrary normal vector field to Φ, we get that $\mathcal{L}\xi = 0$ and the proof ends. □

From Definition 2.10 and Theorem 2.8, if $\Phi : M \to (\overline{M}, g)$ is a minimal immersion and F is a compactly supported variation of Φ, it follows that $V''_F(0) = Q(\xi)$ where ξ is the normal component of the variational field of F. This fact justifies the following definition.

DEFINITION 2.14. Let $\Phi : M \to (\overline{M}, g)$ be a minimal immersion. Φ is called **stable** if $Q(\xi) \geq 0$ for any nonzero $\xi \in \mathfrak{X}_0^{\perp}(M)$. Φ is called **strictly stable** if $Q(\xi) > 0$ for any nonzero $\xi \in \mathfrak{X}_0^{\perp}(M)$.

Definition 2.14 joint with (2.12) and (2.13) give the first interesting result on the stability of minimal immersions.

THEOREM 2.15. *Let $\Phi : M^n \to (\overline{M}^m, g)$ be a totally geodesic immersion of a manifold M.*

1. *If the sectional curvature of $(\overline{M}, g)$ is negative, then Φ is strictly stable.*
2. *If the sectional curvature of $(\overline{M}, g)$ is not positive, then Φ is stable.*
3. *If Φ is a hypersurface $(m = n+1)$ and the Ricci curvature of $(\overline{M}, g)$ is negative, then Φ is strictly stable.*
4. *If Φ is a hypersurface $(m = n+1)$ and the Ricci curvature of $(\overline{M}, g)$ is non-positive, then Φ is stable.*

We conclude this section by introducing the concepts of calibration and calibrated submanifold on a Riemannian manifold and relating them to the stability of minimal submanifolds.

DEFINITION 2.16 ([**HL82**]). Let $(\overline{M}^m, g)$ be a Riemannian manifold. A **calibration** on $\overline{M}$ is an n-form $\eta \in \Omega^n(\overline{M})$ which satisfies the following properties:

(1) η is closed, $\mathrm{d}\eta = 0$.
(2) $\eta(v_1, \ldots, v_n) \leq 1$, for all orthonormal vectors $\{v_1, \ldots, v_n\}$ on $T\overline{M}$.

Under these conditions, an immersion $\Phi : M^n \to \overline{M}^m$ of an oriented manifold M is called a **calibrated immersion** for the calibration η if for any $p \in M$,

$$\eta((\mathrm{d}\Phi)_p(e_1), \ldots, (\mathrm{d}\Phi)_p(e_n)) = 1$$

for an oriented orthonormal basis $\{e_1, \ldots, e_n\}$ of T_pM, i.e., $\Phi^*\eta$ is the volume form on M^n.

PROPOSITION 2.17 ([**HL82**]). *Let $\Phi : M^n \to (\overline{M}^m, g)$ be a calibrated immersion for a calibration $\eta \in \Omega^n(\overline{M})$ of a compact and oriented manifold M. If $\Psi : N^n \to (\overline{M}^m, g)$ is an arbitrary immersion of a compact and oriented manifold N such that $\Phi_*[M] = \Psi_*[N]$ in ${}_\infty H_n(\overline{M}, \mathbb{R})$, then*

$$\mathrm{vol}(M) \leq \mathrm{vol}(N),$$

where $[M]$ and $[N]$ are the fundamental classes of M and N, respectively, in ${}_\infty H_n(M, \mathbb{R})$ and ${}_\infty H_n(N, \mathbb{R})$.

PROOF. From the assumptions we have

$$\int_{\Phi_*[M]} [\eta] = \int_{\Psi_*[N]} [\eta],$$

which implies that

$$\int_M \Phi^*\eta = \int_{[M]} \Phi^*[\eta] = \int_{[N]} \Psi^*[\eta] = \int_N \Psi^*\eta.$$

Since M is a calibrated immersion for the calibration η, then $\Phi^*\eta$ is the volume form on M, and so $\int_M \Phi^*\eta = \mathrm{vol}(M)$. Also, $\Psi^*\eta = h\omega_N$, where ω_N is the volume form on N and, since η is a calibration, $h \leq 1$. This implies that $\int_N \Psi^*\eta \leq \mathrm{vol}(N)$, and this ends the proof. □

COROLLARY 2.18 ([**HL82**]). *Let $\Phi : M^n \to (\overline{M}^m, g)$ be a calibrated immersion for a calibration $\eta \in \Omega^n(\overline{M})$ of a compact and oriented manifold M. Then Φ is a stable minimal immersion. We will say that Φ is* ***volume minimizing*** *in its homology class.*

PROOF. Let $F : (-\epsilon, \epsilon) \times M \to \overline{M}$ any variation of Φ and $V_F(t) = \mathrm{vol}(M, F_t^*g)$ be the associated volume function. For any t, we have that $\Phi = \Phi_0$ is differentiably homotopic to $\Phi_t = F_t$ and so $\Phi_*[M] = (\Phi_t)_*[M]$. Hence, from Proposition 2.17, $V_F(0) \leq V_F(t)$ for any $t \in (-\epsilon, \epsilon)$ and so $t = 0$ is a minimum of $V_F(t)$. Now from the variation formulas it follows the corollary. □

2.5. Index and nullity of compact minimal submanifolds

In this section we will introduce the concepts of index and nullity of a compact minimal submanifold as the index and nullity of the corresponding Jacobi operator (see Definition 2.10). To give a description of the spectrum of the Jacobi operator, we introduce the Sobolev space for sections of the normal bundle.

Let $\Phi : M^n \to (\overline{M}, g)$ be a minimal immersion of a **compact** manifold M. We denote by $L^2(M, T^\perp M)$ the space of measurable sections ξ of the normal bundle of

Φ such that $\|\xi\|^2$ is integrable. This space is a Hilbert space with the product

$$(\xi,\eta)_2 = \int_M g(\xi,\eta)\,dv.$$

So, the index form of Φ can be written as

$$\mathcal{Q}(\xi,\eta) = -(\mathcal{L}\xi,\eta)_2, \quad \forall\,\xi,\eta \in \mathfrak{X}^\perp(M),$$

where $\mathcal{L}$ is the Jacobi operator of Φ.

On the other hand, we denote by $L^2(M, T^*M \otimes T^\perp M)$ the space of measurable sections ω of the vector bundle $T^*M \otimes T^\perp M$ (1-forms on M valuated in the normal bundle) such that $\|\omega\|^2$ is integrable, where the metric on $T^*M \otimes T^\perp M$, also denoted by g, is given by $g(\omega,\nu) = \sum_{i=1}^n g(\omega(e_i),\nu(e_i))$, where $\{e_1,\ldots,e_n\}$ is a local orthonormal reference on M. This space is also a Hilbert space with the product

$$(\omega,\nu)_2 = \int_M g(\omega,\nu)\,dv.$$

If $\xi \in \mathfrak{X}^\perp(M)$, then $\nabla^\perp\xi$ is a smooth section of $T^*M \otimes T^\perp M$. If ω is an arbitrary smooth section of $T^*M \otimes T^\perp M$, then α defined by

$$\alpha(X) = g(\xi,\omega(X)), \quad \forall X \in \mathfrak{X}(M),$$

is a 1-form in $\Omega^1(M)$. It is clear that

$$\delta\alpha = \sum_{i=1}^n \{e_i(g(\xi,\omega(e_i))) - g(\xi,\omega(\nabla_{e_i}e_i))\} = g(\nabla^\perp\xi,\omega) + g(\xi,\delta^\perp\omega),$$

where $\delta^\perp\omega$ defined by

$$\delta^\perp\omega = \sum_{i=1}^n \{\nabla^\perp_{e_i}\omega(e_i) - \omega(\nabla_{e_i}e_i)\},$$

is the codifferential of the 1-form valuated in the normal bundle ω.

So, from Theorem 1.15 it follows that

$$(\nabla^\perp\xi,\omega)_2 + (\xi,\delta^\perp\omega)_2 = 0, \quad \forall\omega \in \Gamma(T^*M \otimes T^\perp M),$$

where $\Gamma(T^*M \otimes T^\perp M)$ denotes the space of smooth sections of $T^*M \otimes T^\perp M$. This expression justifies the next definition.

Definition 2.19. Let $\Phi : M \to (\overline{M}, g)$ be a minimal immersion of a compact manifold M. Given a normal section $\xi \in L^2(M, T^\perp M)$, a section $\nu \in L^2(M, T^*M \otimes T^\perp M)$ is called the **weak derivative** of ξ if ν satisfies

$$0 = g(\nu,\omega)_2 + (\xi,\delta^\perp\omega)_2 \quad \forall\omega \in \Gamma(T^*M \otimes T^\perp M).$$

If ν exists, it is unique and is denoted by $\nabla^\perp\xi$.

The **Sobolev space** $H^{1,2}(M, T^\perp M)$ is defined by

$$H^{1,2}(M, T^\perp M) = \{\xi \in L^2(M, T^\perp M) \mid \xi \text{ has weak derivative}\}.$$

$H^{1,2}(M, T^\perp M)$ is a Hilbert space with the product

$$(\xi,\eta)_{1,2} = (\xi,\eta)_2 + (\nabla^\perp\xi,\nabla^\perp\eta)_2.$$

The space $\mathfrak{X}^\perp(M)$ is dense in the space $(H^{1,2}(M,T^\perp M),(\,,\,)_{1,2})$. It is clear that the index form $\mathcal{Q}$ on $H^{1,2}(M,T^\perp M)$ can be defined by

$$\mathcal{Q}(\xi)=\int_M\{\|\nabla^\perp\xi\|^2-g(\mathcal{A}\xi,\xi)-g(\mathcal{R}\xi,\xi)\}\,dv,\quad \forall\xi\in H^{1,2}(M,T^\perp M).$$

DEFINITION 2.20. Let $\Phi:M\to(\overline{M},g)$ be a minimal immersion of a compact manifold M. If $\xi\in\mathfrak{X}^\perp(M)$ is nonzero and $\lambda\in\mathbb{R}$ satisfy $\mathcal{L}\xi+\lambda\xi=0$, then ξ will be called an **eigensection** of $\mathcal{L}$ with **eigenvalue** λ. The eigenspace associated with the eigenvalue λ is denoted by

$$V_\lambda=\{\xi\in\mathfrak{X}^\perp(M)\mid\mathcal{L}\xi+\lambda\xi=0\}.$$

The eigenspace V_λ is finite dimensional and its dimension will be called the **multiplicity** of the eigenvalue λ. Moreover, eigenspaces corresponding to different eigenvalues are orthogonal in $L^2(M,T^\perp M)$. In the next result we show how the spectrum of the Jacobi operator (the set of its eigenvalues) is defined (compare with Theorem 1.17).

THEOREM 2.21. *Let $\Phi:M\to(\overline{M},g)$ be a minimal immersion of a compact manifold M with Jacobi operator $\mathcal{L}$ and index form $\mathcal{Q}$. Then, there exists a sequence $\{(\lambda_n,\xi_n)\mid n\in\mathbb{N}\}$ where $\{\lambda_n\mid n\in\mathbb{N}\}$ are all the eigenvalues of $\mathcal{L}$ and $\xi_n\in\mathfrak{X}^\perp(M)$ are the corresponding eigensections, $\mathcal{L}\xi_n+\lambda_n\xi_n=0$. This sequence satisfies the following properties:*

(1) $\lambda_1\le\lambda_2\le\dots$ *and* $\{\lambda_n\mid n\in\mathbb{N}\}$ *diverges to* $+\infty$.
(2) $\{\xi_n\mid n\in\mathbb{N}\}$ *is a Hilbert basis of* $L^2(M,T^\perp M)$.
(3) $\lambda_n=\inf\left\{\dfrac{\mathcal{Q}(\xi)}{\|\xi\|_2^2}\;\middle|\;\xi\in H^{1,2}(M,T^\perp M),\ (\xi,\xi_k)_2=0,\ 1\le k\le n-1\right\}$. *Furthermore, if $\xi\in H^{1,2}(M,T^\perp M)$ satisfies $(\xi,\xi_k)_2=0$, $1\le\xi\le n-1$ and $\mathcal{Q}(\xi)=\lambda_n\|\xi\|_2^2$, then $\mathcal{L}\xi+\lambda_n\xi=0$.*

DEFINITION 2.22. Let $\Phi:M\to(\overline{M},g)$ be a minimal immersion of a compact manifold M and $\{\lambda_n\mid n\in\mathbb{N}\}$ the spectrum of the Jacobi operator $\mathcal{L}$. We define the **index** of Φ, and it is denoted by $\mathrm{Ind}(\Phi)$ or $\mathrm{Ind}(M)$, as the index of the Jacobi operator $\mathcal{L}$, that is

$$\mathrm{Ind}(\Phi)=\mathrm{Ind}(M):=\mathrm{Ind}(\mathcal{L})=k,\quad\text{with}\quad\lambda_1\le\lambda_2\le\cdots\le\lambda_k<0,\quad\lambda_{k+1}\ge0.$$

We define the **nullity** of Φ, and it is denoted by $\mathrm{Nul}(\Phi)$ or $\mathrm{Nul}(M)$, as the nullity of the Jacobi operator $\mathcal{L}$, that is

$$\mathrm{Nul}(\Phi)=\mathrm{Nul}(M):=\mathrm{Nul}(\mathcal{L})=\dim V_0,$$

which is the multiplicity of 0 as the eigenvalue of $\mathcal{L}$. Note that the index and the nullity of Φ are respectively the index and the nullity of the index form $\mathcal{Q}$.

Under the above condition (M is compact), $\mathrm{Ind}(\Phi)$ and $\mathrm{Nul}(\Phi)$ are non-negative finite integers. Also, Φ is stable if and only if $\mathrm{Ind}(\Phi)=0$, and Φ is strictly stable if and only if $\mathrm{Ind}(\Phi)=\mathrm{Nul}(\Phi)=0$ (see Definition 2.14).

2.6. Index of two-sided minimal hypersurfaces

In this section we will define the index for two-sided minimal hypersurfaces.

If $\Phi : M^n \to (\overline{M}^{n+1}, g)$ is a two-sided minimal hypersurface, then from Definition 2.1, there exists a unit normal vector field N to Φ which trivializes the normal bundle. This implies that there exists an isomorphism $\mathfrak{X}^{\perp}(M) \to C^{\infty}(M)$ given by

$$\xi \mapsto f, \quad \text{with} \quad \xi = fN.$$

Furthermore, since $\nabla^{\perp} N = 0$, the Jacobi operator of Φ is given by

$$\mathcal{L}(fN) = \left(\Delta f + \|A\|^2 f + \overline{\mathrm{Ricc}}(N) f\right) N.$$

So the Jacobi operator of Φ is actually a Schrödinger operator $\mathcal{L} : C^{\infty}(M) \to C^{\infty}(M)$ given by

$$\mathcal{L} = \Delta + \|A\|^2 + \overline{\mathrm{Ricc}}(N), \tag{2.17}$$

and the index form $\mathcal{Q} : C_0^{\infty}(M) \to \mathbb{R}$ is given by

$$\mathcal{Q}(f) = -\int_M f\mathcal{L}f \, \mathrm{d}v = \int_M \{\|\nabla f\|^2 - (\|A\|^2 + \overline{\mathrm{Ricc}}(N)) f^2\} \, \mathrm{d}v. \tag{2.18}$$

Therefore, if M is compact, the index and the nullity of Φ defined in the previous section, are the index and the nullity of the Schrödinger operator $\mathcal{L} : C^{\infty}(M) \to C^{\infty}(M)$ defined in (2.17), that is the number of negative eigenvalues counted with multiplicity and the multiplicity of zero as an eigenvalue of $\mathcal{L}$, respectively.

This fact justifies to study, in a more general situation, a Schrödinger operator $L = \Delta + q$ on a Riemannian manifold (M, g) which is not necessarily compact. Following Section 1.4, and given a relatively compact open subset Ω of M, we define the **index of L in Ω** as

$$\mathrm{Ind}(L, \Omega) = k, \quad \text{with} \quad \lambda_1 < \lambda_2 \leq \cdots \leq \lambda_k < 0, \quad \lambda_{k+1} \geq 0, \tag{2.19}$$

where $\{\lambda_n\}$ are the eigenvalues of L with Dirichlet condition for Ω (see Theorem 1.17).

We define the **index of the Schrödinger operator** L as

$$\mathrm{Ind}(L) = \sup\{\mathrm{Ind}(L, \Omega) \mid \Omega \text{ is a relatively compact open subset of } M\}. \tag{2.20}$$

If $\Omega \subset \Omega'$ are relatively compact open subsets of M, it is clear that $C_0^{\infty}(\Omega) \subset C_0^{\infty}(\Omega')$ and thus $H_0^{1,2}(\Omega) \subset H_0^{1,2}(\Omega')$. Hence, using the variational characterization of the eigenvalues given in Theorem 1.17, it follows that $\lambda_n(\Omega') \leq \lambda_n(\Omega)$, and hence $\mathrm{Ind}(L, \Omega) \leq \mathrm{Ind}(L, \Omega')$.

This implies that if $\{U_m,\, m \in \mathbb{N}\}$ is a sequence of relatively compact open subsets of M with $\overline{U_m} \subset U_{m+1}$ and $\cup_{n \in \mathbb{N}} U_m = M$, then

$$\mathrm{Ind}(L) = \lim_{m \to \infty} \mathrm{Ind}(L, U_m).$$

It is interesting to remark that when M is not compact, the possible definition of nullity of L is somewhat subtle and could potentially have different meanings.

Definition 2.23. Let $\Phi : M^n \to (\overline{M}^{n+1}, g)$ be a minimal two-sided hypersurface of a manifold M. We define the **index** of Φ, and it will be denoted by $\mathrm{Ind}(\Phi)$ or $\mathrm{Ind}(M)$, as the index of the Schrödinger operator $\mathcal{L} = \Delta + \|A\|^2 + \overline{\mathrm{Ricc}}(N)$.

Remark 2.24. Note that when M is not compact, the index of M can be infinite. Also, if M is compact then M itself is a relatively compact open subset of M, and so $\lim_{m \to \infty} \mathrm{Ind}(\mathcal{L}, U_m) = \mathrm{Ind}(\mathcal{L}, M)$. This means that the two definitions of the index of a compact two-sided hypersurface, given in Definitions 2.22 and 2.23, coincide.

To conclude this section, we will prove some results about the index of a Schrödinger operator on a Riemannian manifold. The relevance of these results is that certain properties of the index will be characterized in terms of suitable functions defined on the whole M.

THEOREM 2.25 ([**FCS80**]). *Let (M, g) be an n-dimensional Riemannian manifold and $L = \Delta + q$ a Schrödinger operator on M, where $q \in C^\infty(M) \cap L^\infty(M)$. Then*

(1) *If there exists a positive smooth function $u : M \to \mathbb{R}$ satisfying $Lu = 0$, then* $\operatorname{Ind}(L) = 0$.
(2) *If M is not compact and* $\operatorname{Ind}(L) = 0$, *then there exists a positive smooth function $u : M \to \mathbb{R}$ satisfying $Lu = 0$.*
(3) *If M is compact and there exists a nonzero smooth function $u : M \to \mathbb{R}$ satisfying $Lu = 0$, then*

$$\operatorname{Ind}(L) \geq m - 1,$$

where m is the number of nodal sets of u, i.e., the number of connected components of the open set $\{p \in M \mid u(p) \neq 0\}$.

REMARK 2.26. Point 2 of the theorem is not true if M is compact. In fact, let (M, g) be a compact Riemannian manifold and $L = \Delta - 1$. Since the first eigenvalue of Δ is zero ([**BGM71**]), then the first eigenvalue of L is 1 and so $\operatorname{Ind}(L) = 0$. But there is no positive smooth function on M satisfying $0 = Lu = \Delta u - u$, because -1 is not an eigenvalue of the Laplacian Δ of M.

PROOF. **Proof of 1.** Let $u \in C^\infty(M)$ be a positive function with $Lu = \Delta u + qu = 0$. If $w = \log u$, then

$$\Delta w = \frac{u\Delta u - \|\nabla u\|^2}{u^2} = -q - \|\nabla w\|^2. \tag{2.21}$$

Let Ω be any relatively compact domain of M and $f \in C_0^\infty(\Omega)$. Considering (2.21) and using that $\int_M \operatorname{div}(f^2 \nabla \omega)\, dv = 0$, it follows that

$$\begin{aligned} \int_\Omega \{f^2(\|\nabla w\|^2 + q)\}\, dv &= -\int_\Omega f^2 \Delta w\, dv \\ &= \int_\Omega g(\nabla f^2, \nabla w)\, dv = 2\int_\Omega f g(\nabla f, \nabla w)\, dv. \end{aligned}$$

But the Schwarz inequality implies that

$$2fg(\nabla f, \nabla w) \leq 2|f||g(\nabla f, \nabla w)| \leq 2|f|\|\nabla f\|\|\nabla w\| \leq f^2\|\nabla w\|^2 + \|\nabla f\|^2.$$

Using this inequality in the last formula we get that $\int_\Omega f^2 q\, dv \leq \int_\Omega \|\nabla f\|^2\, dv$. This means that the quadratic form Q satisfies $Q(f) = \int_\Omega (\|\nabla f\|^2 - qf^2)\, dv \geq 0$ for any $f \in C_0^\infty(\Omega)$, and hence $\operatorname{Ind}(L, \Omega) = 0$. Since Ω is any relatively compact domain of M, we finally get that $\operatorname{Ind}(L) = 0$.

Proof of 2. Since $\operatorname{Ind}(L) = 0$, then $\operatorname{Ind}(L, \Omega) = 0$ for any relatively compact domain Ω in M, and thus the first eigenvalue $\lambda_1(\Omega) \geq 0$.

If Ω is a relatively compact domain with $\lambda_1(\Omega) = 0$, let $u \in C^\infty(\Omega) \cap H^{1,2}(\Omega)$ be a positive function satisfying $Lu = 0$ (see Theorem 1.21).

Since M is not compact, there exists another relatively compact domain Ω' of M such that $\Omega \subset \Omega'$ and $\Omega \neq \Omega'$. Hence $0 = \lambda_1(\Omega) \geq \lambda_1(\Omega')$ and so $\lambda_1(\Omega') = 0$. Then the function $\hat{u}$ on Ω', defined by $\hat{u} = u$ on Ω and $\hat{u} = 0$ on $\Omega' \setminus \Omega$, is a non-negative function on $H^{1,2}(\Omega')$ satisfying $Q(\hat{u}) = 0$. From Theorem 1.17, we have that $L\hat{u} = 0$ and from Proposition 1.20 $\hat{u} = 0$, which is a contradiction. Hence $\lambda_1(\Omega) > 0$ for any relatively compact open subset of M. This implies that the only solution to the problem $Lf = 0$ in Ω and $f = 0$ in $\partial\Omega$ is $f = 0$.

Now, using the Fredhölm alternative, Theorem 1.19,3, there exists a unique solution v to the problem $Lv = -q$ in Ω and $v = 0$ in $\partial\Omega$. Then, the function $\tilde{v} = v + 1$ satisfies $L\tilde{v} = 0$ in Ω and $\tilde{v} = 1$ in $\partial\Omega$.

CLAIM. The function $\tilde{v}$ satisfies that $\tilde{v} > 0$ in Ω.

In fact, if the set

$$\Omega_0 = \{p \in \Omega \mid \tilde{v} < 0\}$$

is not empty, then Ω_0 is a relatively compact domain in M, and $\tilde{v}$ is a solution in Ω_0 of the problem $L\tilde{v} = 0$ with $\tilde{v} = 0$ in $\partial\Omega_0$. This contradicts the above conclusion, and thus $\tilde{v} \geq 0$ in Ω.

Now, as $\tilde{v}$ is not identically zero, Proposition 1.20 implies that $\widetilde{v} > 0$ and we proved the claim.

If p_0 is a point of Ω, then the function $u = \frac{\tilde{v}}{\tilde{v}(p_0)}$ satisfies

$$Lu = 0,\ u > 0 \quad \text{in} \quad \Omega \quad \text{and} \quad u(p_0) = 1.$$

To construct the function u on the whole M, we consider a sequence $\{\Omega_n \mid n \in \mathbb{N}\}$ of relatively compact domains in M, satisfying $\Omega_n \subset \Omega_{n+1}$, $\Omega_n \neq \Omega_{n+1}$ and $\cup_{n\in\mathbb{N}}\Omega_n = M$. Let p_0 a point of M with $p_0 \in \Omega_n$, $\forall n \in \mathbb{N}$ and u_n the positive function on Ω_n with $Lu_n = 0$ in Ω_n and $u_n(p_0) = 1$ constructed above.

If Ω is any relatively compact domain with $p \in \Omega$, then for any n with $\Omega \subset \Omega_n$ and using the Harnack inequality for L, Theorem 1.19,2, it follows that

$$\sup\{u_n(p) \mid p \in \Omega\} \leq C \inf\{u_n(p) \mid p \in \Omega\},$$

where $C > 0$ is a constant. Since $u_n(p_0) = 1$, we get that $\sup\{u_n(p) \mid p \in \Omega\} \leq C$. This means that the family $\{u_n \mid n \in \mathbb{N}\}$ is uniformly bounded on compact subsets of M, and so from [**GT01**], page 90, the family of all the derivatives of u_n are also uniformly bounded on compact subsets of M. This implies that there exists a subsequence $\{u_{n_r}\}$ of $\{u_n\}$ which, together with all its derivatives, converges on compact subsets of M to a function u defined on M, which satisfies $Lu = 0$, $u \geq 0$ and $u(p_0) = 1$. Again using Proposition 1.20 as above, we conclude that $u > 0$. This completes proof of 2.

Proof of 3. If $k = \mathrm{Ind}(L)$, let $\{\lambda_1, \ldots, \lambda_k\}$ be the negative eigenvalues of the operator L and $\{g_1, \ldots, g_k\}$ the corresponding eigenfunctions, i.e., $Lg_i + \lambda_i g_i = 0$, $1 \leq i \leq k$.

If $\{\Omega_1, \ldots, \Omega_m\}$ are the nodal sets of u, we define the functions $u_j : M \to \mathbb{R}$, $1 \leq j \leq m$, by

$$u_j = \begin{cases} u & \text{on } \Omega_j, \\ 0 & \text{on } M \setminus \Omega_j. \end{cases}$$

Then the functions $\{u_1, \ldots, u_m\} \subset H^{1,2}(M) = H_0^{1,2}(M)$ and $\int_M u_i u_j \, dv = 0$ for $i \neq j$. In particular, the functions $\{u_j,\ 1 \leq j \leq m\}$ are linearly independent in $H_0^{1,2}(M)$ and so

$$V = \Big\{ \sum_{j=1}^{m-1} \mu_j u_j \mid \mu_j \in \mathbb{R} \Big\}$$

is an $(m-1)$-dimensional subspace of $H_0^{1,2}(M)$. We define the linear map $F : V \to \mathbb{R}^k$ by

$$F(v) = \left(\int_M vg_1 \, \mathrm{d}v, \ldots, \int_M vg_k \, \mathrm{d}v\right).$$

To prove the result, we will assume that $\mathrm{Ind}(L) < m-1$ and we will get a contradiction. Indeed, if $k < m-1$, there exists a nonzero vector $v = \sum_{i=1}^{m-1} \lambda_i u_i \in \ker F$ and thus $\int_M vg_i \, \mathrm{d}v = 0$ for $1 \leq i \leq k$. From Theorem 1.17 we have that $Q(v) \geq 0$. Hence

$$0 \leq Q(v) = -\sum_{j,k=1}^{m-1} \mu_j \mu_k \int_M u_j L u_k \, \mathrm{d}v = 0,$$

because $Lu_k = 0$. Now, using Theorem 1.17 again, we get that $Lv = 0$ and thus $v \in C^\infty(M)$. But, by construction, $v = 0$ on the non-empty open subset Ω_m. The principle of the unique continuation, [**Pro60**], implies that $v = 0$ on the whole M, which is a contradiction. □

Now we will extend Theorem 2.25,2 to the case where the index of L is finite.

THEOREM 2.27 ([**FC85**]). *Let (M, g) be a non-compact Riemannian manifold and $L = \Delta + q$ a Schrödinger operator on M. If $\mathrm{Ind}(L) < \infty$, then there exists a compact subset K of M such that $\mathrm{Ind}_{M\setminus K}(L) = 0$ and there exists a positive function $u \in C^\infty(M)$ satisfying $Lu = 0$ on $M \setminus K$, where $\mathrm{Ind}_{M\setminus K}(L)$ is the index of L on the Riemannian manifold $(M \setminus K, g)$.*

PROOF. Since $\mathrm{Ind}(L) = k < \infty$, there exists a relatively compact domain Ω in M such that $\mathrm{Ind}(L) = \mathrm{Ind}(L, \Omega) = k$. Hence there exists $\{f_1, \ldots, f_k\} \subset C_0^\infty(\Omega)$ such that $Lf_i + \lambda_i f_i = 0$ with $\lambda_i < 0$, $\forall i \in \{1, \ldots, k\}$.

Now, we will prove that $\mathrm{Ind}_{M\setminus C}(L) = 0$, where C is the compact subset of M given by $C = \overline{\Omega}$. In fact, if $\mathrm{Ind}_{M\setminus C}(L) > 0$, then there exists a relatively compact domain $\Omega' \subset M \setminus C$ such that $\mathrm{Ind}_{M\setminus C}(L, \Omega') > 0$. Hence there exists $f \in C^\infty(\Omega') \cap H^{1,2}(\Omega')$ such that $Lf + \lambda f = 0$, $\lambda < 0$. It is clear that $\mathrm{supp}(f_i) \cap \mathrm{supp}(f) = \emptyset$, and so $\int_M f f_i \, \mathrm{d}v = 0$, $\forall i \in \{1, \ldots, k\}$. This implies that $\{f, f_1, \ldots, f_k\}$ defines a set of $k+1$ linearly independent eigenfunctions of L with negative eigenvalues. Now, if $\widehat{\Omega}$ is some relatively compact domain with $\Omega \cup \Omega' \subset \widehat{\Omega}$, then $\mathrm{Ind}(L, \widehat{\Omega}) \geq k+1$, which contradicts that $\mathrm{Ind}(L) = k$. So we have proved that $\mathrm{Ind}_{M\setminus C}(L) = 0$.

Since $M \setminus C$ is a non-compact manifold and $\mathrm{Ind}_{M\setminus C}(L) = 0$, it follows from Theorem 2.25 that there exists a positive function $v \in C^\infty(M \setminus C)$ such that $Lv = 0$ on $M \setminus C$. Let K be a compact subset of M such that C is contained in the interior $\overset{\circ}{K}$ of K and $\phi \in C^\infty(M)$ is the cut-off function that satisfies $0 \leq \phi \leq 1$, $\phi_{|C} = 1$, and $\mathrm{supp}(\phi) \subset \overset{\circ}{K}$. We define the function $u : M \to \mathbb{R}$ by

$$u = \begin{cases} \phi + (1-\phi)v & \text{on } M \setminus C \\ \phi = 1 & \text{on } C. \end{cases}$$

Then u is a positive smooth function on M and $u_{|M\setminus K} = v_{|M\setminus K}$. So on $M \setminus K$, $Lu = Lv = 0$. Moreover, since $M \setminus K \subset M \setminus C$ and $\mathrm{Ind}_{M\setminus C}(L) = 0$, we have that $\mathrm{Ind}_{M\setminus K}(L) = 0$ too. This concludes the proof. □

To prove the last result of this section, we need to construct a special cut-off function on a complete Riemannian manifold.

LEMMA 2.28. *Let (M,g) be a complete Riemannian manifold and p a point of M. For each real number $R>1$ there exists a function $\varphi^R \in H_0^{1,2}(M)$ such that $\varphi^R = 1$ on $B_p(R)$, $\varphi^R = 0$ on $M \setminus B_p(R^2)$ and $\|\nabla\varphi^R\| \leq \frac{1}{R\log R}$, where $B_p(\rho)$ is the geodesic ball centered at p and radius ρ.*

PROOF. Let $\zeta^R \in H_0^{1,2}(\mathbb{R})$ be the function defined by

$$\zeta^R(t) = \begin{cases} 1 & t < R, \\ 2 - \frac{\log t}{\log R} & t \in [R, R^2], \\ 0 & t > R^2. \end{cases}$$

If f is the function on M given by $f(x) = d(p,x)$, where d is the distance function, then f is Lipschitz on M and thus, from Theorem 1.10,1, $f \in H_0^{1,2}(M)$. Moreover, since ζ^R is piecewise smooth, then $\varphi^R = \zeta^R \circ f \in H_0^{1,2}(M)$ and satisfies $\varphi^R = 1$ on $B_p(R)$ and $\varphi^R = 0$ on $M \setminus B_p(R^2)$. Moreover, since $\|\nabla f\| = 1$ on $M \setminus (\{p\} \cup$ cut locus$(\{p\}))$, we have that for any $x \in \overline{B_p(R^2)} \setminus B_p(R)$, $\|\nabla\varphi^R\|(x) = \frac{1}{f(x)\log R} \leq \frac{1}{R\log R}$. □

The next result is quite important in the study of the index of a Schrödinger operator of a complete Riemannian manifold, because it characterizes the case of finite index in terms of functions of $L^2(M)$. We will give different versions of Theorem 2.29 in Theorems 2.30 and 4.20 and Proposition 4.51.

THEOREM 2.29 ([**FC85**]). *Let (M,g) be a complete n-dimensional Riemannian manifold and $L = \Delta + q$ a Schrödinger operator on M. Then L has finite index k if and only if there exists a k-dimensional subspace W of $L^2(M)$ with an orthonormal basis $\{v_1,\ldots,v_k\}$ with $Lv_i + \lambda_i v_i = 0, \lambda_i < 0$ and $Q(f) \geq 0$ for any function $f \in C_0^\infty(M) \cap W^\perp$, where $W^\perp \subset L^2(M)$ is the L^2-orthogonal complement of W.*

PROOF. If the subspace W exists, then it is easy to check, given suitable cut-off functions φ^R, that, for sufficiently large R, $Q(\varphi^R v) < 0$ for any $v \in W$ and thus $\operatorname{Ind}(L) \geq k$. If $\operatorname{Ind}(L) > k$, from (2.20) there exists a relatively compact open subset $\Omega \subset M$ such that $\operatorname{Ind}(L,\Omega) = l > k$. Let $V \subset C_0^\infty(\Omega)$ be the l-dimensional subspace with $Q(f) < 0$ for any $f \in V$. If $F : V \to W$ is the linear map

$$F(f) = \left(\int_M f v_1 \, dv, \ldots, \int_M f v_k \, dv\right),$$

then there exists $f \in \ker F$ and so we get a function f with $f \in W^\perp$ and $Q(f) < 0$, which contradicts the assertion. Hence $\operatorname{Ind}(L) = k$.

Conversely, if $\operatorname{Ind}(L)$ is finite, then from Theorem 2.27 there exists a compact subset K of M such that $\operatorname{Ind}_{M\setminus K}(L) = 0$. Furthermore, there exists a relatively compact domain $\Omega \subset M$ such that $\operatorname{Ind}(L) = \operatorname{Ind}(L,\Omega)$. Let $R_0 > 1$ be a real number such that $K \cup \Omega \subset B_p(R_0)$, where $B_p(R_0)$ is the geodesic ball of center a point $p \in M$ and radius R_0. So, $\operatorname{Ind}(L) = \operatorname{Ind}(L, B_p(R))$, for any $R \geq R_0$.

If $\phi \in C_0^\infty(M)$, then for any $R \geq R_0$ we consider $\eta^R = 1 - \varphi^R$ (φ^R the function given in Lemma 2.28). Then $\eta^R\phi \in H_0^{1,2}(M)$ and $\operatorname{supp}(\eta^R\phi) \subset M \setminus K$. Since $\operatorname{Ind}_{M\setminus K}(L) = 0$, from Theorem 1.17 we get that $Q(\eta^R\phi) \geq 0$ and so

$$\int_M q(\eta^R\phi)^2 \, dv \leq \int_M \{(\eta^R)^2\|\nabla\phi\|^2 + \phi^2\|\nabla\eta^R\|^2 + 2\eta^R\phi g(\nabla\eta^R, \nabla\phi)\} \, dv. \tag{2.22}$$

Now, using that $2\eta^R\phi g(\nabla\eta^R,\nabla\phi) \leq (\eta^R)^2\|\nabla\phi\|^2 + \phi^2\|\nabla\eta^R\|^2$ and that $Q(\phi) = \int_M\{\|\nabla\phi\|^2 - q\phi^2\}\,dv$, inequality (2.22) becomes

$$\begin{aligned}\int_M (1-(\eta^R)^2)(\|\nabla\phi\|^2 - q\phi^2)\,dv &\leq Q(\phi) + \int_M \{\phi^2\|\nabla\eta^R\|^2 + 2\eta^R\phi g(\nabla\eta^R,\nabla\phi)\}\,dv \\ &\leq Q(\phi) + \int_{B_p(R^2)} \{2\phi^2\|\nabla\eta^R\|^2 + (\eta^R)^2\|\nabla\phi\|^2\}\,dv \\ &\leq 2Q(\phi) + \int_{B_p(R^2)} \{2\phi^2\|\nabla\eta^R\|^2 + q\phi^2\}\,dv,\end{aligned}$$

where in the last inequality, we have used that $(\eta^R)^2 \leq 1$. Now, since $\eta^R = 0$ on $B_p(R)$, from the last inequality we get

$$\begin{aligned}\int_{B_p(R)} \|\nabla\phi\|^2\,dv \leq \int_M (1-(\eta^R)^2)\|\nabla\phi\|^2\,dv &\leq 2Q(\phi) \\ &+ \int_{B_p(R^2)} \{2\phi^2\|\nabla\eta^R\|^2 + (2-(\eta^R)^2)q\phi^2\}\,dv \\ &\leq 2Q(\phi) + 2\Big(\frac{1}{R^2\log^2 R} + \sup_{B_p(R^2)}|q|\Big)\int_{B_p(R^2)}\phi^2\,dv \\ &\leq 2\big(Q(\phi) + C_R\int_M \phi^2\,dv\big),\end{aligned} \tag{2.23}$$

where $C_R = \frac{1}{R^2\log^2 R} + \sup_{B_p(R^2)}|q|$.

Using in (2.23) $R = R_0$, we get that

$$0 \leq Q(\phi) + C_0\int_M \phi^2\,dv, \tag{2.24}$$

where $C_0 = C_{R_0}$ and ϕ is any smooth compact support function on M.

As for any $R \geq R_0$, we have that $k = \operatorname{Ind}(L, B_p(R))$, let $\{\lambda_{1,R},\ldots,\lambda_{k,R}\}$ be the negative eigenvalues of L on $B_p(R)$ and $\{f_{1,R},\ldots,f_{k,R}\} \subset C_0^\infty(B_p(R))$ the corresponding orthonormal eigenfunctions with $\int_M f_{i,R}^2\,dv = 1$, i.e., $Lf_{i,R} + \lambda_{i,R}f_{i,R} = 0$ for $1 \leq i \leq k$. Taken in (2.24), $\phi = f_{i,R}$, we get that $\lambda_{i,R} \geq -C_0$. Also, $\max\{\lambda_{1,R},\ldots,\lambda_{k,R}\}$ is a decreasing function of R, so there exists $\epsilon_0 > 0$ such that $\lambda_{i,R} < -\epsilon_0$. So, finally we get

$$\lambda_{i,R} \in [-C_0, -\epsilon_0], \quad \text{for any} \quad 1 \leq i \leq k \quad \text{and any} \quad R \geq R_0. \tag{2.25}$$

Let $R > R_0$ and $\rho \in [R,\infty)$. Taking in (2.23) $\phi = f_{i,\rho}$ and using that $Q(f_{i,\rho}) < 0$ we get that

$$\int_{B_p(R)} (f_{i,\rho}^2 + \|\nabla f_{i,\rho}\|^2)\,dv < (1+2C_R)\int_M f_{i,\rho}^2\,dv = 1 + 2C_R, \tag{2.26}$$

for any $\rho > R$, $1 \leq i \leq k$. On the other hand, using this $\int_M \operatorname{div}((\eta^R)^2\nabla f_{i,\rho}^2)\,dv = 0$, we have

$$\begin{aligned}2\int_M \eta^R f_{i,\rho} g(\nabla\eta^R,\nabla f_{i,\rho})\,dv = \frac{1}{2}\int_M g(\nabla(\eta^R)^2,\nabla f_{i,\rho}^2)\,dv &= -\frac{1}{2}\int_M (\eta^R)^2\Delta f_{i,\rho}^2\,dv \\ &= \int_M (\eta^R)^2\big(f_{i,\rho}^2(q+\lambda_{i,\rho}) - \|\nabla f_{i,\rho}\|^2\big)\,dv.\end{aligned}$$

Now, using (2.22) in the last equation we get

$$-\lambda_{i,\rho}\int_M (\eta^R)^2 f_{i\rho}^2\,\mathrm{d}v \le \int_M f_{i,\rho}^2\|\nabla\eta^R\|^2\,\mathrm{d}v.$$

Since $-\lambda_{i,\rho}\ge\epsilon_0$ and $\int_M f_{i,\rho}^2\,\mathrm{d}v = 1$, we have for $\rho > R^2$,

$$\begin{aligned}\int_{M\setminus B_p(R^2)} f_{i,\rho}^2\,\mathrm{d}v &= \int_{M\setminus B_p(R^2)} (\eta^R)^2 f_{i,\rho}^2\,\mathrm{d}v \le \int_M (\eta^R)^2 f_{i,\rho}^2\,\mathrm{d}v\\ &\le \frac{1}{\epsilon_0}\int_M f_{i,\rho}^2\|\nabla\eta^R\|^2\,\mathrm{d}v \le \frac{1}{\epsilon_0 R^2\log^2 R}\int_{B_p(R^2)} f_{i,\rho}^2\,\mathrm{d}v\\ &\le \frac{1}{\epsilon_0 R^2\log^2 R}\Big(1-\int_{M\setminus B_p(R^2)} f_{i,\rho}^2\,\mathrm{d}v\Big),\end{aligned}$$

and thus

$$\int_{M\setminus B_p(R^2)} f_{i,\rho}^2\,\mathrm{d}v \le \frac{1}{1+\epsilon_0 R^2\log^2 R},\qquad \text{for any}\quad 1\le i\le k,\ \rho\ge R^2. \tag{2.27}$$

We observe that if $\rho\le R^2$, then $\int_{M\setminus B_p(R^2)} f_{i,\rho}^2\,\mathrm{d}v = 0$.

In summary, we have constructed a family $\{f_{i,\rho}\mid 1\le i\le k,\ \rho\ge R> R_0\}\subset C_0^\infty(M)$ with $Lf_{i,\rho}+\lambda_{i,\rho}f_{i,\rho}=0$, which satisfies (2.25), (2.26) and (2.27).

Now, from (2.26), $A := \{f_{i,\rho},\ 1\le i\le r,\ \rho > R\}$ is bounded in the Sobolev space $H^{1,2}(B_p(R))$, and so, since $H^{1,2}(B_p(R))$ is compactly embedded into $L^2(B_p(R))$, there exists a sequence $\{u_n\}\subset A$ that converges in $L^2(B_p(R))$. Now, we use a Cantor diagonal argument. Repeating the argument with $B_p(R+1)$, we find a subsequence $\{u_{1,n}\}$ of $\{u_n\}$ converging in $L^2(B_p(R+1))$. By an induction argument, for any positive integer m we find a subsequence $\{u_{m,n}\}$ converging in $L^2(B_p(R+m))$. So, we get a sequence $\{f_i,\rho_n\}=\{u_{n,n}\}$ such that

$$\lim_{\rho_n\to\infty} f_{i,\rho_n} = f_i \quad\text{in}\quad L^2(M),\quad 1\le i\le k.$$

From (2.25), the set $\{\lambda_{i,\rho}\}$ is uniformly bounded from below, and so f_i is an eigenfunction of L with eigenvalue $\lambda_i = \lim_{\rho_n\to\infty}\lambda_{i,\rho_n}$.

It also follows from (2.27), that no mass scapes to infinity, i.e., $\int_M f_i^2\,\mathrm{d}v = 1$.

So, we have found an orthonormal $\{f_1,\dots,f_k\}$ in $L^2(M)$ such that $Lf_i+\lambda_i f_i = 0$ and $-C_0\le\lambda_i\le-\epsilon_0$ for any $1\le i\le k$. Let W be its linear span in $L^2(M)$. If $\phi\in C_0^\infty(M)\cap W^\perp$, then for some ρ_n we have

$$\phi = \sum_{i=1}^k a_{i,\rho_n}f_{i,\rho_n} + \phi_{\rho_n},\qquad a_{i,\rho_n} = \int_M \phi f_{i,\rho_n}\,\mathrm{d}v,$$

with ϕ_{ρ_n} orthogonal to $\operatorname{span}\{f_{1,\rho_n},\dots,f_{k,\rho_n}\}$. It is clear that $Q(\phi_{\rho_n})\ge 0$, and so $Q(\phi)\ge\sum_{i=1}^k\lambda_{i,\rho_n}a_{i,\rho_n}^2$. Hence

$$\begin{aligned}Q(\phi) &\ge \sum_{i=1}^k \lim_{\rho_n\to\infty}(\lambda_{i,\rho_n}a_{i,\rho_n}^2) \ge -C_0\sum_{i=1}^n\lim_{\rho_n\to\infty}\Big(\int_M\phi f_{i,\rho_n}\,\mathrm{d}v\Big)^2\\ &= -C_0\sum_{i=1}^k\Big(\int_M \phi f_i\,\mathrm{d}v\Big)^2 = 0.\end{aligned}$$

This concludes the proof. □

2.7. Index of one-sided minimal hypersurfaces

In order to define the index of one-sided minimal hypersurfaces, we start this section by studying the index of a Schrödinger operator $L = \Delta + q$ on a Riemannian manifold (M, g) which admits an isometry τ without fixed points with $\tau^2 = Id$ and $q \circ \tau = q$.

Under the above conditions, the space of smooth functions on M decomposes into

$$C^\infty(M) = C^\infty_+(M) \oplus C^\infty_-(M), \quad C^\infty_\pm(M) = \{f \in C^\infty(M) \mid f \circ \tau = \pm f\}.$$

If $f \in C^\infty_+(M)$ and $g \in C^\infty_-(M)$ are of compact support, then from the change of variable formula, Theorem 1.7,3, it follows that

$$\int_M fg \, dv = \int_M (fg \circ \tau) \, dv = -\int_M fg \, dv, \quad \int_M g \, dv = \int_M (g \circ \tau) \, dv = -\int_M g \, dv,$$

and so f and g are orthogonal in $L^2(M)$ and $\int_M g \, dv = 0$. In particular g changes of sign.

Also, since τ is an isometry, $\Delta(f \circ \tau) = (\Delta f) \circ \tau$, $\forall f \in C^\infty(M)$, and hence $L(f \circ \tau) = (\Delta + q)(f \circ \tau) = (\Delta f) \circ \tau + (q \circ \tau)(f \circ \tau) = Lf \circ \tau$. So we can define two operators

$$L^\pm := L_{|C^\infty_\pm(M)} : C^\infty_\pm(M) \to C^\infty_\pm(M), \quad Lu = L^+u^+ + L^-u^- \quad \text{with} \quad u = u^+ + u^-.$$

So, if Ω is a relatively compact open subset of M and λ is an eigenvalue of L with eigenspace V_λ, then $V_\lambda = V_\lambda^+ \oplus V_\lambda^-$ where

$$V_\lambda^\pm = \{u \in C^\infty_\pm(\Omega) \cap H_0^{1,2}(\Omega) \mid L^\pm u + \lambda u = 0\}.$$

Taking into account Theorem 1.21, and since a function on $C^\infty_-(\Omega)$ cannot be positive, we get that $V_{\lambda_1} = V_{\lambda_1}^+$, where λ_1 is the first eigenvalue of L in Ω.

Now, from (2.19) we have

$$\operatorname{Ind}(L, \Omega) = \operatorname{Ind}(L^+, \Omega) + \operatorname{Ind}(L^-, \Omega),$$

where $\operatorname{Ind}(L^\pm, \Omega)$ is the number of negative eigenvalues of $L^\pm$ counted with multiplicity with the Dirichlet condition for Ω. This implies that

$$\operatorname{Ind}(L) = \operatorname{Ind}(L^+) + \operatorname{Ind}(L^-),$$

where

$$\operatorname{Ind}(L^\pm) = \sup\{\operatorname{Ind}(L^\pm, \Omega) \mid \Omega \text{ is a relatively compact open subset of } M\}. \tag{2.28}$$

In the next result we prove similar results to those given in the last section for the operator L^-.

THEOREM 2.30. *Let (M, g) be an n-dimensional Riemannian manifold and let τ be an isometry without fixed points and with $\tau^2 = Id$. Let $L = \Delta + q$ be a Schrödinger operator on M with $q \circ \tau = q$. Then*

(1) *If $\operatorname{Ind}(L^-) < \infty$, then there exists a compact subset K of M such that $\operatorname{Ind}_{M \setminus K}(L^-) = 0$.*
(2) *If L^- has finite index k, then there exists a k-dimensional subspace W of $L^2_-(M) = \{v \in L^2(M) \mid v \circ \tau = -v\}$ with an orthonormal basis $\{v_1, \ldots, v_k\}$ with $Lv_i + \lambda_i v_i = 0, \lambda_i < 0$ and $Q(f) \geq 0$ for any function $f \in C^\infty_0(M) \cap W^\perp$, where $W^\perp$ is the orthogonal complement of W in $L^2_-(M)$.*

PROOF. The proof of 1 is similar to the proof of the first part of Theorem 2.27.

The proof of 2 is similar to the proof of Theorem 2.29. The only problem is that the test functions we used in the proof must be odd functions with respect to the isometry τ. So, the cut-off functions given in Lemma 2.28 must be even with respect to τ. To do this, we change the geodesic ball $B_p(R)$ in Lemma 2.28 by $C_p(R) = B_p(R) \cup B_{\tau(p)}(R)$. Also, the function $f(x) = d(p,x)$ must be replaced by $f(x) = \min\{d(p,x), d(\tau(p),x)\}$. Since $f \circ \tau = f$, the function $\varphi^R = \zeta^R \circ f$ is even with respect to τ. □

We will use the above arguments to define the index of one-sided minimal hypersurfaces. Let $\Phi : M^n \to (\overline{M}^{n+1}, g)$ be a one-sided minimal hypersurface. Considering Example 2.3 we have three possibilities:

(1) $\overline{M}$ is orientable and M is not.
(2) $\overline{M}$ is not orientable and M is orientable.
(3) $\overline{M}$ and M are not orientable.

CASE 1. $\Phi : M^n \to (\overline{M}^{n+1}, g)$ is a one-sided minimal hypersurface of a nonorientable manifold M into an orientable manifold $\overline{M}$.

Let $\Pi : \widetilde{M} \to M$ be the orientable double covering of M and $\{Id, \tau\}$ the group of automorphisms of the covering. Then τ is an isometry of $\widetilde{M}$ (with respect to the lifted metric) without fixed points and with $\tau^2 = Id$. Using Lemma 2.2 again,

$$\Psi = \Phi \circ \Pi : \widetilde{M} \to \overline{M}^{n+1}$$

is a two-sided minimal hipersurface, satisfying $\Psi \circ \tau = \Psi$. Since Ψ is two-sided, let N be a unit normal vector field to Ψ. It is clear that for any $p \in \widetilde{M}$, the vectors $N_p, N_{\tau(p)} \in T^{\perp}_{\Psi(p)}M$, and so we can define two subsets $\widetilde{M}_+$ and $\widetilde{M}_-$ of $\widetilde{M}$ by

$$\widetilde{M}_{\pm} = \{p \in \widetilde{M} \mid N_{\tau(p)} = \pm N_p\}.$$

It is clear that $\widetilde{M} = \widetilde{M}_+ \cup \widetilde{M}_-$, $\widetilde{M}_+ \cap \widetilde{M}_- = \emptyset$ and that $\widetilde{M}_{\pm}$ are closed subsets of $\widetilde{M}$. Since $\widetilde{M}$ is connected, we have two possibilities: $\widetilde{M} = \widetilde{M}_+$ or $\widetilde{M} = \widetilde{M}_-$. If $\widetilde{M} = \widetilde{M}_+$, then the vector field N projects to a unit normal vector field to M, which is impossible because Φ is one-sided. Hence $\widetilde{M} = \widetilde{M}_-$ and $N \circ \tau = -N$.

If $\xi \in \mathfrak{X}^{\perp}(M)$ is a normal section to Φ and $\widetilde{\xi} \in \mathfrak{X}^{\perp}(\widetilde{M})$ is its lift to $\widetilde{M}$, i.e., $\widetilde{\xi}_p = \xi_{\Pi(p)}$, $\forall p \in \widetilde{M}$, then $\widetilde{\xi}_p = \widetilde{\xi}_{\tau(p)}$. Since Ψ is two-sided, $\widetilde{\xi} = fN$, and since $N_{\tau(p)} = -N_p$, we get that $f \circ \tau = -f$, i.e., $f \in C^{\infty}_{-}(\widetilde{M})$. This gives an isomorphism

$$F : \mathfrak{X}^{\perp}(M) \to C^{\infty}_{-}(\widetilde{M}) = \{f \in C^{\infty}(\widetilde{M}) \mid f \circ \tau = -f\}$$

$$F(\xi) = f, \quad \text{with} \quad \widetilde{\xi} = fN, \quad \widetilde{\xi}_p = \xi_{\Pi(p)}, \forall p \in \widetilde{M}.$$

Now, if $\mathcal{L} : \mathfrak{X}^{\perp}(M) \to \mathfrak{X}^{\perp}(M)$ is the Jacobi operator of the minimal hypersurface Φ and $\widetilde{\mathcal{L}} : C^{\infty}(\widetilde{M}) \to C^{\infty}(\widetilde{M})$ is the Jacobi operator of the two-sided minimal hypersurface Ψ, then the next diagram is commutative:

$$\begin{array}{ccc} \mathfrak{X}^{\perp}(M) & \xrightarrow{\mathcal{L}} & \mathfrak{X}^{\perp}(M) \\ \downarrow F & & \downarrow F \\ C^{\infty}_{-}(\widetilde{M}) & \xrightarrow{\widetilde{\mathcal{L}}^{-}} & C^{\infty}_{-}(\widetilde{M}). \end{array}$$

In fact, if $\widetilde{\xi} = fN$, given that Π is a local isometry, it is easy to check that $\widetilde{\Delta^{\perp}\xi} = (\widetilde{\Delta} f)N$, $\widetilde{\mathcal{A}(\xi)} = \|\widetilde{A}\|^2 fN$, and $\widetilde{\mathcal{R}(\xi)} = \overline{\mathrm{Ricc}}(N)fN$. This implies that $\widetilde{\mathcal{L}(\xi)} = (\widetilde{\mathcal{L}}^{-} f)N$, which proves that the above diagram is commutative.

The above considerations justify the following definition.

DEFINITION 2.31. Let $\Phi : M^n \to \overline{M}^{n+1}$ be a one-sided minimal hypersurface of a nonorientable manifold M into an orientable manifold $\overline{M}^{n+1}$. Let $\Pi : \widetilde{M} \to M$ be the oriented double covering of M and $\Psi = \Phi \circ \Pi : \widetilde{M} \to \overline{M}^{n+1}$ the corresponding two-sided minimal hypersurface. We define the **index** of Φ, and it will be denoted by $\mathrm{Ind}(\Phi)$ or $\mathrm{Ind}(M)$, as

$$\mathrm{Ind}(\Phi) = \mathrm{Ind}(M) := \mathrm{Ind}(\widetilde{\mathcal{L}}^{-}),$$

where $\widetilde{\mathcal{L}} : C^{\infty}(\widetilde{M}) \to C^{\infty}(\widetilde{M})$ is the Jacobi operator of the two-sided minimal hypersurface Ψ and $\mathrm{Ind}(\widetilde{\mathcal{L}}^{-})$ was defined in (2.28).

REMARK 2.32. If M is not compact, then the index of M can be infinite. Also, if M is compact, then $\widetilde{M}$ is also compact, and so from Remark 2.24 we have that $\mathrm{Ind}(\widetilde{\mathcal{L}}) = \mathrm{Ind}(\widetilde{\mathcal{L}}, \widetilde{M})$. So $\mathrm{Ind}(\widetilde{\mathcal{L}}^{-}) = \mathrm{Ind}(\widetilde{\mathcal{L}}^{-}, \widetilde{M})$, and thus the two definitions of the index of M, given in Definitions 2.22 and 2.31, coincide.

CASE 2. $\Phi : M^n \to (\overline{M}^{n+1}, g)$ is a one-sided minimal hypersurface of an orientable manifold M into a nonorientable manifold $\overline{M}$.

Let $\Pi : (\widehat{M}^{n+1}, \widehat{g}) \to (\overline{M}^{n+1}, g)$ be the orientable double covering of $\overline{M}$ and $\{Id, \tau\}$ the group of transformations of the covering. Then τ is an isometry of $(\widehat{M}^{n+1}, \widehat{g})$ without fixed points and with $\tau^2 = Id$.

Now, if Φ admits a lift to $\widehat{M}$, i.e., there exists $\widehat{\Phi} : M \to \widehat{M}$ such that $\Pi \circ \widehat{\Phi} = \Phi$, then $\widehat{\Phi}$ is a hypersurface of an orientable manifold M^n into an orientable manifold $\widehat{M}$, and so, from Lemma 2.2, $\widehat{\Phi}$ is two-sided. This implies that Φ is also two-sided, which contradicts that Φ is one-sided. So, Φ does not admit a lift to $\widehat{M}$, and therefore Π induces on M a non-trivial double covering $\widetilde{\Pi} : \widetilde{M} \to M$ given by

$$\widetilde{M} = \{(p, x) \mid p \in M,\ x \in \Pi^{-1}(\Phi(p))\}, \quad \widetilde{\Pi}(p, x) = p.$$

So we can define $\widetilde{\Phi} : \widetilde{M} \to \widehat{M}$ by $\widetilde{\Phi}(p, x) = x$. If $\widetilde{\tau}$ is the cover transformation of $\widetilde{M}$ with $\widetilde{\tau}^2 = Id$, then $\widetilde{\Phi} \circ \widetilde{\tau} = \tau \circ \widetilde{\Phi}$ and the following diagram is commutative:

$$\begin{array}{ccc} \widetilde{M} & \xrightarrow{\ \widetilde{\Phi}\ } & \widehat{M} \\ \Big\downarrow{\scriptstyle \widetilde{\Pi}} & & \Big\downarrow{\scriptstyle \Pi} \\ M & \xrightarrow{\ \Phi\ } & \overline{M}. \end{array}$$

Since $\widetilde{M}$ is also orientable, $\widetilde{\Phi}$ is a two-sided minimal hypersurface and hence there exists a unit normal vector field N to $\widetilde{\Phi}$. It is clear that for any $\tilde{p} \in \widetilde{M}$, $N_{\widetilde{\tau}(\tilde{p})}$ and $(\mathrm{d}\tau)_{\widetilde{\Phi}(\tilde{p})}(N_{\tilde{p}})$ are unit vectors on the line $T^{\perp}_{\widetilde{\tau}(\tilde{p})}\widetilde{M}$ and hence $(\mathrm{d}\tau)_{\widetilde{\Phi}(\tilde{p})}(N_{\tilde{p}}) = \pm N_{\widetilde{\tau}(\tilde{p})}$ for any $\tilde{p} \in \widetilde{M}$. An argument similar to the one in case 1, shows that either $(\mathrm{d}\tau)_{\widetilde{\Phi}(\tilde{p})}(N_{\tilde{p}}) = N_{\widetilde{\tau}(\tilde{p})}$ for any $\tilde{p} \in \widetilde{M}$ or $(\mathrm{d}\tau)_{\widetilde{\Phi}(\tilde{p})}(N_{\tilde{p}}) = -N_{\widetilde{\tau}(\tilde{p})}$ for any $\tilde{p} \in \widetilde{M}$. If the first assertion holds, then N can be projected to a unit normal vector field on the hypersurface Φ, which is a contradiction because Φ is one-sided. Hence

$$(\mathrm{d}\tau)_{\widetilde{\Phi}(\tilde{p})}(N_{\tilde{p}}) = -N_{\widetilde{\tau}(\tilde{p})}, \quad \forall \tilde{p} \in \widetilde{M}. \tag{2.29}$$

Now if $\xi \in \mathfrak{X}^\perp(M)$, we consider its lift $\widetilde{\xi}$ to $\mathfrak{X}^\perp(\widetilde{M})$ defined by

$$\widetilde{\xi}_{\tilde{p}} = \left((\mathrm{d}\Pi)_{\widetilde{\Phi}(\tilde{p})}\right)^{-1}(\xi_{\widetilde{\Pi}(\tilde{p})}), \qquad \forall \tilde{p} \in \widetilde{M}.$$

Then $\widetilde{\xi} = fN$ for certain $f \in C^\infty(\widetilde{M})$. Since $\widetilde{\xi}$ projects to ξ, we have that $(\mathrm{d}\tau)_{\widetilde{\Phi}(\tilde{p})}(\widetilde{\xi}_{\tilde{p}}) = \widetilde{\xi}_{\tilde{\tau}(\tilde{p})}$, and so, taking into account (2.29), we have that $f \circ \tilde{\tau} = -f$. This gives an isomorphism

$$F : \mathfrak{X}^\perp(M) \to C^\infty_-(\widetilde{M})$$

$$F(\xi) = f, \quad \text{with} \quad \widetilde{\xi} = fN, \quad \widetilde{\xi}_{\tilde{p}} = \left((\mathrm{d}\Pi)_{\widetilde{\Phi}(\tilde{p})}\right)^{-1}(\xi_{\widetilde{\Pi}(\tilde{p})}), \qquad \forall \tilde{p} \in \widetilde{M}.$$

Now if $\mathcal{L} : \mathfrak{X}^\perp(M) \to \mathfrak{X}^\perp(M)$ is the Jacobi operator of the minimal hypersurface Φ and $\widetilde{\mathcal{L}} : C^\infty(\widetilde{M}) \to C^\infty(\widetilde{M})$ is the Jacobi operator of the two-sided minimal hypersurface $\widetilde{\Phi}$, then an argument similar to the one case in 1 proves that the next diagram is commutative:

$$\begin{array}{ccc} \mathfrak{X}^\perp(M) & \xrightarrow{\mathcal{L}} & \mathfrak{X}^\perp(M) \\ \downarrow F & & \downarrow F \\ C^\infty_-(\widetilde{M}) & \xrightarrow{\widetilde{\mathcal{L}}^-} & C^\infty_-(\widetilde{M}). \end{array}$$

The above considerations justify the following definition.

Definition 2.33. Let $\Phi : M^n \to \overline{M}^{n+1}$ be a one-sided minimal hypersurface of an orientable manifold M into a nonorientable manifold $\overline{M}^{n+1}$. Let $\Pi : \widehat{M} \to \overline{M}$ be the oriented double covering of $\overline{M}$ and $\widetilde{\Pi} : \widetilde{M} \to M$ the induced double covering. Then, the lift $\widetilde{\Phi} : \widetilde{M} \to \widehat{M}$ of Φ to $\widehat{M}$ is a two-sided minimal hypersurface satisfying $\Pi \circ \widetilde{\Phi} = \Phi \circ \widetilde{\Pi}$. We define the **index** of Φ, and it will be denoted by $\mathrm{Ind}(\Phi)$ or $\mathrm{Ind}(M)$, as

$$\mathrm{Ind}(\Phi) = \mathrm{Ind}(M) := \mathrm{Ind}(\widetilde{\mathcal{L}}^-),$$

where $\widetilde{\mathcal{L}} : C^\infty(\widetilde{M}) \to C^\infty(\widetilde{M})$ is the Jacobi operator of the hypersurface $\widetilde{\Phi}$ and $\mathrm{Ind}(\widetilde{\mathcal{L}}^-)$ was defined in (2.28).

Remark 2.34. If M is not compact, then the index of M can be infinite. Also, if M is compact, then $\widetilde{M}$ is also compact, and so from Remark 2.24 we have that $\mathrm{Ind}(\widetilde{\mathcal{L}}) = \mathrm{Ind}(\widetilde{\mathcal{L}}, \widetilde{M})$. So $\mathrm{Ind}(\widetilde{\mathcal{L}}^-) = \mathrm{Ind}(\widetilde{\mathcal{L}}^-, \widetilde{M})$, and thus the two definitions of the index of M, given in Definitions 2.22 and 2.33, coincide.

Case 3. $\Phi : M^n \to (\overline{M}^{n+1}, g)$ is a one-sided minimal hypersurface of a nonorientable manifold M into a nonorientable manifold $\overline{M}$.

As in case 2, let $\Pi : (\widehat{M}^{n+1}, \widehat{g}) \to (\overline{M}^{n+1}, g)$ be the orientable double covering of $\overline{M}$ and $\{Id, \tau\}$ the group of transformations of the covering. Then τ is an isometry of $(\widehat{M}^{n+1}, \widehat{g})$ without fixed points and with $\tau^2 = Id$.

Case 3.1. Φ admits a lift to $\widehat{M}$, i.e., there exists $\widehat{\Phi} : M \to \widehat{M}$ such that $\Pi \circ \widehat{\Phi} = \Phi$. Then $\widehat{\Phi}$ is a one-sided minimal hypersurface of a nonorientable manifold M in an orientable manifold $\widehat{M}$. Since Π is a local isometry, then it is easy to check that $F : \mathfrak{X}^\perp_\Phi(M) \to \mathfrak{X}^\perp_{\widehat{\Phi}}(M)$ defined by

$$F(\xi)_p = \left(\mathrm{d}\Pi_{\widehat{\Phi}}\right)^{-1}(\xi_p), \quad \forall p \in M,$$

is an isomorphism and the Jacobi operators $\mathcal{L}$ and $\widehat{\mathcal{L}}$ of Φ and $\widehat{\Phi}$ satisfy $\widehat{\mathcal{L}} \circ F = F \circ \mathcal{L}$, respectively. Using the definition of the index of $\widehat{\Phi}$ given in Definition 2.31, we define the index of Φ as $\operatorname{Ind}(\Phi) := \operatorname{Ind}(\widehat{\Phi})$.

CASE 3.2. Φ does not admit a lift to $\widehat{M}$. Following the arguments of case 2, let $\widetilde{\Pi} : \widetilde{M} \to M$ be the non-trivial induced bundle of Π over M. The transformation $\widetilde{\tau}$ of $\widetilde{M}$ and the minimal immersion $\widetilde{\Phi} : \widetilde{M} \to \widehat{M}$ satisfy $\widetilde{\Phi} \circ \widetilde{\tau} = \tau \circ \widetilde{\Phi}$ and $\Pi \circ \widetilde{\Phi} = \Phi \circ \widetilde{\Pi}$.

Now we have two possibilities for $\widetilde{M}$, to be orientable or not.

If $\widetilde{M}$ is orientable, then $\widetilde{\Phi}$ is two-sided and following case 2, we define the index of Φ as the index of $\widetilde{\mathcal{L}}^-$, where $\mathcal{L} : C^\infty(\widetilde{M}) \to C^\infty(\widetilde{M})$ is the Jacobi operator of $\widetilde{\Phi}$ and $\widetilde{\mathcal{L}}^-$ is the restriction of $\widetilde{\mathcal{L}}$ to $C^\infty_-(\widetilde{M}) = \{f \in C^\infty(\widetilde{M}) \mid f \circ \widetilde{\tau} = -f\}$. Note that although M is not orientable in this case, the arguments from case 2 are still valid.

If $\widetilde{M}$ is not orientable, let $\widehat{\Pi} : \Sigma \to \widetilde{M}$ be the orientable double covering of $\widetilde{M}$, let $\{Id, \widehat{\tau}\}$ be the transformation group of the covering $\widehat{\Pi}$, and let $\Psi : \Sigma \to \widehat{M}$ be the two-sided minimal immersion, so that the following diagrams are commutative:

$$\begin{array}{ccc}
(\Sigma, \widehat{\tau}) & & \\
\downarrow{\scriptstyle \widehat{\Pi}} & \searrow^{\Psi} & \\
(\widetilde{M}, \widetilde{\tau}) & \xrightarrow{\ \widetilde{\Phi}\ } & (\widehat{M}, \tau) \\
\downarrow{\scriptstyle \widetilde{\Pi}} & & \downarrow{\scriptstyle \Pi} \\
M & \xrightarrow{\ \Phi\ } & \overline{M}
\end{array}
\qquad \Psi = \widetilde{\Phi} \circ \widehat{\Pi}, \qquad \Pi \circ \widetilde{\Phi} = \Phi \circ \widetilde{\Pi}.$$

Now, since $\widetilde{\Phi}$ is one-sided, we can decompose $\mathfrak{X}^\perp(\widetilde{M})$ into

$$\mathfrak{X}^\perp(\widetilde{M}) = \mathfrak{X}^\perp_+(\widetilde{M}) \oplus \mathfrak{X}^\perp_-(\widetilde{M}),$$

where

$$\mathfrak{X}^\perp_\pm(\widetilde{M}) = \{\widetilde{\xi} \in \mathfrak{X}^\perp(\widetilde{M}) \mid \widetilde{\xi} \circ \widetilde{\tau} = \pm\widetilde{\xi}\}.$$

Now if $\xi \in \mathfrak{X}^\perp(M)$, we consider its lift $\widetilde{\xi}$ to $\mathfrak{X}^\perp(\widetilde{M})$ defined by $\widetilde{\xi}_{\widetilde{p}} = \xi_{\widetilde{\Pi}(\widetilde{p})}$, $\forall \widetilde{p} \in \widetilde{M}$. It is easy to check that $\widetilde{\xi} \in \mathfrak{X}^\perp_+(\widetilde{M})$ and that the map

$$\begin{aligned} F : \mathfrak{X}^\perp(M) &\to \mathfrak{X}^\perp_+(\widetilde{M}) \\ \xi &\mapsto F(\xi) = \widetilde{\xi} \end{aligned}$$

is an isomorphism. Furthermore, if $\mathcal{L}$ and $\widetilde{\mathcal{L}}$ are the Jacobi operators of the hypersurfaces Φ and $\widetilde{\Phi}$, respectively, then $F \circ \mathcal{L} = \widetilde{\mathcal{L}}^+ \circ F$, where $\widetilde{\mathcal{L}}^+ = \widetilde{\mathcal{L}}_{|\mathfrak{X}^\perp_+(\widetilde{M})}$.

Now we can lift the isometry $\widetilde{\tau}$ to Σ as a map $\tau' : \Sigma \to \Sigma$ that satisfies $\widetilde{\tau} \circ \widehat{\Pi} = \widehat{\Pi} \circ \tau'$ and $\Psi \circ \tau' = \Psi$. Then τ' has two possibilities: either $(\tau')^2 = Id$ or $(\tau')^2 = \widehat{\tau}$. In the first case the group of transformations of the covering $\widetilde{\Pi} \circ \widehat{\Pi}$ is isomorphic to $\mathbb{Z}_2 \times \mathbb{Z}_2$ and in the second one isomorphic to $\mathbb{Z}_4$. But since Ψ is two-sided, let N be a unit normal vector field to Ψ, which trivializes $T^\perp\Sigma$. Since $\widetilde{\Phi}$ is one-sided, $N \circ \widehat{\tau} = -N$. So, $(\tau')^2 = \widehat{\tau}$ is impossible and $(\tau')^2 = Id$. Following the above arguments, we have an isomorphism

$$G : \mathfrak{X}^\perp(M) \to C^\infty_{--}(\Sigma) = \{f \in C^\infty(\Sigma) \mid f \circ \widehat{\tau} = -f,\ f \circ \tau' = -f\}$$

$$G(\xi) = f, \quad \text{with} \quad \widehat{\eta} = fN, \quad \widehat{\eta}_x = \left((\mathrm{d}\Pi)_{\Psi(x)}\right)^{-1}(\xi_{(\widetilde{\Pi} \circ \widehat{\Pi})(x)}), \qquad \forall x \in \Sigma.$$

Now if $\mathcal{L} : \mathfrak{X}^{\perp}(M) \to \mathfrak{X}^{\perp}(M)$ is the Jacobi operator of the minimal hypersurface Φ and $\widehat{\mathcal{L}} : C^{\infty}(\Sigma) \to C^{\infty}(\Sigma)$ is the Jacobi operator of the two-sided minimal hypersurface Ψ, then the next diagram is commutative:

$$\begin{array}{ccc} \mathfrak{X}^{\perp}(M) & \xrightarrow{\mathcal{L}} & \mathfrak{X}^{\perp}(M) \\ \downarrow G & & \downarrow G \\ C^{\infty}_{--}(\Sigma) & \xrightarrow{\widehat{\mathcal{L}}^{--}} & C^{\infty}_{--}(\Sigma), \end{array}$$

where $\widehat{\mathcal{L}}^{--}$ is the restriction of $\widehat{\mathcal{L}}$ to $C^{\infty}_{--}(\Sigma)$. In the latter case we define $\operatorname{Ind}(M) := \operatorname{Ind}(\widehat{\mathcal{L}}^{--})$.

As in cases 1 and 2, the index in this case can be infinite, and if M is compact, the definitions of the index of M are the same as in Definition 2.22

2.8. General properties of the index and nullity

In this section we study some simple properties of the index of a minimal immersion.

In the previous section, we have related the index of a hypersurface to that of some double covering of it. In the first result of this section, we will relate the index of a minimal immersion to that of any covering of it.

PROPOSITION 2.35. *Let $\Phi : M \to (\overline{M}, g)$ be a minimal immersion, $\Pi : \widetilde{M} \to M$ be a covering map, and $\Psi = \Phi \circ \Pi : \widetilde{M} \to (\overline{M}, g)$ be the corresponding minimal immersion.*

(1) *The linear map*

$$F : \mathfrak{X}^{\perp}(M) \to \mathfrak{X}^{\perp}(\widetilde{M}), \quad \textit{given by} \quad F(\xi) = \xi \circ \Pi,$$

is a monomorphism satisfying $\mathcal{L}^{\Psi} \circ F = F \circ \mathcal{L}^{\Phi}$.

(2) *If $\widetilde{M}$ is compact, then $\operatorname{Ind}(\Psi) \geq \operatorname{Ind}(\Phi)$ and $\operatorname{Nul}(\Psi) \geq \operatorname{Nul}(\Phi)$.*

(3) *If Φ is a hypersurface and the cover Π is finite, then $\operatorname{Ind}(\Psi) \geq \operatorname{Ind}(\Phi)$.*

(4) *If Φ is a two-sided stable minimal hypersurface and M is not compact, then Ψ is also stable.*

REMARK 2.36. If the cover Π is not finite, the conclusion of 3 is false, as the following example, due to R. Schoen and pointed out by the referee, proves.

Let $\Pi : \mathbb{H}^2 \to (M, g_0)$ be the universal covering of a hyperbolic compact surface M and $\mathbb{H}^2$ the hyperbolic plane. We take $\overline{M} = (-4, 4) \times M$, endowed with the warped metric $g = dt^2 + f(t)^2 g_0$ with $f(t) = -t^2/16 + 1$. Then $\Phi : (M, g_0) \to (\overline{M}, g)$ defined by $\Phi(p) = (0, p)$ is a two-sided totally geodesic isometric embedding. The corresponding Jacobi eperator is $\mathcal{L} = \Delta + \frac{1}{4}$ and so $\operatorname{Ind}(\Phi) > 0$.

If we consider $\Psi = \Phi \circ \Pi : \mathbb{H}^2 \to \overline{M}$, its Jacobi operator is $\widetilde{\mathcal{L}} = \widetilde{\Delta} + \frac{1}{4}$. But it is well-known that, for any compactly supported open domain $\Omega \subset \mathbb{H}^2$, $\lambda_1(\Omega) \geq \frac{1}{4}$, where λ_1 is the first eigenvalue of the Laplacian with Dirichlet condition for Ω. So, for any $f \in C_0^{\infty}(\mathbb{H}^2)$, we have

$$\widetilde{Q}(f) = \int_{\mathbb{H}^2} \{\|\widetilde{\nabla} f\|^2 - \frac{1}{4} f^2\} \, d\widetilde{v} \geq 0,$$

which implies that $\operatorname{Ind}(\Psi) = 0$.

PROOF. It is clear that F is a monomorphism. Also, since Π is a local isometry, it follows that $\mathcal{L}^{\Psi} \circ F = F \circ \mathcal{L}^{\Phi}$. This proves 1.

Now 2 and 3 follow from 1, taking into account the definitions of index and nullity given in the previous sections and that, if the cover Π is finite, then $F(\mathfrak{X}_0^{\perp}(M)) \subset \mathfrak{X}_0^{\perp}(\widetilde{M})$.

Finally, the minimal immersion Ψ is also two-sided and the Jacobi operators $\mathcal{L}^{\Phi}$ and $\mathcal{L}^{\Psi}$ satisfy $\mathcal{L}^{\Psi}(f \circ \Pi) = \mathcal{L}^{\Phi} f \circ \Pi$ for any $f \in C^{\infty}(M)$. From Theorem 2.25, 2, there exists a positive function u on M such that $\mathcal{L}^{\Phi} u = 0$. So $u \circ \Pi$ is a positive function on $\widetilde{M}$ with $\mathcal{L}^{\Psi}(u \circ \Pi) = 0$. Now 4 follows from Theorem 2.25, 1. □

PROPOSITION 2.37. *Let $\Phi : M \to (M_1, g_1)$ be a minimal immersion and $\Psi : (M_1, g_1) \to (M_2, g_2)$ be a totally geodesic isometric immersion. Then:*

(1) *$\Psi \circ \Phi$ is also a minimal immersion.*
(2) *If M is compact,* $\operatorname{Ind}(\Psi \circ \Phi) \geq \operatorname{Ind}(\Phi)$ *and* $\operatorname{Nul}(\Psi \circ \Phi) \geq \operatorname{Nul}(\Phi)$.
(3) *If M is compact and* $\dim M_1 = \dim M_2$, *i.e., Ψ is a local isometry,* $\operatorname{Ind}(\Psi \circ \Phi) = \operatorname{Ind}(\Phi)$ *and* $\operatorname{Nul}(\Psi \circ \Phi) = \operatorname{Nul}(\Phi)$.
(4) *If $\Phi : M \to M_1$ is a hypersurface and Ψ is a local isometry,* $\operatorname{Ind}(\Psi \circ \Phi) = \operatorname{Ind}(\Phi)$.

PROOF. Since Ψ is a totally geodesic immersion and Φ is a minimal immersion, $\Psi \circ \Phi$ is also a minimal immersion.

We relate the spaces of normal sections of both minimal immersions by the linear map

$$F : \mathfrak{X}_{\Phi}^{\perp}(M) \to \mathfrak{X}_{\Psi\circ\Phi}^{\perp}(M)$$
$$F(\xi)_p = \mathrm{d}\Psi_{\Phi(p)}(\xi_p), \quad \forall p \in M,$$

which is clearly a monomorphism.

Given that Ψ is a totally geodesic immersion together with (2.1) and (2.2), it is not difficult to prove that the Jacobi operators $\mathcal{L}^{\Phi}$ and $\mathcal{L}^{\Psi\circ\Phi}$ of the minimal immersions Φ and $\Psi \circ \Phi$ satisfy

$$\mathcal{L}^{\Psi\circ\Phi} \circ F = F \circ \mathcal{L}^{\Phi}.$$

Since F is a monomorphism, the above equation proves 2.

If Ψ is a local isometry, F is an isomorphism with

$$F^{-1} : \mathfrak{X}_{\Psi\circ\Phi}^{\perp}(M) \to \mathfrak{X}_{\Phi}^{\perp}(M)$$
$$F^{-1}(\eta)_p = \left(\mathrm{d}\Psi_{\Phi(p)}\right)^{-1}(\eta_p), \quad \forall p \in M,$$

and so the above relation proves 3 and 4. □

In the following result we analyze the effect on the index and nullity of the homotheties of the ambient space.

PROPOSITION 2.38. *Let $\Phi : M \to (\overline{M}, g)$ be a minimal immersion and $\overline{g} = \lambda g$, $\lambda \in \mathbb{R}, \lambda > 0$, be a metric on $\overline{M}$ homothetic to g. If we denote by $\Phi_\lambda : M \to (\overline{M}, \overline{g})$ the immersion Φ itself, then Φ_λ is also a minimal immersion and*

(1) *If M is compact, then* $\operatorname{Ind}(\Phi) = \operatorname{Ind}(\Phi_\lambda)$ *and* $\operatorname{Nul}(\Phi) = \operatorname{Nul}(\Phi_\lambda)$.
(2) *If Φ is a hypersurface, then* $\operatorname{Ind}(\Phi) = \operatorname{Ind}(\Phi_\lambda)$.

PROOF. From Proposition 1.5, the Levi-Civita connections on $\overline{M}$ associated to $\overline{g}$ and g are the same. As the normal bundles of Φ and Φ_λ are the same, it follows that the second fundamental forms, as well as the normal connections, of both immersions are the same, and thus, as the induced metrics on M are homothetic, the Jacobi operators $\mathcal{L}^\lambda$ and $\mathcal{L}$ of the immersions Φ_λ and Φ are related by $\mathcal{L}^\lambda = \lambda^{-1}\mathcal{L}$, and the proof follows. □

PROPOSITION 2.39 (Graphs). *Let (M^n, g) be a Riemannian manifold and $f : M^n \to \mathbb{R}$ be a smooth map satisfying*

$$\operatorname{div}\left(\frac{\nabla f}{\sqrt{1+\|\nabla f\|^2}}\right) = 0.$$

Then the graph of f

$$\Phi : M^n \to (M^n \times \mathbb{R}, \overline{g} = g \times \mathrm{d}t^2)$$
$$\Phi(p) = (p, f(p))$$

is a two-sided stable minimal embedded hypersurface. Moreover, if M is orientable, then Φ is a calibrated immersion for some calibration on $(M^n \times \mathbb{R}, \overline{g})$. In particular, if M is compact, M is volume minimizing in its homology class.

PROOF. It is clear that Φ is an embedding and the metric induced by $\overline{g}$ on M is $h = g + \mathrm{d}f \otimes \mathrm{d}f$. Also the vector field N given by

$$N = \frac{1}{\sqrt{1+\|\nabla f\|^2}}(-\nabla f, 1),$$

is a unit normal vector field on Φ, and hence Φ is a two-sided hypersurface.

Then the Weingarten endomorphism A of Φ is given by

$$h(Av, w) = \frac{(\nabla^2 f)(v, w)}{\sqrt{1+\|\nabla f\|^2}}, \quad v, w \in TM.$$

Now, if $\{e_1, \dots, e_n\}$ is a local orthonormal reference on (M, g), then $h_{ij} = \delta_{ij} + A_{ij}$ for any $1 \leq i, j \leq n$, where $A_{ij} = \mathrm{d}f(e_i)\mathrm{d}f(e_j)$. Then, the inverse matrix of h_{ij} is given by $h^{ij} = \delta_{ij} - (1 + \|\nabla f\|^2)^{-1} A_{ij}$. This implies, using the above expression for A, that the mean curvature H of Φ is given by

$$\begin{aligned} nH &= \sum_{i,j=1}^n h^{ij} h(Ae_i, e_j) = \frac{\Delta f}{\sqrt{1+\|\nabla f\|^2}} - \sum_{i,j=1}^n \frac{A_{ij}\,(\nabla^2 f)(e_i, e_j)}{(1+\|\nabla f\|^2)^{3/2}} \\ &= \frac{\Delta f}{\sqrt{1+\|\nabla f\|^2}} - \frac{(\nabla^2 f)(\nabla f, \nabla f)}{(1+\|\nabla f\|^2)^{3/2}} = \operatorname{div}\left(\frac{\nabla f}{\sqrt{1+\|\nabla f\|^2}}\right) = 0, \end{aligned}$$

and so Φ is a minimal immersion.

On the other hand, the vector field on $(M \times \mathbb{R}, \overline{g})$ defined by $X = (0, 1)$ is parallel and so it is a Killing vector field. From Proposition 2.12, the normal component of the restriction of X to M is a Jacobi field, and so $u = \overline{g}(N, X) = \frac{1}{\sqrt{1+\|\nabla f\|^2}}$ is a positive Jacobi function on M, i.e., $\mathcal{L}u = 0$, where $\mathcal{L}$ is the Jacobi operator of Φ. Now, Theorem 2.25, 1) proves that Φ is stable.

Assuming that M is orientable, we will define a calibration on $(M \times \mathbb{R}, \overline{g})$. We take the orientable manifold $\overline{M} = M \times \mathbb{R}$ with the orientation given by the basis $\{N_p, \mathrm{d}\Phi_p(e_1), \dots, \mathrm{d}\Phi_p(e_n)\}$ of $T_{(p,t)}\overline{M}$, where $\{e_1, \dots, e_n\}$ is an oriented basis of T_pM. Let α be the corresponding volume form of $(M \times \mathbb{R}, \overline{g})$.

Let $X \in \mathfrak{X}(M \times \mathbb{R})$ be the unit vector field given by

$$X_{(p,t)} = \frac{1}{\sqrt{1+\|\nabla f\|^2}}(-\nabla f, 1).$$

We observe that $N = X \circ \Phi$. We define $\eta \in \Omega^n(M \times \mathbb{R})$ by

$$\eta(X_1, \dots, X_n) = \alpha(X, X_1, \dots, X_n), \quad X_i \in \mathfrak{X}(M \times \mathbb{R}), \quad 1 \le i \le n,$$

i.e., η is the interior product of α with X, $\eta = i_X\alpha$. Then (see [**War83**]), we have that

$$\mathrm{d}\eta = \mathrm{d}i_X\alpha = L_X\alpha - i_X\mathrm{d}\alpha = L_X\alpha = (\operatorname{div} X)\alpha,$$

where $L_X\alpha$ is the Lie derivative of α with respect to the vector field X and $\operatorname{div} X$ is the divergence of X in $(M \times \mathbb{R}, \overline{g})$. Now it is easy to check that

$$(\operatorname{div} X)(p,t) = -\operatorname{div}\left(\frac{\nabla f}{\sqrt{1+\|\nabla f\|^2}}\right)(p) = 0,$$

and so η is a closed n-form on $M \times \mathbb{R}$.

Also, if $\{v_1, \dots, v_n\}$ are orthonormal vectors in $T_{(p,f(p))}\overline{M}$, and v_{n+1} is a unit vector such that $\{v_{n+1}, v_1 \dots, v_n\}$ is a positively oriented orthonormal basis of $T_{(p,f(p))}\overline{M}$, then

$$\eta_{(p,t)}(v_1, \dots, v_n) = \alpha_{(p,t)}(X_{(p,t)}, v_1, \dots, v_n) = \overline{g}(X, v_{n+1}) \le 1. \tag{2.30}$$

Thus, the n-form η defines a calibration on $(M \times \mathbb{R}, \overline{g})$ (Definition 2.16).

Now, for any $p \in M$, if $\{e_1, \dots, e_n\}$ is a positively oriented orthonormal basis of T_pM, then

$$\eta_{(p,f(p))}(\mathrm{d}\Phi_p(e_1), \dots, \mathrm{d}\Phi_p(e_n)) = \alpha_{(p,f(p))}(N_p, (e_1, \mathrm{d}f_p(e_1)), \dots, (e_n, \mathrm{d}f_p(e_n))) = 1,$$

because $\{N_p, (e_1, e_1(f)), \dots, (e_n, e_n(f))\}$ is a positively oriented orthonormal basis of $T_{(p,f(p))}\overline{M}$. This implies that Φ is a calibrated immersion with respect to the calibration η on $(\overline{M}, \overline{g})$. □

PROPOSITION 2.40 (Slices). *Let $\{(M_i, g_i),\ i = 1, 2\}$ be Riemannian manifolds. For each $a \in M_2$, the map $\Phi : (M_1, g_1) \to (M_1 \times M_2, g_1 \times g_2)$ defined by*

$$\Phi(p) = (p, a), \quad \forall p \in M_1,$$

is a stable totally geodesic isometric embedding. Furthermore, if M_1 is compact, then $\operatorname{Nul}(\Phi) = \dim M_2$.

PROOF. First, it is clear that Φ is a totally geodesic isometric embedding. Also, $T_p^\perp M_1 \equiv T_a M_2$ for any $p \in M_1$ and so, taking into account the expression for the curvature $\overline{R}$ of the product manifold, we have that $\overline{R}(\nu, v)w = 0$ for any $\nu \in T_p^\perp M_1$ and $v, w \in T_pM_1$. Hence $\mathcal{R}\xi = 0$, $\forall \xi \in \mathfrak{X}^\perp(M_1)$ and so the Jacobi operator of Φ is given by $\mathcal{L} = \Delta^\perp$. From here, the index form is

$$\mathcal{Q}(\xi) = \int_{M_1} \|\nabla^\perp \xi\|^2 \,\mathrm{d}v_1 \ge 0, \quad \forall \xi \in \mathfrak{X}_0^\perp(M_1),$$

and so Φ is stable.

Since $\mathfrak{X}^\perp(M_1) \equiv C^\infty(M_1, T_aM_2)$, if n is the dimension of M_2 and $\{v_1, \dots, v_n\}$ is an orthonormal basis of T_aM_2, then any section $\xi \in \mathfrak{X}^\perp(M_1)$ is given by

$$\xi = \sum_{i=1}^n f_i v_i, \quad f_i \in C^\infty(M_1).$$

If M_1 is compact, $\mathcal{L}(\xi) = \sum_{i=1}^{n}(\Delta f_i)v_i = 0$ if and only if $\Delta f_i = 0$ for all $1 \leq i \leq n$. This means that all f_i are constant and therefore $\mathrm{Nul}(\Phi) = n$. □

PROPOSITION 2.41 (Product immersion). *Let $\Phi_i : M_i \to (\overline{M}_i, g_i)$, $i = 1, 2$ be minimal immersions of compact manifolds M_i and $\Phi = \Phi_1 \times \Phi_2 : M_1 \times M_2 \to (\overline{M}_1 \times \overline{M}_2, g_1 \times g_2)$ the product immersion. Then Φ is a minimal immersion and $M_1 \times M_2$ is stable if and only if M_i, $i = 1, 2$, are stable.*

PROOF. Since $T^{\perp}_{(p,q)} M_1 \times M_2 \equiv T^{\perp}_p M_1 \times T^{\perp}_q M_2$ for any $(p, q) \in M_1 \times M_2$, the second fundamental form σ of Φ is given by

$$\sigma(v, w) = \big(\sigma_1(v_1, w_1), \sigma_2(v_2, w_2)\big), \quad \forall\, v = (v_1, v_2), w = (w_1, w_2) \in T_{(p,q)} M_1 \times M_2,$$

where σ_i is the second fundamental form of Φ_i, $i = 1, 2$. From this we see that Φ is a minimal immersion.

First we assume that Φ is stable. Then, given $\xi \in \mathfrak{X}^{\perp}(M_1)$, we consider $(\xi, 0) \in \mathfrak{X}^{\perp}(M_1 \times M_2)$. If $\mathcal{L}$ and $\mathcal{L}^1$ are the Jacobi operators of Φ and Φ_1, it is easy to check that $\mathcal{L}(\xi, 0) = (\mathcal{L}^1\xi, 0)$ and thus, using Fubini's theorem (Theorem 1.7), the corresponding index forms satisfy

$$\mathcal{Q}((\xi, 0)) = \mathcal{Q}_1(\xi)\,\mathrm{vol}(M_2, g_2).$$

This implies that $\mathcal{Q}_1(\xi) \geq 0$, $\forall \xi \in \mathfrak{X}^{\perp}(M_1)$ and so Φ_1 is stable. In the same way we would prove that Φ_2 is also stable.

Now we assume that Φ_i, $i = 1, 2$ are stable. If $\xi \in \mathfrak{X}^{\perp}(M_1 \times M_2)$, then

$$\xi_{(p,q)} = \big(\xi^1_{(p,q)}, \xi^2_{(p,q)}\big) \in T^{\perp}_p M_1 \times T^{\perp}_q M_2, \quad \forall\, (p, q) \in M_1 \times M_2.$$

We define two families of normal vector fields

$$\{\eta_q \mid q \in M_2\} \subset \mathfrak{X}^{\perp}(M_1) \quad \text{and} \quad \{\mu_p \mid p \in M_1\} \subset \mathfrak{X}^{\perp}(M_2)$$

as follows: $(\eta_q)_p = \xi^1_{(p,q)}$ for each $p \in M_1$ and $(\mu_p)_q = \xi^2_{(p,q)}$ for each $q \in M_2$.

Now it is not difficult to check that for any $(p, q) \in M_1 \times M_2$ we have

$$\mathcal{L}\xi_{(p,q)} = \big((\mathcal{L}^1\eta_q)_p, (\mathcal{L}^2\mu_p)_q\big),$$

and so, again using Fubini's theorem

$$\mathcal{Q}(\xi) = \int_{M_2} \mathcal{Q}_1(\eta_q)\,\mathrm{d}v_2 + \int_{M_1} \mathcal{Q}_2(\mu_p)\,\mathrm{d}v_1.$$

Since Φ_i, $i = 1, 2$ are stable, the last expression implies that $\mathcal{Q}(\xi) \geq 0$, $\forall \xi \in \mathfrak{X}^{\perp}(M_1 \times M_2)$, which means that Φ is stable. □

We end this section with two technical results that will be used throughout the book. In both results we assume that the Riemannian ambient manifold admits an isometric immersion in the Euclidean space, which allows to find good test eigensections to study the index and nullity of its minimal submanifolds.

The first result is inspired by the ideas developed by H. B. Lawson Jr. and J. Simons in [**LS73**].

PROPOSITION 2.42. *Let $\Phi : M^n \to (\overline{M}^m, g)$ be a minimal immersion of a compact manifold M. Suppose $\overline{M}$ admits an isometric immersion $\Psi : (\overline{M}, g) \to (\mathbb{R}^p, \langle , \rangle)$ in an Euclidean space. Then, if $\{a_1, \ldots, a_p\}$ is an orthonormal basis of*

$\mathbb{R}^p$, $a_\mu^\perp$ is the normal component in $T_\Phi^\perp M$ of the vector a_μ and $\mathcal{Q}$ the index form of Φ, we have that

$$\sum_{\mu=1}^{p} \mathcal{Q}(a_\mu^\perp) = \sum_{i=1}^{n}\sum_{r=1}^{m-n} \int_M \{-\langle\overline{\sigma}(e_i,e_i),\overline{\sigma}(\xi_r,\xi_r)\rangle + 2\|\overline{\sigma}(e_i,\xi_r)\|^2\}\, dv,$$

where $\{e_1,\ldots,e_n\}$ (respectively $\{\xi_1,\ldots,\xi_r\}$) is a local orthonormal reference on TM (respectively on $T_\Phi^\perp M$) and $\overline{\sigma}$ is the second fundamental form of the immersion Ψ.

PROOF. Given an arbitrary vector $a \in \mathbb{R}^p$, we decompose it as

$$a = a^\top + a^\perp + a^N,$$

where $a^\top$ is the tangential component of a and $a^\perp$ (respectively a^N) is the normal component in $T_\Phi^\perp M$ of a (respectively the normal component in $T^\perp\overline{M}$ of a). Then, using (2.1) we have that

$$\nabla_v a^\top = A_{a^\perp} v + \sum_{j=1}^{n}\langle\overline{\sigma}(v,e_j),a\rangle e_j,$$

$$\nabla_v^\perp a^\perp = -\sigma(v,a^\top) + \sum_{r=1}^{m-n}\langle\overline{\sigma}(v,\xi_r),a\rangle\xi_r.$$

From the last equations and using (2.1) and (2.2), we have that

$$\Delta^\perp a^\perp = -\mathcal{A}(a^\perp) + \sum_{i=1}^{n}\left(\overline{R}(a^\top,e_i)e_i\right)^\perp - \sum_{i,j=1}^{n}\langle\overline{\sigma}(e_i,e_j),a\rangle\sigma(e_i,e_j)$$
$$+\sum_{i=1}^{n}\sum_{r=1}^{m-n}\{\langle -\overline{A}_{\overline{\sigma}(e_i,\xi_r)}e_i + \overline{\sigma}(e_i,\nabla_{e_i}^\perp\xi_r) + (\nabla\overline{\sigma})(e_i,\xi_r,e_i) - \overline{\sigma}(e_i,A_{\xi_r}e_i),a\rangle\xi_r\}$$
$$+\sum_{i=1}^{n}\sum_{r=1}^{m-n}\langle\overline{\sigma}(e_i,\xi_r),a\rangle\nabla_{e_i}^\perp\xi_r,$$

where $\overline{A}$ is the Weingarten endomorphism of the immersion Ψ. Thus

$$\mathcal{L}a^\perp = \sum_{i=1}^{n}\left(\overline{R}(a^\top + a^\perp,e_i)e_i\right)^\perp - \sum_{i,j=1}^{n}\langle\overline{\sigma}(e_i,e_j),a\rangle\sigma(e_i,e_j) \tag{2.31}$$
$$+\sum_{i=1}^{n}\sum_{r=1}^{m-n}\{\langle -\overline{A}_{\overline{\sigma}(e_i,\xi_r)}e_i + \overline{\sigma}(e_i,\nabla_{e_i}^\perp\xi_r) + (\nabla\overline{\sigma})(e_i,\xi_r,e_i) - \overline{\sigma}(e_i,A_{\xi_r}e_i),a\rangle\xi_r\}$$
$$+\sum_{i=1}^{n}\sum_{r=1}^{m-n}\langle\overline{\sigma}(e_i,\xi_r),a\rangle\nabla_{e_i}^\perp\xi_r.$$

From (2.31) we get

$$\begin{aligned} g(\mathcal{L}a^\perp, a^\perp) &= \sum_{i=1}^n \overline{R}(a^\top + a^\perp, e_i, e_i, a^\perp) - \sum_{i,j=1}^n \langle \overline{\sigma}(e_i, e_j), a\rangle \langle \sigma(e_i, e_j), a\rangle \\ &+ \sum_{i=1}^n \{ \langle -\overline{A}_{\overline{\sigma}(e_i, a^\perp)} e_i) + (\nabla \overline{\sigma})(e_i, a^\perp, e_i) - \overline{\sigma}(e_i, A_{a^\perp} e_i), a\rangle \} \\ &+ \sum_{i=1}^n \sum_{r=1}^{m-n} \{ \langle \overline{\sigma}(e_i, \nabla^\perp_{e_i}\xi_r), a\rangle \langle \xi_r, a\rangle + \langle \overline{\sigma}(e_i, \xi_r), a\rangle \langle \nabla^\perp_{e_i}\xi, a\rangle \}. \end{aligned}$$

Now, considering that $\{a_1, \ldots, a_p\}$ is an orthonormal basis of the ambient space $\mathbb{R}^p$, we get from the last equation and the Gauss equation of Ψ

$$\begin{aligned} \sum_{\mu=1}^p g(\mathcal{L}a_\mu^\perp, a_\mu^\perp) &= \sum_{i=1}^n \sum_{\mu=1}^p \overline{R}(a_\mu^\perp, e_i, e_i, a_\mu^\perp) - \sum_{i=1}^n \sum_{\mu=1}^p \|\overline{\sigma}(e_i, a_\mu^\perp)\|^2 \\ &= \sum_{i=1}^n \sum_{r=1}^{m-n} \{ \overline{R}(\xi_r, e_i, e_i, \xi_r) - \|\overline{\sigma}(e_i, \xi_r)\|^2 \} \\ &= \sum_{i=1}^n \sum_{r=1}^{m-n} \{ \langle \overline{\sigma}(e_i, e_i), \overline{\sigma}(\xi_r, \xi_r)\rangle - 2\|\overline{\sigma}(e_i, \xi_r)\|^2 \} \end{aligned}$$

and the proposition follows. □

This second result is inspired by the ideas developed by A. Savo in [**Sav10**].

PROPOSITION 2.43. *Let $\Phi : M^n \to (\overline{M}^{n+1}, g)$ be a minimal hypersurface of a compact manifold M. Suppose $\overline{M}$ admits an isometric immersion $\Psi : (\overline{M}, g) \to (\mathbb{R}^p, \langle , \rangle)$ into an Euclidean space. Given a harmonic vector field X on M and an orthonormal basis $\{a_1, \ldots, a_p\}$ of $\mathbb{R}^p$, the normal sections*

$$\xi^X_{\lambda\mu} = \langle X, a_\mu\rangle a_\lambda^\perp - \langle X, a_\lambda\rangle a_\mu^\perp \in \mathfrak{X}^\perp_\Phi(M), \quad 1 \le \lambda, \mu \le p,$$

where $a_\mu^\perp$ is the normal component in $T^\perp_\Phi M$ of the vector a_μ, satisfy

$$\begin{aligned} \sum_{1\le\lambda<\mu\le p} \mathcal{Q}(\xi^X_{\lambda\mu}) &= 2\sum_{i=1}^n \int_M \{ \|\overline{\sigma}(X, e_i)\|^2 + \|X\|^2 \|\overline{\sigma}(N, e_i)\|^2 \} \, dv \\ &- \sum_{i=1}^n \int_M \{ \langle \overline{\sigma}(X, X) + \|X\|^2 \overline{\sigma}(N, N), \overline{\sigma}(e_i, e_i)\rangle \} \, dv, \end{aligned}$$

where $\overline{\sigma}$ is the second fundamental form of Ψ, $\{e_1, \ldots, e_n\}$ is a local orthonormal reference on TM, and N is a local unit vector field on $T^\perp M$.

REMARK 2.44. Although a similar calculation could be done in codimension greater than one, the result has applications only in the case of hypersurfaces.

PROOF. Let $a, b \in \mathbb{R}^p$ with $a \neq b$ and $\xi^X_{ab} \in \mathfrak{X}^\perp_\Phi(M)$ be the normal section

$$\xi^X_{ab} = \langle X, b\rangle a^\perp - \langle X, a\rangle b^\perp.$$

Then, from (2.1) we have

$$\begin{aligned} \nabla^\perp_{e_i}\xi^X_{ab} &= \langle \nabla_{e_i}X + \sigma(e_i, X) + \overline{\sigma}(e_i, X), b\rangle a^\perp + \langle X, b\rangle \nabla^\perp_{e_i} a^\perp \\ &- \langle \nabla_{e_i}X + \sigma(e_i, X) + \overline{\sigma}(e_i, X), a\rangle b^\perp - \langle X, a\rangle \nabla^\perp_{e_i} b^\perp. \end{aligned}$$

From the last equation and using (2.1) and (2.2) we get that

$$\begin{aligned}\mathcal{L}\xi_{ab}^X &= \langle \Delta X, b\rangle a^\perp + \langle X, b\rangle \mathcal{L}a^\perp - \langle \Delta X, a\rangle b^\perp - \langle X, a\rangle \mathcal{L}b^\perp \\ &\quad + 2\sum_{i=1}^n \langle \sigma(e_i, \nabla_{e_i}X) + \overline{\sigma}(e_i, \nabla_{e_i}X) + \overline{\sigma}(e_i, \sigma(X, e_i)), b\rangle a^\perp \\ &+ \sum_{i=1}^n \langle -A_{\sigma(e_i,X)}e_i - \overline{A}_{\overline{\sigma}(e_i,X)}e_i + (\nabla\sigma)(e_i, X, e_i) + (\overline{\nabla}\overline{\sigma})(e_i, X, e_i), b\rangle a^\perp \\ &\quad + 2\sum_{i=1}^n \langle \nabla_{e_i}X + \sigma(e_i, X) + \overline{\sigma}(e_i, X), b\rangle \nabla_{e_i}^\perp a^\perp \\ &\quad - 2\sum_{i=1}^n \langle \sigma(e_i, \nabla_{e_i}X) + \overline{\sigma}(e_i, \nabla_{e_i}X) + \overline{\sigma}(e_i, \sigma(X, e_i)), a\rangle b^\perp \\ &- \sum_{i=1}^n \langle -A_{\sigma(e_i,X)}e_i - \overline{A}_{\overline{\sigma}(e_i,X)}e_i + (\nabla\sigma)(e_i, X, e_i) + (\overline{\nabla}\overline{\sigma})(e_i, X, e_i), a\rangle b^\perp \\ &\quad - 2\sum_{i=1}^n \langle \nabla_{e_i}X + \sigma(e_i, X) + \overline{\sigma}(e_i, X), a\rangle \nabla_{e_i}^\perp b^\perp,\end{aligned}$$

where $\overline{A}$ is the Weingarten endomorphism of the immersion Ψ.

Now, since $\xi_{aa}^X = 0$ and $\xi_{ba}^X = -\xi_{ab}^X$, from the last equation we get

$$\begin{aligned}(2.32)\quad 2\sum_{1\le\lambda<\mu\le p} \langle \xi_{\lambda\mu}^X, \mathcal{L}\xi_{\lambda\mu}^X\rangle &= \sum_{1\le\lambda,\mu\le p} \langle \xi_{\lambda\mu}^X, \mathcal{L}\xi_{\lambda\mu}^X\rangle = 2\|X\|^2 \sum_{\lambda=1}^p \langle a_\lambda^\perp, \mathcal{L}a_\lambda^\perp\rangle \\ &+ 2\left(\langle \Delta X, X\rangle - \sum_{i=1}^n \{\|\sigma(e_i, X)\|^2 + \|\overline{\sigma}(e_i, X)\|^2\}\right)\left(\sum_{\lambda=1}^p \|a_\lambda^\perp\|^2\right) \\ &+ 4\sum_{i=1}^n \langle \nabla_{e_i}X, X\rangle \left(\sum_{\lambda=1}^p \langle \nabla_{e_i}^\perp a_\lambda^\perp, a_\lambda^\perp\rangle\right) - 4\sum_{i,\lambda} \langle X, a_\lambda\rangle \langle \nabla_{e_i}^\perp a_\lambda^\perp, \sigma(e_i, X)\rangle.\end{aligned}$$

We are going to compute the different terms of (2.32). From Proposition 2.42 we get

$$\sum_{\lambda=1}^p \langle a_\lambda^\perp, \mathcal{L}a_\lambda^\perp\rangle = \sum_{i=1}^n \{\langle \overline{\sigma}(e_i, e_i), \overline{\sigma}(N, N)\rangle - 2\|\overline{\sigma}(e_i, N)\|^2\}.$$

Also, as $\langle \Delta X, X\rangle = \mathrm{Ricc}(X)$, from the Gauss equation we have

$$\langle \Delta X, X\rangle = \sum_{i=1}^n \{\langle \overline{\sigma}(X, X), \overline{\sigma}(e_i, e_i)\rangle - \|\sigma(X, e_i)\|^2 - \|\overline{\sigma}(X, e_i)\|^2\}.$$

Since it is clear that $\sum_{\lambda=1}^p \|a_\lambda^\perp\|^2 = 1$, then

$$\sum_{\lambda=1}^p \langle \nabla_{e_i}^\perp a_\lambda^\perp, a_\lambda^\perp\rangle = \frac{1}{2} e_i\Big(\sum_{\lambda=1}^p \|a_\lambda^\perp\|^2\Big) = 0.$$

Finally, from Proposition 2.42,

$$\nabla_{e_i}^\perp a_\lambda^\perp = -\sigma(e_i, a_\lambda^\top) + \langle \overline{\sigma}(e_i, N), a_\lambda\rangle N,$$

where $a_\lambda^\top$ is the tangent component of a_λ. So

$$\sum_{i,\lambda}\langle X,a_\lambda\rangle\langle\nabla^\perp_{e_i}a^\perp_\lambda,\sigma(e_i,X)\rangle=-\sum_{i=1}^n\|\sigma(X,e_i)\|^2.$$

Putting all this information on (2.32) we get

$$\begin{aligned}\sum_{1\le\lambda<\mu\le p}\langle\xi^X_{\lambda\mu},\mathcal{L}\xi^X_{\lambda\mu}\rangle=\|X\|^2\sum_{i=1}^n\{\langle\overline{\sigma}(N,N),\overline{\sigma}(e_i,e_i)\rangle-2\|\overline{\sigma}(N,e_i)\|^2\}\\+\sum_{i=1}^n\{\langle\overline{\sigma}(X,X),\overline{\sigma}(e_i,e_i)\rangle-2\|\overline{\sigma}(X,e_i)\|^2\},\end{aligned}$$

which proves the result using the definition of the index form. □

2.9. Examples of minimal submanifolds

In this section we show some important and simple examples of minimal submanifolds, computing their indices and nullities. Many of them will be characterized in the following chapters. Throughout the book, a large number of examples of minimal submanifolds are presented, whose computation of index and nullity requires more sophisticated techniques.

EXAMPLE 2.45. *The totally geodesic embedding $\Psi:\mathbb{R}^m\to\mathbb{R}^p$ given by $\Psi(x)=(x,0)$ is stable. Moreover, if $\Phi:M^n\to\mathbb{R}^m$ is a stable minimal immersion, then $\Psi\circ\Phi:M^n\to\mathbb{R}^p$ is also a stable minimal immersion.*

PROOF. By Theorem 2.15, the minimal embedding Ψ is stable.

On the other hand, it is clear that $\{\xi_i\mid 1\le i\le p-m\}$ defined by

$$\xi_i=(0,\dots,0,\overset{m+i}{1},0\dots,0),\quad 1\le i\le p-m$$

is a global orthonormal reference on the normal bundle of Ψ, and thus any section $\nu\in\mathfrak{X}^\perp(\mathbb{R}^m)$ can be written as $\nu=\sum_{i=1}^{p-m}f_i\xi_i$, with $f_i\in C^\infty(\mathbb{R}^m)$. So if $\xi\in\mathfrak{X}^\perp_{\Psi\circ\Phi}(M)$, then ξ can be written as

$$\xi=\eta+\sum_{i=1}^{p-m}f_i\xi_i,\quad\text{with}\quad\eta\in\mathfrak{X}^\perp_\Phi(M)\quad\text{and}\quad f_i\in C^\infty(M).$$

Since the vector fields ξ_i are parallel and Ψ is totally geodesic, it is easy to check that

$$\mathcal{L}^{\Psi\circ\Phi}\xi=\mathcal{L}^\Phi\eta+\sum_{i=1}^{p-m}(\Delta f_i)\xi_i,$$

which implies that if ξ has compact support

$$\mathcal{Q}^{\Psi\circ\Phi}(\xi)=\mathcal{Q}^\Phi(\eta)+\sum_{i=1}^{p-m}\int_M\|\nabla f_i\|^2\,dv\ge\mathcal{Q}^\Phi(\eta).$$

This inequality proves the second statement in the example. □

EXAMPLE 2.46. *The totally geodesic embedding* $\Psi : \mathbb{S}^m \to \mathbb{S}^p$ *given by* $\Psi(x) = (x, 0)$ *has index* $p - m$ *and nullity* $(m+1)(p-m)$*. Furthermore, if* $\Phi : M^n \to \mathbb{S}^m$ *is a minimal immersion of a compact manifold* M*, then*

$$\mathrm{Ind}(\Psi \circ \Phi) = \mathrm{Ind}(\Phi) + (p-m) \sum_{\lambda_k < n} \mu(\lambda_k) \geq \mathrm{Ind}(\Phi) + (p-m),$$

$$\mathrm{Nul}(\Psi \circ \Phi) = \mathrm{Nul}(\Phi) + (p-m)\mu(n),$$

where λ_k *are the eigenvalues of* Δ *acting on* $C^\infty(M)$ *and* $\mu(\lambda_k)$ *their respective multiplicities.*

PROOF. First, since $\mathbb{S}^p$ has constant curvature 1, we have that $\mathcal{R}(\xi) = m\xi$ for any $\xi \in \mathfrak{X}^\perp(\mathbb{S}^m)$. Also, since Ψ is totally geodesic, $\mathcal{A} = 0$, and hence the Jacobi operator of Ψ is $\mathcal{L} = \Delta^\perp + mId$.

Considering the spheres embedded in the corresponding Euclidean spaces, it is clear that $\{\xi_i \mid 1 \leq i \leq p - m\}$ defined by

$$\xi_i = (0, \ldots, 0, \overset{m+1+i}{1}, 0 \ldots, 0), \quad 1 \leq i \leq p - m$$

is a global orthonormal reference on the normal bundle of Ψ, so any section $\xi \in \mathfrak{X}^\perp(\mathbb{S}^m)$ can be written as $\xi = \sum_{i=1}^{p-m} f_i \xi_i$, with $f_i \in C^\infty(\mathbb{S}^m)$. As $\nabla^\perp \xi_i = 0$, we get

$$\mathcal{L}\xi = \sum_{i=1}^{p-m} (\Delta f_i + m f_i)\xi_i.$$

Hence, λ is an eigenvalue of $\mathcal{L}$, i.e., $\mathcal{L}\xi + \lambda\xi = 0$ for certain nonzero $\xi \in \mathfrak{X}^\perp(\mathbb{S}^m)$ if and only if $\Delta f_i + (m+\lambda) f_i = 0$ for all $1 \leq i \leq p-m$. Now, from Theorem 1.18, the first two eigenvalues of the Laplacian of $\mathbb{S}^m$ with its canonical metric of constant curvature 1 are 0 (with multiplicity 1) and m (with multiplicity $m+1$). So we get that the first two eigenvalues of $\mathcal{L}$ are $-m$ (with multiplicity $p - m$) and 0 (with multiplicity $(m+1)(p-m)$). This proves the first statement of the example.

To prove the second one, we follow the argument of Example 2.45. If $\xi \in \mathfrak{X}^\perp_{\Psi\circ\Phi}$ then ξ can be written as

$$\xi = \eta + \sum_{i=1}^{p-m} f_i \xi_i, \quad \text{with} \quad \eta \in \mathfrak{X}^\perp_\Phi(M) \quad \text{and} \quad f_i \in C^\infty(M).$$

Since the vector fields ξ_i are parallel and Ψ is totally geodesic, it is easy to check that

$$\mathcal{L}^{\Psi\circ\Phi}\xi = \mathcal{L}^\Phi \eta + \sum_{i=1}^{p-m} (\Delta f_i + n f_i)\xi_i,$$

which implies that

$$\mathcal{L}^{\Psi\circ\Phi}\xi + \lambda\xi = 0 \Leftrightarrow \mathcal{L}^\Phi \eta + \lambda \eta = 0 \quad \text{and} \quad \Delta f_i + (n+\lambda) f_i = 0, \quad 1 \leq i \leq p - m.$$

This leads to the second statement, taking into account that 0 is always an eigenvalue of Δ and n is an eigenvalue of Δ if M^n is a minimal submanifold of a sphere, since $\Delta(\Psi \circ \Phi) + n(\Psi \circ \Phi) = 0$. □

The following result can be proved using arguments similar to those used in the last two examples.

EXAMPLE 2.47. *The totally geodesic embedding $\Psi : \mathbb{H}^m \to \mathbb{H}^p$ given by $\Psi(x) = (0, x)$ is strictly stable. Moreover, if $\Phi : M^n \to \mathbb{H}^m$ is a stable minimal immersion, then $\Psi \circ \Phi : M^n \to \mathbb{H}^p$ is also a stable minimal immersion.*

Unlike what happens in Euclidean space $\mathbb{R}^3$, where the only complete two-sided stable minimal surfaces are planes (see Theorem 4.2), in the case of the hyperbolic space $\mathbb{H}^3$ there are families of complete two-sided stable minimal surfaces that are not totally geodesic. We will present examples of minimal revolution surfaces of $\mathbb{H}^3$ given by H. Mori [**Mor81**] and M. do Carmo and M. Dacjzer [**dCD83**].

EXAMPLE 2.48 (Spherical catenoids [**Mor81**,**dCD83**]). *For each constant $a > 1/2$, let $\phi : \mathbb{R} \to \mathbb{R}$ be the function given by*

$$\phi(s) = \sqrt{a^2 - 1/4} \int_0^s \frac{\sqrt{a\cosh(2t) - 1/2}}{a\cosh(2t) + 1/2}\, \mathrm{d}t.$$

We define $\Phi_a : \mathbb{R} \times \mathbb{S}^1 \to \mathbb{H}^3 \subset \mathbb{R}^4_1$ by

$$\begin{aligned}\Phi_a(s, e^{i\theta}) = \Big(&\sqrt{a\cosh(2s) - 1/2}\,(\cos\theta, \sin\theta),\\ &\sqrt{a\cosh(2s) + 1/2}\,(\sinh\phi(s), \cosh\phi(s))\Big).\end{aligned}$$

*Then it is easy to check that the induced metric on $\mathbb{R} \times \mathbb{S}^1$ is complete and Φ_a is a two-sided minimal surface. Furthermore, in [**dCD83**] it was proved that Φ_a is an embedding and the family of embeddings $\{\Phi_a \mid a > 1/2\}$ are called **spherical catenoids**.*

PROPOSITION 2.49. *The family $\{\Phi_a : \mathbb{R} \times \mathbb{S}^1 \to \mathbb{H}^3,\, a \geq 17/2\}$ of spherical catenoids of $\mathbb{H}^3$ is stable.*

PROOF. The Jacobi operator of Φ_a is given by $\mathcal{L} = \Delta + \|A\|^2 - 2$ and it is easy to prove (see [**Mor81**]) that

$$\|A\|^2(s, e^{i\theta}) = \frac{2(a^2 - 1/4)}{(a\cosh 2s - 1/2)^2} \leq \frac{2(a^2 - 1/4)}{(a - 1/2)^2} = \frac{2(a + 1/2)}{a - 1/2}.$$

If Ω is a relatively compact domain of $\mathbb{R} \times \mathbb{S}^1$, then there exists a domain $\Omega' = (a, b) \times \mathbb{S}^1$ with $\overline{\Omega} \subset \Omega'$ and so the first eigenvalues of the Laplacian with Dirichlet conditions on the boundary satisfy $\lambda_1(\Omega') \leq \lambda_1(\Omega)$. Now using [**Che70a**], $\lambda_1(\Omega') \geq \big(h(\Omega')/2\big)^2$, where $h(\Omega')$ is the Cheeger constant of Ω' defined as

$$h(\Omega') = \inf_{E \subset \overline{\Omega'}} \frac{P(E)}{\mathrm{Area}(E)},$$

where $P(E)$ is the perimeter of E. As Ω' is a doubly connected domain and the Gauss curvature of $\mathbb{R} \times \mathbb{S}^1$ satisfies $K = -1 - \|A\|^2/2 \leq -1$, using a result of R. Osserman ([**Oss79**], Theorem 8,A) we obtain that $h(\Omega') \geq 1$. Hence $\lambda_1(\Omega) \geq 1/4$.

Now, if $a \geq 17/2$, for any function $f \in C_0^\infty(\Omega)$ we have

$$\mathcal{Q}(f) = \int_M \{\|\nabla f\|^2 - (\|A\|^2 - 2)f^2\}\, \mathrm{d}v \geq \int_M \left(\frac{1}{4} - \frac{2(a + 1/2)}{a - 1/2} + 2\right) f^2\, \mathrm{d}v \geq 0.$$

Since Ω is any relatively compact domain of $\mathbb{R} \times \mathbb{S}^1$, the above inequality proves that Φ_a is stable when $a \geq 17/2$. □

EXAMPLE 2.50 (Clifford hypersurfaces). *Given two integers* $n, m \geq 1$, *we consider the canonical embedding* $\Phi : \mathbb{S}^n(r_1) \times \mathbb{S}^m(r_2) \to \mathbb{S}^{n+m+1}$ *where* $r_1 = \sqrt{n/n+m}$ *and* $r_2 = \sqrt{m/n+m}$ *are the radii of the spheres. Then* Φ *is a two-sided minimal hypersurface with* $\mathrm{Ind}(\Phi) = n+m+3$ *and* $\mathrm{Nul}(\Phi) = (n+1)(m+1)$.

PROOF. First, from Lemma 2.2 Φ is a two-sided hypersurface and $N_{(p,q)} = (\sqrt{\frac{m}{n}}p, -\sqrt{\frac{n}{m}}q)$, $\forall p \in \mathbb{S}^n(r_1)$, $\forall q \in \mathbb{S}^m(r_2)$, is a unit normal vector field to Φ. It is also not difficult to check that the Weingarten endomorphism A associated to N is given by

$$A(v,0) = -\sqrt{\frac{m}{n}}(v,0), \quad A(0,w) = \sqrt{\frac{n}{m}}(0,w) \quad \forall v \in T\mathbb{S}^n, \forall w \in T\mathbb{S}^m.$$

So the trace of A is zero, i.e., Φ is a minimal hypersurface and $\|A\|^2 = n+m$. Since $\overline{\mathrm{Ricc}}(N) = n+m$, from (2.17), the Jacobi operator $\mathcal{L}$ of Φ is given by $\mathcal{L} = \Delta + 2(n+m)$.

From Theorem 1.18, it follows that the eigenvalues of the Laplacian of $\mathbb{S}^n(r_1) \times \mathbb{S}^m(r_2)$ with $r_1 = \sqrt{\frac{n}{n+m}}$ and $r_2 = \sqrt{\frac{m}{n+m}}$, are $\{0, n+m, 2(n+m), \dots\}$ with multiplicities $\{1, n+m+2, (n+1)(m+1), \dots\}$. This implies that the corresponding eigenvalues of the Jacobi operator $\mathcal{L}$ are $\{-2(n+m), -(n+m), 0, \dots\}$ with multiplicities $\{1, n+m+2, (n+1)(m+1), \dots\}$. Thus, the index and the nullity of the Clifford hypersurfaces are $n+m+3$ and $(n+1)(m+1)$, respectively. □

REMARK 2.51. The hypersurfaces studied in Examples 2.46 and 2.50 are two-sided compact minimal isoparametric (i.e. with constant principal curvatures) hypersurfaces with one or two constant principal curvatures. B. Solomon, [**Sol90a, Sol90b**], computed the index and nullity of two-sided compact minimal isoparametric hypersurfaces of the sphere with three constant principal curvatures. In Table 1 we show the list of the above hypersurfaces as well as their indices, nullities, and multiplicities of the principal curvatures,

TABLE 1. Minimal hypersurfaces of the sphere with three constant principal curvatures

Hypersurface	Dimension	Multiplicity	Index	Nullity
$\mathrm{SO}(3)/\mathbb{Z}_2 \times \mathbb{Z}_2$	3	1	20	7
$\mathrm{SU}(3)/T^2$	6	2	44	20
$\mathrm{Sp}(3)/\mathrm{Sp}(1)^3$	12	4	119	70
$\mathrm{F}_4/\mathrm{Spin}(8)$	24	8	377	273

where $\mathrm{Sp}(n)$ is the symplectic group, $\mathrm{Spin}(n)$ is the spinorial group and F_4 is one of the five exceptional compact simple Lie groups.

In the following two examples we introduce some two-sided hypersurfaces of the product of two spheres, which will be characterized by their indices in Section 5.5.

EXAMPLE 2.52. *Let* $\Phi : \mathbb{S}^{n_1-1} \times \mathbb{S}^{n_2} \to \mathbb{S}^{n_1} \times \mathbb{S}^{n_2}$ $(n_1 \geq 2)$ *be the two-sided totally geodesic embedding* $\Phi(p,q) = ((p,0),q)$, $\forall p \in \mathbb{S}^{n_1-1}$ *and* $\forall q \in \mathbb{S}^{n_2}$.

(1) *If* $n_2 \geq n_1 - 1$, *then*

$$\operatorname{Ind}(\Phi) = 1, \qquad \operatorname{Nul}(\Phi) = \begin{cases} n_1 & \text{if } n_2 > n_1 - 1, \\ n_1 + n_2 + 1 & \text{if } n_2 = n_1 - 1. \end{cases}$$

(2) *If* $n_2 < n_1 - 1$, *then*

$$\operatorname{Ind}(\Phi) = \sum_{\lambda_k < n_1 - 1} \mu(\lambda_k) \geq n_2 + 2, \qquad \operatorname{Nul}(\Phi) = n_1 + \mu(n_1 - 1),$$

where λ_k *are the eigenvalues of the Laplacian of* $\mathbb{S}^{n_2}$ *and* $\mu(\lambda_k)$ *are the corresponding multiplicities, where* $\mu(n_1 - 1) = 0$ *if* $n_1 - 1$ *is not an eigenvalue of* $\mathbb{S}^{n_2}$.

PROOF. From Lemma 2.2, Φ is a two-sided hypersurface and N defined by $N_{(p,q)} = ((0,1),0)$, $\forall (p,q) \in \mathbb{S}^{n_1-1} \times \mathbb{S}^{n_2}$, is a unit normal vector field on Φ. This implies that $\overline{\operatorname{Ricc}}(N) = n_1 - 1$ and hence the Jacobi operator of the embedding Φ is given by $\mathcal{L} = \Delta + n_1 - 1$.

From Theorem 1.18, the eigenvalues of the Laplacian of $\mathbb{S}^{n_1-1} \times \mathbb{S}^{n_2}$ are $\{\lambda_{kl} = \lambda_k + \mu_l, \lambda, l \geq 1\}$, where

$$\{\lambda_k = (k-1)(k+n_1-3) \mid k \geq 1\} \quad \text{and} \quad \{\mu_l = (l-1)(l+n_2-2) \mid l \geq 1\}$$

are the eigenvalues of the Laplacian of $\mathbb{S}^{n_1-1}$ and $\mathbb{S}^{n_2}$ respectively. So the eigenvalues of the Jacobi operator are $\{-(n_1-1) + \lambda_{kl} \mid k, l \geq 1\}$.

If $n_2 \geq n_1 - 1$, the corresponding eigenvalue of the Jacobi operator is negative only for $k = l = 1$. Since its multiplicity is one, we get that $\operatorname{Ind}(\Phi) = 1$. Also, only for $k = 2$ and $l = 1$ the corresponding eigenvalue of the Jacobi operator is zero, unless $n_2 = n_1 - 1$, in which case $k = 1$ and $l = 2$ also gives the eigenvalue zero. In the first case $\operatorname{Nul}(\Phi) = n_1$ and in the second case (when $n_2 = n_1 - 1$) $\operatorname{Nul}(\Phi) = n_1 + n_2 + 1$.

If $n_2 < n_1 - 1$, the corresponding eigenvalue of the Jacobi operator is negative not only for $k = l = 1$, but also for $k = 1$ and any l such that $\mu_l = (l-1)(l+n_2-2) < n_1 - 1$. From here the result follows. □

Similarly, we can prove the following result.

EXAMPLE 2.53. *Let* $\Phi : \mathbb{S}^{n_1} \times \mathbb{S}^{n_2-1} \to \mathbb{S}^{n_1} \times \mathbb{S}^{n_2}$ $(n_2 \geq 2)$ *be the two-sided totally geodesic embedding* $(p,q) \mapsto (p,(q,0))$ *of* $\mathbb{S}^{n_1} \times \mathbb{S}^{n_2-1}$ *into* $\mathbb{S}^{n_1} \times \mathbb{S}^{n_2}$.

(1) *If* $n_1 \geq n_2 - 1$, *then*

$$\operatorname{Ind}(\Phi) = 1, \qquad \operatorname{Nul}(\Phi) = \begin{cases} n_2 & \text{if } n_1 > n_2 - 1, \\ n_1 + n_2 + 1 & \text{if } n_1 = n_2 - 1. \end{cases}$$

(2) *If* $n_1 < n_2 - 1$, *then*

$$\operatorname{Ind}(\Phi) = \sum_{\lambda_k < n_2 - 1} \mu(\lambda_k) \geq n_1 + 2, \qquad \operatorname{Nul}(\Phi) = n_2 + \mu(n_2 - 1)$$

where λ_k *are the eigenvalues of the Laplacian of* $\mathbb{S}^{n_1}$ *and* $\mu(\lambda_k)$ *are the corresponding multiplicities, where* $\mu(n_2 - 1) = 0$ *if* $n_2 - 1$ *is not an eigenvalue of* $\mathbb{S}^{n_1}$.

In the next example, we will give a way to compute the index and the nullity of a minimal submanifold of the real projective space.

LEMMA 2.54. *Let $\Phi : M^n \to \mathbb{RP}^m$ be an immersion of a manifold M into the real projective space. We consider the canonical projection $\Pi : \mathbb{S}^m \to \mathbb{RP}^m$ induced by the antipodal map $A : \mathbb{S}^m \to \mathbb{S}^m$ defined by $A(p) = -p$. Then the double covering $\widetilde{\Pi} : \widetilde{M} \to M$ induced from Π by Φ is trivial if and only if Φ admits a lift to $\mathbb{S}^m$.*

PROOF. The induced covering $\widetilde{\Pi}$ is defined by $\widetilde{M} = \{(p,x) \mid p \in M, x \in \Pi^{-1}(\Phi(p)\}$ and $\widetilde{\Pi}(p,x) = p$.

If $\widetilde{\Pi}$ is trivial, then there exists a smooth map $F : M \to \widetilde{M}$ such that $\widetilde{\Pi} \circ F = Id$. So F must be given by $F(p) = (p, \widetilde{\Phi}(p))$ with $\widetilde{\Phi}(p) \in \Pi^{-1}(\Phi(p))$. It is clear that $\widetilde{\Phi}$ defines an immersion $\widetilde{\Phi} : M \to \mathbb{S}^m$ with $\Pi \circ \widetilde{\Phi} = \Phi$. So $\widetilde{\Phi}$ is a lift of Φ to $\mathbb{S}^m$.

Conversely, if $\widetilde{\Phi} : M \to \mathbb{S}^m$ is a lift of Φ to $\mathbb{S}^m$, then the map $F : M \to \widetilde{M}$ defined by $F(p) = (p, \widetilde{\Phi}(p))$ is a trivialization of the covering $\widetilde{\Pi}$. □

Now suppose that $\Phi : M^n \to \mathbb{RP}^m$ is a minimal immersion which does not admit a lift to $\mathbb{S}^m$. Then the induced covering $(\widetilde{M}, \widetilde{\Pi})$ over M is non-trivial and we can define a map $\widetilde{\Phi} : \widetilde{M} \to \mathbb{S}^m$ by $\widetilde{\Phi}(p,x) = x$, for any $(p,x) \in \widetilde{M}$. It is clear that $\Pi \circ \widetilde{\Phi} = \Phi \circ \widetilde{\Pi}$ and that $\widetilde{\Phi}$ is also a minimal immersion. Then the space of normal sections of $\widetilde{\Phi}$ can be decomposed as

$$\mathfrak{X}^\perp(\widetilde{M}) = \mathfrak{X}^\perp_+(\widetilde{M}) \oplus \mathfrak{X}^\perp_-(\widetilde{M}), \qquad \mathfrak{X}^\perp_\pm(\widetilde{M}) = \{\xi \in \mathfrak{X}^\perp(\widetilde{M}) \mid \xi_{\tau(\tilde{p})} = \pm\xi_{\tilde{p}}, \forall \tilde{p} \in \widetilde{M}\},$$

where τ is the deck transformation of $\widetilde{M}$ with $\tau^2 = Id$. In fact, $\mathfrak{X}^\perp_\pm(\widetilde{M})$ is well defined because $T_{\widetilde{\Phi}(\tau(\tilde{p}))}\mathbb{S}^m = T_{\widetilde{\Phi}(\tilde{p})}\mathbb{S}^m$ for any $\tilde{p} \in \widetilde{M}$. It is clear that if $\xi \in \mathfrak{X}^\perp_+(\widetilde{M}) \cap \mathfrak{X}^\perp_-(\widetilde{M})$ then $\xi = 0$. Finally, any normal section ξ can be written as $\xi = \xi^+ + \xi^-$ where $\xi^\pm = \frac{\xi \pm \xi\circ\tau}{2} \in \mathfrak{X}^\perp_\pm(\widetilde{M})$.

Since the antipodal map is an isometry of the sphere, we have that $\widetilde{\Delta}^\perp(\xi \circ A) = (\widetilde{\Delta}^\perp \xi) \circ A$ and thus the Jacobi operator $\widetilde{\mathcal{L}}$ of $\widetilde{\Phi}$ satisfies $\widetilde{\mathcal{L}}(\xi \circ A) = \widetilde{\mathcal{L}}\xi \circ A$. So $\widetilde{\mathcal{L}}(\mathfrak{X}^\perp_\pm(\widetilde{M})) \subset \mathfrak{X}^\perp_\pm(\widetilde{M})$ and thus $\widetilde{\mathcal{L}}$ splits into $\widetilde{\mathcal{L}} = \widetilde{\mathcal{L}}^+ + \widetilde{\mathcal{L}}^-$ where $\widetilde{\mathcal{L}}^\pm = \widetilde{\mathcal{L}}_{|\mathfrak{X}^\perp_\pm(\widetilde{M})}$.

Now, we define a linear map $F : \mathfrak{X}^\perp(M) \to \mathfrak{X}^\perp_-(\widetilde{M})$ by

$$F(\xi)_{\tilde{p}} = \left((\mathrm{d}\Pi)_{\widetilde{\Phi}(\tilde{p})}\right)^{-1}(\xi_{\widetilde{\Pi}(\tilde{p})}), \quad \forall \tilde{p} \in \widetilde{M}, \quad \forall \xi \in \mathfrak{X}^\perp(M).$$

Note that $F(\xi)_{\tau(\tilde{p})} = -F(\xi)_{\tilde{p}}$, $\forall \tilde{p} \in \widetilde{M}$, and hence $F(\xi) \in \mathfrak{X}^\perp_-(\widetilde{M})$. It is clear that $\ker F = 0$ and that if $\eta \in \mathfrak{X}^\perp_-(\widetilde{M})$, then $\xi \in \mathfrak{X}^\perp(M)$ defined by

$$\xi_p = \mathrm{d}\Pi_{\widetilde{\Phi}(\tilde{p})}(\eta_{\tilde{p}}), \quad \forall p \in M, \quad \text{where} \quad \widetilde{\Pi}(\tilde{p}) = p,$$

is a well-defined vector field in $\mathfrak{X}^\perp(M)$ with $F(\xi) = \eta$. This means that F is an isomorphism.

Now, if $\mathcal{L}$ denotes the Jacobi operator of Φ, since the projection Π is a local isometry, it is easy to check that

$$F \circ \mathcal{L} = \widetilde{\mathcal{L}}^- \circ F. \tag{2.33}$$

We can summarize the above reasoning as follows:

PROPOSITION 2.55. *Let $\Phi : M^n \to \mathbb{RP}^m$ be a minimal immersion of a compact manifold M and $\Pi : \mathbb{S}^m \to \mathbb{RP}^m$ be the canonical projection induced by the antipodal map A of $\mathbb{S}^m$. Then,*

(1) *If Φ admits a lift on $\mathbb{S}^m$, i.e., there exist an immersion $\widetilde{\Phi} : M \to \mathbb{S}^m$ with $\Pi \circ \widetilde{\Phi} = \Phi$, then $\widetilde{\Phi}$ is minimal,* $\mathrm{Ind}(\Phi) = \mathrm{Ind}(\widetilde{\Phi})$ *and* $\mathrm{Nul}(\Phi) = \mathrm{Nul}(\widetilde{\Phi})$.

(2) *If* Φ *does not admit a lift to* $\mathbb{S}^m$, $(\widetilde{M}, \widetilde{\Pi})$ *is the double covering on* M *induced by* Π *and* $\widetilde{\Phi} : \widetilde{M} \to \mathbb{S}^m$ *the minimal immersion satisfying* $\Pi \circ \widetilde{\Phi} = \Phi \circ \widetilde{\Pi}$, *then* $\mathrm{Ind}(\Phi) = \mathrm{Ind}(\widetilde{\mathcal{L}}^-)$ *and* $\mathrm{Nul}(\Phi) = \mathrm{Nul}(\widetilde{\mathcal{L}}^-)$, *where*

$$\widetilde{\mathcal{L}}^- = \widetilde{\mathcal{L}}_{|\mathfrak{X}^\perp_-(\widetilde{M})}, \quad \mathfrak{X}^\perp_-(\widetilde{M}) = \{\xi \in \mathfrak{X}^\perp(\widetilde{M}) \mid \xi_{\tau(\tilde{p})} = -\xi_{\tilde{p}}, \forall \tilde{p} \in \widetilde{M}\},$$

where $\widetilde{\mathcal{L}}$ *is the Jacobi operator of* $\widetilde{\Phi}$ *and* τ *is the deck transformation of* $\widetilde{M}$ *with* $\tau^2 = Id$.

PROOF. (1) follows from Proposition 2.37,2. It follows from (2.33) that the eigenvalues of $\mathcal{L}$ and $\widetilde{\mathcal{L}}^-$ are equal, which proves (2). □

REMARK 2.56. If in the last result, $m = n + 1$, i.e., Φ is a hypersurface, then the conclusions about the index hold even without the compactness assumption.

As a consequence of the above proposition we get the following special cases.

EXAMPLE 2.57. *Let* $\Phi : \mathbb{RP}^n \to \mathbb{RP}^m$ *be the totally geodesic embedding, which makes the following diagram commutative:*

$$\begin{array}{ccc} \mathbb{S}^n & \xrightarrow{\widetilde{\Phi}} & \mathbb{S}^m \\ \downarrow{\scriptstyle \widetilde{\Pi}} & & \downarrow{\scriptstyle \Pi} \\ \mathbb{RP}^n & \xrightarrow{\Phi} & \mathbb{RP}^m, \end{array}$$

where $\widetilde{\Phi}$ *is the embedding of Example* 2.46. *Then* $\mathbb{RP}^n$ *is stable and* $\mathrm{Nul}(\Phi) = (n+1)(m-n)$.

PROOF. From Proposition 2.55, $\mathrm{Ind}(\Phi) = \mathrm{Ind}(\widetilde{\mathcal{L}}^-)$ and $\mathrm{Nul}(\Phi) = \mathrm{Nul}(\widetilde{\mathcal{L}}^-)$, where $\widetilde{\mathcal{L}}$ is the Jacobi operator of the totally geodesic embedding $\widetilde{\Phi}$.

According to Example 2.46, the non-positive eigenvalues of the Jacobi operator $\widetilde{\mathcal{L}}$ of $\widetilde{\Phi}$ are $-n$ and 0 with the respective multiplicities $(m-n)$ and $(m-n)(n+1)$. But the eigensections of $-n$ are $\xi = \sum_{i=1}^{m-n} a_i \xi_i$, $a_i \in \mathbb{R}$. It is clear that these sections belong to $\mathfrak{X}^\perp_+(\mathbb{S}^n)$. So $-n$ is not an eigenvalue of $\widetilde{\mathcal{L}}^-$ and so $\mathrm{Ind}(\Phi) = 0$. Also, the eigensections of 0 are $\xi = \sum_{i=1}^{m-n} f_i \xi_i$, where $\Delta f_i + n f_i = 0$. In this case these sections belong to $\mathfrak{X}^\perp_-(\mathbb{S}^n)$ and so 0 is an eigenvalue of $\widetilde{\mathcal{L}}^-$. This proves that $\mathrm{Nul}(\Phi) = (n+1)(m-n)$. □

From Proposition 2.55,1 it follows the next result.

EXAMPLE 2.58. *Let* $\Phi : \mathbb{S}^n \to \mathbb{RP}^m$ *be the totally geodesic immersion, which makes the following diagram commutative:*

$$\begin{array}{ccc} & & \mathbb{S}^m \\ & \overset{\widetilde{\Phi}}{\nearrow} & \downarrow{\scriptstyle \Pi} \\ \mathbb{S}^n & \xrightarrow{\Phi} & \mathbb{RP}^m, \end{array}$$

where $\widetilde{\Phi}$ *is the embedding of Example* 2.46. *Then* $\mathrm{Ind}(\Phi) = m - n$ *and* $\mathrm{Nul}(\Phi) = (n+1)(m-n)$.

EXAMPLE 2.59 (Clifford hypersurfaces of the real projective space). *The Clifford hypersurface* $\widetilde{\Phi} : \mathbb{S}^n(r_1) \times \mathbb{S}^m(r_2) \to \mathbb{S}^{n+m+1}$, $r_1 = \sqrt{n/n+m}$ *and* $r_2 = \sqrt{m/n+m}$ *(Example* 2.50*) is invariant under the antipodal map of* $\mathbb{S}^{n+m+1}$, *and* *so $\widetilde{\Phi}$ projects into a minimal hypersurface $\Phi : (\mathbb{S}^n(r_1) \times \mathbb{S}^m(r_2))/\mathbb{Z}_2 \to \mathbb{RP}^{n+m+1}$, so the following diagram is commutative:*

$$\begin{array}{ccc} \mathbb{S}^n(r_1) \times \mathbb{S}^m(r_2) & \xrightarrow{\widetilde{\Phi}} & \mathbb{S}^{n+m+1} \\ \big\downarrow \widetilde{\Pi} & & \big\downarrow \Pi \\ \frac{\mathbb{S}^n(r_1)\times\mathbb{S}^m(r_2)}{\mathbb{Z}_2} & \xrightarrow{\Phi} & \mathbb{RP}^{n+m+1}. \end{array}$$

Then, $\mathrm{Ind}(\Phi) = 1$ *and* $\mathrm{Nul}(\Phi) = (n+1)(m+1)$.

PROOF. It is clear that Φ is a minimal hypersurface, which makes the above diagram commutative.

Also, from Proposition 2.55, $\mathrm{Ind}(\Phi) = \mathrm{Ind}(\widetilde{\mathcal{L}}^-)$ and $\mathrm{Nul}(\Phi) = \mathrm{Nul}(\widetilde{\mathcal{L}}^-)$, where $\widetilde{\mathcal{L}}$ is the Jacobi operator of the embedding $\widetilde{\Phi}$. Since $\widetilde{\Phi}$ is a two-sided hypersurface, the operator $\widetilde{\mathcal{L}}$ reduces to an operator acting on functions.

Following Example 2.50, $N_{(p,q)} = (\sqrt{\frac{m}{n}}p, -\sqrt{\frac{n}{m}}q)$ is a unit normal vector field on $\widetilde{\Phi}$ satisfying $N_{(-p,-q)} = -N_{(p,q)}$. Hence $N \in \mathfrak{X}^\perp_-(\mathbb{S}^{n_1}(r_1) \times \mathbb{S}^{n_2}(r_2))$. This implies that any section $\xi \in \mathfrak{X}^\perp_-(\mathbb{S}^{n_1}(r_1) \times \mathbb{S}^{n_2}(r_2))$ can be written as $\xi = fN$, where $f(-p,-q) = f(p,q)$, $\forall (p,q) \in \mathbb{S}^{n_1}(r_1) \times \mathbb{S}^{n_2}(r_2)$, i.e., $f \in C^\infty_+(\mathbb{S}^{n_1}(r_1) \times \mathbb{S}^{n_2}(r_2))$. Hence, $\mathrm{Ind}(\Phi) = \mathrm{Ind}(\widetilde{\mathcal{L}}^+)$ and $\mathrm{Nul}(\Phi) = \mathrm{Nul}(\widetilde{\mathcal{L}}^+)$, where we now consider $\widetilde{\mathcal{L}} : C^\infty(\mathbb{S}^{n_1}(r_1) \times \mathbb{S}^{n_2}(r_2)) \to C^\infty(\mathbb{S}^{n_1}(r_1) \times \mathbb{S}^{n_2}(r_2))$.

Following Example 2.50, the non-positive eigenvalues of the Jacobi operator $\widetilde{\mathcal{L}} : C^\infty(\mathbb{S}^{n_1}(r_1) \times \mathbb{S}^{n_2}(r_2)) \to C^\infty(\mathbb{S}^{n_1}(r_1) \times \mathbb{S}^{n_2}(r_2))$ of $\widetilde{\Phi}$ are $-2(n+m)$, $-(n+m)$ and 0, with respective multiplicities 1, $n+m+2$ and $(n+1)(m+1)$. But it is not difficult to check that the eigenfunctions corresponding to the eigenvalues $-2(n+m)$ and 0 are in $C^\infty_+(\mathbb{S}^{n_1}(r_1) \times \mathbb{S}^{n_2}(r_2))$ and the eigenfunctions corresponding to the eigenvalue $-(n+m)$ lie in $C^\infty_-(\mathbb{S}^{n_1}(r_1) \times \mathbb{S}^{n_2}(r_2))$. Hence $\mathrm{Ind}(\Phi) = 1$ and $\mathrm{Nul}(\Phi) = (n+1)(m+1)$. □

Now we compute the index and nullity of two minimal submanifolds of the complex projective space.

EXAMPLE 2.60. *Let $\Phi : \mathbb{CP}^n \to \mathbb{CP}^m$ be the totally geodesic embedding, which makes the following diagram commutative:*

$$\begin{array}{ccc} \mathbb{S}^{2n+1} & \xrightarrow{\widetilde{\Phi}} & \mathbb{S}^{2m+1} \\ \big\downarrow \widetilde{\Pi} & & \big\downarrow \Pi \\ \mathbb{CP}^n & \xrightarrow{\Phi} & \mathbb{CP}^m, \end{array}$$

where $\widetilde{\Phi}$ is the embedding of the Example 2.46 *and* Π *and* $\widetilde{\Pi}$ *are the Hopf fibrations. Then $\mathbb{CP}^n$ is stable and* $\mathrm{Nul}(\Phi) = 2(n+1)(m-n)$.

PROOF. According to the definition of $\mathbb{CP}^m$ given in Section 1.6, the Hopf fibration $\Pi : \mathbb{S}^{2m+1} \to \mathbb{CP}^m$ is a Riemannian submersion and for each point $p \in \mathbb{S}^{2m+1}$, $\ker d\Pi_p = \langle ip \rangle$. So if H_p is the orthogonal complement of $\langle ip \rangle$ in $T_p\mathbb{S}^{2m+1}$, then $d\Pi_p : H_p \to T_{\Pi(p)}\mathbb{CP}^m$ is an isomorphism.

If $\xi \in \mathfrak{X}^\perp(\mathbb{CP}^n)$, let $\widetilde{\xi} \in \mathfrak{X}^\perp(\mathbb{S}^{2n+1})$ be its horizontal lift, defined by

$$\widetilde{\xi}_p = (d\Pi_{\widetilde{\Phi}(p)})^{-1}(\xi_{\widetilde{\Pi}(p)}) \in H_{\widetilde{\Phi}(p)}, \quad \forall p \in \mathbb{S}^{2n+1}.$$

For each $t \in \mathbb{R}$, let F_t be the isometry of $\mathbb{S}^{2m+1}$ (respectively G_t the isometry of $\mathbb{S}^{2n+1}$) defined by $F_t(p) = e^{it}p$, $\forall p \in \mathbb{S}^{2m+1}$ (respectively $G_t(p) = e^{it}p$, $\forall p \in \mathbb{S}^{2n+1}$). Then $\Pi \circ F_t = \Pi$ (respectively $\widetilde{\Pi} \circ G_t = \widetilde{\Pi}$) and so

$$\widetilde{\xi}_{G_t(p)} = (\mathrm{d}F_t)_{\widetilde{\Phi}(p)}(\widetilde{\xi}_p), \quad \forall p \in \mathbb{S}^{2n+1}, \quad \forall t \in \mathbb{R}.$$

Now, if

$$\mathfrak{X}^{\perp}_{+}(\mathbb{S}^{2n+1}) = \{\eta \in \mathfrak{X}^{\perp}(\mathbb{S}^{2n+1}) \mid \eta_{G_t(p)} = (\mathrm{d}F_t)_{\widetilde{\Phi}(p)}(\eta_p), \quad \forall p \in \mathbb{S}^{2n+1}, \quad \forall t \in \mathbb{R}\},$$

then $F : \mathfrak{X}^{\perp}(\mathbb{CP}^n) \to \mathfrak{X}^{\perp}_{+}(\mathbb{S}^{2n+1})$ defined by $F(\xi) = \widetilde{\xi}$ is an isomorphism.

Since G_t and F_t are isometries, the Jacobi operator $\widetilde{\mathcal{L}} = \widetilde{\Delta}^{\perp} + (2n+1)Id$ of $\widetilde{\Phi}$ satisfies that $\widetilde{\mathcal{L}}(\mathfrak{X}^{\perp}_{+}(\mathbb{S}^{2n+1})) \subset \mathfrak{X}^{\perp}_{+}(\mathbb{S}^{2n+1})$ and so, we can define an operator

$$\widetilde{\mathcal{L}}^{+} = \widetilde{\mathcal{L}}_{|\mathfrak{X}^{\perp}_{+}(\mathbb{S}^{2n+1})} : \mathfrak{X}^{\perp}_{+}(\mathbb{S}^{2n+1}) \to \mathfrak{X}^{\perp}_{+}(\mathbb{S}^{2n+1}).$$

As Φ is totally geodesic, the Jacobi operator of Φ is given by $\mathcal{L} = \Delta^{\perp} + 2nId$. If $\{e_1, \ldots, e_{2n}\}$ is a local orthonormal reference on $\mathbb{CP}^n$, then $\{\widetilde{e}_1, \ldots, \widetilde{e}_{2n}, V\}$ is an orthonormal reference on $\mathbb{S}^{2n+1}$, where $\widetilde{e}_i$ is the horizontal lift of e_i and $V_p = ip$ for any $p \in \mathbb{S}^{2n+1}$. Then for any $\xi \in \mathfrak{X}^{\perp}(\mathbb{CP}^n)$ and using the equations of a Riemannian submersion, we have

$$\begin{aligned}\widetilde{\mathcal{L}}^{+}(F\xi) &= \sum_{i=1}^{2n} (\widetilde{\nabla}^{\perp}_{\widetilde{e}_i}\widetilde{\nabla}^{\perp}_{\widetilde{e}_i} - \widetilde{\nabla}^{\perp}_{\widetilde{\nabla}_{\widetilde{e}_i}\widetilde{e}_i})(\widetilde{\xi}) + (\widetilde{\nabla}^{\perp}_{V}\widetilde{\nabla}^{\perp}_{V} - \widetilde{\nabla}^{\perp}_{\widetilde{\nabla}_V V})(\widetilde{\xi}) + (2n+1)F\xi \\ &= F(\Delta^{\perp}\xi) - F\xi + (2n+1)F\xi = F(\mathcal{L}\xi),\end{aligned}$$

and so the eigenvalues of $\mathcal{L}$ and $\widetilde{\mathcal{L}}^{+}$ are the same. Following Example 2.46, the first two eigenvalues of $\widetilde{\mathcal{L}}$ are $-(2n+1)$ with multiplicity $2(m-n)$ and 0 with multiplicity $4(n+1)(m-n)$.

Considering Example 2.46, if $\{\xi_j, i\xi_j \mid 1 \leq j \leq m-n\}$ is the global orthonormal reference of the normal bundle of $\widetilde{\Phi}$, then a normal section $\xi = \sum_{j=1}^{m-n}\{f_j\xi_j + g_j i\xi_j\}$ is an eigensection of the eigenvalue $-(2n+1)$ if f_j and g_j are constant. But then, ξ is not a section of $\mathfrak{X}^{\perp}_{+}(\mathbb{S}^{2n+1})$ and so $-(2n+1)$ is not an eigenvalue of $\widetilde{\mathcal{L}}^{+}$. This implies that $\mathbb{CP}^n$ is stable. Also, ξ is an eigensection of the eigenvalue 0 if $\xi = \sum_{j=1}^{m-n}\{\langle p, a_j\rangle\xi_j + \langle p, b_j\rangle i\xi_j\}$, where a_j, b_j are vectors of $\mathbb{R}^{2n+2}$. Now, it is easy to prove that $\xi \in \mathfrak{X}^{\perp}_{+}(\mathbb{S}^{2n+1})$ if and only if $b_j = ia_j$, $\forall 1 \leq j \leq m-n$. This implies that the multiplicity of the eigenvalue 0 is $2(n+1)(m-n)$, which proves the result. □

Example 2.61 (Clifford hypersurfaces of the complex projective space). *The Clifford hypersurface* $\widetilde{\Phi} : \widetilde{M} = \mathbb{S}^{2p+1}(r_1) \times \mathbb{S}^{2q+1}(r_2) \to \mathbb{S}^{2n+1}$, $n = p+q+1$, $r_1 = \sqrt{2p+1/2n}$, $r_2 = \sqrt{2q+1/2n}$ *(Example* 2.50*) is invariant under the Hopf fibration* $\Pi : \mathbb{S}^{2n+1} \to \mathbb{CP}^n$, *and so* $\widetilde{\Phi}$ *projects to a two-sided minimal hypersurface* $\Phi : M_{p,q} := \widetilde{M}/\mathbb{S}^1 \to \mathbb{CP}^n$, *which makes the following diagram commutative:*

$$\begin{array}{ccc} \widetilde{M} & \xrightarrow{\ \widetilde{\Phi}\ } & \mathbb{S}^{2n+1} \\ \Big\downarrow{\scriptstyle \widetilde{\Pi}} & & \Big\downarrow{\scriptstyle \Pi} \\ M_{p,q} & \xrightarrow{\ \Phi\ } & \mathbb{CP}^n. \end{array}$$

Also, $\mathrm{Ind}(\Phi) = 1$ *and* $\mathrm{Nul}(\Phi) = 2(p+1)(q+1)$.

REMARK 2.62. The hypersurfaces $M_{p,q}$ of the complex projective space $\mathbb{CP}^n$ described in Example 2.61 have been studied and characterized by several authors from different points of view (see for example [**Law70b, Tak75**]). It is also interesting to note that $M_{0,n-1}$ and $M_{n-1,0}$ are geodesic spheres of the complex projective space $\mathbb{CP}^n$.

PROOF. The Clifford hypersurface $\widetilde{M}$ of $\mathbb{S}^{2n+1}$ (Example 2.50) is invariant under the action of $\mathbb{S}^1$ on $\mathbb{S}^{2n+1}$, and so the Hopf fibration Π defines a projection $\widetilde{\Pi}:\widetilde{M}\to M_{p,q}$, and the embedding $\widetilde{\Phi}$ projects into an embedding $\Phi: M_{p,q}\to\mathbb{CP}^n$ satisfying $\Phi\circ\widetilde{\Pi}=\Pi\circ\widetilde{\Phi}$.

Following the notation of Example 2.60, for each $t\in\mathbb{R}$, let F_t be the isometry of $\mathbb{S}^{2n+1}$ (respectively G_t the isometry of $\widetilde{M}$) defined by $F_t(p)=e^{it}p$, $\forall p\in\mathbb{S}^{2n+1}$ $\big($respectively $G_t((p,q))=(e^{it}p,e^{it}q)$, $\forall(p,q)\in\widetilde{M}\big)$.

The normal vector field $\widetilde{N}$ to $\widetilde{\Phi}$ (see Example 2.50) satisfies

$$\widetilde{N}_{G_t(p,q)}=(\mathrm{d}F_t)_{\widetilde{\Phi}(p,q)}(\widetilde{N}_{(p,q)}),\quad \forall\,(p,q)\in\widetilde{M},$$

and so $\widetilde{N}$ projects to a unit normal vector field N to Φ. This means that Φ is a two-sided embedding.

The principal curvatures of the hypersurface $M_{p,q}$ are $\{-\lambda,\frac{1}{\lambda},\frac{1-\lambda^2}{\lambda}\}$ with respective multiplicities $\{2p,2q,1\}$, where $\lambda=\sqrt{\frac{2q+1}{2p+1}}$. So Φ is a minimal immersion and $\|A\|^2=2(n-1)$. Since $\overline{\mathrm{Ricc}}(N)=2n+2$, we get that the Jacobi operator $\mathcal{L}$ of the immersion Φ is $\mathcal{L}=\Delta+4n$, where Δ is the Laplacian of $M_{p,q}$. Also, from Example 2.50, the Jacobi operator of the immersion $\widetilde{\Phi}$ is $\widetilde{\mathcal{L}}=\widetilde{\Delta}+4n$.

On the other hand, if

$$C^\infty_+(\widetilde{M})=\{f\in\widetilde{M}\mid f\circ G_t=f,\,\forall t\in\mathbb{R}\},$$

then $F: C^\infty(M_{p,q})\to C^\infty_+(\widetilde{M})$ defined by $F(f)=f\circ\widetilde{\Pi}$ is an isomorphism. Moreover, as $\widetilde{\mathcal{L}}(C^\infty_+(\widetilde{M}))\subset C^\infty_+(\widetilde{M})$, we can define the operator $\widetilde{\mathcal{L}}^+=\widetilde{\mathcal{L}}_{|C^\infty_+(\widetilde{M})}$: $C^\infty_+(\widetilde{M})\to C^\infty_+(\widetilde{M})$. Using a similar reasoning as in Example 2.61, we prove that $F\circ\mathcal{L}=\widetilde{\mathcal{L}}^+\circ F$, and so $\mathcal{L}$ and $\widetilde{\mathcal{L}}^+$ have the same eigenvalues. Following Example 2.50, the non-positive eigenvalues of $\widetilde{\mathcal{L}}$ are $-4n$ with multiplicity 1, $-2n$ with multiplicity $2(n+1)$, and 0 with multiplicity $4(p+1)(q+1)$.

It is clear that $-4n$ is an eigenvalue of $\widetilde{\mathcal{L}}^+$ because the constant functions are in $C^\infty_+(\widetilde{M})$. Also, $-2n$ is not an eigenvalue of $\widetilde{\mathcal{L}}^+$ because the functions $(p,q)\mapsto\langle p,a\rangle$ and $(p,q)\mapsto\langle q,b\rangle$ with $a\in\mathbb{R}^{2p+2}$ and $b\in\mathbb{R}^{2q+2}$ do not belong to $C^\infty_+(\widetilde{M})$. This implies that $\mathrm{Ind}(M_{p,q})=1$.

Finally, if $\{a_1,\dots,a_{p+1},ia_1,\dots,ia_{p+1}\}$ $\big($resp. $\{b_1,\dots,b_{p+1},ib_1,\dots,ib_{p+1}\}\big)$ is an orthonormal basis of $\mathbb{R}^{2p+2}$ (resp. $\mathbb{R}^{2q+2}$), the eigenspace of $\widetilde{\mathcal{L}}$ associated to the eigenvalue 0 is given by

$$\widetilde{V}_0=\left\{h:=\sum_{j=1}^{p+1}\sum_{k=1}^{q+1}\{\lambda_{jk}f_jg_k+\lambda_{j^*k^*}f_{j^*}g_{k^*}+\lambda_{jk^*}f_jg_{k^*}+\lambda_{j^*k}f_{j^*}g_k\}\right\},$$

where the $\lambda's$ are real numbers and $f_j(p,q)=\langle p,a_j\rangle$, $f_{j^*}(p,q)=\langle p,ia_j\rangle$, $g_k(p,q)=\langle q,b_k\rangle$ and $g_{k^*}(p,q)=\langle q,ib_k\rangle$. Now it is easy to prove that $h\in C^\infty_+(\widetilde{M})$ if and only if $\lambda_{jk} = \lambda_{j^*k^*}$ and $\lambda_{jk^*} = -\lambda_{j^*k}$ for all j, k. This implies that 0 is an eigenvalue of $\widetilde{\mathcal{L}}^+$ with multiplicity $2(p+1)(q+1)$, which means that $\mathrm{Nul}(\Phi) = 2(p+1)(q+1)$. □

To conclude this chapter, we will introduce a one-parameter family of Riemannian manifolds, called Berger spheres, which are a deformation of the standard sphere. In contrast to the sphere case, in some of them we will give examples of stable compact minimal submanifolds.

Let $\mathbb{C}^{m+1}$ be the complex Euclidean space of dimension $m+1$, and let

$$\mathbb{S}^{2m+1} = \left\{(z_1, \dots, z_{m+1}) \in \mathbb{C}^{m+1} \mid \|z_1\|^2 + \cdots + \|z_{m+1}\|^2 = 1\right\}$$

be the $(2m+1)$-dimensional unit sphere. If g denotes the standard metric on $\mathbb{S}^{2m+1}$, we define a one-parameter family of metrics $\{g_\tau \mid \tau \in (0,1]\}$ on $\mathbb{S}^{2m+1}$ by

$$g_\tau(v, w) = g(v, w) + (\tau^2 - 1)g(v, iz)g(w, iz), \quad v, w \in T_z\mathbb{S}^{2m+1}.$$

The **Berger spheres** are the family of Riemannian manifolds

$$\left\{(\mathbb{S}^{2m+1}, g_\tau) \mid \tau \in (0,1]\right\},$$

denoted simply by $\mathbb{S}^{2m+1}_\tau$. Note that $\mathbb{S}^{2m+1}_1$ is the standard sphere $\mathbb{S}^{2m+1}$. This deformation of the standard metric g can also be defined for $\tau \in (1, \infty)$, but the geometry of these Berger spheres is quite different from the geometry of $\mathbb{S}^{2m+1}_\tau$, $\tau \in (0,1)$, which makes it very difficult to study the stability of their minimal submanifolds.

It is clear that the unitary group $\mathrm{U}(m+1)$ acts transitively by isometries on $\mathbb{S}^{2m+1}_\tau$ and thus $\mathbb{S}^{2m+1}_\tau$ is a homogeneous Riemannian manifold.

According to Section 1.6 and Example 2.60, if $\Pi : \mathbb{S}^{2m+1} \to \mathbb{CP}^m$ is the Hopf fibration and $T_z\mathbb{S}^{2m+1} = H_z \oplus \langle Jz \rangle$ is the decomposition of $T_z\mathbb{S}^{2m+1}$ in horizontal and vertical subspaces, then

$$g_\tau(v, w) = g(v, w), \quad g_\tau(iz, iz) = \tau^2, \quad v, w \in H_z.$$

Hence $\Pi : \mathbb{S}^{2m+1}_\tau \to \mathbb{CP}^m$ is a Riemannian submersion and the length of the fiber of Π is $2\pi\tau$. Hence $\lim_{\tau\to 0} \mathrm{vol}\,(\mathbb{S}^{2m+1}_\tau) = 0$ and the injectivity radius of $\mathbb{S}^{2m+1}_\tau$ is $\pi\tau$. Thus the family of metrics $\{g_\tau\}$ is an example of a Cheeger-Gromov collapse, [**CG86**], and the metrics $\{g_\tau\}$ collapse $\mathbb{S}^{2m+1}$ into the limit space $\mathbb{CP}^m$.

From now on, and to simplify the notation, the Berger's metric g_τ will be denoted by $\langle , \rangle$.

Just as the unit sphere $\mathbb{S}^{2m+1}$ is a geodesic sphere of the Euclidean space $\mathbb{C}^{m+1}$, the Berger spheres $\mathbb{S}^{2m+1}_\tau$, $0 < \tau < 1$ are geodesic spheres of the complex projective space and those corresponding to $1 < \tau < \infty$ are geodesic spheres of the complex hyperbolic space. That is why the former are called elliptic, while the latter are called hyperbolic.

In fact, for any $0 < \tau < 1$ the map $F : \mathbb{S}^{2m+1}_\tau \to \mathbb{CP}^{m+1}(4(1-\tau^2))$ given by

$$F(z) = \left[\left(\frac{\tau}{\sqrt{1-\tau^2}}, z\right)\right]$$

is an isometric embedding and $F(\mathbb{S}^{2m+1}_\tau)$ is the geodesic sphere of $\mathbb{CP}^{m+1}(4(1-\tau^2))$ with radius $\frac{\arccos\tau}{\sqrt{1-\tau^2}}$ and center $[(\frac{1}{\sqrt{1-\tau^2}}, 0, \dots, 0)]$. Furthermore, the restriction of complex structures of $\mathbb{C}^{m+1}$ and $\mathbb{CP}^{m+1}(4(1-\tau^2))$ to $\mathbb{S}^{2m+1}_\tau$ coincide and the unit vector field $\xi_z = \frac{1}{\tau}iz$ for $z \in \mathbb{S}^{2m+1}_\tau \subset \mathbb{C}^{m+1}$ verifies that $J\xi$ is normal to F, where

J is the complex structure of $\mathbb{CP}^{m+1}(4(1-\tau^2))$. The second fundamental form $\widehat{\sigma}$ of F is given by

$$\widehat{\sigma}(X,Y) = \left(\tau\langle X,Y\rangle - \frac{1-\tau^2}{\tau}\langle X,\xi\rangle\langle Y,\xi\rangle\right) J\xi, \quad \forall X,Y \in \mathfrak{X}(\mathbb{S}^{2m+1}_\tau). \tag{2.34}$$

The proof of the above statements is straightforward and is left as an exercise for the reader.

Using the Gauss equation of the embedding F and (2.34) one can check that the curvature tensor of $\mathbb{S}^{2m+1}_\tau$ is given by

$$\begin{aligned}\overline{R}(X,Y)Z &= \langle Y,Z\rangle X - \langle X,Z\rangle Y \\ &\quad + (1-\tau^2)\big[\langle JY,Z\rangle JX - \langle JX,Z\rangle JY - 2\langle JX,Y\rangle JZ\big] \\ &\quad + (1-\tau^2)\langle Z,\xi\rangle\big[\langle X,\xi\rangle Y - \langle Y\xi\rangle X\big] \\ &\quad + (1-\tau^2)\big[\langle Y,\xi\rangle\langle X,Z\rangle - \langle X,\xi\rangle\langle Y,Z\rangle\big]\xi,\end{aligned} \tag{2.35}$$

for each $X,Y,Z \in \mathfrak{X}(\mathbb{S}^{2m+1}_\tau)$. From (2.35) it follows that the sectional curvature of $\mathbb{S}^{2m+1}_\tau$ satisfies $K \geq \tau^2$ and that the scalar curvature is $2m[2(m+1)-\tau^2]$.

Finally, it is interesting to make explicit the relation between the Levi-Civita connection $\overline{\nabla}$ of $\mathbb{S}^{2m+1}_\tau$ and the Levi-Civita connection ${}^g\overline{\nabla}$ of the unit sphere $\mathbb{S}^{2m+1}$. In fact, using (1.1) and the relation between the two metrics, it is easy to get that

$${}^g\overline{\nabla}_X Y = \overline{\nabla}_X Y + \frac{1-\tau^2}{\tau}\left[\langle Y,\xi\rangle(JX)^\top + \langle X,\xi\rangle(JY)^\top\right], \; \forall X,Y \in \mathfrak{X}(\mathbb{S}^{2m+1}_\tau), \tag{2.36}$$

where $\top$ stands for the tangential component in $\mathbb{S}^{2m+1}_\tau$. Now, since ${}^g\overline{\nabla}_X\xi = \frac{1}{\tau}(JX)^\top$, from the last equation we get

$$\overline{\nabla}_X\xi = \tau(JX)^\top, \qquad \forall X \in \mathfrak{X}(\mathbb{S}^{2m+1}_\tau). \tag{2.37}$$

EXAMPLE 2.63 ([**TU22**]). *Let $\Phi : \mathbb{S}^{2n+1} \to \mathbb{S}^{2m+1}_\tau$ be the embedding defined by*

$$\Phi(z_1,\ldots,z_{n+1}) = (z_1,\ldots,z_{n+1},0\ldots,0).$$

Then Φ is a totally geodesic isometric embedding of $\mathbb{S}^{2n+1}_\tau$ in $\mathbb{S}^{2m+1}_\tau$ whose index is given by

$$\operatorname{Ind}(\mathbb{S}^{2n+1}_\tau) = \begin{cases} 0 & \textit{if } 0 < \tau^2 \leq \frac{1}{2n+2}, \\ 2(m-n) & \textit{if } \frac{1}{2n+2} < \tau^2 \leq 1. \end{cases}$$

PROOF. First, Φ is a totally geodesic embedding because $\mathbb{S}^{2n+1}$ is the fixed point set of the isometry $A = \begin{pmatrix} \mathrm{Id}_{n+1} & 0 \\ 0 & -\mathrm{Id}_{m-n}\end{pmatrix} \in U(m+1)$. Second, it is clear that the induced metric on $\mathbb{S}^{2n+1}$ is the Berger metric associated to τ.

Furthermore, using (2.35) it follows that

$$\mathcal{R}\eta = \sum_{i=1}^{2n+1}\left(\overline{R}(\eta,e_i)e_i\right)^\perp = (2m+\tau^2)\eta, \qquad \forall \eta \in \mathfrak{X}^\perp(\mathbb{S}^{2n+1}_\tau),$$

which, given that Φ is totally geodesic, implies that the Jacobi operator of Φ is

$$\mathcal{L}\eta = \Delta^\perp\eta + (2m+\tau^2)\eta, \qquad \forall \eta \in \mathfrak{X}^\perp(\mathbb{S}^{2n+1}_\tau).$$

On the other hand, the unit vector field $\xi \in \mathfrak{X}(\mathbb{S}^{2m+1}_\tau)$ is tangent to the submanifold $\mathbb{S}^{2n+1}_\tau$, and the normal bundle of Φ is invariant under J. Thus it is clear that

$\{a_j, ia_j : 1 \leq j \leq m-n\}$ defined by

$$a_j = (0, \ldots, 0, \overset{n+1+j}{1}, 0, \ldots, 0) \in \mathbb{C}^{m+1}$$

is a global orthonormal reference of the normal bundle of Φ. If $\mathcal{D}_j = \operatorname{span}\{a_j, ia_j\}$, then the normal bundle splits as $T^\perp \mathbb{S}^{2n+1}_\tau = \mathcal{D}_1 \oplus \cdots \oplus \mathcal{D}_{m-n}$. So any normal section $\eta \in \mathfrak{X}^\perp(\mathbb{S}^{2n+1}_\tau)$ can be written as $\eta = \sum_{j=1}^{m-n}\{f_j a_j + g_j\, ia_j\}$, with $f_j, g_j \in C^\infty(\mathbb{S}^{2n+1}_\tau)$.

But since a_j is parallel with respect to the standard connection on $\mathbb{S}^{2m+1}$, then using (2.36) we get for every $u \in T\mathbb{S}^{2n+1}_\tau$ that

$$\nabla^\perp_u a_j = \overline{\nabla}_u a_j = -\frac{1-\tau^2}{\tau}\langle u, \xi\rangle ia_j, \quad \nabla^\perp_u ia_j = \overline{\nabla}_u ia_j = \frac{1-\tau^2}{\tau}\langle u, \xi\rangle a_j,$$

and so

$$\Delta^\perp a_j = -\frac{(1-\tau^2)^2}{\tau^2}a_j, \quad \Delta^\perp ia_j = -\frac{(1-\tau^2)^2}{\tau^2}ia_j.$$

Finally, if $\eta = \sum_{j=1}^{m-n}\{f_j a_j + g_j\, ia_j\}$ is an arbitrary normal section of Φ, then

$$\begin{aligned}\mathcal{L}\eta = &\sum_{j=1}^{m-n}\left[\Delta f_j + (2m+2-\tfrac{1}{\tau^2})f_j + 2\tfrac{1-\tau^2}{\tau}L_\xi g_j\right]a_j \\ &+ \sum_{j=1}^{m-n}\left[\Delta g_j + (2m+2-\tfrac{1}{\tau^2})g_j - 2\tfrac{1-\tau^2}{\tau}L_\xi f_j\right]ia_j,\end{aligned} \tag{2.38}$$

where L_ξ is the Lie derivative with respect to ξ.

As a consequence, if $\Gamma(\mathcal{D}_j)$ denotes the space of sections of the subbundle $\mathcal{D}_j$, then $\mathcal{L}(\Gamma(\mathcal{D}_j)) \subset \Gamma(\mathcal{D}_j)$, and so $\operatorname{Ind}(\mathcal{L}) = \sum_{j=1}^{m-n}\operatorname{Ind}(\mathcal{L}_j)$, where $\mathcal{L}_j$ is the restriction of $\mathcal{L}$ to $\Gamma(\mathcal{D}_j)$.

We will compute the index of $\mathcal{L}_j$ and prove that these numbers are independent of the sub-bundles $\mathcal{D}_j$. To simplify the notation, let $\mathcal{D} = \operatorname{span}\{a, ia\}$ and $\mathcal{L} : \Gamma(\mathcal{D}) \to \Gamma(\mathcal{D})$.

Let $\eta = fa + gia$ be an eigensection of $\mathcal{L}$ associated to the eigenvalue ρ, i.e. $\mathcal{L}\eta + \rho\eta = 0$. Then we get from (2.38)

$$\begin{aligned}\Delta f + [\rho + (2m+2-\tfrac{1}{\tau^2})]f &= -2\tfrac{1-\tau^2}{\tau}L_\xi g, \\ \Delta g + [\rho + (2m+2-\tfrac{1}{\tau^2})]g &= 2\tfrac{1-\tau^2}{\tau}L_\xi f.\end{aligned} \tag{2.39}$$

We need a result by S. Tanno in which he calculates the spectrum of the Berger spheres.

THEOREM 2.64 ([**Tan79**]). *The spectrum of the Laplacian operator Δ of $\mathbb{S}^{2n+1}_\tau$ is contained in the set*

$$\left\{\mu_{k,p} = \lambda_k + \tfrac{1-\tau^2}{\tau^2}(k-2p)^2 \mid k \in \mathbb{Z},\, k \geq 0,\, p \in \mathbb{Z},\, 0 \leq p \leq \lfloor \tfrac{k}{2} \rfloor\right\},$$

where $\{\lambda_k = k(2n+k),\, k \in \mathbb{Z},\, k \geq 0\}$ is the spectrum of $\mathbb{S}^{2n+1}$ and $\lfloor\cdot\rfloor$ is the integer part. Furthermore, $\Delta \circ L_\xi = L_\xi \circ \Delta$ and the eigenspace $V(\lambda_k)$ of the eigenvalue λ_k decomposes into

$$V(\lambda_k) = V(\mu_{k,0}) \oplus \cdots \oplus V(\mu_{k,[\frac{k}{2}]}),$$

where some of $V(\mu_{k,p})$ may be trivial. Moreover, for every $\varphi \in V(\mu_{k,p})$ we have $(L_\xi)^2\varphi + \frac{1}{\tau^2}(k-2p)^2\varphi = 0$.

Let $f = \sum_{k,p} f_{k,p}$ and $g = \sum_{k,p} g_{k,p}$ be the decompositions of f and g into eigenfunctions of the Laplacian Δ of $\mathbb{S}^{2n+1}_\tau$ with $f_{k,p}, g_{k,p} \in V(\mu_{k,p})$. More precisely,

$$\Delta f_{k,p} + \mu_{k,p} f_{k,p} = 0, \quad \Delta g_{k,p} + \mu_{k,p} g_{k,p} = 0, \quad \mu_{k,p} = k(2m+k) + \tfrac{1-\tau^2}{\tau^2}(k-2p)^2,$$

where k and p are integers satisfying $k \geq 0$ and $0 \leq p \leq \lfloor \frac{k}{2} \rfloor$.

Since $(L_\xi)^2 f_{k,p} = -\frac{1}{\tau^2}(k-2p)^2 f_{k,p}$, it follows from (2.39) and Theorem 2.64 that

$$\begin{aligned} &[(-\mu_{k,p} + \rho + (2n+2-\tfrac{1}{\tau^2}))^2 - 4\tfrac{(1-\tau^2)^2}{\tau^4}(k-2p)^2] f_{k,p} = 0, \\ &[(-\mu_{k,p} + \rho + (2n+2-\tfrac{1}{\tau^2}))^2 - 4\tfrac{(1-\tau^2)^2}{\tau^4}(k-2p)^2] g_{k,p} = 0, \end{aligned} \tag{2.40}$$

for any $k \geq 0$ and $0 \leq p \leq \lfloor \frac{k}{2} \rfloor$, and so, the eigenvalue ρ of $\mathcal{L}$ is of the form

$$\rho = (2n+1+k)(k-1) + \tfrac{1-\tau^2}{\tau^2}(k-2p\pm 1)^2. \tag{2.41}$$

for some $k \geq 0$ and $0 \leq p \leq \lfloor \frac{k}{2} \rfloor$.

Since $\rho \geq 0$ for $k \geq 1$, we only need to analyze the case $k = 0$. In this case, $p = 0$ and $\rho = \frac{1}{\tau^2} - (2n+2)$. So $\rho \geq 0$ when $\tau^2 \leq \frac{1}{2n+2}$ and so $\operatorname{Ind}(\mathcal{L}) = 0$ for $\tau^2 \leq \frac{1}{2n+2}$. If $\tau^2 > \frac{1}{2n+2}$, ρ is negative and from (2.40) we deduce that f and g are constant functions and so the eigensections associated to ρ are of the form $\alpha a + \beta i a$, with $\alpha, \beta \in \mathbb{R}$. It follows that for $\tau^2 > \frac{1}{2n+2}$, $\operatorname{Ind}(\mathcal{L})$ is twice the multiplicity of $\mu_{0,0}$, which is one. So, in this case, $\operatorname{Ind}(\mathcal{L}) = 2$. Considering the above decomposition of the normal bundle, we prove the result. $\square$

CHAPTER 3

Complete Minimal Submanifolds of Non-negatively Curved Riemannian Manifolds

In this chapter we will present results of different nature on the stability and index of complete minimal submanifolds of Riemannian manifolds. The common denominator will be that the curvature of the ambient manifold is non-negative and that we will not consider concrete examples as ambient manifolds. In addition, in Sections 3.1 and 3.2, we will study abstract Schrödinger operators which reflect certain Jacobi operators of minimal hypersurfaces.

3.1. Index of Schrödinger operators on complete surfaces

The motivation for studying the Schrödinger operators that we will consider in this section is that they generalize the Jacobi operators of two-sided minimal surfaces in 3-manifolds with non-negative scalar curvature.

THEOREM 3.1 ([**FC85**]). *Let (M^2, g) be a complete surface and $L = \Delta - K + q$ a Schrödinger operator on M (see Section 1.4), where K is the Gauss curvature of (M, g) and q is a non-negative function. Then*

(1) *If $\mathrm{Ind}(L) < \infty$ and M is not compact, then*
   (a) *The Gauss curvature K^* of $(M, g^* = u^2 g)$ satisfies $K^* \geq 0$ on $M \backslash C$,*
   (b) *(M, g^*) is complete,*
   where C and u are respectively the compact subset of M and the positive function on M with $Lu = 0$ on $M \setminus C$ given in Theorem 2.27.

(2) *If $q \geq c > 0$ on M for some real number c, then $\mathrm{Ind}(L) < \infty$ if and only if M is compact.*

REMARK 3.2. Considering the proof of Theorem 2.27, the assumption $\mathrm{Ind}(L) < \infty$ in Theorem 3.1 can be weakened by there exists a compact $C \subset M$ with $\mathrm{Ind}_{M\backslash C}(L) = 0$, because the existence of the function u is assured.

PROOF. **Proof of 1**. Since $g^* = u^2 g$, it follows from Proposition 1.5,4 that the Gauss curvature K^* of the metric g^* is given by

$$K^* = \frac{K - \Delta \log u}{u^2}, \tag{3.1}$$

where Δ is the Laplacian associated to the metric g and K is the Gauss curvature of the metric g. Since $Lu = 0$ on $M \setminus C$, $q \geq 0$ and $u > 0$, we get that $\Delta u \leq Ku$ on $M \setminus C$, which implies that

$$\Delta \log u = \frac{u\Delta u - \|\nabla u\|^2}{u^2} \leq K - \frac{\|\nabla u\|^2}{u^2}, \quad \text{on } M \setminus C. \tag{3.2}$$

From (3.1) and (3.2) we now get that

$$K^* \geq \frac{\|\nabla u\|^2}{u^4} \geq 0, \quad \text{on} \quad M \setminus C, \tag{3.3}$$

and so a) is proved.

Now, we will prove b), i.e., that the metric g^* on M is complete. From Theorem 1.2,4, it suffices to prove that the length of any divergence curve in (M, g^*) is infinite.

Let $\alpha : [0, \infty) \to M$ be a divergent curve, $p = \alpha(0)$ and $\{B_p(R),\, R > 0\}$ geodesic balls of (M, g) centered at p with radius R. The completeness of (M, g) (Theorem 1.2,3) ensures that $\overline{B_p(R)}$ are compact subsets of M and so, since α is a divergent curve, $\forall R > 0$, there exists a real number $t_R > 0$ such that $\alpha(t_R) \in \partial B_p(R)$ and $\alpha(t) \in M \setminus \overline{B_p(R)}$ for any $t > t_R$.

Now, for any $R > 0$, let $\eta_R : M \to \mathbb{R}$ be a cut-off function satisfying $0 \leq \eta_R \leq 1$, $\eta_R = 0$ on $\overline{B_p(R)}$ and $\eta_R = 1$ on $M \setminus \overline{B_p(R+1)}$. Then, there exists a constant μ such that the function $(u + \eta_R)^2 \geq \mu > 0$ on M. In fact, on $\overline{B_p(R+1)}$ we have that $(u + \eta_R)^2 \geq u^2 \geq \lambda > 0$, where λ is the minimum of the positive function u^2 over the compact subset $\overline{B_p(R+1)}$, and on $M \setminus \overline{B_p(R+1)}$ we have that $(u + \eta_R)^2 > \eta_R^2 = 1$. Now, it is enough to take $\mu = \min\{\lambda, 1\}$. This implies that the metric $g_R = (u + \eta_R)^2 g$ is complete, since $(u + \eta_R)^2 \geq \mu > 0$ and the metric μg is complete.

If d_R is the distance associated to the metric g_R, let $q_R \in \partial B_p(R)$ be the minimum of the continuous function $d_R(p, -) : \partial B_p(R) \to \mathbb{R}$ defined on the compact subset $\partial B_p(R)$. The completeness of g_R (Theorem 1.2) implies that there exists a geodesic $\gamma_R : [0, \infty) \to M$ in (M, g_R) such that $\gamma_R(0) = p$, $\gamma_R(s_R) = q_R$ and $d_R(p, q_R) = s_R$. Moreover, by definition of q_R, we have that $\gamma_R([0, s_R)) \subset B_p(R)$.

But on $\overline{B_p(R)}$ the function $\eta_R = 0$, and so $g_R = g^*$ on $\overline{B_p(R)}$. So for any $R > 0$, there exists a geodesic $\gamma_R : [0, s_R] \to \overline{B_p(R)}$ in (M, g^*), such that $\gamma_R(0) = p$, $\gamma_R(s_R) = q_R \in \partial B_p(R)$ and $L^*(\gamma_R) = d^*(p, \partial B_p(R)) = s_R$.

Now, we reparametrize the geodesics γ_R to curves β_R with $g(\beta_R', \beta_R') = 1$ and, taking a subsequence if necessary, $\{\beta_R'(0)\} \to v \in T_pM$ (as $R \mapsto \infty$) with $g(v, v) = 1$. Using ODE theory, the curves $\{\beta_R\}$ converge, on compact subsets of $[0, \infty)$, to a limit curve $\beta : [0, \infty) \to M$ that minimizes the distance d^* between any two points and $g(\beta', \beta') = 1$. So, in particular, we have that

$$L^*(\beta) = \int_0^\infty \|\beta'\|_* \, \mathrm{d}s = \int_0^\infty (u \circ \beta)(s) \, \mathrm{d}s.$$

On the other hand, by construction, we have that

$$L^*(\alpha) \geq L^*(\alpha_{|[0,t_R]}) \geq L^*(\gamma_{R|[0,s_R]}) \quad \forall R > 0.$$

Now, taking limits when $R \mapsto \infty$, we finally get that

$$L^*(\alpha) \geq L^*(\beta) = \int_0^\infty (u \circ \beta)(s) \, \mathrm{d}s.$$

We may assume that $\beta \subset M \setminus C$, because, as C is compact, $\beta \cap C \subset \beta([0, a])$ for some $a > 0$, and so we can replace β with $\beta_{|[a,\infty)}$.

So to prove that g^* is a complete metric on M, we only need to prove that

$$\int_0^\infty (u \circ \beta)(s) \, \mathrm{d}s = \infty.$$

Since β minimizes d^* between any two points, the second variation formula of the lenght implies that

$$\int_0^\infty \left(\frac{(\phi')^2}{u\circ\beta} - (K^*\circ\beta)(u\circ\beta)\phi^2\right) \mathrm{d}s \geq 0.$$

for any piecewise smooth function $\phi : [0,\infty) \to \mathbb{R}$ such that $\phi_{|[0,\infty)\setminus(a,b)} = 0$, with $a < b$ (see [**Cha06**]).

Using (3.3), the last inequality gives the following one

$$\int_0^\infty \left(\frac{(\phi')^2}{u\circ\beta} - \frac{\|(\nabla u)\circ\beta\|^2\phi^2}{(u\circ\beta)^3}\right) \mathrm{d}s \geq 0,$$

where the gradient of u and its length are computed with respect to the metric g. Since $g(\beta',\beta') = 1$, we have that

$$\|(\nabla u)\circ\beta\|^2 \geq g((\nabla u)\circ\beta,\beta')^2 g(\beta',\beta') = g((\nabla u)\circ\beta,\beta')^2 = ((u\circ\beta)')^2.$$

From here, the last integral inequality is written as

$$\int_0^\infty \left(\frac{(\phi')^2}{u\circ\beta} - \frac{((u\circ\beta)')^2\phi^2}{(u\circ\beta)^3}\right) \mathrm{d}s \geq 0.$$

If $\psi = \frac{\phi}{u\circ\beta}$, a simple calculation transforms the last inequality into

$$0 \leq \int_0^\infty \left((u\circ\beta)(\psi')^2 + 2(u\circ\beta)'\psi\psi'\right) \mathrm{d}s,$$

and integrating by parts we finally get

$$\int_0^\infty (u\circ\beta)\left((\psi')^2 + 2\psi\psi''\right) \mathrm{d}s \leq 0, \tag{3.4}$$

for any piecewise smooth function $\psi : [0,\infty) \to \mathbb{R}$ such that $\psi_{|[0,\infty)\setminus(c,d)} = 0$ for certain real numbers $c < d$.

Now, for any $a > 0$, let $\psi_a : [0,\infty) \to \mathbb{R}$ be the function given by

$$\psi_a(s) = \begin{cases} \frac{s}{a} & s \in [0,a], \\ \frac{4}{3} - \frac{s^2}{3a^2} & s \in [a,2a], \\ 0 & s \in [2a,\infty). \end{cases}$$

Then ψ_a is a piecewise smooth function such that $(\psi_a)_{|\mathbb{R}\setminus(0,2a)} = 0$. Using (3.4) for this function ψ_a and that $u > 0$, we get that

$$\begin{aligned} 0 &\geq \int_0^\infty (u\circ\beta)\left(((\psi_a)')^2 + 2\psi_a\psi_a''\right) \mathrm{d}s \\ &= \int_0^a \frac{u\circ\beta}{a^2}\,\mathrm{d}s + \int_a^{2a} (u\circ\beta)\left(\frac{8s^2}{9a^4} - \frac{16}{9a^2}\right) \mathrm{d}s \\ &\geq \int_0^a \frac{u\circ\beta}{a^2}\,\mathrm{d}s - \frac{16}{9}\int_a^{2a} \frac{u\circ\beta}{a^2}\,\mathrm{d}s \geq \int_0^a \frac{u\circ\beta}{a^2}\,\mathrm{d}s - \frac{16}{9}\int_a^\infty \frac{u\circ\beta}{a^2}\,\mathrm{d}s, \end{aligned}$$

and thus,

$$\int_0^a (u\circ\beta)\,\mathrm{d}s \leq \frac{16}{9}\int_a^\infty (u\circ\beta)\,\mathrm{d}s, \quad \forall a > 0,$$

which is impossible unless $\int_0^\infty (u \circ \beta)\, ds = \infty$. This proves that the metric g^* is complete.

Proof of 2. We assume that M is not compact and we will find to a contradiction. If M is not compact, according to the proof of part 1, for any $a > 0$, the function $\psi_a : [0, \infty) \to \mathbb{R}$ defined by

$$\psi_a(s) = \begin{cases} \sin \frac{\pi s}{a} & s \in [0, a] \\ 0 & s \in [a, \infty) \end{cases}$$

is picewise smooth and $(\psi_a)_{|[0,\infty)\setminus(0,a)} = 0$. Using (3.4) for these functions ψ_a we get that

$$0 \geq \int_0^\infty \left(\frac{\pi^2}{a^2} \cos^2 \frac{\pi s}{a} + \left(c - \frac{2\pi^2}{a^2} \right) \sin^2 \frac{\pi s}{a} \right) (u \circ \beta)(s)\, ds.$$

If a is large enough to make $c - \frac{2\pi^2}{a^2} > 0$, the last inequality gives a contradiction. □

COROLLARY 3.3 ([**FC85**]). *Let (M, g) be a complete non-compact orientable surface and $L = \Delta - K + q$ a Schrödinger operator on M, where K is the Gauss curvature of (M, g) and q a non-negative function. If $\operatorname{Ind}(L) < \infty$, then*

(1) *M is conformally equivalent to a compact Riemann surface with a finite number of points removed,*
(2) *$\int_M q\, dv < \infty$.*

PROOF. To prove the corollary, we will make use of the classical results of Huber and Conh-Vossen.

THEOREM 3.4 ([**CV35, Hub57**]). *Let (M^2, g) be a complete orientable surface with the negative part of its Gauss curvature K integrable, i.e., $\int_M |K^-|\, dv < \infty$, where*

$$K^- = \begin{cases} 0 & \text{if} \quad K > 0, \\ K & \text{if} \quad K \leq 0. \end{cases}$$

Then M is conformally equivalent to a compact Riemann surface with a finite number of points removed, and

$$\int_M K\, dv \leq 2\pi\chi(M),$$

where $\chi(M)$ is the Euler characteristic of M.

From Theorem 3.1, the surface (M, g^*) is complete and there exists a compact subset C of M such that $K^* \geq 0$ on $M \setminus C$. So, the set $A = \{p \in M \mid K^*(p) < 0\} \subset C$ and

$$\int_M |(K^*)^-|\, dv^* = \int_A |(K^*)^-|\, dv^* \leq \int_C |(K^*)^-|\, dv^* < \infty,$$

because C is compact. Now 1 is a consequence of Theorem 3.4.

On the other hand, from Theorem 3.1, there exists a positive function u on M such that $0 = Lu = \Delta u - Ku + qu$ on $M \setminus C$. From (3.1) we get that

$u^2K^* = q + \frac{\|\nabla u\|^2}{u^2} \geq q$ on $M \setminus C$ and so

$$\int_M q\,\mathrm{d}v = \int_C q\,\mathrm{d}v + \int_{M\setminus C} q\,\mathrm{d}v \leq \int_C q\,\mathrm{d}v + \int_{M\setminus C} u^2K^*\,\mathrm{d}v$$
$$= \int_C q\,\mathrm{d}v + \int_{M\setminus C} K^*\,\mathrm{d}v^* \leq \int_C q\,\mathrm{d}v - \int_C K^*\mathrm{d}v^* + 2\pi\chi(M) < \infty,$$

where we have used that C is compact, that $\mathrm{d}v^* = u^2\mathrm{d}v$ and Theorem 3.4. □

Using the above results, we study the special case $\mathrm{Ind}(L) = 0$.

THEOREM 3.5 ([**FC85**]). *Let (M, g) be a complete surface and $L = \Delta - K + q$ a Schrödinger operator on M, where K is the Gauss curvature of (M, g) and q is a non-negative function. If $\mathrm{Ind}(L) = 0$, then one of the following things happens:*

(1) *M is conformally equivalent to the sphere $\mathbb{S}^2$.*
(2) *M is diffeomorphic to the real projective plane $\mathbb{RP}^2$.*
(3) *M is a flat torus and q vanishes identically.*
(4) *M is a flat Klein bottle and q vanishes identically.*
(5) *M is conformally equivalent to the complex plane $\mathbb{C}$.*
(6) *M is a flat cylinder and q vanishes identically.*
(7) *M is a flat Möbius strip and q vanishes identically.*

Furthermore, if $\mathrm{Ind}(L) = 0$ and $q \geq c > 0$, then M is either conformally equivalent to the sphere $\mathbb{S}^2$ or diffeomorphic to the real projective plane $\mathbb{RP}^2$.

REMARK 3.6. Not all the results of Theorem 3.5 appear in [**FC85**]. The new contributions are unpublished and are due to A. Ros and F. Urbano.

PROOF. First, assume that M is compact. Then from Theorem 1.17 and using the constant function 1 as a test function we have

$$0 \leq Q(1) = \int_M (K - q)\,\mathrm{d}v = 2\pi\chi(M) - \int_M q\,\mathrm{d}v \leq 2\pi\chi(M), \tag{3.5}$$

where $\chi(M)$ is the Euler characteristic of M and we have used the Gauss-Bonnet theorem.

We have only three possibilities: either $\chi(M) = 2$ and M is diffeomorphic to $\mathbb{S}^2$, or $\chi(M) = 1$ and M is diffeomorphic to $\mathbb{RP}^2$ or $\chi(M) = 0$ and M is diffeomorphic to either a torus or a Klein bottle. In the first case, from Theorem 1.22, M is conformally diffeomorphic to $\mathbb{S}^2$ and we get 1). In the second case, we get 2). In the third case, we get from (3.5) that $\int_M q\,\mathrm{d}v = 0$ and so, since $q \geq 0$, we get $q = 0$. Also, since $Q(1) = 0$, Theorem 1.17 implies that $0 = L1 = -K$. So M is flat and we get 3) and 4).

Now, we assume that M is not compact. From Theorem 2.25,2 and Theorem 3.1 it follows that there exists a positive smooth function u on M such that $Lu = 0$ and the metric $g^* = u^2g$ is complete with non-negative Gauss curvature. We distinguish the cases depending on whether M is orientable or not.

If M is orientable, by Corollary 3.3, M is conformally diffeomorphic to a compact Riemann surface Σ with a finite number of points $\{p_1, \ldots, p_k\}$ removed. Since M is not compact, $k \geq 1$. Now, the Euler characteristics of M and Σ are related by $\chi(M) = \chi(\Sigma) - k$, and, if h is the genus of Σ, then the last equality becomes $\chi(M) = 2 - 2h - k$. Thus, since $K^* \geq 0$, Theorem 3.4 gives that

$$0 \leq \int_M K^*\mathrm{d}v^* \leq 2\pi\chi(M) = 2\pi(2 - 2h - k).$$

Since $h \geq 0$ and $k \geq 1$, the last inequality implies that there are only two possibilities: either $\{h = 0,\ k = 1\}$ or $\{h = 0,\ k = 2\}$.

In the first case, $h = 0$, $k = 1$, M is conformally equivalent to a sphere with one point removed. This means that M is conformally equivalent to the complex plane $\mathbb{C}$. Hence we get 5).

In the second case, $h = 0$, $k = 2$, M is diffeomorphic to a cylinder. Moreover, since $2 - 2h - k = 0$, we get from the last inequality that the curvature $K^* = 0$. But, since $g^* = u^2 g$ and $0 = Lu = \Delta u - Ku + qu$, it follows that

$$0 = u^2 K^* = K - \Delta \log u = K - \frac{\Delta u}{u} + \frac{\|\nabla u\|^2}{u^2} = q + \frac{\|\nabla u\|^2}{u^2}.$$

This implies that $q = 0$ and u is a constant function. So $K = 0$ and (M, g) is a flat cylinder. So we get 6).

Now we assume that M is not compact and not orientable. Let $\widetilde{M}$ be the orientable double covering of M and $\widetilde{g}$ be the covering metric on $\widetilde{M}$. Then $(\widetilde{M}, \widetilde{g})$ is a complete orientable surface and, on it, we consider the Schrödinger operator $\widetilde{L} = \widetilde{\Delta} - \widetilde{K} + \widetilde{q}$, where $\widetilde{\Delta}$ is the Laplacian of $\widetilde{g}$, $\widetilde{K}$ is the Gauss curvature of $\widetilde{g}$, and $\widetilde{q} = q \circ \Pi$, where $\Pi : \widetilde{M} \to M$ is the covering projection. Since Π is a local isometry, we have that $\widetilde{K} = K \circ \Pi$ and $\widetilde{\Delta}(f \circ \Pi) = \Delta f \circ \Pi$. Now we proceed as in the proof of Proposition 2.35,4. Since $\operatorname{Ind}(L) = 0$ and M is not compact, it follows from Theorem 2.25,2 that there exists a positive function $u \in C^\infty(M)$ with $Lu = 0$. Then the function $\widetilde{u} = u \circ \Pi \in C^\infty(\widetilde{M})$ is positive and $\widetilde{L}\widetilde{u} = 0$. So, from Theorem 2.25,1, it follows that $\operatorname{Ind}(\widetilde{L}) = 0$. Thus, using the study made in the orientable case, we have that either $\widetilde{M}$ is conformally equivalent to $\mathbb{C}$ or $\widetilde{M}$ is a flat cylinder and $\widetilde{q}$ vanishes identically on $\widetilde{M}$. The first case is impossible because $\mathbb{C}$ cannot be the orientable double covering of a nonorientable surface. In the second case, M is a flat Möbius strip and $q = 0$, and we get 8).

Finally, if $q \geq c > 0$ and $\operatorname{Ind}(L) = 0$, from Theorem 3.1, M is compact and, since q cannot vanish, M is conformally equivalent to $\mathbb{S}^2$ or diffeomorphic to $\mathbb{RP}^2$. This concludes the proof. □

3.2. Index and nullity of Schrödinger operators on a compact Riemann surface associated to a meromorphic function

In this section we will study the index of certain Schrödinger operators on a compact Riemann surface coming from holomorphic functions on such a Riemann surface. The main motivation for studying these Schrödinger operators is that they are a generalization of the Jacobi operator of a two-sided minimal surface of Euclidean space of dimension three. The content of this section is based on the results of S. Montiel and A. Ros, [**MR91**].

Let M be a compact Riemann surface and $\phi : M \to \mathbb{S}^2$ a holomorphic function. If g is an arbitrary conformal metric on M (see Section 1.5) and Δ is the Laplacian of (M, g), we consider on M the Schrödinger operator L_ϕ defined by

$$L_\phi = \Delta + \|\nabla \phi\|^2, \quad \text{with} \quad \|\nabla \phi\|^2(p) = \sum_{i=1}^{2} \|\mathrm{d}\phi_p(e_i)\|^2, \quad \forall p \in M,$$

where $\{e_1, e_2\}$ is an orthonormal basis of T_pM.

If g^* is another conformal metric on M, then $g^* = e^{2u} g$ for certain $u \in C^\infty(M)$, and considering Proposition 1.5, the corresponding Schrödinger operator L_ϕ^* is given by $L_\phi^* = e^{-2u}L_\phi$. If Q_ϕ and Q_ϕ^* are the quadratic forms associated with L_ϕ and L_ϕ^* respectively, then for any $f \in C^\infty(M)$ we have

$$Q_\phi^*(f) = -\int_M fL_\phi^* f \, dv^* = -\int_M fL_\phi f \, dv = Q_\phi(f),$$

and, in particular, L_ϕ and L_ϕ^* have the same index and nullity. Thus such numbers depend only on the Riemann surface M and of the holomorphic map ϕ and not on the chosen conformal metric in the conformal structure of M.

DEFINITION 3.7. Let M be a compact Riemann surface and $\phi : M \to \mathbb{S}^2$ a holomorphic map. We will denote by $\text{Ind}(M, \phi)$ and $\text{Nul}(M, \phi)$ the index and nullity of the operator $L_\phi = \Delta + \|\nabla\phi\|^2$ associated to any conformal metric g of the conformal structure of M.

Now we study some important properties of $\text{Ind}(M, \phi)$ and $\text{Nul}(M, \phi)$.

Since $\phi : M \to \mathbb{S}^2$ is holomorphic, then given a conformal metric g of the conformal structure of M, the map $\phi : (M, g) \to (\mathbb{S}^2, \langle,\rangle)$ is harmonic, where $\langle,\rangle$ is the canonical metric on both $\mathbb{S}^2$ and $\mathbb{R}^3$. This means that

$$\Delta\phi + \|\nabla\phi\|^2\phi = 0.$$

So, for any $a \in \mathbb{R}^3$, if $\phi_a = \langle\phi, a\rangle$, we have that

$$L_\phi(\phi_a) = 0, \quad \forall a \in \mathbb{R}^3.$$

Moreover, if ϕ is not constant, the kernel of the linear map

$$a \in \mathbb{R}^3 \mapsto \phi_a \in C^\infty(M),$$

is zero, and so $\{\phi_a, | \, a \in \mathbb{R}^3\}$ is a 3-dimensional space of eigenfunctions of L_ϕ with corresponding eigenvalue 0.

On the other hand, given $a \in \mathbb{R}^3$, let $n(a)$ be the number of nodal sets of ϕ_a and $\{\Omega_1, \ldots, \Omega_{n(a)}\}$ the corresponding nodal sets of ϕ_a, i.e., the connected components of the open set $\{p \in M \mid \phi_a(p) \neq 0\}$. Note that if $\mathbb{S}_a^2(+) = \{q \in \mathbb{S}^2 \mid \langle q, a\rangle > 0\}$ and $\mathbb{S}_a^2(-) = \{q \in \mathbb{S}^2 \mid \langle q, a\rangle < 0\}$, then the map ϕ applies every nodal set in $\mathbb{S}_a^2(\pm)$. In particular, the restriction of ϕ_a to a nodal set is either positive or negative. Thus, from Theorem 2.25 we get:

PROPOSITION 3.8. *Let M be a compact Riemann surface and $\phi : M \to \mathbb{S}^2$ a non-constant holomorphic map. Then* $\text{Nul}(M, \phi) \geq 3$ *and for any $a \in \mathbb{R}^3$,* $\text{Ind}(M, \phi) \geq n(a) - 1$, *where $n(a)$ is the number of nodal sets of $\phi_a = \langle\phi, a\rangle$.*

COROLLARY 3.9. *Let M be a compact Riemann surface and $\phi : M \to \mathbb{S}^2$ a holomorphic map. Then* $\text{Ind}(M, \phi) = 0$ *if and only if ϕ is constant.*

PROOF. If ϕ is constant, we have that $L_\phi = \Delta$, and its index is zero. So $\text{Ind}(M, \phi) = 0$.

Conversely, if $\text{Ind}(M, \phi) = 0$ and ϕ is not constant, then for any $a \in \mathbb{R}^3$, it follows that $\Delta\phi_a + \|\nabla\phi\|^2\phi_a = 0$, and so, using Theorem 1.7,5, we obtain that

$$0 = \int_M (\Delta\phi_a + \|\nabla\phi\|^2\phi_a) \, dv = \int_M \|\nabla\phi\|^2\phi_a \, dv.$$

Since ϕ is not constant, this implies that the function ϕ_a changes sign and thus the number of its nodal sets $n(a) \geq 2$. Proposition 3.8 gives that $\text{Ind}(M, \phi) \geq 1$, which is a contradiction. This concludes the proof. □

The following result computes the index and nullity of L_ϕ in terms of the degree of ϕ, for a certain family of holomorphic maps.

THEOREM 3.10 ([**MR91**]). *Let M be a compact Riemann surface and $\phi : M \to \mathbb{S}^2$ a non-constant holomorphic map. If all the ramification values of ϕ lie in an equator of $\mathbb{S}^2$, i.e., there exists a vector $a \in \mathbb{R}^3$ such that $\phi_a(p) = \langle \phi(p), a\rangle = 0$ for any ramification point p of ϕ, then* $\operatorname{Ind}(M, \phi) = 2d - 1$ *and* $\operatorname{Nul}(M, \phi) = 3$, *where d is the degree of the map ϕ.*

PROOF. First, we will see that the number of nodal sets of ϕ_a is $n(a) = 2d$. In fact, if $M^+ = \{p \in M \mid \phi_a(p) > 0\}$ and $M^- = \{p \in M \mid \phi_a(p) < 0\}$, then, from the hypothesis about the ramification points, $\phi : M^+ \to \mathbb{S}^2_a(+)$ and $\phi : M^- \to \mathbb{S}^2_a(-)$ are regular maps and thus d-sheeted coverings. Since $\mathbb{S}^2_a(\pm)$ are simply connected we get that M^+ and M^- have d connected components and hence $n(a) = 2d$.

From Proposition 3.8, we get that $\operatorname{Ind}(M, \phi) \geq n(a) - 1 = 2d - 1$. To finish the proof, we just need to prove that $\operatorname{Ind}(M, \phi) > 2d - 1$ is not possible.

Indeed, if $m = \operatorname{Ind}(M, \phi) > 2d-1$, let $\{\lambda_1, \dots, \lambda_m\}$ be the negative eigenvalues of the operator L_ϕ and $\{h_1, \dots, h_m\}$ the corresponding eigenfunctions, i.e., $L_\phi h_i + \lambda_i h_i = 0$, $1 \leq i \leq m$. Since $L_\phi \phi_a = 0$, it follows that $\int_M \phi_a h_i \, dv = 0$ for $1 \leq i \leq m$, and so $\{h_1, \dots, h_m, \phi_a\}$ spans a subspace V of $C^\infty(M)$ with $\dim V = m + 1 > 2d$. For simplicity, we denote $\phi_a = h_{m+1}$, and so V is generated by $\{h_1, \dots, h_{m+1}\}$.

If $\{\Omega_1, \dots, \Omega_{2d}\}$ are the nodal sets of ϕ_a, then the linear equation system

$$\sum_{j=1}^{m+1} \mu_j \int_{\Omega_i} h_j \|\nabla\phi\|^2 \, dv = 0, \quad 1 \leq i \leq 2d,$$

has a non-trivial solution since $m + 1 > 2d$. Hence, there exist real numbers $\mu_1, \dots, \mu_{m+1}$ such that $h = \sum_{j=1}^{m+1} \mu_j h_j$ is a nonzero function satisfying

$$\int_{\Omega_i} h \|\nabla\phi\|^2 \, dv = 0, \quad 1 \leq i \leq 2d. \tag{3.6}$$

Now, we calculate $Q_\phi(h)$. Using that $L_\phi h_i + \lambda_i h_i = 0$, $1 \leq i \leq m$ and that $L_\phi h_{m+1} = L_\phi \phi_a = 0$, it follows that

$$\begin{aligned} Q_\phi(h) &= -\int_M h L_\phi h \, dv = -\sum_{j,k=1}^{m+1} \mu_j \mu_k \int_M h_j L_\phi h_k \, dv \\ &= \sum_{j=1}^{m+1} \sum_{k=1}^{m} \mu_j \mu_k \lambda_k \int_M h_j h_k \, dv = \sum_{k=1}^{m} \mu_k^2 \lambda_k \int_M h_k^2 \, dv \leq 0. \end{aligned} \tag{3.7}$$

On the other hand, we can also estimate $Q_\phi(h)$ in the following way. Since the complement in M of the nodal set of ϕ_a has measure zero, we have that

$$Q_\phi(h) = \int_M \left(\|\nabla h\|^2 - \|\nabla\phi\|^2 h^2\right) dv = \sum_{i=1}^{2d} \int_{\Omega_i} \left(\|\nabla h\|^2 - \|\nabla\phi\|^2 h^2\right) dv.$$

But $\phi_i := \phi_{|\Omega_i} : \Omega_i \to \mathbb{S}^2_a(\pm)$ is a conformal diffeomorphism and so, considering the independence of Q_ϕ of the metric in the conformal class, if g_0 is the canonical metric in $\mathbb{S}^2_a(\pm)$ and $g_i = (\phi_i)^* g_0$, we have that

$$\int_{\Omega_i} \left(\|\nabla h\|^2 - \|\nabla\phi\|^2 h^2\right) dv = \int_{\Omega_i} \left(\|\nabla_i h\|_i^2 - \|\nabla_i \phi\|_i^2 h^2\right) dv_i,$$

where in the second term of the equality everything is computed with respect to the metric g_i. Now, since $\phi_i : (\Omega_i, g_i) \to (\mathbb{S}^2_a(\pm), g_0)$ is an isometry, if $\psi_i = h \circ \phi_i^{-1}$, we get that

$$\int_{\Omega_i} (\|\nabla_i h\|_i^2 - \|\nabla_i \phi\|_i^2 h^2)\, dv_i = \int_{\mathbb{S}^2_a(\pm)} (\|\nabla_0 \psi_i\|_0^2 - \|\nabla_0(\phi \circ \phi_i^{-1})\|_0^2 \psi_i^2)\, dv_0$$
$$= \int_{\mathbb{S}^2_a(\pm)} (\|\nabla_0 \psi_i\|_0^2 - 2\psi_i^2)\, dv_0.$$

Finally we have obtained that

$$Q_\phi(h) = \sum_{i=1}^{2d} \int_{\mathbb{S}^2_a(\pm)} (\|\nabla_0 \psi_i\|_0^2 - 2\psi_i^2)\, dv_0.$$

Now, from the change of variable formula (Theorem 1.7,3), for the diffeomorphism $\phi_i : (\Omega_i, g) \to (S^2_\pm, g_0)$ and (3.6) it follows that $\int_{\mathbb{S}^2_a(\pm)} \psi_i\, dv_0 = \frac{1}{2}\int_{\Omega_i} h\|\nabla\phi\|^2 dv = 0$.

The first two eigenvalues of the Laplacian in $(\mathbb{S}^2(\pm), g_0)$ with Neumann boundary conditions are 0 and 2, and the constant functions are the eigenfunctions corresponding to the eigenvalue 0. As $\int_{\mathbb{S}^2_a(\pm)} \psi_i\, dv_0 = 0$, the functions ψ_i are $L^2(\mathbb{S}^2(\pm))$-orthogonal to the constant functions and thus $\int_{\mathbb{S}^2_a \pm)} \|\nabla_0 \psi_i\|_0^2 dv_0 \geq 2\int_{\mathbb{S}^2(\pm)} \psi_i^2 dv_0$. Considering the last formula, this implies that $Q_\phi(h) \geq 0$.

Since from (3.7) $Q_\phi(h) \leq 0$, we finally get that $Q_\phi(h) = 0$, and thus $\mu_k = 0$ for $k = 1, \ldots, m$. This implies that $h = \mu_{m+1} h_{m+1} = \mu_{m+1}\phi_a$ with $\mu_{m+1} \neq 0$. But we know that $0 = \int_{\Omega_i} h\|\nabla\phi\|^2 dv = \mu_{m+1}\int_{\Omega_i} \phi_a\|\nabla\phi\|^2 dv$ for any $i \in \{1, \ldots, 2d\}$. This is impossible because the function ϕ_a is either positive or negative on every nodal set Ω_i and ϕ is not constant. So the assumption $\mathrm{Ind}(M, \phi) > 2d - 1$ implies a contradiction and so $\mathrm{Ind}(M, \phi) = 2d - 1$.

Now suppose $\mathrm{Nul}(M, \phi) > 3$. Then there exists a function h satisfying $L_\phi h = 0$ and h is $L^2(M)$-orthogonal to the space $\{\phi_b = \langle \phi, b\rangle \mid b \in \mathbb{R}^3\}$. According to the above notation, on each nodal set Ω_i we have $g_i = e^{2u_i} g$ and so, $L^i_\phi h = e^{-2u_i} L_\phi h = 0$. So $(\Delta_0 + 2)(\psi_i) = 0$ and ψ_i is an eigenfunction of Δ_0 with eigenvalue 2. From Theorem 1.18 it follows that $\psi_i(p) = \langle p, b_i\rangle$ for some $b_i \in \mathbb{R}^3$, and since $h = \psi_i \circ \phi_i$ on Ω_i, we have that $h = \phi_{b_i}$ on Ω_i. Hence $L_\phi(h - \phi_{b_i}) = 0$ on M and $h - \phi_{b_i}$ vanishes on Ω_i. Again using again the principle of unique continuation, $h = \phi_{b_i}$ on M, which is a contradiction. □

COROLLARY 3.11 ([**MR91**]). *Let $\phi : M \to \mathbb{S}^2$ be a non-constant holomorphic map on a compact Riemann surface M.*

(1) *If the degree of ϕ is $d = 1$, then* $\mathrm{Ind}(M, \phi) = 1$ *and* $\mathrm{Nul}(M, \phi) = 3$.
(2) *If the degree of ϕ is $d = 2$ and the genus of M is $h = 0$, then* $\mathrm{Ind}(M, \phi) = \mathrm{Nul}(M, \phi) = 3$.
(3) *If the genus of M is $h = 0$, then* $\mathrm{Ind}(M, \phi) \neq 2$.

PROOF. **Proof of 1**. From the definition of degree, if $d = 1$, ϕ has not ramification points and thus ϕ has not ramification values. The proof follows from Theorem 3.10.

Proof of 2. If $d = 2$ and $h = 0$, from the Riemann-Hurwitz formula, Theorem 1.23, the total ramification of ϕ is $b_\phi = 2$. Thus, ϕ has either two simple ramification points or one ramification point with ramification order 2. In either case, the ramification values of ϕ lie on an equator of $\mathbb{S}^2$, and the proof follows from Theorem 3.10.

Proof of 3. From Corollary 3.9 and Corollary 3.11 we can assume that the degree of ϕ is $d \geq 2$. Now, Theorem 1.23 gives that the total number of ramification of ϕ is $b_\phi \geq 2$. This means that Φ has at least two simple ramification points $\{p_1, p_2\}$ or one ramification point q of order 2. In both cases, there exists $a \in \mathbb{S}^2$ such that $\langle \phi(p_i), a\rangle = 0$, $i = 1, 2$ or $\langle \phi(q), a\rangle = 0$. This implies that either the points $\{p_1, p_2\}$ or the point q are in the nodal lines of ϕ_a, and so they are corners of the nodal lines.

Under these conditions, if we prove that the number of nodal sets of ϕ_a is $n(a) \geq 4$, it follows from Proposition 3.8 that $\operatorname{Ind}(M, \phi) \geq 3$ and the proof is finished.

If $\{\Omega_1, \dots, \Omega_{n(a)}\}$ are the nodal sets of ϕ_a, then from the Gauss-Bonnet theorem for the domain Ω_i, [**Cha06**], we have that

$$\int_{\Omega_i} K \, \mathrm{d}v + \int_{\partial\Omega_i} k_g^i(s) \, \mathrm{d}s + \sum_{j=0}^{k_i} \theta_j^i = 2\pi\chi(\Omega_i), \quad 1 \leq i \leq n(a),$$

where K is the Gauss curvature of a conformal metric on M, k_g^i is the geodesic curvature of the nodal line $\partial\Omega_i$, $\{\theta_j^i, 0 \leq j \leq k_i\}$ are the exterior angles at the vertices of $\partial\Omega_i$ and $\chi(\Omega_i)$ is the Euler characteristic of Ω_i. From the last relation we now get

$$\int_M K \, \mathrm{d}v + \sum_{i=1}^{n(a)} \sum_{j=0}^{k_i} \theta_j^i = 2\pi \sum_{i=1}^{n(a)} \chi(\Omega_i),$$

where we have used that $\sum_{i=1}^{n(a)} \int_{\partial\Omega_i} k_g^i(s) \, \mathrm{d}s = 0$. Since Ω_i is connected, orientable and open in M, we have that $\chi(\Omega_i) = \beta_0(\Omega_i) - \beta_1(\Omega_i) + \beta_2(\Omega_i) = 1 - \beta_1(\Omega_i) \leq 1$, where $\beta_j(\Omega_i)$ is the j-th Betti number of Ω_i. Moreover, since M has genus zero, $\int_M K \, \mathrm{d}v = 4\pi$ and hence

$$\sum_{i=1}^{n(a)} \sum_{j=0}^{k_i} \theta_j^i \leq 2\pi(n(a) - 2).$$

Now, if ϕ has two simple ramification points $\{p_1, p_2\}$, there are four nodal sets locally around each one. So the sum of the exterior angles at each p_i is $\sum_{i=1}^{4}(\pi - \alpha_i) = 2\pi$. If there is a point q with ramification order 2, then there are 6 nodal sets locally around q. So the sum of the exterior angles at q is $\sum_{i=1}^{6}(\pi - \alpha_i) = 4\pi$. So in any case $\sum_{i=1}^{n(a)} \sum_{j=0}^{k_i} \theta_j^i \geq 4\pi$ and we finally get that $2\pi(n(a) - 2) \geq 4\pi$, which implies that $n(a) \geq 4$. This concludes the proof. $\square$

To conclude this section, we will study the case $\operatorname{Ind}(M, \phi) = 1$.

THEOREM 3.12 ([**MR91**]). *Let M be a compact Riemann surface and $\phi : M \to \mathbb{S}^2$ a holomorphic map with* $\operatorname{Ind}(M, \phi) = 1$. *Then for any non-constant holomorphic map $\psi : M \to \mathbb{S}^2$ we have that* $\deg \psi \geq \deg \phi$, *and if the equality holds then $\psi = G \circ \phi$ for some conformal transformation G of $\mathbb{S}^2$.*

PROOF. Let λ_1 be the first eigenvalue of $L_\phi = \Delta + \|\nabla\phi\|^2$ and u be a positive eigenfunction, $L_\phi u + \lambda_1 u = 0$ (Theorem 1.21,2). From Proposition 1.8,2, given a non-constant holomorphic map $\psi : M \to \mathbb{S}^2$, there exists $a \in \mathbb{R}^3$ such that the conformal transformation $\psi_a : \mathbb{S}^2 \to \mathbb{S}^2$ given in (1.3) satisfies

$$\int_M u(\psi_a \circ \psi)\, dv = 0.$$

Thus, since $\operatorname{Ind}(M, \phi) = 1$, using the vector valued function $\psi_a \circ \psi$ as a test function, and using the conformal invariance of the Dirichlet integral (1.4), we have that

$$0 \leq \int_M \left(\|\nabla(\psi_a \circ \psi)\|^2 - \|\nabla\phi\|^2 \|\psi_a \circ \psi\|^2\right) dv = \int_M \left(\|\nabla\psi\|^2 - \|\nabla\phi\|^2\right) dv.$$

If ω is the volume 2-form on $\mathbb{S}^2$, we have that $\int_M \psi^*\omega = 4\pi \deg \psi$. Since ψ is conformal, then $\psi^*\omega = \frac{\|\nabla\psi\|^2}{2}\eta$, where η is the volume 2-form on M. So, $\int_M \|\nabla\psi\|^2 dv = 8\pi \deg \psi$. Similarly, $\int_M \|\nabla\phi\|^2\, dv = 8\pi \deg \phi$, and from the last inequality we get that $\deg \psi \geq \deg \phi$ for any non-constant holomorphic map ψ.

If the equality holds, i.e., $\deg \psi = \deg \phi$, then $L_\phi(\psi_a \circ \psi) = 0$ and so $\Delta(\psi_a \circ \psi) + \|\nabla\phi\|^2(\psi_a \circ \psi) = 0$. Since $\psi_a \circ \psi$ is a holomorphic map, we have that $\Delta(\psi_a \circ \psi) + \|\nabla(\psi_a \circ \psi)\|^2(\psi_a \circ \psi) = 0$, and hence $\|\nabla\phi\|^2 = \|\nabla(\psi_a \circ \psi)\|^2$.

We write $\widehat{\psi} = \psi_a \circ \psi$, which is also a non-constant holomorphic map. We will prove that $\|\nabla\phi\|^2 = \|\nabla\widehat{\psi}\|^2$ implies that there exists an isometry A of $\mathbb{S}^2$ such that $A \circ \widehat{\psi} = \phi$. Taking $G = \psi_a^{-1} \circ A^{-1}$ we finish the proof.

To prove the above statement, we use a complex coordinate z in M. Then the fact that ϕ and $\widehat{\psi}$ are holomorphic means that

$$\langle \phi_z, \phi_z \rangle = 0, \quad \langle \widehat{\psi}_z, \widehat{\psi}_z \rangle = 0,$$

where $\langle , \rangle$ is the usual bilinear complex product on $\mathbb{C}^3$. From the above property together with $\langle \phi, \phi \rangle = \langle \widehat{\psi}, \widehat{\psi} \rangle = 1$ and $\|\phi_z\|^2 = \|\widehat{\psi}_z\|^2$ we get that

$$\begin{aligned} &\langle \phi, \phi_z \rangle = 0, \quad \phi_{z\bar{z}} + \|\phi_z\|^2 \phi = 0, \quad \phi_{zz} = (\log \|\phi_z\|^2)_z \phi_z, \\ &\langle \widehat{\psi}, \widehat{\psi}_z \rangle = 0, \quad \widehat{\psi}_{z\bar{z}} + \|\phi_z\|^2 \widehat{\psi} = 0, \quad \widehat{\psi}_{zz} = (\log \|\phi_z\|^2)_z \widehat{\psi}_z. \end{aligned} \tag{3.8}$$

If $\{p_1, \ldots, p_r\}$ are the ramification points of ϕ and b is any vector in $\mathbb{R}^3$, we define $A(b) : M - \{p_1, \ldots, p_r\} \to \mathbb{R}^3$ by

$$A(b) = \langle \widehat{\psi}, b \rangle \phi + \frac{1}{\|\phi_z\|^2}\left(\langle \widehat{\psi}_z, b \rangle \phi_{\bar{z}} + \langle \widehat{\psi}_{\bar{z}}, b \rangle \phi_z\right), \tag{3.9}$$

which is independent of the complex coordinate z on M. Differentiating $A(b)$ with respect to z and considering (3.8) we get that $A(b)_z = 0$ and hence $A(b)$ is a constant vector of $\mathbb{R}^3$. This defines a map $A : \mathbb{R}^3 \to \mathbb{R}^3$. From (3.9) and (3.8) it follows that

$$\langle \phi, A(b) \rangle = \langle \widehat{\psi}, b \rangle, \quad \langle \phi_z, A(b) \rangle = \langle \widehat{\psi}_z, b \rangle, \quad \langle \phi_{\bar{z}}, A(b) \rangle = \langle \widehat{\psi}_{\bar{z}}, b \rangle,$$

and so

$$\|A(b)\|^2 = \langle \phi, A(b) \rangle^2 + 2\frac{\|\langle \phi_z, A(b) \rangle\|^2}{\|\phi_z\|^2} = \langle \widehat{\psi}, b \rangle^2 + 2\frac{\|\langle \widehat{\psi}_z, b \rangle\|^2}{\|\widehat{\psi}_z\|^2} = \|b\|^2.$$

So A defines an isometry of $\mathbb{S}^2$ and from (3.9), $A \circ \widehat{\psi} = \phi$. □

COROLLARY 3.13. *Let M be a compact Riemann surface of genus h and $\phi : M \to \mathbb{S}^2$ be a non-constant holomorphic map of degree d. If* $\operatorname{Ind}(M, \phi) = 1$, *then $d \leq \lfloor \frac{h}{2} \rfloor + 1$, where $\lfloor \cdot \rfloor$ stands for integer part. In particular,*

(1) *If* $h = 0$, *then* $d = 1$.
(2) $h \neq 1$.
(3) *If* $h = 2, 3$ *then* $d = 2$.
(4) *If* $h = 4, 5$ *then* $d \leq 3$.

PROOF. From Bill-Noether theory, see [**GH78**], there exists a non-constant holomorphic map $\psi : M \to \mathbb{S}^2$ such that $\deg \psi \leq \lfloor \frac{h+1}{2} \rfloor + 1$, and so Theorem 3.12 gives that $d = \deg \phi \leq \deg \psi \leq \lfloor \frac{h+1}{2} \rfloor + 1$. If h is even, then $\lfloor \frac{h+1}{2} \rfloor = \lfloor \frac{h}{2} \rfloor$, and so $d \leq \lfloor \frac{h}{2} \rfloor + 1$. If h is odd, then $\lfloor \frac{h+1}{2} \rfloor + 1 = \lfloor \frac{h}{2} \rfloor + 2$. If $\deg \psi < \lfloor \frac{h}{2} \rfloor + 2$, then $d \leq \lfloor \frac{h}{2} \rfloor + 1$. If $\deg \psi = \lfloor \frac{h}{2} \rfloor + 2 = \lfloor \frac{h+1}{2} \rfloor + 1$, Brill-Noether theory ensures that, up to conformal transformations of $\mathbb{S}^2$, ψ is not unique. Thus, Theorem 3.12 gives that $\deg \phi < \deg \psi$ and hence $d \leq \lfloor \frac{h}{2} \rfloor + 1$. This concludes the proof. □

REMARK 3.14. In Theorem 4.26 we will prove that, under the conditions of the previous corollary, if the genus of M is $h = 4$, then the degree of the holomorphic map ϕ is $d = 2$. The reason for not including the proof in this section is that it uses the theory of minimal surfaces of $\mathbb{R}^3$, which will be developed in Chapter 4.

3.3. Index of complete minimal hypersurfaces in Riemannian manifolds with non-negative curvature

First, we will use the results of Section 3.1 to study the index of minimal surfaces of 3-manifolds with non-negative Ricci or scalar curvature. Later we will consider the higher dimensional case, where we will study interesting properties of stable hypersurfaces.

Let $\Phi : M \to (\overline{M}^3, g)$ be a two-sided minimal immersion of a surface M in a 3-dimensional Riemannian manifold and N be a unit normal vector field to Φ. From (2.17), the Jacobi operator $\mathcal{L} : C^\infty(M) \to C^\infty(M)$ associated to Φ is given by
$$\mathcal{L} = \Delta + \|A\|^2 + \overline{\text{Ricc}}(N).$$
If ρ is the scalar curvature of $(\overline{M}^3, g)$, we have that $\rho = 2\overline{\text{Ricc}}(N) + 2\overline{K}$, where $\overline{K}$ is the extrinsic curvature of Φ, i.e., $\overline{K} = \overline{R}(e_1, e_2, e_2, e_1)$, where $\{e_1, e_2\}$ is an orthonormal reference on M. The Gauss equation (2.2) of Φ says that $K = \overline{K} - \|A\|^2/2$, where K is the Gauss curvature of M. Thus, the Jacobi operator of Φ can also be written as
$$\mathcal{L} = \Delta - K + \frac{\rho + \|A\|^2}{2}.$$
This means that, in this case, the Jacobi operator of Φ is a Schrödinger operator studied in Section 3.1 with $q = (\rho + \|A\|^2)/2$. Thus, we can particularize all the results of that section to our setting.

As a consequence of Theorem 3.1 and Corollary 3.3 we get

COROLLARY 3.15 ([**FCS80, FC85**]). *Let* $\Phi : M \to (\overline{M}, g)$ *be a two-sided minimal immersion of a complete surface into a 3-dimensional Riemannian manifold* $\overline{M}$ *whose scalar curvature* ρ *is non-negative with* $\text{Ind}(M) < \infty$.

(1) *If* M *is not compact, then*
  (a) *There exists a compact subset* C *of* M *and a positive function* $u \in C^\infty(M)$ *such that* $M \setminus C$ *is stable and* $\mathcal{L}u = 0$ *on* $M \setminus C$.
  (b) $(M, g^* = u^2 g)$ *is a complete surface with Gauss curvature* $K^* \geq 0$ *on* $M \setminus C$.

(2) *If M is orientable and not compact,*
   (a) *M is conformally equivalent to a compact Riemann surface with a finite number of points removed.*
   (b) $\int_M(\|A\|^2+\rho)\,\mathrm{d}v<\infty$.
(3) *If the scalar curvature $\rho \geq c > 0$, then M is compact.*

Some statements in the Corollary 3.15 are also true for one-sided minimal immersions, but under certain conditions. In fact, we can prove the following result:

COROLLARY 3.16. *Let $\Phi : M \to (\overline{M}, g)$ be a one-sided minimal immersion of a complete surface M in a 3-dimensional orientable Riemannian manifold $\overline{M}$ with scalar curvature $\rho \geq c > 0$. If the genus of the orientable double covering of M is finite and* $\mathrm{Ind}(M) < \infty$*, then M is compact.*

REMARK 3.17. Since Φ is one-sided and $\overline{M}$ is orientable, from Lemma 2.2 M is not orientable, and so it admits a non-trivial oriented double covering. This Corollary is an unplubished result due to F. Urbano.

PROOF. Let $\Pi : \widetilde{M} \to M$ be the oriented double covering of M and $\{Id, \tau\}$ be the transformation group of the covering.

CLAIM. If the genus of $\widetilde{M}$ is finite, then there exists a compact subset K of M such that $M \setminus K$ is orientable.

PROOF OF THE CLAIM. Since M is not orientable, there exists a Möbius strip embedded in M. Let γ_1 be the middle Jordan curve of the Möbius strip and $\widetilde{\gamma}_1 = \Pi^{-1}(\gamma_1)$. Then $\widetilde{\gamma}_1$ is a connected Jordan curve and its complement $\widetilde{M} \setminus \widetilde{\gamma}_1$ is connected or has two connected components. Since $\tau(\widetilde{\gamma}_1) = \widetilde{\gamma}_1$, if $\widetilde{M} \setminus \widetilde{\gamma}_1$ has two connected components D_1 and D_2, then $\tau(D_1) = D_2$ and thus $\Pi : D_1 \to M \setminus \gamma_1$ is a homeomorphism. Since D_1 is orientable, $M \setminus \gamma_1$ is also orientable.

Now, if $M \setminus \gamma_1$ is not orientable, repeating the argument, there exists a Jordan curve γ_2 in $M \setminus \gamma_1$ such that $\widetilde{M} \setminus (\widetilde{\gamma}_1 \cup \widetilde{\gamma}_2)$ is connected or $M \setminus (\gamma_1 \cup \gamma_2)$ is orientable. If g is the genus of $\widetilde{M}$, repeating the argument, and since $\widetilde{M} \setminus (\widetilde{\gamma}_1 \cup \cdots \cup \widetilde{\gamma}_{g+1})$ is not connected, we get that $M \setminus (\gamma_1 \cup \cdots \cup \gamma_{g+1})$ is orientable. This completes the proof of the claim, taking $K = \gamma_1 \cup \cdots \cup \gamma_{g+1}$.

Since $\overline{M}$ is orientable, $\Phi : M \setminus K \to \overline{M}$ is a two-sided minimal immersion, and so the Jacobi operator on $M \setminus K$ is $\mathcal{L} = \Delta - K + (\rho + \|A\|^2)/2$. If $\mathrm{Ind}(M)$ is finite, then $\mathrm{Ind}(M \setminus K) = \mathrm{Ind}_{M\setminus K}(\mathcal{L})$ is also finite. Thus, from Theorem 2.27 there exists a compact subset $C \subset M \setminus K$, such that $\mathrm{Ind}_{M\setminus(C\cup K)}(\mathcal{L}) = 0$. Hence M is a complete surface and the Schrödinger operator $\mathcal{L} = \Delta - K + (\rho + \|A\|^2)/2$ on M satisfies that $\mathrm{Ind}_{M\setminus(C\cup K)}(\mathcal{L}) = 0$. Theorem 3.1 and Remark 3.2 prove the corollary. □

Similarly, it follows from Theorem 3.5 that:

COROLLARY 3.18 ([**FCS80, FC85**]). *Let $\Phi : M \to (\overline{M}, g)$ be a two-sided minimal immersion of a complete surface into a 3-dimensional Riemannian manifold $\overline{M}$ whose scalar curvature ρ is non-negative. If M is stable, then one of the following possibilities holds:*

(1) *M is conformally equivalent to $\mathbb{S}^2$.*
(2) *M is diffeomorphic to $\mathbb{RP}^2$.*

(3) *M is a totally geodesic flat torus and* $\rho_{|\Phi(M)} = 0$.
(4) *M is a totally geodesic flat Klein bottle and* $\rho_{|\Phi(M)} = 0$.
(5) *M is conformally equivalent to* $\mathbb{C}$.
(6) *M is a totally geodesic flat cylinder and* $\rho_{|\Phi(M)} = 0$.
(7) *M is a totally geodesic flat Möbius strip and* $\rho_{|\Phi(M)} = 0$.

If $\rho \geq c > 0$ for some constant c and M is stable, then M is either conformally equivalent to $\mathbb{S}^2$ or diffeomorphic to $\mathbb{RP}^2$.

If the Ricci curvature of the Riemannian manifold $\overline{M}$ is non-negative (in particular its scalar curvature is also non-negative), we can improve the results of Corollary 3.18.

COROLLARY 3.19 ([**FCS80, FC85**]). *Let $\Phi : M \to (\overline{M}, g)$ be a two-sided minimal immersion of a complete surface into a 3-dimensional Riemannian manifold $\overline{M}$ with non-negative Ricci curvature. If M is stable, then Φ is a totally geodesic immersion, and one of the following possibilities holds:*

(1) *M is conformally equivalent to $\mathbb{S}^2$, $\overline{\mathrm{Ricc}}(N) = 0$ and $K \geq 0$.*
(2) *M is diffeomorphic to $\mathbb{RP}^2$, $\overline{\mathrm{Ricc}}(N) = 0$ and $K \geq 0$.*
(3) *M is a flat torus and $\overline{\mathrm{Ricc}}_{|\Phi(M)} = 0$.*
(4) *M is a flat Klein bottle and $\overline{\mathrm{Ricc}}_{|\Phi(M)} = 0$.*
(5) *M is conformally equivalent to $\mathbb{C}$, $\overline{\mathrm{Ricc}}(N) = 0$ and $K \geq 0$.*
(6) *M is a flat cylinder and $\overline{\mathrm{Ricc}}_{|\Phi(M)} = 0$.*
(7) *M is a flat Möbius strip and $\overline{\mathrm{Ricc}}_{|\Phi(M)} = 0$.*

where N is a unit normal field to Φ and K is the Gauss curvature of M.

PROOF. Since the Ricci curvature of $\overline{M}$ is non-negative, the scalar curvature is also non-negative and so we can apply Corollary 3.18. Since $\overline{M}$ is of dimension three, if $\overline{\mathrm{Ricc}} \geq 0$ and $\rho(p) = 0$ then $\overline{\mathrm{Ricc}}(p) = 0$. So it is clear that 3, 4, 6 and 7 follow directly from the Corollary 3.18.

To prove 1 and 2, and since in both cases M is compact, the stability of Φ implies that

$$0 \leq \mathcal{Q}(1) = -\int_M (\|A\|^2 + \overline{\mathrm{Ricc}}(N))\, dv \leq 0,$$

and then Φ is totally geodesic and $\overline{\mathrm{Ricc}}(N) = 0$. The Gauss equation gives that $0 \leq \rho = 2\overline{K} = 2K$. Now, 1 and 2 follow from Corollary 3.18.

Finally, we only have to consider the case M is conformally equivalent to $\mathbb{C}$. In this case, since M is not compact, it follows from Theorem 2.25,2 that there exists a positive smooth function u over M such that $\mathcal{L}u = 0$. This means that $\Delta u = -(\|A\|^2 + \overline{\mathrm{Ricc}}(N))u \leq 0$ and thus u is a positive superharmonic function on M. Liouville's theorem says that u is constant and hence Φ is totally geodesic and $\overline{\mathrm{Ricc}}(N) = 0$. An argument similar to that used in 1 and 2 would prove that $K \geq 0$, and so 5 is proved. □

REMARK 3.20. According to Proposition 2.40, if (M, g) is a complete Riemannian surface, we consider the 3-manifold $M \times \mathbb{R}$, endowed with the product metric $g \times dt^2$. Then $p \in M \mapsto (p, 0) \in M \times \mathbb{R}$ defines a totally geodesic stable two-sided embedding $\Phi : M \to (M \times \mathbb{R}, g + dt^2)$ with $N = (0, 1)$ and $\overline{\mathrm{Ricc}}(N) = 0$. If the Gauss curvature of M is $K \geq 0$, then the Ricci curvature of $\overline{M}$ is also non-negative and we are under the hypothesis of Corollary 3.19. Now it is clear that all cases of Corollary 3.19 can happen.

COROLLARY 3.21. *A 3-dimensional Riemannian manifold with positive Ricci curvature does not admit two-sided complete stable minimal surfaces.*

REMARK 3.22. It is interesting to note that there may be examples of one-sided complete stable minimal surfaces in a 3-dimensional manifold with positive Ricci curvature.

We end this section by studying the case $\mathrm{Ind}(M) = 1$.

THEOREM 3.23 ([**Ros06**]). *Let $\Phi : M \to (\overline{M}, g)$ be a two-sided minimal immersion of a compact orientable surface into a 3-dimensional Riemannian manifold $\overline{M}$ with non-negative Ricci curvature. If $\mathrm{Ind}(M) = 1$, then the genus of M is $h \leq 3$.*

PROOF. First, the Jacobi operator $\mathcal{L}$ can also be written as $\mathcal{L} = \Delta - 2K + \overline{\mathrm{Ricc}}(e_1) + \overline{\mathrm{Ricc}}(e_2)$, where $\{e_1, e_2\}$ is an orthonormal reference on M.

Let f be a function with $\mathcal{L}f + \lambda_1 f = 0$, where λ_1 is the negative eigenvalue of $\mathcal{L}$. From Brill-Noether theory, see [**GH78**], there exists a non-constant holomorphic map $\phi : M \to \mathbb{S}^2$ such that $d := \deg \phi \leq \lfloor \frac{h+1}{2} \rfloor + 1$, where h is the genus of M and $\lfloor \cdot \rfloor$ stands for the integer part. Moreover, from Proposition 1.8,2 there exists a conformal transformation ψ of $\mathbb{S}^2$ such that $\int_M f(\psi \circ \phi) dv = 0$. Since the second eigenvalue of $\mathcal{L}$ is non-negative, it follows from Theorem 1.17, the Gauss-Bonnet theorem, and an argument as in the proof of Theorem 3.12 that

$$
\begin{aligned}
0 &\leq \int_M \left\{ \|\nabla(\psi \circ \phi)\|^2 + \left(2K - \overline{\mathrm{Ricc}}(e_1) - \overline{\mathrm{Ricc}}(e_2)\right) \|\psi \circ \phi\|^2 \right\} dv \\
&= 8\pi(d + 1 - h) - \int_M (\overline{\mathrm{Ricc}}(e_1) + \overline{\mathrm{Ricc}}(e_2))\, dv \leq 8\pi\left(\lfloor \tfrac{h+1}{2} \rfloor + 2 - h\right),
\end{aligned}
\tag{3.10}
$$

which implies that $h \leq 5$.

Suppose that $h = 4$ or 5. Then $\lfloor \frac{h+1}{2} \rfloor + 2 - h = 0$ and we get the equality in (3.10). This implies that $\overline{\mathrm{Ricc}}(e_1) + \overline{\mathrm{Ricc}}(e_2) = 0$, $d+1 = h$ and $\mathcal{L}(\psi \circ \phi) = 0$. So the Jacobi operator is given by $\mathcal{L} = \Delta - 2K$ and thus $\Delta(\psi \circ \phi) - 2K(\psi \circ \phi) = 0$. But since $\psi \circ \phi : M \to \mathbb{S}^2$ is a holomorphic map, then $\Delta(\psi \circ \phi) + \|\nabla(\psi \circ \phi)\|^2 (\psi \circ \phi) = 0$, which implies that $\|\nabla(\psi \circ \phi)\|^2 = -2K$. Finally, we get that $\mathcal{L} = \Delta + \|\nabla(\psi \circ \phi)\|^2 = L_{\psi \circ \phi}$ and $\mathrm{Ind}(M) = \mathrm{Ind}(M, \psi \circ \phi)$ (see Section 3.2).

As $\mathrm{Ind}(M, \psi \circ \phi) = 1$, from Corollary 3.13,4, $d \leq 3$. Since $d + 1 = h$, we get that h cannot be 5, and so $h = 4$ and $d = 3$. But from Theorem 4.26 (see also Remark 3.14) this is impossible. So $h \leq 3$ and the result is proved. □

REMARK 3.24. Recently, A. Ros (see [**Ros25**]) constructed a metric on the 3-dimensional real projective space with positive sectional curvature and an embedded minimal compact orientable surface of index one and genus three in such a Riemanian 3-manifold.

The study of stability for hypersurfaces of dimension at least three is much more complicated and a generalization of the results for surfaces is not to be expected. With the following example we prove that Corollary 3.15,3 is false in higher dimension.

EXAMPLE 3.25 ([**CLS22**]). *Let $\overline{M} = \mathbb{R}^2 \times \mathbb{S}^{n-1}$, $n \geq 3$, be endowed with the product metric. This manifold has constant scalar curvature $\rho = (n-1)(n-2) > 0$. Now consider the two-sided complete totally geodesic embedding $\Phi : \mathbb{R} \times \mathbb{S}^{n-1} \to \mathbb{R}^2 \times \mathbb{S}^{n-1}$ given by $\Phi(t, p) = ((t, 0), p)$. Then Proposition 2.41 gives that the immersion Φ is stable.*

In the following example we prove that Corollary 3.21 is false in higher dimensions, even under the stronger condition that the ambient manifold has positive sectional curvature.

EXAMPLE 3.26 ([**CLS22**]). *For $\alpha \in (0,1]$ let $\rho_\alpha : [0,\infty) \to \mathbb{R}$ be the function given by*

$$\rho_\alpha(r) = \alpha r + (1-\alpha)\int_0^r e^{-s^2}\mathrm{d}s.$$

We consider on $(0,\infty)\times\mathbb{S}^n \equiv \mathbb{R}^{n+1}\setminus\{0\}$, $n \geq 3$, the warped metric

$$g_\alpha = \mathrm{d}r^2 + \rho_\alpha(r)^2\langle,\rangle,$$

where $\langle,\rangle$ is the standard metric on $\mathbb{S}^n$. Since $\rho'_\alpha(0) = 1$ and $\rho_\alpha^{(even)}(0) = 0$, the metric g_α extends to $[0,\infty)\times\mathbb{S}^n \equiv \mathbb{R}^{n+1}$ as a complete metric. If $\alpha = 1$, g_1 is the Euclidean metric $\langle,\rangle$ on $\mathbb{R}^{n+1}$. Moreover, since $\alpha r \leq \rho_\alpha(r) \leq r$, we have that the measures $\mathrm{d}v$ and $\mathrm{d}v_0$ associated to the metrics g_α and the Euclidean metric $\langle,\rangle$ are related by $\alpha^{n-1}\mathrm{d}v_0 \leq \mathrm{d}v \leq \mathrm{d}v_0$. Also, the sectional curvatures of g_α are pinched between

$$-\frac{\rho''_\alpha(r)}{\rho_\alpha(r)} = \frac{2(1-\alpha)re^{-r^2}}{\rho_\alpha(r)} \quad \text{and} \quad \frac{1-\rho'_\alpha(r)^2}{\rho_\alpha(r)^2} = \frac{1-\left(\alpha+(1-\alpha)e^{-r^2}\right)^2}{\rho_\alpha(r)^2},$$

and so $(\mathbb{R}^{n+1}, g_\alpha)$, $\alpha \in (0,1)$, has positive sectional curvature and its Ricci curvatures satisfy

$$\overline{\mathrm{Ricc}}(v) \leq \frac{n(1-\alpha^2)}{\alpha^2 r^2}, \quad \text{for any unit vector } v.$$

Now let $\Phi_\alpha : \mathbb{R}^n \equiv [0,\infty)\times\mathbb{S}^{n-1} \to (\mathbb{R}^{n+1} \equiv [0,\infty)\times\mathbb{S}^n, g_\alpha)$ be the embedding given by $\Phi_\alpha(r,p) = (r,(p,0))$. Then it is clear that Φ is a two-sided complete totally geodesic embedding and so the Jacobi operator of Φ_α is given by $\mathcal{L} = \Delta + \overline{\mathrm{Ricc}}(N)$, where N is a unit normal vector field to Φ_α.

We will prove that Φ_α is stable. Indeed, if $f \in C_0^\infty(\mathbb{R}^n)$ and the subindex $_0$ stands for the Euclidean metric, then

$$\begin{aligned}
\mathcal{Q}(f) &= \int_{\mathbb{R}^n}\{\|\nabla f\|^2 - \overline{\mathrm{Ricc}}(N)f^2\}\mathrm{d}v \geq \int_{\mathbb{R}^n}\{\|\nabla_0 f\|_0^2 - \overline{\mathrm{Ricc}}(N)f^2\}\mathrm{d}v \\
&\geq \alpha^{n-1}\int_{\mathbb{R}^n}\|\nabla_0 f\|_0^2\,\mathrm{d}v_0 - \tfrac{n(1-\alpha^2)}{\alpha^2}\int_{\mathbb{R}^n}\frac{f^2}{r^2}\mathrm{d}v \\
&\geq \tfrac{\alpha^{n-1}(n-2)^2}{4}\int_{\mathbb{R}^n}\frac{f^2}{r^2}\mathrm{d}v_0 - \tfrac{n(1-\alpha^2)}{\alpha^2}\int_{\mathbb{R}^n}\frac{f^2}{r^2}\mathrm{d}v \\
&\geq \left(\tfrac{\alpha^{n-1}(n-2)^2}{4} - \tfrac{n(1-\alpha^2)}{\alpha^2}\right)\int_{\mathbb{R}^n}\frac{f^2}{r^2}\mathrm{d}v,
\end{aligned}$$

where we have used the Hardy's inequality, the relation between the measures and the upper bound of the Ricci curvature. It is clear that there exists $\alpha_0 \in (0,1)$ sufficiently close to 1 such that $\mathcal{Q}(f) \geq 0$. Thus Φ_{α_0} is stable.

Note that the above arguments are true for $n = 2$, except for Hardy's inequality.

To study the stability of higher dimensional minimal hypersurfaces, we will need the notion of parabolicity of a Riemannian manifold, as well as the notion of parabolicity of ends.

DEFINITION 3.27. A complete Riemannian manifold (M, g) is said to be **parabolic** if any positive superharmonic function f on M, i.e. $\Delta f \leq 0$, is constant. Otherwise, M is said to be **non-parabolic**.

DEFINITION 3.28. Let (M^n, g) be a complete Riemannian manifold, K a compact subset of M and E an unbounded connected component of $M \setminus K$ with smooth boundary. We say that E is parabolic if it does not admit a positive harmonic function f satisfying

$$f_{|\partial E} = 1 \quad \text{and} \quad f_{|E} < 1.$$

Otherwise E is non-parabolic and a function f satisfying the above conditions will be called a barrier function of E.

REMARK 3.29. P. Li, ([**Li04**], Lemma 3.6), proved that if E is not parabolic, there exists a minimal barrier function f on E, i.e. $f \leq g$ on E for any barrier function g, and furthermore f has finite energy, that is $\int_E \|\nabla f\|^2 \mathrm{d}v < \infty$.

PROPOSITION 3.30. *Let (M^n, g) be a complete non-compact Riemannian manifold, and let $K \subset \widetilde{K}$ be compact subsets of M with smooth boundary. If E is a non-parabolic unbounded connected component of $M \setminus K$, then there exists a non-parabolic unbounded conected component $\widetilde{E}$ of $M \setminus \widetilde{K}$ with $\widetilde{E} \subset E$.*

PROOF. From Definition 3.28 it follows that there exists a positive harmonic function f on E such that $f_{|\partial E} = 1$ and $f_{|E} < 1$. If $p \in K$ and $R > 0$, we can assume that

$$\lim_{R \to \infty} \inf_{\partial B_p(R) \cap E} f = 0,$$

because if $\lim_{R \to \infty} \inf_{\partial B_p(R) \cap E} f = c > 0$, we replace f by $g = \frac{f-c}{1-c}$.

Under these conditions, there exists an unbounded connected component $\widetilde{E}$ of $M \setminus \widetilde{K}$ such that $\widetilde{E} \subset E$ and $\lim_{R \to \infty} \inf_{\partial B_p(R) \cap \widetilde{E}} f = 0$.

To prove that $\widetilde{E}$ is non-parabolic, we consider a sequence $R_i \mapsto \infty$ such that $\widetilde{K} \subset B_p(R_i)$ for any i. Minimizing the energy functional $\int_{\widetilde{E} \cap B_p(R_i)} \|\nabla u\|^2 \mathrm{d}v$ among the functions $u \in H_0^{1,2}(\widetilde{E} \cap B_p(R_i))$ with $u = 1$ on $\partial \widetilde{E}$ and $u = 0$ on $\partial B_p(R_i)$, we get that the minimizer u_i is the unique solution of the Dirichlet problem on $\widetilde{E} \cap B_p(R_i)$

$$\begin{cases} \Delta u_i = 0 & \text{on } \widetilde{E} \cap B_p(R_i) \\ u_i = 1 & \text{on } \partial \widetilde{E} \\ u_i = 0 & \text{on } \widetilde{E} \cap \partial B_p(R_i). \end{cases}$$

According to the maximum principle, $0 \leq u_i \leq 1$. Moreover the sequence u_i converges uniformly on compact subsets to a harmonic function u on $\widetilde{E}$ such that $0 < u \leq 1$, $u_{|\partial \widetilde{E}} = 1$.

If $a = \inf_{\partial \widetilde{E}} f > 0$, then $u_i \leq a^{-1} f$ on $\partial \widetilde{E} \cup \big(\widetilde{E} \cap \partial B_p(R_i)\big)$, and so the maximum principle gives that $u_i \leq a^{-1} f$ on $\widetilde{E} \cap B_p(R_i)$. This implies that $u \leq a^{-1} f$ on $\widetilde{E}$ and so

$$\lim_{R \to \infty} \inf_{\partial B_p(R) \cap \widetilde{E}} u = 0.$$

This means that u is not the constant function 1 and using the maximum principle again we get that $u_{|\widetilde{E}} < 1$. This proves that $\widetilde{E}$ is parabolic. □

The above proposition allows us to define the notion of non-parabolic end.

DEFINITION 3.31. Let (M^n, g) be a complete non-compact Riemannian manifold and let $\{E_i \mid i \in \mathbb{N}\}$ be an end of M associated to the sequence of compact sets $\{K_i \mid i \in \mathbb{N}\}$ (see Definition 1.3). The end $\{E_i\}$ is called **non-parabolic** if for each $i \in \mathbb{N}$ there exists a connected open subset $F_i \subset E_i$ with smooth boundary such that $E_i \setminus K_{i+1} \subset F_i$ and F_i is non-parabolic for i sufficiently large.

Note that if the compact sets K_i have smooth boundary, we can take $F_i = E_i$ in the above definition. Also, for i sufficiently large we have that F_i is unbounded.

PROPOSITION 3.32. *Let (M^n, g) be a complete non-compact Riemannian manifold.*

(1) *If M is non-parabolic, then M has at least a non-parabolic end.*
(2) *If M has at least two non-parabolic ends, then there exists a non-constant harmonic function u on M with finite energy, i.e. $\int_M \|\nabla u\|^2 dv < \infty$.*

PROOF. **Proof of 1.** If M is non-parabolic, there exists a non-constant positive superharmonic function f on M. Obviously, we can assume that $\inf_M f = 0$. Then there exists a compact set $K \subset M$ with smooth boundary and an unbounded connected component E of $M \setminus K$ such that $\inf_E f = 0$. Let $g = f_{|E}$. Then, by the strong maximum principle, g is a positive superharmonic function on E satisfying

$$\inf_E g = 0 \quad \text{and} \quad \inf_{\partial E} g = b > 0,$$

for some constant b. Using $b^{-1}g$ as a barrier function, we can solve the Dirichlet problem on E, and so there exists a harmonic function u on E such that

$$u_{|\partial E} = 1 \quad \text{and} \quad u_{|E} < 1.$$

This proves that E is non-parabolic and so, M has at least a non-parabolic end.

Proof of 2. We assume that M has at least two non-parabolic ends. Then there exists a compact set $K \subset M$ with smooth boundary such that $M \setminus K$ has at least two unbounded connected components E and F which are non-parabolic. Let f and g be the minimal barrier functions on E and F respectively given in Definition 3.28. It is clear that we can assume that $\inf_E f = 0$ and $\inf_F g = 0$. Then the harmonic function h on $M \setminus K$ is defined by

$$\begin{cases} h = f & \text{on } E \\ h = 1 - g & \text{on } F \\ h = 0 & \text{on the other connected components of } M \setminus K, \end{cases}$$

satisfies $0 \leq h \leq 1$, $h_{\partial E} = 1$, $h_{|\partial F} = 0$, $\inf_E h = 0$ and $\sup_F h = 1$. Moreover as

$$\int_{M\setminus K} \|\nabla h\|^2 dv = \int_E \|\nabla f\|^2 dv + \int_F \|\nabla g\|^2 dv,$$

from Remark 3.29 we get that h has finite energy, i.e., $\int_{M\setminus K} \|\nabla h\|^2 dv < \infty$.

Let p be a point of K, $\{R_i \mid i \in \mathbb{N}\}$ be a sequence with $R_i < R_{i+1}$, $\lim_{i\to\infty} R_i = \infty$ such that $K \subset B_p(R_1)$. For each i let f_i be a harmonic function on $B_p(R_i)$ such that $(f_i)_{|\partial B_p(R_i)} = h_{|\partial B_p(R_i)}$. From the maximum principle and the properties of h it follows that $0 \leq f_i \leq 1$, $f_i \leq f$ on $E \cap B_p(R_i)$ and $f_i \leq 1 - g$ on $F \cap B_p(R_i)$. Then there exists a subsequence, which we also denote by $\{f_i\}$, which converges uniformly on compact subsets to a harmonic function u on M, which is non-constant by construction.

Moreover, since f_i is a minimizer of the functional $\int_{B_p(R_i)} \|\nabla v\|^2 dv$, we have for $i > 1$

$$\int_{B_p(R_i)} \|\nabla f_i\|^2 dv \leq \int_{B_p(R_1)} \|\nabla f_1\|^2 dv + \int_{B_p(R_i)\setminus B_p(R_1)} \|\nabla h\|^2 dv.$$

Then

$$\int_M \|\nabla u\|^2 dv = \lim_{i\to\infty} \int_{B_p(R_i)} \|\nabla f_i\|^2 dv \leq \int_{B_p(R_1)} \|\nabla f_1\|^2 dv + \int_{M\setminus K} \|\nabla h\|^2 dv < \infty.$$

□

We now begin to study the stability of minimal hypersurfaces of a Riemannian manifold with non-negative curvature in the simplest case where the hypersurface is parabolic. This includes the case where the hypersurface is compact. It is interesting to compare the following two results with the case of surfaces studied in Corollaries 3.19 and 3.21.

PROPOSITION 3.33. *Let $\Phi : M^n \to (\overline{M}^{n+1}, g)$ be a two-sided minimal hypersurface of a complete parabolic manifold M^n into a Riemannian manifold $\overline{M}$ with non-negative Ricci curvature.*

1. *If Φ is stable, then Φ is totally geodesic, $\overline{\text{Ricc}}(N)$ vanishes on M, and so M has non-negative scalar curvature (N is a unit normal vector field to Φ).*
2. *If $\overline{M}$ has positive Ricci curvature, then Φ is unstable.*

PROOF. First assume that M is compact. Then using the constant function 1 as a test function in (2.18) and as Φ is stable,

$$0 \geq \int_M \{\|A\|^2 + \overline{\text{Ricc}}(N)\} dv.$$

From the above inequality and the Gauss equation (2.2), 1 and 2 follow easily.

If M is not compact, then it follows from Theorem 2.25,2 that there exists a positive function $u \in C^\infty(M)$ such that $\mathcal{L}(u) = 0$. So $0 = \Delta u + \big(\|A\|^2 + \overline{\text{Ricc}}(N)\big)u \geq \Delta u$. This means that u is a positive superharmonic function on a parabolic manifold and so u is constant. Now 1 and 2 follow as in the compact case. □

COROLLARY 3.34. *Let $\Phi : M^n \to (\overline{M}^{n+1}, g)$ be a two-sided orientable minimal hypersurface of a complete non-compact parabolic manifold M^n into a Riemannian manifold $\overline{M}$ with non-negative sectional curvature. If M is stable, then either*

1. *M has only one end, or*
2. *M has two ends and it is isometric to the Riemannian product $\mathbb{R} \times P$, where P is a compact $(n-1)$-dimensional Riemannian manifold with non-negative sectional curvature.*

PROOF. From Proposition 3.33 it follows that Φ is totally geodesic, and so, from the Gauss equation (2.2), M also has non-negative sectional curvature. Now we use a result of J. Cheeger and D. Gromoll (Theorem 2.2, [**CG72**]), to conclude that M is diffeomorphic to the normal bundle of the soul $S(M) \subset M$, which is a compact totally geodesic submanifold of M. If the codimension of $S(M)$ in M is greater than one, then M has only one end. If the codimension is one, and since M

is orientable, Theorem 4.2 in [**CG72**] says that M is isometric to the Riemannian product $S(M) \times \mathbb{R}$. This proves the result. □

In the following result we collect important properties of the minimal stable complete hypersurfaces of a Riemannian manifold with non-negative sectional curvature.

THEOREM 3.35 ([**SY76**]). *Let $\Phi : M^n \to (\overline{M}^{n+1}, g)$ be a two-sided minimal hypersurface of a complete non-compact manifold M^n into a Riemannian manifold $\overline{M}$ with non-negative sectional curvature. If Φ is stable, then*

(1) *The volume of M is infinite.*
(2) *If u is a harmonic function on M with finite energy, i.e., $\int_M \|\nabla u\|^2 \mathrm{d}v < \infty$, then u is constant.*
(3) *If M is non-parabolic, then M has exactly one non-parabolic end.*

PROOF. **Proof of 1**. For $R > 1$, let $\varphi^R \in H_0^{1,2}(M)$ be the cut-off function given in Lemma 2.28. Since Φ is stable, it follows from (2.18) that

$$\int_M \|\nabla\varphi^R\|^2 \,\mathrm{d}v \geq \int_M \{\|A\|^2 + \overline{\mathrm{Ricc}}(N)\}(\varphi^R)^2\}\,\mathrm{d}v \geq \int_M \|A\|^2(\varphi^R)^2\,\mathrm{d}v,$$

where N is a unit normal vector field to Φ and we have used that $\overline{\mathrm{Ricc}}(N) \geq 0$. Since $\|\nabla\varphi^R\| \leq \frac{1}{R\log R}$ and $\varphi^R = 1$ on the geodesic ball $B_p(R)$, the above equation becomes

$$\frac{\mathrm{vol}(M)}{R^2\log^2 R} \geq \int_{B(R)} \|A\|^2 \mathrm{d}v.$$

If the volume of M is finite, by letting $R \mapsto \infty$ in the above inequality, we get that Φ is totally geodesic and then the Gauss equation implies that M also has non-negative sectional curvature. This fact contradicts a result of S.T. Yau ([**Yau74a**], Corollary) and so M has infinite volume. This proves 1.

Proof of 2. Given a harmonic map u on M, we will show the following inequality:

$$\int_M \|\nabla\varphi\|^2\|\nabla u\|^2\mathrm{d}v \geq \frac{1}{n}\int_M \varphi^2\|A\|^2\|\nabla u\|^2\mathrm{d}v + \frac{1}{n-1}\int_M \varphi^2\|\nabla\|\nabla u\|\|^2\mathrm{d}v, \tag{3.11}$$

where φ is an arbitrary non-negative function with compact support in $H_0^{1,2}(M)$.

To prove inequality (3.11), we consider for each $\delta > 0$ the function $a_\delta := \sqrt{\|\nabla u\|^2 + \delta}$. Taking as the test function $a_\delta\varphi \in H_0^{1,2}(M)$ in (2.18), the stability of Φ together with the curvature assumption on $\overline{M}$, implies that

$$\int_M \|\nabla(a_\delta\varphi)\|^2\mathrm{d}v \geq \int_M \varphi^2 a_\delta^2\big(\|A\|^2 + \overline{\mathrm{Ricc}}(N)\big)\mathrm{d}v \geq \int_M \varphi^2 a_\delta^2\|A\|^2\mathrm{d}v.$$

Since $\nabla a_\delta = \frac{\nabla\|\nabla u\|^2}{2a_\delta}$, it follows that

$$\|\nabla(a_\delta\varphi)\|^2 = \frac{\varphi^2\|\nabla\|\nabla u\|^2\|^2}{4a_\delta^2} + a_\delta^2\|\nabla\varphi\|^2 + \varphi g(\nabla\|\nabla u\|^2, \nabla\varphi),$$
$$\mathrm{div}\,(\varphi^2\nabla\|\nabla u\|^2) = \varphi^2\Delta\|\nabla u\|^2 + 2\varphi g(\nabla\|\nabla u\|^2, \nabla\varphi),$$

and so, using Theorem 1.7,5, the above stability inequality becomes

$$\int_M \left\{\frac{\varphi^2\|\nabla\|\nabla u\|^2\|^2}{4a_\delta^2} + a_\delta^2\|\nabla\varphi\|^2\right\}\mathrm{d}v \geq \int_M \{\frac{1}{2}\varphi^2\Delta\|\nabla u\|^2 + \varphi^2 a_\delta^2\|A\|^2\}\mathrm{d}v.$$

Since ∇u is a harmonic vector field, from Lemma 1.16, it follows

$$\frac{n}{4(n-1)}\|\nabla\|\nabla u\|^2\|^2 \leq \|\nabla u\|^2\|\nabla\nabla u\|^2 \leq a_\delta^2\|\nabla\nabla u\|^2$$
$$= a_\delta^2\big(\frac{1}{2}\Delta\|\nabla u\|^2 - S(\nabla u, \nabla u)\big),$$

and so, the above inequality becomes

$$\begin{aligned}\int_M a_\delta^2\|\nabla\varphi\|^2 dv &\geq \int_M \left\{\varphi^2 S(\nabla u, \nabla u) + \frac{\varphi^2\|\nabla\|\nabla u\|^2\|^2}{4(n-1)a_\delta^2} + \varphi^2 a_\delta^2\|A\|^2\right\} dv \\ &= \int_M \left\{\varphi^2 S(\nabla u, \nabla u) + \frac{\varphi^2\|\nabla a_\delta\|^2}{(n-1)} + \varphi^2 a_\delta^2\|A\|^2\right\} dv.\end{aligned} \tag{3.12}$$

On the other hand, since M is a minimal immersion, we have

$$\begin{aligned}\|A\|^2 &= \sum_{i,j=1}^n g(Ae_i, e_j)^2 \geq g(Ae_1, e_1)^2 + \sum_{i=2}^n g(Ae_i, e_i)^2 + 2\sum_{j=2}^n g(Ae_1, e_j)^2 \\ &\geq g(Ae_1, e_1)^2 + \frac{\left(\sum_{i=2}^n g(Ae_i, e_i)\right)^2}{n-1} + 2\sum_{j=2}^n g(Ae_1, e_j)^2 \\ &\geq \frac{n}{n-1}\sum_{j=1}^n g(Ae_1, e_j)^2.\end{aligned}$$

It is clear that we can work on the open subset $\{p \in M \mid \nabla u_p \neq 0\}$. On such an open subset we can choose an orthonormal reference $\{e_1, \ldots, e_n\}$ such that $\nabla u = \|\nabla u\| e_1$, and so, using the Gauss equation (2.2), the above inequality and the curvature assumption we get

$$\begin{aligned}S(\nabla u, \nabla u) &= \|\nabla u\|^2 S(e_1, e_1) = \|\nabla u\|^2 \sum_{i=1}^n \{\overline{R}(e_1, e_i, e_i, e_1) - g(Ae_1, e_i)^2\} \\ &\geq -\frac{n-1}{n}\|\nabla u\|^2\|A\|^2 \geq -\frac{n-1}{n}a_\delta^2\|A\|^2.\end{aligned}$$

Using the above equation, the inequality (3.12) becomes

$$\int_M \|\nabla\varphi\|^2 a_\delta^2\, dv \geq \frac{1}{n}\int_M \varphi^2\|A\|^2\|\nabla u\|^2 dv + \frac{1}{n-1}\int_M \varphi^2\|\nabla a_\delta\|^2 dv.$$

Letting $\delta \mapsto 0$ in the above inequality and using Theorem 1.7 we get the inequality (3.11).

To prove 2, we now consider for $R > 1$ the cut-off function $\varphi^R \in H_0^{1,2}(M)$, which is given in Lemma 2.28. Then inequality (3.11) and the facts that $\|\nabla\varphi^R\| \leq \frac{1}{R\log R}$ and $\varphi^R = 1$ on the geodesic ball $B_p(R)$, give

$$\begin{aligned}\int_{B_p(R)} \|\nabla\|\nabla u\|\|^2 dv &= \int_{B_p(R)} (\varphi^R)^2\|\nabla\|\nabla u\|\|^2 dv \\ &\leq \int_M (\varphi^R)^2\|\nabla\|\nabla u\|\|^2 dv \leq (n-1)\int_M \|\nabla u\|^2\|\nabla\varphi^R\|^2 dv \\ &\leq \frac{n-1}{R^2\log^2 R}\int_M \|\nabla u\|^2 dv.\end{aligned}$$

Since $\int_M \|\nabla u\|^2 \mathrm{d}v < \infty$ and $R > 1$ is an arbitrary number, we get that $\|\nabla u\|$ is a constant λ. If λ is positive then $\lambda^2 \operatorname{vol}(M) = \int_M \|\nabla u\|^2 \mathrm{d}v < \infty$, which contradicts 1. So $\lambda = 0$ and so u is a constant function. This proves 2.

Proof of 3. Statement 3 is a consequence of 2 and Proposition 3.32. □

The next result of P. Li and J. Wang, [**LW04**], is a generalization of Corollary 3.34 to the general case, i.e., the assumption of non-parabolicity of the hypersurface. To do this, the authors must impose certain global properties on the ambient manifold to ensure the existence of Busemann functions on it.

Let $(\overline{M}^{n+1}, g)$ be a complete Riemannian manifold and $\gamma : [0, \infty) \to \overline{M}$ a geodesic ray. For each $t \geq 0$ we associate a function $\beta_t : \overline{M} \to \mathbb{R}$ by $\beta_t(p) = t - d(p, \gamma(t))$, where d is the metric distance on $\overline{M}$. The function β_t is continuous and the family $\{\beta_t \mid t \geq 0\}$ is uniformly equicontinuous. Moreover, when $t \mapsto \infty$, β_t converges uniformly on compact sets to a continuous function $\beta_\gamma(p) = \lim_{t\to\infty} \beta_t(p)$. The function β_γ is called the **Busemann** function associated to the ray γ. In [**CG72**], J. Cheeger and D. Gromoll proved that if $\overline{M}$ has non-negative Ricci curvature (or non-negative sectional curvature), then for any geodesic ray γ in $\overline{M}$, the Busemann function β_γ is subharmonic (or convex). Furthermore, the Busemann function satisfies $\|\overline{\nabla}\beta_\gamma\| = 1$.

If $\Phi : M \to (\overline{M}^{n+1}, g)$ is a two-sided minimal hypersurface and $\overline{M}$ has non-negative sectional curvature, then for any geodesic ray γ on $\overline{M}$ the function $f = \beta_\gamma \circ \Phi$ is a subarmonic function on M. In fact, since β_γ is convex, the hessian of β_γ is positive semidefinite and so, for any vector v in TM we have

$$0 \leq g(\overline{\nabla}_v \overline{\nabla}\beta_\gamma, v) = g(\nabla_v \nabla f, v) - g(\overline{\nabla}\beta_\gamma, N) g(Av, v),$$

where N is a unit normal vector field to Φ. By summing the above inequality along an orthonormal reference of TM, we prove that f is subarmonic.

THEOREM 3.36 ([**LW04**]). *Let $\Phi : M^n \to (\overline{M}^{n+1}, g)$ be a two-sided properly immersed minimal hypersurface of a complete non-compact orientable manifold M^n in a complete non-compact Riemannian manifold $\overline{M}$ with non-negative sectional curvature. If Φ is stable, then either*

(1) *M has only one end, or*
(2) *Φ is totally geodesic and M is isometric to the Riemannian product $\mathbb{R} \times P$, where P is a compact Riemannian manifold with non-negative sectional curvature.*

PROOF. From Corollary 3.34 we can assume that M is non-parabolic. In this case Theorem 3.35,3, ensures that M has exactly one non-parabolic end. To prove the result, we will assume that M has at least two ends and we will prove 2.

Let $K \subset M$ be a compact set with smooth boundary such that $M \setminus K$ has at least two unbounded connected components E and F with E being non-parabolic. Let v be a minimal barrier function on E given in Definition 3.28. It is clear that we can assume that $\inf_E v = 0$.

Since Φ is proper, the set $\Phi(F)$ is unbounded in $\overline{M}$. Therefore, there exists a geodesic ray on $\overline{M}$ whose associated Busemann function β satisfies that $f = \beta \circ \Phi$ is unbounded on F. Since $\|\nabla f\| \leq \|\overline{\nabla}\beta\| = 1$, we have that f has at most linear growth.

Let $\{R_i \mid i \in \mathbb{N}\}$ be a sequence with $R_i < R_{i+1}$, $\lim_{i\to\infty} R_i = \infty$ such that $K = B_p(R_1)$. For each i let f_i be a harmonic function on $F(R_i) := B_p(R_i) \cap F$

such that $f_i = 0$ on ∂F and $f_i = 1$ on $\partial B_p(R_i) \cap F$. Since f_i is not constant, we define positive numbers c_i by

$$c_i \int_{F(R_i)} \|\nabla f_i\|^2 dv = 1.$$

Then the function $u_i = c_i f_i$ is also harmonic on $F(R_i)$ and satisfies $u_i = 0$ on ∂F and $u_i = c_i$ on $\partial B_p(R_i) \cap F$.

For any positive real number R, let $s(R) = \sup_{\partial B_p(R) \cap F} f$, where we recall that f is a subharmonic function on M. Now we will prove that

$$\limsup_{i\to\infty} \frac{c_i}{s(R_i)} = C, \quad \text{for some constant} \quad 0 < C < \infty. \tag{3.13}$$

In fact, on $\partial F(R_i) = (\partial B_p(R_1) \cap F) \cup (\partial B_p(R_i) \cap F)$ we have the inequality

$$\frac{c_i}{s(R_i)} \big(f(x) - s(1)\big) \leq u_i(x),$$

and thus, by the maximum principle, the inequality is satisfied on $F(R_i)$. Since f is unbounded on F, there exists R_{i_0} such that $s(R_{i_0}) > s(1)$. Let $x_0 \in \partial B_p(R_{i_0}) \cap F$ such that $s(R_{i_0}) = f(x_0)$. For $i \geq i_0$, $x_0 \in F(R_i)$ and so, from the last inequality we have

$$\frac{c_i}{s(R_i)} \big(s(R_{i_0}) - s(1)\big) = \frac{c_i}{s(R_i)} \big(f(x_0) - s(1)\big) \leq u_i(x_0), \quad i \geq i_0.$$

If we let $i \mapsto \infty$, then

$$\lim_{i\to\infty} \frac{c_i}{s(R_i)} \leq \frac{1}{s(R_{i_0}) - s(1)} \lim_{i\to\infty} u_i(x_0) = \frac{u(x_0)}{s(R_{i_0}) - s(1)} = C < \infty,$$

which proves (3.13).

Thus, taking into account that f has at most linear growth, we obtain

$$c_i \leq Cs(R_i) \leq CR_i + B \quad \text{for } i \text{ sufficiently large},$$

where B is a constant.

Now for each i let k_i be the function on $B_p(R_i)$ such that

$$\begin{cases} k_i = u_i & \text{on } F(R_i) \\ k_i = v & \text{on } E(R_i) := B_p(R_i) \cap E \\ k_i = 0 & \text{on } B_p(R_i) \setminus (E(R_i) \cup F(R_i)). \end{cases}$$

Let w_i be a harmonic function on $B_p(R_i)$ such that $w_i = k_i$ on $\partial B_p(R_i)$. Since w_i minimizes the Dirichlet integral, we have that

$$\begin{aligned} \int_{B_p(R_i)} \|\nabla w_i\|^2 dv &\leq \int_{B_p(R_i)} \|\nabla k_i\|^2 dv \\ &= \int_{E(R_i)} \|\nabla v\|^2 dv + \int_{F(R_i)} \|\nabla u_i\|^2 dv \leq CR_i + \widehat{B}. \end{aligned}$$

The sequence w_i converges, up to a subsequence, to a harmonic map w on M.

We consider the non-negative function φ on $H_0^{1,2}(B_p(R_i))$ defined by

$$\varphi(x) = \begin{cases} 1 & \text{on } B_p(R_1) \\ \frac{r_i - d(p,x)}{R_i - R_1} & \text{on } B_p(R_i) \setminus B_p(R_1). \end{cases}$$

Since w_i is a harmonic function on $B_p(R_i)$ and $B_p(R_i)$ is stable, the stability inequality (3.11) implies that

$$\begin{aligned}&\frac{1}{n}\int_{B_p(R_1)}\|\nabla w_i\|^2\|A\|^2 dv+\frac{1}{n-1}\int_{B_p(R_1)}\|\nabla\|\nabla w_i\|\|^2 dv\\&=\frac{1}{n}\int_{B_p(R_1)}\varphi^2\|\nabla w_i\|^2\|A\|^2 dv+\frac{1}{n-1}\int_{B_p(R_1)}\varphi^2\|\nabla\|\nabla w_i\|\|^2 dv\\&\le\frac{1}{n}\int_{B_p(R_i)}\varphi^2\|\nabla w_i\|^2\|A\|^2 dv+\frac{1}{n-1}\int_{B_p(R_i)}\varphi^2\|\nabla\|\nabla w_i\|\|^2 dv\\&\le\int_{B_p(r_i)}\|\nabla\varphi\|^2\|\nabla w_i\|^2 dv\le\frac{1}{(R_i-R_1)^2}\int_{B_p(R_i)}\|\nabla w_i\|^2 dv\\&\le\frac{CR_i+\widehat{B}}{(R_i-R_1)^2}.\end{aligned}$$

If we let $i\mapsto\infty$, the last inequality implies that

$$\frac{1}{n}\int_{B_p(R_1)}\|\nabla w\|^2\|A\|^2 dv+\frac{1}{n-1}\int_{B_p(R_1)}\|\nabla\|\nabla w\|\|^2 dv=0,$$

and so, since r_1 is arbitrary and w is a non-constant harmonic map, we conclude that $A=0$ and thus M is totally geodesic. Since $\overline{M}$ has non-negative sectional curvature, M also has non-negative sectional curvature, and since M has at least two ends, Theorem 4.2 in [**CG72**] says that M is isometric to the Riemannian product of $\mathbb{R}\times P$, where P is a compact Riemannian manifold with non-negative sectional curvature. This proves 2 and ends the proof. □

Some results obtained in Corollaries 3.18, 3.19 and 3.21 for minimal surfaces in 3-dimensional Riemannian manifolds were extended to minimal parabolic hypersurfaces in Proposition 3.33. Little progress has been made in this direction without the strong parabolicity hypothesis. Recently O. Chodosh, C. Li and D. Stryker have generalized some results to minimal hypersurfaces of 4-dimensional Riemannian manifolds with stricter conditions on the geometry of the ambient manifold. In particular, they proved the following result.

THEOREM 3.37 ([**CLS22**]). *Let $\Phi : M^3 \to (\overline{M}^4, g)$ be a two-sided complete minimal hypersurface in a complete 4-dimensional Riemannian manifold $\overline{M}^4$ with weakly bounded geometry, non-negative sectional curvature and scalar curvature $\rho\ge c>0$. If M is stable, then Φ is totally geodesic and $\overline{\mathrm{Ricc}}(N)=0$, where N is a unit normal vector field to Φ.*

The concept of weakly bounded geometry can be found in [**CLS22**], Definition 2.3. We will not prove the above theorem, mainly because it uses the theory of μ-bubles, which would mean going into Gromov's theory, [**Gro96**], which is beyond the scope of this book. In the above theorem, it is clear that we can assume that the hypersurface is non-parabolic, and therefore it follows from Theorem 3.35 that it has only one non-parabolic end. The authors construct an appropriate test function to be used in the stability inequality, by decomposing the hypersurface into blocks along the non-parabolic end.

Also very recently, G. Catino, P. Mastrolia and A. Roncoroni weakly generalized the result of Corollary 3.15,3 to minimal hypersurfaces of dimension $n\le 5$. They use a similar idea as D. Fischer-Colbrie in the case of surfaces.

THEOREM 3.38 ([**CMR24**]). *Let $\Phi : M^n \to (\overline{M}^{n+1}, g)$, $n \leq 5$, be a two-sided complete minimal hypersurface in an $(n+1)$-dimensional Riemannian manifold $\overline{M}^{n+1}$ with non-negative sectional curvature and Ricci curvature $\overline{\mathrm{Ricc}} \geq c > 0$. If M has finite index, then M is compact.*

PROOF. We follow the idea and the notation of Theorem 3.1.

We assume that M is not compact, and we will reach a contradiction. Since M has finite index, from Theorem 2.27, there exists a compact set $K \subset M$ and a positive function $u \in C^\infty(M)$ such that the Jacobi operator $\mathcal{L}u = \left(\Delta + \|A\|^2 + \overline{\mathrm{Ricc}}(N)\right)u = 0$ on $M \setminus K$. We consider the metric $g^* = u^{2k}g$ with $\frac{n-1}{n} \leq k < \frac{4}{n-1}$. Here we are using that $n \leq 5$.

If $\alpha : [0, \infty) \to M$ is a divergent curve, then reasoning as in Theorem 3.1, we conclude that there exists a curve $\beta : [0, \infty) \to M$ with $g(\beta', \beta') = 1$ which minimizes the distance d^* between any two points and that $L^*(\alpha) \geq L^*(\beta)$. Thus,

$$L^*(\beta) = \int_0^\infty (u \circ \beta)^k(s) \mathrm{d}s.$$

We can assume that $\beta \subset M \setminus K$, because since K is compact, $\beta \cap K \subset \beta([0, a])$ for some $a > 0$, and so we can replace β by $\beta_{|[a,\infty)}$.

Since β minimizes d^* between any two points, if $\phi : [0, \infty) \to \mathbb{R}$ is a piecewise smooth function such that $\phi_{|[0,\infty)\setminus(a,b)} = 0$, with $a < b$, then from the second variation formula of length (see [**Cha06**]) we get

$$\int_0^\infty \frac{1}{(u \circ \beta)^k} \left((n-1)(\phi')^2 - S^*(\beta', \beta')\phi^2\right) \mathrm{d}s \geq 0, \tag{3.14}$$

where S^* is the Ricci tensor of (M, g^*).

From Proposition 1.5 and considering that $u^{2k} = e^{2f}$, it follows that

$$\begin{aligned} S^*(\beta', \beta') = S(\beta', \beta') - \left(k\left(\frac{\Delta u}{u} \circ \beta\right) + k\big(k(n-2) - 1\big)\left(\frac{\|\nabla u\|^2}{u^2} \circ \beta\right)\right) \\ - k(n-2)\frac{(\nabla^2 u)(\beta', \beta')}{u \circ \beta} + k(k+1)(n-2)\left(\frac{(u \circ \beta)'}{u \circ \beta}\right)^2. \end{aligned} \tag{3.15}$$

Since u is a Jacobi function and $\overline{\mathrm{Ricc}} \geq c$, we get

$$-\frac{\Delta u}{u} \circ \beta = \left(\|A\|^2 + \overline{\mathrm{Ricc}}(N)\right) \circ \beta \geq \left(\|A\|^2 \circ \beta\right) + c.$$

Also, if $\{e_1, \ldots, e_n\}$ is a local orthonormal reference such that $Ae_i = \lambda_i e_i$, then

$$n\lambda_i^2 = (n-1)\lambda_i^2 + \Big(-\sum_{j \neq i} \lambda_j\Big)^2 \leq (n-1)\lambda_i^2 + (n-1)\sum_{j \neq i} \lambda_j^2 = (n-1)\|A\|^2,$$

and so, from the Gauss equation (2.2) and using that $\overline{M}$ has non-negative sectional curvature we get

$$\begin{aligned} S(\beta', \beta') &\geq -\|A\beta'\|^2 = -\sum_{i=1}^n g(\beta', e_i)^2 \lambda_i^2 \\ &\geq -\frac{n-1}{n}(\|A\|^2 \circ \beta)\sum_{i=1}^n g(\beta', e_i)^2 = -\frac{n-1}{n}(\|A\|^2 \circ \beta). \end{aligned}$$

Using the above information and the fact that $k-\frac{n-1}{n}\geq 0$, equation (3.15) becomes

$$\begin{aligned} S^*(\beta',\beta') \geq kc - k\big(k(n-2)-1\big)\left(\frac{\|\nabla u\|^2}{u^2}\circ\beta\right) \\ - k(n-2)\frac{(\nabla^2 u)(\beta',\beta')}{u\circ\beta} + k(k+1)(n-2)\left(\frac{(u\circ\beta)'}{u\circ\beta}\right)^2. \end{aligned} \tag{3.16}$$

On the other hand, since β minimize d^*, then the first variation formula of length implies

$$\nabla^*_{\beta'}\beta' = k\frac{(u\circ\beta)'}{u\circ\beta}\beta',$$

and so, from Proposition 1.5, we get

$$\begin{aligned} k\frac{(u\circ\beta)'}{u\circ\beta} &= g(\nabla^*_{\beta'}\beta',\nabla u) = g(\nabla_{\beta'}\beta',\nabla u) + 2k\frac{((u\circ\beta)')^2}{u\circ\beta} - k\left(\frac{\|\nabla u\|^2}{u}\circ\beta\right) \\ &= (u\circ\beta)'' - (\nabla^2 u)(\beta',\beta') + 2k\frac{((u\circ\beta)')^2}{u\circ\beta} - k\left(\frac{\|\nabla u\|^2}{u}\circ\beta\right). \end{aligned}$$

So, (3.16) becomes

$$\begin{aligned} S^*(\beta',\beta') &\geq kc - k(n-2)\frac{(u\circ\beta)''}{u\circ\beta} + k\left(\frac{\|\nabla u\|^2}{u^2}\circ\beta\right) + k(n-2)\left(\frac{(u\circ\beta)'}{u\circ\beta}\right)^2 \\ &\geq k\left(c-(n-2)\frac{(u\circ\beta)''}{u\circ\beta} + (n-1)\left(\frac{(u\circ\beta)'}{u\circ\beta}\right)^2\right), \end{aligned}$$

where we used $\|\nabla u\|^2\circ\beta \geq \big((u\circ\beta)'\big)^2$ and $k>0$.

Using the above inequality in (3.14) we get

$$\begin{aligned} &\int_0^\infty \frac{1}{(u\circ\beta)^k}\left((n-1)(\phi')^2 - ck\phi^2 + k(n-2)\frac{(u\circ\beta)''\phi^2}{u\circ\beta}\right)\mathrm{d}s \\ &- k(n-1)\int_0^\infty \frac{\big((u\circ\beta)'\phi\big)^2}{(u\circ\beta)^{k+2}}\mathrm{d}s \geq 0. \end{aligned}$$

Now, by calling $\psi = (u\circ\beta)^{-\frac{k}{2}}\phi$, the last inequality becomes

$$\begin{aligned} &\int_0^\infty \left((n-1)(\psi')^2 + k(n-2)\frac{(u\circ\beta)''\psi^2}{u\circ\beta} - ck\psi^2\right)\mathrm{d}s \\ &+ k(n-1)\int_0^\infty \left(\frac{(u\circ\beta)'\psi\psi'}{u\circ\beta} + (\frac{k}{4}-1)\left(\frac{(u\circ\beta)'\psi}{u\circ\beta}\right)^2\right)\mathrm{d}s \geq 0. \end{aligned}$$

Integrating by parts, the last inequality becomes

$$\begin{aligned} 0 \leq &\int_0^\infty \left((n-1)(\psi')^2 - k(n-3)\frac{(u\circ\beta)'\psi\psi'}{u\circ\beta} - ck\psi^2\right)\mathrm{d}s \\ &+ k\left(\frac{k(n-1)}{4}-1\right)\int_0^\infty \left(\frac{(u\circ\beta)'\psi}{u\circ\beta}\right)^2\mathrm{d}s. \end{aligned} \tag{3.17}$$

Now for $\epsilon > 0$ we have

$$0 \leq \left\| \sqrt{\epsilon}\frac{(u\circ\beta)'\psi}{u\circ\beta} + \frac{1}{\sqrt{\epsilon}}\psi' \right\|^2 = \epsilon\left(\frac{(u\circ\beta)'\psi}{u\circ\beta}\right)^2 + \frac{1}{\epsilon}(\psi')^2 + 2\frac{(u\circ\beta)'\psi\psi'}{u\circ\beta},$$

and since $\frac{k(n-1)}{4} - 1 < 0$, inequality (3.17) becomes

$$\begin{aligned} 0 \leq & \int_0^\infty \left(\left(n-1+\frac{k(n-3)}{2\epsilon}\right)(\psi')^2 - ck\psi^2\right) \mathrm{d}s \\ & + k\left(\frac{k(n-1)}{4} - 1 + (n-3)\frac{\epsilon}{2}\right)\int_0^\infty \left(\frac{(u\circ\beta)'\psi}{u\circ\beta}\right)^2 \mathrm{d}s \\ \leq & \int_0^\infty \left(\left(n-1+\frac{k(n-3)}{2\epsilon}\right)(\psi')^2 - ck\psi^2\right) \mathrm{d}s, \end{aligned}$$

because for ϵ small enough, $\frac{k(n-1)}{4} - 1 + (n-3)\frac{\epsilon}{2} \leq 0$. Integrating by parts again, the last inequality gives

$$\int_0^\infty (ck\psi^2 + B\psi\psi'')\mathrm{d}s \leq 0,$$

for some positive constant B.

Reasoning as in Theorem 3.1,2, for any $a > 0$, the function $\psi_a : [0,\infty) \to \mathbb{R}$ defined by

$$\psi_a(s) = \begin{cases} \sin\frac{\pi s}{a} & s \in [0,a] \\ 0 & s \in [a,\infty) \end{cases}$$

is piecewise smooth and $(\psi_a)_{|[0,\infty)\setminus(0,a)} = 0$. So, using ψ_a in the last inequality we get

$$(ck - \frac{B\pi^2}{a^2})\int_0^a \sin^2\frac{\pi s}{a}\mathrm{d}s \leq 0.$$

If a is large enough to make $ck - \frac{B\pi^2}{a^2} > 0$, the last inequality gives a contradiction. □

COROLLARY 3.39 ([**CLS22**]). *An $(n+1)$-dimensional, $n \leq 5$, Riemannian manifold with non-negative sectional curvature and Ricci curvature $\overline{\mathrm{Ricc}} \geq c > 0$ does not admit two-sided complete stable minimal hypersurfaces.*

REMARK 3.40. Corollary 3.39 extends Corollary 3.21 for surfaces. Also, if the ambient manifold is compact with positive sectional curvature, then it satisfies the assumptions of Corollary 3.39, and so it does not admit two-sided complete stable minimal hypersurfaces. This result was proved in [**CLS22**] for $n = 3$.

PROOF. From Theorem 3.38, M is compact and thus parabolic. The result follows from Proposition 3.33,2. □

Regarding Corollary 3.39 and Remark 3.40, when the dimension n is very large, little is known about the existence or not of two-sided complete stable minimal hypersurfaces. The study of these seems to be a complicated problem and even in simple examples, such as the sphere $\mathbb{S}^{n+1}$ or the real projective space $\mathbb{RP}^{n+1}$, neither non-compact examples nor obstructions to their existence are known.

3.4. Index of compact orientable minimal surfaces in orientable Riemannian 4-manifolds

In this last section of the chapter we will study the index and nullity of orientable surfaces in 4-dimensional orientable Riemannian manifolds. In this case the normal bundle, on whose sections the Schrödinger operator acts, acquires a holomorphic line bundle structure on the surface. This allows us to use its holomorphic sections as test sections to estimate the index and nullity of these surfaces. In Chapters 4, 5 and 6, we will study similar problems when the ambient manifold is $\mathbb{R}^4$, $\mathbb{S}^4$, $\mathbb{CP}^2$, and $\mathbb{S}^2 \times \mathbb{S}^2$. The content of this section is based on the results of M.J. Micallef and J.G. Wolfson, [**MW93**] and S. Montiel and F. Urbano, [**MU97**].

We begin by studying some geometric properties of 4-dimensional orientable Riemannian manifolds, as well as their minimal surfaces.

Let $(\overline{M}, g)$ be a 4-dimensional Riemannian manifold. The **Weyl tensor** of $(\overline{M}, g)$ is the $(0,4)$-tensor W, given in local coordinates by

$$W_{ijkl} = R_{ijkl} - \frac{1}{2}(S_{il}g_{jk} + S_{jk}g_{il} - S_{jl}g_{ik} - S_{ik}g_{jl}) + \frac{\rho}{6}(g_{il}g_{jk} - g_{ik}g_{jl}),$$

where R is the Riemannian curvature tensor, S is the Ricci tensor, and ρ is the scalar curvature of $(\overline{M}, g)$. The Weyl tensor W acts on 2-forms of $\overline{M}$ by

$$W : \Omega^2(\overline{M}) \to \Omega^2(\overline{M})$$

$$(W\alpha)(X,Y) = \frac{1}{2}\sum_{i,j=1}^{4} W(e_i, e_j, Y, X)\alpha(e_i, e_j),$$

for each $\alpha \in \Omega^2(\overline{M})$ and $X, Y \in \mathfrak{X}(\overline{M})$, where $\{e_1, e_2, e_3, e_4\}$ is a local orthonormal reference on $\overline{M}$.

On the other hand, for each $x \in \overline{M}$ we consider the complexified tangent space $T_x\overline{M} \otimes \mathbb{C}$. We will also denote by g the complex linear extension of the metric g to $T_x\overline{M} \otimes \mathbb{C}$ and $\overline{R}$ will denote the complex linear extension of the Riemannian curvature tensor R to the complexified $T_x\overline{M} \otimes \mathbb{C}$. If $\{z, w\}$ is a basis of a complex two-plane $V \subset T_x\overline{M} \otimes \mathbb{C}$, then the real number given by

$$\overline{K}(V) = \frac{\overline{R}(z, w, \overline{w}, \overline{z})}{\|z \wedge w\|^2},$$

depends only on the plane V and is called the **complex sectional curvature** of the complex plane V.

A complex two-plane $V \subset T_x\overline{M} \otimes \mathbb{C}$ is said to be **totally isotropic** if $g(z,z) = 0$, $\forall z \in V$. This means that if $z = a + ib$ then $\|a\|^2 = \|b\|^2$ and $g(a,b) = 0$. In [**MM88**] it was proved that a complex two-plane V in $T_x\overline{M} \otimes \mathbb{C}$ is totally isotropic if and only if V has a basis $\{z = e_1 + ie_2, w = e_3 + ie_4\}$ where $\{e_1, e_2, e_3, e_4\}$ is an orthonormal basis of $T_x\overline{M}$. This basis is called standard. Given this basis, it is easy to check that

$$\begin{aligned}\overline{K}(V) = {}& \overline{R}(e_1, e_3, e_3, e_1) + \overline{R}(e_1, e_4, e_4, e_1) + \overline{R}(e_2, e_3, e_3, e_2) \\ & + \overline{R}(e_2, e_4, e_4, e_2) + 2\overline{R}(e_1, e_2, e_3, e_4).\end{aligned} \tag{3.18}$$

If $\{\theta_1, \theta_2, \theta_3, \theta_4\}$ is the dual basis in $T_x^*\overline{M}$ of $\{e_1, e_2, e_3, e_4\}$ and $\alpha = \theta_1 \wedge \theta_2 + \theta_3 \wedge \theta_4$, it is easy to check, using the definition of W, that

$$\overline{K}(V) = g\left(\left(\frac{\rho}{6} - W\right)(\alpha), \alpha\right) \tag{3.19}$$

where $g(\theta_i \wedge \theta_j, \theta_k \wedge \theta_l) = \delta_{ik}\delta_{jl} - \delta_{il}\delta_{jk}$.

In [**MM88**] the authors studied Riemannian manifolds with positive curvature on totally isotropic planes, i.e., $\overline{K}(V) > 0$ for any totally isotropic plane V, and considering (3.19), a necessary and sufficient condition for positive curvature on totally isotropic two-planes is that $\frac{\rho}{6} - W$ be definite positive acting on 2-forms.

From now on we assume that $(\overline{M}, g)$ is an oriented manifold. In this case, the star operator (Section 1.3) acts on 2-forms $\star : \Omega^2(\overline{M}) \to \Omega^2(\overline{M})$ with $\star\star = Id$. Thus its eigenvalues are ± 1 and the corresponding 3-dimensional eigenspaces are denoted by

$$\Omega^2_\pm(\overline{M}) = \{\alpha \in \Omega^2(\overline{M}) \mid \star\alpha = \pm\alpha\}.$$

The elements of $\Omega^2_+(\overline{M})$ are called **self-dual 2-forms** and those of $\Omega^2_-(\overline{M})$ are called **anti-self-dual 2-forms**.

In [**ST69**] the authors proved that $\star W = W\star$, and thus the Weyl tensor W decomposes into $W^\pm : \Omega^2_\pm(\overline{M}) \to \Omega^2_\pm(\overline{M})$. Under these conditions, the oriented four manifold $(\overline{M}, g)$ is called **self-dual** if $W^- = 0$ and it is called **anti-self-dual** if $W^+ = 0$. It is interesting to note that a change of orientation on $\overline{M}$ exchanges the self-dual and the anti-self-dual conditions. If we now consider a Kähler surface $(\overline{M}, g, J)$ with its canonical orientation ω^2, where ω is the Kähler 2-form, then $\omega \in \Omega^2_+(\overline{M})$ holds.

Under the above conditions, a totally isotropic plane $V \subset T_x\overline{M} \otimes \mathbb{C}$ is called positive if $\{e_1, e_2, e_3, e_4\}$ is a positively oriented basis, where $\{z = e_1 + ie_2, w = e_3 + ie_4\}$ is a standard basis of V. It is clear that this condition is satisfied for any standard basis of V and thus it depends only of the plane V. If $\{e_1, e_2, e_3, e_4\}$ is a negatively oriented basis, the totally isotropic plane V will be called negative.

If $\{\theta_1, \theta_2, \theta_3, \theta_4\}$ is its dual basis on $T^*_x M$, then

$$\{e_1, e_2, e_3, e_4\} \text{ is positively oriented if and only if } \theta_1 \wedge \theta_2 + \theta_3 \wedge \theta_4 \in \Lambda^2_+ T^*_x\overline{M},$$
$$\{e_1, e_2, e_3, e_4\} \text{ is negatively oriented if and only if } \theta_1 \wedge \theta_2 + \theta_3 \wedge \theta_4 \in \Lambda^2_- T^*_x\overline{M}.$$

LEMMA 3.41 ([**MW93**]). *Let $(\overline{M}, g)$ be an oriented Riemannian 4-manifold.*

(1) *If V is a positive (or negative) totally isotropic complex plane with standard basis $\{z = e_1 + ie_2, w = e_3 + ie_4\}$ then*

$$\overline{K}(V) = g\left(\left(\frac{\rho}{6} - W^+\right)(\alpha), \alpha\right) \quad \left(or \;\; \overline{K}(V) = \left(\left(\frac{\rho}{6} - W^-\right)(\alpha), \alpha\right)\right),$$

where $\alpha = \theta_1 \wedge \theta_2 + \theta_3 \wedge \theta_4$ lies in $\Lambda^2_+ T^*_x\overline{M}$ (or $\Lambda^2_- T^*_x\overline{M}$) and $\{\theta_1, \theta_2, \theta_3, \theta_4\}$ is the dual basis of $\{e_1, e_2, e_3, e_4\}$.

(2) *If $(\overline{M}, g, J)$ is a Kähler surface, with Kähler 2-form ω, endowed with the canonical orientation ω^2, then if V is a positive totally isotropic complex plane,*

$$\overline{K}(V) = \left(\left(\frac{\rho}{6} - W^+\right)(\alpha), \alpha\right) = \frac{\rho}{2}\left(1 - \frac{g(\alpha, \omega)^2}{4}\right).$$

PROOF. 1 is a direct consequence of (3.19). To prove 2, we take a positive orthonormal basis $\{v_1, v_{1*} = Jv_1, v_2, v_{2*} = Jv_2\}$ with dual basis $\{\eta_1, \eta_{1*}, \eta_2, \eta_{2*}\}$. Then

$$\{w = \eta_1 \wedge \eta_{1*} + \eta_2 \wedge \eta_{2*},\ \theta = \eta_1 \wedge \eta_2 + \eta_{2*} \wedge \eta_{1*},\ \nu = \eta_1 \wedge \eta_{2*} + \eta_{1*} \wedge \eta_2\}$$

is a basis of $\Lambda^2_+ T^*_x M$ and so $\alpha = \frac{1}{2}\{g(\alpha,\omega)\omega + g(\alpha,\theta)\theta + g(\alpha,\nu)\nu\}$. Then,

$$\begin{aligned} g\left(\left(\frac{\rho}{6} - W^+\right)(\alpha),\alpha\right) &= \frac{g(\alpha,\theta)^2}{4} g\left(\left(\frac{\rho}{6} - W^+\right)(\theta),\theta\right) \\ &\quad + \frac{g(\alpha,\nu)^2}{4} g\left(\left(\frac{\rho}{6} - W^+\right)(\nu),\nu\right) = \frac{\rho}{2}\left(\frac{g(\alpha,\theta)^2}{4} + \frac{g(\alpha,\nu)^2}{4}\right) \\ &= \frac{\rho}{2}\left(1 - \frac{g(\alpha,\omega)^2}{4}\right). \end{aligned}$$

□

To conclude this intoduction, we will characterize self-duality and anti-self-duality in terms of twistor bundles. Indeed, given a point $x \in \overline{M}$, let $\mathcal{P}^\pm_x$ be the set of almost Hermitian structures $J^\pm_x$ over $T_x\overline{M}$ such that if $\omega^\pm(u,v) = g(J^\pm_x u, v)$, then $\pm\omega^\pm \wedge \omega^\pm$ is the orientation on $T_x\overline{M}$ from $\overline{M}$. Then $\mathcal{P}^\pm = \bigcup_{x\in\overline{M}} \mathcal{P}^\pm_x$ are $\mathbb{CP}^1$-fiber bundles over $\overline{M}$ called the **twistor bundles** of $\overline{M}$. Atiyah, Hitchin and Singer [**AHS78**] defined almost complex structures $\mathcal{J}^\pm$ on $\mathcal{P}^\pm$ and proved the following central result: *$(\mathcal{P}^+, \mathcal{J}^+)$ is a complex manifold if and only if the Riemannian metric of $\overline{M}$ is anti-self-dual, and $(\mathcal{P}^-, \mathcal{J}^-)$ is a complex manifold if and only if the Riemannian metric of $\overline{M}$ is self-dual.*

Now let $\Phi : M \to \overline{M}$ be an immersion of an oriented surface M. If $\{e_1, e_2, e_3, e_4\}$ is an oriented orthonormal local reference on $\Phi^*T\overline{M}$ such that $\{e_1, e_2\}$ is an oriented reference on TM, then we define two almost complex structures $J^\pm$ on $\Phi^*T\overline{M}$ by

$$J^\pm(e_1) = e_2, \quad J^\pm e_3 = \pm e_4. \tag{3.20}$$

We note that $J^\pm$, restricted to TM, is the complex structure on M compatible with the given orientation, and that, on $T^\perp M$, one has $\nabla^\perp J^\pm = 0$. This allows to use the Koszul-Malgrange theorem [**KM58**] which gives two unique holomorphic line bundle structures on M, denoted by $N^\pm$. In fact, the complexified normal bundle decomposes as $T^\perp M \otimes \mathbb{C} = T^\perp_\pm M^{1,0} \oplus T^\perp_\pm M^{0,1}$ where

$$T^\perp_\pm M^{1,0} = \{v - iJ^\pm v \mid v \in T^\perp M\}, \quad T^\perp_\pm M^{0,1} = \{v + iJ^\pm v \mid v \in T^\perp M\}.$$

Then $N^\pm \equiv T^\perp_\pm M^{1,0}$. We note that the dual bundle $(N^+)^*$ is biholomorphic to N^- by

$$F : (N^+)^* \to N^-, \quad F(a(\theta_3 + i\theta_4)) = a(e_3 + ie_4), \quad \forall a \in \mathbb{C}.$$

Now, given $\xi \in \mathfrak{X}^\perp(M)$, we define

$$D^\pm_v \xi = \nabla^\perp_{J^\pm v}\xi - J^\pm \nabla^\perp \xi, \quad \forall v \in TM. \tag{3.21}$$

Then, for any section $\xi - iJ^\pm\xi$ of $N^\pm$, we have that

$$\nabla^\perp_{e_1+ie_2}(\xi - i\,\mathrm{J}^\pm\,\xi) = i\big(D^\pm_{e_1}\xi - i\,\mathrm{J}^\pm\, D^\pm_{e_1}\xi\big),$$

and so, a section $\xi - iJ^\pm\xi$ of $N^\pm$ is holomorphic if and only if $D^\pm\xi = 0$. This property defines an isomorphism

$$\begin{aligned} Z^\pm = \{\xi \in \mathfrak{X}^\perp(M) \mid D^\pm\xi = 0\} &\to H^0(M, \mathcal{O}(N^\pm)) \\ \xi &\mapsto \xi - iJ^\pm\xi, \end{aligned} \tag{3.22}$$

from the complex linear space $Z^\pm$, $((a+ib)\xi = a\xi + bJ^\pm\xi)$, to the complex linear space of holomorphic sections of $N^\pm$.

Under the above conditions, we define the **normal curvature** $K^{\perp}$ of the immersion Φ as

$$K^{\perp} = R^{\perp}(e_1, e_2, e_3, e_4), \tag{3.23}$$

and the Euler characteristic of $T^{\perp}M$ by

$$\chi^{\perp} = \frac{1}{2\pi}\int_M K^{\perp} dv.$$

It is clear that $K^{\perp}$ is well defined and that it changes sign when we change the orientation on the surface M. In this context, the degrees of the holomorphic line bundles $N^{\pm}$ are given by $\deg(N^{\pm}) = \mp\chi^{\perp}$ (see Definition 1.25).

Let $V^{\pm}$ be the complex planes with basis $\{z^+ = e_1 + ie_2, w^+ = e_3 + ie_4\}$ and $\{z^- = e_1 + ie_2, w^- = e_4 + ie_3\}$. Then V^+ (or V^-) is a positive (or negative) complex totally isotropic plane and we will denote the corresponding complex sectional curvature by $\overline{K}^{\pm}$.

Now suppose that Φ is a minimal immersion. Then it is interesting to define two $T^{\perp}M$-valued bilinear forms $\sigma^{\pm}$ on TM by

$$\sigma^{\pm}(v, w) = \sigma(J^{\pm}v, w) - J^{\pm}\sigma(v, w), \quad \forall v, w \in TM. \tag{3.24}$$

Using the Ricci equation (2.2), it is easy to check that

$$K^{\perp} = \overline{R}(e_1, e_2, e_3, e_4) \pm \frac{\|\sigma^{\pm}\|^2 - 2\|\sigma\|^2}{4}. \tag{3.25}$$

DEFINITION 3.42. An immersion $\Phi : M \to (\overline{M}^4, g)$ of an oriented surface M into an oriented manifold $\overline{M}$ is said to be **superminimal with positive or negative spin** if $\sigma^{\pm} = 0$, i.e., $\sigma(J^{\pm}v, w) = J^{\pm}\sigma(v, w)$ for any $v, w \in TM$.

From the definition we have that a superminimal immersion is, in particular, a minimal immersion, because if v is a unit tangent vector to M, then

$$0 = \sigma^{\pm}(v, v) + \sigma^{\pm}(J^{\pm}v, J^{\pm}v) = -2J^{\pm}\vec{H}.$$

Now we will give a geometric interpretation of superminimality.

The almost complex structures $J^{\pm}$ on $\Phi^*T\overline{M}$ give to $\Phi^*T\overline{M}$ a structure of complex vector bundle over M. In this way we have associated to the immersion Φ two almost Hermitian structures $J^{\pm}_{\Phi(p)}$ on $T_{\Phi(p)}\overline{M}$ such that $J^{\pm}_{\Phi(p)} \in \mathcal{P}^{\pm}_{\Phi(p)}$ for each point $p \in M$. So one can define the **twistor liftings** $\widetilde{\Phi}^{\pm} : M \to \mathcal{P}^{\pm}$ by

$$\widetilde{\Phi}^{\pm}(p) = J^{\pm}_{\Phi(p)}.$$

Although not explicitely stated, the following result was proved in [**Fri84**].

PROPOSITION 3.43. *Let $\Phi : M \to \overline{M}$ be an immersion of an oriented surface M into an oriented 4-dimensional Riemannian manifold. Then the following assertions are equivalent:*

(1) *Φ is superminimal with positive or negative spin.*
(2) *Φ is a minimal immersion and the twistor liftings map of Φ,*

$$\widetilde{\Phi}^{\pm} : (M, J) \to (\mathcal{P}^{\pm}, \mathcal{J}^{\pm})$$

are horizontal.
(3) *The almost Hermitian structures $J^{\pm}$ on $\Phi^*T\overline{M}$ are parallel with respect to the induced connection $\overline{\nabla}$, i.e., $\overline{\nabla}J^{\pm} = 0$.*

Finally, to conclude with the preliminaries, we consider the case in which the ambient manifold is a **Kähler surface** $(\overline{M}, g, J)$, where J is the complex structure endowed with the canonical orientation ω^2, where ω is the Kähler 2-form. If $\Phi : M \to \overline{M}$ is an immersion of an oriented surface, then the **Kähler function** C on M is defined by $\Phi^*\omega = C\,\mathrm{d}A$, where $\mathrm{d}A$ is the area 2-form on M. It is clear that the Kähler function satisfies $-1 \leq C \leq 1$. Surfaces with $C = 1$ (or $C = -1$) are called **holomorphic** (or **anti-holomorphic**). **Complex** surface will be synonymous with either holomorphic or anti-holomorphic surface. The surface is called **totally real** if $C^2 < 1$.

There is a relation between J and the almost complex structures $J^{\pm}$ defined in (3.20), namely

$$(Jv)^{\top} = CJ^{\pm}v, \quad (J\xi)^{\perp} = CJ^{+}\xi = -CJ^{-}\xi, \quad J^{\pm}(Jv)^{\perp} \pm (JJ^{\pm}v)^{\perp} = 0 \tag{3.26}$$

for any $v \in TM$ and $\xi \in T^{\perp}M$ and where $\top$ and $\perp$ stand for tangent and normal components, respectively.

Once the preliminares are over, we begin by studying the index form of an oriented minimal surface of an oriented 4-manifold $\overline{M}$.

PROPOSITION 3.44. *Let $\Phi : M \to (\overline{M}^4, g)$ be a minimal immersion of an oriented surface M into an oriented 4-dimensional Riemannian manifold $\overline{M}$. Then, for any $\xi \in \mathfrak{X}_0^{\perp}(M)$ we have*

$$\mathcal{Q}(\xi) + \mathcal{Q}(J^{\pm}\xi) = \int_M \left\{ \|D^{\pm}\xi\|^2 - \left(\frac{\|\sigma^{\pm}\|^2}{2} + \overline{K}^{\pm} \right) \|\xi\|^2 \right\} \mathrm{d}v,$$

where $J^{\pm}$, $D^{\pm}$, $\sigma^{\pm}$ and $\overline{K}^{\pm}$ were previously defined.

REMARK 3.45. Since $J^{-}\xi = -J^{+}\xi$ and $\mathcal{Q}$ is a quadratic form, then $\mathcal{Q}(\xi) + \mathcal{Q}(J^{+}\xi) = \mathcal{Q}(\xi) + \mathcal{Q}(J^{-}\xi)$.

PROOF. From (2.12) we have that

$$\mathcal{Q}(\xi) = \int_M \|\nabla^{\perp}\xi\|^2 \mathrm{d}v - \int_M \|A_{\xi}\|^2 \mathrm{d}v - \sum_{i=1}^{2} \int_M \overline{R}(\xi, e_i, e_i, \xi)\, \mathrm{d}v, \tag{3.27}$$

where $\{e_1, e_2\}$ is a local oriented orthonormal reference on TM. From (3.21) we now get

$$\frac{1}{2}\|D^{\pm}\xi\|^2 = \|\nabla^{\perp}\xi\|^2 - \sum_{i=1}^{2} g(\nabla^{\perp}_{J^{\pm}e_i}\xi, \nabla^{\perp}_{e_i}J^{\pm}\xi).$$

If α is the 1-form on M defined by $\alpha(v) = g(\nabla^{\perp}_{J^{\pm}v}\xi, J^{\pm}\xi)$, then from Theorem 1.15 it follows that

$$0 = \int_M \delta\alpha\, \mathrm{d}v = \sum_{i=1}^{2} \int_M g(\nabla^{\perp}_{J^{\pm}e_i}\xi, \nabla^{\perp}_{e_i}J^{\pm}\xi)\, \mathrm{d}v + \sum_{i=1}^{2} \int_M g(\nabla^{\perp}_{e_i}\nabla^{\perp}_{J^{\pm}e_i}\xi, J^{\pm}\xi)\, \mathrm{d}v.$$

But

$$\sum_{i=1}^{2} g(\nabla^{\perp}_{e_i}\nabla^{\perp}_{J^{\pm}e_i}\xi, J^{\pm}\xi) = \frac{1}{2}\sum_{i=1}^{2} R^{\perp}(e_i, J^{\pm}e_i, \xi, J^{\pm}\xi) = \pm K^{\perp}\|\xi\|^2.$$

The two expressions above together with (3.27) give

$$
\begin{aligned}
\mathcal{Q}(\xi) = {} & \frac{1}{2}\int_M \|D^{\pm}\xi\|^2 \mathrm{d}v - \int_M \pm K^{\perp}\|\xi\|^2 \mathrm{d}v - \int_M \|A_\xi\|^2 \mathrm{d}v \\
& - \sum_{i=1}^{2}\int_M \{\overline{R}(\xi, e_i, e_i, \xi)\mathrm{d}v.
\end{aligned}
\tag{3.28}
$$

Doing the same computations for $J^{\pm}\xi$ and as $D^{\pm}J^{\pm}\xi = J^{\pm}D^{\pm}\xi$ we get

$$
\begin{aligned}
\mathcal{Q}(J^{\pm}\xi) = {} & \frac{1}{2}\int_M \|D^{\pm}\xi\|^2 \mathrm{d}v - \int_M \pm K^{\perp}\|\xi\|^2 \mathrm{d}v - \int_M \|A_{J^{\pm}\xi}\|^2 \mathrm{d}v \\
& - \sum_{i=1}^{2}\int_M \{\overline{R}(J^{\pm}\xi, e_i, e_i, J^{\pm}\xi)\mathrm{d}v.
\end{aligned}
\tag{3.29}
$$

Now as

$$\|A_\xi\|^2 + \|A_{J^{\pm}\xi}\|^2 = \|\sigma\|^2\|\xi\|^2$$

and

$$\overline{R}(\xi, e_i, e_i, \xi) + \overline{R}(J^{\pm}\xi, e_i, e_i, J^{\pm}\xi) = \sum_{j=3}^{4}\overline{R}(e_j, e_i, e_i, e_j)\|\xi\|^2,$$

from (3.25),(3.28) and (3.29) it follows that

$$
\begin{aligned}
\mathcal{Q}(\xi) + \mathcal{Q}(J^{\pm}\xi) &= \int_M \|D^{\pm}\xi\|^2 \mathrm{d}v - \frac{1}{2}\int_M \|\sigma^{\pm}\|^2\|\xi\|^2 \mathrm{d}v \\
&\quad - \int_M \left\{ \pm 2\overline{R}(e_1, e_2, e_3, e_4) + \sum_{i=1}^{2}\sum_{j=3}^{4}\overline{R}(e_j, e_i, e_i, e_j) \right\} \|\xi\|^2 \mathrm{d}v \\
&= \int_M \left\{ \|D^{\pm}\xi\|^2 - \left(\frac{\|\sigma^{\pm}\|^2}{2} + \overline{K}^{\pm} \right) \|\xi\|^2 \right\} \mathrm{d}v.
\end{aligned}
$$

□

In the following theorem we collect the main results obtained in [**MW93**].

THEOREM 3.46 ([**MW93**]). *Let $\Phi : M \to (\overline{M}^4, g)$ be a minimal immersion of an oriented compact surface M of genus h in an oriented Riemannian 4-manifold $\overline{M}$ with scalar curvature ρ. If $W^{\pm}$ is the self-dual part (the anti-self-dual part) of the Weil tensor W of $\overline{M}$ and $\chi^{\perp}$ is the Euler characteristic of the normal bundle, then*

1. *If $\frac{\rho}{6} - W^+ \geq 0$ $\left(\text{or } \frac{\rho}{6} - W^- \geq 0\right)$, then $\mathrm{Ind}(M) + \mathrm{Nul}(M) \geq 1 - h - \chi^{\perp}$ $\left(\text{or } \mathrm{Ind}(M) + \mathrm{Nul}(M) \geq 1 - h + \chi^{\perp}\right)$.*
2. *If $\frac{\rho}{6} - W^+ > 0$ $\left(\text{or } \frac{\rho}{6} - W_- > 0\right)$, then $\mathrm{Ind}(M) \geq 1 - h - \chi^{\perp}$ $\left(\text{or } \mathrm{Ind}(M) \geq 1 - h + \chi^{\perp}\right)$.*
3. *If $\frac{\rho}{6} - W \geq 0$, then $\mathrm{Ind}(M) + \mathrm{Nul}(M) \geq 1 - h + |\chi^{\perp}|$.*
4. *If $\frac{\rho}{6} - W > 0$, then $\mathrm{Ind}(M) \geq 1 - h + |\chi^{\perp}|$. In particular, if M is a sphere, then M is unstable.*
5. *If $\overline{M}$ is a Kähler surface endowed with the standard orientation, $\rho > 0$, and Φ is stable with $\chi^{\perp} \leq 1 - h$, then Φ is a complex surface.*
6. *If $\overline{M}$ is a Kähler surface endowed with the standard orientation, $\rho > 0$ and Φ is stable and totally real, then $h = 0$, i.e., M is a sphere.*

(7) *If $\overline{M}$ is a hyperKähler surface, i.e., on $\overline{M}$ there are three structures of Kähler surface $\{J_1, J_2, J_3\}$ satisfying $J_1J_2 = -J_2J_1 = J_3$, and Φ is stable with $\chi^\perp \leq 3 - h$, then Φ is complex with respect to one of the complex structures $\{aJ_1 + bJ_2 + cJ_3 \mid a, b, c \in \mathbb{R}, a^2 + b^2 + c^2 = 1\}$ of $\overline{M}$.*

(8) *If $\overline{M}$ is Einstein with $\frac{\rho}{6} - W^+ \leq 0$ (or $\frac{\rho}{6} - W^- \leq 0$) and Φ is superminimal with positive spin (or negative spin), then Φ is stable.*

PROOF. **Proof of 1 and 2.** From Lemma 3.41,1, $\frac{\rho}{6} - W^\pm \geq 0$ (or $\frac{\rho}{6} - W^\pm > 0$) if and only if $\overline{K}^\pm \geq 0$ (or $\overline{K}^\pm > 0$). Thus, under the assumptions of 1 and 2, for any section $\xi \in \mathfrak{X}^\perp(M)$, Proposition 3.44 implies that

$$\mathcal{Q}(\xi) + \mathcal{Q}(J^\pm\xi) \leq \int_M \|D^\pm\xi\|^2 dv, \quad \Big(\text{or} \quad \mathcal{Q}(\xi) + \mathcal{Q}(J^\pm\xi) < \int_M \|D^\pm\xi\|^2 dv\Big).$$

So, if ξ is a holomorphic section of the holomorphic line bundle $N^\pm$, then

$$Q(\xi) + Q(J^+\xi) \leq 0 \quad \big(\text{or} \quad Q(\xi) + Q(J^+\xi) < 0\big).$$

Thus, taking into account (3.22), $\operatorname{Ind}(M) + \operatorname{Nul}(M) \geq \dim H^0(M, \mathcal{O}(N^\pm))$ if $\frac{\rho}{6} - W^\pm \geq 0$ (or $\operatorname{Ind}(M) \geq \dim H^0(M, \mathcal{O}(N^\pm))$ if $\frac{\rho}{6} - W^\pm > 0$), where $H^0(M, \mathcal{O}(N^\pm))$ is the complex space of holomorphic sections of the holomorphic line bundle $N^\pm$. Now the Riemann-Roch theorem, Theorem 1.24, tells us that

$$\dim H^0(M, \mathcal{O}(N^\pm)) = 1 - h + \deg(N^\pm) + \dim H^0(M, \mathcal{O}(\kappa \otimes (N^\pm)^*)) \geq 1 - h \mp \chi^\perp,$$

and so we prove 1 and 2.

Proof of 3 and 4. These are simple consequences of 1 and 2.

Proof of 5. Let J be the complex structure of the Kähler surface $\overline{M}$. Then we define a 1-form β on M valued in $T^\perp M$ by

$$\beta(v) = (Jv)^\perp, \quad \forall v \in TM,$$

where $\perp$ is the normal component.

CLAIM. β defines a holomorphic section of the holomorphic line bundle $\kappa \otimes (N^+)^* \equiv \kappa \otimes N^-$.

We begin by proving that

$$(\nabla^\perp_{J^+V}\beta)W = (\nabla^\perp_V\beta)(J^+W), \quad \forall V, W \in \mathfrak{X}(M). \tag{3.30}$$

where $\nabla\beta$ was defined in (1.5). In fact, since $\overline{M}$ is Kähler, taking normal components in the equality $\overline{\nabla}_V JW = J\overline{\nabla}_V W$ and using (2.1), (3.26) and (1.5), we get

$$\begin{aligned}(\nabla^\perp_V\beta)W &= \nabla^\perp_V(JW)^\perp - (J\nabla_V W)^\perp \\ &= (J\sigma(V,W))^\perp - \sigma(V,(JW)^\top) = CJ^+\sigma(V,W) - C\sigma(V,J^+W),\end{aligned}$$

where $\top$ is the tangent component. Since Φ is minimal, $\sigma(J^+V, W) = \sigma(V, J^+W)$ and so, from the last equation we have that

$$\begin{aligned}(\nabla^\perp_{J^+V}\beta)W - (\nabla^\perp_V\beta)(J^+W) &= CJ^+\sigma(J^+V,W) - C\sigma(J^+V,J^+W) \\ &\quad - CJ^+\sigma(V,J^+W) - C\sigma(V,W) = 0,\end{aligned}$$

and (3.30) is proved.

Now, we define a section of $\kappa \otimes N^-$, also called β,

$$\beta(v - iJ^+v) := (J(v - iJ^+v))^\perp = (Jv)^\perp + iJ^+(Jv)^\perp = (Jv)^\perp - iJ^-(Jv)^\perp,$$

where we used (3.26). Taking into account (3.30), we get

$$\begin{aligned}(\nabla^{\perp}_{e_1+ie_2}\beta)(W - iJ^+W) &= (\nabla^{\perp}_{e_1}\beta)W + (\nabla^{\perp}_{e_2}\beta)(J^+W)\\ &\quad - i\{(\nabla^{\perp}_{e_1}\beta)(J^+W) - (\nabla^{\perp}_{e_2}\beta)W\} = 0,\end{aligned}$$

which gives that β is a holomorphic section of $\kappa \otimes (N^+)^* \equiv \kappa \otimes N^-$. This proves the claim.

Now suppose that Φ is not a complex immersion. Then, since $|\beta|^2 = 2(1-C^2)$, where C is the Kähler function of M, we get that β is a non-trivial holomorphic section of $\kappa \otimes (N^+)^*$ and hence $\dim H^0(M, \kappa \otimes (N^+)^*) \geq 1$. This fact, together with Theorem 1.24 and the fact that $\chi^{\perp} \leq 1-h$, implies that $\dim H^0(M, \mathcal{O}(N^+) \geq 2 - h - \chi^{\perp} \geq 1$. Then, from (3.22), we can take a non-trivial holomorphic section ξ in the holomorphic line bundle N^+ and, from Proposition 3.44 and Lemma 3.41 it follows that

$$Q(\xi) + Q(J^+\xi) \leq -\int_M \frac{\rho}{2}\left(1 - \frac{g(\alpha,\omega)^2}{4}\right)\|\xi\|^2\,dv = -\int_M \frac{\rho}{2}(1 - C^2)\|\xi\|^2 dv < 0.$$

This means that M is unstable and we finish the proof of 5.

Proof of 6. Since Φ is a totally real immersion, we have two consequences: $C^2 < 1$ and $\chi^{\perp} = \chi(M) = 2(1-h)$. Thus, the 1-form β defined above is also, in this case, a non-trivial holomorphic section of $\kappa \otimes (N^+)^*$ and thus, as in 5, $\dim H^0(M, \kappa \otimes (N^+)^*) \geq 1$. This implies that $\dim H^0(M, \mathcal{O}(N^+) \geq 2 - h - \chi^{\perp} = 2 - h - 2(1-h) = h$. If $h \geq 1$, we use a similar argument as in the proof of 5 to get that M is unstable. So $h = 0$ and 6 is proved.

Proof of 7. Suppose that $\overline{M}$ is a hyperkähler manifold and let $V = \{aJ_1 + bJ_2 + cJ_3 \mid a, b, c \in \mathbb{R}, a^2 + b^2 + c^2 = 1\}$ be the three-dimensional space of Kähler structures on $\overline{M}$ generated by $\{J_1, J_2, J_3\}$. In a similar way as above, we can define a linear map

$$\begin{aligned}V &\to H^0(M, \mathcal{O}(\kappa \otimes (N^+)^*))\\ J &\mapsto \beta_J(v) = (Jv)^{\perp}, \quad \forall v \in TM.\end{aligned}$$

If M is not a complex surface with respect to any of of the Kähler structures of V, then reasoning as in the above cases, we deduce that $\dim H^0(M, \mathcal{O}(\kappa \otimes (N^+)^*)) \geq 3$. This implies that $\dim H^0(M, \mathcal{O}(N^+)) \geq 4 - h - \chi^{\perp} \geq 1$. Reasoning as in the proof of 5, we get that M is unstable. So, M is complex with respect to some structure of V.

Proof of 8. Since $\overline{M}$ is Einstein, the Ricci tensor $\overline{S} = \frac{\rho}{4}g$, and thus,

$$\sum_{i=1}^{2} \overline{R}(\xi, e_i, e_i, \xi) = \left(\frac{\rho}{4} - \overline{R}(e_3, e_4, e_4, e_3)\right)\|\xi\|^2, \quad \forall \xi \in \mathfrak{X}^{\perp}(M).$$

If Φ is superminimal with positive spin, then we have that $\sigma^+ = 0$ and so, (3.25), becomes

$$K^{\perp}\|\xi\|^2 = \left(\overline{R}(e_1, e_2, e_3, e_4) - \frac{\|\sigma\|^2}{2}\right)\|\xi\|^2.$$

Now, since $\frac{\rho}{6} - W^+ \leq 0$, (3.18) says that

$$\begin{aligned}0 \geq \overline{K}^+ &= \overline{\mathrm{Ricc}}(e_3) + \overline{\mathrm{Ricc}}(e_4) - 2\overline{R}(e_3, e_4, e_4, e_3) + 2\overline{R}(e_1, e_2, e_3, e_4)\\ &= \frac{\rho}{2} - 2\overline{R}(e_3, e_4, e_4, e_3) + 2\overline{R}(e_1, e_2, e_3, e_4).\end{aligned}$$

From the last three equations we get

$$K^{\perp}\|\xi\|^2 + \sum_{i=1}^{2} \overline{R}(\xi, e_i, e_i, \xi) \leq -\frac{\|\sigma\|^2}{2}\|\xi\|^2,$$

and so the equation (3.28) says that

$$\mathcal{Q}(\xi) \geq \int_M \left(\frac{\|\sigma\|^2}{2}\|\xi\|^2 - \|A_\xi\|^2\right) dv.$$

Since Φ is a superminimal immersion with positive spin, it is not difficult to check that $\|A_{e_3}\|^2 = \|A_{e_4}\|^2$ and $g(A_{e_3}, A_{e_4}) = 0$, which implies that $\|\sigma\|^2\|\xi\|^2 = 2\|A_\xi\|^2$. So, we have proved that $\mathcal{Q}(\xi) \geq 0$ for any $\xi \in \mathfrak{X}^{\perp}(M)$ and hence M is stable.

The proof for the case $\frac{\rho}{6} - W^- \leq 0$ is completely analogous. This proves 8. □

From now on we assume that the oriented manifold $(\overline{M}, g)$ is *Einstein and self-dual*, i.e., $W^- = 0$, and that $\Phi : M \to \overline{M}$ is a *superminimal immersion with negative spin* of an oriented surface M. In this case the Jacobi operator of the immersion Φ can be written in a very nice form.

LEMMA 3.47. *Let $\Phi : M \to \overline{M}$ be a superminimal immersion with negative spin of an oriented surface into an oriented Einstein self-dual four manifold $\overline{M}$. Then, the Jacobi operator of Φ is given by*

$$\mathcal{L}\xi = \Delta^{\perp}\xi - K\xi + \frac{\rho}{4}\xi, \quad \forall \xi \in \mathfrak{X}^{\perp}(M),$$

where K is the Gauss curvature of M and ρ is the scalar curvature of $\overline{M}$.

PROOF. If $\{e_1, e_2, e_3, e_4\}$ is an oriented orthonormal local reference on $\Phi^*T\overline{M}$, with $\{e_1, e_2\}$ an oriented orthonormal reference on TM, then as $\overline{M}$ is Einstein, from (3.18) and Lemma 3.41,1 it follows that

$$\overline{R}(e_1, e_2, e_2, e_1) + \overline{R}(e_1, e_2, e_3, e_4) = \overline{R}(e_3, e_4, e_4, e_3) + \overline{R}(e_1, e_2, e_3, e_4) = \frac{\rho}{12}. \tag{3.31}$$

Since $\overline{M}$ is Einstein, we have that $g(\mathcal{R}\xi, J^-\xi) = 0$ and

$$g(\mathcal{R}\xi, \xi) = \left(\frac{\rho}{4} - \overline{R}(e_3, e_4, e_4, e_3)\right)\|\xi\|^2,$$

for any $\xi \in \mathfrak{X}^{\perp}(M)$. So from (3.31) we obtain

$$\mathcal{R}\xi = \frac{\rho}{4}\xi - \overline{R}(e_1, e_2, e_2, e_1)\xi, \quad \forall \xi \in \mathfrak{X}^{\perp}(M). \tag{3.32}$$

Since Φ is a superminimal immersion with negative spin, from (3.24) we also get that $g(\mathcal{A}\xi, J^-\xi) = 0$ and that

$$\mathcal{A}\xi = \frac{\|\sigma\|^2}{2}\xi, \quad \forall \xi \in \mathfrak{X}^{\perp}(M). \tag{3.33}$$

Using the Gauss equation of Φ, $K = \overline{R}(e_1, e_2, e_2, e_1) - \frac{\|\sigma\|^2}{2}$, together with (3.32) and (3.33), we get that

$$\mathcal{L}\xi = \Delta^{\perp}\xi + \mathcal{A}\xi + \mathcal{R}\xi = \Delta^{\perp}\xi - K\xi + \frac{\rho}{4}\xi, \quad \forall \xi \in \mathfrak{X}^{\perp}(M).$$

□

A direct consequence of this expression obtained for the Jacobi operator is that this operator commutes with the almost Hermitian structure J^-. Thus, all its eigenspaces are invariant with respect to J^- and hence *the multiplicity of all its eigenvalues is even*. In particular, $\operatorname{Ind}(\Phi)$ and $\operatorname{Nul}(\Phi)$ are even. We begin by proving the following result:

THEOREM 3.48 ([**MU97**]). *Let $\Phi : M \to \overline{M}$ be a superminimal immersion with negative spin of a compact orientable surface M of genus h into a self-dual Einstein 4-dimensional manifold M with scalar curvature ρ and let λ_1 be the first eigenvalue of the Jacobi operator $\mathcal{L}$ of Φ. Then $\lambda_1 \geq -\rho/6$. In addition,*

(1) *Suppose that $\rho = 0$. If $h \geq 2$, then $\lambda_1 = 0$ with multiplicity $2h$ or $2h-2$. If $h = 1$, then $\lambda_1 = 0$ with multiplicity 2 or $\lambda_1 > 0$ depending on whether the normal bundle N is trivial or not. If $h = 0$, then $\lambda_1 > 0$.*
(2) *If $\rho > 0$, then either $h = 0$ and Φ is a stable immersion with $\operatorname{Area}(M) = \frac{24\pi}{\rho}$ or $\lambda_1 = -\frac{\rho}{6}$ with multiplicity $\frac{\rho}{12\pi}\operatorname{Area}(M) + 2h - 2$.*

REMARK 3.49. We observe that if $\rho \leq 0$, then M is stable because the first eigenvalue of $\mathcal{L}$ is non-negative. This result was proved in Theorem 3.46,8 because $\frac{\rho}{6} - W^- = \frac{\rho}{6} \leq 0$.

PROOF. Let ξ be any normal section to Φ. Then from (3.28), (3.25), (3.32) and considering (3.31), we get

$$\mathcal{Q}(\xi) = \int_M \left\{ \frac{1}{2}\|D^-\xi\|^2 + \frac{\|\sigma\|^2}{2}\|\xi\|^2 - \|A_\xi\|^2 - \frac{\rho}{6}\|\xi\|^2 \right\} dv.$$

Since Φ is a superminimal immersion with negative spin, then $2\|A_\xi\|^2 = \|\sigma\|^2\|\xi\|^2$ (see proof of Theorem 3.46,8) and so, the last equation becomes

$$\mathcal{Q}(\xi) = \int_M \left\{ \frac{1}{2}\|D^-\xi\|^2 - \frac{\rho}{6}\|\xi\|^2 \right\} dv \geq -\frac{\rho}{6}\int_M \|\xi\|^2 dv.$$

So, the first eigenvalue of $\mathcal{L}$ is greater than or equal to $-\frac{\rho}{6}$ and it is exactly this quantity if and only if $\dim H^0(M, \mathcal{O}(N^-))$ is positive (see (3.22)). Now the Riemann-Roch theorem (Theorem 1.24) together with $\deg(N^-) = \chi^\perp$ tell us that

$$\dim H^0(M, \mathcal{O}(N^-)) = 1 - h + \chi^\perp + \dim H^0(M, \mathcal{O}(\kappa \otimes (N^-)^*)). \tag{3.34}$$

But from (3.31) and considering (3.25) and the Gauss equation of Φ (2.2), we get that $K + K^\perp = \frac{\rho}{12}$, and hence

$$\frac{\rho}{24\pi}\operatorname{Area}(M) = 2(1-h) + \chi^\perp.$$

Then, (3.34) becomes

$$\dim H^0(M, \mathcal{O}(N^-)) = h - 1 + \frac{\rho}{24\pi}\operatorname{Area}(M) + \dim H^0(M, \mathcal{O}(\kappa \otimes (N^-)^*)). \tag{3.35}$$

But we have $\deg(\kappa \otimes (N^-)^*) = \deg(\kappa) - \deg(N^-) = -\frac{\rho}{24\pi}\operatorname{Area}(M)$.

So if $\rho = 0$, then $\dim H^0(M, \mathcal{O}(\kappa \otimes (N^-)^*))$ is either one or zero, depending on whether the line bundle $\kappa \otimes (N^-)^*$ is trivial or not. With (3.35), our statement 1 is proved.

If ρ is positive, then $\kappa \otimes (N^-)^*$ is a negative line bundle and that has no global sections on M. So in this case the formula (3.35) gives us

$$\dim H^0(M, \mathcal{O}(N^-)) = \frac{\rho}{24\pi}\operatorname{Area}(M) + h - 1.$$

From here, the only case where $\dim H^0(M,\mathcal{O}(N^-))$ is zero corresponds to $h=0$ and $\operatorname{Area}(M)=\frac{24\pi}{\rho}$. In this case, Theorem 3.50 below tells us that the immersion is stable. □

THEOREM 3.50 ([**MU97**]). *Let $\Phi : M \to \overline{M}$ be a superminimal immersion with negative spin of an oriented compact surface M of genus h into a self-dual Einstein oriented 4-dimensional manifold with positive scalar curvature ρ. Then,*

$$\operatorname{Ind}(M)=\frac{\rho}{12\pi}\operatorname{Area}(M)+2h-2 \quad \textit{and} \quad \operatorname{Nul}(M)\geq\frac{\rho}{12\pi}\operatorname{Area}(M)+2-2h.$$

Furthermore, if $\rho\operatorname{Area}(M)>48\pi(h-1)$, then we have the equality

$$\operatorname{Nul}(M)=\frac{\rho}{12\pi}\operatorname{Area}(M)+2-2h.$$

REMARK 3.51. The bound given in Theorem 3.46,2, applied to this case $\left(\frac{\rho}{6}-W^-=\frac{\rho}{6}>0\right)$, says that

$$\operatorname{Ind}(M)\geq\frac{\rho}{24\pi}\operatorname{Area}(M)+h-1.$$

PROOF. Let $\xi\in\mathfrak{X}^\perp(M)$ be an eigensection of the Jacobi operator $\mathcal{L}$ corresponding to the eigenvalue λ, i.e., $\mathcal{L}\xi+\lambda\xi=0$. From Lemma 3.47 and as $K+K^\perp=\frac{\rho}{12}$, the Jacobi operator is given by $\mathcal{L}=\Delta^\perp+K^\perp+\frac{\rho}{6}$. Thus, the normal vector field ξ satisfies

$$(\Delta^\perp+K^\perp)\xi=-\left(\lambda+\frac{\rho}{6}\right)\xi. \tag{3.36}$$

We associate to the normal field $\xi\in\mathfrak{X}^\perp(M)$ a 1-form α_ξ on M, valued in the normal bundle $T^\perp M$, by the relation

$$\alpha_\xi(v)=D^-_v\xi=\nabla^\perp_{J^-v}\xi-J^-\nabla^\perp_v\xi,\quad \forall v\in TM. \tag{3.37}$$

We define the codifferential of α_ξ, as the normal vector field $\delta\alpha_\xi$ given by

$$\delta\alpha_\xi=\sum_{i=1}^2(\nabla^\perp_{e_i}\alpha_\xi)e_i=\sum_{i=1}^2\{\nabla^\perp_{e_i}\alpha_\xi(e_i)-\alpha_\xi(\nabla_{e_i}e_i)\},$$

where $\{e_1,e_2\}$ is a local orthonormal reference on TM. Then, using (3.23) and (3.36) we get that

$$\begin{aligned}\delta\alpha_\xi&=\sum_{i=1}^2\{\nabla^\perp_{e_i}\nabla^\perp_{J^-e_i}\xi-J^-\nabla^\perp_{e_i}\nabla^\perp_{e_i}\xi-\nabla^\perp_{J^-\nabla_{e_i}e_i}\xi+J^-\nabla^\perp_{\nabla_{e_i}e_i}\xi\}\\&=-J^-\Delta^\perp\xi-K^\perp J^-\xi=\left(\lambda+\frac{\rho}{6}\right)J^-\xi.\end{aligned} \tag{3.38}$$

Now we associate to ξ a bilinear form T_ξ on M, valued in the normal bundle $T^\perp M$, defined by

$$T_\xi(v,w)=(\nabla^\perp_{J^-v}\alpha_\xi)J^-w-(\nabla^\perp_v\alpha_\xi)w. \tag{3.39}$$

It is clear that $T_\xi(J^-v,J^-w)=-T_\xi(v,w)$ and so, T_ξ is a trace-free symmetric tensor. We define the codifferential of T_ξ, as the 1-form δT_ξ on M valued in the normal bundle given by

$$\delta T_\xi(V)=\sum_{i=1}^2(\nabla^\perp_{e_i}T_\xi)(e_i,V)=\sum_{i=1}^2\{\nabla^\perp_{e_i}T_\xi(e_i,V)-T_\xi(\nabla_{e_i}e_i,V)-T_\xi(e_i,\nabla_{e_i}V)\},$$

where $\{e_1, e_2\}$ is a local orthonormal reference on TM. Using the Ricci identity, the symmetry of T_ξ and definition (3.39), we have

$$\begin{aligned}\delta T_\xi(v) &= \sum_{i=1}^{2}(\nabla^\perp_{e_i}T_\xi)(e_i,v) = \sum_{i=1}^{2}(\nabla^\perp_{e_i}T_\xi)(v,e_i)\\ &= \sum_{i=1}^{2}\{(\nabla^\perp_{e_i}\nabla^\perp_{J^-v}\alpha_\xi)J^-e_i - (\nabla^\perp_{e_i}\nabla^\perp_{v}\alpha_\xi)e_i\}\\ &= \sum_{i=1}^{2}\{(\nabla^\perp_{J^-v}\nabla^\perp_{e_i}\alpha_\xi)J^-e_i + R^\perp(e_i,J^-v)\alpha_\xi(J^-e_i) - \alpha_\xi(R(e_i,J^-v)J^-e_i)\\ &\quad - (\nabla^\perp_{v}\nabla^\perp_{e_i}\alpha_\xi)e_i - R^\perp(e_i,v)\alpha_\xi(e_i) + \alpha_\xi(R(e_i,v)e_i)\}.\end{aligned}$$

Now, using that $T_\xi(J^- -, J^- -) = -T_\xi(-,-)$ and the definitions of Gauss and normal curvature, we get

$$\delta T_\xi(v) = -\sum_{i=1}^{2}\{J^-(\nabla^\perp_{J^-v}\nabla^\perp_{e_i}\alpha_\xi)e_i + (\nabla^\perp_{v}\nabla^\perp_{e_i}\alpha_\xi)e_i\} - 2(K+K^\perp)\alpha_\xi(v).$$

But

$$\begin{aligned}\sum_{i=1}^{2}(\nabla^\perp_{v}\nabla^\perp_{e_i}\alpha_\xi)e_i &= \nabla^\perp_v\delta\alpha_\xi - \sum_{i=1}^{2}\{(\nabla^\perp_{\nabla_v e_i}\alpha_\xi)e_i + (\nabla^\perp_{e_i}\alpha_\xi)\nabla_v e_i\}\\ &= \left(\lambda+\frac{\rho}{6}\right)\nabla^\perp_v J^-\xi - \sum_{i,j=1}^{2} g(\nabla_v e_i, e_j)\{(\nabla^\perp_{e_j}\alpha_\xi)e_i + (\nabla^\perp_{e_i}\alpha_\xi)e_j\}\\ &= \left(\lambda+\frac{\rho}{6}\right)\nabla^\perp_v J^-\xi.\end{aligned}$$

So, from (3.37) and as $K+K^\perp = \frac{\rho}{12}$, we finally get $\delta T_\xi = \lambda\alpha_\xi$.

Now we define a 1-form γ_ξ on M by

$$\gamma_\xi(v) = \sum_{i=1}^{2} g(\alpha_\xi(e_i), T_\xi(v,e_i)), \quad \forall v\in TM.$$

Then, using that $T_\xi(J^- -, J^- -) = -T_\xi(-,-)$, (3.39) and that $\delta T_\xi = \lambda\alpha_\xi$, we get that the codifferential of γ_ξ is given by

$$\begin{aligned}\delta\gamma_\xi &= \sum_{j=1}^{2}(\nabla_{e_j}\gamma_\xi)e_j = \sum_{i,j=1}^{2}\{g((\nabla^\perp_{e_j}\alpha_\xi)e_i, T_\xi(e_j,e_i)) + g(\alpha_\xi(e_i),(\nabla^\perp_{e_j}T_\xi)(e_j,e_i))\}\\ &\quad + \sum_{i,j=1}^{2}\{g(\alpha_\xi(\nabla_{e_j}e_i), T_\xi(e_j,e_i)) + g(\alpha_\xi(e_i), T_\xi(e_j,\nabla_{e_j}e_i))\}\\ &= \frac{1}{2}\sum_{i,j=1}^{2}\{g((\nabla^\perp_{J^-e_j}\alpha_\xi)J^-e_i, T_\xi(J^-e_j,J^-e_i)) + g((\nabla^\perp_{e_j}\alpha_\xi)e_i, T_\xi(e_j,e_i))\}\\ &\quad + \lambda\|\alpha_\xi\|^2 + \sum_{i,j,k=1}^{2} g(\nabla_{e_i}e_j,e_k)\{g(\alpha_\xi(e_k),T_\xi(e_j,e_i)) + g(\alpha_\xi(e_i),T_\xi(e_j,e_k))\}\\ &= -\frac{1}{2}\|T_\xi\|^2 + \lambda\|\alpha_\xi\|^2.\end{aligned}$$

Now Theorem 1.15 gives

$$\int_M \|T_\xi\|^2 dv = 2\lambda \int_M \|\alpha_\xi\|^2 dv.$$

Thus, in the case where the eigenvalue λ corresponding to the eigenfield ξ is negative, we have shown that $\alpha_\xi = 0$, i.e., the field ξ defines a holomorphic section of the line bundle N^-. Then, the proof of the Theorem 3.48,2, tells us that $\lambda = -\frac{\rho}{6}$ and

$$\operatorname{Ind}(M) = 2 \dim H^0(M, \mathcal{O}(N^-)) = \frac{\rho}{12\pi}\operatorname{Area}(M) + 2h - 2.$$

On the other hand, if $\lambda = 0$, that is $\mathcal{L}\xi = 0$, we have from the same integral formula that $T_\xi = 0$.

CLAIM. $T_\xi = 0$ implies that α_ξ defines a holomorphic section of the holomorphic line bundle $\kappa^* \otimes N^-$.

We define a section of the holomorphic line bundle $\kappa^* \otimes N^-$, which will also be denoted by α_ξ, by

$$\alpha_\xi(v + iJ^- v) = \alpha_\xi(v) + i\alpha_\xi(J^- v) = \alpha_\xi(v) - iJ^- \alpha_\xi(v),$$

where we have used (3.37) and the identification between κ and the conjugate of κ^*. Considering (3.39) and that $T_\xi = 0$, we have

$$\begin{aligned}(\nabla^\perp_{e_1+ie_2}\alpha_\xi)(V + iJ^- V) &= \{(\nabla^\perp_{e_1}\alpha_\xi)V - (\nabla^\perp_{J^- e_1}\alpha_\xi)J^- V\} \\ &\quad + i\{(\nabla^\perp_{e_1}\alpha_\xi)J^- V + (\nabla^\perp_{J^- e_1}\alpha_\xi)V\} \\ &= -T_\xi(e_1, V) - iT_\xi(Je_1, v) = 0,\end{aligned}$$

which means that α_ξ is a holomorphic section of the holomorphic line bundle $\kappa^* \otimes N^-$.

Thus, if $\mathcal{J}(M) = \{\xi \in \mathfrak{X}^\perp(M) \mid \mathcal{L}\xi = 0\}$ is the space of Jacobi fields, we have defined a complex linear map

$$\begin{aligned}\alpha : \mathcal{J}(M) &\to H^0(M, \mathcal{O}(\kappa^* \otimes N^-)) \\ \xi &\mapsto \alpha_\xi\end{aligned}$$

from the complex linear space $\mathcal{J}(M)$ $\big((a+ib)\xi = a\xi + bJ^-\xi\big)$ to the complex linear space of holomorphic sections of $\kappa^* \otimes N^-$.

Our goal is to see that this map α is an isomorphism. Indeed, if $\alpha_\xi = 0$ for some $\xi \in \mathcal{J}(M)$, then from (3.37), $\xi \in H^0(M, \mathcal{O}(N^-))$. But in the proof of Theorem 3.48, we have shown that all the holomorphic sections of N^- are eigensections of $\mathcal{L}$ corresponding to the eigenvalue $-\frac{\rho}{6}$. We conclude that $\xi = 0$, and so, α is a monomorphism. Now, take a holomorphic section $\beta \in H^0(M, \mathcal{O}(\kappa^* \otimes N^-))$ and, following the argument above, identify β with a 1-form on M valued in the normal bundle satisfying $\beta \circ J^- = -J^- \circ \beta$ and with the corresponding $T = 0$ (see (3.39)). Define the normal field ξ to be $-\frac{6}{\rho}J^-\delta\beta$. Using (3.38), we have

$$\mathcal{L}\xi = \Delta^\perp\xi + K^\perp\xi + \frac{\rho}{6}\xi = J^-\delta\beta + \frac{\rho}{6}\xi = 0,$$

and so, $\xi \in \mathcal{J}(M)$. Using definition (3.37), we have

$$\alpha_\xi(v) = -\frac{6}{\rho}\sum_{i=1}^{2}\{J^-(\nabla^2\beta)(J^- v, e_i, e_i) + (\nabla^2\beta)(v, e_i, e_i)\}$$

for any $v \in TM$. Now, from the Ricci identity, the holomorphicity of β, and the fact that $K + K^\perp = \frac{\rho}{12}$, we get

$$\begin{aligned}\alpha_\xi(v) &= \frac{6}{\rho}\sum_{i=1}^{2}\{J^-(\nabla^2\beta)(e_i, J^-v, J^-e_i) - (\nabla^2\beta)(e_i, v, e_i) + R^\perp(J^-v, e_i)\beta(Je_i)\\ &\quad - \beta(R(J^-v, e_i)J^-e_i) - R^\perp(v, e_i)\beta(e_i) + \beta(R(v, e_i)e_i)\}\\ &= \frac{12}{\rho}(K + K^\perp)\beta(v) = \beta(v).\end{aligned}$$

So the map α is surjective and we get the isomorphism

$$\mathcal{J}(M) \equiv H^0(M, \mathcal{O}(\kappa^* \otimes N^-)).$$

Thus, the Riemann-Roch theorem (Theorem 1.24) allows us to prove that the nullity of M is given by

$$\mathrm{Nul}(M) = \frac{\rho}{12\pi}\mathrm{Area}(M) + 2 - 2h + 2\dim H^0(M, \mathcal{O}(\kappa^2 \otimes (N^-)^*)).$$

The inequality about the nullity follows directly from this. Also, as $\frac{\rho}{24\pi}\mathrm{Area}(M) = 2(1-h) + \chi^\perp$ (see proof of Theorem 3.48),

$$\deg(\kappa^2 \otimes (N^-)^*) = -\frac{\rho}{24\pi}\mathrm{Area}(M) + 2h - 2,$$

and so, the holomorphic line bundle $\kappa^2 \otimes (N^-)^*$ is negative if $\rho\mathrm{Area}(M) > 48\pi(h - 1)$, and in this case we have the equality. $\square$

CHAPTER 4

Index of Minimal Submanifolds of Euclidean Spaces

This chapter is devoted to the study of the second variation of the volume of minimal submanifolds of Euclidean space, with special emphasis on the hypersurfaces. The case of surfaces is treated in the first two sections, and the case where the codimension is greater than one is treated in the last section. However, in Chapter 6 we will study a special class of stable minimal submanifolds of complex Euclidean space, the so-called special Lagrangian submanifolds, which are also calibrated submanifolds for the special Lagrangian calibration.

If $\Phi : M^n \to (\mathbb{R}^m, \langle , \rangle)$ is a minimal immersion, then by Definition 2.10, the Jacobi operator of Φ is given by

$$\mathcal{L}\xi = \Delta^{\perp}\xi + \mathcal{A}\xi, \quad \forall \xi \in \mathfrak{X}^{\perp}(M).$$

If M is a two-sided minimal hypersurface, from (2.17), the Jacobi operator is the Schrödinger operator

$$\mathcal{L}f = \Delta f + \|A\|^2 f, \quad \forall f \in C^{\infty}(M),$$

where A is the Weingarten endomorphism associated to an orientation N on M.

It is worth recalling an important property of minimal submanifolds of Euclidean space. It is clear that

$$\Delta\Phi = n\vec{H} = 0,$$

i.e., the coordinate functions are harmonic, which implies the following result.

PROPOSITION 4.1. *If M is a minimal submanifold of $\mathbb{R}^m$, then M is not compact.*

4.1. Orientable minimal surfaces in $\mathbb{R}^3$

Let $\Phi : M \to \mathbb{R}^3$ be a minimal immersion of an orientable surface M. Since $\mathbb{R}^3$ is orientable, Φ is two-sided (Lemma 2.2). The Gauss equation of Φ, (2.2), gives that the Gauss curvature of M is given by $K = -\|A\|^2/2$, and thus $K \leq 0$. Also, if $N : M \to \mathbb{S}^2$ is the Gauss map of Φ, then it is easy to check that N is a holomorphic map, considering the natural Riemann surface structure on M, and that $\|\nabla N\|^2 = -2K$. Thus the Jacobi operator of Φ can also be written as

$$\mathcal{L} = \Delta + \|A\|^2 = \Delta - 2K = \Delta + \|\nabla N\|^2. \tag{4.1}$$

In this section interesting examples of minimal surfaces of $\mathbb{R}^3$ will appear in relation to the behavior of their indices. We have chosen not to introduce them in detail and we will always refer to the texts of T.H. Colding and W.P. Minicozzi, [**CM11**], F.J. López and F. Martín, [**LM99**] and W. Meeks and J. Pérez, [**MP12**], where the main properties and historical references of these examples can be found.

The first result about the index of orientable minimal surfaces of $\mathbb{R}^3$ is a consequence of Corollaries 3.19 and 3.15.

THEOREM 4.2 ([**dCP79, Pog81, FCS80**]). *Let $\Phi : M \to \mathbb{R}^3$ be a minimal immersion of an orientable complete surface M in $\mathbb{R}^3$.*

(1) *If M is stable, then $\Phi(M)$ is a plane of $\mathbb{R}^3$.*
(2) *If $\mathrm{Ind}(M)$ is finite, then M is conformally equivalent to a compact Riemann surface with a finite number of points removed and*

$$\int_M \|A\|^2 \,\mathrm{d}v < \infty.$$

REMARK 4.3. The proof of the above theorem given here is that of D. Fischer-Colbrie and R. Schoen, [**FCS80**]. The proofs of M. do Carmo and C.K. Peng, [**dCP79**], and of A.V. Pogorelov, [**Pog81**], are quite different and are not a consequence of a more general theory. M. do Carmo and C.K. Peng, using the uniformization theorem (Theorem 1.22), prove that if the surface is conformally equivalent to the disk, then it is unstable, and if the surface is conformally equivalent to $\mathbb{C}$ and stable, then it is a plane. In contrast, Pogorelov's proof goes on to show that stability ensures that geodesic balls of the surface have area quadratic growth.

The integrability of $\|A\|^2$ is equivalent to $\int_M |K| \,\mathrm{d}v < \infty$. These complete minimal surfaces of $\mathbb{R}^3$ are called of **finite total curvature** and they have been widely studied. In the next result we have summarized some known results about these minimal surfaces which will be used along this section.

THEOREM 4.4 ([**Oss69**]). *Let $\Phi : M \to \mathbb{R}^3$ be a minimal immersion of an orientable complete surface M in $\mathbb{R}^3$ with finite total curvature. Then*

(1) *There exists a compact Riemann surface Σ and a conformal diffeomorphism $F : M \to \Sigma \setminus \{p_1, \dots, p_n\}$.*
(2) *There exists a holomorphic map $\phi : \Sigma \to \mathbb{S}^2$ such that $N_p = \phi_{F(p)}$ for any $p \in M$, being N the Gauss map of M.*
(3) *If d is the degree of ϕ and $\widehat{g}$ is an arbitrary conformal metric on Σ, then*

$$4\pi d = \int_\Sigma \frac{\|\widehat{\nabla}\phi\|^2_{\widehat{g}}}{2} \,\mathrm{d}\widehat{v} = \int_M \frac{\|\nabla N\|^2}{2} \,\mathrm{d}v = -\int_M K \,\mathrm{d}v.$$

*$\int_M K \,\mathrm{d}v$ will be called the **total curvature** of M.*
(4) *$d = 0$ if and only if $\Phi(M)$ is a plane.*
(5) *$d = 1$ if and only if $\Phi(M)$ is either a catenoid or the Enneper surface (see Figure 1).*

*The points $\{p_1, \dots, p_n\}$ represent the ends of M. Also, the **genus** of M is the genus of the compact Riemann surface Σ.*

To compute the index of some remarkable minimal surfaces of $\mathbb{R}^3$ as well as to understand some of their properties, we need to introduce the Weiertrass representation of a minimal surface of $\mathbb{R}^3$.

If $\Phi : M \to \mathbb{R}^3$ is a minimal immersion of an oriented surface M, we consider the associated structure of Riemann surface on M and z will denote a complex parameter in M. Since Φ is minimal, $\Phi_{z\bar{z}} = 0$, and hence $\omega(z) = \Phi_z \mathrm{d}z$ is a global holomorphic 1-form on M. If $\Phi = (\Phi_1, \Phi_2, \Phi_3)$, then we write $\omega = (\omega_1, \omega_2, \omega_3)$,

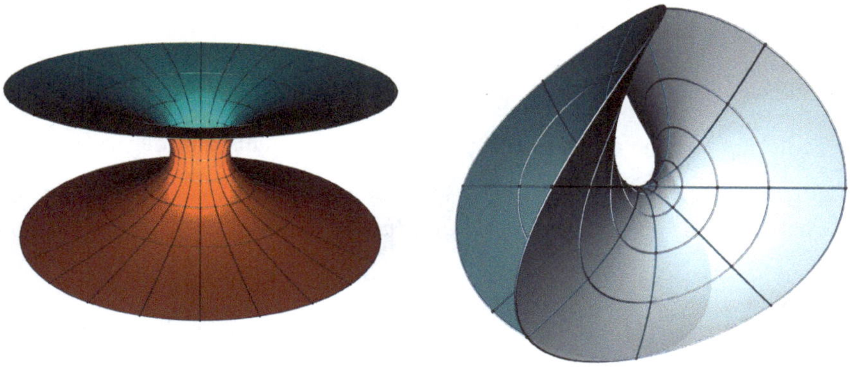

FIGURE 1. Catenoid surface (left), and Enneper surface (right). Creative Commons (cc)(i)(=) Matthias Weber, `http://minimalsurfaces.blog`

with $\omega_i(z) = (\Phi_i)_z \mathrm{d}z$, $i = 1, 2, 3$. Since Φ is a conformal immersion, $\langle \Phi_z, \Phi_z \rangle = 0$ and so

$$\omega_1^2 + \omega_2^2 + \omega_3^2 \equiv 0.$$

The induced metric $\mathrm{d}s^2$ on M is given by $\mathrm{d}s^2 = (\|\omega_1\|^2 + \|\omega_2\|^2 + \|\omega_3\|^2)\|\mathrm{d}z\|^2$, and since Φ is an immersion, $\|\omega_1\|^2 + \|\omega_2\|^2 + \|\omega_3\|^2$ has no zeros.

Under these conditions, the immersion Φ is given by

$$\Phi(z) - \Phi(z_0) = \Re \int_\gamma \omega,$$

where γ is any smooth curve in M from z_0 to z. If

$$g = \frac{\omega_3}{\omega_1 - i\omega_2}, \qquad \eta = \omega_3,$$

then g is a meromorphic map and η is a holomorphic 1-form with zeros exactly at the poles of g. Now, ω and the Gauss map N of Φ are given by

(4.2)
$$\omega = \left(\frac{1}{2}(g^{-1} - g), \frac{i}{2}(g^{-1} + g), 1\right)\eta, \quad N = \left(\frac{2\Re(g)}{1 + \|g\|^2}, \frac{2\Im(g)}{1 + \|g\|^2}), \frac{\|g\|^2 - 1}{\|g\|^2 + 1}\right).$$

The second formula means that the meromorphic function g is the stereographic projection, from the point $(0, 0, 1)$, of the Gauss map N of Φ.

DEFINITION 4.5. If $\Phi : M \to \mathbb{R}^3$ is a minimal immersion of an oriented surface M, then (M, ω) or (M, g, η) is called the **Weierstrass representation** of Φ. The induced metric and the Gauss curvature K of M are given by

$$\mathrm{d}s^2 = \frac{1}{4}\|\eta\|^2 \left(\|g\|^{-1} + \|g\|\right)^2 \|\mathrm{d}z\|^2, \qquad K = -\left(\frac{4\|g\|\|g'\|}{\|\eta\|(1 + \|g\|^2)^2}\right)^2. \tag{4.3}$$

Now suppose that $\Phi : M \to \mathbb{R}^3$ is a minimal immersion of a complete orientable surface M of finite total curvature. From Theorem 4.4, $M = \Sigma \setminus \{p_1, \ldots, p_n\}$ (we identify M with $F(M)$), where Σ is a compact Riemann surface. The Gauss map N extends to a holomorphic map ϕ on Σ, and the Weiertrass data $\{g, \eta, \omega\}$ extends meromorphically to Σ.

The geometry of the ends $\{p_1, \ldots, p_n\}$ is controlled by the order of the poles of ω at p_i. L. Jorge and W. Meeks, [**JM83**], defined the **multiplicity** d_i of the end p_i as

$$d_i + 1 = \max\{\operatorname{ord}(\omega_j, p_i),\ j = 1, 2, 3\},$$

where $\operatorname{ord}(\omega_j, p_i)$ is the order of the pole of ω_j at p_i. This number satisfies $d_i \geq 1$.

THEOREM 4.6 ([**JM83**]). *Let $\Phi : M \to \mathbb{R}^3$ be a minimal immersion of a complete orientable surface of finite total curvature of genus h and n ends $\{p_1, \ldots, p_n\}$. Then*

(1) *Φ is a proper map.*
(2) *For $R > 0$, if $\Gamma_R = \Phi(M) \cap \mathbb{S}^2(R)$, then $\{\frac{1}{R} q \mid q \in \Gamma_R\}$ consist of n closed curves $\{\Gamma_1, \ldots, \Gamma_n\}$ in $\mathbb{S}^2$ which converge smoothly as R goes to infinity to great circles $\{\gamma_1, \ldots, \gamma_n\}$ of $\mathbb{S}^2$, with multiplicities $\{d_1, \ldots, d_n\}$.*
(3) *The total curvature of Φ is given by*

$$\int_M K \, \mathrm{d}v = -4\pi\Big(h - 1 + \frac{n}{2} + \frac{1}{2}\sum_{i=1}^{n} d_i\Big).$$

We say that the end p_i is embedded if there exists a neighborhood D_i of p_i in Σ such that $\Phi_{|D_i \setminus \{p_i\}} : D_i \setminus \{p_i\} \to \mathbb{R}^3$ is an embedding. This is equivalent to the fact that $d_i = 1$.

The Enneper surface (see Figure 1) is a complete orientable minimal surface of finite total curvature with an end of multiplicity three. It is a basic example, because many minimal surfaces have ends Enneper type.

The following result by D. Fischer-Colbrie characterizes the complete orientable minimal surfaces of $\mathbb{R}^3$ of finite index.

THEOREM 4.7 ([**FC85**]). *Let $\Phi : M \to \mathbb{R}^3$ be a minimal immersion of an orientable complete surface M in $\mathbb{R}^3$. Then $\operatorname{Ind}(M)$ is finite if and only if M has finite total curvature, i.e., $\int_M |K| \, \mathrm{d}v < \infty$. In such case,*

$$\operatorname{Ind}(M) = \operatorname{Ind}(\Sigma, \phi),$$

where ϕ and Σ were given in Theorem 4.4 and $\operatorname{Ind}(\Sigma, \phi)$ was defined in Definition 3.7.

PROOF. If M has finite index, then Theorem 4.2 says that M has finite total curvature.

Conversely, if M has finite total curvature, then, from Theorem 4.4, there exists a conformal diffeomorphism $F : M \to \widetilde{\Sigma} = \Sigma \setminus \{p_1, \ldots, p_n\}$ and the Gauss map $N : M \to \mathbb{S}^2$ extends to a holomorphic map $\phi : \Sigma \to \mathbb{S}^2$ making commutative the following diagram:

$$\begin{array}{ccccc} M & \xrightarrow{F} & \widetilde{\Sigma} & \xrightarrow{i} & \Sigma \\ & {\scriptstyle N}\searrow & \downarrow{\scriptstyle \widetilde{\phi}} & \swarrow{\scriptstyle \phi} & \\ & & \mathbb{S}^2, & & \end{array}$$

where i is the inclusion and $\widetilde{\phi} = \phi_{|\widetilde{\Sigma}}$.

Let $\widetilde{g}$ be the metric on $\widetilde{\Sigma}$ that becomes $F : (M, g) \to (\widetilde{\Sigma}, \widetilde{g})$ in an isometry, and $\widehat{g}$ any conformal metric on Σ. It is clear that $\widetilde{g}$ and $\widehat{g}_{|\widetilde{\Sigma}}$ are conformal metrics.

From (4.1), the Jacobi operator of M is given by $\mathcal{L} = \Delta + \|\nabla N\|^2$. We define the Schrodinger operators $\widetilde{L}$ and $\widehat{L}$ on the Riemannian surfaces $(\widetilde{\Sigma}, \widetilde{g})$ and $(\Sigma, \widehat{g})$, respectively by

$$\widetilde{L} = \widetilde{\Delta} + \|\widetilde{\nabla}\widetilde{\phi}\|^2_{\widetilde{g}}, \qquad \widehat{L} = \widehat{\Delta} + \|\widehat{\nabla}\phi\|^2_{\widehat{g}}.$$

Since F is an isometry, we have that $\mathrm{Ind}(M) := \mathrm{Ind}(\mathcal{L}) = \mathrm{Ind}(\widetilde{L})$. Also, $\mathrm{Ind}(\widetilde{L}) = \mathrm{Ind}_{\widetilde{\Sigma}}(\widehat{L})$, because, using the conformal invariance of the Dirichlet integral (1.4), we have

$$\int_{\widetilde{\Sigma}} (\|\widetilde{\nabla} f\|^2_{\widetilde{g}} - \|\widetilde{\nabla}\widetilde{\phi}\|^2_{\widetilde{g}} f^2) \mathrm{d}\widetilde{v} = \int_{\widetilde{\Sigma}} (\|\widehat{\nabla} f\|^2_{\widehat{g}} - \|\widehat{\nabla}\phi\|^2_{\widehat{g}} f^2) \mathrm{d}\widehat{v}, \quad \forall f \in C_0^\infty(\widetilde{\Sigma}).$$

So $\mathrm{Ind}(M) = \mathrm{Ind}_{\widetilde{\Sigma}}(\widehat{L}) \leq \mathrm{Ind}(\widehat{L}) < \infty$, since Σ is compact. This proves the first part of the result.

Assuming that M has finite index, we have proved that $\mathrm{Ind}(M) = \mathrm{Ind}(\widetilde{L}) \leq \mathrm{Ind}(\widehat{L})$. Now we will prove that $\mathrm{Ind}(\widehat{L}) \leq \mathrm{Ind}(\widetilde{L})$.

First, for any $\epsilon > 0$ small enough, we construct a function $h^\epsilon \in H^{1,2}(\Sigma)$ such that

$$\lim_{\epsilon \to 0} \int_\Sigma \|\widehat{\nabla} h^\epsilon\|^2_{\widehat{g}} \mathrm{d}\widehat{v} = 0. \tag{4.4}$$

To do this, let $\eta^\epsilon \in H_0^{1,2}(\mathbb{R})$ be the function defined by

$$\eta^\epsilon(t) = \begin{cases} 0 & t < \epsilon^2, \\ 2 - \frac{\log t}{\log \epsilon} & t \in [\epsilon^2, \epsilon], \\ 1 & t > \epsilon. \end{cases}$$

As $\widetilde{\Sigma} = \Sigma \setminus \{p_1, \ldots, p_n\}$, let $z_i : U_i \to B(1) \subset \mathbb{C}$ be complex coordinates on Σ such that $p_i \in U_i$ and $U_i \cap U_j = \emptyset$ for any $1 \leq i, j \leq n$ with $i \neq j$. Then we define $h^\epsilon \in H^{1,2}(\Sigma)$ by

$$h^\epsilon = \begin{cases} 1 & \text{on } \Sigma - \cup_{i=1}^n U_i \\ \eta^\epsilon(\|z_i\|) & \text{on } U_i, \quad 1 \leq i \leq n. \end{cases}$$

For each U_i, we have $\widehat{g} = e^{2u_i}\|\mathrm{d}z_i\|^2$, and so from the conformal invariance of the Dirichlet integral (1.4), we have

$$\int_{U_i} \|\widehat{\nabla} h^\epsilon\|^2_{\widehat{g}} \mathrm{d}\widehat{v} = \int_{U_i} \|\nabla_0 h^\epsilon\|^2 \mathrm{d}z = \frac{1}{(\log \epsilon)^2} \int_{A_\epsilon} \frac{\mathrm{d}z}{\|z\|^2} = -\frac{2\pi}{\log \epsilon},$$

where $A_\epsilon = \{z_i \in U_i \mid \epsilon^2 \leq \|z_i\| \leq \epsilon\}$ and ∇_0 denotes the Levi-Civita connection of the metric $\|\mathrm{d}z\|^2$. Hence,

$$\lim_{\epsilon \to 0} \int_\Sigma \|\widehat{\nabla} h^\epsilon\|^2_{\widehat{g}} \mathrm{d}\widehat{v} = \lim_{\epsilon \to 0} \sum_{i=1}^n \int_{U_i} \|\widehat{\nabla} h^\epsilon\|^2_{\widehat{g}} \mathrm{d}\widehat{v} = \lim_{\epsilon \to 0} (-\frac{2\pi n}{\log \epsilon}) = 0.$$

Suppose that $\mathrm{Ind}(\widehat{L}) = k$ and let $\{f_1, \ldots, f_k\} \subset C^\infty(\Sigma)$ be eigenfunctions of $\widehat{L}$ corresponding to the negative eigenvalues $\lambda_1 < \lambda_2 \leq \cdots \leq \lambda_k < 0$.

For each small $\epsilon > 0$ we consider $g_i^\epsilon = h^\epsilon f_i \in H^{1,2}(\Sigma)$. Then,

$$\widehat{L} g_i^\epsilon = -\lambda_i g_i^\epsilon + f_i \widehat{\Delta} h^\epsilon + 2\widehat{g}(\widehat{\nabla} h^\epsilon, \widehat{\nabla} f_i),$$

and thus,

$$\widehat{\mathcal{Q}}(g_i^\epsilon, g_j^\epsilon) = -\int_\Sigma g_j^\epsilon \widehat{L} g_i^\epsilon \, \mathrm{d}\widehat{v} = \lambda_i \int_\Sigma g_i^\epsilon g_j^\epsilon \mathrm{d}\widehat{v} - \int_\Sigma (f_i f_j h^\epsilon \widehat{\Delta} h^\epsilon + 2 f_j h^\epsilon \widehat{g}(\widehat{\nabla} h^\epsilon, \widehat{\nabla} f_i) \mathrm{d}\widehat{v}.$$

But

$\widehat{\mathrm{div}}\,(f_i f_j h^\epsilon \widehat{\nabla} h^\epsilon) = f_i f_j h^\epsilon \widehat{\Delta} h^\epsilon + f_j h^\epsilon \widehat{g}(\widehat{\nabla} f_i, \widehat{\nabla} h^\epsilon) + f_i h^\epsilon \widehat{g}(\widehat{\nabla} f_j, \widehat{\nabla} h^\epsilon) + f_i f_j \|\widehat{\nabla} h^\epsilon\|^2_{\widehat{g}},$

and so, using that $\widehat{Q}$ is symmetric, we get

$$\widehat{\mathcal{Q}}(g_i^\epsilon, g_j^\epsilon) = \frac{\lambda_i + \lambda_j}{2} \int_\Sigma g_i^\epsilon g_j^\epsilon \, d\widehat{v} + \int_\Sigma f_i f_j \|\widehat{\nabla} h^\epsilon\|^2_{\widehat{g}} d\widehat{v}.$$

Using (4.4), the definition of h^ϵ and the fact that the functions f_i are bounded, we get that

$$\text{(4.5)} \quad \lim_{\epsilon \to 0} \widehat{\mathcal{Q}}(g_i^\epsilon) = \lambda_i \int_\Sigma f_i^2 d\widehat{v} < 0, \quad \lim_{\epsilon \to 0} \widehat{\mathcal{Q}}(g_i^\epsilon, g_j^\epsilon) = \lim_{\epsilon \to 0} \int_\Sigma f_i f_j \|\widehat{\nabla} h^\epsilon\|^2_{\widehat{g}} d\widehat{v} \leq 0,$$

when $i \neq j$.

If V is the subspace of $C^\infty(\Sigma)$ generated by $\{f_1, \ldots, f_k\}$ and V^ϵ is the subspace of $H^{1,2}(\Sigma)$ generated by $\{g_1^\epsilon, \ldots, g_k^\epsilon\}$, then $V^\epsilon = \{h^\epsilon f \mid f \in V\}$. From (4.5) we see that there exists ϵ_0 small enough so that $\widehat{\mathcal{Q}}$ is definite negative on V^{ϵ_0}. Now, if $W = \{(g^{\epsilon_0})_{|\widetilde{\Sigma}} \mid g^{\epsilon_0} \in V^{\epsilon_0}\}$, then we have that W is a k-dimensional subspace of $H_0^{1,2}(\widetilde{\Sigma})$ and that

$$\widetilde{\mathcal{Q}}(g^{\epsilon_0}_{|\widetilde{\Sigma}}) = \widehat{\mathcal{Q}}(g^{\epsilon_0}) < 0, \quad \forall\, g^{\epsilon_0}_{|\widetilde{\Sigma}} \in W.$$

Hence $\mathrm{Ind}(\widetilde{L}) \geq k = \mathrm{Ind}(\widehat{L})$, and hence $\mathrm{Ind}(M) = \mathrm{Ind}(\widetilde{L}) = \mathrm{Ind}(\widehat{L})$.

Finally, from Definition 3.7, it follows that $\mathrm{Ind}(M) = \mathrm{Ind}(\Sigma, \phi)$. □

Theorem 4.7, together with the results obtained in Section 3.2 about $\mathrm{Ind}(\Sigma, \phi)$, allow us to obtain interesting results about the index of complete orientable minimal surfaces of $\mathbb{R}^3$.

COROLLARY 4.8 ([**Cho90, MR91**]). *Let $\Phi : M \to \mathbb{R}^3$ be a minimal immersion of a complete orientable surface M of finite index. Then*

(1) *If M is stable, then $\Phi(M)$ is a plane of $\mathbb{R}^3$.*
(2) *If the genus of M is $h = 0$, then* $\mathrm{Ind}(M) \neq 2$.
(3) *If the total curvature of M is -8π, then* $\mathrm{Ind}(M) = 3$.
(4) *If M is a Jorge-Meeks n-noid, with $n \geq 2$* ([**JM83**]) *(see Figure 2), then* $\mathrm{Ind}(M) = 2n - 3$.

REMARK 4.9. The complete minimal orientable surfaces of $\mathbb{R}^3$ with finite total curvature -8π were classified by F.J. López in [**Ló92**]. In his classification he showed five families of such surfaces with genus zero (see Figure 3, left) and one example (the Chen-Gackstatter surface [**CG82**]) with genus one (see Figure 3, right). Among the examples of genus zero, the Richmond surface is one of them. Its index was computed by A. Tuzhilin in [**Tuz91**].

Also, the Jorge-Meeks 2-noid is the catenoid and the Jorge-Meeks 3-noid is the trinoid, which is one of the examples of the above five families (see Figure 2).

PROOF. From Theorem 4.7, we have $\mathrm{Ind}(M) = \mathrm{Ind}(\Sigma, \phi)$, where Σ is the compactification of M and $\phi : \Sigma \to \mathbb{S}^2$ the holomorphic map given in Theorem 4.4.

If M is a stable complete minimal orientable surface of $\mathbb{R}^3$, then $\mathrm{Ind}(\Sigma, \phi) = 0$ and so, Corollary 3.9 gives that ϕ is constant. This implies that the Gauss map N is constant and hence, M is totally geodesic and thus a plane.

Point 2 follows from the Corollary 3.11.

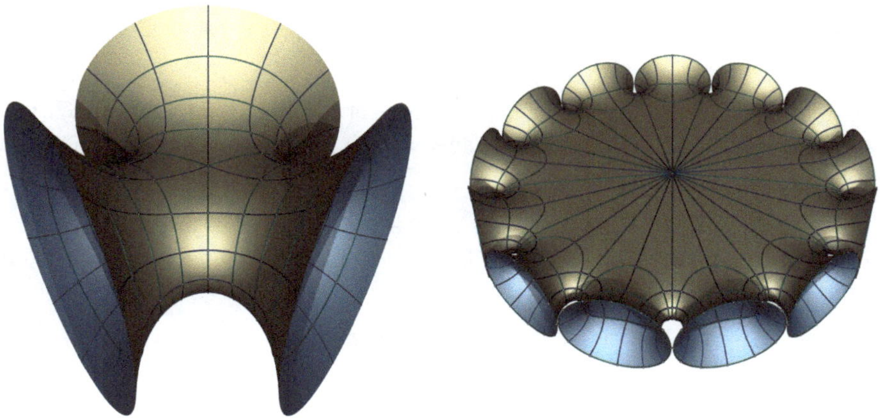

FIGURE 2. Trinoid surface (left), and 13-noid surface (right). Creative Commons 🅭🅯🄎 Matthias Weber, `http://minimalsurfaces.blog`

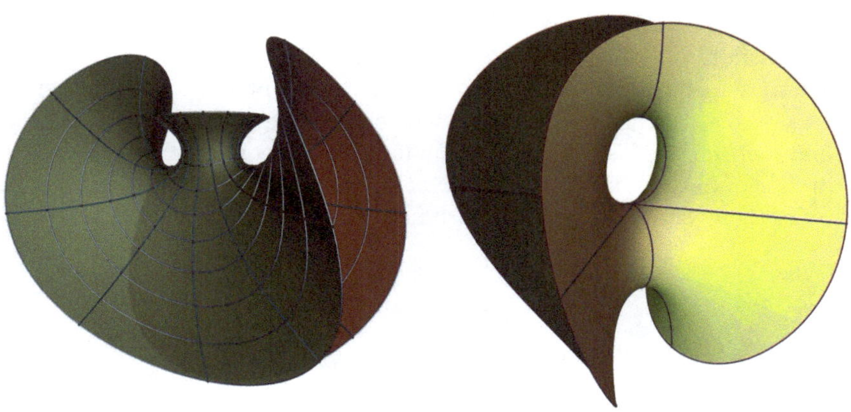

FIGURE 3. López surface (left), and Chen-Gackstatter surface (right). Creative Commons 🅭🅯🄎 Matthias Weber, `http://minimalsurfaces.blog`

If M has finite total curvature -8π then, as we said in Remark 4.9, the genus of M is either 0 or 1. If the genus is zero, then 2 is a consequence of Corollary 3.11. If the genus is one, then M is the Chen-Gackstatter surface. The corresponding torus Σ is $\mathbb{C}/\Lambda$, where Λ is the lattice in $\mathbb{C}$ generated by $\{1, i\}$ and the map $\phi : \mathbb{C}/\Lambda \to \overline{\mathbb{C}}$ is $\phi = a\mathcal{P}'/\mathcal{P}$, where a is a positive constant and $\mathcal{P}$ is the Weierstrass function associated to the lattice Λ. It is not difficult to check (see [**Ló92**]) that ϕ has four different ramification points and also four different ramification values, all of them belonging to the same equator of $\overline{\mathbb{C}} \equiv \mathbb{S}^2$. Now it follows from Theorem 3.10 that $\text{Ind}(M) = \text{Ind}(\Sigma, \phi) = 2d - 1 = 3$.

Finally, to prove 4, let M be a Jorge-Meeks n-noid with $(n \geq 2)$. Then Σ is a sphere and the holomorphic map ϕ is given by $\phi(z) = z^{n-1}$. Then the degree

$d = n - 1$, and $\{0, \infty\}$ are the ramification points of ϕ. Thus, the ramification values of ϕ lie in an equator of $\mathbb{S}^2$. From Theorem 3.10 it follows that $\mathrm{Ind}(M) = \mathrm{Ind}(\Sigma, \phi) = 2d - 1 = 2n - 3$. □

Although the following result is a consequence of Theorem 4.21, which will be proved later, we have thought it is convenient to give its original proof now since the argument used is interesting.

THEOREM 4.10 ([**LR89**]). *Let $\Phi : M \to \mathbb{R}^3$ be a minimal immersion of a complete orientable surface M. Then* $\mathrm{Ind}(M) = 1$ *if and only if Φ is congruent to the catenoid or the Enneper surface.*

PROOF. From Theorem 4.4,4, the degree d of the extended Gauss map of the catenoid and the Enneper surface is $d = 1$. Now, Corollary 3.11 gives that $\mathrm{Ind}(\Sigma, \phi) = 1$ and so, Theorem 4.7 proves that the catenoid and the Enneper surface have index one.

Conversely, if $\mathrm{Ind}(M) = 1$, then from Theorems 4.7 and 4.4, M is conformally equivalent to a compact Riemann surface Σ with a finite number of points removed, the Gauss map N of M extends to a holomorphic map $\phi : \Sigma \to \mathbb{S}^2$ and $\mathrm{Ind}(\Sigma, \phi) = 1$.

From Theorem 3.12, for any non-constant holomorphic map $\psi : \Sigma \to \mathbb{S}^2$ we have $\deg \psi \geq \deg \phi$ and if the equality holds, then $\psi = G \circ \phi$, for a some conformal diffeomorphism G of $\mathbb{S}^2$.

Since M is not totally geodesic, ϕ is not a constant map and so, there exists a point $p_0 \in \Sigma$ such that $(\widehat{\nabla}\phi)_{p_0} \neq 0$, where $\widehat{\nabla}$ is the gradient with respect to a conformal metric $\widehat{g}$ on Σ. From the Riemann-Roch theorem (Theorem 1.24), if D is the divisor of Σ given by $D(p) = 0$ for any $p \neq p_0$ and $D(p_0) = h$, where h is the genus of Σ, then there exists a holomorphic map $\psi_0 : \Sigma \to \mathbb{S}^2$ such that

$$b_{\psi_0}(p) \begin{cases} = 0 & \text{for } p \neq p_0 \\ \leq h & \text{for } p = p_0, \end{cases}$$

where $b_{\psi_0}(q)$ is the ramification of ψ_0 at q. So, if $q_0 = \psi_0(p_0)$,

$$\deg \psi_0 = \sum_{p \in \psi_0^{-1}(q_0)} (b_{\psi_0}(p) + 1) \leq h + 1.$$

From Theorem 4.6 and since M is not totally geodesic, we have that $\deg \phi \geq h + 1$ and so, $\deg \phi \geq \deg \psi_0$. As $\deg \psi_0 \geq \deg \phi$, we have that $\deg \phi = \deg \psi_0$ and thus, $\psi_0 = G \circ \phi$ for certain conformal diffeomorphism G of $\mathbb{S}^2$. Since $(\widehat{\nabla}\phi)_p \neq 0$, then $(\widehat{\nabla}\psi_0)_p \neq 0$ and so, $0 = b_{\psi_0}(p)$. This implies that $\deg \phi = \deg \psi_0 = 1$ and the proof follows from Theorem 4.4,5. □

J. Choe, [**Cho90**], defined the vision number of a surface and related this notion to the index.

DEFINITION 4.11 ([**Cho90**]). If $\Phi : M \to \mathbb{R}^3$ is an immersion of an orientable surface M and $X \in \mathfrak{X}(\mathbb{R}^3)$, then the **vision number** of M with respect to X is the number of connected components of the open set $\{p \in M \mid \langle N, X\rangle(p) \neq 0\}$, where N is the Gauss map of Φ.

REMARK 4.12. If $\Phi : M \to \mathbb{R}^3$ is a minimal immersion of finite total curvature, given a constant vector $a \in \mathbb{R}^3$, then the function $\phi_a = \langle \phi, a\rangle$ is an extension to Σ

of the function $\langle N, a\rangle : M \to \mathbb{R}^3$, i.e., $\phi_a \circ F = \langle N, a\rangle$. Since a is a Killing vector field in $\mathbb{R}^3$, it follows from Proposition 2.12 that $\mathcal{L}\langle N, a\rangle = 0$ and thus, the number of nodal sets of ϕ_a (see Proposition 3.8) is exactly the vision number of M with respect to the constant field a of $\mathbb{R}^3$.

In [**Sch83**] R. Schoen proved that if $\Phi : M \to \mathbb{R}^3$ is a complete minimal surface of finite total curvature and each end of M is embedded, then each end E_i is the graph of a function u_i with bounded slope over the exterior of a bounded region in some plane Π_i. Furthermore, if $\{x, y\}$ are coordinates on Π_i, then the functions u_i satisfy

$$u_i(x, y) = a \log \sqrt{x^2 + y^2} + b + \frac{c_1 x}{x^2 + y^2} + \frac{c_2 y}{x^2 + y^2} + \mathcal{O}\left(\frac{1}{x^2 + y^2}\right),$$

where a, b, c_1 and c_2 are constants.

If $N : M \to \mathbb{S}^2$ is the Gauss map of M and $\phi : \Sigma \to \mathbb{S}^2$ its holomorphic extension (see Theorem 4.4) to the compact Riemann surface Σ, we also assume that the ends of M are parallel, i.e., without loss of generality $\phi(p_i) = \pm(0, 0, 1) \in \mathbb{S}^2$, for any $1 \leq i \leq n$. Let X be the Killing vector field on $\mathbb{R}^3$, given by

$$X_{(x_1,x_2,x_3)} = (-x_2, x_1, 0), \quad \forall\, (x_1, x_2, x_3) \in \mathbb{R}^3,$$

whose associated isometry group is the group of rotations around $0\vec{x}_3$. Then, from Proposition 2.12 $f = \langle N, X\rangle$ is a Jacobi function on M, i.e., $\mathcal{L}\langle N, X\rangle = 0$. Since at each end E_i the vector field X is given by $X = (-y, x, 0)$ and $N = (-\frac{\partial u_i}{\partial x}, -\frac{\partial u_i}{\partial y}, 1)$, the Jacobi function f is given at the end by $f = y\frac{\partial u_i}{\partial x} - x\frac{\partial u_i}{\partial y}$. Using the expression of u_i, the function f extends smoothly to a function $\hat{f} : \Sigma \to \mathbb{R}$ with $\hat{f}(p_i) = 0$ for each end p_i, $1 \leq i \leq n$. If $\widehat{L}$ is the corresponding Schrödinger operator on Σ (see Theorem 4.7), then $\widehat{L}\hat{f} = 0$. It is clear that the number of nodal sets of $\hat{f}$ is the number of connected components of the set $\{p \in M \mid \langle N, X\rangle(p) \neq 0\}$. Now, taking into account Definition 4.11 and Theorem 2.25, we get the following result.

PROPOSITION 4.13 ([**Cho90**]). *Let $\Phi : M \to \mathbb{R}^3$ be a minimal immersion of a complete orientable surface of finite total curvature with embedded ends. If the normal vectors at the ends are all parallel to the axis $0\vec{x}_3$, then*

$$\mathrm{Ind}(M) \geq \nu(M, X) - 1,$$

where $\nu(M, X)$ is the vision number of M with respect to the vector field $X \in \mathfrak{X}(\mathbb{R}^3)$ given by

$$X_{(x_1,x_2,x_3)} = (-x_2, x_1, 0) \quad \forall\, (x_1, x_2, x_3) \in \mathbb{R}^3.$$

We will use Proposition 4.13 to estimate the index of an important family of minimal surfaces of $\mathbb{R}^3$, the so-called Costa-Hoffman-Meeks surfaces (see Figure 4). We begin by defining this family of surfaces.

C. Costa ([**Cos84**]) and D. Hoffman and W. Meeks ([**HM90**]) constructed for each integer $h \geq 1$ (Costa for $h = 1$ and Hoffman and Meeks for $h \geq 2$), a complete embedded minimal surface $M(h)$ in $\mathbb{R}^3$ of finite total curvature $-4\pi(h+2)$ of genus h and three parallel ends. Up to rotations, the normal vectors at the ends can be taken parallel to the axis $0\vec{x}_3$, and moreover the surface $M(h)$ is symmetric with respect to the $h + 1$ planes $\{\Pi_1, \ldots, \Pi_{h+1}\}$ which meet each other at equal angles $\frac{\pi}{h+1}$ along $0\vec{x}_3 = \cap_{i=1}^{h+1}\Pi_i$.

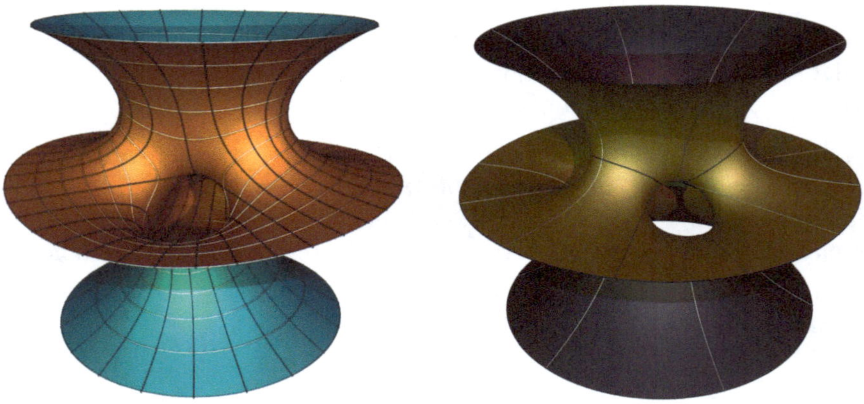

FIGURE 4. Costa surface (left), and Hoffman-Meeks surface (right). Creative Commons ©①⑨ Matthias Weber, `http://minimalsurfaces.blog`

Since $M(h)$ is symmetric with respect to Π_i, if $p \in M(h) \cap \Pi_i$, then the normal vector $N_p \in \Pi_i$ and hence $\langle N, X\rangle(p) = 0$, where X is the vector field of $\mathbb{R}^3$ given in Proposition 4.13. Hence

$$M(h) \cap \left(\cup_{i=1}^{h+1} \Pi_i\right) \subset \{p \in M(h) \mid \langle N, X\rangle(p) = 0\}.$$

Since $\mathbb{R}^3 \setminus \left(\cup_{i=1}^{h+1} \Pi_i\right)$ has $2h+2$ connected components, the number of connected components of $\{p \in M(h) \mid \langle N, X\rangle(p) \neq 0\}$ is greater than or equal to $2h+2$, i.e., the vision number $\nu(M(h), X) \geq 2h+2$. From Proposition 4.13 we get:

PROPOSITION 4.14 ([**Cho90**]). *If $M(h)$ is the Costa-Hoffman-Meeks surface of genus $h \geq 1$, then* $\mathrm{Ind}(M(h)) \geq 2h+1$.

S. Nayatani for $h \leq 37$ ([**Nay93**] for $h = 1$, [**Nay92**] for $2 \leq h \leq 37$) and F. Morabito for $h \geq 38$ ([**Mor09**]) proved that $\mathrm{Ind}(M(h)) = 2h+3 = 2d-1$, where d is the degree of the Gauss map. If Σ denotes the compact Riemann surface of genus h and $\phi : \Sigma \to \mathbb{S}^2$ the holomorphic map given in Theorem 4.4, the idea of their proofs is to construct a deformation $\phi_t : \Sigma \to \mathbb{S}^2$, $t \in (0, \infty)$, by holomorphic maps of the map $\phi = \phi_1$ and to study the behavior of $\mathrm{Ind}(\Sigma, \phi_t)$ with respect to t.

Finally, J. Choe ([**Cho90**]) conjectured that the index of any complete orientable minimal surface M of $\mathbb{R}^3$ with finite total curvature $-4\pi d$ satisfies $\mathrm{Ind}(M) \leq 2d-1$. An answer to this conjecture was independently given by S. Montiel and A. Ros ([**MR91**]) and N. Ejiri and M. Kotani ([**EK93**]) in the case where the genus of M is zero. In fact, Theorem 23 in [**MR91**] or Therorem A in [**EK93**], ensure that the index of an orientable complete minimal surface M of $\mathbb{R}^3$ with finite total curvature $-4\pi d$ and genus zero satisfies $\mathrm{Ind}(M) \leq 2d-1$. Moreover, it follows from their results that, generically, the index is $2d-1$. In the special case of finite total curvature -12π, i.e., $d = 3$, it follows from Theorem 25 in [**MR91**] that $\mathrm{Ind}(M) = 4 < 5 = 2d-1$ if and only if the four ramification points of the holomorphic map $\phi : \mathbb{S}^2 \to \mathbb{S}^2$ form an equianharmonic quadruple. Otherwise, the index is five. Their proofs use complex geometry and the fact that the genus is zero.

R. Kusner in [**Kus87**] constructed a complete orientable minimal surface of $\mathbb{R}^3$ with finite total curvature -12π and four embedded planar ends. Its compactification in $\mathbb{S}^3$ produces a Willmore immersion of $\mathbb{S}^2$ into $\mathbb{S}^3$ with Willmore energy 12π, [**Bry84**]. This example has four ramifications points and it is not difficult to check that they form an equianharmonic quadruple, and so its index is 4.

The rest of this section is devoted to obtaining a lower bound of the index of complete minimal orientable surfaces of $\mathbb{R}^3$ in terms of the genus, the number of ends and their multiplicities (Theorem 4.21). This result is due to O. Chodosh and D. Maximo, although weaker results had been obtained earlier. To prove this we start with some preliminary results.

Let $\Phi : M \to \mathbb{R}^3$ be a minimal immersion of an oriented surface M. If $\{e_1, e_2\}$ is an orthonormal reference on M compatible with the orientation of M, let N be the Gauss map of Φ such that $\{e_1, e_2, N\}$ is positively oriented in $\mathbb{R}^3$. The complex structure J on M is defined by $Jv = N \wedge v$, for any $v \in TM$, where $\wedge$ is the cross product in $\mathbb{R}^3$.

Following the notation of Section 1.3, let $\mathcal{H}(M)$ be the space of harmonic vector fields on M. If $a \in \mathbb{R}^3$, then the tangential component $a^\top$ of a satisfies

$$g(\nabla_v a^\top, w) = \langle N, a\rangle g(Av, w), \quad \text{for any} \quad v, w \in TM,$$

where A is the Weingarten endomorphism associated to N and g is the induced metric on M. Then $g(\nabla_- a^\top, -)$ is symmetric and, since Φ is minimal, $\operatorname{div}(a^\top) = 0$. This implies that $a^\top$ is a harmonic vector field on M, and so we have a linear map $a \in \mathbb{R}^3 \mapsto a^\top \in \mathcal{H}(M)$. If M is not totally geodesic, then its kernel is zero and so

$$\mathcal{V}(M) = \{a^\top \mid a \in \mathbb{R}^3\}$$

is a linear subspace of dimension three of $\mathcal{H}(M)$. It is clear that if X is a harmonic vector field, then JX is also harmonic and so

$$\mathcal{V}^*(M) = \{Ja^\top \mid a \in \mathbb{R}^3\} = \{N \wedge a^\top \mid a \in \mathbb{R}^3\} = \{N \wedge a \mid a \in \mathbb{R}^3\}$$

is also a linear subspace of dimension three of $\mathcal{H}(M)$, provided that M is not totally geodesic.

LEMMA 4.15 ([**Ros06**]). *Let $\Phi : M \to \mathbb{R}^3$ be a non-totally geodesic minimal immersion of an oriented surface M. If X is a harmonic vector field on M, then*

$$\mathcal{L}\langle X, a\rangle = 2g(\nabla X, A)\langle N, a\rangle \quad \textit{for any} \quad a \in \mathbb{R}^3. \tag{4.6}$$

Furthermore, $g(\nabla X, A) = 0$ if and only if $X \in \mathcal{V}^*(M)$.

PROOF. Using the properties of X and (1.6) we have that

$$\begin{aligned}
\Delta\langle X, a\rangle &= \langle \Delta X, a\rangle + 2\sum_{i=1}^{2} g(\nabla_{e_i} X, Ae_i)\langle N, a\rangle - \langle A^2 X, a\rangle \\
&= K\langle X, a\rangle + 2g(\nabla X, A)\langle N, a\rangle - \frac{\|A\|^2}{2}\langle X, a\rangle \\
&= 2g(\nabla X, A)\langle N, a\rangle - \|A\|^2\langle X, a\rangle,
\end{aligned}$$

and (4.6) follows.

Suppose now that $X \in \mathcal{V}^*(M)$. Then there exists $a \in \mathbb{R}^3$ such that $X = N \wedge a$ and so

$$g(\nabla X, A) = \sum_{i=1}^{2} g(\nabla_{e_i} X, Ae_i) = -\sum_{i=1}^{2}\langle Ae_i \wedge a, Ae_i\rangle = 0.$$

Conversely, if $g(\nabla X, A) = 0$, then $\nabla X = hJA$, for some function h, because $\{A, JA\}$ is a reference in the linear bundle over M of the traze-free symmetric bilinear maps.

Derivating $\nabla X = hJA$ and using the minimality of M, we get that $\Delta X = JA\nabla h$ and since X is a harmonic vector field we get $KJX = -A\nabla h$. Hence $KAJX = -A^2\nabla h = K\nabla h$. Since K has only isolated zeros, we get that $\nabla h = AJX$.

Now, the map $-JX + hN : M \to \mathbb{R}^3$ is constant, because considering that $\nabla X = hJA$ and that $\nabla h = AJX$, we have, for any $v \in TM$, that

$$\overline{\nabla}_v(-JX + hN) = -J\nabla_v X - g(AJX, v)N + g(\nabla h, v)N - hAv = 0.$$

So, there exists $a \in \mathbb{R}^3$ such that $-JX + hN = a$, which implies that $N \wedge a = -N \wedge JX = X$. Hence $X \in \mathcal{V}^*(M)$ and the lemma is proved. $\square$

LEMMA 4.16. *Let $\Phi : M \to \mathbb{R}^3$ be a complete minimal immersion of an orientable surface of finite total curvature. Let $p \in \Sigma$ (see Theorem 4.4) be an end of Φ with multiplicity d. Then, there exist constants $C_1, C_2 > 0$ and an integer $n \geq d$ such that:*

(1) $|\langle N, \Phi\rangle| \leq C_1\|\Phi\|^{-\frac{n+1}{d}+1}$ *in $U \setminus \{p\}$, where U is an open subset of Σ with $p \in U$.*
(2) $|K| \leq C_2\|\Phi\|^{-2-\frac{2(n+1)}{d}}$ *in $V \setminus \{p\}$, where V is an open subset of Σ with $p \in V$.*

PROOF. **Proof of 1.** Since the Gauss map N extends to Σ, up to a rotation in $\mathbb{R}^3$, we can assume that $N(p) = (0, 0, -1)$.

Using the Weierstrass representation of M, let z be a complex coordinate in a neighborhood U of p in Σ with $z(p) = 0$. Then, considering that $g(0) = 0$, the definition of the multiplicity of the end, and that the order of the pole 0 of ω_1 and ω_2 is $d + 1$, we can write

$$\omega_1(z) = \left(\frac{A}{z^{d+1}} + \frac{B_1(z)}{z^d}\right)\mathrm{d}z$$
$$\omega_2(z) = \left(\frac{iA}{z^{d+1}} + \frac{B_2(z)}{z^d}\right)\mathrm{d}z$$
$$\eta(z) = \omega_3(z) = \frac{B_3(z)}{z^d}\mathrm{d}z,$$

with B_1, B_2 and B_3 holomorphic functions on U and $A \in \mathbb{C}^*$. Integrating these equations we get

$$\Phi(z) = \Re\left\{\left(-\frac{A}{d}z^{-d}, -\frac{iA}{d}z^{-d}, 0\right) + \mathcal{O}(z^{-d+1})\right\}, \tag{4.7}$$

and so on U

$$\|\Phi(z)\| \simeq \|z\|^{-d}. \tag{4.8}$$

Also, if $B_3(z) = bz^n + \mathcal{O}(z^{n+1})$, with $b \in \mathbb{C}^*$ and $n \geq 0$, it follows that $\eta(z) = \left(bz^{n-d} + \mathcal{O}(z^{n+1-d})\right)\mathrm{d}z$ and, since η is holomorphic, $n \geq d$. This implies

$$g(z) = \frac{\eta(z)}{\omega_1(z) - i\omega_2(z)} = \frac{bz^{n-d}}{2Az^{-(d+1)}} + \mathcal{O}(z^{n+2}) = \frac{b}{2A}z^{n+1} + \mathcal{O}(z^{n+2}). \tag{4.9}$$

Given this expression for g, it is easy to see that

$$N(z) = \left(2\Re\left(\frac{b}{2A}z^{n+1}\right) + \mathcal{O}(z^{n+2}), 2\Im\left(\frac{b}{2A}z^{n+1}\right) + \mathcal{O}(z^{n+2}), -1 + \mathcal{O}(z^{2n+2})\right),$$

and so from (4.7) we get

$$\langle N, \Phi\rangle(z) = -\frac{1}{d}\Re(bz^{n+1-d}) + \mathcal{O}(z^{n+2-d}).$$

With (4.8) we finally get

$$|\langle N, \Phi\rangle|(z) \simeq \|z\|^{n+1-d} \simeq \|\Phi\|^{1-\frac{n+1}{d}}(z),$$

which proves 1.

Proof of 2. From (4.9) it follows that $g'(z) = \frac{(n+1)b}{2A}z^n + \mathcal{O}(z^{n+1})$, and so, using (4.9), from equation (4.3) we get

$$K(z) = -\left(\frac{(n+1)\|b\|}{\|A\|^2}\|z\|^{n+1+d} + \mathcal{O}(z^{n+d+2})\right)^2.$$

So, using (4.8)

$$|K(z)| \simeq \|z\|^{2(n+1+d)} \simeq \|\Phi\|^{-\dfrac{2(n+1+d)}{d}}(z),$$

which proves 2. □

The following result is a refinement of the log cutoff trick.

LEMMA 4.17. *Let $\Phi : M \to \mathbb{R}^3$ be a minimal immersion of an orientable surface M. Then for each $R > 1$ there exists a function $\varphi^R \in H_0^{1,2}(M)$ such that $\varphi^R(x) = 1$ if $\|\Phi(x)\| \leq R$, $\varphi^R(x) = 0$ if $\|\Phi(x)\| \geq R^2$, and*

$$\|\nabla\varphi^R\| \leq \frac{1}{\|\Phi\|\log R}, \quad |\Delta\varphi^R| \leq \frac{2\langle N, \Phi\rangle^2}{\|\Phi\|^4 \log R}.$$

PROOF. To do this, let $\zeta^R \in H_0^{1,2}(\mathbb{R})$ be the function defined by

$$\zeta^R(t) = \begin{cases} 1 & t < R, \\ 2 - \frac{\log t}{\log R} & t \in [R, R^2], \\ 0 & t > R^2. \end{cases}$$

If f is the function on M given by $f(x) = \|\Phi(x)\|$, then $\varphi^R = \zeta^R \circ f \in H_0^{1,2}(M)$ satisfies $\varphi^R(x) = 1$ if $\|\Phi(x)\| < R$, $\varphi^R(x) = 0$ if $\|\Phi(x)\| > R^2$.

Now we can check directly that for $R \leq \|\Phi\| \leq R^2$, we have $\nabla\varphi^R = -\frac{\Phi^\top}{\|\Phi\|^2 \log R}$, where $\Phi^\top$ is the tangent component of Φ. Thus

$$\|\nabla\varphi^R\|^2 = \frac{\|\Phi^\top\|^2}{\|\Phi\|^4 \log^2 R} \leq \frac{1}{\|\Phi\|^2 \log^2 R},$$

which proves the bound of $\|\nabla\varphi^R\|$.

Now, using the above expression of $\nabla\varphi^R$, if $\{e_1, e_2\}$ is an orthonormal reference on M, we have for $R \leq \|\Phi\| \leq R^2$

$$\Delta\varphi^R = \sum_{i=1}^{2}\langle\nabla_{e_i}\nabla\varphi^R, e_i\rangle = \frac{2\left(\|\Phi^\top\|^2 - \|\Phi\|^2\right)}{\|\Phi\|^4 \log R} = -\frac{2\langle N, \Phi\rangle^2}{\|\Phi\|^4 \log R}.$$ □

If $\Phi : M \to \mathbb{R}^3$ is a minimal immersion of an orientable surface, we consider on M the **weight function** $\mathcal{P} : M \to \mathbb{R}$ given by

$$\mathcal{P} = \frac{1}{(1 + \|\Phi\|^2)\log(2 + \|\Phi\|)},$$

which satisfies $0 < \mathcal{P} \leq (\log 2)^{-1}$. Then we consider the space

$$L^2_*(M) = \{\mathcal{P}^{-1/2} f \mid f \in L^2(M)\},$$

endowed with the norm

$$\|g\|^2_{L^2_*(M)} = \int_M g^2 \mathcal{P}\, \mathrm{d}v, \qquad \forall g \in L^2_*(M).$$

Then, $\left(L^2_*(M), \|.\|_{L^2_*(M)}\right)$ is a Banach space. In the next technical lemma we will prove some properties that will simplify the proof of the next results.

LEMMA 4.18 ([**CM23**]). *Let $\Phi : M \to \mathbb{R}^3$ be a complete minimal immersion of an orientable surface of finite total curvature.*

(1) *$L^2(M) \subset L^2_*(M)$ and if $M_\rho = \{x \in M \mid \|\Phi(x)\| < \rho\}$, then $L^2(M_\rho) = L^2_*(M_\rho)$.*
(2) *If $R > 1$ and $M_{R,R^2} = \{x \in M \mid R < \|\Phi(x)\| < R^2\}$, then for any $f \in L^2_*(M)$*

$$\lim_{R\to\infty} \int_{M_{R,R^2}} \frac{f^2}{\|\Phi\|^2 \log^2 \|\Phi\|} \mathrm{d}v = 0.$$

(3) *If $f \in L^2_*(M)$ and φ^R is the cut-off function given in Lemma 4.17, then*

$$\lim_{R\to\infty} \int_M \|\nabla\varphi^R\|^2 f^2 \mathrm{d}v = 0 \quad \text{and} \quad \lim_{R\to\infty} \int_M \varphi^R |\Delta\varphi^R| f^2 \mathrm{d}v = 0.$$

PROOF. **Proof of 1**. If $f \in L^2(M)$, then

$$\int_M f^2 \mathcal{P}\, \mathrm{d}v \leq \frac{1}{\log 2} \int_M f^2 \mathrm{d}v < \infty,$$

which implies that $f \in L^2_*(M)$. We also note that if $f \in L^2_*(M_\rho)$,

$$\int_{M_\rho} f^2 \mathrm{d}v \leq (1 + \rho^2)\log(2 + \rho) \int_{M_\rho} f^2 \mathcal{P}\, \mathrm{d}v < \infty,$$

and so, $f \in L^2(M_\rho)$. This proves 1.
Proof of 2. Using $R < \|\Phi\| < R^2$ and that $f \in L^2_*(M)$, we get

$$\begin{aligned}
\lim_{R\to\infty} \int_{M_{R,R^2}} \frac{f^2}{\|\Phi\|^2 \log^2 \|\Phi\|} \mathrm{d}v &= \lim_{R\to\infty} \int_{M_{R,R^2}} \left(1 + \frac{1}{\|\Phi\|^2}\right) \frac{\log(2 + \|\Phi\|)}{\log^2 \|\Phi\|} f^2 \mathcal{P}\, \mathrm{d}v \\
&\leq \lim_{R\to\infty} \left(1 + \frac{1}{R^2}\right) \frac{\log(2 + R^2)}{\log^2 R} \int_M f^2 \mathcal{P}\, \mathrm{d}v \\
&= \lim_{R\to\infty} \frac{\log(2 + R^2)}{\log^2 R} \|f\|^2_{L^2_*(M)} = 0,
\end{aligned}$$

which proves 2.

Proof of 3. From Lemma 4.17 and 2 it follows that

$$\lim_{R\to\infty}\int_M \|\nabla\varphi^R\|^2 f^2 dv \le \lim_{R\to\infty}\int_{M_{R,R^2}} \frac{f^2}{\|\Phi\|^2\log^2 R}dv$$

$$\le \lim_{R\to\infty}\int_{M_{R,R^2}} \frac{4f^2}{\|\Phi\|^2\log^2\|\Phi\|}dv = 0.$$

Finally, it follows from Lemma 4.17, Lemma 4.16 and 2 that

$$\lim_{R\to\infty}\int_M \varphi^R|\Delta\varphi^R| f^2 dv \le \lim_{R\to\infty}\int_{M_{R,R^2}} \frac{2\langle\Phi,N\rangle^2 f^2}{\|\Phi\|^4\log R}dv$$

$$= \sum_{i=1}^n \lim_{R\to\infty}\int_{M_{R,R^2}\cap U_i} \frac{2C_i f^2}{\|\Phi\|^{2+\frac{2(n_i+1)}{d_i}}\log R}dv$$

$$\le \sum_{i=1}^n \lim_{R\to\infty} 2C_i\int_{M_{R,R^2}\cap U_i} \frac{f^2}{\|\Phi\|^2\log^2\|\Phi\|}\frac{\log\|\Phi\|}{\|\Phi\|^{\frac{2(n_i+1)}{d_i}}}dv$$

$$\le \sum_{i=1}^n \lim_{R\to\infty}\frac{4C_i\log R}{R^{\frac{2(n_i+1)}{d_i}}}\int_{M_{R,R^2}\cap U_i} \frac{f^2}{\|\Phi\|^2\log^2\|\Phi\|}dv = 0,$$

because from 2,

$$\lim_{R\to\infty}\int_{M_{R,R^2}\cap U_i} \frac{f^2}{\|\Phi\|^2\log^2\|\Phi\|}dv \le \lim_{R\to\infty}\int_{M_{R,R^2}} \frac{f^2}{\|\Phi\|^2\log^2\|\Phi\|}dv = 0$$

and, since $n_i \ge d_i$, $\lim_{R\to\infty}\frac{\log R}{R^{\frac{2(n_i+1)}{d_i}}} = 0$. □

PROPOSITION 4.19 ([**CM23**]). *Let $\Phi : M \to \mathbb{R}^3$ be a complete minimal immersion of an orientable surface of finite total curvature. Let $\{p_1, \dots, p_n\} \subset \Sigma$ be the ends of Φ and $\{d_1 \dots, d_n\}$ their respective multiplicities. Then:*

(1) *The dimension of the space*

$$\mathcal{H}_*(M) := \{X \in \mathcal{H}(M) \mid \|X\| \in L^2_*(M)\}$$

is $2h - 2 + 2\sum_{i=1}^n(1 + d_i)$, where h is the genus of M.

(2) *If $f \in L^2_*(M) \cap C^\infty(M)$ satisfies the weighted eigenvalue equation*

$$\mathcal{L}f + \lambda\mathcal{P}f = 0, \qquad \lambda \in \mathbb{R},$$

then $\|\nabla f\| \in L^2(M)$. The functions that satisfy the above equation are called $\mathcal{P}$-eigenfunctions of the Jacobi operator.

(3) *If $X \in \mathcal{H}_*(M)$, then $\|\nabla X\| \in L^2(M)$.*

PROOF. **Proof of 1.** On the compact Riemann surface Σ we consider the divisor $\mathcal{D}$ defined by

$$\mathcal{D} = p_1^{-(d_1+1)} \cdot \dots \cdot p_n^{-(d_n+1)}.$$

The Riemann-Roch theorem, [**FK92**], implies that the space of meromorphic 1-forms on Σ with poles only at the points $\{p_1, \dots, p_n\}$ and of order $l \le d_i + 1$, has dimension

$$h - 1 - \deg(\mathcal{D}) = h - 1 + \sum_{i=1}^n (d_i + 1).$$

The restriction of these 1-forms to M defines a space V of holomorphic 1-forms on M, and the real and imaginary parts of these 1-forms define nonzero harmonic 1-forms. Considering that $\mathcal{H}^1(M)$ and $\mathcal{H}(M)$ are isomorphic, $\dim \mathcal{H}(M) = \dim \mathcal{H}^1(M) \geq 2h - 2 + 2\sum_{i=1}^n (1 + d_i)$. Now we prove that every holomorphic 1-form on V satisfies $\|\alpha\| \in L^2_*(M)$, and so $\dim \mathcal{H}_*(M) \geq 2h - 2 + 2\sum_{i=1}^n (1 + d_i)$.

By Lemma 4.16,2, if P is the compact subset of M given by $P = M \setminus \cup_{i=1}^n (U_i \setminus \{p_i\})$ y $\alpha \in V$ then

$$\|\alpha\|^2_{L^2_*(M)} = \int_P \|\alpha\|^2 \mathcal{P}\, \mathrm{d}v + \sum_{i=1}^n \int_{U_i \setminus \{p_i\}} \|\alpha\|^2 \mathcal{P}\, \mathrm{d}v.$$

The first integral is finite, and we will check that each integral of the second term is finite. Let z be a complex coordinate in U_i in Σ with $z(p_i) = 0$. Then the metric $g = e^{2u}\|\mathrm{d}z\|^2$ and we have that $\|\alpha\|^2 \mathrm{d}v = \|\alpha\|_0^2 \mathrm{d}z$, where $\|\,\|_0$ stands for the Euclidean metric. Now,

$$\alpha(z) = \left(\frac{A}{z^l} + \frac{B(z)}{z^{l-1}}\right) \mathrm{d}z, \quad l \leq d_i + 1,$$

with $A \in \mathbb{C}^*$ and B a holomorphic function on $U_i \setminus \{p_i\}$. Considering (4.8) we have

$$\begin{aligned}\int_{U_i \setminus \{p_i\}} \|\alpha\|^2 \mathcal{P}\, \mathrm{d}v &\leq C \int_{U_i \setminus \{p_i\}} \|z\|^{-2l}(1 + \|\Phi\|^2)^{-1}(\log(2 + \|\Phi\|)^{-1} \mathrm{d}z \\ &\leq C' \int_{U_i \setminus \{p_i\}} \frac{\|z\|^{2d_i - 2l}}{(1 + \|z\|^{2d_i}) \log(2 + \|z\|^{-d_i})} \mathrm{d}z = (I).\end{aligned}$$

Now, since $l \leq d_i + 1$ and $1 \leq d_i$, we have

$$\begin{aligned}(I) &\leq C' \int_{U_i \setminus \{p_i\}} \frac{\|z\|^{-2}}{(1 + \|z\|^{2d_i}) \log(2 + \|z\|^{-1})} \mathrm{d}z \\ &= C' \int_0^1 \frac{\mathrm{d}r}{r(1 + r^{2d_i}) \log(2 + r^{-1})} \leq C' \int_0^1 \frac{\mathrm{d}r}{r \log(2 + r^{-1})} < \infty.\end{aligned}$$

Hence $\|\alpha\| \in L^2_*(M)$.

Conversely, if $\omega \in \mathcal{H}^1_*(M) = \{\omega \in \mathcal{H}^1(M) \mid \|\omega\| \in L^2_*(M)\}$, then $\alpha = \omega + i \star \omega$ is a holomorphic 1-form on M, where $\star$ is the star operator on M (see Section 1.3). Using that $\|\omega\| \in L^2_*(M)$, the definition of the weight function $\mathcal{P}$ and the behavior of the immersion Φ at the ends, which is given in (4.8), it is not difficult to check that α extends meromorphically to the compact Riemann surface Σ with poles of order $l \leq d_j + 1$ at the ends. This concludes Proof 1.

Proof of 2. We consider the functions φ^R given in Lemma 4.17. Then

$$\{(\varphi^R)^2 \|\nabla f\|^2 \mid R > 1\}$$

is a pointwise non-decreasing set of non-negative functions, such that

$$\lim_{R \to \infty} (\varphi^R)^2 \|\nabla f\|^2 = \|\nabla f\|^2.$$

Then the monotone convergence theorem (Theorem 1.7) implies that

$$\int_M \|\nabla f\|^2 \mathrm{d}v = \lim_{R \to \infty} \int_M (\varphi^R)^2 \|\nabla f\|^2 \mathrm{d}v. \tag{4.10}$$

Let us check that the above limit is finite. Using that $(\varphi^R)^2 f\nabla f$ has compact support, the divergence theorem, and the equation that f satisfies, we have that

$$(4.11)\quad \int_M (\varphi^R)^2\|\nabla f\|^2 \mathrm{d}v = -\int_M (\varphi^R)^2 f\Delta f\,\mathrm{d}v - \frac{1}{2}\int_M g(\nabla(\varphi^R)^2, \nabla f^2)\mathrm{d}v$$
$$= -\int_M 2(\varphi^R)^2 K f^2 \mathrm{d}v + \lambda\int_M (\varphi^R)^2 f^2 \mathcal{P}\mathrm{d}v - \frac{1}{2}\int_M g(\nabla(\varphi^R)^2, \nabla f^2)\mathrm{d}v.$$

By Lemma 4.16,2, if P is the compact subset of M given by $P = M\setminus\cup_{i=1}^n(V_i\setminus\{p_i\})$, then

$$\int_M (\varphi^R)^2 K f^2 \mathrm{d}v = \int_P (\varphi^R)^2 K f^2\mathrm{d}v + \sum_{i=1}^n \int_{V_i\setminus\{p_i\}} (\varphi^R)^2 K f^2 \mathrm{d}v.$$

It is clear that the first term is finite. Also, using Lemma 4.16,2 again, there exist constants $C_i > 0$ such that

$$-\int_{V_i\setminus\{p_i\}} (\varphi^R)^2 K f^2\mathrm{d}v \le C_i \int_{V_i\setminus\{p_i\}} \frac{(\varphi^R)^2 f^2}{\|\Phi\|^{2+\frac{2(n_i+1)}{d_i}}}\mathrm{d}v$$
$$= C_i\int_{V_i\setminus\{p_i\}} \frac{(\varphi^R)^2(1+\|\Phi\|^2)\log(2+\|\Phi\|)}{\|\Phi\|^{2+\frac{2(n_i+1)}{d_i}}} f^2\mathcal{P}\mathrm{d}v$$

and the limit when $R\to\infty$ of the above integral is finite.

Also, since $f \in L^2_*(M)$, and $0 \le \varphi^R \le 1$, we have, using the dominate convergence theorem (Theorem 1.7), that

$$\lim_{R\to\infty}\int_M (\varphi^R)^2 f^2\mathcal{P}\mathrm{d}v = \|f\|_{L^2_*(M)} < \infty.$$

So from (4.11), we only need to check that $\lim_{R\to\infty}\int_M g(\nabla(\varphi^R)^2, \nabla f^2)\,\mathrm{d}v < \infty$. Using the fact that $f^2\nabla(\varphi^R)^2$ has compact support and the divergence theorem (Theorem 1.7), we have

$$-\frac{1}{2}\int_M g(\nabla(\varphi^R)^2, \nabla f^2)\,\mathrm{d}v = \frac{1}{2}\int_M f^2\Delta(\varphi^R)^2\mathrm{d}v$$
$$= \int_M f^2\varphi^R\Delta\varphi^R\mathrm{d}v + \int_M f^2\|\nabla\varphi^R\|^2\mathrm{d}v.$$

Now the result follows from Lemma 4.18.

Proof of 3. From Lemma 1.16,1,

$$\Delta\|X\|^2 = 2K\|X\|^2 + 2\|\nabla X\|^2.$$

Since $(\varphi^R)^2\nabla\|X\|^2$ has compact support, from the divergence theorem and the above equation we get

$$\int_M (\varphi^R)^2\|\nabla X\|^2\mathrm{d}v = -\int_M (\varphi^R)^2 K\|X\|^2\mathrm{d}v - \frac{1}{2}\int_M g(\nabla\|X\|^2, \nabla(\varphi^R)^2)\,\mathrm{d}v.$$

This formula is similar to (4.11) where f^2 is replaced by $\|X\|^2$. So, using a similar reasoning as in part 2, we finish the proof of the proposition. □

To finish the preliminaries of Theorem 4.21, let us prove a weighted version of Theorem 2.29. This version is inspired by the work of B. Simon, [**Sim81**], and references therein, which proves that the index of a quadratic form associated to a

Schrödinger operator taken with respect to weighted and unweighted L^2-spaces is the same.

THEOREM 4.20 ([**CM23**]). *Let $\Phi : M \to \mathbb{R}^3$ be a minimal immersion of a complete orientable surface M with finite total curvature. If $\mathrm{Ind}(M) = k$, then there exists a k-dimensional subspace W of $L^2_*(M)$ with an L^2_*-orthonormal basis $\{v_1, \dots, v_k\}$ with $\mathcal{L}v_i + \lambda_i \mathcal{P} v_i = 0$, $\lambda_i < 0$, and $\mathcal{Q}(f) \geq 0$ for any function $f \in C_0^\infty(M) \cap W^\perp$, where $W^\perp \subset L^2_*(M)$ is the L^2_*-orthogonal complement of W.*

PROOF. We will follow the proof of Theorem 2.29. If $\mathrm{Ind}(M)$ is finite, then there exists a compact subset C of M such that $\mathrm{Ind}_{M\setminus C}(\mathcal{L}) = 0$, where $\mathcal{L} = \Delta - 2K$. Moreover there exists a relatively compact domain Ω in M such that $\mathrm{Ind}(M) = \mathrm{Ind}(\mathcal{L}, \Omega)$. Since Φ is proper, let $R_0 > 1$ such that $C, \Omega \subset M_{R_0} = \Phi^{-1}(B(R_0))$. Hence $\mathrm{Ind}(M) = \mathrm{Ind}(\mathcal{L}, M_R)$ for any $R \geq R_0$.

If $\phi \in C_0^\infty(M)$, then for any $R \geq R_0$ we consider $\eta^R = 1 - \varphi^R$ (φ^R is the function given in Lemma 4.17). Then, $\eta^R\phi \in H_0^{1,2}(M)$ and $\mathrm{supp}(\eta^R\phi) \subset M \setminus C$. The stability of $M \setminus C$ implies that $\mathcal{Q}(\eta^R\phi) \geq 0$ and so,

$$\int_M -2(\eta^R\phi)^2 K \,\mathrm{d}v \leq \int_M \{(\eta^R)^2\|\nabla\phi\|^2 + \phi^2\|\nabla\eta\|^2 + 2\eta^R\phi g(\nabla\eta^R, \nabla\phi)\} \,\mathrm{d}v. \tag{4.12}$$

Now, adding $\mathcal{Q}(\phi)$ to both sides of the inequality and using that $2\eta^R\phi g(\nabla\eta^R, \nabla\phi) \leq (\eta^R)^2\|\nabla\phi\|^2 + \phi^2\|\nabla\eta^R\|^2$, we obtain

$$\begin{aligned}
\int_M (1 - (\eta^R)^2) \left(\|\nabla\phi\|^2 + 2K\phi^2\right) \mathrm{d}v &\leq \\
&\leq \mathcal{Q}(\phi) + \int_M \{\phi^2\|\nabla\eta^R\|^2 + 2\eta^R\phi g(\nabla\eta^R, \nabla\phi)\} \,\mathrm{d}v \\
&\leq \mathcal{Q}(\phi) + \int_{M_{R^2}} \{2\phi^2\|\nabla\eta^R\|^2 + (\eta^R)^2\|\nabla\phi\|^2\} \,\mathrm{d}v \\
&\leq 2\mathcal{Q}(\phi) + \int_{M_{R^2}} \{2\phi^2\|\nabla\eta^R\|^2 - 2K\phi^2\} \,\mathrm{d}v,
\end{aligned}$$

where, in the last inequality, we have used that $(\eta^R)^2 \leq 1$. Now, since $\eta^R = 0$ on M_R, we get from the last inequality

$$\begin{aligned}
\int_{M_{R^2}} \|\nabla\phi\|^2 \mathrm{d}v &\leq \int_M (1 - (\eta^R)^2)\|\nabla\phi\|^2 \mathrm{d}v \\
&\leq 2\mathcal{Q}(\phi) + 2\int_{M_{R^2}} \{\phi^2\|\nabla\eta^R\|^2 + (1 + (\eta^R)^2 K)\phi^2\} \mathrm{d}v \\
&\leq 2\mathcal{Q}(\phi) + 2\left(\frac{1}{R^2\log^2 R} + 2\sup_{M_{R^2}}|K|\right)\int_{M_{R^2}} \phi^2 \mathrm{d}v \\
&\leq 2\left(\mathcal{Q}(\phi) + C_R\int_M \phi^2 \mathrm{d}v\right),
\end{aligned} \tag{4.13}$$

where $C_R = \frac{1}{R^2\log^2 R} + 2\sup_{M_{R^2}}|K|$.

Using (4.13) with $R = R_0$, we get that

$$0 \leq Q(\phi) + C_0\int_M \phi^2 \mathrm{d}v, \tag{4.14}$$

where $C_0 = C_{R_0}$ and ϕ is any smooth compactly supported function on M.

As for any $R \geq R_0$, we have that $k = \mathrm{Ind}(\mathcal{L}, M_R)$ and from Lemma 4.18, $L^2_*(M_R) = L^2(M_R)$, let $\{\lambda_{1,R}, \dots, \lambda_{k,R}\}$ be the negative $L^2_*(M_R)$-weighted eigenvalues and $\{f_{1,R}, \dots, f_{k,R}\} \subset C_0^\infty(M_R)$ the corresponding orthonormal eigenfunctions, i.e., $\mathcal{L}f_{i,R} + \lambda_{i,R}\mathcal{P}f_{i,R} = 0$ for $1 \leq i \leq k$ and $R \geq R_0$ and $\int_M f_{i,R}f_{j,R}\mathcal{P}\mathrm{d}v = \delta_{ij}$. Taken in (2.24), $\phi = f_{i,R}$, we obtain that $\lambda_{i,R} \geq -C_0$. Also, $\max\{\lambda_{1,R}, \dots, \lambda_{k,R}\}$ is a decreasing function of R and hence there exists $\epsilon_0 > 0$ such that $\lambda_{i,R} < -\epsilon$. So, finally we get

$$\lambda_{i,R} \in [-C_0, -\epsilon_0], \quad \text{for any} \quad 1 \leq i \leq k \quad \text{and any} \quad R \geq R_0. \tag{4.15}$$

Let $R_0 < R$ and $\rho \in [R, \infty)$. Using (4.13) with $\phi = f_{i,\rho}$, we get that

$$\int_{M_R} (f_{i,\rho}^2 + \|\nabla f_{i,\rho}\|^2)\mathrm{d}v < (1 + 2C_R)\int_M f_{i,\rho}^2\mathrm{d}v = 1 + 2C_R, \tag{4.16}$$

for any $\rho > R$, $1 \leq i \leq r$.

Also, integrating by parts, we have

$$\begin{aligned} 2\int_M \eta^R f_{i,\rho} g(\nabla\eta^R, \nabla f_{i,\rho})\,\mathrm{d}v &= \frac{1}{2}\int_M g(\nabla(\eta^R)^2, \nabla f_{i,\rho}^2), \mathrm{d}v = -\frac{1}{2}\int_M (\eta^R)^2\Delta f_{i,\rho}^2\mathrm{d}v \\ &= \int_M (\eta^R)^2 (f_{i,\rho}^2(-2K + \lambda_{i,\rho}\mathcal{P}) - \|\nabla f_{i,\rho}\|^2)\,\mathrm{d}v. \end{aligned}$$

Now, using (4.12) in the last equation we get

$$-\lambda_{i,\rho}\int_M (\eta^R)^2 f_{i\rho}^2\mathcal{P}\mathrm{d}v \leq \int_M f_{i,\rho}^2\|\nabla\eta^R\|^2\mathrm{d}v.$$

Since $-\lambda_{i,\rho} \geq \epsilon_0$ and $\int_M f_{i,\rho}^2\mathcal{P}\mathrm{d}v = 1$, we have for $\rho > R^2$,

$$\begin{aligned} \int_{M\setminus M_{R^2}} f_{i,\rho}^2\mathcal{P}\mathrm{d}v &= \int_{M\setminus M_{R^2}} (\eta^R)^2 f_{i,\rho}^2\mathcal{P}\mathrm{d}v \leq \int_M (\eta^R)^2 f_{i,\rho}^2\mathcal{P}\mathrm{d}v \\ &\leq \frac{1}{\epsilon_0}\int_M f_{i,\rho}^2\|\nabla\eta^R\|^2\mathrm{d}v \leq \frac{1}{\epsilon_0 R^2\log^2 R}\int_{M_{R^2}} f_{i,\rho}^2\mathrm{d}v \\ &\leq \frac{(1+R^4)\log^2(2+R^2)}{\epsilon_0 R^2\log^2 R}\int_{M_{R^2}} f_{i,\rho}^2\mathcal{P}\mathrm{d}v \\ &\leq \frac{(1+R^4)\log^2(2+R^2)}{\epsilon_0 R^2\log^2 R}\left(1 - \int_{M\setminus M_{R^2}} f_{i,\rho}^2\mathcal{P}\mathrm{d}v\right), \end{aligned}$$

and thus

$$\int_{M\setminus M_{R^2}} f_{i,\rho}^2\mathrm{d}v \leq \frac{1}{1 + \frac{\epsilon_0 R^2\log^2 R}{(1+R^4)\log^2(2+R^2)}}, \quad \text{for any} \quad 1 \leq i \leq k, \ \rho \geq R^2. \tag{4.17}$$

We observe that if $\rho \leq R^2$, then $\int_{M\setminus B(R^2)} f_{i,\rho}^2\mathcal{P}\mathrm{d}v = 0$.

Finally we have a family $\{f_{i,\rho} \mid 1 \leq i \leq k, \ \rho \geq R_0\} \subset C_0^\infty(M)$ with $\mathcal{L}f_{i,\rho} + \lambda_{i,\rho}\mathcal{P}f_{i,\rho} = 0$, which satisfies (4.15), (4.16) and (4.17).

Now, from (4.16), $A := \{f_{i,\rho}, 1 \leq i \leq r, \rho > R\}$ is bounded in the Sobolev space $H^{1,2}(M_R)$, and so, as $H^{1,2}(M_R)$ is compactly embedded into $L^2_*(M_R)$, there exists a sequence $\{u_n\} \subset A$ that converges in $L^2_*(M_R)$. Now, we use a Cantor diagonal argument. Repeating the argument with M_{R+1}, we find a subsequence $\{u_{1,n}\}$ of $\{u_n\}$ converging in $L^2_*(M_{R+1})$. Making an induction argument, for any positive integer m we find a subsequence $\{u_{m,n}\}$ converging in $L^2_*(M_{R+m})$. So we get a sequence $\{f_i, \rho_n\} = \{u_{n,n}\}$ such that

$$\lim_{\rho_n\to\infty} f_{i,\rho_n} = f_i \quad \text{in} \quad L^2_*(M), \quad 1 \le i \le k.$$

From (4.15), the set $\{\lambda_{i,\rho}\}$ is uniformly bounded from below, and so f_i is a weighted eigenfuction of $\mathcal{L}$ with eigenvalue $\lambda_i = \lim_{\rho_n\to\infty} \lambda_{i,\rho_n}$.

It also follows from (4.17), that no mass scapes to infinity, i.e., $\int_M f_i^2 \mathcal{P} dv = 1$.

So we have found an orthonormal $\{f_1, \dots, f_k\}$ in $L^2_*(M)$ such that $\mathcal{L}f_i + \lambda_i \mathcal{P} f_i = 0$ and $-C_0 \le \lambda_i \le -\epsilon_0$ for any $1 \le i \le k$. Let W be their linear span in $L^2_*(M)$. If $\phi \in C_0^\infty(M) \cap W^\perp$, then for any ρ_n we have

$$\phi = \sum_{i=1}^k a_{i,\rho_n} f_{i,\rho_n} + \phi_{\rho_n}, \qquad a_{i,\rho_n} = \int_M \phi f_{i,\rho_n} \mathcal{P} dv,$$

with ϕ_{ρ_n} L^2_*-orthogonal to $\operatorname{span}\{f_{1,\rho_n}, \dots, f_{k,\rho_n}\}$. It is clear that $\mathcal{Q}(\phi_{\rho_n}) \ge 0$, and so $\mathcal{Q}(\phi) \ge \sum_{i=1}^k \lambda_{i,\rho_n} a_{i,\rho_n}^2$. Hence,

$$\begin{aligned}\mathcal{Q}(\phi) &\ge \sum_{i=1}^k \lim_{\rho_n\to\infty} (\lambda_{i,\rho_n} a_{i,\rho_n}^2) \ge -C_0 \sum_{i=1}^n \lim_{\rho_n\to\infty} \left(\int_M \phi f_{i,\rho_n} \mathcal{P} dv\right)^2 \\ &= -C_0 \sum_{i=1}^k \left(\int_M \phi f_i \mathcal{P} dv\right)^2 = 0.\end{aligned}$$

This conclude the proof. □

THEOREM 4.21 ([**CM23**]). *Let $\Phi : M \to \mathbb{R}^3$ be a minimal immersion of a complete orientable surface of finite total curvature of genus h and n ends $\{p_1, \dots, p_n\}$ with multiplicities $\{d_1, \dots, d_n\}$ respectively. Then*

$$\operatorname{Ind}(M) \ge \frac{1}{3}\left(2h - 5 + 2\sum_{i=1}^n (1 + d_i)\right).$$

REMARK 4.22. A. Ros in [**Ros06**] proved that, under the assumptions of Theorem 4.21, $\operatorname{Ind}(M) \ge 2h/3$. O. Chodosh and D. Maximo, in [**CM16**] proved that $\operatorname{Ind}(M) \ge \frac{2}{3}(h+n) - 1$. The idea to obtain these bounds comes from [**Ros06**] and consists of using square integrable harmonic 1-forms on the surface as test functions. Later, in [**CM16, CM23**], the authors consider weighted square integrable harmonic 1-forms as test functions.

PROOF. Suppose that $\operatorname{Ind}(M) = k$. From Theorem 4.20 there exists a k-dimensional subspace $W \subset L^2_*(M)$ with an L^2_*-orthonormal basis $\{v_1, \dots, v_k\}$ with

$$\mathcal{L}v_i + \lambda_i \mathcal{P} v_i = 0, \quad \lambda_i < 0.$$

We consider the linear map $F : \mathcal{H}_*(M) \to \mathbb{R}^{3k}$ defined by

$$F(X) = \left(\int_M v_1 \mathcal{P} X \, dv, \dots, \int_M v_k \mathcal{P} X \, dv\right).$$

Then, from Proposition 4.19,1, $\dim \ker F \ge 2h + 2\sum_{i=1}^n (d_i + 1) - 2 - 3k$. We will prove that $\dim \ker F \le 3$, and the theorem will follow.

CLAIM. If $X \in \ker F$, then $\mathcal{Q}(X,Y) = \sum_{i=1}^{3} \mathcal{Q}(\langle X, e_i\rangle, \langle Y, e_i\rangle) = 0$ for any compact support smooth map $Y : M \to \mathbb{R}^3$ with $Y \in L^2_*(M) \cap W^{\perp}$, i.e., $\langle Y, e_i\rangle \in L^2_*(M) \cap W^{\perp}$, $1 \le i \le 3$, where $\{e_1, e_2, e_3\}$ is an orthonormal basis of $\mathbb{R}^3$.

PROOF OF THE CLAIM. We choose R large enough so that $\{x \in M \mid \|\Phi(x)\| < R\}$ contains the support of Y, and φ^R the cut-off function given in Lemma 4.17. Now, for each $t \in \mathbb{R}$ we define

$$X_t = \varphi^R\big(X + tY + v_1 a_1 + \cdots + v_k a_k\big),$$

where $\{a_1, \dots, a_k\}$ are vectors of $\mathbb{R}^3$ depending on R. It is possible to choose the vectors a_i so that $X_t \in W^{\perp}$. In fact, as $\int_M t\varphi^R v_i \mathcal{P} Y \, \mathrm{d}v = \int_M t v_i \mathcal{P} Y \, \mathrm{d}v = 0$, the vectors a_i are solution of the following linear system

$$\sum_{j=1}^{k} \left(\int_M \varphi^R v_i v_j \mathcal{P} \, \mathrm{d}v \right) a_j = - \int_M \varphi^R v_i \mathcal{P} X \, \mathrm{d}v \quad 1 \le i \le k.$$

By the dominate convergence theorem (Theorem 1.7) $\lim_{R\to\infty} \big(\int_M \varphi^R v_i v_j \mathcal{P} \, \mathrm{d}v\big)$ is the identity matrix, and thus, for sufficiently large R, the above system has a solution. We observe that, again by the dominate covergence theorem,

$$\lim_{R\to\infty} \int_M \varphi^R v_i \mathcal{P} X \, \mathrm{d}v = \int_M v_i \mathcal{P} X = 0,$$

and so $\lim_{R\to\infty} a_i = 0$.

Since X_t is a smooth vector field with compact support in $W^{\perp}$, it follows from Theorem 4.20 that $\mathcal{Q}(X_t) \ge 0$. Thus

$$\begin{aligned} 0 \le \mathcal{Q}(X_t) &= t^2 \mathcal{Q}(Y) + 2t\mathcal{Q}(\varphi^R X, Y) + \sum_{i=1}^{k} \mathcal{Q}(Y, \varphi^R v_i a_i) + \mathcal{Q}(\varphi^R X) \\ &\quad + \mathcal{Q}(\varphi^R \sum_{i=1}^{k} v_i a_i) + 2 \sum_{i=1}^{k} \mathcal{Q}(\varphi^R X, \varphi^R v_i a_i) \\ &= t^2 \mathcal{Q}(Y) + 2t\mathcal{Q}(X,Y) + \mathcal{Q}(\varphi^R X) \\ &\quad + \mathcal{Q}\left(\varphi^R \sum_{i=1}^{k} v_i a_i\right) + 2 \sum_{i=1}^{k} \mathcal{Q}(\varphi^R X, \varphi^R v_i a_i), \end{aligned}$$

where we have used that

$$\mathcal{Q}(\varphi^R X, Y) = \mathcal{Q}(X, \varphi^R Y) = \mathcal{Q}(X,Y)$$

and that

$$\mathcal{Q}(Y, \varphi^R v_i a_i) = -\lambda_i \int_M v_i \mathcal{P} \langle Y, a_i\rangle \mathrm{d}v = 0.$$

From the last inequality, it follows

(4.18)

$$\mathcal{Q}(X,Y)^2 \le \mathcal{Q}(Y) \left(\mathcal{Q}(\varphi^R X) + \mathcal{Q}\left(\varphi^R \sum_{i=1}^{k} v_i a_i \right) + 2 \sum_{i=1}^{k} \mathcal{Q}(\varphi^R X, \varphi^R v_i a_i) \right).$$

We will prove that when $R \mapsto \infty$, the limit of the three terms between parenthesis is zero. In such a case, as from Theorem 4.20 $\mathcal{Q}(Y) \ge 0$, we will obtain that $\mathcal{Q}(X,Y) = 0$ and the claim follows.

In fact, using Lemma 4.15 and the divergence theorem we have

$$\begin{aligned}\mathcal{Q}(\varphi^R X) &= -\int_M \{\varphi^R g(\Delta(\varphi^R X), X) - 2(\varphi^R)^2 K\|X\|^2\}\mathrm{d}v \\ &= \int_M \{-\varphi^R \Delta\varphi^R \|X\|^2 - 2\varphi^R g(\nabla_{\nabla\varphi^R} X, X)\}\mathrm{d}v \\ &= \int_M \{-\varphi^R \Delta\varphi^R \|X\|^2 - \frac{1}{2} g(\nabla(\varphi^R)^2, \nabla\|X\|^2)\}\mathrm{d}v \\ &= \int_M \|\nabla\varphi^R\|^2 \|X\|^2 \mathrm{d}v.\end{aligned}$$

Now from Lemma 4.18,3, we have $\lim_{R\to\infty} \mathcal{Q}(\varphi^R X) = 0$.

Also,

$$\mathcal{Q}(\varphi^R \sum_{i=1}^k v_i a_i) = \sum_{i,j=1}^k \langle a_i, a_j\rangle \mathcal{Q}(\varphi^R v_i, \varphi^R v_j).$$

Using the divergence theorem for the compactly supported vector field $v_i v_j \varphi^R \nabla\varphi^R$, we get

$$\begin{aligned}\mathcal{Q}(\varphi^R v_i, \varphi^R v_j) &= -\frac{1}{2}\int_M \{\varphi^R v_j L(\varphi^R v_i) + \varphi^R v_i L(\varphi^R v_j)\}\mathrm{d}v \\ &= \frac{\lambda_i + \lambda_j}{2}\int_M (\varphi^R)^2 v_i v_j \mathcal{P}\,\mathrm{d}v - \int_M \{v_i v_j \varphi^R \Delta\varphi^R + \varphi^R g(\nabla\varphi^R, \nabla(v_i v_j))\}\mathrm{d}v \\ &= \frac{\lambda_i + \lambda_j}{2}\int_M (\varphi^R)^2 v_i v_j \mathcal{P}\,\mathrm{d}v \int_M \|\nabla\varphi^R\|^2 v_i v_j \mathrm{d}v.\end{aligned}$$

So

$$\begin{aligned}\lim_{R\to\infty} \langle a_i, a_j\rangle \mathcal{Q}(\varphi^R v_i, \varphi^R v_j) &= \frac{\lambda_i + \lambda_j}{2}\lim_{R\to\infty} \langle a_i, a_j\rangle \int_M (\varphi^R)^2 v_i v_j \mathcal{P}\,\mathrm{d}v \\ &\quad + \lim_{R\to\infty} \langle a_i, a_j\rangle \int_M \|\nabla\varphi^R\|^2 v_i v_j \mathrm{d}v = 0,\end{aligned}$$

because when $R \to \infty$, $a(R) \to 0$, and $\lim_{R\to\infty} \int_M (\varphi^R)^2 v_i v_j \mathcal{P}\,\mathrm{d}v < \infty$ by the dominate convergence theorem and $\lim_{R\to\infty} \int_M \|\nabla\varphi^R\|^2 v_i v_j \mathrm{d}v < \infty$ by Proposition 4.19,2.

Finally,

(4.19)

$$\begin{aligned}\mathcal{Q}(\varphi^R X, \varphi^R v_i a_i) &= -\int_M \varphi^R \mathcal{L}(\varphi^R v_i)\langle X, a_i\rangle \mathrm{d}v = \lambda_i \left\langle \int_M (\varphi^R)^2 v_i \mathcal{P} X, a_i \right\rangle \mathrm{d}v \\ &\quad - \int_M \varphi^R (v_i \Delta\varphi^R + 2g(\nabla\varphi^R, \nabla v_i))\langle X, a_i\rangle \mathrm{d}v.\end{aligned}$$

Since v_i and X belong to $L^2_*(M)$, by the dominate convergence theorem we have

$$\lim_{R\to\infty} \int_M (\varphi^R)^2 v_i \mathcal{P} X\,\mathrm{d}v \in \mathbb{R}^3.$$

Since a_i converges to 0, the first term of (4.19) converges to zero when $R \to \infty$.

Also,

$$\left|\int_M \varphi^R \Delta\varphi^R v_i \langle X, a_i\rangle\,\mathrm{d}v\right| \le \|a_i\| \int_M \varphi^R |\Delta\varphi^R| |v_i| \|X\| \mathrm{d}v,$$

and thus, the second term of (4.19) also converges to 0 when $R \to \infty$ using Lemma 4.18 and that $a_i \to 0$.

Finally, using the Schwarz inequality

$$\left| \int_M \varphi^R g(\nabla \varphi^R, \nabla v_i) \langle X, a_i \rangle \, dv \right| \le \|a_i\| \int_M \|\nabla \varphi^R\| \|X\| \|\nabla v_i\| \, dv$$
$$\le \|a_i\| \left(\int_M \|\nabla \varphi^R\|^2 \|X\|^2 dv \right)^{1/2} \left(\int_M \|\nabla v_i\|^2 dv \right)^{1/2}.$$

So from Lemma 4.18 and Proposition 4.19, it follows that the third term of (4.19) converges to 0 when $R \to \infty$, using that $a_i \to 0$. This completes the proof of the claim.

Now, given a compactly supported smooth map $Y : M \to \mathbb{R}^3$ with $Y \in L^2_*(M) \cap W^\perp$, and φ^R the cut-off function given in Lemma 4.17, we define

$$Y_R = \varphi^R (Y + v_1 b_1 + \cdots + v_k b_k),$$

where $\{b_1, \ldots, b_k\}$ are vectors of $\mathbb{R}^3$ depending on R. As in the proof of the claim, it is possible to choose the vectors b_i such that $Y_R \in W^\perp$ and $\lim_{R\to\infty} b_i = 0$, for all $1 \le i \le k$. So, since Y_R has compact support, from Lemma 4.15 and the last claim, we get

$$0 = \mathcal{Q}(X, Y_R) = -2 \int_M g(\nabla X, A) \langle N, Y_R \rangle \, dv.$$

Now, since $\lim_{R\to\infty} b_i = 0$, we have that

$$\lim_{R\to\infty} g(\nabla X, A) \langle N, Y_R \rangle = g(\nabla X, A) \langle N, Y \rangle.$$

Furthermore, it follows from Proposition 4.19,3 that $\|\nabla X\| \in L^2(M)$ and from Lemma 4.16 that in a neighborhood of each end, $\|A\|^2 = 2|K| \le C \|\phi\|^{-2-\frac{2(m+1)}{d}}$, where $m \ge d$ is an integer and d is the multiplicity of the end. So, we can apply the dominated convergence theorem (Theorem 1.7) and conclude that

$$\int_M g(\nabla X, A) \langle N, Y \rangle \, dv = 0, \tag{4.20}$$

for any compactly supported smooth map $Y : M \to \mathbb{R}^3$ with $Y \in L^2_*(M) \cap W^\perp$.

On the other hand, if b is any vector of $\mathbb{R}^3$, then for any $1 \le i \le k$ we have

$$\mathcal{Q}(X, v_i b) = \sum_{j=1}^{3} \mathcal{Q}(\langle X, e_j \rangle, v_i \langle b, e_j \rangle) = -\sum_{j=1}^{3} \int_M \langle X, e_j \rangle \langle b, e_j \rangle \mathcal{L}(v_i) \, dv$$
$$= \lambda_i \int_M \mathcal{P} \langle X, v_i b \rangle = 0,$$

because $X \in W^\perp \subset L^2_*(M)$ and $v_i b \in W \subset L^2_*(M)$. Thus, from Lemma 4.15 we have

$$0 = \mathcal{Q}(X, v_i b) = -2 \int_M g(\nabla X, A) \langle N, v_i b \rangle \, dv, \tag{4.21}$$

for any $b \in \mathbb{R}^3$ and any $1 \le i \le k$. Now it follows from (4.20) and (4.21) that $g(\nabla X, A) = 0$ and Lemma 4.15 implies that $X \in \mathcal{V}^*(M)$. So, $\ker F \subset \mathcal{V}^*(M)$ and hence, $\dim \ker F \le 3$. This concludes the proof. $\square$

COROLLARY 4.23. *Let $\Phi : M \to \mathbb{R}^3$ be a minimal immersion of a complete orientable surface M. Then:*

1. *M is stable if and only if M is a plane.*
2. $\mathrm{Ind}(M) = 1$ *if and only if Φ is congruent to the catenoid or the Enneper surface.*
3. $\mathrm{Ind}(M) \neq 2$.
4. *If Φ is an embedding,* $\mathrm{Ind}(M) \neq 3$.

REMARK 4.24. This gives a new proof of Theorem 4.2. Also, in [**CM16**], 3 was proved under the embedeness assumption.

PROOF. Suppose that $\mathrm{Ind}(M) \leq 2$. Then from Theorem 4.21 we have

$$5 \geq h + \sum_{i=1}^{n}(1 + d_i) \geq h + 2,$$

and so $h \leq 3$.

If $h = 3$, then $n = 1$ and $d_1 = 1$, hence the minimal surface has only one end, which is embedded. This implies that $\Phi(M)$ is a plane of $\mathbb{R}^3$ with $h = 0$ and so, we get a contradiction.

If $h = 2$, the last formula gives $\sum_{i=1}^{n}(1 + d_i) \leq 3$, and so we have two possibilities: $n = 1$, $d_1 = 1$ or $n = 1$, $d_1 = 2$. We know that the first one is impossible, and if the second one happens, from Theorems 4.6 and 4.4, the degree of the Gauss map is $d = 5/2$, which is a contradiction.

If $h = 1$, the above formula gives $\sum_{i=1}^{n}(1 + d_i) \leq 4$, and so we have four possibilities: $n = 1$ and $d_1 = 1, 2, 3$ or $n = 2$, $n_1 = n_2 = 1$. Using the similar arguments as before, we only need to study the cases $n = 1$, $d_1 = 3$ and $n = 2$, $d_1 = d_2 = 1$. In both cases, Theorems 4.6 and 4.4 imply that $d = 2$, which means that M has finite total curvature -8π. This fact contradicts Corollary 4.8.

If $h = 0$, then by the same argument $d = -1 + \frac{1}{2}\sum_{i=1}^{n}(1 + d_i) \leq -1 + \frac{5}{2} = \frac{3}{2}$. So $d \leq 1$ and Theorem 4.4 finishes the proof of 1,2 and 3.

Now suppose that Φ is an embedding and $\mathrm{Ind}(M) = 3$. Then from Theorem 4.21 and using that $d_i = 1$ for any end,

$$7 \geq h + 2n.$$

From the main result in [**Sch83**], if $n \leq 2$, then M is either the plane or the catenoid, whose indices are 0 and 1 respectively. So $n = 3$ and $h = 0, 1$. If $h = 0$, from [**LR91**] it follows that M is either the plane or the catenoid. Hence we get that $h = 1$ and $n = 3$. Using Costa's classification of embedded minimal tori with three ends of $\mathbb{R}^3$, [**Cos84, Cos91**], M must be an element of the deformation family of the Costa surface ([**Cos91, HM90**]). The proof finishes with the next result, in which we prove that the index of any surface in the deformation of the Costa surface is at least 4. □

The deformation of the Costa surface $M(1)$ is a family of minimal surfaces $\{M_t, t \in [1, \infty)\}$ ($M(1) = M_1$, see [**Cos84, Cos91**] for details) with finite total curvature, genus one and three ends. The compact Riemann surface Σ_t given in Theorem 4.4 is the torus $\Sigma_t = \mathbb{C}/\Lambda_t$ where Λ_t is the lattice $\Lambda_t = \{m + int \mid m, n \in \mathbb{Z}\}$ and the ends are given by $\{\Pi(0), \Pi(1/2), \Pi(it/2)\}$ where $\Pi : \mathbb{C} \to \Sigma_t$ is the projection. These surfaces are invariant under the isometries of $\mathbb{R}^3$:

$$(x_1, x_2, x_3) \mapsto (-x_1, x_2, x_3), \qquad (x_1, x_2, x_3) \mapsto (x_1, -x_2, x_3),$$

which descend to isometries $\{\tau_i,\ 1 \le i \le 2\}$ of Σ_t given by

$$\tau_1(\Pi(x+iy)) = \Pi(1-x+iy), \qquad \tau_2(\Pi(x+iy)) = \Pi(x+i(t-y)),$$

where $x+iy \in \mathbb{C}$.

THEOREM 4.25 ([**CM23**]). *If M_t with $t \in [1,\infty)$ is the deformation of the Costa surface $M(1) = M_1$, then* $\mathrm{Ind}(M_t) \ge 4$ *for all* $t \ge 1$.

PROOF. Following the same argument as in Proposition 4.14, the vision number $\nu(M_t, X) \ge 4$ and thus $\mathrm{Ind}(M_t) \ge 3$. The proof consists of assuming that $\mathrm{Ind}(M_t) = 3$ for some $t \in [1,\infty)$ and to obtain a contradiction.

If $\mathrm{Ind}(M_t) = 3$ for some $t \in [1,\infty)$, then $\nu(M_t, X) = 4$ and so, $M_t \setminus (\Pi_1 \cup \Pi_2)$ has 4 connected components, where the Jacobi function $\langle N, X\rangle$ has no zeros, being Π_1 and Π_2 the planes $x_2 = 0$ and $x_1 = 0$ respectively. Then, Theorem 2.25 implies that every connected component is stable.

According to Theorem 4.20, if $\{v_1, v_2, v_3\}$ is a basis of the 3-dimensional space $W \subset L^2_*(M_t)$ with $\mathcal{L}v_i + \lambda_i \mathcal{P} v_i = 0$, $\lambda_i < 0$ and some v_i satisfies $v_i \circ \tau_1 = -v_i$ and $v_i \circ \tau_2 = -v_i$, then v_i vanishes on the boundary of the 4 connected components, and since each connected component is stable, the restriction of v_i to each connected component vanishes. This implies that $v_i = 0$ on M_t. So we get that

$$W^{--} := W \cap C^{--} = \{0\}, \quad \text{with} \quad C^{--} = \{f \in C^\infty(M_t) \mid f \circ \tau_1 = -f,\ f \circ \tau_2 = -f\},$$

and thus

$$W = W^{++} \oplus W^{+-} \oplus W^{-+},$$

where W^{++}, W^{+-} and W^{--} are the other eigenspaces of τ_1 and τ_2.

Now, using another argument we will arrive at a contradiction.

According to the proof of the Theorem 4.21, the space

$$\mathcal{H}_*(M_t) = \{X \in \mathcal{H}(M_t) \mid \|X\| \in L^2_*(M_t)\}$$

has dimension 12.

CLAIM. If

$$\mathcal{H}_*(M_t) = \mathcal{H}_*^{++}(M_t) \oplus \mathcal{H}_*^{+-}(M_t) \oplus \mathcal{H}_*^{-+}(M_t) \oplus \mathcal{H}_*^{--}(M_t)$$

is the decomposition of $\mathcal{H}_*(M_t)$ into eigenspaces of τ_1, τ_2, then

$$\dim \mathcal{H}_*^{+-}(M_t) = \dim \mathcal{H}_*^{-+}(M_t) = 4, \quad \dim \mathcal{H}_*^{++}(M_t) = \dim \mathcal{H}_*^{--}(M_t) = 2.$$

Assuming the claim holds, we will get the desired contradiction. If $F^{+-} = F_{|\mathcal{H}_*^{+-}(M_t)} : \mathcal{H}_*^{+-}(M_t) \to \mathbb{R}^9$, with

$$F(X) = \left(\int_{M_t} v_1 \mathcal{P} X \,\mathrm{dv}, \int_{M_t} v_2 \mathcal{P} X \,\mathrm{dv}, \int_{M_t} v_3 \mathcal{P} X \,\mathrm{dv} \right),$$

then, following the proof of Theorem 4.21, we have $\ker F^{+-} = \ker F \cap \mathcal{H}_*^{+-}(M_t) \subset \mathcal{V}^*(M_t) \cap \mathcal{H}_*^{+-}(M_t)$, and so, $\dim \ker F^{+-} \le 1$, because the basis $\{N \wedge e_1, N \wedge e_2, N \wedge e_3\}$ of $\mathcal{V}^*(M_t)$ satisfies $N \wedge e_1 \in \mathcal{H}_*^{+-}(M_t)$, $N \wedge e_2 \in \mathcal{H}_*^{-+}(M_t)$ and $N \wedge e_3 \in \mathcal{H}_*^{--}(M_t)$. So it follows from the claim that $4 \le 1 + \dim \operatorname{img} F^{+-} \le 1 + \dim W^{++} + \dim W^{+-}$, because if $X \in \mathcal{H}_*^{+-}(M_t)$ then $\langle X, e_1\rangle \in C^{--}$, $\langle X, e_2\rangle \in C^{++}$, $\langle X, e_3\rangle \in C^{+-}$ and $W^{--} = \{0\}$, where

$$C^\infty(M_t) = C^{++} \oplus C^{+-} \oplus C^{-+} \oplus C^{--}$$

is the decomposition of $C^\infty(M_t)$ into eigenspaces of τ_1, τ_2.

Thus we have $3 \leq \dim W^{++} + \dim W^{+-}$. Similarly, we would get that $3 \leq \dim W^{++} + \dim W^{-+}$ and $1 \leq \dim W^{+-} + \dim W^{-+}$. Since $3 = \dim W = \dim W^{++} + \dim W^{+-} + \dim W^{-+}$, we get a contradiction.

PROOF OF THE CLAIM. As we did in Proposition 4.19, since $\mathcal{H}(M_t)$ is isomorphic to the corresponding space of harmonic 1-forms $\mathcal{H}^1(M_t)$, we will prove the statement with the decomposition

$$\mathcal{H}^1_*(M_t) = H^{++} \oplus H^{+-} \oplus H^{-+} \oplus H^{--}.$$

It is easy to check that the basic 1-forms dx and dy on Σ_t are of type $-+$ and $+-$ respectively, and that they span $\mathcal{H}^1_*(M_t)$. Using this that fact, if $\star : \mathcal{H}^1(M_t) \to \mathcal{H}^1(M_t)$ is the star operator on M_t (see Section 1.3), it is easy to check that $\star : H^{++} \to H^{--}$ and $\star : H^{+-} \to H^{-+}$ are isomorphisms. Therefore, to prove the claim, we only need to check that $\dim H^{++} = 2$.

Taking into account that the basic 1-forms dx and dy are of type $-+$ and $+-$ respectively, it is easy to check that

(4.22)
$$\Re\left(\frac{dz}{z}\right) \in H^{++}, \quad \Im\left(\frac{dz}{z}\right) \in H^{--}, \quad \Re\left(\frac{dz}{z^2}\right) \in H^{-+}, \quad \Im\left(\frac{dz}{z^2}\right) \in H^{+-}.$$

If $\omega \in H^{++}$, then $\alpha = \omega + i \star \omega$ is a meromorphic 1-form on M_t with poles at the points $\{\Pi(0), \Pi(1/2), \Pi(it/2)\}$ of order $l \leq 2$. We will prove that the order of α at the three poles is $l \leq 1$ and that the residues of α are real.

If $\omega \in H^{++}$ and $\alpha = \omega + i \star \omega$, then around the end $\Pi(0)$, $\alpha = \left(\sum_{n=-2}^{\infty} a_n z^n\right) dz$ and so, using (4.22), $\Re(\alpha) \in H^{++}$ implies that $a_{-2} = 0$ and $a_{-1} \in \mathbb{R}$. So, the order of the pole $\Pi(0)$ is at most 1 and $\operatorname{Res}_{\Pi(0)} \alpha \in \mathbb{R}$. The same proof applies to the other two ends.

Since the sum of the residues of α is equal zero, we can define a linear map

$$\begin{aligned} G : H^{++} &\to \{(x_1, x_2, x_3) \in \mathbb{R}^3 \mid x_1 + x_2 + x_3 = 0\} \\ \omega &\mapsto G(\omega) = \left(\operatorname{Res}_{\Pi(0)} \alpha, \operatorname{Res}_{\Pi(1/2)} \alpha, \operatorname{Res}_{\Pi(it/2)} \alpha\right). \end{aligned}$$

If $\omega \in \ker G$, then α is a meromorphic 1-form on M_t without poles, and so α extends to a holomorphic 1-form $\widetilde{\alpha}$ on the torus Σ_t. This implies that $\widetilde{\alpha} = a dz$ with $a \in \mathbb{C}$. Since the real part of α is invariant under τ_1 and τ_2, we get that $a = 0$. Hence $\ker G = \{0\}$. Also, if $\{a_1, a_2, a_3\}$ are real numbers with $a_1 + a_2 + a_3 = 0$, then from Theorem II.5.3 in [**FK92**], there exists a meromorphic 1-form α on M_t with only simple poles at $\Pi(0), \Pi(1/2)$ and $\Pi(it/2)$ and $\operatorname{Res}_{\Pi(0)} \alpha = a_1$, $\operatorname{Res}_{\Pi(1/2)} \alpha = a_2$ and $\operatorname{Res}_{\Pi(it/2)} \alpha = a_3$. Using (4.22), it is easy to prove that $\omega = \Re(\alpha) \in H^{++}$. So G is an isomorphism and hence $\dim H^{++} = 2$. □

Considering the above results, it seems natural to conjecture that:

The only orientable complete minimal surfaces embedded in $\mathbb{R}^3$ with index ≤ 5 are the plane, the catenoid and the Costa surface.

We end this section with an appendix proving a result mentioned in Remark 3.14, for which we need the theory of minimal surfaces of $\mathbb{R}^3$.

THEOREM 4.26 ([**Ros06**]). *Let M be a compact Riemann surface of genus 4 and let $\phi : M \to \mathbb{S}^2$ be a non-constant holomorphic map of degree d. If $\operatorname{Ind}(M, \phi) = 1$, then $d = 2$.*

PROOF. According to Corollary 3.13 it is sufficient to exclude the case $d = 3$. From Theorem III.8.7 in [**FK92**], given a compact Riemann surface of genus 4, there are three possibilities: either M is hyperelliptic, or M admits, up to conformal transformations of $\mathbb{S}^2$, a unique holomorphic map $\psi : M \to \mathbb{S}^2$ of degree 3, or M admits two holomorphic maps of degree 3 which are not conformal transformations of each other. Since $\mathrm{Ind}(M, \phi) = 1$, Theorem 3.12 implies that the first and third possibilities cannot happen. So, the second possibility occurs, and from the proposition on page 114 of [**FK92**], in this case there exists a basis $\{\omega_1, \omega_2, \omega_3, \omega_4\}$ of the complex space of the holomorphic 1-forms on M such that $\omega_1^2 + \omega_2^2 + \omega_3^2 = 0$. Furthermore, the unique holomorphic map of degree 3 on M, up to conformal transformations of $\mathbb{S}^2$, is given by $g = \omega_3/(\omega_1 - i\omega_2) : M \to \overline{\mathbb{C}} \equiv \mathbb{S}^2$. If $\eta := \omega_3$, then we get that

$$\omega_1 = \frac{1}{2}(g^{-1} - g)\eta, \quad \omega_2 = \frac{i}{2}(g^{-1} + g)\eta, \quad \omega_3 = \eta.$$

Since g has three poles and η has six zeros, it follows that the zeros of η coincide with the poles of g and the order of the zero is twice the order of the pole at these points.

It is clear that since g and ϕ are conformal transformations of each other, we can modify appropriately η so that $\{\eta, \phi\}$ satisfy the same properties. Now, if $\Pi : \mathbb{D} \to M$ is the holomorphic universal covering, then $\{\Pi^*\eta, \phi \circ \Pi\}$ are the Weierstrass data of a minimal immersion $\Phi : \mathbb{D} \to \mathbb{R}^3$ (see [**Oss69**]). By conformal modification of the metric on M, the projection Π becomes in a local isometry and we know that the index does not change.

Since M has genus 4, the space $\mathcal{H}(M)$ of harmonic vector fields has dimension 8, and if $X \in \mathcal{H}(M)$, then its lift $\widetilde{X}$ to $\mathbb{D}$ is also a harmonic vector field, which can be written $\widetilde{X} = (\widetilde{X}_1, \widetilde{X}_2, \widetilde{X}_3)$, with $\widetilde{X}_i \in C^\infty(\mathbb{D})$. It is clear that there exist functions $X_i \in C^\infty(M)$, $1 \leq i \leq 3$, such that $\widetilde{X}_i = X_i \circ \Pi$. If $\mathcal{L} = \widetilde{\Delta} + \|\widetilde{\nabla} N\|^2$ is the Jacobi operator of Φ, then $\mathcal{L}(\widetilde{X_i}) = \widetilde{LX_i}$ and so, using Lemma 4.15, we get

$$\sum_{i=1}^{3} \langle LX_i, X_i \rangle = \sum_{i=1}^{3} \langle \mathcal{L}\widetilde{X}_i, \widetilde{X}_i \rangle = -\langle \widetilde{\nabla}\widetilde{X}, A \rangle \sum_{i=1}^{3} \langle N, \widetilde{X} \rangle = 0.$$

Since $\mathrm{Ind}(M, \phi) = 1$, let λ_1 be the unique negative eigenvalue of $L = \Delta + \|\nabla\phi\|^2$ and $f : M \to \mathbb{R}$ be a function satisfying $Lf + \lambda_1 f = 0$. We define a linear map $F : \mathcal{H}(M) \to \mathbb{R}^3$ by

$$F(X) = \left(\int_M f X_1 \, dv, \int_M f X_2 \, dv, \int_M f X_3 \, dv \right).$$

If $X \in \ker F$, then $Q(X_i) \geq 0$ and so

$$0 \leq \sum_{i=1}^{3} Q(X_i) = -\sum_{i=1}^{3} \int_M \langle LX_i, X_i \rangle \, dv = 0.$$

So $LX_i = 0$ for all i and so $\mathcal{L}\widetilde{X} = 0$. Again using Lemma 4.15, $\widetilde{X} \in \mathcal{V}^*(\mathbb{D})$. Hence we have a monomorphism

$$X \in \ker F \mapsto \widetilde{X} \in \mathcal{V}^*(\mathbb{D}),$$

from a vector space of dimension at least 5 into a 3-dimensional space. This is a contradiction and the second possibility does not happen. Thus, $d = 2$. □

4.2. Nonorientable minimal surfaces in $\mathbb{R}^3$

Let $\Phi : M \to \mathbb{R}^3$ be a minimal immersion of a nonorientable surface M. If $\Pi : \widetilde{M} \to M$ is the orientable double covering of M and τ the isometry of $\widetilde{M}$ which generates the transformation group of the covering, then the Gauss map of Φ is the map $G : M \to \mathbb{RP}^2$ which makes the following diagram commutative

$$\begin{array}{ccc} \widetilde{M} & \xrightarrow{N} & \mathbb{S}^2 \\ \downarrow{\scriptstyle \Pi} & & \downarrow{\scriptstyle \Pi_0} \\ M & \xrightarrow{G} & \mathbb{RP}^2, \end{array}$$

where N is the Gauss map of the two-sided minimal immersion $\widetilde{\Phi} = \Phi \circ \Pi : \widetilde{M} \to \mathbb{R}^3$ and Π_0 is the standard projection.

The minimal surface M has finite total curvature if $\int_M |K|\, dv < \infty$. In the next result we have summarized some well-known results that will be used along this section.

THEOREM 4.27 ([**LM99**]). *Let $\Phi : M \to \mathbb{R}^3$ be a minimal immersion of a nonorientable complete surface M in $\mathbb{R}^3$ with finite total curvature. Then:*

(1) *There exists a nonorientable compact surface Σ and a diffeomorphism $F : M \to \Sigma \setminus \{p_1, \dots, p_n\}$. The points $\{p_1, \dots, p_n\}$ are called the ends of M.*
(2) *There exists a map $\phi : \Sigma \to \mathbb{RP}^2$ such that $G_p = \phi_{F(p)}$ for any $p \in M$, where G is the Gauss map of M.*
(3) *If d is the degree of ϕ and d_i the multiplicity of the end p_i, then*
$$-\int_M K\,\mathrm{d}v = 2\pi d = 2\pi\big(-\chi(\Sigma) + \sum_{i=1}^{n}(1 + d_i)\big),$$
where $\chi(\Sigma)$ is the Euler characteristic of Σ.
(4) *$d \geq 3$ and the equality holds if and only if M is the Meeks minimal Möbius strip ([**Mee81**]).*

REMARK 4.28. If $\widetilde{M}$ is the orientable double covering of M and $\widetilde{K}$ is its Gauss curvature, then $\int_{\widetilde{M}} \widetilde{K}\, \mathrm{d}\widetilde{v} = 2\int_M K\, \mathrm{d}v$. Also, the minimal immersion $\widetilde{\Phi} : \widetilde{M} \to \mathbb{R}^3$ has $2n$ ends given by $\{q_1, \dots, q_n, \tau(q_1), \dots, \tau(q_n)\}$, $\Pi(q_i) = p_i$, with multiplicities $d(q_i) = d(\tau(q_i)) = d_i$, for any $1 \leq i \leq n$.

Now, given a minimal immersion $\Phi : M \to \mathbb{R}^3$ of a nonorientable surface M, from Definition 2.31, if $\widetilde{\mathcal{L}}$ is the Jacobi operator of $\widetilde{M}$, then

$$\mathrm{Ind}(M) = \mathrm{Ind}(\widetilde{\mathcal{L}}^-),$$

where $\widetilde{\mathcal{L}}^-$ is the restriction of $\widetilde{\mathcal{L}}$ to $C^\infty_-(\widetilde{M}) = \{f \in C^\infty(\widetilde{M}) \mid f \circ \tau = -f\}$.

THEOREM 4.29 ([**Ros06**]). *Let $\Phi : M \to \mathbb{R}^3$ be a minimal immersion of a nonorientable complete surface M in $\mathbb{R}^3$. Then,* $\mathrm{Ind}(M)$ *is finite if and only if M has finite total curvature.*

PROOF. If M has finite total curvature, then $\widetilde{M}$ also has finite total curvature (see Remark 4.28), and Theorem 4.7 says that $\mathrm{Ind}(\widetilde{M})$ is finite. Since $\mathrm{Ind}(M) = \mathrm{Ind}(\widetilde{\mathcal{L}}^-) \leq \mathrm{Ind}(\widetilde{\mathcal{L}}) = \mathrm{Ind}(\widetilde{M})$, we get that $\mathrm{Ind}(M)$ is finite.

Conversely, we assume that $\mathrm{Ind}(M) < \infty$. If the genus of $\widetilde{M}$ is finite, then by Corollary 3.16, M has finite total curvature.

Now suppose that $\widetilde{M}$ has infinite genus. Then if $\mathrm{Ind}(M) = \mathrm{Ind}(\widetilde{\mathcal{L}}^-) = k$, from Theorem 2.30,2, there exist $L^2(\widetilde{M})$-functions $\{v_1, \dots, v_k\}$ with $v_i \circ \tau = -v_i$ and $\widetilde{\mathcal{L}}v_i + \lambda_i v_i = 0$, $\lambda_i < 0$, and such that, if $f \in C_0^\infty(\widetilde{M})$ satisfies $f \circ \tau = -f$ and is L^2-orthogonal to v_i, $1 \le i \le k$, then

$$\widetilde{\mathcal{Q}}(f) \ge 0.$$

Since the genus of $\widetilde{M}$ is infinite, the space of L^2-harmonic vector fields on $\widetilde{M}$ has also infinite dimension. We consider the linear map $F : \mathcal{H}(\widetilde{M}) \cap L^2(M) \to \mathbb{R}^{3k}$ defined by

$$F(X) = \left(\int_{\widetilde{M}} v_1 X \, d\tilde{v}, \dots, \int_{\widetilde{M}} v_k X \, d\tilde{v} \right),$$

and let $X \in \ker F$. Let $Y : \widetilde{M} \to \mathbb{R}^3$ be any compactly supported smooth map with $Y \circ \tau = -Y$ and $\int_{\widetilde{M}} v_i Y \, d\tilde{v} = 0$ for any $1 \le i \le k$. Choose R large enough in Lemma 4.17 so that $\{x \in M \mid \|\Phi(x)\| < R\}$ contains the support of Y, and let φ^R be the cut-off function defined in Lemma 4.17. Now, for each $t \in \mathbb{R}$ we define

$$X_t = \varphi^R\big(X + tY + v_1 a_1 + \dots + v_k a_k\big),$$

where $\{a_1, \dots, a_k\}$ are vectors of $\mathbb{R}^3$. It is possible to choose the vectors a_i such that X_t is $L^2(\widetilde{M})$-orthonormal to v_i for any $1 \le i \le k$. In fact, since $\int_{\widetilde{M}} t\varphi^R v_Y \, d\tilde{v} = \int_{\widetilde{M}} t v_i Y \, d\tilde{v} = 0$, the vectors a_i are solution of the following linear system

$$\sum_{j=1}^k \left(\int_{\widetilde{M}} \varphi^R v_i v_j \, d\tilde{v} \right) a_j = - \int_{\widetilde{M}} \varphi^R v_i X \, d\tilde{v}, \quad 1 \le i \le k.$$

By the dominate convergence theorem, $\lim_{R\to\infty} \big(\int_{\widetilde{M}} \varphi^R v_i v_j \, d\tilde{v} \big)$ is the identity matrix, and thus, for sufficiently large R, the above system has a solution. We observe that, again by the dominate covergence theorem, $\lim_{R\to\infty} \int_{\widetilde{M}} \varphi^R v_i X \, d\tilde{v} = \int_{\widetilde{M}} v_i X \, d\tilde{v} = 0$, and so, $\lim_{R\to\infty} a_i = 0$.

Since X_t is a compactly supported smooth vector field $L^2(\widetilde{M})$-orthogonal to v_i, $1 \le i \le k$, it follows that $\widetilde{\mathcal{Q}}(X_t) \ge 0$. So,

$$\begin{aligned} 0 &\le t^2 \widetilde{\mathcal{Q}}(Y) + 2t\widetilde{\mathcal{Q}}(\varphi^R X, Y) + \sum_{i=1}^k \widetilde{\mathcal{Q}}(Y, \varphi^R v_i a_i) + \widetilde{\mathcal{Q}}(\varphi^R X) \\ &\quad + \widetilde{\mathcal{Q}}(\varphi^R \sum_{i=1}^k v_i a_i) + 2 \sum_{i=1}^k \widetilde{\mathcal{Q}}(\varphi^R X, \varphi^R v_i a_i) \\ &= t^2 \widetilde{\mathcal{Q}}(Y) + 2t\widetilde{\mathcal{Q}}(X, Y) + \widetilde{\mathcal{Q}}(\varphi^R X) + \widetilde{\mathcal{Q}}(\varphi^R \sum_{i=1}^k v_i a_i) + 2 \sum_{i=1}^k \widetilde{\mathcal{Q}}(\varphi^R X, \varphi^R v_i a_i), \end{aligned}$$

where we have used that

$$\widetilde{\mathcal{Q}}(\varphi^R X, Y) = \widetilde{\mathcal{Q}}(X, \varphi^R Y) = \widetilde{\mathcal{Q}}(X, Y)$$

and that

$$\widetilde{\mathcal{Q}}(Y, \varphi^R v_i a_i) = -\lambda_i \int_{\widetilde{M}} v_i \langle Y, a_i \rangle \, d\tilde{v} = 0.$$

From the last inequality it follows that

$$\widetilde{\mathcal{Q}}(X,Y)^2 \leq \widetilde{\mathcal{Q}}(Y)\left(\widetilde{\mathcal{Q}}(\varphi^R X) + \widetilde{\mathcal{Q}}(\varphi^R \sum_{i=1}^{k} v_i a_i) + 2\sum_{i=1}^{k} \widetilde{\mathcal{Q}}(\varphi^R X, \varphi^R v_i a_i)\right). \tag{4.23}$$

Following a similar reasoning as in the proof of Theorem 4.21, we conclude that $\widetilde{\mathcal{Q}}(X,Y) = 0$. But from Lemma 4.15, it follows that

$$\widetilde{\mathcal{Q}}(X,Y) = -2\int_{\widetilde{M}} g(\widetilde{\nabla} X, A)\langle \widetilde{N}, Y\rangle \, d\tilde{v}.$$

Hence, $g(\widetilde{\nabla} X, A) = 0$ and from Lemma 4.15 again, $X \in \mathcal{V}(\widetilde{M})$. So, $\dim(\mathcal{H}(\widetilde{M}) \cap L^2(M)) \leq 3k + 3$, which is a contradiction. □

THEOREM 4.30 ([**CM23**]). *Let $\Phi : M \to \mathbb{R}^3$ be a minimal immersion of a complete nonorientable surface of finite total curvature with n ends $\{p_1, \ldots, p_n\}$ with multiplicities $\{d_1, \ldots, d_n\}$ respectively. If h is the genus of the orientable double covering $\widetilde{M}$ of M, then*

$$\operatorname{Ind}(M) \geq \frac{1}{3}\Big(h - 4 + 2\sum_{i=1}^{n}(1 + d_i)\Big).$$

PROOF. The proof will follow the same scheme of Theorem 4.21 with the appropriate modifications due to the non-orientability of the minimal surface.

As we indicated above, if $\Pi : \widetilde{M} \to M$ is the orientable double covering of M and $\widetilde{\Phi} = \Phi \circ \Pi : \widetilde{M} \to \mathbb{R}^3$ the corresponding two-sided minimal surface, then $\operatorname{Ind}(M) = \operatorname{Ind}(\widetilde{\mathcal{L}}^-)$, where $\widetilde{\mathcal{L}}^-$ is the restriction of $\widetilde{\mathcal{L}}$ to $C^\infty_-(\widetilde{M}) = \{f \in C^\infty(\widetilde{M}) \mid f \circ \tau = -f\}$, and τ is the isometry of $\widetilde{M}$ which generates the transformation group of the covering.

The space $\mathcal{H}(\widetilde{M})$ admits a decomposition $\mathcal{H}(\widetilde{M}) = \mathcal{H}_+(\widetilde{M}) \oplus \mathcal{H}_-(\widetilde{M})$ where

$$\mathcal{H}_\pm(\widetilde{M}) = \{X \in (\widetilde{M}) \mid \tau_* X = \pm X\}.$$

Moreover, if J is the complex structure of $\widetilde{M}$, since τ changes orientation on $\widetilde{M}$, we have that $\tau_* \circ J = -J \circ \tau_*$ and so, $J : \mathcal{H}_+(\widetilde{M}) \to \mathcal{H}_-(\widetilde{M})$ is an isomorphism. Hence, as $\mathcal{H}_-(\widetilde{M}) \equiv \mathcal{H}(M)$, from Proposition 4.19, the dimension of $\mathcal{H}^-_*(\widetilde{M}) := \mathcal{H}_-(\widetilde{M}) \cap L^2_*(\widetilde{M})$ is at least $h - 1 + 2\sum_{i=1}^n (1 + d_i)$.

In the same way that a version of Theorem 2.29 was obtained for Schrödinger operators acting on anti-invariant functions (Theorem 2.30), we can obtain a version of Theorem 4.20 for the $\widetilde{\mathcal{L}}^-$ operator. Thus, if $\operatorname{Ind}(M) = \operatorname{Ind}(\widetilde{\mathcal{L}}^-) = k$, then there exists a k-dimensional subspace $W \subset L^2_*(\widetilde{M})$ with an L^2_* orthonormal basis $\{v_1, \ldots, v_k\}$ with

$$\widetilde{\mathcal{L}}^- v_i + \lambda_i \mathcal{P} v_i = 0, \quad \lambda_i < 0.$$

We consider the linear map $F : \mathcal{H}^-_*(\widetilde{M}) \to \mathbb{R}^{3k}$ defined by

$$F(X) = \left(\int_{\widetilde{M}} v_1 \mathcal{P} X \, d\tilde{v}, \ldots, \int_{\widetilde{M}} v_k \mathcal{P} X \, d\tilde{v}\right).$$

Then, $\dim \ker F \geq h + 2\sum_{i=1}^n (d_i + 1) - 1 - 3k$. We will prove that $\dim \ker F \leq 3$, and the theorem will follow.

CLAIM. If $X \in \ker F$, then $\widetilde{Q}(X,Y) = \sum_{i=1}^{3} \widetilde{Q}(\langle X, e_i\rangle, \langle Y, e_i\rangle) = 0$ for any compactly supported smooth map $Y : \widetilde{M} \to \mathbb{R}^3$ with $Y \in L^2_*(M) \cap W^{\perp}$ and $Y \circ \tau = -Y$, where $\{e_1, e_2, e_3\}$ is an orthonormal basis of $\mathbb{R}^3$.

The proof of the claim is similar to that of Theorem 4.21, taking into account that the cut-off function φ^R satisfies $\varphi^R \circ \tau = \varphi^R$ and so, $X_t = \varphi^r(X + tY + v_1 a_1 + \cdots + v_k a_k)$ is anti-invariant, i.e., $X_t \circ \tau = -X_t$.

The rest of the proof is similar to that of the Theorem 4.21. □

COROLLARY 4.31. *Let $\Phi : M \to \mathbb{R}^3$ be a minimal immersion of a complete nonorientable surface M. Then,* $\mathrm{Ind}(M) \geq 2$.

REMARK 4.32. A. Ros proved in [**Ros06**] that, under the assumptions of Corollary 4.31, $\mathrm{Ind}(M) \geq 1$.

PROOF. If $\mathrm{Ind}(M) \leq 1$, then Theorem 4.30 states that

$$h + 2\sum_{i=1}^{n}(1 + d_i) \leq 7.$$

But since $d_i \geq 1$ for any $1 \leq i \leq n$, we have that M has only one end ($n = 1$) and d_1 is either 1 or 2.

If $d_1 = 1$, then the orientable double covering $\widetilde{M}$ of M is a minimal surface with finite total curvature in $\mathbb{R}^3$ with two embedded ends. By [**Sch83**], $\widetilde{M}$ is a catenoid, which is embedded and so, it cannot be an orientable double covering of M.

If $d_1 = 2$, then the orientable double covering $\widetilde{M}$ of M is a minimal surface with finite total curvature in $\mathbb{R}^3$ with two ends $\{\widetilde{p}_1, \widetilde{p}_2\}$ of multiplicities $\widetilde{d}_i = 2$, $1 \leq i \leq 2$. Then, $\sum_{i=1}^{2} \widetilde{d}_i = 4$. But using Proposition 3.1 of [**HK97**], for any point $q \in \mathbb{R}^3$, the number of points of $\widetilde{\Phi}^{-1}(q)$ is at most $(\sum_{i=1}^{2} \widetilde{d}_i) - 1 = 3$, where $\widetilde{\Phi} = \Phi \circ \Pi : \widetilde{M} \to \mathbb{R}^3$. Since $\widetilde{\Phi}^{-1}(q)$ is even, we get that for any $q \in \Phi(M)$, $\widetilde{\Phi}^{-1}(q)$ has two points and thus Φ is an embedding. This contradicts the one-sidedness of Φ, since a proper embedding in $\mathbb{R}^3$ divides $\mathbb{R}^3$ in two connected components which implies that Φ is two-sided. □

4.3. Minimal hypersurfaces in $\mathbb{R}^{n+1}$

In this section we will study complete stable orientable hypersurfaces of Euclidean space as well as those of finite index. The general results obtained in Section 3.9 will be used, and we will culminate with the resolution of R. Schoen's conjecture ([**CM11**], Conjecture 2.12), which claimed that the hyperplanes of $\mathbb{R}^4$ are the only minimal and stable complete orientable hypersurfaces. Surprisingly, three different proofs have been given in a short period of time, two of them by O. Chodosh and C. Li ([**CL24, CL23**]) and one by G. Catino, P. Mastrolia and A. Roncoroni ([**CMR24**]). Although many of the results in this section are valid for surfaces, we should focus on the case where the dimension of the hypersurface is $n \geq 3$.

First, we will prove some interesting properties about the ends of a minimal hypersurface of $\mathbb{R}^{n+1}$.

THEOREM 4.33 ([**CSZ97, LW02**]). *Let $\Phi : M^n \to \mathbb{R}^{n+1}$ be a two-sided complete minimal immersion. Then:*

(1) *Each end of M has infinite volume.*

(2) *If $n \geq 3$, each end of M is non-parabolic.*
(3) *If M is stable, then M has only one end.*
(4) *If M is stable and $n \geq 3$, then M is non-parabolic.*

PROOF. **Proof of 1**. Following to S.T. Yau, [**Yau75**], we will study the volume of geodesic balls of M. If p is any point of M, we consider the function $f : M \to \mathbb{R}$ given by

$$f(x) = \|\Phi(x) - \Phi(p)\|^2.$$

Then, it is easy to see that the gradient and the Laplacian of f are given by

$$\nabla f = 2\big(\Phi(x) - \Phi(p)\big)^\top, \qquad \Delta f = 2n,$$

where $b^\top$ is the tangential component of $b \in \mathbb{R}^{n+1}$. Integrating $\Delta f = 2n$ over the geodesic ball $B_p(R)$ and using Theorem 1.11, we have that

$$\begin{aligned} 2n \operatorname{vol}(B_p(R)) &= \int_{\partial B_p(R)} \langle \nabla f, \nu \rangle \, \mathrm{d}v = 2 \int_{\partial B_p(R)} \langle \Phi(x) - \Phi(p), \nu \rangle \, \mathrm{d}v \\ &\leq 2 \int_{\partial B_p(R)} \|\Phi(x) - \Phi(p)\| \, \mathrm{d}v \leq 2 \int_{\partial B_p(R)} d(x,p) \, \mathrm{d}v \\ &\leq 2R \operatorname{vol}(\partial B_p(R)), \end{aligned}$$

where ν is the outward normal vector to $\partial B_p(R)$ and d is the distance on M. Then,

$$\frac{\mathrm{d}}{\mathrm{d}R}\left(\frac{\operatorname{vol}(B_p(R))}{R^n}\right) = \frac{1}{R^{n+1}}\big(R \operatorname{vol}(\partial B_p(R)) - n \operatorname{vol}(B_p(R))\big) \geq 0,$$

where we have used that $(\operatorname{vol}(B_p(R))' = \operatorname{vol}(\partial B_p(R))$ and the above inequality. So, $R^{-n} \operatorname{vol}(B_p(R))$ is not decreasing and

$$\frac{\operatorname{vol}(B_p(R))}{R^n} \geq \lim_{R \to 0} \frac{\operatorname{vol}(B_p(R))}{R^n} = \operatorname{vol}(B(1)) \qquad \forall R > 0,$$

where $B(1)$ is the unit ball in $\mathbb{R}^n$, and the last equality is easy to prove using polar coordinates on $B_p(R)$.

Now suppose K is a compact subset of M and E is an unbounded connected component of $M \setminus K$ with $\operatorname{vol}(E) < \infty$. Then, there exists $R_0 > 0$ such that $\operatorname{vol}(E) < (R_0)^n \operatorname{vol}(B(1))$. We choose a point p in E such that $d(p, \partial E) > R_0$, which implies that $B_p(R_0) \subset E$. Then, considering the last inequality, we get the following contradiction:

$$\operatorname{vol}(E) \geq \operatorname{vol}(B_p(R_0)) \geq (R_0)^n \operatorname{vol}(B(1)) > \operatorname{vol}(E).$$

Proof of 2. Let E be an unbounded connected component of $M \setminus K$, where K is a compact subset of M with smooth boundary. We will prove that there exists a non-constant harmonic function f on E such that $f = 1$ on ∂E and $0 \leq f \leq 1$. So, from Definition 3.28, E will be non-parabolic.

Let p be a point of K and $R > 0$ such that $K \subset B_p(R)$. Let f_R be the solution to the Dirichlet problem

$$\begin{cases} \Delta f_R = 0 & \text{on } E_R := E \cap B_p(R) \\ f_R = 1 & \text{on } \partial E \\ f_R = 0 & \text{on } E \cap \partial B_p(R). \end{cases}$$

By the maximum principle, $0 \leq f_R \leq 1$, and following a standard argument used earlier in the book, $\{f_R\}$ converges uniformly as $R \mapsto \infty$ on compact subsets of E

to a harmonic function f on E such that $f = 1$ on ∂E and $0 \leq f \leq 1$. Now we will prove that f is not constant.

Let $R_0 > 0$ with $K \subset B_p(R_0)$ and for $R > R_0$, let φ a cut-off function on E_R such that $\varphi = 1$ on $E_R \setminus E_{R_0}$, $\varphi = 0$ on ∂E and $\|\nabla\varphi\| \leq c$ for some constant c. Then, as $n \geq 3$, the Sobolev inequality of Michael-Simon, [**MS73**], implies that

$$\begin{aligned}\left(\int_{E_R} (\varphi f_R)^{\frac{2n}{n-2}}\,\mathrm{d}v\right)^{\frac{n-2}{n}} &\leq c(n)\int_{E_R} \|\nabla(\varphi f_R)\|^2\,\mathrm{d}v \\ &= c(n)\int_{E_R}\{f_R^2\|\nabla\varphi\|^2 + \varphi^2\|\nabla f_R\|^2 + \frac{1}{2}g(\nabla\varphi^2, \nabla f_R^2)\}\,\mathrm{d}v \\ &= c(n)\int_{E_R}\{f_R^2\|\nabla\varphi\|^2 + \frac{1}{2}\operatorname{div}(\varphi^2\nabla f_R^2)\}\,\mathrm{d}v = c(n)\int_{E_R} f_R^2\|\nabla\varphi\|^2\,\mathrm{d}v,\end{aligned}$$

where $c(n)$ is a positive constant depending only of n and where we have used that f_R is harmonic and Theorem 1.7.

Now, if $R_0 < R_1 < R$, then

$$\begin{aligned}\left(\int_{E_{R_1}\setminus E_{R_0}} (f_R)^{\frac{2n}{n-2}}\,\mathrm{d}v\right)^{\frac{n-2}{n}} &= \left(\int_{E_{R_1}\setminus E_{R_0}} (\varphi f_R)^{\frac{2n}{n-2}}\,\mathrm{d}v\right)^{\frac{n-2}{n}} \\ &\leq c(n)\int_{E_R} f_R^2\|\nabla\varphi\|^2\,\mathrm{d}v \leq c(n)c\int_{E_{R_0}} f_R^2\,\mathrm{d}v.\end{aligned}$$

If the limit function f were constant, then $f = 1$, and so letting $R \to \infty$ in the last inequality we arrive to

$$\big(\operatorname{vol}(E_{R_1}) - \operatorname{vol}(E_{R_0})\big)^{\frac{n-2}{n}} \leq c(n)c\operatorname{vol}(E_{R_0}).$$

Since $R_1 > R_0$ is arbitrary, letting $R_1 \mapsto \infty$ we get that $\operatorname{vol}(E)$ is finite, which contradicts 1.

Proof of 3. Suppose that M has at least two ends that are non-parabolic. Hence, Proposition 3.32 says that there exists a non-constant harmonic function on M with finite energy. This fact contradicts Theorem 3.35,2.

Proof of 4. From Proposition 3.33, if M is parabolic then M is totally geodesic and so M is a hyperplane isometric to $\mathbb{R}^3$. Considering that $\mathbb{R}^n$ is parabolic if and only if $n = 2$, we get a contradiction. □

Besides hyperplanes, the catenoid is perhaps one of the most important examples of a minimal hypersurface in Euclidean space. Now we will introduce the higher dimensional catenoid and prove that its index is 1. Following the work of M. do Carmo and M. Dajczer, [**dCD83**], we know that there exists a 1-parameter family of higher dimensional catenoids in $\mathbb{R}^{n+1}$, $n \geq 3$, any two of which differ in a homothety of $\mathbb{R}^{n+1}$. Since the index is invariant under homotheties (see Proposition 2.38), we will work in a normalized situation.

EXAMPLE 4.34 ([**dCD83, TZ09**]). *Let $h(t)$ be the solution of*

$$h' = \left(h^{2(n-1)} - 1\right)^{\frac{1}{2}}, \qquad h(0) = 1,\ n \geq 3.$$

Then, extending h as an even function, h is a positive function defined on $(-a, a)$, where

$$a = \int_1^\infty \frac{\mathrm{d}r}{\left(r^{2(n-1)} - 1\right)^{\frac{1}{2}}} < \infty.$$

We define the **higher dimensional catenoid** *of* $\mathbb{R}^{n+1}$ $(n \geq 3)$ *as the two-sided embedding*

$$\Phi : M = (-a, a) \times \mathbb{S}^{n-1} \to \mathbb{R}^{n+1}$$
$$(t, p) \mapsto \Phi(t, p) = (h(t)p, t).$$

Φ has two different principal curvatures, namely h^{-n} with multiplicity $n-1$ and $-(n-1)h^{-n}$ with multiplicity 1. Thus, Φ is a minimal embedding with $\|A\|^2 = n(n-1)h^{-2n}$ and the induced metric $h^{2(n-1)}dt^2 + h^2 g$ is complete, where g is the standard metric on $\mathbb{S}^{n-1}$.

THEOREM 4.35 ([**TZ09**]). *The index of the higher dimensional catenoid is* 1.

PROOF. Since the catenoid has two ends, we deduce from Theorem 4.33,3 that the catenoid is unstable. Now we will prove that its index is at most 1. To do this, for any R with $0 < R < a$, we consider the relatively compact domain Ω_R of M given by $\Omega_R = (-R, R) \times \mathbb{S}^{n-1}$. Since $\Omega_R \subset \Omega_{R'}$ for $R < R'$ and $M = \cup_{0<R<a} \Omega_R$, following Section 2.6, it suffices to prove that $\mathrm{Ind}(\mathcal{L}, \Omega_R) \leq 1$ for R near a.

If there exists $0 < R_0 < a$ such that $\mathrm{Ind}(\mathcal{L}, \Omega_{R_0}) \geq 2$, then it follows from Theorem 1.21 that there exists a non-trivial function f on Ω_{R_0} and a real number $\lambda < 0$ satisfying

$$\begin{cases} \mathcal{L}f + \lambda f = 0, \\ f_{|\partial\Omega_{R_0}} = 0 \\ f \quad \text{changes sign on } \Omega_{R_0}. \end{cases}$$

Given $v \in \mathbb{S}^{n-1}$, let $\sigma_v : \mathbb{S}^{n-1} \to \mathbb{S}^{n-1}$ be the reflection with respect to the hyperplane of $\mathbb{R}^n$ perpendicular to v, i.e., $\sigma_v(p) = p - 2\langle p, v\rangle v$. Then, the map $F_v : M \to M$ given by $F(t,p) = (t, \sigma_v(p))$ is an isometry of M. Hence $\mathcal{L}(f \circ F_v) + \lambda(f \circ F_v) = 0$, for any $v \in \mathbb{S}^{n-1}$.

Now, for any $v \in \mathbb{S}^{n-1}$, the function $f_v = f - (f \circ F_v)$ satisfies

$$\begin{cases} \mathcal{L}f_v + \lambda f_v = 0, \\ (f_v)_{|\partial\Omega_{R_0}} = 0 \\ f_v = 0 \quad \text{on} \quad (-R_0, R_0) \times (\mathbb{S}^{n-1} \cap \langle v\rangle^\perp). \end{cases}$$

If $\Pi_v^+ = \{p \in \mathbb{R}^n \mid \langle p, v\rangle > 0\}$ and $\Omega_{R_0}^+ = (-R_0, R_0) \times (\mathbb{S}^{n-1} \cap \Pi_V^+)$, then Φ is a graph over $\Omega_{R_0}^+$ and $f_v = 0$ on $\partial\Omega_{R_0}^+$. Since a graph is stable, this implies that $f_v = 0$ on $\Omega_{R_0}^+$. Doing the same for $\Omega_{R_0}^-$, we conclude that $f_v = 0$ for any $v \in \mathbb{S}^{n-1}$. This implies that $f = f \circ F_v$ and thus $f(t, v) = f(t, Av)$ for any $A \in \mathrm{O}(n)$. Hence $f(t, p) = f(t)$ for any $p \in \mathbb{S}^{n-1}$ and $t \in (-R_0, R_0)$. Since f changes sign, there exists $t_0 \in (-R_0, R_0)$ such that $f(t_0) = 0$. Assuming that $t_0 \geq 0$, then the domain $\Omega = (t_0, R_0) \times \mathbb{S}^{n-1}$ is a graph and the function f vanishes on $\partial\Omega$. The stability of the graph Ω implies that $f = 0$ on Ω, which is a contradiction. □

From now on, following the relevant article by R. Schoen, L. Simon and S.T. Yau, [**SSY75**], and using Simons' formula, we will obtain sufficient conditions for a complete and stable hypersurface of Euclidean space to be a hyperplane. This will allow us to prove Bernstein's famous problem when the dimension of the hypersurface is less than or equal to five.

LEMMA 4.36. *Let $\Phi : M^n \to \mathbb{R}^{n+1}$ be a two-sided minimal hypersurface. Then its second fundamental form A satisfies*

(1) $\frac{1}{2}\Delta\|A\|^2 + \|A\|^4 = \|\nabla A\|^2$, (Simon's formula)

(2) $\frac{n+2}{4n}\|\nabla\|A\|^2\|^2 \le \|A\|^2\|\nabla A\|^2$. (Simon's inequality)

PROOF. First, we will prove the Simons' formula. If $\{e_1, \dots, e_n\}$ is a local orthonormal reference on M, then $\|A\|^2 = \sum_{i=1}^n \|Ae_i\|^2$. So, using Codazzi's equation (2.2),

$$g(\nabla\|A\|^2, v) = 2\sum_{i=1}^n g((\nabla A)(v, e_i), Ae_i) = 2\sum_{i=1}^n g((\nabla A)(e_i, v), Ae_i), \quad \forall v \in TM.$$

So, using the Ricci's identity we get

$$\begin{aligned}\frac{1}{2}\Delta\|A\|^2 &= \sum_{i,j=1}^n \{g((\nabla^2 A)(e_j, e_i, e_j), Ae_i) + \|(\nabla A)(e_i, e_j)\|^2\}\\ &= -\sum_{i,j=1}^n \{R(e_j, e_i, e_j, A^2 e_i) + R(e_j, e_i, Ae_i, Ae_j)\} + \|\nabla A\|^2\\ &= -\|A\|^4 + \|\nabla A\|^2,\end{aligned}$$

which proves 1.

Next we estimate $\|\nabla\|A\|^2\|^2$. To do this, let $\{e_1, \dots, e_n\}$ be a local orthonormal reference on M that diagonalizes A, i.e., $Ae_i = \lambda_i e_i$, $1 \le i \le n$. Then $\|A\|^2 = \sum_{i=1}^n \lambda_i^2$ and so, using Schwarz inequality for the vectors $(\lambda_1, \dots, \lambda_n)$ and $(e_j(\lambda_1), \dots, e_j(\lambda_n))$ and the minimality of M, we have

$$\begin{aligned}\|\nabla\|A\|^2\|^2 &= 4\sum_{j=1}^n \left(\sum_{i=1}^n \lambda_i e_j(\lambda_i)\right)^2 \le 4\|A\|^2 \sum_{i,j=1}^n (e_j(\lambda_i))^2\\ &= 4\|A\|^2 \left\{\sum_{j=1}^n \left(e_j(-\sum_{k\neq j}\lambda_k)\right)^2 + \sum_{j\neq i}(e_j(\lambda_i))^2\right\}\\ &\le 4\|A\|^2\left\{(n-1)\sum_{k\neq j}(e_j(\lambda_k))^2 + \sum_{j\neq i}(e_j(\lambda_i))^2\right\}\\ &= 4n\|A\|^2\sum_{i\neq j}(e_j(\lambda_i))^2.\end{aligned}$$

From the last inequalities we now get

$$\|\nabla\|A\|^2\|^2 \le 4\|A\|^2\sum_{i,j}(e_j(\lambda_i))^2, \qquad \|\nabla\|A\|^2\|^2 \le 4n\|A\|^2\sum_{i\neq j}(e_j(\lambda_i))^2.$$

Using Codazzi's equation (2.2), we have

$$\begin{aligned}\frac{1}{4n}\|\nabla\|A\|^2\|^2 &\le \|A\|^2\sum_{i\neq j} g((\nabla A)(e_j, e_i), e_i)^2\\ &= \frac{\|A\|^2}{2}\sum_{i\neq j}\{g((\nabla A)(e_i, e_j), e_i)^2 + g((\nabla A)(e_i, e_i), e_j)^2\}\end{aligned}$$

and thus,

$$\frac{n+2}{4n}\|\nabla\|A\|^2\|^2$$
$$\leq\|A\|^2\left(\sum_{i,j}g((\nabla A)(e_j,e_i),e_i)^2+\sum_{i\neq j}\{g((\nabla A)(e_i,e_j),e_i)^2+g((\nabla A)(e_i,e_i),e_j)^2\}\right)$$
$$\leq\|A\|^2\|\nabla A\|^2,$$

which proves 2. □

The following result is fundamentally due to R. Schoen, L. Simon and S.T. Yau. The other cited papers are merely small extensions of their method.

THEOREM 4.37 ([**CL24, dCP82, SSY75**]). *Let $\Phi : M^n \to \mathbb{R}^{n+1}$ be a two-sided stable minimal hypersurface. For every non-negative Lipschitz function f with compact support, we have*

(1) $\int_M \|A\|^p f^{p-2}\,\mathrm{d}v \leq c_1 \int_M \|A\|^2\|\nabla f\|^{p-2}\,\mathrm{d}v,$ *for* $p\in[4,4+\sqrt{8/n})$,

(2) $\int_M \|A\|^p f^p\,\mathrm{d}v \leq c_2 \int_M \|\nabla f\|^p\,\mathrm{d}v,$ *for* $p\in[4,4+\sqrt{8/n})$,

(3) $\int_M \|A\|^p f^p\,\mathrm{d}v \leq c_3 \int_M \|\nabla f\|^p\,\mathrm{d}v,$ *for* $p\in[3,4)$ *and* $n<8$,

where c_1, c_2 and c_3 are constants depending only of n and p.

PROOF. Since M is stable and the Jacobi operator is $\mathcal{L}=\Delta+\|A\|^2$, then for any $h\in H_0^{1,2}(M)$ we have

$$\int_M h^2\|A\|^2\,\mathrm{d}v \leq \int_M \|\nabla h\|^2\,\mathrm{d}v.$$

Let q be a positive real number. For any $\delta>0$ we define $a_\delta=(\|A\|^2+\delta)^q$. Using the above inequality for the function $h=fa_\delta$, and using that $\nabla a_\delta = q(a_\delta)^{\frac{q-1}{q}}\nabla\|A\|^2$, we have

$$\begin{aligned}\int_M f^2a_\delta^2\|A\|^2\,\mathrm{d}v &\leq q^2\int_M f^2(a_\delta)^{\frac{2(q-1)}{q}}\|\nabla\|A\|^2\|^2\,\mathrm{d}v+\int_M a_\delta^2\|\nabla f\|^2\,\mathrm{d}v\\ &\quad+2q\int_M f(a_\delta)^{\frac{2q-1}{q}}g(\nabla\|A\|^2,\nabla f)\,\mathrm{d}v.\end{aligned}\tag{4.24}$$

If we multiply the Simons identity (Lemma 4.36,1) by $f^2(a_\delta)^{\frac{2q-1}{q}}$ and integrate over M, taking into account that

$$\begin{aligned}\operatorname{div}\left(f^2(a_\delta)^{\frac{2q-1}{q}}\nabla\|A\|^2\right) &= f^2(a_\delta)^{\frac{2q-1}{q}}\Delta\|A\|^2+2f(a_\delta)^{\frac{2q-1}{q}}g(\nabla\|A\|^2,\nabla f)\\ &\quad+(2q-1)f^2(a_\delta)^{\frac{2(q-1)}{q}}\|\nabla\|A\|^2\|^2,\end{aligned}$$

we get

$$\begin{aligned}\int_M f^2(a_\delta)^{\frac{2q-1}{q}}\|\nabla A\|^2\,\mathrm{d}v &= -\int_M f(a_\delta)^{\frac{2q-1}{q}}g(\nabla\|A\|^2,\nabla f)\,\mathrm{d}v\\ &\quad-\frac{2q-1}{2}\int_M f^2(a_\delta)^{\frac{2(q-1)}{q}}\|\nabla\|A\|^2\|^2\,\mathrm{d}v+\int_M f^2(a_\delta)^{\frac{2q-1}{q}}\|A\|^4\,\mathrm{d}v.\end{aligned}$$

From Lemma 4.36,2, it follows that

$$\frac{n+2}{4n}\|\nabla\|A\|^2\|^2 \le \|A\|^2\|\nabla A\|^2 \le (a_\delta)^{\frac{1}{q}}\|\nabla A\|^2,$$

and so, the last equality becomes

$$\left(q+\frac{1}{2n}-\frac{1}{4}\right)\int_M f^2(a_\delta)^{\frac{2(q-1)}{q}}\|\nabla\|A\|^2\|^2\,dv \le \int_M f^2a_\delta^2\|A\|^2\,dv - \int_M f(a_\delta)^{\frac{2q-1}{q}}g(\nabla\|A\|^2,\nabla f)\,dv. \tag{4.25}$$

Adding (4.24) and (4.25), we obtain

$$\left(\frac{1}{2n}-\frac{(2q-1)^2}{4}\right)\int_M f^2(a_\delta)^{\frac{2(q-1)}{q}}\|\nabla\|A\|^2\|^2\,dv \le (2q-1)\int_M f(a_\delta)^{\frac{2q-1}{q}}g(\nabla\|A\|^2,\nabla f)\,dv + \int_M a_\delta^2\|\nabla f\|^2\,dv.$$

Using the inequality

$$2g\left(f(a_\delta)^{\frac{q-1}{q}}\nabla\|A\|^2, a_\delta\nabla f\right) \le \epsilon f^2(a_\delta)^{\frac{2(q-1)}{q}}\|\nabla\|A\|^2\|^2 + \frac{1}{\epsilon}a_\delta^2\|\nabla f\|^2$$

in the last inequality, we get

$$\begin{aligned}&\left(\frac{1}{2n}-\frac{(2q-1)^2}{4}-\epsilon\frac{|2q-1|}{2}\right)\int_M f^2(a_\delta)^{\frac{2(q-1)}{q}}\|\nabla\|A\|^2\|^2\,dv\\ &\qquad\le\left(1+\frac{|2q-1|}{2\epsilon}\right)\int_M a_\delta^2\|\nabla f\|^2\,dv,\end{aligned} \tag{4.26}$$

where ϵ is a positive number.

Applying the Schwarz inequality

$$2g\left(a_\delta\nabla f, qf(a_\delta)^{\frac{q-1}{q}}\nabla\|A\|^2\right) \le a_\delta^2\|\nabla f\|^2 + q^2f^2(a_\delta)^{\frac{2(q-1)}{q}}\|\nabla\|A\|^2\|^2$$

in (4.24) we have

$$\int_M f^2a_\delta^2\|A\|^2\,dv \le 2\int_M a_\delta^2\|\nabla f\|^2\,dv + 2q^2\int_M f^2(a_\delta)^{\frac{2(q-1)}{q}}\|\nabla\|A\|^2\|^2\,dv.$$

To prove 1, 2 and 3, we take $q=\frac{p-2}{4}$. Then, if $p\in[4,4+\sqrt{8/n})$, we have that $2q-1\in[0,\sqrt{2/n})$. So $\frac{1}{2n}-\frac{(2q-1)^2}{4}>0$. If $p\in[3,4)$, we have that $2q-1\in[-\frac{1}{2},0)$ and again $\frac{1}{2n}-\frac{(2q-1)^2}{4}>0$ if $n<8$. So, in both cases, there exists $\epsilon>0$ such that $\frac{1}{2n}-\frac{(2q-1)^2}{4}-\frac{\epsilon(|2q-1|)}{2}>0$.

Now, from (4.26) and the last inequality, we get

$$\int_M f^2a_\delta^2\|A\|^2\,dv \le \left(\frac{2q^2(1+\frac{|2q-1|}{\epsilon})}{\frac{1}{2n}-\frac{(2q-1)^2}{4}-\frac{\epsilon(|2q-1|)}{2}}+2\right)\int_M a_\delta^2\|\nabla f\|^2\,dv. \tag{4.27}$$

Since $p=4q+2$, taking the limit as $\delta\to 0$ in (4.27) and using the dominated convergence theorem (Theorem 1.7), we have

$$\int_M f^2\|A\|^p\,dv \le c\int_M \|A\|^{p-2}\|\nabla f\|^2\,dv, \tag{4.28}$$

for some positive constant c.

Proof of 1. We observe that if $p = 4$, then (4.28) proves statement 1. So, from now on we can assume that $p > 4$.

Now we want to manipulate (4.28) so that the right integrand contains only $\|A\|^2$. To do this, we use the Young's inequality

$$ab \leq \frac{\alpha^s a^s}{s} + \frac{\alpha^{-t} b^t}{t}, \quad \frac{1}{t} + \frac{1}{s} = 1,$$

where $\alpha > 0$ is arbitrary and $1 < s, t < \infty$. We take

$$s = \frac{p-2}{p-4}, \quad t = \frac{p-2}{2}, \quad a = \|A\|^{\frac{p(p-4)}{p-2}} \quad \text{and} \quad b = \|A\|^{\frac{4}{p-2}} \|\nabla f\|^2 f^{-2},$$

and then Young's inequality implies

$$\|A\|^{p-2}\|\nabla f\|^2 = f^2 ab \leq \frac{\alpha^s}{s} f^2 \|A\|^{\frac{p-2}{2}} + \frac{\alpha^{-t}}{t} f^{2-p}\|A\|^2\|\nabla f\|^{p-2}.$$

Using the above inequality and making α sufficiently small, (4.28) becomes

$$\int_M f^2 \|A\|^p \, dv \leq b_2 \int_M f^{4-p}\|A\|^2\|\nabla f\|^{p-2} \, dv.$$

Finally, replacing f by $f^{\frac{p-2}{2}}$, we prove 1.

Proof of 2 and 3. Now, replacing f by $f^{\frac{p}{2}}$ in (4.28), we get

$$\int_M f^p \|A\|^p \, dv \leq c \int_M f^{p-2}\|A\|^{p-2}\|\nabla f\|^2 \, dv,$$

for some positive constant c. Hölder's inequality gives

$$\int_M f^p \|A\|^p \, dv \leq c \left(\int_M (f^{p-2}\|A\|^{p-2})^{\frac{p}{p-2}} \, dv \right)^{\frac{p-2}{p}} \left(\int_M \|\nabla f\|^p \, dv \right)^{\frac{2}{p}},$$

which proves 2 and 3. $\square$

THEOREM 4.38 (**[dCP82]**). *Let $\Phi : M^n \to \mathbb{R}^{n+1}$ be a two-sided stable minimal hypersurface of a complete manifold M. If q is a point of M, $B_q(R)$ is the geodesic ball at q with radius R, and $\|A\|^2$ satisfies*

$$\lim_{R \to \infty} \frac{\int_{B_q(R)} \|A\|^2 dv}{R^{p-2}} = 0, \qquad p \in [4, 4 + \sqrt{8/n}),$$

then $\Phi(M)$ is a hyperplane. In particular, if M satisfies $\int_M \|A\|^2 \, dv < \infty$, then $\Phi(M)$ is a hyperplane.

PROOF. We fix $0 < a < 1$ and we define a non-negative function $f \in H_0^{1,2}(M)$ by

$$f(x) = \begin{cases} 1 & \text{for } d(x,q) \leq aR \\ \frac{R - d(x,q)}{(1-a)R} & \text{for } aR \leq d(x,q) \leq R \\ 0 & \text{for } d(x,q) \geq R, \end{cases}$$

where d is the distance in M. Then, $\|\nabla f\| \leq \frac{1}{(1-a)R}$ and so, from Theorem 4.37,1 we have

$$\int_{B_q(aR)} \|A\|^p \, dv \leq \int_M \|A\|^p f^{p-2} \, dv \leq c_1 \int_M \|A\|^2 \|\nabla f\|^{p-2} \, dv$$
$$\leq \frac{c_1}{(1-a)^{p-2}} \frac{\int_{B_q(R)} \|A\|^2 \, dv}{R^{p-2}}.$$

By letting $R \to \infty$ we get that $\int_M \|A\|^p \,dv = 0$, which implies that M is totally geodesic. □

THEOREM 4.39 ([**SSY75**]). *Let $\Phi : M^n \to \mathbb{R}^{n+1}$ be a two-sided complete stable minimal hypersurface of Euclidean space with $n \leq 5$. If $B_q(R)$ denotes the geodesic ball at q of radius R, and there exists $V < \infty$ such that*

$$\sup_{R>0} \frac{\mathrm{vol}(B_q(R))}{R^n} \leq V,$$

then $\Phi(M)$ is a hyperplane.

PROOF. We define the function $f \in H_0^{1,2}(M)$ by

$$f(x) = \begin{cases} 1 & \text{for } d(x,q) \leq R, \\ 2 - \frac{d(x,q)}{R} & \text{for } R \leq d(x,q) \leq 2R, \\ 0 & \text{for } d(x,q) \geq 2R. \end{cases}$$

From the inequality 2 of Theorem 4.37, we get

$$\begin{aligned} \int_{B_q(R)} \|A\|^p \,dv \leq \int_M f^p \|A\|^p \,dv \leq c_2 \int_M \|\nabla f\|^p \,dv \\ = c_2 \int_{B_q(2R)} \frac{dv}{R^p} \leq \frac{2^n c_2 V R^n}{R^p}. \end{aligned}$$

Since $n \leq 5$, there exists $p \in [4, 4 + \sqrt{8/n})$ with $n < p$. For such p and letting $R \to \infty$ we have $\int_M \|A\|^p \,dv = 0$ and the result follows. □

COROLLARY 4.40 ([**SSY75**]). *Let $\Phi : M^n \to \mathbb{R}^{n+1}$ be a two-sided complete stable minimal hypersurface of Euclidean space with $n \leq 5$. If $B(R)$ denotes the ball in $\mathbb{R}^{n+1}$ of radius R, and there exists $V < \infty$ such that*

$$\sup_{R>0} \frac{\mathrm{vol}(\Phi^{-1}(B(R)))}{R^n} \leq V,$$

then $\Phi(M)$ is a hyperplane.

PROOF. If q is a point of M with $\Phi(q) = 0$, then it is clear that $B_q(R) \subset \Phi^{-1}(B(R))$. So $\mathrm{vol}(B_q(R)) \leq \mathrm{vol}(\Phi^{-1}(B(R)))$ and the corollary follows. □

REMARK 4.41. Very recently, in [**Bel25**], C. Belletini proves Corollary 4.40 for $n = 6$, using different ideas from those used in the proof of Corollary 4.40.

If we consider in Proposition 2.39 the case where the Riemannian manifold (M^n, g) is the Euclidean space $\mathbb{R}^n$, then it is clear that every affine function $f : \mathbb{R}^n \to \mathbb{R}$ is a solution of the equation

$$\mathrm{div}\left(\frac{\nabla f}{\sqrt{1 + \|\nabla f\|^2}}\right) = 0,$$

which gives a hyperplane of $\mathbb{R}^{n+1}$ as an associated graph . The **Bernstein problem** consists of proving that the only solutions of the above equation defined on all $\mathbb{R}^n$ are the affine functions. That is, if the only entire graphs in $\mathbb{R}^{n+1}$ are hyperplanes. In the next result we give a positive answer to the Bernstein problem for $n \leq 6$.

THEOREM 4.42 ([**Ber27**,**DG61**,**Alm66**,**Sim68**]). *If $f:\mathbb{R}^n\to\mathbb{R}$ is a solution of the equation*

$$\operatorname{div}\left(\frac{\nabla f}{\sqrt{1+\|\nabla f\|^2}}\right)=0$$

and $n\le 6$, then f is an affine function.

REMARK 4.43. The result was proved by S. Bernstein for $n=2$, by E. De Giorgi for $n=3$, by F. J. Almgren, Jr. for $n=4$ and by J. Simons for $n=5,6,7$. Also, E. Bombieri, E. De Giorgi and E. Giusti, [**BDGG69**], proved that for $n\ge 8$ there exist non-affine solutions $f:\mathbb{R}^n\to\mathbb{R}$ of the above equation, that thus, there exist complete minimal graphs in $\mathbb{R}^{n+1}$ which are not hyperplanes.

PROOF. By Proposition 2.39, if f is a solution of the above equation, then its graph, Graph(f), is a two-sided stable embedded complete minimal hypersurface of $\mathbb{R}^{n+1}$, which is a calibrated submanifold with respect to a calibration $\eta\in\Omega^n(\mathbb{R}^{n+1})$.

For almost all sufficiently large R, the boundary $\partial B(R)=\mathbb{S}^n(R)$ intersects Graph(f) in a connected $(n-1)$-dimensional submanifold N of $\mathbb{S}^n(R)$. Thus $\mathbb{S}^n(R)\setminus N$ has two connected components and at least one of them has volume at most equal to $\operatorname{vol}(\mathbb{S}^n(R))/2=R^n\operatorname{vol}(\mathbb{S}^n)/2$. If Σ is such a connected component, then it follows from Theorem 1.11 that

$$\int_{\mathrm{Graph}(f)\cap B(R)}\eta=\int_\Sigma\eta.$$

Since Graph(f) is calibrated with respect to η, we get that

$$\operatorname{vol}(\mathrm{Graph}(f)\cap B(R))=\int_{\mathrm{Graph}(f)\cap B(R)}\eta=\int_\Sigma\eta\le\operatorname{vol}(\Sigma)\le\frac{R^n\operatorname{vol}(\mathbb{S}^n)}{2}.$$

Finally, it follows from Corollary 4.40 and Remak 4.41 that Graph(f) is a hyperplane and thus, f is an affine function. □

Now we begin to study complete minimal hypersurfaces of $\mathbb{R}^{n+1}$ with finite index. In this study, as in the case of surfaces (see Sections 4.1 and 4.2), the space of harmonic fields on the hypersurface will play an important role. We begin by proving a result of B. Palmer which opens the door to such a study.

THEOREM 4.44 ([**Pal91**]). *Let $\Phi:M^n\to\mathbb{R}^{n+1}$ be a two-sided complete minimal hypersurface. If M is stable, then the space*

$$\mathcal{H}(M)\cap L^2(M,TM)=\{X\in\mathfrak{X}(M)\mid X\text{ is harmonic and }\int_M\|X\|^2\,dv<\infty\}$$

is trivial.

PROOF. Let X be a nonzero harmonic vector field on M with $\int_M\|X\|^2\,dv<\infty$.

From the Gauss equation (2.2) and if $\{e_i\mid 1\le i\le n\}$ is a local reference diagonalizing A, i.e., $Ae_i=\lambda_ie_i$, we have that

$$\begin{aligned}-S(X,X)=\|AX\|^2=\sum_{i=1}^n\lambda_i^2\langle X,e_i\rangle^2&=\|A\|^2\|X\|^2\\&-\sum_{i\ne j}\lambda_i^2\langle X,e_j\rangle^2=\|A\|^2\|X\|^2-W.\end{aligned}$$

Note that $W \geq 0$ and $W = 0$ if and only if $A = 0$. So, it follows from Lemma 1.16 that

$$\Delta\|X\|^2 = 2S(X,X) + 2\|\nabla X\|^2 = -2\|A\|^2\|X\|^2 + 2W + 2\|\nabla X\|^2. \tag{4.29}$$

Now we assume that M is stable and we come to a contradiction. For any $\delta > 0$, consider the following regularization of $\|X\|$:

$$e_\delta = (\|X\|^2 + \delta)^{1/2}.$$

It is clear that

$$\nabla e_\delta = \frac{1}{2}e_\delta^{-1}\nabla\|X\|^2, \quad \Delta e_\delta = \frac{1}{2}e_\delta^{-1}\Delta\|X\|^2 - \frac{1}{4}e_\delta^{-3}\|\nabla\|X\|^2\|^2.$$

If M is stable, then for any $f \in H_0^{1,2}(M)$ of compact support and using $fe_\delta \in H_0^{1,2}(M)$ as a test function, we have that

$$\begin{aligned}
0 \leq & -\int_M \{fe_\delta\Delta(fe_\delta) + \|A\|^2 f^2 e_\delta^2\}\,dv \\
= & -\int_M \{fe_\delta^2\Delta f + \frac{1}{2}f^2\Delta\|X\|^2 - \frac{1}{4}f^2e_\delta^{-2}\|\nabla\|X\|^2\|^2 + fg(\nabla f, \nabla\|X\|^2)\}\,dv \\
& -\int_M \|A\|^2 f^2 e_\delta^2\,dv = -\int_M \{fe_\delta^2\Delta f + f^2W + f^2\|\nabla X\|^2\}\,dv \\
& +\int_M \{\frac{1}{4}f^2e_\delta^{-2}\|\nabla\|X\|^2\|^2 - fg(\nabla f, \nabla\|X\|^2) - \|A\|^2f^2(e_\delta^2 - \|X\|^2)\}\,dv,
\end{aligned}$$

where we have taken into account (4.29) and the expressions of ∇e_δ and Δe_δ.

As $e_\delta^2 \geq \|X\|^2$, then from Lemma 1.16

$$\|\nabla X\|^2 = e_\delta^{-2}e_\delta^2\|\nabla X\|^2 \geq e_\delta^{-2}\|X\|^2\|\nabla X\|^2 \geq \frac{n}{2(n-1)}e_\delta^{-2}\|\nabla\|X\|^2\|^2. \tag{4.30}$$

Using this information in the stability inequality, we get

(4.31)

$$\begin{aligned}
0 \leq & -\int_M \left\{fe_\delta^2\Delta f + f^2W + \left(\frac{n}{2(n-1)} - \frac{1}{4}\right)f^2e_\delta^{-2}\|\nabla\|X\|^2\|^2\right\}\,dv \\
& -\int_M fg(\nabla f, \nabla\|X\|^2)\,dv \leq -\int_M \{fe_\delta^2\Delta f + f^2W + fg(\nabla f, \nabla\|X\|^2)\}\,dv,
\end{aligned}$$

considering that $\frac{n}{2(n-1)} - \frac{1}{4} > 0$. Taking the limit as $\delta \to 0$ in the last inequality and using the dominated convergence theorem (Theorem 1.7), we finally obtain

$$\int_M f^2W\,dv \leq -\int_M \{f\|X\|^2\Delta f + fg(\nabla f, \nabla\|X\|^2)\}\,dv.$$

Now, since

$$f\|X\|^2\Delta f + \|X\|^2\|\nabla f\|^2 + fg(\nabla f, \nabla\|X\|^2) = \frac{1}{2}\operatorname{div}\left(\|X\|^2\nabla f^2\right),$$

from Theorem 1.7, the last inequality gives

$$\int_M \|X\|^2\|\nabla f\|^2\,dv \geq \int_M f^2W\,dv.$$

Finally, taking f as the cut-off function given in Lemma 2.28, it follows that

$$\frac{1}{R^2 \log^2 R} \int_M \|X\|^2 \, dv \geq \int_{B_p(R)} W \, dv,$$

where $B_p(R)$ is a geodesic ball of radius R centered at $p \in M$. Letting $R \to \infty$ and as $\int_M \|X\|^2 \, dv < \infty$, we get that $W = 0$. This implies that $A = 0$ and therefore, $\Phi(M)$ is a hyperplane. The contradiction comes from the fact that $\mathbb{R}^n$ with the Euclidean metric does not have harmonic vector fields X with $\int_{\mathbb{R}^n} \|X\|^2 \, dv < \infty$. □

From Theorems 4.2 and 4.4 it follows, among other things, that if a complete and orientable minimal surface of $\mathbb{R}^3$ has finite index, then the number of ends is also finite. The following result by P. Li and J. Wang extends the above property to the case of hypersurfaces $\mathbb{R}^{n+1}$.

THEOREM 4.45 ([**LW02**]). *Let $\Phi : M^n \to \mathbb{R}^{n+1}$, $n \geq 3$, be a two-sided complete minimal hypersurface. If M has finite index, then M must have finitely many ends.*

PROOF. To prove the theorem we will use the following result of P. Li and L. F. Tam, [**LT92**]:

If M is a complete Riemannian manifold, then the number of non-parabolic ends of M is at most the dimension of the space of bounded harmonic functions with finite Dirichlet integral.

But if f is a harmonic function with finite Dirichlet integral, then its gradient ∇f is a harmonic vector field with $\nabla f \in L^2(M, TM)$. Since the kernel of the gradient map $C^\infty(M) \to \mathfrak{X}(M)$ consists of the constant functions, we get that the number of non-parabolic ends of M is at most $1 + \dim\big(\mathcal{H}(M) \cap L^2(M, TM)\big)$.

Since M is a minimal hypersurface of $\mathbb{R}^{n+1}$, Theorem 4.33,2 states that all the ends of M are non-parabolic, so the proof of the result reduces to prove that $\dim\big(\mathcal{H}(M) \cap L^2(M, TM)\big)$ is finite.

Since M has finite index, it follows from Theorem 2.27 that there exists a compact subset $K \subset M$ such that $M \setminus K$ is a stable minimal hypersurface of $\mathbb{R}^{n+1}$. It is clear that we can take $K = B_p(R_0)$, for some $p \in M$ and $R_0 > 0$.

Now we follow the notation and calculations of Theorem 4.44. If $X \in \mathcal{H}(M) \cap L^2(M, TM)$, for any $\delta > 0$ we define $e_\delta = (\|X\|^2 + \delta)^{1/2}$.

For any $f \in H_0^{1,2}(M \setminus B_p(R_0))$ and using $f e_\delta \in H_0^{1,2}(M)$ as a test function, the stability of $M \setminus B_p(R_0)$ together with the first inequality of (4.31) and that $W \geq 0$ imply that

$$\begin{aligned} 0 \leq & - \int_{M \setminus B_p(R_0)} \left\{ f e_\delta^2 \Delta f + \left(\frac{n}{2(n-1)} - \frac{1}{4} \right) f^2 e_\delta^{-2} \|\nabla \|X\|^2\|^2 \right\} dv \\ & - \int_{M \setminus B_p(R_0)} f g(\nabla f, \nabla \|X\|^2) \, dv. \end{aligned}$$

As $\operatorname{div}(fe_\delta^2\nabla f) = fe_\delta^2\Delta f + e_\delta^2\|\nabla f\|^2 + fg(\nabla f, \nabla\|X\|^2)$, the above inequality becomes, together with Theorem 1.7,

$$\begin{aligned} 0 &\leq \int_{M\setminus B_p(R_0)} \left\{ e_\delta^2\|\nabla f\|^2 - \frac{n+1}{2(n-1)} f^2 e_\delta^{-2}\|\nabla\|X\|^2\|^2 \right\} dv \\ &= \int_{M\setminus B_p(R_0)} \left\{ e_\delta^2\|\nabla f\|^2 - \frac{n+1}{(n-1)} f^2\|\nabla e_\delta\|^2 \right\} dv. \end{aligned}$$

Now, combining the above inequality with the Sobolev inequality for minimal hypersurfaces of $\mathbb{R}^{n+1}$, [**MS73**], we get

$$\begin{aligned} \left(\int_{M\setminus B_p(R_0)} (fe_\delta)^{\frac{2n}{n-2}} dv \right)^{\frac{n-2}{n}} &\leq c(n) \int_{M\setminus B_p(R_0)} \|\nabla(fe_\delta)\|^2 dv \\ &= c(n) \int_{M\setminus B_p(R_0)} \{ f^2\|\nabla e_\delta\|^2 + e_\delta^2\|\nabla f\|^2 + 2fe_\delta g(\nabla f, \nabla e_\delta) \} dv \\ &\leq 2c(n) \int_{M\setminus B_p(R_0)} \{ f^2\|\nabla e_\delta\|^2 + e_\delta^2\|\nabla f\|^2 \} dv \\ &\leq \frac{4n}{n+1} c(n) \int_{M\setminus B_p(R_0)} e_\delta^2\|\nabla f\|^2 dv, \end{aligned}$$

where $c(n)$ is a positive constant depending only of n and where we have used the inequality $\|f\nabla e_\delta - e_\delta\nabla f\|^2 \geq 0$. Taking the limit as $\delta \to 0$ in the last inequality, we finally obtain

$$\left(\int_{M\setminus B_p(R_0)} (f\|X\|)^{\frac{2n}{n-2}} dv \right)^{\frac{n-2}{n}} \leq \frac{4n}{n+1} c(n) \int_{M\setminus B_p(R_0)} \|X\|^2\|\nabla f\|^2 dv.$$

Now, for any $R > R_0 + 1$ we choose a cut-off function f with the following properties:

$$\begin{cases} f = 0 & \text{on } B_p(R_0), \\ \|\nabla f\| \leq c_1 & \text{on } B_p(R_0+1) \setminus B_p(R_0), \\ f = 1 & \text{on } B_p(R) \setminus B_p(R_0+1), \\ \|\nabla f\| \leq c_2 R^{-1} & \text{on } B_p(2R) \setminus B_p(R), \\ f = 0 & \text{on } M \setminus B_p(2R), \end{cases}$$

for some constants c_1 and c_2. If we pluge the above function f in the last inequality, we get

$$\begin{aligned} \left(\int_{B_p(R)\setminus B_p(R_0+1)} \|X\|^{\frac{2n}{n-2}} dv \right)^{\frac{n-2}{n}} &\leq \left(\int_{M\setminus B_p(R_0)} (f\|X\|)^{\frac{2n}{n-2}} dv \right)^{\frac{n-2}{n}} \\ &\leq c_3 \int_{B_p(R_0+1)\setminus B_p(R_0)} \|X\|^2 dv + \frac{c_4}{R^2} \int_{B_p(2R)\setminus B_p(R)} \|X\|^2 dv \\ &\leq c_3 \int_{B_p(R_0+1)\setminus B_p(R_0)} \|X\|^2 dv + \frac{c_4}{R^2} \int_M \|X\|^2 dv, \end{aligned}$$

for some positive constants c_3 and c_4. As $X \in L^2(M,TM)$, letting $R \to \infty$ in the last inequality we have

$$\left(\int_{M\setminus B_p(R_0+1)} \|X\|^{\frac{2n}{n-2}}\, dv\right)^{\frac{n-2}{n}} \le c_3 \int_{B_p(R_0+1)\setminus B_p(R_0)} \|X\|^2\, dv. \tag{4.32}$$

On the other hand, Hölder's inequality implies that

$$\int_{B_p(R_0+2)\setminus B_p(R_0+1)} \|X\|^2\, dv \le \lambda \left(\int_{B_p(R_0+2)\setminus B_p(R_0+1)} \|X\|^{\frac{2n}{n-2}}\, dv\right)^{\frac{n-2}{n}},$$

where $\lambda = \big(\mathrm{vol}\,(B_p(R_0+2)\setminus B_p(R_0+1))\big)^{\frac{2}{n}}$. Using (4.32) in the above inequality we finally get

(4.33)

$$\begin{aligned}
\int_{B_p(R_0+2)} \|X\|^2\, dv &= \int_{B_p(R_0+1)} \|X\|^2\, dv + \int_{B_p(R_0+2)\setminus B_p(R_0+1)} \|X\|^2\, dv \\
&\le \int_{B_p(R_0+1)} \|X\|^2\, dv + \lambda \left(\int_{B_p(R_0+2)\setminus B_p(R_0+1)} \|X\|^{\frac{2n}{n-2}}\, dv\right)^{\frac{n-2}{n}} \\
&\le \int_{B_p(R_0+1)} \|X\|^2\, dv + c_3\lambda \int_{B_p(R_0+1)\setminus B_p(R_0)} \|X\|^2\, dv \\
&\le c_5 \int_{B_p(R_0+1)} \|X\|^2\, dv,
\end{aligned}$$

for some positive constant c_5.

On the other hand, since the function $\|X\|$ satisfies the differential inequality given in Lemma 1.16, then the Moser iteration argument, [**Li93**], can be applied to obtain that

$$\|X\|^2(x) \le c_6 \int_{B_x(1)} \|X\|^2\, dv, \qquad \forall x \in M,$$

for some positive constant c_6. Let x be a point of $\overline{B_p(R_0+1)}$ such that

$$\|X\|^2(x) = \sup\{\|X\|^2(y) \mid y \in \overline{B_p(R_0+1)}\}.$$

As $B_x(1) \subset B_p(R_0+2)$, we get that

$$\sup\{\|X\|^2(y) \mid y \in \overline{B_p(R_0+1)}\} \le c_6 \int_{B_p(R_0+2)} \|X\|^2\, dv.$$

This inequality together with (4.33) gives

$$\sup\{\|X\|^2(y) \mid y \in \overline{B_p(R_0+1)}\} \le c_7 \int_{B_p(R_0+1)} \|X\|^2\, dv, \tag{4.34}$$

for some positive constant c_7.

Now, in order to prove that $\dim\big(\mathcal{H}(M)\cap L^2(M,TM)\big)$ is finite, we will show that the dimension v of any finite dimensional subspace $V \subset \big(\mathcal{H}(M)\cap L^2(M,TM)\big)$ is bounded by a constant independent of V.

Let $F : \mathcal{H}(M)\cap L^2(M,TM) \to \mathcal{H}(B_p(R_0+1))\cap L^2(B_p(R_0+1), TB_p(R_0+1))$ be the linear map $F(X) = X_{|B_p(R_0+1)}$.

By the unique continuation property, $\ker F = 0$ and thus $F(V)$ is a subspace isomorphic to V and of dimension v. Applying Lemma 11 of [**Li80**] to the finite dimensional subspace $F(V)$ of $\mathcal{H}(B_p(R_0+1))\cap L^2(B_p(R_0+1),TB_p(R_0+1))$, there exists $F(X)\in F(V)$ such that

$$v\int_{B_p(R_0+1)}\|X\|^2\,\mathrm{d}v\leq \min\{n,v\}\operatorname{vol}(B_p(R_0+1))\sup\{\|X\|^2(y)\mid y\in\overline{B_p(R_0+1)}\}.$$

Now using inequality (4.34), we finally get

$$v\int_{B_p(R_0+1)}\|X\|^2\,\mathrm{d}v\leq c_8\,\min\{n,v\}\int_{B_p(R_0+1)}\|X\|^2\,\mathrm{d}v,$$

for some positive constant c_8. This inequality implies that $v\leq\min\{n,nc_8\}$, which is independent of V. This concludes the proof. □

The generalization of minimal surfaces of $\mathbb{R}^3$ with finite total curvature to hypersurfaces of $\mathbb{R}^{n+1}$ was studied by M. Anderson in [**And86**], where he proved many interesting results about them. A two-sided complete minimal hypersurface M of $\mathbb{R}^{n+1}$ is said to be of **finite total curvature** if $\int_M\|A\|^n\,\mathrm{d}v<\infty$. This means that, if ρ is the scalar curvature of M, then $\int_M|\rho|^{\frac{n}{2}}\,\mathrm{d}v<\infty$. In the next result we show some interesting properties of this kind of minimal hypersurfaces. (Compare with Theorem 4.4).

THEOREM 4.46 ([**And86, Tys89, Sch83**]). *Let $\Phi: M^n\to\mathbb{R}^{n+1}$, $n\geq 3$, be a two-sided complete minimal hypersurface of finite total curvature. Then:*

(1) *There exists a diffeomorphism $F: M\to\widehat{M}\setminus\{p_1,\dots,p_r\}$, where $\widehat{M}$ is a compact manifold and the metric on M extends conformally to a Riemannian metric on $\widehat{M}$.*
(2) *If $\{E_1,\dots,E_r\}$ are unbounded connected components of the complement of a compact subset of M representing to the ends of M, then each E_i is diffeomorphic $\mathbb{S}^{n-1}\times[0,\infty)$ and its multiplicitiy $m_i\in\mathbb{Z}^+$, given by*

$$m_i=\lim_{R\to\infty}\frac{\operatorname{vol}\left(\Phi^{-1}(B(R))\cap E_i\right)}{R^{n-1}\operatorname{vol}(\mathbb{S}^{n-1})},$$

is $m_i=1$ for any $1\leq i\leq r$.
(3) *M is regular at infinity, that is, each E_i can be chosen to be the graph of a function u_i over a hyperplane Π_i of $\mathbb{R}^{n+1}$, such that for any $x\in\Pi_i$, we have*

$$\|x\|^{n-3}|u(x)|+\|x\|^{n-2}\|(\nabla u)(x)\|+\|x\|^{n-1}\|(\nabla^2u)(x)\|\leq c,$$

for some positive constant c.
(4) *If M has only an end, then $\Phi(M)$ is a hyperplane.*
(5) *If M has two ends, then $\Phi(M)$ is the higher dimensional catenoid.*

REMARK 4.47. Item 5 was proved by R. Schoen ([**Sch83**]) by assuming instead of the finiteness of the total scalar curvature, the weaker condition that M is regular at infinity. Also, if M is regular at infinity, then, using the expression of the second fundamental form of a graph given in Proposition 2.39, $\|A\|$ is bounded on every E_i and hence it is bounded on M.

COROLLARY 4.48 ([**SZ98**]). *Let $\Phi: M^n\to\mathbb{R}^{n+1}$ be a two-sided stable minimal hypersurface of a complete manifold M. If M has finite total curvature then $\Phi(M)$ is a hyperplane.*

REMARK 4.49. Corollary 4.48 was proved by P. Bérard, [**Bé91**] for $n\leq 5$.

PROOF. The result follows from Theorems 4.33,3 and 4.46. □

In [**LY83**], P. Li and S.T. Yau obtained, for a Schrödinger operator on $\mathbb{R}^n, n \geq 3$ with non-negative potential, an upper bound on the number of eigenvalues of such an operator less than a certain number arbitrarily prescribed. In particular, they obtained an upper bound on the index of this operator. As they note in the paper, their method is valid on any complete Riemannian manifold as long as a Sobolev inequality exists on it. Tysk, in [**Tys89**], using the Sobolev inequality which has any orientable minimal and complete hypersurface of $\mathbb{R}^{n+1}$, obtained the following result:

THEOREM 4.50 ([**LY83, Tys89**]). *Let $\Phi : M^n \to \mathbb{R}^{n+1}$ $(n \geq 3)$ be a complete orientable minimal hypersurface. Then*

$$\mathrm{Ind}(M) \leq c(n) \int_M \|A\|^n \,\mathrm{d}v,$$

where $c(n)$ is a positive constant depending only on n. In particular, if M has finite total curvature, then $\mathrm{Ind}(M)$ *is finite.*

By Theorem 4.50, we know that the complete minimal hypersurfaces of $\mathbb{R}^{n+1}$ with finite total curvature have finite index. Next we will give a result of C. Li in which we obtain a lower bound on the index for the above family. This idea will also be used in Chapter 5 when the ambient manifold is a sphere, the real projective space or the product of spheres and the minimal hypersurfaces are compact. (Theorems 5.2, 5.32 and 5.35).

The non-compactness of the minimal hypersurfaces of $\mathbb{R}^{n+1}$ requires that we first prove a modified version of Theorem 2.29.

PROPOSITION 4.51 ([**Li17**]). *Let $\Phi : M^n \to \mathbb{R}^{n+1}$ be a two-sided complete minimal hypersurface with index k that is regular at infinity. Let $W \subset L^2(M)$ be the k-dimensional subspace with an orthonormal basis $\{v_1, \dots, v_k\}$ with $\mathcal{L}v_i + \lambda_i v_i = 0$, $\lambda_i < 0$ given in Theorem 2.29. Then for any function $f \in C^\infty(M) \cap H_0^{1,2}(M) \cap W^\perp$ we have $\mathcal{Q}(f) \geq 0$, and if $\mathcal{Q}(f) = 0$ then $\mathcal{L}(f) = 0$, where $W^\perp$ is the $L^2(M)$-orthogonal complement of W.*

PROOF. A first observation is that the functions $v_i \in H_0^{1,2}(M)$. In fact, from Remark 4.47, the function $\|A\|^2$ is bounded. Also, since M is regular at infinity, the Laplacian Δ on M is a uniformly elliptic operator. Therefore, as $\Delta v_i + \|A\|^2 v_i + \lambda_i v_i = 0$, then using a covering argument and elliptic estimates, we have that $\|\nabla v_i\|_{L^2(M)} \leq C\|v_i\|_{L^2(M)} < \infty$.

If $f \in C^\infty(M) \cap H_0^{1,2}(M) \cap W^\perp$, we choose φ^R to be the cut-off function given in Lemma 4.17. Now we define

$$f^R = \varphi^R(f + c_1 v_1 + \cdots + c_k v_k),$$

where $\{c_1, \dots, c_k\}$ are constants depending on R. Then $f^R \in W^\perp$ if the constants c_i are solutions of the following linear system

$$\sum_{j=1}^{k} \left(\int_M \varphi^R v_i v_j \,\mathrm{d}v \right) c_j = - \int_M \varphi^R f v_i \,\mathrm{d}v, \quad 1 \leq i \leq k.$$

By the dominate convergence theorem (Theorem 1.7), $\lim_{R\to\infty} \left(\int_M \varphi^R v_i v_j \,\mathrm{d}v \right)$ is the identity matrix, and thus, for sufficiently large R, the above system has a solution. We note that, again by the dominate covergence theorem,

$$\lim_{R\to\infty}\int_M \varphi^{\frac{Rfv_i}{}}\, dv = \int_M fv_i\, dv = 0,$$

and so, $\lim_{R\to\infty} c_i = 0$.

Now, since $f^R \in C_0^\infty(M)\cap W^\perp$, from Theorem 2.29 we have that $\mathcal{Q}(f^R)\geq 0$ for R large enough. Hence,

$$0 \leq \mathcal{Q}(f^R) = \mathcal{Q}(\varphi^R f) + 2\sum_{i=1}^r c_i \mathcal{Q}(\varphi^R f, \varphi^R v_i) + \sum_{i,j=1}^r c_i c_j \mathcal{Q}(\varphi^R v_i, \varphi^r v_j).$$

If we let $R\to\infty$ and use that $\lim_{R\to\infty} c_i = 0$, we get that $\mathcal{Q}(f)\geq 0$. This proves the first part of the proposition.

Now suppose that $\mathcal{Q}(f) = 0$. We will prove that $\mathcal{Q}(f,h) = 0$ for any $h \in H_0^{1,2}(M)$, which would demonstrate that $\mathcal{L}f = 0$.

First we assume that $h \in C_0^\infty(M)\cap W^\perp$. We do a reasoning quite similar to the one above. We choose φ^R as the cut-off function given in Lemma 4.17, so that the support of φ^R is contained in $B_p(R)$. For each $t\in\mathbb{R}$, we define

$$f_t^R = \varphi^R(f + th + c_1 v_1 + \cdots + c_k v_k),$$

where $\{c_1,\ldots,c_k\}$ are constants depending of R. Then $f_t^R \in W^\perp$ if the constant c_i are solutions of the following linear system

$$\sum_{j=1}^k \left(\int_M \varphi^R v_i v_j\, \mathrm{d}\right) c_j = -\int_M \varphi^R f v_i, \quad 1\leq i\leq k,$$

where we have taken into account that $\int_M \varphi^R h v_i\, dv = \int_M h v_i\, dv = 0$. Reasoning as before, Theorem 2.29 implies that $\mathcal{Q}(f_t^R)\geq 0$ for any t and R large enough. Since $\mathcal{Q}(\varphi^R h, \varphi^R v_i) = 0$, we have

$$\begin{aligned} 0 \leq \mathcal{Q}(f_t^R) &= t^2\mathcal{Q}(\varphi^R h) + 2t\mathcal{Q}(\varphi^R f, \varphi^R h) \\ &\quad + \left(\mathcal{Q}(\varphi^R f) + 2\sum_{i=1}^k c_i \mathcal{Q}(\varphi^R f, \varphi^R v_i) + \sum_{i,j=1}^k c_i c_j \mathcal{Q}(\varphi^R v_i, \varphi^R v_j)\right). \end{aligned}$$

Since the above expression is satisfied for every $t\in\mathbb{R}$, we obtain

$$\mathcal{Q}(\varphi^R f, \varphi^R h)^2 \leq \mathcal{Q}(\varphi^R h)\left(\mathcal{Q}(\varphi^R f) + 2\sum_{i=1}^k c_i \mathcal{Q}(\varphi^R f, \varphi^r v_i) + \sum_{i,j=1}^k c_i c_j \mathcal{Q}(\varphi^R v_i, \varphi^R v_j)\right).$$

Letting $R\to\infty$ we have that $c_i\to 0$, $\mathcal{Q}(\varphi^R f)\to\mathcal{Q}(f) = 0$ and $\mathcal{Q}(\varphi^R f, \varphi^R h))\to \mathcal{Q}(f,h)$. So we get $\mathcal{Q}(f,h)^2\leq 0$, hence $\mathcal{Q}(f,h) = 0$.

Second we consider $h\in C^\infty(M)\cap H_0^{1,2}(M)\cap W^\perp$. In this case, h can be approximated by compactly supported smooth functions on $W^\perp$. Using the first case and standard arguments we deduce that $\mathcal{Q}(f,h) = 0$.

Finally, we need to prove that $\mathcal{Q}(f, v_i) = 0$ for any $1\leq i\leq k$. From the proof of Theorem 2.29, we know that the functions v_i are weak limits of a sequence of eigenfunctions of $\mathcal{L}$ on $B_p(R_i)$. Repeating a cut-off argument similar to the one before, we get $\mathcal{Q}(f, v_i) = 0$ and thus $\mathcal{Q}(f,h) = 0$ for $h\in W$. This concludes the proof. □

THEOREM 4.52 ([**Li17**]). *Let $\Phi : M^n \to \mathbb{R}^{n+1}$ be a two-sided complete minimal hypersurface of finite total curvature, with $n \geq 3$. Then,*

$$\operatorname{Ind}(M) + N(M) \geq \frac{2}{n(n+1)}\left(r + \beta_1(\widehat{M}) - 1\right),$$

where r is the number of ends of M, $\beta_1(\widehat{M})$ is the first Betti number of $\widehat{M}$ (see Theorem 4.46) and $N(M) = \dim\{f \in L^2(M) \cap C^\infty(M) \mid \mathcal{L}f = 0\}$.

PROOF. We follow the notation and calculations of Proposition 2.43. Taking in that proposition $\Psi : (\mathbb{R}^{n+1}, \langle,\rangle) \to (\mathbb{R}^{n+1}, \langle,\rangle)$ as the identity map, for each harmonic vector field X on M we can define the functions $\{f^X_{\lambda,\mu} \mid 1 \leq \lambda < \mu \leq n+1\}$ on the hypersurface M by

$$f^X_{\lambda,\mu} = \langle N, a_\lambda\rangle\langle X, a_\mu\rangle - \langle N, a_\mu\rangle\langle X, a_\lambda\rangle,$$

where $\{a_1, \dots a_{n+1}\}$ is an orthonormal basis of $\mathbb{R}^{n+1}$. Following Proposition 2.43 again, and as $\overline{\sigma} = 0$, we obtain that

(4.35)

$$\begin{aligned}
\nabla f^X_{\lambda,\mu} &= -\langle X, a_\mu\rangle A(a_\lambda)^T + \langle X, a_\lambda\rangle A(a_\mu)^T + \langle N, a_\lambda\rangle \nabla_{(a_\mu)^T} X - \langle N, a_\mu\rangle \nabla_{(a_\lambda)^T} X,\\
\mathcal{L} f^X_{\lambda,\mu} &= -2g(\nabla_{(a_\mu)^T} X, A(a_\lambda)^T) + 2g(\nabla_{(a_\lambda)^T} X, A(a_\mu)^T).
\end{aligned}$$

Now we prove that if $X \in \mathcal{H}(M) \cap L^2(M, TM)$, then the functions $f^X_{\lambda,\mu} \in H^{1,2}_0(M)$. Since M is complete, it suffices to prove, thanks to Theorem 1.10, that $f^X_{\lambda,\mu} \in L^2(M)$ and $\nabla f^X_{\lambda,\mu} \in L^2(M, TM)$. Since $|f^X_{\lambda,\mu}| \leq 2\|X\|$ and $X \in L^2(M, TM)$, it follows that $f^X_{\lambda,\mu} \in L^2(M)$.

To prove that $\nabla f^X_{\lambda,\mu} \in L^2(M)$, we proceed as follows. To simplify the notation, we denote f by $f^X_{\lambda,\mu}$ and $h = \Delta f$. Considering the above expression of $\mathcal{L} f^X_{\lambda,\mu}$, we have that $h = -\|A\|^2 f - 2g(\nabla_{(a_\mu)^T} X, A(a_\lambda)^T) + 2g(\nabla_{(a_\lambda)^T} X, A(a_\mu)^T)$. Since, from Remark 4.47, the function $\|A\|$ is bounded, then $h \in L^2(M)$. We fix a point $p \in M$ and we denote by B_1 and B_2 the balls of $\mathbb{R}^{n+1}$ of radius 1 and 2 centered at $\Phi(p)$. We take a cut-off function φ in $\mathbb{R}^{n+1}$ such that $0 \leq \varphi \leq 1$, $\varphi = 1$ on B_1, its support is contained on B_2 and its gradient satisfies $\|\nabla\varphi\| \leq 2$. If $\eta = \varphi \circ \Phi$, then

$$\begin{aligned}
&\int_{\Phi^{-1}(B_1)} \|\nabla f\|^2 \, dv \leq \int_{\Phi^{-1}(B_2)} \eta \|\nabla f\|^2 \, dv\\
&= \int_{\Phi^{-1}(B_2)} \{\operatorname{div}(f\eta\nabla f) - f\eta\Delta f - fg(\nabla f, \nabla\eta)\} \, dv\\
&= -\int_{\Phi^{-1}(B_2)} \{f\eta h + fg(\nabla f, \nabla\eta)\} \, dv\\
&\leq \|f\|_{L^2(\Phi^{-1}(B_2))}\|h\|_{L^2(\Phi^{-1}(B_2))} + \int_{\Phi^{-1}(B_2 \setminus B_1)} fg(\nabla f, \nabla\eta) \, dv\\
&\leq \|f\|_{L^2(\Phi^{-1}(B_2))}\|h\|_{L^2(\Phi^{-1}(B_2))} + \frac{1}{2}\int_{\Phi^{-1}(B_2 \setminus B_1)} \{\epsilon\|\nabla f\|^2 + \frac{1}{\epsilon} f^2 \|\nabla\eta\|^2\} \, dv\\
&\leq \|f\|_{L^2(\Phi^{-1}(B_2))}\|h\|_{L^2(\Phi^{-1}(B_2))} + \frac{1}{2}\int_{\Phi^{-1}(B_2 \setminus B_1)} \{\epsilon\|\nabla f\|^2 + \frac{4}{\epsilon} f^2\} \, dv,
\end{aligned}$$

where $\epsilon > 0$ is a small number and we have used Theorem 1.7 and the arithmetic-geometric mean inequality.

Let C be the cube inscribed in B_1 with center $\Phi(p)$. If we represent by $3C$ the cube of center $\Phi(p)$ under the homothety by a factor of 3, then $B_2 \subset 3C$ and the last inequality becomes

$$\int_{\Phi^{-1}(C)} \|\nabla f\|^2 \, dv \leq \|f\|_{L^2(\Phi^{-1}(3C))} \|h\|_{L^2(\Phi^{-1}(3C))} + \frac{1}{2} \int_{\Phi^{-1}(3C \setminus C)} \left\{ \epsilon \|\nabla f\|^2 + \frac{4}{\epsilon} f^2 \right\} dv.$$

Next cover $\mathbb{R}^{n+1}$ with parallel cubes with no common interiors and equal sice C. Adding the above inequality for each cube of the cover, and since each point is covered by the cubes $3C$ at most 3^{n+1} times, we get

$$\int_M \|\nabla f\|^2 \, dv \leq 3^{2(n+1)} \|f\|_{L^2(M)} \|h\|_{L^2(M)} + \int_M \left\{ \epsilon \frac{3^{n+1}}{2} \|\nabla f\|^2 + 2 \frac{3^{n+1}}{\epsilon} f^2 \right\} dv.$$

If we take $\epsilon = 3^{-(n+1)}$, then we finally obtain

$$\frac{1}{2} \int_M \|\nabla f\|^2 \, dv \leq 3^{2(n+1)} \left[\|f\|_{L^2(M)} \|h\|_{L^2(M)} + 2 \int_M f^2 \, dv \right] < \infty.$$

This proves that $f^X_{\lambda,\mu} \in H^{1,2}_0(M)$.

Now, for each $X \in \mathcal{H}(M) \cap L^2(M, TM)$ and using (4.35), the properties of the harmonic vector field X and the Gauss equation (2.2), we deduce

(4.36)

$$\begin{aligned}
\sum_{1 \leq \lambda < \mu \leq n+1} \mathcal{Q}(f^X_{\lambda,\mu}) &= \sum_{1 \leq \lambda < \mu \leq n+1} \int_M \{ \|\nabla f^X_{\lambda,\mu}\|^2 - \|A\|^2 (f^X_{\lambda,\mu})^2 \} \, dv \\
&= \frac{1}{2} \sum_{\lambda,\mu=1}^{n+1} \int_M \{ \|\nabla f^X_{\lambda,\mu}\|^2 - \|A\|^2 (f^X_{\lambda,\mu})^2 \} \, dv \\
&= \int_M \left\{ \|X\|^2 \|A\|^2 + \|\nabla X\|^2 - \|AX\|^2 - \frac{1}{2} \|A\|^2 \sum_{\lambda,\mu=1}^{n+1} (f^X_{\lambda,\mu})^2 \right\} dv \\
&= \int_M \{ \|\nabla X\|^2 - \|AX\|^2 \} \, dv = - \int_M \{ \langle \Delta X, X \rangle + \|AX\|^2 \} \, dv \\
&= - \int_M \{ S(X,X) + \|AX\|^2 \} \, dv = 0.
\end{aligned}$$

If $\operatorname{Ind}(M) = k < \infty$. Let $\{v_1, \ldots, v_k\}$ be the functions given in Proposition 4.51 and F the linear map

$$F : \mathcal{H}(M) \cap L^2(M, TM) \to \mathbb{R}^{n(n+1)k/2}$$

$$X \mapsto F(X) = \left(\int_M f^X_{\lambda,\mu} v_i \, dv \right)_{1 \leq \lambda < \mu \leq n+1, \, 1 \leq i \leq k}.$$

Therefore,

$$\dim \left(\mathcal{H}(M) \cap L^2(M, TM) \right) \leq \dim \ker F + \frac{n(n+1)k}{2}. \tag{4.37}$$

First we prove that $\dim \left(\mathcal{H}(M) \cap L^2(M, TM) \right) \geq r + \beta_1(\widehat{M}) - 1$. From Theorem 1.14,5, $\beta_1(\widehat{M}) = \dim \mathcal{H}^1(\widehat{M}) = \dim \mathcal{H}(\widehat{M})$, and so, there exist $\beta_1(\widehat{M})$ linearly independent harmonic vector fields on $\widehat{M}$. It is clear that their restrictions on M give $\beta_1(\widehat{M})$ linear independent non-exact harmonic vector fields on M which belong to $L^2(M, TM)$, where non-exact means that they are not gradients of harmonic functions.

On the other hand, in the proof of Theorem 4.45, it was proved that there exist $r-1$ linearly independent exact harmonic vector fields on M which belong to $L^2(M, TM)$. Thus our claim is proved.

Second, we prove that $\dim \ker F \leq \frac{n(n+1)}{2} N(M)$. To do this, we define the linear map

$$G : \ker F \to \bigoplus_{i=1}^{n(n+1)/2} C^\infty(M)$$
$$X \mapsto G(X) = \left(f^X_{\lambda,\mu}\right)_{1\leq\lambda<\mu\leq n+1}.$$

If $X \in \ker G$, then $f^X_{\lambda,\mu} = 0$ for $1 \leq \lambda, \mu \leq n+1$. So $\langle N, a_\lambda\rangle X = \langle X, a_\lambda\rangle N$. As N has no zeros and N and X are orthogonal, then $X = 0$. So $\ker G = \{0\}$ and therefore, $\dim \ker F = \dim \operatorname{img} G$. But if $X \in \ker F$, then $f^X_{\lambda,\mu}$ is L^2-orthogonal to the subspace W generated by $\{v_1, \ldots, v_k\}$, and Proposition 4.51 gives that $\mathcal{Q}(f^X_{\lambda,\mu}) \geq 0$ for any $1 \leq \lambda < \mu \leq n+1$. Now from (4.36),

$$\sum_{1\leq\lambda<\mu\leq n+1} \mathcal{Q}(f^X_{\lambda,\mu}) = 0,$$

and so $\mathcal{Q}(f^X_{\lambda,\mu}) = 0$ for all $1 \leq \lambda < \mu \leq n+1$. Using Proposition 4.51 again, $\mathcal{L}(f^X_{\lambda,\mu}) = 0$ and thus, $\dim \ker F = \dim \operatorname{img} G \leq \frac{n(n+1)}{2} N(M)$. Now the result follows from (4.37). □

At the end of this section we will prove R. Schoen's conjecture ([**CM11**], Conjecture 2.12), which claims that the hyperplanes of $\mathbb{R}^4$ are the only minimal and stable complete orientable hypersurfaces. As we indicated at the beginning of this section, there are three different proofs of this. In this text we will explain the one due to G. Catino, P. Mastrolia and A. Roncoroni ([**CMR24**]), because we believe that it is the most appropriate in the context of the book.

The first proof was given by O. Chodosh and C. Li, [**CL23**]. Since by Theorem 4.33,4 M is non-parabolic, for any point $p \in M$ there exists the corresponding Green's function u for the Laplacian defined on $M \setminus \{p\}$. Given a suitable cut-off function f, the authors use inequality 3 of Theorem 4.37 for the test function fu. Assuming that the second fundamental form is uniformly bounded, they manage to get control over the integral of $\|\nabla(fu)\|^3$ and conclude that the hypersurface is totally geodesic. Finally, using a standard point-picking rescaling argument, they prove the result without any condition on the second fundamental form.

The second proof of O. Chodosh and C. Li, [**CL23**], consists of proving that the volume growth of geodesic balls is cubic and then using Theorem 4.39. A.V. Pogorelov [**Pog81**] also proved the result for surfaces in $\mathbb{R}^3$ following a similar argument. Fundamental to the control of the volume of geodesic balls is the μ-bubble technique, first developed by M. Gromov, [**Gro96**].

The proof given in [**CMR24**] follows the ideas of D. Fischer-Colbrie and R. Schoen (Theorem 3.1) and R. Schoen, L. Simon and S.T. Yau (Theore 4.37). We begin by proving a result using arguments very similar to those of Theorem 3.38.

LEMMA 4.53 ([**CMR24**]). *Let $\Phi : M^3 \to \mathbb{R}^4$ be a two-sided complete stable minimal immersion and let $u \in C^\infty(M)$ be the positive function satisfying $\Delta u + \|A\|^2 u = 0$ given in Theorem 2.25,2. Then,*

(1) *The Ricci tensor S^* of $(M, g^* = u^{\frac{4}{3}} g)$ satisfies*
$$S^* \geq -\frac{2}{3u}\nabla^2 u + \frac{4}{3u^2}\mathrm{d}u \otimes \mathrm{d}u,$$
in the sense of quadratic forms, where g is the metric on M induced by Φ.

(2) *The 2-Bakry-Emery-Ricci tensor on $(M, g^* = u^{\frac{4}{3}} g)$ given by*
$$S^* + (\nabla^*)^2 f - \frac{1}{2}\mathrm{d}f \otimes \mathrm{d}f \geq 0$$
in the sense of quadratic forms, with $e^{2f} = u^{\frac{4}{3}}$.

(3) *The metric $g^* = u^{\frac{4}{3}} g$ is complete.*

PROOF. **Proof of 1**. From Proposition 1.5 and considering that $u^{\frac{4}{3}} = e^{2f}$, we have
$$\begin{aligned} S^* &= S - \left(\frac{2}{3u}\Delta u - \frac{2}{9u^2}\|\nabla u\|^2\right) g - \frac{2}{3}\nabla^2 \log u + \frac{4}{9u^2}\mathrm{d}u \otimes \mathrm{d}u \\ &= -A^2 + \left(\frac{2}{3}\|A\|^2 + \frac{2}{9u^2}\|\nabla u\|^2\right) g - \frac{2}{3u}\nabla^2 u + \frac{10}{9u^2}\mathrm{d}u \otimes \mathrm{d}u \\ &\geq -\frac{2}{3u}\nabla^2 u + \frac{4}{3u^2}\mathrm{d}u \otimes \mathrm{d}u, \end{aligned}$$
where we have used the Gauss equation, that u is a Jacobi function, i.e., $\Delta u + \|A\|^2 u = 0$, and the inequalities
$$\|\nabla u\|^2 g \geq \mathrm{d}u \otimes \mathrm{d}u, \quad A^2 \leq \frac{2}{3}\|A\|^2 g.$$
Proof of 2. We use Proposition 1.5 again to get
$$\begin{aligned} (\nabla^*)^2 f - \frac{1}{2} df \otimes df &= \nabla^2 f - \frac{5}{2} df \otimes df + \|\nabla f\|^2 g \\ &= \frac{2}{3u}\nabla^2 u - \frac{16}{9u^2}\mathrm{d}u \otimes \mathrm{d}u + \frac{4}{9u^2}\|\nabla u\|^2 g \\ &\geq \frac{2}{3u}\nabla^2 u - \frac{4}{3u^2}\mathrm{d}u \otimes \mathrm{d}u, \end{aligned}$$
where we again used again the inequality $\|\nabla u\|^2 g \geq \mathrm{d}u \otimes \mathrm{d}u$. The above inequality together with part 1 prove 2.

Proof of 3. We follow the idea and the notation of Theorem 3.1, taking into account that in this case M is not a surface but a manifold of dimension three.

From Theorem 1.2 it suffices to prove that the length of any divergent curve in (M, g^*) is infinite. If $\alpha : [0, \infty) \to M$ is a divergent curve, then by reasoning as in Theorem 3.1, there exists a curve $\beta : [0, \infty) \to M$ with $g(\beta', \beta') = 1$ which minimizes the distance d^* between any two points and that $L^*(\alpha) \geq L^*(\beta)$.

Since β minimizes d^* between any two points, the first variation formula of the lenght implies that
$$\nabla^*_{\beta'}\beta' = \frac{2(u \circ \beta)'}{3(u \circ \beta)}\beta',$$

and the second variation formula of the lenght implies that

$$\int_0^\infty \frac{1}{(u\circ\beta)^{\frac{2}{3}}}\left(2(\phi')^2 - S^*(\beta',\beta')\phi^2\right) \mathrm{d}s \geq 0. \tag{4.38}$$

for any piecewise smooth function $\phi : [0,\infty) \to \mathbb{R}$ such that $\phi_{|[0,\infty)\setminus(a,b)} = 0$, with $a < b$.

From Lemma 4.53, it follows that

$$\begin{aligned} S^*(\beta',\beta') &\geq -\frac{2}{3(u\circ\beta)}(\nabla^2 u)(\beta',\beta') + \frac{4}{3(u\circ\beta)^2}((u\circ\beta)')^2 \\ &= -\frac{2}{3(u\circ\beta)}(u\circ\beta)'' + \frac{2}{3(u\circ\beta)}g(\nabla u, \nabla_{\beta'}\beta') + \frac{4}{3(u\circ\beta)^2}((u\circ\beta)')^2. \end{aligned}$$

But, using Lemma 4.53 again, it follows that

$$\begin{aligned} \frac{2((u\circ\beta)')^2}{3(u\circ\beta)} &= g(\nabla^*_{\beta'}\beta', \nabla u) = g(\nabla_{\beta'}\beta', \nabla u) + \frac{4((u\circ\beta)')^2}{3(u\circ\beta)} - \frac{2\|\nabla u\|^2}{3(u\circ\beta)} \\ &\leq g(\nabla_{\beta'}\beta', \nabla u) + \frac{2((u\circ\beta)')^2}{3(u\circ\beta)}, \end{aligned}$$

where we again use that $\|\nabla u\|^2 g \geq \mathrm{d}u \otimes \mathrm{d}u$. So,

$$S^*(\beta',\beta') \geq -\frac{2}{3(u\circ\beta)}(u\circ\beta)'' + \frac{4}{3(u\circ\beta)^2}((u\circ\beta)')^2.$$

Using the above inequality in (4.38), we obtain

$$\int_0^\infty \frac{1}{(u\circ\beta)^{\frac{2}{3}}}\left((\phi')^2 + \frac{1}{3(u\circ\beta)}(u\circ\beta)''\phi^2 - \frac{2}{3(u\circ\beta)^2}((u\circ\beta)')^2\phi^2\right) \mathrm{d}s \geq 0.$$

Now, by calling $\psi = (u\circ\beta)^{-\frac{2}{3}}\phi$, the last inequality becomes

$$\begin{aligned} \int_0^\infty &\left((u\circ\beta)^{\frac{2}{3}}(\psi')^2 + \frac{1}{3}(u\circ\beta)^{-\frac{1}{3}}(u\circ\beta)''\psi^2 + \frac{4}{3}(u\circ\beta)^{-\frac{1}{3}}(u\circ\beta)'\psi\psi'\right) \mathrm{d}s \\ &- \int_0^\infty \frac{2}{9}(u\circ\beta)^{-\frac{4}{3}}((u\circ\beta)')^2\psi^2 \,\mathrm{d}s \geq 0. \end{aligned}$$

Integrating by parts, the last inequality implies that

$$\int_0^\infty \left((u\circ\beta)^{\frac{2}{3}}(\psi')^2 + \frac{2}{3}(u\circ\beta)^{-\frac{1}{3}}(u\circ\beta)'\psi\psi' - \frac{1}{9}(u\circ\beta)^{-\frac{4}{3}}((u\circ\beta)')^2\psi^2\right) \mathrm{d}s \geq 0. \tag{4.39}$$

Now we will deal with $I := \int_0^\infty (u\circ\beta)^{-\frac{1}{3}}(u\circ\beta)'\psi\psi' \,\mathrm{d}s$.

On the one hand, integration by parts gives

$$I = \frac{3}{2}\int_0^\infty \left((u\circ\beta)^{\frac{2}{3}}\right)'\psi\psi' \,\mathrm{d}s = -\frac{3}{2}\int_0^\infty (u\circ\beta)^{\frac{2}{3}}(\psi')^2 \,\mathrm{d}s - \frac{3}{2}\int_0^\infty (u\circ\beta)^{\frac{2}{3}}\psi\psi'' \,\mathrm{d}s.$$

On the other hand, it follows from $\|\sqrt{\frac{2}{3}}(u\circ\beta)^{-\frac{2}{3}}(u\circ\beta)'\psi + \sqrt{\frac{3}{2}}(u\circ\beta)^{\frac{1}{3}}\psi'\|^2 \geq 0$ that

$$-I \leq \frac{1}{3}\int_0^\infty (u\circ\beta)^{-\frac{4}{3}}((u\circ\beta)')^2\psi^2 \,\mathrm{d}s + \frac{3}{4}\int_0^\infty (u\circ\beta)^{\frac{2}{3}}(\psi')^2 \,\mathrm{d}s.$$

So,

$$\begin{aligned}I &= \frac{3}{2}I - \frac{1}{2}I\\ &\leq \int_0^\infty \left(-\frac{15}{8}(u\circ\beta)^{\frac{2}{3}}(\psi')^2 - \frac{9}{4}(u\circ\beta)^{\frac{2}{3}}\psi\psi'' + \frac{1}{6}(u\circ\beta)^{-\frac{4}{3}}((u\circ\beta)')^2\psi^2\right) ds\end{aligned}$$

and (4.39) becomes

$$\int_0^\infty (u\circ\beta)^{\frac{2}{3}}\left((\psi')^2 + 6\psi\psi''\right) ds \leq 0.$$

Now, comparing the above inequality with (3.4), the proof follows the same reasoning as in Theorem 3.1. □

LEMMA 4.54 ([**CMR24**]). *Under the same assumptions of Lemma 4.53, there exist $c > 0$ and $0 < \lambda < \frac{1}{25}$ such that*

$$\int_M \psi^{5+\lambda} u^{-2(1+\frac{\lambda}{3})}\|A\|^{5+\lambda}\, dv \leq c \int_M u^{-2(1+\frac{\lambda}{3})}\|\nabla\psi\|^{5+\lambda}\, dv,$$

for any non-negative Lipschitz function ψ of compact support on M, where $u \in C^\infty(M)$ is the positive function satisfying $\Delta u + \|A\|^2 u = 0$ given in Theorem 2.25,2.

PROOF. In the proof of the lemma, we will follow the notation and ideas of Theorem 4.37.

From (4.27) we have that

$$\int_M f^2 a_\delta^{\frac{2}{3}}\|A\|^2\, dv \leq c_1 \int_M a_\delta^{\frac{2}{3}}\|\nabla f\|^2\, dv,$$

for any non-negative Lipschitz funcion with compact support f, where c_1 is a positive constant, $a_\delta = (\|A\|^2 + \delta)^{\frac{p-2}{4}}$ and $p \in [4, 4 + \sqrt{8/3})$.

Given $\alpha < -1$, we define ψ by $f = u^\alpha\psi$. Then

$$\begin{aligned}\|\nabla f\|^2 &= u^{2\alpha}\|\nabla\psi\|^2 + \psi^2\|\nabla u^\alpha\|^2 + 2u^\alpha\psi g(\nabla u^\alpha, \nabla\psi)\\ &\leq 2\left(u^{2\alpha}\|\nabla\psi\|^2 + \psi^2\|\nabla u^\alpha\|^2\right),\end{aligned}$$

and the last integral inequality becomes

$$\int_M u^{2\alpha}a_\delta^2\psi^2\|A\|^2\, dv \leq 2c_1 \int_M \left(u^{2\alpha}a_\delta^2\|\nabla\psi\|^2 + a_\delta^2\psi^2\|\nabla u^\alpha\|^2\right) dv. \tag{4.40}$$

Now we are going to deal with the last summand of (4.40) using that $\Delta u + \|A\|^2 u = 0$. In fact, since

$$\begin{aligned}\operatorname{div}(u^\alpha a_\delta^2\psi^2\nabla u^\alpha) &= a_\delta^2\psi^2\|\nabla u^\alpha\|^2 + 2u^\alpha a_\delta^2\psi g(\nabla u^\alpha, \nabla\psi)\\ &\quad + \frac{p-2}{2}u^\alpha a_\delta^{\frac{2(p-4)}{p-2}}\psi^2 g(\nabla u, \nabla\|A\|^2) + u^\alpha a_\delta^2\psi^2\Delta u^\alpha,\end{aligned}$$

and

$$\Delta u^\alpha = \alpha u^{\alpha-1}\Delta u + \alpha(\alpha-1)u^{\alpha-2}\|\nabla u\|^2 = -\alpha u^\alpha\|A\|^2 + \frac{\alpha-1}{\alpha}u^{-\alpha}\|\nabla u^\alpha\|^2,$$

we get that

$$\begin{aligned}\operatorname{div}(u^\alpha a_\delta^2\psi^2\nabla u^\alpha) &= (2 - \frac{1}{\alpha})a_\delta^2\psi^2\|\nabla u^\alpha\|^2 + 2u^\alpha a_\delta^2\psi g(\nabla u^\alpha, \nabla\psi)\\ &\quad + \frac{p-2}{2}u^\alpha a_\delta^{\frac{2(p-4)}{p-2}}\psi^2 g(\nabla u, \nabla\|A\|^2) - \alpha u^{2\alpha}a_\delta^2\psi^2\|A\|^2.\end{aligned}$$

Using the following inequality in the above equation

$$-2u^\alpha\psi g(\nabla u^\alpha,\nabla\psi)\le\epsilon\psi^2\|\nabla u^\alpha\|^2+\frac{1}{\epsilon}u^{2\alpha}\|\nabla\psi\|^2,\quad\text{for}\quad\epsilon>0,$$

and using the Theorem 1.7, we obtain

$$\begin{aligned}(4.41)\qquad &(2-\frac{1}{\alpha}-\epsilon)\int_M a_\delta^2\psi^2\|\nabla u^\alpha\|^2\,dv\le\frac{1}{\epsilon}\int_M u^{2\alpha}a_\delta^2\|\nabla\psi\|^2\,dv\\ &-\frac{p-2}{2}\int_M u^\alpha a_\delta^{\frac{2(p-4)}{p-2}}\psi^2 g(\nabla u^\alpha,\nabla\|A\|^2)\,dv+\alpha\int_M u^{2\alpha}a_\delta^2\psi^2\|A\|^2\,dv.\end{aligned}$$

Now with

$$-2u^\alpha a_\delta^{\frac{2(p-4)}{p-2}}\psi^2 g(\nabla u^\alpha,\nabla\|A\|^2)\le\frac{1}{\eta}a_\delta^2\psi^2\|\nabla u^\alpha\|^2+\eta u^{2\alpha}a_\delta^{\frac{2(p-6)}{p-2}}\psi^2\|\nabla\|A\|^2\|^2,$$

for $\eta>0$, in (4.41) and that $\alpha<-1$, we deduce

$$\begin{aligned}(4.42)\qquad &(2-\frac{1}{\alpha}-\epsilon-\frac{(p-2)}{4\eta})\int_M a_\delta^2\psi^2\|\nabla u^\alpha\|^2\,dv\\ &\le\frac{1}{\epsilon}\int_M u^{2\alpha}a_\delta^2\|\nabla\psi\|^2\,dv-\int_M u^{2\alpha}a_\delta^2\psi^2\|A\|^2\,dv\\ &+\frac{(p-2)\eta}{4}u^{2\alpha}a_\delta^{\frac{2(p-6)}{p-2}}\psi^2\|\nabla\|A\|^2\|^2.\end{aligned}$$

We next deal with the first summand of (4.40). To do this, we use (4.25) with $q=\frac{p-2}{4}$ and $n=3$, and we get

$$\begin{aligned}(4.43)\qquad &(\frac{3p-7}{12})\int_M f^2 a_\delta^{\frac{2(p-6)}{p-2}}\|\nabla\|A\|^2\|^2\,dv\\ &\le\int_M f^2a_\delta^2\|A\|^2\,dv-\int_M f a_\delta^{\frac{2(p-4)}{p-2}}g(\nabla\|A\|^2,\nabla f)\,dv.\end{aligned}$$

Again using that

$$-2fa_\delta^{\frac{2(p-4)}{p-2}}g(\nabla\|A\|^2,\nabla f)\le\frac{1}{2}f^2a_\delta^{\frac{2(p-6)}{p-2}}\|\nabla\|A\|^2\|^2+2a_\delta^2\|\nabla f\|^2,$$

the inequality (4.43) becomes

$$(4.44)\quad (\frac{3p-10}{12})\int_M f^2a_\delta^{\frac{2(p-6)}{p-2}}\|\nabla\|A\|^2\|^2\,dv\le\int_M f^2a_\delta^2\|A\|^2\,dv+\int_M a_\delta^2\|\nabla f\|^2\,dv.$$

Since $f=u^\alpha\psi$, we have

$$\begin{aligned}\|\nabla f\|^2&=u^{2\alpha}\|\nabla\psi\|^2+\psi^2\|\nabla u^\alpha\|^2+2u^\alpha\psi g(\nabla u^\alpha,\nabla\psi)\\ &\le(1+\frac{1}{\epsilon})u^{2\alpha}\|\nabla\psi\|^2+(1+\epsilon)\psi^2\|\nabla u^\alpha\|^2),\end{aligned}$$

and so the inequality (4.44) gives

$$\begin{aligned}(4.45)\qquad &\left(\frac{3p-10}{12}\right)\int_M u^{2\alpha}a_\delta^{\frac{2(p-6)}{p-2}}\psi^2\|\nabla\|A\|^2\|^2,dfv\le\int_M u^{2\alpha}a_\delta^2\psi^2\|A\|^2\,dv\\ &+\left(1+\frac{1}{\epsilon}\right)\int_M u^{2\alpha}a_\delta^2\|\nabla\psi\|^2\,dv+(1+\epsilon)\int_M a_\delta^2\psi^2\|\nabla u^\alpha\|^2\,dv.\end{aligned}$$

Adding (4.42) and (4.45) and taking $p = 5+\lambda \in [4, 4+\sqrt{8/3})$ we have

$$\begin{aligned}(1 - \frac{1}{\alpha} - 2\epsilon - \frac{3+\lambda}{4\eta}) \int_M a_\delta^2 \psi^2 \|\nabla u^\alpha\|^2\, dv \\ \leq (\frac{\eta(3+\lambda)}{4} - \frac{5+3\lambda}{12}) \int_M u^{2\alpha} a_\delta^{\frac{2(\lambda-1)}{3+\lambda}} \psi^2 \|\nabla\|A\|^2\|^2\, dv \\ +(1+\frac{2}{\epsilon}) \int_M u^{2\alpha} a_\delta^2 \|\nabla\psi\|^2\, dv.\end{aligned}$$

Now, we take $\eta = \frac{5+3\lambda}{9+3\lambda} > 0$ and $\alpha = -(1+\frac{\lambda}{3}) < -1$ and so, the last inequality gives us

$$\begin{aligned}(\frac{6+\lambda}{3+\lambda} - \frac{3(3+\lambda)^2}{4(5+3\lambda)} - 2\epsilon) \int_M a_\delta^2 \psi^2 \|\nabla u^{-(1+\frac{\lambda}{3})}\|^2\, dv \\ \leq (1+\frac{2}{\epsilon}) \int_M u^{-2(1+\frac{\lambda}{3})} a_\delta^2 \|\nabla\psi\|^2\, dv.\end{aligned}$$

Since $\frac{6+\lambda}{3+\lambda} - \frac{3(3+\lambda)^2}{4(5+3\lambda)} > 0$, there exists $\epsilon > 0$ such that $\frac{6+\lambda}{3+\lambda} - \frac{3(3+\lambda)^2}{4(5+3\lambda)} - 2\epsilon > 0$. So from the last inequality we deduce

$$\int_M a_\delta^2 \psi^2 \|\nabla u^{-(1+\frac{\lambda}{3})}\|^2\, dv \leq c_2 \int_M u^{-2(1+\frac{\lambda}{3})} a_\delta^2 \|\nabla\psi\|^2\, dv,$$

where c_2 is a positive constant. Now, using this inequality in (4.40), we get

$$\int_M u^{-2(1+\frac{\lambda}{3})} a_\delta^2 \psi^2 \|A\|^2\, dv \leq c_3 \int_M u^{-2(1+\frac{\lambda}{3})} a_\delta^2 \|\nabla\psi\|^2\, dv,$$

where c_3 is a positive constant. Letting $\delta \to 0$ and using the dominated convergence theorem 1.7, we finally have

$$\int_M u^{-2(1+\frac{\lambda}{3})} \psi^2 \|A\|^{5+\lambda}\, dv \leq c_3 \int_M u^{-2(1+\frac{\lambda}{3})} \|A\|^{3+\lambda} \|\nabla\psi\|^2\, dv.$$

Next we use Young's inequality

$$ab \leq \frac{\beta^t a^t}{t} + \frac{\beta^{-s} b^s}{s}, \quad \frac{1}{t} + \frac{1}{s} = 1, \quad t, s > 1, \quad \beta > 0,$$

for $a = \|A\|^{3+\lambda} \psi^{\frac{2(3+\lambda)}{5+\lambda}}$, $b = \|\nabla\psi\|^2 \psi^{\frac{-2(3+\lambda)}{5+\lambda}}$, $t = \frac{5+\lambda}{3+\lambda}$ and $s = \frac{5+\lambda}{2}$, so that

$$\|A\|^{3+\lambda} \|\nabla\psi\|^2 \leq \frac{\beta^t}{t} \|A\|^{5+\lambda} \psi^2 + \frac{\beta^{-s}}{s} \|\nabla\psi\|^{5+\lambda} \psi^{-(3+\lambda)}.$$

Taking β small enough to be $\frac{\beta^t}{t} < 1$, we finally get from the last two inequalities that

$$\int_M u^{-2(1+\frac{\lambda}{3})} \psi^2 \|A\|^{5+\lambda}\, dv \leq c_4 \int_M u^{-2(1+\frac{\lambda}{3})} \psi^{-(3+\lambda)} \|\nabla\psi\|^{5+\lambda}\, dv,$$

where c_4 is a positive constant. Since $\nabla\psi^{\frac{2}{5+\lambda}} = \frac{2}{5+\lambda} \psi^{-\frac{3+\lambda}{5+\lambda}} \nabla\psi$, the last inequality becomes

$$\int_M u^{-2(1+\frac{\lambda}{3})} \psi^2 \|A\|^{5+\lambda}\, dv \leq c \int_M u^{-2(1+\frac{\lambda}{3})} \|\nabla\psi^{\frac{2}{5+\lambda}}\|^{5+\lambda}\, dv.$$

To finish the proof, we replace $\psi^{\frac{2}{5+\lambda}}$ by ψ in the above inequality. □

THEOREM 4.55 ([**CMR24**]). *Let $\Phi : M^3 \to \mathbb{R}^4$ be a two-sided complete minimal hypersurface of $\mathbb{R}^4$. If M is stable, then $\Phi(M)$ is a hyperplane.*

PROOF. From Lemma 4.53, the manifold $(M^3, g^* = u^{\frac{4}{3}} g)$ is complete and the 2-Bakry-Emery-Ricci tensor $S^* + (\nabla^*)^2 f - \frac{1}{2} df \otimes df$ is non-negative semidefinite, with $e^{2f} = u^{\frac{4}{3}}$. Now, we can use the weighted Bishop's theorem proved by G. Wei and W. Wylie, [**WW09**], Theorem A.1, which states that for any point $p \in M$ and any $R > 0$,

$$\int_{B_p^*(R)} e^{-f} \, dv^* \leq C R^5,$$

where C is a positive constant, $B_p^*(R)$ is the geodesic ball centered at p and radius R of $(M, g^* = u^{\frac{4}{3}} g)$, and dv^* is the measure associated to g^*. Since $dv^* = e^{3f} dv$, we get

$$\int_{B_p^*(R)} u^{\frac{4}{3}} \, dv \leq C R^5. \tag{4.46}$$

Now, given $p \in M$ and $R > 0$, we define the non-negative Lipschitz function ψ with compact support

$$\psi(x) = \begin{cases} 1 & \text{on } B_p^*(R), \\ \frac{2R - d^*(x,p)}{R} & \text{on } B_p^*(2R) \setminus B_p^*(R), \\ 0 & \text{on } M \setminus B_p^*(2R), \end{cases}$$

where d^* is the distance associated to g^*. Then, $\|\nabla^* \psi\|_* \leq \frac{1}{R}$ and using Lemma 4.54 and (4.46), we have

$$\begin{aligned} \int_{B_p^*(R)} u^{-2(1+\frac{\lambda}{3})} \|A\|^{5+\lambda} \, dv &= \int_{B_p^*(R)} \psi^{5+\lambda} u^{-2(1+\frac{\lambda}{3})} \|A\|^{5+\lambda} \, dv \\ &\leq \int_M \psi^{5+\lambda} u^{-2(1+\frac{\lambda}{3})} \|A\|^{5+\lambda} \, dv \leq c \int_M u^{-2(1+\frac{\lambda}{3})} \|\nabla \psi\|^{5+\lambda} \, dv \\ &= c \int_M u^{\frac{4}{3}} \|\nabla^* \psi\|_*^{5+\lambda} \, dv = c \int_{B_p^*(2R)} u^{\frac{4}{3}} \|\nabla^* \psi\|_*^{5+\lambda} \, dv \\ &\leq \frac{c}{R^{5+\lambda}} \int_{B_p^*(2R)} u^{\frac{4}{3}} \, dv \leq \frac{\hat{c}}{R^\lambda}. \end{aligned}$$

Finally letting $R \to \infty$ in the above inequality, we get that $A = 0$ and so $\Phi(M)$ is a hyperplane of $\mathbb{R}^4$. $\square$

After the first draft was written and during the revision of the text by the reviewers, important progress has been made in the study of two-sided complete minimal stable hypersurfaces of Euclidean space. The first result, due to O. Chodosh, C. Li, P. Minter and D. Stryker, is the following:

THEOREM 4.56 ([**CLMS24**]). *Let $\Phi : M^4 \to \mathbb{R}^5$ be a two-sided complete minimal hypersurface of $\mathbb{R}^5$. If M is stable, then $\Phi(M)$ is a hyperplane.*

The ideas for the proof of Theorem 4.56 are a generalization of the ones given by O. Chodosh and C. Li, [**CL23**], for the case of hypersurfaces of $\mathbb{R}^4$. The difficulty of the proof in this case is much higher and requires the development of new and important ideas in the μ-buble technique.

THEOREM 4.57 ([**Maz24**]). *Let $\Phi : M^5 \to \mathbb{R}^6$ be a two-sided complete minimal hypersurface of $\mathbb{R}^6$. If M is stable, then $\Phi(M)$ is a hyperplane.*

In the case of Theorem 4.57, L. Mazet introduces some clever and subtle ideas to generalize the arguments in the proof of Theorem 4.56.

Also, E. Bombieri, E. De Giorgi and E. Giusti, [**BDGG69**], proved that for $n \geq 7$ there exist two-sided complete stable minimal hypersurfaces of $\mathbb{R}^{n+1}$ which are not hyperplanes. Thus the open problem is:

Classify the two-sided complete stable minimal hypersurfaces of $\mathbb{R}^7$. Are they hyperplanes?

Since Theorem 4.39 has not been proved for $n = 6$, there is an added difficulty in proving, using a similar method to the one used in the proof of Theorems 4.56 and 4.57, that the only two-sided complete stable minimal hypersurfaces of $\mathbb{R}^7$ are the hyperplanes.

4.4. Minimal surfaces in $\mathbb{R}^4$

The study of the stability of minimal surfaces of Euclidean space with codimension greater than one is a difficult problem, and only few results are known. Complex immersions of Riemann surfaces in complex Euclidean $\mathbb{C}^n$ are stable (see Theorem 6.4) and most of the results are aimed at proving that the stability of minimal immersions implies that they are holomorphic with respect to some orthogonal complex structure of $\mathbb{R}^{2n}$. However, C. Arezzo, M.J. Micallef and G.P. Pirola, [**AMP02**], proved that there are examples of stable minimal complete surfaces in $\mathbb{R}^n$, $n \geq 5$, which are not complex curves.

In this section we will present an important result due to M.J. Micallef, [**Mic84**], in which he classified, assuming that the surface is parabolic, the complete and orientable minimal stable surfaces of $\mathbb{R}^4$, proving that they are holomorphic with respect to some orthogonal complex structure of $\mathbb{R}^4$. Anyway, it seems natural to conjecture that the result is true without the parabolicity assumption, i.e.,

Conjecture: Every orientable complete minimal stable surface of $\mathbb{R}^4$ is holomorphic with respect to some orthogonal complex structure of $\mathbb{R}^4$.

In codimension greater than two, some positive results in this direction are due to M.J. Micallef, [**Mic84**], and A. Fraser and R. Schoen, [**FS23**], under the hypothesis of finite total curvature and genus 0 or 1 respectively.

THEOREM 4.58 ([**Mic84**]). *Let $\Phi : M^2 \to \mathbb{R}^4$ be a stable minimal immersion of a complete oriented surface M. If M is parabolic, then Φ is holomorphic with respect to some orthogonal complex structure of $\mathbb{R}^4$.*

PROOF. To prove the result, we will use the theory developed in Section 3.4. Following the notation of that section, we consider the complex structure J^+ defined by

$$J^+(e_1) = e_2, \quad J^+e_3 = e_4,$$

where $\{e_1, e_2, e_3, e_4\}$ is an oriented orthonormal reference on $\Phi^*T\mathbb{R}^4$ such that $\{e_1, e_2\}$ is an oriented reference on TM.

If $\xi \in \mathfrak{X}_0^\perp(M)$, then we have from Proposition 3.44 that

$$\mathcal{Q}(\xi) + \mathcal{Q}(J^+\xi) = \int_M \|D^+\xi\|^2 \, dv - \frac{1}{2}\int_M \|\sigma^+\|^2\|\xi\|^2 \, dv, \tag{4.47}$$

where σ^+ was defined in (3.24) and D^+ was defined in (3.21).

Now, if $f \in C_0^\infty(M)$ and $a \in \mathbb{R}^4$ with $\|a\| = 1$, then we consider the normal section $\xi \in \mathfrak{X}_0^\perp(M)$ defined by $\xi = fa^\perp$, where $a^\perp$ is the normal component of a. If $a = a^\top + a^\perp$, where $a^\top$ is the tangential component of a, then by derivating the above expression we get

$$\nabla_v^\perp a^\perp = -\sigma(v, a^\top), \quad \nabla_v a^\top = A_{a^\perp} v, \quad v \in TM, \tag{4.48}$$

and so,

$$D_v^+\xi = \nabla_{J^+v}^\perp(fa^\perp) - J^+\nabla_v^\perp(fa^\perp) = g(\nabla f, J^+v)a^\perp - g(\nabla f, v)J^+a^\perp - f\sigma^+(v, a^\top).$$

It follows that

$$\|D^+\xi\|^2 = 2\|\nabla f\|^2\|a^\perp\|^2 + f^2\sum_{i=1}^2 \|\sigma^+(e_i, a^\top)\|^2 + 4f\langle\sigma^+(\nabla f, a^\top), J^+a^\perp\rangle. \tag{4.49}$$

Now, if div is the divergence operator on M, then

$$\begin{aligned}\operatorname{div}\left(f^2(A_{J^+a^\perp}J^+a^\top - A_{a^\perp}a^\top)\right) = f^2 \operatorname{div}\left(A_{J^+a^\perp}J^+a^\top - A_{a^\perp}a^\top\right)\\ + 2f\langle\sigma^+(a^\top, \nabla f), J^+a^\perp\rangle,\end{aligned}$$

hence, using (4.49), we have that (4.47) becomes

$$\mathcal{Q}(\xi) + \mathcal{Q}(J^+\xi) = 2\int_M \|\nabla f\|^2\|a^\perp\|^2 \, dv$$

$$+ \int_M f^2 \left\{\sum_{i=1}^2 \|\sigma^+(e_i, a^\top)\|^2 - \frac{\|\sigma^+\|^2}{2}\|a^\perp\|^2 - 2\operatorname{div}\left(A_{J^+a^\perp}J^+a^\top - A_{a^\perp}a^\top\right)\right\} dv.$$

The stability of M implies that $\mathcal{Q}(\xi) + \mathcal{Q}(J^+\xi) \geq 0$ and so, the last equation together with $\|a^\perp\|^2 \leq \|a\|^2 = 1$ implies that

$$\int_M \|\nabla f\|^2 \, dv \geq \int_M q_a f^2 \, dv, \quad \text{for any} \quad f \in C_0^\infty(M), \tag{4.50}$$

where q_a is the smooth function on M given by

$$q_a = -\frac{1}{2}\sum_{i=1}^2 \|\sigma^+(e_i, a^\top)\|^2 + \frac{\|\sigma^+\|^2}{4}\|a^\perp\|^2 + \operatorname{div}\left(A_{J^+a^\perp}J^+a^\top - A_{a^\perp}a^\top\right).$$

From Theorem 2.25,2, the inequality (4.50) implies that there exists a positive function $u_a \in C^\infty(M)$ such that $\Delta u_a + q_a u_a = 0$. If $w_a = \log u_a$, then the above equation implies that

$$\Delta w_a = -q_a - \|\nabla w_a\|^2, \tag{4.51}$$

and thus,

$$\operatorname{div}(f^2\nabla w_a) = f^2\Delta w_a + 2fg(\nabla f, \nabla w_a).$$

As $f \in C_0^\infty(M)$, using (4.51) and Theorem 1.7, we get

$$2\int_M fg(\nabla f, \nabla w_a) \, dv = \int_M f^2(q_a + \|\nabla w_a\|^2) \, dv, \quad \text{for any} \quad f \in C_0^\infty(M).$$

Also, since

$$2|fg(\nabla f, \nabla w_a)| \leq 2|f|\|\nabla f\|\|\nabla w_a\| \leq 2\|\nabla f\|^2 + \frac{1}{2}f^2\|\nabla w_a\|^2,$$

from the last equation we have

$$\int_M \|\nabla f\|^2 \,dv \geq \int_M \left(\frac{2q_a + \|\nabla w_a\|^2}{4}\right) f^2 \,dv, \quad \text{for any} \quad f \in C_0^\infty(M). \tag{4.52}$$

Now, if $\{a_1, a_2, a_3, a_4\}$ is an orthonormal basis of $\mathbb{R}^4$, then from the definition of the function q_a, it follows

$$\begin{aligned}
\sum_{\alpha=1}^{4} q_{a_\alpha} = & -\frac{1}{2}\sum_{\alpha=1}^{4}\sum_{i,j,k=1}^{2} \langle a_\alpha, e_j\rangle\langle a_\alpha, e_k\rangle\langle\sigma^+(e_i,e_j),\sigma^+(e_i,e_k)\rangle \\
& + \frac{\|\sigma^+\|^2}{4}\sum_{\alpha=1}^{4}\sum_{i=3}^{4}\langle a_\alpha, e_i\rangle^2 \\
& + \operatorname{div}\left\{\sum_{\alpha=1}^{4}\sum_{i=1}^{2}\sum_{j=3}^{4}(\langle a_\alpha, Je_j\rangle\langle a_\alpha, Je_i\rangle - \langle a_\alpha, e_j\rangle\langle a_\alpha, e_i\rangle)A_{e_j}e_i\right\} \\
= & -\frac{1}{2}\|\sigma^+\|^2 + \frac{1}{2}\|\sigma^+\|^2 = 0.
\end{aligned}$$

So from (4.52), we finally deduce that

$$\int_M \|\nabla f\|^2 \,dv \geq \int_M \widehat{q} f^2 \,dv, \quad \text{for any} \quad f \in C^\infty(M),$$

being $\widehat{q} = \frac{\sum_{\alpha=1}^4 \|\nabla w_{a_\alpha}\|^2}{16}$. Again using Theorem 2.25, from the last inequality we obtain a positive function $\widehat{u} \in C^\infty(M)$ such that $\Delta\widehat{u} + \widehat{q}\widehat{u} = 0$. Since $\widehat{q} \geq 0$, the function $\widehat{u}$ is superharmonic and the parabolicity of M implies that $\widehat{u}$ is constant. Thus, since $\widehat{u} > 0$, we get that $\widehat{q} = 0$, which gives that w_{a_α} is constant for any $1 \leq \alpha \leq 4$. Now, from (4.51) we finally get that $q_{a_\alpha} = 0$ for any α, and so,

(4.53)
$$0 = q_a = -\frac{1}{2}\sum_{i=1}^{2}\|\sigma^+(e_i, a^\top)\|^2 + \frac{\|\sigma^+\|^2}{4}\|a^\perp\|^2 + \operatorname{div}\left(A_{J^+a^\perp}J^+a^\top - A_{a^\perp}a^\top\right),$$

for any $a \in \mathbb{R}^4$ with $\|a\| = 1$. We will next use the complex structure of M and the holomorphic 4-differential on M associated to the minimal immersion Φ. Let $z = x + iy$ be a local complex coordinate on M and let $\partial_z = \frac{1}{2}(\frac{\partial}{\partial x} - i\frac{\partial}{\partial y})$ and $\partial_{\overline{z}} = \frac{1}{2}(\frac{\partial}{\partial x} + i\frac{\partial}{\partial y})$ be the Cauchy-Riemann operators. If the induced metric is given by $g = e^{2u}\|dz\|^2$, then we have that

$$\begin{aligned}
\Phi_{zz} &= 2u_z\Phi_z + \sigma(\partial_z, \partial_z), \\
\Phi_{z\overline{z}} &= -\frac{e^{2u}}{2}\Phi,
\end{aligned} \tag{4.54}$$

where we have extended to the complexified bundles all the operators associated to the immersion Φ. If we denote by $\Omega = \sigma(\partial_z, \partial_z)$, then we define on M a 4-differential $\Theta(z) = g(\Omega, \Omega) \otimes (dz)^4$. Using the Codazzi equation and (4.54), we have that

$$\nabla^\perp_{\frac{\partial}{\partial \overline{z}}}\Omega = (\nabla\sigma)(\partial_z, \partial_z, \partial_{\overline{z}}) = \nabla^\perp_{\frac{\partial}{\partial z}}\sigma(\partial_z, \partial_{\overline{z}}) - 2u_z\sigma(\partial_z, \partial_{\overline{z}}) = 0, \tag{4.55}$$

and thus $\partial_{\overline{z}} g(\Omega, \Omega) = 0$. This means that Θ is a holomorphic 4-differential on M. We have only two possibilities: either $\Theta \equiv 0$ or Θ has only isolated zeros.

In the first case, $\Theta = 0$, we have that

$$\|\sigma(e_1, e_1)\| = \|\sigma(e_1, e_2)\|, \quad \langle \sigma(e_1, e_1), \sigma(e_1, e_2)\rangle = 0,$$

where $\{e_1, e_2\}$ is the orthonormal reference $e_1 = e^{-u}\partial_x$ and $e_2 = e^{-u}\partial_y$. This implies that either σ^+ or σ^- vanish identically and so, the immersion Φ is superminimal with positive or negative spin (see Definition 3.42).

The twistor spaces of $\mathbb{R}^4$ are $\mathbb{R}^4 \times \mathbb{S}^2$ with the appropriate complex structure. From Proposition 3.43, the twistor lifting of Φ is horizontal and so, Φ is complex with respect to some orthogonal complex structure of $\mathbb{R}^4$.

Finally, we will see that the second case, Θ nonzero cannot happen. In this second case, let p be a point of M where $\Theta(p) \neq 0$. Then, with a suitable change of the complex parameter in a neighborhood U of the point p, we can assume that $\Theta = (dz)^4$ and thus $\|\sigma(e_1, e_1)\|^2 = 4e^{-4u} + \|\sigma(e_1, e_2)\|^2$ and $\langle \sigma(e_1, e_1), \sigma(e_1, e_2)\rangle = 0$ in U.

Under these conditions, we take $a = \frac{\sigma(e_1,e_1)}{\|\sigma(e_1,e_1)\|}(p)$. Then $a^\top(p) = 0$ and $a^\perp(p) = a$. If $\lambda = \|\sigma(e_1, e_1)\|(p)$ and $\mu = \frac{\langle \sigma(e_1,e_2), J^+\sigma(e_1,e_1)\rangle}{\|\sigma(e_1,e_1)\|}(p)$, the above conditions give that at the point p,

$$A_a = \begin{pmatrix} \lambda & 0 \\ 0 & -\lambda \end{pmatrix}, \qquad A_{J^+a} = \begin{pmatrix} 0 & \mu \\ \mu & 0 \end{pmatrix}, \tag{4.56}$$

Since $q_a = 0$, we have

$$0 = q_a(p) = \frac{\|\sigma^+\|^2(p)}{4} + \operatorname{div}\left(A_{J^+a^\perp} J^+ a^\top - A_{a^\perp} a^\top\right)(p). \tag{4.57}$$

Using that $a^\top(p) = 0$, Codazzi's equation, (4.48) and (4.56), we get

$$\begin{aligned}
\operatorname{div}&\left(A_{J^+a^\perp} J^+ a^\top - A_{a^\perp} a^\top\right)(p) \\
&= \operatorname{div}\left\{\sum_{i=1}^{2} (\langle \sigma(J^+a^\top, e_i), J^+a^\perp\rangle - \langle \sigma(e_i, a^\top), a^\perp\rangle) e_i\right\}(p) \\
&= \sum_{i=1}^{2} e_i(\langle \sigma(J^+a^\top, e_i), J^+a^\perp\rangle - \langle \sigma(e_i, a^\top), a^\perp\rangle)(p) \\
&= \sum_{i=1}^{2} (\langle \sigma(\nabla_{e_i} J^+a^\top, e_i), J^+a^\perp\rangle - \langle \sigma(e_i, \nabla_{e_i} a^\top), a^\perp\rangle)(p) \\
&= \sum_{i=1}^{2} (\langle A_{J^+a} e_i, J^+ A_a e_i\rangle - \langle A_a e_i, A_a e_i\rangle)(p) = 2\lambda(\mu - \lambda),
\end{aligned}$$

where we have used that $\operatorname{div}(e^{-u}\partial_x) = \operatorname{div}(e^{-u}\partial_y) = 0$.

So, since $\|\sigma^+\|^2(p) = 8\lambda(\lambda - \mu - 2e^{-4u(p)})$, equation (4.57) becomes $0 = -4\lambda e^{-4u(p)}$, which is impossible because $\lambda > 0$. So, the second case cannot happen and the theorem is proved. ☐

COROLLARY 4.59 ([**Mic84**]). *Let $\Phi : M^2 \to \mathbb{R}^4$ be a stable minimal immersion of a complete oriented surface M. If one of the following conditions holds:*

(1) *Φ is an entire graph,*
(2) *M has finite total curvature,*
(3) *Φ has quadratic area growth,*

then Φ is holomorphic with respect to some orthogonal complex structure on $\mathbb{R}^4$.

PROOF. In [**Oss69**] it was proved that if Φ is minimal and an entire graph, then M is conformally equivalent to the complex plane, which is parabolic. This proves 1.

In [**CO67**] it was proved that if M is minimal with finite total curvature, then M is conformally equivalent to a compact Riemann surface with finitely many punctures, which is also parabolic. This proves 2.

In [**CY75**] it was proved that if M is complete and Φ has quadratic area growth, then M is parabolic. This proves 3. □

THEOREM 4.60 ([**Mic84**]). *Let $\Phi : M^2 \to T^4$ be a stable minimal immersion of a compact oriented surface M into a flat torus $T = \mathbb{R}^4/\mathbb{Z}^4$. Then, Φ is holomorphic with respect to some orthogonal complex structure of T^4.*

PROOF. The proof is the same as in the Theorem 4.58 taking into account the following facts. First, given a vector $a \in \mathbb{R}^4$, then a projects into a parallel vector field of T^4. Second, we have that M is parabolic, because it is compact. Also, in some parts of the proof we can simplify the arguments of the Theorem 4.58 by using constant functions instead of compactly supported functions. □

CHAPTER 5

Index of Minimal Submanifolds of Spheres, Real Projective Spaces, Product of Spheres and Elliptic Berger Spheres

In this chapter we will present the main results on the index and nullity of compact minimal submanifolds of some remarkable Riemannian manifolds of non-negative curvature. In particular, we will consider as ambient manifolds the sphere $\mathbb{S}^m$, the real projective space $\mathbb{RP}^m$, the product of two spheres $\mathbb{S}^n \times \mathbb{S}^m$ and the elliptic Berger spheres.

5.1. Minimal submanifolds of the sphere

If $\Phi : M^n \to \mathbb{S}^m$ is a minimal immersion of a compact manifold M, we will denote by $\langle , \rangle$ the Euclidean metric on $\mathbb{R}^{m+1}$ as well as the induced metric on $\mathbb{S}^m$. The induced metric on M is denoted by g. From Definition 2.10, the Jacobi operator $\mathcal{L}$ of Φ is given by

$$\mathcal{L}\xi = \Delta^{\perp}\xi + \mathcal{A}\xi + n\xi, \qquad \forall \xi \in \mathfrak{X}^{\perp}(M).$$

In Examples 2.46 and 2.50 the index and the nullity of the totally geodesic embedding $\mathbb{S}^n \subset \mathbb{S}^m$ and the Clifford hypersurfaces of $\mathbb{S}^m$ were computed.

We begin this chapter with a classical and important result of J. Simons, in which we obtain a lower bound on the index and nullity of any compact minimal submanifold of the sphere, and prove that the totally geodesic one attains both lower bounds. It is interesting to note that there are no stable ones and, moreover, that $-n$ is always an eigenvalue of the Jacobi operator $\mathcal{L}$.

THEOREM 5.1 ([**Sim68**]). *Let $\Phi : M^n \to \mathbb{S}^m$ be a minimal immersion of a compact manifold M. Then*

(1) *$-n$ is an eigenvalue of the Jacobi operator $\mathcal{L}$ with multiplicity either $m-n$ (and $\Phi(M)$ is the totally geodesic sphere $\mathbb{S}^n$ (Example 2.46)) or $\geq m+1$.*
(2) *$\mathrm{Ind}(M) \geq m-n$, and the equality holds if and only if Φ is an embedding and $\Phi(M)$ is the totally geodesic sphere $\mathbb{S}^n$. Furthermore, if Φ is not totally geodesic, then $\mathrm{Ind}(M) \geq m+1$.*
(3) *$\mathrm{Nul}(M) \geq (m-n)(n+1)$, and the equality holds if and only if Φ is an embedding and $\Phi(M)$ is the totally geodesic sphere $\mathbb{S}^n$.*

PROOF. **Proof of 1**. We consider the embedding $\mathbb{S}^m \subset \mathbb{R}^{m+1}$, whose second fundamental form is $\overline{\sigma}(v,w) = -g(v,w)p$, $\forall v,w \in T_p\mathbb{S}^m$. Now, from the proof of Proposition 2.42 we have that for any $a \in \mathbb{R}^{m+1}$ the Jacobi operator $\mathcal{L}$ of Φ satisfies

$$\mathcal{L}a^{\perp} - na^{\perp} = 0, \quad \text{where } a^{\perp} \text{ is the normal component of a in } T^{\perp}M.$$

We will study the multiplicity of $-n$ as an eigenvalue of $\mathcal{L}$. For this we define the linear map

$$F:\mathbb{R}^{m+1}\to\mathfrak{X}^\perp(M),\qquad F(a)=a^\perp,\ \forall a\in\mathbb{R}^{m+1}.$$

It is clear that the multiplicity ν of $-n$ satisfies $\nu\geq\dim\operatorname{img}(F)=m+1-\dim\ker(F)$. If $\dim\ker(F)\geq 1$, then there exists a nonzero $a\in\mathbb{R}^{m+1}$ with $a^\perp=0$. So $0\neq a=a^\top+\langle\Phi,a\rangle\Phi$, where $a^\top$ is the tangential component of a in TM. If $\langle\Phi,a\rangle=0$, then its derivative is also zero and so, $a^\top=0$, which is impossible because $a\neq 0$. So $\langle\Phi,a\rangle$ is a nonzero function. Moreover, it is easy to check that its Hessian satisfies that $\nabla^2\langle\Phi,a\rangle-\langle\Phi,a\rangle g=0$, and then, from Obata's theorem, [**Oba62**], we conclude that M is isometric to the sphere $\mathbb{S}^n$ with its standard metric of constant curvature 1. Since M is minimal, from the Gauss equation (2.2) we see that Φ is the totally geodesic embedding of $\mathbb{S}^n$ and hence $\dim\ker F=n+1$. Hence $\dim\ker(F)$ is either $n+1$ and Φ is the totally geodesic $\mathbb{S}^n$ or 0 and $\nu\geq m+1$. This proves 1.

Proof of 2. This is a direct consequence of 1.

Proof of 3. First, from Example 2.46, the nullity of the totally geodesic embedding is $(m-n)(n+1)$.

Now, let $\mathcal{K}(\mathbb{S}^m)$ be the space of Killing vector fields on $\mathbb{S}^m$. Its dimension is $\frac{m(m+1)}{2}$ and for each skew symmetric $(m+1)$-matrix B we have that the corresponding Killing vector field X_B is defined by $X_B(p)=Bp$, $\forall p\in\mathbb{S}^m$. Let

$$\mathcal{K}_M(\mathbb{S}^m)=\{(X_B)_{|M}\mid X_B\in\mathcal{K}(\mathbb{S}^m)\}$$

be the space of restrictions on M of the Killing vector fields on $\mathbb{S}^m$. To have some control over its dimension, we define the linear map

$$F:\mathcal{K}(\mathbb{S}^m)\to\mathcal{K}_M(\mathbb{S}^m),\qquad F(X_B)=(X_B)_{|M},\ \forall B\in so(m+1).$$

Then $\dim\mathcal{K}_M(\mathbb{S}^m)=\frac{m(m+1)}{2}-\dim\ker(F)$. If $X_B\in\ker(F)$, then $B\Phi(p)=0$, $\forall p\in M$, and hence $Bv=0$, $\forall v\in TM$. Taking derivatives again, we get that $B\sigma(v,w)=0$, $\forall v,w\in TM$. Thus, for any point $p\in M$, if $\{v_1,\dots,v_n\}$ (respectively $\{\xi_1,\dots,\xi_{m-n}\}$) is an orthonormal basis of T_pM (respectively of $T_p^\perp M$), then, with respect to the orthonormal basis $\{\Phi(p),v_1,\dots,v_n,\xi_1,\dots,\xi_{m-n}\}$ of $\mathbb{R}^{m+1}$, the skew-symmetric matrix B is given by $\left(\begin{smallmatrix}0&0\\0&C\end{smallmatrix}\right)$, where $C\in so(m-n)$. So

$$\dim\ker(F)\leq\dim so(m-n)=\frac{(m-n)(m-n-1)}{2}.$$

If the immersion is not totally geodesic, then there exists a point $p\in M$ such that the first normal space at p

$$N_1(p)=\{\sigma(v,w)\mid v,w\in T_pM\}$$

has positive dimension. Since we know that $B\sigma(v,w)=0$, $\forall v,w\in T_pM$, this means that any matrix B in $\ker(F)$ is given by $\left(\begin{smallmatrix}0&0\\0&D\end{smallmatrix}\right)$, where $D\in so(k)$, with $k<m-n$, which means that $\dim\ker(F)<\frac{(m-n)(m-n-1)}{2}$. As a consequence, we get that $\dim\mathcal{K}_M(\mathbb{S}^m)\geq\frac{(2m-n)(n+1)}{2}$ and, if the equality holds, then Φ is totally geodesic.

Now, we consider the linear map $G:\mathcal{K}_M(\mathbb{S}^m)\to\mathcal{J}(M)$ given by $G((X_B)_{|M})=X_B^\perp$, discussed in Proposition 2.12, where $\mathcal{J}(M)$ is the space of Jacobi fields on M

(Definition 2.11). Then

$$\begin{aligned}\mathrm{Nul}(M) &\geq \dim \mathcal{J}(M) \geq \dim \mathcal{K}_M(\mathbb{S}^m) - \dim \ker(G) \\ &\geq \frac{(2m-n)(n+1)}{2} - \dim \ker(G).\end{aligned}$$

But if $X_B \in \ker(G)$, then the tangential component of X_B in M is also a Killing vector field on M, and so, using Theorem 1.6, $\dim \ker(G) \leq \dim \mathrm{Iso}(M) \leq \frac{n(n+1)}{2}$. Finally we get that $\mathrm{Nul}(M) \geq (m-n)(n+1)$ and if the equality holds, then Φ is the totally geodesic embedding of $\mathbb{S}^n$. This concludes the proof. □

If we consider hypersurfaces of $\mathbb{S}^m$, we can improve the results of the last theorem.

THEOREM 5.2 ([**Sim68, Per01, Per02, Sav10**]). *Let $\Phi : M^n \to \mathbb{S}^{n+1}$ be a non-totally geodesic minimal hypersurface of an n-dimensional compact manifold M. Then:*

(1) *If Φ is two-sided (M is orientable), then* $\mathrm{Ind}(M) \geq n+3$. *Furthermore, if* $\mathrm{Ind}(M) = n+3$, *then the first nonzero eigenvalue of the Laplacian Δ of M is n and its multiplicity is $n+2$.*
(2) *If Φ is two-sided, then the first eigenvalue λ_1 of the Jacobi operator satisfies $\lambda_1 \leq -2n$ and the equality holds if and only if Φ is an embedding and $\Phi(M)$ is a Clifford hypersurface (Example* 2.50*).*
(3) *If Φ is two-sided and invariant under the antipodal map of $\mathbb{S}^{n+1}$, i.e., there exists an isometry τ of M such that $\tau^2 = Id$ and $\Phi \circ \tau = -\Phi$, then* $\mathrm{Ind}(M) = n+3$ *if and only if Φ is an embedding and $\Phi(M)$ is a Clifford hypersurface.*
(4) *If $n \geq 3$, then* $\mathrm{Ind}(M) \geq \frac{2}{(n+2)(n+1)}\beta_1(M) + n + 2$, *where $\beta_1(M)$ is the first Betti number of M.*

REMARK 5.3. Two-sided compact minimal hypersurfaces of the unit sphere $\mathbb{S}^{n+1}$ whose first nonzero eigenvalue of the Laplacian is n and whose multiplicity is $n+2$ are called standard minimal hypersurfaces. S.-T. Yau, [**Yau82**], conjectured in 1982 that the first nonzero eigenvalue of any compact embedded minimal hypersurface of the sphere S^{n+1} is $\lambda_1 = n$. Several results have been proved concerning this conjecture, perhaps the most interesting being that of H.I. Choi and A.N. Wang, [**CW83**], in which they prove that $\lambda_1 \geq n/2$.

REMARK 5.4. Following the proof of part 4 and taking m as the number of negative eigenvalues of the Jacobi operator, we would conclude that $\mathrm{Ind}(M) \geq \frac{2}{(n+2)(n+1)}\beta_1(M)$, even for $n = 2$.

PROOF. **Proof of 1.** Since the immersion Φ is two-sided, the Jacobi operator transforms into a Schrödinger operator $\mathcal{L} : C^\infty(M) \to C^\infty(M)$ given by $\mathcal{L} = \Delta + \|A\|^2 + n$ (see Section 2.6). From Theorem 1.21 it follows that the first eigenvalue of $\mathcal{L}$ has multiplicity 1 and the corresponding eigenfunctions have constant sign. Since M is not totally geodesic, from the last theorem, the first eigenvalue is not $-n$. So the first eigenvalue is less than $-n$ and thus, $\mathrm{Ind}(M) \geq n+3$.

To prove the last statement of 1, we consider the spectrum $\{0 = \lambda_0 < \lambda_1 < \cdots < \lambda_{k-1} < \lambda_k = n < \dots\}$ of the Laplacian Δ of M. We recall that n is an eigenvalue of Δ and its eigenspace $V_{\lambda_k} \supset W = \{\langle \Phi, a\rangle \mid a \in \mathbb{R}^{n+2}\}$. Since Φ is not totally geodesic, dim $V_{\lambda_k} \geq \dim(W) = n+2$ (see proof of 1 in Theorem 5.1). Now we consider the space

$$V = V_{\lambda_0} \oplus V_{\lambda_1} \oplus \cdots \oplus V_{\lambda_k} \subset C^\infty(M),$$

where V_{λ_i} is the eigenspace corresponding to the eigenvalue λ_i. If $f \in V$ is a nonzero function, the f is written as $f = \sum_{i=0}^k f_i$, with $f_i \in V_{\lambda_i}$. Then

$$\mathcal{L}f = \sum_{i=0}^{k-1}(\|A\|^2 + n - \lambda_i)f_i + \|A\|^2 f_k = \sum_{i=0}^{k-1}(n-\lambda_i)f_i + \|A\|^2 f,$$

and so, since $\int_M f_i f_j \, dv = 0$, $i \neq j$, we get

$$\mathcal{Q}(f) = -\sum_{i=0}^{k-1}\int_M (n-\lambda_i)f_i^2 \, dv - \int_M \|A\|^2 f^2 \, dv \leq 0,$$

and the equality holds if and only if $f = f_k$ and $\|A\|^2 f^2 = 0$.

Now let us suppose that $\mathrm{Ind}(M) = n+3$. Then M is not totally geodesic and so $O = \{p \in M \mid A(p) \neq 0\} \neq \emptyset$. If the equality occurs in the last inequality, the function $f = f_k$ vanishes at O. But $\Delta f + nf = 0$, and so, by the unique continuation property, $f = 0$ on M, which is a contradiction. Hence $Q(f) < 0$ for every nonzero $f \in V$. In particular, $n+3 = \mathrm{Ind}(M) \geq 1 + \sum_{i=1}^{k-1} \dim V_i + n + 2 \geq n+3$ and so, $n = \lambda_1$ and its multiplicity is $n+2$. This proves 1.

Proof of 2. To prove 2, we will use as test functions the family of functions $\{a_\delta \mid \delta > 0\}$ where $a_\delta = (\|A\|^2 + \delta)^{\frac{1}{2}}$.

First, it is clear that

$$\Delta a_\delta = \frac{\Delta\|A\|^2}{2a_\delta} - \frac{\|\nabla\|A\|^2\|^2}{4a_\delta^3}.$$

But Simons' formula ([**Sim68**]) says that

$$\frac{1}{2}\Delta\|A\|^2 = \|\nabla A\|^2 + \|A\|^2(n - \|A\|^2),$$

and so,

$$\mathcal{L}a_\delta = \frac{\|\nabla A\|^2}{a_\delta} + \frac{\|A\|^2(n-\|A\|^2)}{a_\delta} - \frac{\|\nabla\|A\|^2\|^2}{4a_\delta^3} + (\|A\|^2 + n)a_\delta.$$

Thus the first eigenvalue λ_1 satisfies

(5.1)
$$\lambda_1 \int_M a_\delta^2 \, dv \leq \mathcal{Q}(a_\delta) = \int_M \left\{ -\|\nabla A\|^2 - 2n\|A\|^2 + \frac{\|\nabla\|A\|^2\|^2}{4a_\delta^2} - \delta(\|A\|^2 + n) \right\} dv.$$

Now, if $\{e_1, \ldots, e_n\}$ is a local orthonormal reference on M and N a unit normal vector field to Φ, we have

(5.2)
$$\|\nabla\|A\|^2\|^2 = \sum_{k=1}^n \left(e_k(\|A\|^2)\right)^2 = 4\sum_{k=1}^n \left(\sum_{i,j=1}^n g(Ae_i, e_j) g((\nabla A)(e_k, e_i), e_j) \right)^2$$

$$= 4\sum_{k=1}^n \left(\operatorname{trace} AB^k\right)^2 \leq 4\sum_{k=1}^n \operatorname{trace} A^2 \operatorname{trace}(B^k)^2 = 4\|A\|^2\|\nabla A\|^2 \leq 4a_\delta^2\|\nabla A\|^2,$$

where A and B^k are the symmetric matrices

$$A_{ij} = g(Ae_i, e_j), \quad B^k_{ij} = g((\nabla A)(e_k, e_i), e_j).$$

Therefore, the inequality (5.1) becomes

$$\lambda_1 \int_M a_\delta^2 \, dv \le -(2n+\delta) \int_M \|A\|^2 \, dv - \delta n \operatorname{vol}(M).$$

If we let $\delta \mapsto 0$, we get that $(\lambda_1 + 2n) \int_M \|A\|^2 \, dv \le 0$. Since Φ is not totally geodesic, then $\int_M \|A\|^2 \, dv > 0$, and thus $\lambda_1 \le -2n$.

From Example 2.50 it follows that $\lambda_1 = -2n$ for the Clifford hypersurfaces.

Suppose now that $\lambda_1 = -2n$. Let $\rho > 0$ be an eigenfunction of $\mathcal{L}$ corresponding to $\lambda_1 = -2n$. Then

$$\Delta\rho + \|A\|^2\rho - n\rho = 0. \tag{5.3}$$

We define $f = \rho^{-1}\|A\|$. Let $p \in M$ be a maximum of f and, since M is not totally geodesic, let Ω be an open subset of M containing p where f is greater than a positive constant.

CLAIM. On Ω we have that $\Delta\|A\| \ge n\|A\| - \|A\|^3$ and the equality holds if and only if $\|A\|$ is constant on Ω.

Using the claim and (5.3) we get on Ω that

$$\begin{aligned}\Delta f &= \rho^{-1}\Delta\|A\| + \|A\|\big(-\rho^{-2}\Delta\rho + 2\rho^{-3}\|\nabla\rho\|^2\big) - 2\rho^{-2}g(\nabla\rho, \nabla\|A\|)\\ &\ge 2\rho^{-3}\|A\|\|\nabla\rho\|^2 - 2\rho^{-2}g(\nabla\rho, \nabla\|A\|) = -2\rho^{-1}g(\nabla\rho, \nabla f).\end{aligned}$$

Since f attains its maximum at the interior of Ω and $\Delta f + 2\rho^{-1}g(\nabla\rho, \nabla f) \ge 0$ on Ω, the maximum principle says that f is constant on Ω and thus $\|A\| = c\rho$ on Ω for some positive constant c. From (5.3) $\Delta\|A\| = n\|A\| - \|A\|^3$ on Ω and so from the claim, $\|A\|^2 = n$ on Ω. Hence ρ is also constant in Ω and by the unique continuation principle, ρ is constant on M. Now (5.3) says that $\|A\|^2 = n$ on M and from [**CdCK70**] the result follows.

PROOF OF THE CLAIM. Using the Simons formula and the inequality $\|\nabla\|A\|^2\|^2 \ge 4\|A\|^2\|\nabla A\|^2$ given in (5.2), we get on Ω

$$\begin{aligned}\|A\|\Delta\|A\| + \|\nabla\|A\|\|^2 = \frac{1}{2}\Delta\|A\|^2 &\ge \frac{\|\nabla\|A\|^2\|^2}{4\|A\|^2} + n\|A\|^2 - \|A\|^4\\ &= \|\nabla\|A\|\|^2 + n\|A\|^2 - \|A\|^4,\end{aligned}$$

hich implies

$$\Delta\|A\| \ge n\|A\| - \|A\|^3 \quad \text{on } \Omega. \tag{5.4}$$

Also, the equality in (5.4) holds if and only if $\|\nabla\|A\|^2\|^2 = 4\|A\|^2\|\nabla A\|^2$. From (5.2), the equality holds if and only if $B^k = \alpha(e_k)A$, for any $k \in \{1, \dots, n\}$, where α is a 1-form on Ω. Thus, the equality in (5.4) happens if and only if

$$(\nabla A)(v, w) = \alpha(v)\, Aw, \quad \forall v, w \in T\Omega.$$

Since (∇A) is symmetric and A is nonzero on Ω, we deduce that $\nabla A = 0$ and hence, $\|A\|^2$ is constant on Ω.

Proof of 3. From Example 2.50 we see that the index of any Clifford hypersurface is $n+3$.

Conversely, suppose now that there exists an isometry τ of M such that $\tau^2 = Id$, $\Phi\circ\tau = -\Phi$ and that the index of M is $n+3$. Then, the space $C^\infty(M)$ decomposes as $C^\infty(M) = C^\infty_+(M)\oplus C^\infty_-(M)$ with $C^\infty_\pm(M) = \{f\in C^\infty(M)\mid f\circ\tau = \pm f\}$. Moreover, since $\|A\|^2\circ\tau = \|A\|^2$, the Jacobi operator $\mathcal{L} = \Delta + \|A\|^2 + n$ also decomposes into two operators $\mathcal{L}^\pm = \mathcal{L}_{|C^\infty_\pm} : C^\infty_\pm \to C^\infty_\pm$, where $\mathcal{L}u = \mathcal{L}^+u^+ + \mathcal{L}^-u^-$ and $u = u^+ + u^-$. Let f be an eigenfunction corresponding to the first eigenvalue of $\mathcal{L}$, whose multiplicity is 1. Then f has sign and so, $f\in C^\infty_+(M)$. Hence, $f^2\in C^\infty_+(M)$. If we look at the vector valued function $\Phi : M\to\mathbb{R}^{n+2}$, then $\Phi\in C^\infty_-(M,\mathbb{R}^{n+2})$ and so, as τ is an isometry

$$\int_M f^2\Phi\,dv = \int_M (f^2\Phi\circ\tau)\,dv = -\int_M f^2\Phi\,dv,$$

which means $\int_M f^2\Phi\,dv = 0$. Thus, the vector valued function $f\Phi$ is L^2-orthogonal to the first eigenspace of $\mathcal{L}$ and, since $\operatorname{Ind}(M) = n+3$, $\mathcal{L}$ has only two negative eigenvalues, namely λ_1 of multiplicity 1 and $-n$ of multiplicity $n+2$. So, we get

$$\mathcal{Q}(f\Phi) \geq -n\int_M \langle f\Phi, f\Phi\rangle\,dv = -n\int_M f^2\,dv. \tag{5.5}$$

We will compute $\mathcal{Q}(f\Phi)$. In fact, it is easy to see that $\Delta(f\Phi) = (\Delta f)\Phi - nf\Phi + 2\nabla f$. Then, $\mathcal{L}(f\Phi) = (\mathcal{L}f)\Phi - nf\Phi + 2\nabla f = -(\lambda_1+n)f\Phi + 2\nabla f$. Since $\langle\nabla f,\Phi\rangle = 0$, we get that $\mathcal{Q}(f\Phi) = (\lambda_1+n)\int_M f^2\,dv$. Thus (5.5) together with $\int_M f^2\,dv > 0$ implies that $\lambda_1\geq -2n$. Now the second part of the theorem says that $\lambda_1\leq -2n$ and so we get that $\lambda_1 = -2n$. Therefore the proof of 3 follows from part 2.

Proof of 4. We use Proposition 2.43 when $\overline{M}^{n+1} = \mathbb{S}^{n+1}$. As in this case $\mathbb{R}^p = \mathbb{R}^{n+2}$ and $\overline{\sigma} = -g\Phi$, if X is a harmonic vector field on M and $\{a_1,\dots,a_{n+2}\}$ is an orthonormal basis of $\mathbb{R}^{n+2}$, then

$$\sum_{1\leq\lambda<\mu\leq n+2}\mathcal{Q}(\xi^X_{\lambda\mu}) = -2(n-1)\int_M \|X\|^2\,dv, \tag{5.6}$$

where $\xi^X_{\lambda\mu} : M\to\mathbb{R}$ are the normal sections to Φ given by $\xi^X_{\lambda\mu} = \langle X, a_\mu\rangle a_\lambda^\perp - \langle X, a_\lambda\rangle a_\mu^\perp$.

We know that $-n$ is an eigenvalue of the Jacobi operator $\mathcal{L}$ (Theorem 5.1). Suppose that m is the number of eigenvalues of the Jacobi operator less than $-n$. If $\lambda_1\leq\lambda_2\leq\lambda_3\leq\cdots\leq\lambda_m$ are these eigenvalues and $\{\eta_1,\dots,\eta_m\}$ the corresponding eigensections, we define a map $F:\mathcal{H}(M)\to\mathbb{R}^{\frac{(n+2)(n+1)m}{2}}$ by

$$F(X) = \left(\int_M \langle\xi^X_{\lambda\mu},\eta_i\rangle\,dv\right)_{1\leq\lambda<\mu\leq n+2,\,1\leq i\leq m},$$

where $\mathcal{H}(M)$ is the space of harmonic vector fields on M. Obviously, F is a linear map. Moreover, if $X\in\ker F$, then every $\xi^X_{\lambda\mu}$, $\lambda<\mu$, is L^2-orthogonal to η_i, $1\leq i\leq m$, and so, $\mathcal{Q}(\xi^X_{\lambda\mu})\geq -n\int_M\|\xi^X_{\lambda\mu}\|^2\,dv$. From (5.6) it follows that

$$-2(n-1)\int_M\|X\|^2\,dv \geq -n\sum_{1\leq\lambda<\mu\leq n+2}\int_M\|\xi^X_{\lambda\mu}\|^2\,dv = -n\int_M\|X\|^2\,dv.$$

This implies that $(n-2)\int_M\|X\|^2\,dv\leq 0$ and as $n\geq 3$, we get that $X=0$. Thus, $\ker F = 0$ and $\beta_1(M) = \dim\mathcal{H}\leq\frac{(n+2)(n+1)m}{2}$. Now, since M is not totally geodesic, from Theorem 5.1, the multiplicity of $-n$ is at least $n+2$, and it follows that $\mathrm{Ind}(M) \geq \frac{2\beta_1(M)}{(n+2)(n+1)} + n + 2$, which proves 4. □

THEOREM 5.5 ([**Sav10, Urb90, Urb13**]). *Let $\Phi : M \to \mathbb{S}^3$ be a minimal immersion of a complete surface M.*

(1) *If M is not compact, then* $\mathrm{Ind}(M) = \infty$.
(2) *If M is compact and Φ is two-sided, then* $\mathrm{Ind}(M) \leq 5$ *if and only if either* $\mathrm{Ind}(M) = 1$, *Φ is an embedding and $\Phi(M)$ is the totally geodesic sphere $\mathbb{S}^2$ (Example* 2.46*) or* $\mathrm{Ind}(M) = 5$, *Φ is an embedding and $\Phi(M)$ is the Clifford torus (Example* 2.50*).*
(3) *If M is compact and Φ is two-sided and not totally geodesic, then*

$$\mathrm{Ind}(M) \geq \frac{h}{2} + 4,$$

where h is the genus of M.
(4) *If M is compact and Φ is one-sided, then* $\mathrm{Ind}(M) \geq 5$.

Considering Theorems 5.2 and 5.5, an interesting open problem for two-sided compact minimal hypersurfaces of $\mathbb{S}^{n+1}$, $n \geq 3$, is:

Classify the two-sided compact minimal hypersurfaces of $\mathbb{S}^{n+1}$ ($n \geq 3$), with index $n+3$. Are they the Clifford hypersurfaces?

REMARK 5.6. 1 was proved in [**Urb13**], 2 in [**Urb90**] and 3 in [**Sav10**]. 4 is an unpublished result of F. Urbano.

REMARK 5.7. In [**KW20**], N. Kapouleas and D. Wiygul have computed the index of the Lawson's minimal surfaces ξ_{1h} of $\mathbb{S}^3$ (see [**Law70a**]), where $h \geq 2$ is the genus of the orientable compact surface ξ_{1h}. They prove that $\mathrm{Ind}(\xi_{1h}) = 2h+3$, and thus $\mathrm{Ind}(\xi_{12}) = 7$. It is reasonable to conjecture that:

ξ_{12} is the only two-sided compact minimal surface of $\mathbb{S}^3$ with index 7, and that there are none of index 6.

PROOF. We begin the proof by obtaining the value of the Jacobi operator when acting on a harmonic vector field of the surface, if Φ is two-sided, i.e., M is orientable. This calculation is similar to the one done in Lemma 4.15 for surfaces of Euclidean space of dimension three. It will be used twice in the proof.

Given a harmonic vector field $X \in \mathcal{H}(M)$, then it is not difficult to check that

$$\begin{aligned}\Delta\langle X, a\rangle &= \langle \Delta X, a\rangle - \langle X, a\rangle - \langle A^2 X, a\rangle + 2\sum_{i=1}^{n} g(\nabla_{e_i} X, Ae_i)\langle N, a\rangle \\ &= \langle \Delta X, a\rangle - \langle X, a\rangle - \frac{\|A\|^2}{2}\langle X, a\rangle + 2g(\nabla X, A)\langle N, a\rangle,\end{aligned}$$

where $\{e_1, e_2\}$ is an orthonormal reference on M, N is a unit normal vector field on M and $a \in \mathbb{R}^4$. But since X is a harmonic vector field on M, from (1.6) $\Delta X = KX$, and so the above formula, together with the Gauss one, give $\Delta\langle X, a\rangle = -\|A\|^2\langle X, a\rangle + 2g(\nabla X, A)\langle N, a\rangle$. From here, it follows that

$$\mathcal{L}\langle X, a\rangle = 2\langle X, a\rangle + 2g(\nabla X, A)\langle N, a\rangle. \tag{5.7}$$

Proof of 1. Suppose that $\mathrm{Ind}(M)$ is finite. If Φ is two-sided (M is orientable), from Corollary 3.15,6, M is compact and 1 follows. If Φ is one-sided (M is not orientable) and the orientable double covering $\widetilde{M}$ of M is of finite genus, from

Corollary 3.16, M is compact and 1 follows. So, we only need to analyze the case where Φ is one-sided and the orientable double covering $\widetilde{M}$ of M has infinite genus.

Let τ be the change of sheet in $\widetilde{M}$ and let $\Pi:\widetilde{M}\to M$ be the projection. Then, $\Psi=\Phi\circ\Pi$ is also a minimal immersion which is two-sided. If $\mathcal{L}:C^\infty(\widetilde{M})\to C^\infty(\widetilde{M})$ is the Jacobi operator of Ψ, from Definition 2.31 and Remark 2.32, we have that $\operatorname{Ind}(M)=\operatorname{Ind}(\mathcal{L}^-)$, which is the restriction of $\mathcal{L}$ to the space $C^\infty_-(\widetilde{M})=\{f\in C^\infty(\widetilde{M})\mid f\circ\tau=-f\}$.

If $\operatorname{Ind}(M)=\operatorname{Ind}(\mathcal{L}^-)=k$, from Theorem 2.30,2, there exist $L^2(\widetilde{M})$-functions $\{v_1,\dots,v_k\}$ with $v_i\circ\tau=-v_i$ and $\mathcal{L}v_i+\lambda_iv_i=0$, $\lambda_i<0$, and such that, if $f\in C^\infty_0(\widetilde{M})$ satisfies $f\circ\tau=-f$ and is L^2-orthogonal to v_i, $1\le i\le k$, then $\mathcal{Q}(f)\ge 0$.

Since the genus of $\widetilde{M}$ is infinite, the space $\mathcal{H}(\widetilde{M})\cap L^2(\widetilde{M},T\widetilde{M})$ of L^2-harmonic vector fields on $\widetilde{M}$ has also infinite dimension. So let V be an l-dimensional ($l>4k$) subspace of $\mathcal{H}^-(\widetilde{M})\cap L^2(\widetilde{M},T\widetilde{M})$ where $\mathcal{H}^-(\widetilde{M})=\{X\in\mathcal{H}(\widetilde{M})\mid X\circ\tau=-X\}$.

Let $\{B_p(n)\mid n\in\mathbb{N}\}$ be an exhaustion of the complete surface $\widetilde{M}$ with p a point of $\widetilde{M}$ and $\{\varphi_n\mid n\in\mathbb{N}\}$ cut-off functions on $\widetilde{M}$ with $\varphi_n=1$ on $B_p(n)$, $\operatorname{supp}(\varphi_n)\subset B_p(2n)$, $\|\widetilde{\nabla}\varphi_n\|^2\le 1$ and $\varphi_n\circ\tau=\varphi_n$.

For each $n\in\mathbb{N}$, let $F_n:V\to\mathbb{R}^{4k}$ be the linear map given by

$$F_n(X)=\left(\int_{\widetilde{M}}\varphi_nv_1X\,\mathrm{d}\tilde{v},\dots,\int_{\widetilde{M}}\varphi_nv_kX\,\mathrm{d}\tilde{v}\right).$$

Since $l>4k$, let $X_n\in\ker F_n$ with $\int_{\widetilde{M}}\|X_n\|^2\,\mathrm{d}\tilde{v}=1$. Thus, for any n, the function φ_nX_n has compact support, is L^2-orthogonal to v_i, $1\le i\le k$ and, since $X_n\in\mathcal{H}^-(\widetilde{M})\cap L^2(\widetilde{M},T\widetilde{M})$, we have that $\varphi_nX_n\circ\tau=-\varphi_nX_n$. So $\mathcal{Q}(\varphi_nX_n)\ge 0$, for all $n\in\mathbb{N}$.

Now,

$$\begin{aligned}\mathcal{Q}(\varphi_nX_n)=&-\int_{\widetilde{M}}\varphi_n(\widetilde{\Delta}\varphi_n)\|X_n\|^2\,\mathrm{d}\tilde{v}-\frac{1}{2}\int_{\widetilde{M}}g(\widetilde{\nabla}\varphi_n^2,\widetilde{\nabla}\|X_n\|^2)\,\mathrm{d}\tilde{v}\\&-\int_{\widetilde{M}}\varphi_n^2\langle\mathcal{L}X_n,X_n\rangle\,\mathrm{d}\tilde{v}.\end{aligned}$$

As $\operatorname{div}(\|X_n\|^2\widetilde{\nabla}\varphi_n^2)=g(\widetilde{\nabla}\varphi_n^2,\widetilde{\nabla}\|X_n\|^2)+2(\varphi_n\widetilde{\Delta}\varphi_n+\|\widetilde{\nabla}\varphi_n\|^2)\|X_n\|^2$, we finally get that

$$\int_{\widetilde{M}}\varphi_n^2\langle\mathcal{L}X_n,X_n\rangle\,\mathrm{d}\tilde{v}\le\int_{\widetilde{M}}\|\widetilde{\nabla}\varphi_n\|^2\|X_n\|^2\,\mathrm{d}\tilde{v},\quad\forall n\in\mathbb{N}.$$

Now, from (5.7), we get that

$$\mathcal{L}X_n=2X_n+2g(\widetilde{\nabla}X_n,\widetilde{A})N,$$

where N is a unit normal vector field to Ψ. So we get that

$$\int_{\widetilde{M}}\|\widetilde{\nabla}\varphi_n\|^2\|X_n\|^2\,\mathrm{d}\tilde{v}\ge 2\int_{\widetilde{M}}\varphi_n^2\|X_n\|^2\,\mathrm{d}\tilde{v}.\tag{5.8}$$

It is clear that the sequence of harmonic vector fields X_n, up to the extraction of a subsequence, converges in V to a harmonic vector field X with $\int_{\widetilde{M}}\|X\|^2\,\mathrm{d}\tilde{v}=1$. Furthermore, if $\{V^1,\dots,V^l\}$ is an L^2-orthonormal basis of V, then the function h

on $\widetilde{M}$ defined by $h = k^2 \max\{\tilde{g}(V_i, V_j)\,,\, 1 \leq i, j \leq l\}$ is integrable and $\|X_n\|^2 \leq h$ for any $n \in \mathbb{N}$. Hence from Theorem 1.7, we have that

$$\lim_{n\to\infty} \int_{\widetilde{M}} \|\widetilde{\nabla}\varphi_n\|^2 \|X_n\|^2 \, d\tilde{v} = \int_{\widetilde{M}} \lim_{n\to\infty} \|\widetilde{\nabla}\varphi_n\|^2 \|X_n\|^2 \, d\tilde{v} = 0,$$

because $\|\widetilde{\nabla}\varphi_n\|^2 \to 0$.

So, from (5.8) we get that

$$0 = \lim_{n\to\infty} \int_{\widetilde{M}} \varphi_n^2 \|X_n\|^2 \, d\tilde{v} = \int_{\widetilde{M}} \lim_{n\to\infty} \varphi_n^2 \|X_n\|^2 \, d\tilde{v} = \int_{\widetilde{M}} \|X\|^2 \, d\tilde{v} = 1,$$

which is a contradiction.

Proof of 2. We know that the Clifford torus in $\mathbb{S}^3$ has index 5 (Example 2.50). We also know from Theorem 5.2 that if Φ is not totally geodesic, then $\operatorname{Ind}(M) \geq 5$ and the first eigenvalue of the Jacobi operator of M satisfies $\lambda_1 \leq -4$.

CLAIM. If $\lambda_2 = -2$, then M is the Clifford torus.

In fact, from Theorem 1.21, the eigenfunctions of λ_1 have constant sign, and so we can take f with $\mathcal{L}f + \lambda_1 f = 0$ and $f > 0$. By Proposition 1.8, there exists a vector $a \in \mathbb{R}^4$ with $\|a\| < 1$ such that

$$\int_M f(\psi_a \circ \Phi) \, dv = 0,$$

where ψ_a is the conformal transformation of $\mathbb{S}^3$ given by

$$\psi_a(p) = a + \frac{1 - \|a\|^2}{\|p + a\|^2}(p + a), \quad p \in \mathbb{S}^3.$$

So, the vector valued function $\psi_a \circ \Phi$ is L^2-orthogonal to f and then,

$$\begin{aligned} \mathcal{Q}(\psi_a \circ \Phi) &= \int_M \|\nabla(\psi_a \circ \Phi)\|^2 \, dv - \int_M (\|A\|^2 + 2)\|\psi_a \circ \Phi\|^2 \, dv \\ &\geq -2 \int_M \|\psi_a \circ \Phi\|^2 \, dv = -2\text{Area}\,(M). \end{aligned}$$

This implies, using the Gauss formula $K = 1 - \frac{\|A\|^2}{2}$ and the Gauss-Bonnet theorem, that

$$\int_M \|\nabla(\psi_a \circ \Phi)\|^2 \, dv \geq \int_M \|A\|^2 \, dv = 2 \int_M (1 - K) \, dv = 2\text{Area}\,(M) - 4\pi(1 - h),$$

where h is the genus of M.

From the definition of ψ_a it follows that

$$\|\nabla(\psi_a \circ \Phi)\|^2 = \frac{2(1 - \|a\|^2)^2}{\|\Phi + a\|^4},$$

and thus the last inequality becomes

$$\int_M \frac{(1 - \|a\|^2)^2}{\|\Phi + a\|^4} \, dv \geq \text{Area}\,(M) - 2\pi(1 - h). \tag{5.9}$$

Now, we compute the Laplacian of the function $\log\|\Phi+a\|^2$ and we get

$$\begin{aligned}\Delta\log\|\Phi+a\|^2 &= -\frac{4}{\|\Phi+a\|^4}\left(\|\Phi+a\|^2\langle\Phi,a\rangle+\|a^T\|^2\right)\\ &= -\frac{4}{\|\Phi+a\|^4}\left(\langle\Phi,a\rangle+\|a\|^2\langle\Phi,a\rangle+\langle\Phi,a\rangle^2+\|a\|^2-\langle N,a\rangle^2\right),\end{aligned}$$

where we have used that $\|a\|^2=\|a^T\|^2+\langle N,a\rangle^2+\langle\Phi,a\rangle^2$. From this it follows that

$$\Delta\log\|\Phi+a\|^2=\left(\frac{(1-\|a\|^2)^2}{\|\Phi+a\|^4}-1\right)+\frac{4\langle N,a\rangle^2}{\|\Phi+a\|^4}\geq\left(\frac{(1-\|a\|^2)^2}{\|\Phi+a\|^4}-1\right),$$

and the equality holds if and only if $\langle N,a\rangle=0$. Now integrating the above inequality, (5.9) becomes in $1-h\geq 0$. If $h=0$, then it follows from [**Law70a**] that M is totally geodesic, which is impossible by hypothesis. The other possibility is that $h=1$. In this case we have equality and so, $\langle N,a\rangle=0$. This implies that the function $\langle\Phi,a\rangle$ satisfies the equation $\nabla^2\langle\Phi,a\rangle+\langle\Phi,a\rangle\,g=0$. If $\langle\Phi,a\rangle$ is not trivial, then Obata's theorem, [**Oba62**], says that M is a sphere of constant curvature one. This means that Φ is totally geodesic which is impossible. Hence $\langle\Phi,a\rangle=0$. Now, if $a\neq 0$, then $\Phi(M)$ lies on the equator of $\mathbb{S}^3$ orthogonal to a, and again Φ is totally geodesic, which is impossible. Finally we get that $a=0$ and so, the test function $\psi_a\circ\Phi=\Phi$. Thus, $\mathcal{L}\Phi-2\Phi=0$. But it is easy to check that $\mathcal{L}\Phi=\|A\|^2\Phi$, and hence $\|A\|^2=2$. The Gauss equation says that M is flat, and again using [**Law70a**], M is a finite covering of the Clifford torus. Since $\mathrm{Ind}(M)=5$, we get that M is the Clifford torus. This proves the claim.

Now, if $\mathrm{Ind}(M)=5$, then $\lambda_2=-2$ and so, M is the Clifford torus.

Proof of 3. From part 2 of this theorem, we can assume that $h\geq 3$. If $X\in\mathcal{H}(M)$ is a harmonic vector field on M, we get from (5.7) that

$$\mathcal{Q}(X)=-\int_M\langle\mathcal{L}X,X\rangle\,\mathrm{d}v=-2\int_M\|X\|^2\,\mathrm{d}v. \tag{5.10}$$

Now let $\alpha=\lfloor\frac{h+1}{2}\rfloor$ and $\{f_1,\ldots,f_{\alpha-1}\}$ be the first $\alpha-1$ eigenfunctions of $\mathcal{L}$. We define a linear map $F:\mathcal{H}(M)\to\mathbb{R}^{4(\alpha-1)}$ by

$$F(X)=\left(\int_M f_1X\,\mathrm{d}v,\ldots,\int_M f_{\alpha-1}X\,\mathrm{d}v\right).$$

Then $\dim\ker F\geq 2h-4(\alpha-1)\geq 2$. Moreover, if $X\in\ker F$, then $\mathcal{Q}(X)\geq\lambda_\alpha\int_M\|X\|^2\,\mathrm{d}v$ and thus (5.10) says that $\lambda_\alpha\leq -2$.

Next we will prove that $\lambda_\alpha<-2$. In fact, if $\lambda_\alpha=-2$, then for any $X\in\ker F$ we have that $\mathcal{L}X-2X=0$ and hence, from (5.7) it follows that $g(\nabla X,A)=0$ for any $X\in\ker F$.

Claim. If X is a harmonic vector field on M such that $g(\nabla X,A)=0$, then $g(\nabla\|A\|^2,AX)=0$.

Since $\dim\ker F\geq 2$, we get that $A\nabla\|A\|^2=0$ and since A has only isolated zeros, $\|A\|^2$ is constant. Since this is impossible because $h\geq 3$, we get that $\lambda_\alpha<-2$ and so, from Theorem 5.1.1, we get

$$\mathrm{Ind}(M)\geq\alpha+4\geq\frac{h}{2}+4,$$

which proves part 3 of the theorem.

PROOF OF THE CLAIM. The zeros of X are isolated, as are the zeros of A. We work outside these zeros, and in this open subset O of M we take an orthonormal reference $\{e_1, e_2\}$ which diagonalizes the second fundamental form. Then, since $g(\nabla X, A) = 0$ and X is harmonic, we have that with respect to the reference $\{e_1, e_2\}$,

$$A = \frac{\|A\|}{\sqrt{2}}\begin{pmatrix}1 & 0\\ 0 & -1\end{pmatrix}, \qquad \nabla X = \begin{pmatrix}0 & f\\ f & 0\end{pmatrix}$$

for certain smooth function f on O. Also, since

$$g(\nabla X, A) = \sum_{i=1}^{2} g(\nabla_{e_i} X, Ae_i) = \sum_{i=1}^{n} e_i(g(X, Ae_i)) = \sum_{i=1}^{n} e_i(g(AX, e_i)) = \operatorname{div}(AX),$$

we have that $\operatorname{div}(AX) = 0$.

On the other hand, the connection ∇ on O is given by

$$\nabla_{e_1} e_1 = xe_2,\ \nabla_{e_1} e_2 = -xe_1,\ \nabla_{e_2} e_1 = -ye_2,\ \nabla_{e_2} e_2 = ye_1,$$

for functions x, y on O. Now, the Codazzi equation $(\nabla A)(e_1, e_2) = (\nabla A)(e_2, e_1)$ implies that

$$\nabla\|A\| = 2\|A\|(y\,e_1 + x\,e_2). \tag{5.11}$$

It is also easy to check that

$$\Delta X = (g(\nabla f, e_2) - 2fx)e_1 + (g(\nabla f, e_1) - 2fy))e_2,$$

and so, as $\Delta X = KX$, we get that

$$\nabla f = \big(2fy + Kg(X, e_2)\big)e_1 + \big(2fx + Kg(X, e_1)\big)e_2. \tag{5.12}$$

If J denotes the complex structure of the Riemann surface M with $Je_1 = e_2$, it follows from (5.11) and (5.12) that

$$\begin{aligned} J\nabla\left(\frac{f}{\|A\|}\right) &= \frac{K}{\|A\|}\big(g(X, e_2)e_2 - g(X, e_1)e_1\big) \\ &= -\frac{\sqrt{2}K}{\|A\|^2}\big(g(X, e_2)Ae_2 + g(X, e_1)Ae_1\big) = -\frac{\sqrt{2}K}{\|A\|^2}AX. \end{aligned}$$

Finally, since $\operatorname{div} J\nabla k = 0$, for any smooth function k, and we know that $\operatorname{div} AX = 0$ and that $2K = 2 - \|A\|^2$, we get from the last formula

$$0 = \operatorname{div}\left(\frac{2 - \|A\|^2}{\|A\|^2}AX\right) = -\frac{2}{\|A\|^4}g(AX, \nabla\|A\|^2),$$

which proves the claim.

Proof of 4. From Theorem 5.1 we know that $\operatorname{Ind}(M) \geq 4$ and that -2 is an eigenvalue of the Jacobi operator of M with multiplicity at least 4. We will prove that the index of M cannot be 4.

Since Φ is one-sided, from Definition 2.31 and Remark 2.32, if $\Pi : \widetilde{M} \to M$ is the orientable double covering of M and $\Psi = \Phi \circ \Pi : \widetilde{M} \to \mathbb{S}^3$ the corresponding two-sided minimal immersion, then $\operatorname{Ind}(M) = \operatorname{Ind}(\mathcal{L}^-)$, where $\mathcal{L}^-$ is the restriction of the Jacobi operator $\mathcal{L}$ of Ψ to the space $C^\infty_-(\widetilde{M}) = \{f \in C^\infty(\widetilde{M}) \mid f \circ \tau = -f\}$, where $\{I, \tau\}$ is the automorphism group of $\widetilde{M}$.

If $\operatorname{Ind}(M) = 4$, then the first eigenvalue of $\mathcal{L}^-$ is $\lambda_1^- = -2$. Thus from Brill-Noether theory, there exists a holomorphic map $F : \widetilde{M} \to \mathbb{S}^2$ such that $F \circ \tau = -F$

and $\deg F \leq h+1$, where h is the genus of $\widetilde{M}$. So, using F as the vector valued test function, we have

$$-\int_{\widetilde{M}}\langle\mathcal{L}^-F,F\rangle\,\mathrm{d}\tilde{v}\geq -2\int_{\widetilde{M}}\|F\|^2\,\mathrm{d}\tilde{v}=-2\mathrm{Area}(\widetilde{M}). \tag{5.13}$$

But

$$\begin{aligned}-\int_{\widetilde{M}}\langle\mathcal{L}^-F,F\rangle\,\mathrm{d}\tilde{v}&=\int_{\widetilde{M}}\{\|\nabla F\|^2-\|A\|^2-2\}\,\mathrm{d}\tilde{v}\\&\leq 8\pi\deg F-2\mathrm{Area}(\widetilde{M})-\int_{\widetilde{M}}\|A\|^2\,\mathrm{d}\tilde{v},\end{aligned}$$

which, together with (5.13), implies that $\int_{\widetilde{M}}\|A\|^2\,\mathrm{d}\tilde{v}\leq 8\pi(1+h)$. From the Gauss equation and the Gauss-Bonnet theorem we get $\int_{\widetilde{M}}\|A\|^2\,\mathrm{d}\tilde{v}=2\mathrm{Area}(\widetilde{M})-8\pi(1-h)$, and so, we get $\mathrm{Area}(\widetilde{M})\leq 8\pi$.

Now we use a result of P. Li and S.T. Yau, [**LY82**], which says that if $\vec{H}$ is the mean curvature vector of $\Psi:\widetilde{M}\to\mathbb{R}^4$ and μ is the maximum multiplicity of Ψ, then $\int_{\widetilde{M}}\|\vec{H}\|^2\,\mathrm{d}\tilde{v}\geq 4\pi\mu$. Since Ψ is minimal in $\mathbb{S}^3$, $\|\vec{H}\|^2=1$. Also, since $\widetilde{M}$ is the orientable double covering of M, $\mu\geq 2$. So $\mathrm{Area}(\widetilde{M})\geq 8\pi$, which together with the above inequality for the area of $\widetilde{M}$ says that $\mathrm{Area}(\widetilde{M})=8\pi$. This means that the equality holds in (5.13). So, $\mathcal{L}^-F-2F=0$ and $\deg F=h+1$. Finally, we have a holomorphic map F such that $\Delta F+\|A\|^2F=0$ with $\deg F=h+1$. But, since F is holomorphic, $\Delta F+\|\nabla F\|^2F=0$, and so, $\|A\|^2=\|\nabla F\|^2$.

From [**Law70a**], the numbers of zeros of A is $4h-4$. Also, the number of zeros of ∇F is the total number of ramification of F, which, using the Riemann-Hurwitz formula (Theorem 1.23), is $b(F)=4h$. This is a contradiction, and so 4 is proved. □

A particularly interesting case is the consideration of compact minimal surfaces of $\mathbb{S}^4$. Considering Section 3.4, since $\mathbb{S}^4$ is a self-dual and anti-self-dual Riemannian manifold, the twistor spaces $(\mathcal{P}^+,\mathcal{J}^+)$ and $(\mathcal{P}^-,\mathcal{J}^-)$ are complex manifolds, and it is known that they are biholomorphic to the complex projective space $\mathbb{CP}^3$ with its standard complex structure. Also the twistor projections $\pi^\pm:\mathbb{CP}^3\to\mathbb{S}^4$ are related by $\pi^-=A\circ\pi^+$, where A is the antipodal map on $\mathbb{S}^4$, and thus the negative spin superminimal surfaces of $\mathbb{S}^4$ are the images by the antipodal map of the positive spin superminimal surfaces. So, in this case we simply speak of superminimal surfaces.

Let $\Phi:M\to\mathbb{S}^4$ be a full superminimal immersion (Definition 3.42) of a compact orientable surface. Then, its twistor lifting $\widetilde{\Phi}$ can be viewed as a holomorphic immersion of M into $\mathbb{CP}^3$, whose image lies on the contact horizontal distribution. Furthermore, as a horizontal complex curve in $\mathbb{CP}^3$, which is not linearly full, must be a line, which is the twistor lifting of the totally geodesic immersion in $\mathbb{S}^4$, we have that $\widetilde{\Phi}$ is linearly full in $\mathbb{CP}^3$. As a consequence, the space of full superminimal immersions in $\mathbb{S}^4$ can be constructed by projecting a certain class of linearly full complex curves of the complex projective space $\mathbb{CP}^3$. The following properties are known:

1) Every minimal immersion of a two-sphere in $\mathbb{S}^4$ is superminimal ([**Bry82**, **Cal67**], see also proof of Corollary 5.9).

2) The area Area(M) of a full superminimal immersion Φ is $4\pi \deg \widetilde{\Phi}$. So Area$(M) \geq 12\pi$, and the equality holds if and only if $\widetilde{\Phi}$ is a horizontal rational normal curve in $\mathbb{CP}^3$ ([**Cal67, Bar75, Gau87**]).

3) The non-compact Lie group $SO(5,\mathbb{C})$ of complex contact transformations of $\mathbb{CP}^3$ acts naturally on the space of full superminimal immersions in $\mathbb{S}^4$, keeping the area fixed. This action is transitive on the subspace of the smallest area Area$(M) = 12\pi$ because any two rational normal curves are always projectively equivalent ([**Bar75, GH78**]).

4) The **Veronese immersion** $\Phi : \mathbb{S}^2(\sqrt{3}) \to \mathbb{S}^4$ given by

$$\Phi(x,y,z) = \frac{1}{\sqrt{3}}\left(xy, xz, yz, \frac{1}{2}(x^2-y^2), \frac{1}{2\sqrt{3}}(x^2+y^2-2z^2)\right)$$

gives us a full superminimal immersion of the two-sphere into $\mathbb{S}^4$, which has parallel second fundamental form and induced metric with constant Gauss curvature $\frac{1}{3}$. So, this Veronese map is a full superminimal immersion with area 12π. Thus we get all superminimal immersions in $\mathbb{S}^4$ with area 12π by deforming it by means of the action of the group $SO(5,\mathbb{C})$. We will refer to this family of superminimal immersions as the **twistor deformations of the Veronese immersion.**

THEOREM 5.8 ([**MU97, KW24**]). *Let $\Phi : M \to \mathbb{S}^4$ be a minimal immersion of a compact orientable surface M.*

(1) *If Φ is superminimal and full, then* Ind$(M) \geq 10$ *and* Nul$(M) \geq 14$. *Furthermore, the equality in some of these inequalities holds if and only if Φ is the Veronese immersion or one of its twistor deformations.*
(2) *If the genus of M is $h = 1$, then* Ind$(M) \geq 6$ *and the equality holds if and only if Φ is congruent to the Clifford torus (Example* 2.50*) in some totally geodesic sphere $\mathbb{S}^3 \subset \mathbb{S}^4$.*

PROOF. **Proof of 1.** From Theorem 3.50 and since the scalar curvature of $\mathbb{S}^4$ is 12, we have that

$$\operatorname{Ind}(M) = \frac{\operatorname{Area}(M)}{\pi} + 2h - 2, \quad \operatorname{Nul}(M) \geq \frac{\operatorname{Area}(M)}{\pi} + 2 - 2h,$$

where h is the genus of M. If $\widetilde{\Phi}$ is the twistor lifting of Φ to $\mathbb{CP}^3$, since $\widetilde{\Phi}$ is linearly full, from the Brill-Noether theorem, ([**GH78**]), it follows that

$$\frac{\operatorname{Area}(M)}{4\pi} = \deg \widetilde{\Phi} \geq \lfloor \tfrac{3+3h}{4} \rfloor + 3.$$

Now, last two formulas imply that Ind$(M) \geq 10$ and Nul$(M) \geq 14$. Moreover, if the equality occurs in some of the above inequalities, then $h = 0$ and Area$(M) = 12\pi$. This implies, considering the above properties, that M is the Veronese immersion or some of its twistor deformations.

Finally, the above properties prove that the index and the nullity of the Veronese immersion are 10 and 14, respectively.

Proof of 2. We will use the complex structure of M and the holomorphic 4-differential on M associated to the minimal immersion Φ. Since M is a torus, let $z = x + iy$ be a global complex coordinate on M and let $\partial_z = \frac{1}{2}(\frac{\partial}{\partial x} - i\frac{\partial}{\partial y})$ and $\partial_{\bar{z}} = \frac{1}{2}(\frac{\partial}{\partial x} + i\frac{\partial}{\partial y})$ be the Cauchy-Riemann operators. If the induced metric is given by $\langle,\rangle = e^{2u}\|dz\|^2$, then we have that

$$\begin{aligned}\Phi_{zz} &= 2u_z\Phi_z + \sigma(\partial_z,\partial_z)\\ \Phi_{z\bar z} &= -\frac{e^{2u}}{2}\Phi,\end{aligned} \tag{5.14}$$

where we have extended to the complexified bundles all the operators associated to the immersion Φ. If we denote by $\Omega = \sigma(\partial_z,\partial_z)$, then we define on M a 4-differential $\Theta(z) = \langle\Omega,\Omega\rangle \otimes (dz)^4$. Using the Codazzi equation and (5.14) we have that

$$\nabla^\perp_{\bar z}\Omega = (\nabla\sigma)(\partial_z,\partial_z,\partial_{\bar z}) = \nabla^\perp_z\sigma(\partial_z,\partial_{\bar z}) - 2u_z\sigma(\partial_z,\partial_{\bar z}) = 0, \tag{5.15}$$

and thus $\partial_{\bar z}\langle\Omega,\Omega\rangle = 0$. This means that Θ is a holomorphic 4-differential on the torus M. We have only two possibilities: either $\Theta\equiv 0$ or Θ has no zeros, and in this case, with a suitable change of the complex parameter, we can assume that $\Theta = (dz)^4$, i.e., $\langle\Omega,\Omega\rangle = 1$.

If $\Omega = \Omega_1 + i\Omega_2$, i.e., $\Omega_1 = \frac{1}{2}\sigma(\partial_x,\partial_x)$ and $\Omega_2 = -\frac{1}{2}\sigma(\partial_x,\partial_y)$, then $\langle\Omega,\Omega\rangle = \langle\Omega_1,\Omega_1\rangle - \langle\Omega_2,\Omega_2\rangle + 2i\langle\Omega_1,\Omega_2\rangle$.

If $\Theta = 0$, then, since M is not totally geodesic, we have that $\|\sigma(\partial_x,\partial_x)\|^2 = \|\sigma(\partial_x,\partial_y)\|^2 > 0$ and $\langle\sigma(\partial_x,\partial_x),\sigma(\partial_x,\partial_y)\rangle = 0$. This means that Φ is a superminimal immersion, and so, from part 1, $\mathrm{Ind}(M)\ge 10 > 6$.

If $\Theta = (dz)^4$, then $\langle\Omega_1,\Omega_1\rangle = 1 + \langle\Omega_2,\Omega_2\rangle$ and $\langle\Omega_1,\Omega_2\rangle = 0$. Using (5.14) and (5.15) we have:

$$\begin{aligned}\Delta^\perp\Omega &= 2e^{-2u}(\nabla^\perp_z\nabla^\perp_{\bar z}\Omega + \nabla^\perp_{\bar z}\nabla^\perp_z\Omega)\\ &= 2e^{-2u}\nabla^\perp_{\bar z}\{(\nabla\sigma)(\partial_z,\partial_z,\partial_z) + 4u_z\Omega\} = 2e^{-2u}\{(\nabla^2\sigma)(\partial_{\bar z},\partial_z,\partial_z,\partial_z) + 4u_{z\bar z}\Omega\}\\ &= 2e^{-2u}\{R^\perp(\partial_{\bar z},\partial_z)\Omega - 2\sigma(R(\partial_{\bar z},\partial_z)\partial_z,\partial_z)\} + 8e^{-2u}u_{z\bar z}\Omega\\ &= 2e^{-2u}\{\sigma(A_\Omega\partial_z,\partial_{\bar z}) - \sigma(A_\Omega\partial_{\bar z},\partial_z) + (4u_{z\bar z} + e^{2u}K)\Omega\} = 4e^{-4u}(\overline{\Omega} - \|\Omega\|^2\Omega),\end{aligned}$$

where we have used the Ricci identity and the fact that $4u_{z\bar z} + e^{2u}K = 0$. From this equation, and taking the real and imaginary parts, it follows that

$$\Delta^\perp\Omega_1 = -8e^{-4u}\|\Omega_2\|^2\Omega_1, \quad \Delta^\perp\Omega_2 = -8e^{-4u}\|\Omega_1\|^2\Omega_2. \tag{5.16}$$

On the other hand, if $\mathcal{A}$ is the operator appearing in the Jacobi operator, we have that

$$\mathcal{A}(\Omega_i) = 4e^{-2u}\mathcal{R}\{\sigma(\partial_z, A_{\Omega_i}\partial_{\bar z})\} = 8e^{-4u}\sum_{j=1}^{2}\langle\Omega_i,\Omega_j\rangle\Omega_j,$$

which implies that

$$\mathcal{A}(\Omega_i) = 8e^{-4u}\|\Omega_i\|^2\Omega_i, \quad i = 1,2. \tag{5.17}$$

Since Ω_1 is a non-null normal section to Φ, from (5.16) and (5.17), we have that

$$\mathcal{Q}(\Omega_1) = -2\int_M (4e^{-4u}+1)\|\Omega_1\|^2\,dv < -2\int_M \|\Omega_1\|^2\,dv,$$

which proves that the first eigenvalue of the Jacobi operator of Φ satisfies $\lambda_1 < -2$.

Also from Theorem 5.1, 2, and since Φ is not totally geodesic, -2 is an eigenvalue of the Jacobi operator with multiplicity at least 5, which proves that the index of M is $\mathrm{Ind}(M)\ge 6$.

From Theorem 1.18, the first nonzero eigenvalue of the Laplacian of the Clifford torus is 2. Thus, Examples 2.46 and 2.50 imply that the index of the Clifford torus in $\mathbb{S}^4$ is 6.

We assume that $\text{Ind}(M) = 6$. Then, the multiplicity of -2 is five.

CLAIM. If the multiplicity $\mu(-2) = 5$, then Φ is not full, i.e., there exists a non-null vector $a \in \mathbb{R}^5$ such that $\langle \Phi, a \rangle = 0$.

In such a case, $\Phi(M)$ lies in a totally geodesic $\mathbb{S}^3$ of $\mathbb{S}^4$. Hence from Example 2.46 and Theorem 5.5 we get that $6 = \text{Ind}(M) = \text{Ind}(\Phi') + 1 \geq 6$, where Φ' is the immersion $\Phi : M \to \mathbb{S}^3$. Thus, from Theorem 5.5 we see that M is the Clifford torus.

PROOF OF THE CLAIM. We will use the normal sections $J\Omega_i$, $i = 1, 2$ as test sections, where J is one of the two complex structures on the normal bundle of Φ (see Section 3.4). If $\mathcal{A}$ is the operator appearing in the Jacobi operator, we have that

$$\mathcal{A}(J\Omega_i) = 4e^{-2u}\mathcal{R}\left\{\sigma(\partial_z, A_{J\Omega_i}\partial_{\bar{z}})\right\} = 8e^{-4u}\sum_{j=1}^{2} g(J\Omega_i, \Omega_j)\Omega_j,$$

which implies that

$$\mathcal{A}(J\Omega_1) = 8e^{-4u}g(J\Omega_1, \Omega_2)\Omega_2, \quad \mathcal{A}(J\Omega_2) = 8e^{-4u}g(J\Omega_2, \Omega_1)\Omega_1. \tag{5.18}$$

Moreover, since $\nabla^\perp J = J\nabla^\perp$, we have that $\Delta^\perp J = J\Delta^\perp$, and so, from (5.16) and (5.17) it follows that

$$\mathcal{L}(J\Omega_i) - 2(J\Omega_i) = 0, \quad i = 1, 2.$$

Since the multiplicity of -2 is 5 and from Theorem 5.1, 2 it follows that $\mathcal{L}a^\perp - 2a^\perp = 0$, $\forall a \in \mathbb{R}^5$ and $\dim\{a^\perp \mid a \in \mathbb{R}^5\} = 5$, we get that

$$J\,\Omega_i = a_i^\perp, \quad i = 1, 2, \tag{5.19}$$

for certain $a_i \in \mathbb{R}^5$.

Considering that $\langle \Omega_1, \Omega_1 \rangle = 1 + \langle \Omega_2, \Omega_2 \rangle$ and $\langle \Omega_1, \Omega_2 \rangle = 0$, we have that there exists a function θ on the torus M and a global unit normal sections $\{e_3, e_4\}$ such that $\Omega_1 = (\cosh\theta)\, e_3$ and $\Omega_2 = (\sinh\theta)\, e_4$. Without loss of generality, the complex structure J on the normal bundle of Φ is given by

$$Je_3 = e_4, \quad Je_4 = -e_3.$$

Now the equations (5.19) become

$$a_1^\perp = (\cosh\theta)e_4, \quad a_2^\perp = -(\sinh\theta)e_3. \tag{5.20}$$

From (5.15), (5.19) and (5.20) it follows that $\nabla_z^\perp e_3 = i\theta_z e_4$ and $\nabla_z^\perp e_4 = -i\theta_z e_3$, and so, taking the derivative in (5.20), we get

$$\begin{aligned}
\theta_z(\sinh\theta\, e_4 - i\cosh\theta\, e_3) &= \nabla_z^\perp a_1^\perp = -\sigma(\partial_z, a_1^T)\\
&= -2e^{-2u}\langle \partial_{\bar{z}}, a_1 \rangle(\cosh\theta\, e_3 + i\sinh\theta\, e_4),\\
-\theta_z(i\sinh\theta\, e_4 + \cosh\theta\, e_3) &= \nabla_z^\perp a_2^\perp = -\sigma(\partial_z, a_2^T)\\
&= -2e^{-2u}\langle \partial_{\bar{z}}, a_2 \rangle(\cosh\theta\, e_3 + i\sinh\theta\, e_4).
\end{aligned}$$

From the above equations we finally get that $2e^{-2u}\langle \partial_{\bar{z}}, a_2 \rangle = \theta_z = -2ie^{-2u}\langle \partial_{\bar{z}}, a_1 \rangle$, and taking derivatives with respect to z and using (5.14) we get that $-\langle \Phi, a_2 \rangle =$

$i\langle\Phi, a_1\rangle$. This means that $\langle\Phi, a_1\rangle = 0$ and, since $a_1 \neq 0$ because $a_1^\perp = \cosh\theta\, e_3 \neq 0$, the immersion is not full. □

COROLLARY 5.9. *Let $\Phi : M \to \mathbb{S}^4$ be a full minimal immersion of a compact orientable surface M of genus zero. Then,* $\mathrm{Ind}(M) \geq 10$ *and* $\mathrm{Nul}(M) \geq 14$. *Furthermore, the equality in some of these inequalities holds if and only if Φ is the Veronese immersion or one of its twistor deformations.*

PROOF. By the proof of Theorem 5.8,2, since M is conformal to a sphere, the holomorphic 4-differential $\Theta \equiv 0$ and hence Φ is a superminimal immersion. □

Now, as in the last corollary, we consider minimal immersions of a compact and orientable surface of genus zero, but in a sphere of arbitrary dimension. The aim is to stimate the index, and for this purpose we recall some classical results on this type of immersions studied by E. Calabi and S.S. Chern, [**Cal67, Che70b**], as well as the examples known as the Boruvka spheres.

For any positive integer n, the set of spherical harmonics of degree n of three variables is a $(2n+1)$-dimensional vector space. Considering the unit sphere $\mathbb{S}^{2n}$ in this vector space, we get an isometric minimal immersion

$$\Phi : \mathbb{S}^2\left(\sqrt{\tfrac{n(n+1)}{2}}\right) \to \mathbb{S}^{2n}$$

called the **Boruvka sphere**.

We observe that $\Phi = Id$ for $n = 1$ and Φ is congruent to the Veronese immersion for $n = 2$.

THEOREM 5.10 ([**Cal67, Che70b**]). *Let $\Phi : M \to \mathbb{S}^m$ be a full minimal immersion of a compact orientable surface M of genus zero. Then:*

(1) *m is even, $m = 2n$.*
(2) *The area of M is an integer multiple of 2π and* $\mathrm{Area}(M) \geq 2\pi n(n+1)$.
(3) *If M has constant curvature c, then $c = \frac{2}{n(n+1)}$ and $\Phi(M)$ is congruent to the Boruvka sphere.*

THEOREM 5.11 ([**Eji83**]). *Let $\Phi : M \to \mathbb{S}^{2n}$ be a full minimal immersion of a compact orientable surface M of genus zero. Then:*

(1) *The first eigenvalue of the Jacobi operator of Φ is $\lambda_1 = -2$ and its multiplicity is* $\frac{1}{\pi}\mathrm{Area}(M) + 2(n-3)$.
(2) $\mathrm{Ind}(M) \geq \frac{1}{\pi}\mathrm{Area}(M) + 2(n-3) \geq 2(n-1)(n+3)$.

REMARK 5.12. When $n = 2$, part 1 of the theorem is a consequence of Theorem 3.48, and part 2 is contained in Corollary 5.9.

PROOF. Following the proof of Theorem 5.8, if we consider a local complex coordinate $z = x + iy$ on M, then the 4-differential $\Theta(z) = \langle\sigma(\partial_z, \partial_z), \sigma(\partial_z, \partial_z)\rangle \otimes (dz)^4$ is holomorphic, and since M is a sphere, then $\Theta \equiv 0$. This means that there exists an orthonormal reference $\{e_1, e_2\}$ on M, compatible with the complex structure J of M, $Je_1 = e_2$, satisfying

$$\|\sigma(e_1, e_1)\| = \|\sigma(e_1, e_2)\| = k_1, \quad \langle\sigma(e_1, e_1), \sigma(e_1, e_2)\rangle = 0.$$

Moreover, [**Che70b**], the set $M_1 = \{p \in M \mid k_1(p) \neq 0\}$ is open and $M - M_1$ is finite. On M_1 we can define an orthonormal reference $\{e_3, e_4\}$ of the so-called first normal space $N_1(M_1)$ by

$$e_3 = \frac{\sigma(e_1, e_1)}{k_1}, \quad e_4 = \frac{\sigma(e_1, e_2)}{k_1}.$$

Now, on M_1, we consider the third fundamental form σ_3 defined by

$$\sigma_3(v, w, x) = \text{the component of } (\nabla\sigma)(v, w, x) \text{ orthogonal to } N_1(M_1).$$

Then σ_3 is a symmetric tensor, and using that the holomorphic 6-differential $\Theta_2(z) = \langle \sigma_3(\partial_z, \partial_z, \partial_z), \sigma_3(\partial_z, \partial_z, \partial_z)\rangle \otimes (\mathrm{d}z)^6$ vanishes on the sphere M, we have that

$$\|\sigma_3(e_1, e_1, e_1)\| = \|\sigma_3(e_1, e_1, e_2)\| = k_2, \quad \langle \sigma_3(e_1, e_1, e_1), \sigma_3(e_1, e_1, e_2)\rangle = 0.$$

Moreover, **[Che70b]**, the set $M_2 = \{p \in M_1 \mid k_2(p) \neq 0\}$ is open and $M - M_2$ is finite. On M_2 we can define an orthonormal reference $\{e_5, e_6\}$ of the second normal space $N_2(M_2)$ by

$$e_5 = \frac{\sigma_3(e_1, e_1, e_1)}{k_2}, \quad e_6 = \frac{\sigma_3(e_1, e_1, e_2)}{k_2}.$$

Repeating the argument, and since the normal bundle has rank $2(n-1)$, we can define the higher fundamental forms σ_s, for $s = 2, 3, \ldots, n$, $(\sigma_2 = \sigma)$, satisfying that they are symmetric tensors and that

$$\|\sigma_s(e_1, \ldots, e_1, e_1)\| = \|\sigma_s(e_1, \ldots, e_1, e_2)\| = k_{s-1},$$
$$\langle \sigma_s(e_1, \ldots, e_1, e_1), \sigma_s(e_1, \ldots, e_1, e_2)\rangle = 0.$$

Moreover, **[Che70b]**, the set $M_{s-1} = \{p \in M_{s-2} \mid k_{s-1}(p) \neq 0\}$ is open and $M - M_{s-1}$ is finite. On M_{s-1} we can define an orthonormal reference $\{e_{2s-1}, e_{2s}\}$ of the normal space $N_{s-1}(M_{s-1})$ by

$$e_{2s-1} = \frac{\sigma_s(e_1, \ldots, e_1, e_1)}{k_{s-1}}, \quad e_{2s} = \frac{\sigma_s(e_1, \ldots, e_1, e_2)}{k_{s-1}}.$$

So we have a Whitney decomposition

$$T^\perp M_{n-1} = N_1(M_{n-1}) \oplus \cdots \oplus N_{n-1}(M_{n-1}).$$

Now we can define an almost complex structure J on the normal bundle $T^\perp M_{n-1}$ by

$$Je_{2s-1} = e_{2s}, \quad Je_{2s} = -e_{2s-1}, \qquad 2 \leq s \leq n.$$

It is easy to check that J is independent of the reference $\{e_1, e_2\}$ which is compatible with the complex structure of M and so J is well defined.

Next we will prove that $\nabla^\perp J = 0$ on M_{n-1}. Since $\{e_3, \ldots, e_{2n}\}$ is an orthonormal reference of $T^\perp M_{n-1}$, then we have that $\nabla^\perp J = 0$ if and only if $\nabla^\perp Je_s = J\nabla^\perp e_s$ for $3 \leq s \leq 2n$. We will check this property for $s = 3$, because for $s > 3$ the proof is similar. Given the definitions of e_3 and $e_4 = Je_3$, the proof that $\nabla^\perp Je_3 = J\nabla^\perp e_3$, is equivalent to seeing that

$$\nabla^\perp_v \sigma(e_1, e_2) = J\nabla^\perp_v \sigma(e_1, e_1),$$

for any tangent vector v to M_{n-1}. The last equation is equivalent to

$$(\nabla\sigma)(v, e_1, e_2) - 2\alpha(v)\sigma(e_1, e_1) = J(\nabla\sigma)(v, e_1, e_1) + 2\alpha(v)J\sigma(e_1, e_2),$$

where α is the 1-form on M given by $\alpha(v) = \langle \nabla_v e_1, e_2\rangle$. From the definition of J, proving the last equality is equivalent to proving that

$$(\nabla\sigma)(v, e_1, e_2) = J(\nabla\sigma)(v, e_1, e_1), \quad \forall v \in TM_{n-1}. \tag{5.21}$$

Since we have from the definition of J that $\sigma_3(v,e_1,e_2)=J\sigma_3(v,e_1,e_1)$, the proof of equality (5.21) is equivalent to prove the following one:

$$\sum_{s=3}^{4}\langle(\nabla\sigma)(v,e_1,e_2),e_s\rangle e_s=\sum_{s=3}^{4}\langle(\nabla\sigma)(v,e_1,e_1),e_s\rangle Je_s. \tag{5.22}$$

To prove (5.22), we use that the 4-differential $\Theta\equiv 0$. Taking derivatives with respect to z we get that

$$\langle(\nabla\sigma)(\partial_z,\partial_z,\partial_z),\sigma(\partial_z,\partial_z)\rangle=0,$$

and so,

$$\langle(\nabla\sigma)(e_1,e_1,e_1),e_3\rangle=\langle(\nabla\sigma)(e_1,e_1,e_2),e_4\rangle,$$
$$\langle(\nabla\sigma)(e_1,e_1,e_1),e_4\rangle+\langle(\nabla\sigma)(e_1,e_1,e_2),e_3\rangle=0.$$

These last equations prove equality (5.22) and thus $\nabla^\perp J=0$ on $T^\perp M_{n-1}$.

Now, we will extend the almost complex structure J to $T^\perp M$. In fact, let $p\in M-M_{n-1}$ and let $\{\xi_3,\dots,\xi_{2n}\}$ be an orthonormal basis of $T_p^\perp M$. If $\gamma:(-\epsilon,\epsilon)\to M$ is a smooth curve with $\alpha(0)=p$ and $\operatorname{img}(\gamma)\cap(M-M_1)=\{p\}$, we consider the orthonormal parallel sections (with respect to $\nabla^\perp$) along γ: $\{\xi_3(t),\dots,\xi_{2n}(t)\}$, satisfying $\xi_s(0)=\xi_s$ for any $3\le s\le 2n$. Then,

$$J\xi_s(t)=\sum_{r=3}^{2n}a_{sr}(t)\gamma_r(t),\quad \forall t\in(-\epsilon,\epsilon)-\{0\}.$$

Since the sections are parallel along γ and $\nabla^\perp J=0$ on $T^\perp M_{n-1}$, the last equations imply that the functions $a_{sr}(t)$ are constant on $(-\epsilon,\epsilon)-\{0\}$. Hence we define $J\xi_s=\sum_{r=3}^{2n}a_{sr}\xi_r$. It is clear that the definition of J at the point p is independent of the orthonormal basis of $T_p^\perp M$ and of the curve γ. Thus, J defines an almost complex structure on $T^\perp M$, which is parallel with respect to the normal connection $\nabla^\perp$. So, we deduce that $T^\perp M$ is a holomorphic bundle over M, and that a section $\xi\in\mathfrak{X}^\perp(M)$ is holomorphic if $D\xi=0$, where

$$D_v\xi=\nabla^\perp_{Jv}\xi-J\nabla^\perp_v\xi,\quad \forall\, v\in TM. \tag{5.23}$$

Now, given any $a\in\mathbb{R}^{2n+1}$, we prove that the normal section $a^\perp$ satisfies that $Da^\perp=0$. In fact, by derivating $a=a^\top+a^\perp+\langle\Phi,a\rangle\Phi$ with respect to any tangent vector v, we have that $\nabla^\perp a^\perp=-\sigma(a^\top,v)$. So,

$$D_va^\perp=-\sigma(Jv,a^\top)+J\sigma(v,a^\top)=\sum_{i,j=1}^{2}\langle a,e_i\rangle g(v,e_j)(-\sigma(Je_j,e_i)+J\sigma(e_j,e_i))=0,$$

considering only the definition of J. In particular, for any $a\in\mathbb{R}^{2n+1}$ the normal section $a^\perp$ is holomorphic. From Theorem 5.1, $La^\perp-2a^\perp=0$ and so, it is natural to ask about the relationship between the holomorphic sections of the normal bundle and the eigensections corresponding to the eigenvalue -2. To do this we proceed in a similar way as in the proof of Proposition 3.44. From (5.23) it follows, for any $\xi\in\mathfrak{X}^\perp M$, that

$$\frac{1}{2}\|D\xi\|^2=\|\nabla^\perp\xi\|^2-\sum_{i=1}^{2}\langle\nabla^\perp_{Je_i}\xi,\nabla^\perp_{e_i}J\xi\rangle.$$

Integrating by parts, we have that

$$-\int_M \sum_{i=1}^{2} \langle \nabla^\perp_{Je_i}\xi, \nabla^\perp_{e_i} J\xi\rangle \, dv = \int_M \sum_{i=1}^{2} \langle \nabla^\perp_{e_i}\nabla^\perp_{Je_i}\xi, J\xi\rangle \, dv.$$

But, from the Ricci equation and the properties of the second fundamental form, we have

$$\begin{aligned}\sum_{i=1}^{2} \langle \nabla^\perp_{e_i}\nabla^\perp_{Je_i}\xi, J\xi\rangle &= \frac{1}{2}\sum_{i=1}^{2} R^\perp(e_i, Je_i, \xi, J\xi) \\ &= -2\left(\langle\sigma(e_1,e_1),\xi\rangle^2 + \langle\sigma(e_1,e_2),\xi\rangle^2\right) = -\langle \mathcal{A}\xi, \xi\rangle.\end{aligned}$$

Thus, considering the above equations, the index form $\mathcal{Q}$ satisfies

$$\mathcal{Q}(\xi) = \int_M \left(\frac{1}{2}\|D\xi\|^2 - 2\|\xi\|^2\right) dv.$$

From this we see that the first eigenvalue of the Jacobi operator satisfies $\lambda_1 \geq -2$. But from Theorem 5.1, -2 is an eigenvalue of the Jacobi operator. So, $\lambda_1 = -2$ and the corresponding eigenspace is the space of holomorphic sections of $T^\perp M$. Hence its multiplicity is $m(-2) = 2\dim H^0(M, \mathcal{O}(T^\perp M))$.

Now, the Riemann-Roch Theorem (Theorem 1.24) gives that

$$\begin{aligned}\dim\, H^0(M, \mathcal{O}(T^\perp M)) &= c_1(T^\perp M) + \operatorname{rank}(T^\perp M)(1-h) + \dim\, H^1(M, \mathcal{O}(T^\perp M)) \\ &= c_1(T^\perp M) + n - 1 + \dim\, H^1(M, \mathcal{O}(T^\perp M)),\end{aligned}$$

where $c_1(T^\perp M)$ is the first Chern class of $T^\perp M$. Using Gauss and Ricci equations (2.2), we get

$$c_1(T^\perp M) = \frac{1}{2\pi}\int_M R^\perp(e_1, e_2, e_4, e_3)\, dv = \frac{1}{2\pi}\int_M \frac{\|\sigma\|^2}{2}\, dv = \frac{\operatorname{Area}(M)}{2\pi} - 2.$$

On the other hand, using [**Ser51**], the holomorphic bundle $T^\perp M$ over the sphere M is a Whitney sum of holomorphic line bundles

$$T^\perp M = E_1 \oplus \cdots \oplus E_{n-1}.$$

Moreover, $\dim\, H^1(M, \mathcal{O}(E_i)) = \dim\, H^0(M, \mathcal{O}(\kappa \otimes E_i^*))$, where κ is the canonical bundle over M and E_i^* the dual bundle of E_i. But $c_1(\kappa \otimes E_i^*) = -2 - c_1(E_i^*) \leq -2$, because $c_1(E_i^*) \geq 0$. Since the first Chern class of $\kappa\otimes E_i^*$ is negative, such line bundle has no holomorphic sections and so, $\dim\, H^1(M, \mathcal{O}(E_i)) = 0$ for any $1 \leq i \leq n-1$. Hence, $\dim\, H^1(M, \mathcal{O}(T^\perp M)) = 0$ and $\dim\, H^0(M, \mathcal{O}(T^\perp M)) = \frac{\operatorname{Area}(M)}{2\pi} + n - 3$, which proves part 1 of the theorem.

The proof of part 2 follows from part 1 and Theorem 5.10. □

5.2. Legendrian minimal submanifolds of the sphere

In this section we will study the index and nullity of a special class of minimal submanifolds of odd-dimensional spheres, the so-called Legendrian submanifolds.

Let $\mathbb{C}^{n+1}$ be the $(n+1)$-dimensional complex Euclidean space, $\langle,\rangle$ the Euclidean metric and ω the Kähler 2-form on $\mathbb{C}^{n+1}$. Then $\omega = d\Lambda$, where Λ is the Liouville 1-form given by

$$2\Lambda(v) = \langle v, Jp\rangle,$$

where $v \in T_p\mathbb{C}^{n+1}$, $p \in \mathbb{C}^{n+1}$ and J is the complex structure on $\mathbb{C}^{n+1}$. We consider the $(2n+1)$-dimensional sphere $\mathbb{S}^{2n+1} \subset \mathbb{C}^{n+1}$ and also denote by Λ its restriction to $\mathbb{S}^{2n+1}$.

DEFINITION 5.13. An immersion $\Phi : M^n \to \mathbb{S}^{2n+1}$ of an n-dimensional manifold M is called **Legendrian** if $\Phi^*\Lambda = 0$, i.e., $J\Phi \in \mathfrak{X}^\perp(M)$.

In particular, a Legendrian immersion satisfies that $\Phi^*\omega = 0$, and then M is an isotropic submanifold of $\mathbb{C}^{n+1}$. If $\mathbb{CP}^n$ is the n-dimensional complex projective space and $\Pi : \mathbb{S}^{2n+1} \to \mathbb{CP}^n$ the Hopf fibration, then $\Phi : M^n \to \mathbb{S}^{2n+1}$ is a horizontal immersion with respect to Π and so, $\Pi \circ \Phi : M^n \to \mathbb{CP}^n$ is a Lagrangian immersion (see Definition 6.6).

Let $\Phi : M^n \to \mathbb{S}^{2n+1}$ be a Legendrian immersion. Since $J\Phi$ is a normal vector field to Φ, the normal bundle of Φ decomposes as $T^\perp M = J(TM) \oplus \langle J\Phi\rangle$, where $\langle J\Phi\rangle$ is the trivial line bundle spanned by $J\Phi$. So, a section $\xi \in \mathfrak{X}^\perp(M)$ can be decomposed as

$$\xi = JX + fJ\Phi,$$

where $X \in \mathfrak{X}(M)$ and $f \in C^\infty(M)$.

Under these conditions, the Gauss and Weingarten formulas (2.1) say that

$$\begin{aligned} JA_{JX}Y &= \sigma(X,Y), \\ \nabla^\perp_X JY &= J\nabla_X Y - \langle X,Y\rangle J\Phi, \\ A_{J\Phi} = 0, &\qquad \nabla^\perp_X J\Phi = JX, \end{aligned} \tag{5.24}$$

where X, Y are vector fields tangent to M.

Now suppose that $\Phi : M^n \to \mathbb{S}^{2n+1}$ is a minimal Legendrian immersion. Using (5.24) and writing $\xi = JX + fJ\Phi$, we have that

$$\Delta^\perp\xi = J\Delta X - JX + 2J\nabla f + (\Delta f - nf - 2\operatorname{div} X)J\Phi,$$

div is the divergence operator, ∇f is the gradient of f, and Δ is the rough Laplacian acting on vector fields of M. Also, from (5.24) we get

$$\mathcal{A}(\xi) = \sum_{i=1}^n JA_{\sigma(X,e_i)}e_i,$$

where $\{e_1, \dots, e_n\}$ is an orthonormal reference on M. Now, using the Gauss equation (2.2) of Φ we finally get

$$\mathcal{L}\xi = J\Delta X - J\Big(\sum_{i=1}^n R(X,e_i)e_i\Big) + 2(n-1)JX + 2J\nabla f + (\Delta f - 2\operatorname{div} X)J\Phi.$$

Since the tangent vector field X can be identified with its dual 1-form $X^\flat$, we will consider the following identification:

$$\begin{aligned} \mathfrak{X}^\perp(M) &\equiv \Omega^1(M) \oplus C^\infty(M) \\ \xi &\equiv (X^\flat, f), \end{aligned}$$

i.e., $X^\flat(v) = \langle v, X\rangle = \langle Jv, \xi\rangle$ for any v tangent to M. Now, taking into account the relation between Δ and the Hodge-Laplacian acting on 1-forms, one finally gets the expression of the Jacobi operator $\mathcal{L}$ of a minimal Legendrian immersion $\Phi : M^n \to \mathbb{S}^{2n+1}$:

$$\mathcal{L} : \Omega^1(M) \oplus C^\infty(M) \to \Omega^1(M) \oplus C^\infty(M) \tag{5.25}$$
$$\mathcal{L}(\alpha, f) = (\Delta\alpha + 2(n-1)\alpha + 2\mathrm{d}f, \Delta f - 2\delta\alpha),$$

where δ is the codifferential operator acting on 1-forms.

In Theorem 5.1 it was proved that $-n$ and 0 are eigenvalues of the Jacobi operator of any minimal submanifold of dimension n of a sphere. The normal vector fields $a^\perp$, with $a \in \mathbb{R}^{2n+2}$, satisfy $\mathcal{L}a^\perp - na^\perp = 0$ and the Jacobi fields ξ coming from the Killing vector fields of the sphere satisfy $\mathcal{L}\xi = 0$. It is interesting to identify these eigensections in terms of 1-forms and functions.

The vector space of Killing vector fields on $\mathbb{S}^{2n+1}$ can be identified with the space of skew-symmetric matrices $so(2n+2)$ in such a way that the Killing field corresponding to a matrix $B \in so(2n+2)$ is given by $X_B(p) = Bp$ for any $p \in \mathbb{S}^{2n+1}$. This space can be decomposed as $so(2n+2) = \langle J \rangle \oplus so^+(2n+2) \oplus so^-(2n+2)$, where $\langle J \rangle$ is the linear line spanned by the skew-symmetric matrix J and

$$so^+(2n+2) = \{B \in so(2n+2) \,/\, BJ = JB \text{ and } \operatorname{trace} BJ = 0\} \text{ and}$$
$$so^-(2n+2) = \{B \in so(2n+2) \,/\, BJ = -JB\}.$$

This decomposition of $so(2n+2)$ is orthogonal with respect to the inner product $\langle B, C \rangle = -\operatorname{trace} BC$, $\forall B, C \in so(2n+2)$. It is clear that $\dim so^+(2n+2) = (n+1)^2 - 1$ and $\dim so^-(2n+2) = n(n+1)$. Now it is easy to check the following statements:

(1) If $a \in \mathbb{C}^{n+1}$, then the eigensection $a^\perp$ corresponding to the eigenvalue $-n$ of $\mathcal{L}$ is identified with
$$a^\perp \equiv (\mathrm{d}f, f), \text{ with } f = \langle J\Phi, a \rangle \qquad (\Delta f + nf = 0).$$
(2) The Jacobi field $X_J^\perp$ is described by
$$X_J^\perp \equiv (0, f), \text{ with } f(p) = 1 \quad \forall p \in M.$$
(3) If $B \in so^+(2n+2)$, then the Jacobi field $X_B^\perp$ is identified with
$$X_B^\perp \equiv (\mathrm{d}g, 2g), \text{ with } g = \langle B\Phi, J\Phi \rangle \qquad (\Delta g + 2(n+1)g = 0). \tag{5.26}$$
(4) If $B \in so^-(2n+2)$, then the Jacobi field $X_B^\perp$ is identified with
$$X_B^\perp \equiv (\alpha, 0), \text{ with } \alpha_p(v) = \langle B\Phi(p), Jv \rangle \quad \forall p \in M, \forall v \in T_pM. \tag{5.27}$$
The 1-form α satisfies $\Delta\alpha + 2(n-1)\alpha = 0$ and $\alpha = \delta\omega$ where ω is the 2-form $\omega_p(v, w) = \langle Bv, Jw \rangle$, $\forall p \in M$, $\forall v \in T_pM$.

THEOREM 5.14 ([**Urb03**]). *Let $\Phi : M^n \to \mathbb{S}^{2n+1}$ be a minimal Legendrian immersion of a compact manifold M. Then:*

$$\operatorname{Ind}(M) = \beta_1(M) + \sum_{0<\lambda_k<2(n+1)} m(\lambda_k) + \sum_{0<\mu_k<2(n-1)} m(\mu_k),$$
$$\operatorname{Nul}(M) = 1 + m(2(n+1)) + m(2(n-1)),$$

where λ_k (or μ_k) are the eigenvalues of Δ acting on $C^\infty(M)$ (or on $\delta\Omega^2(M)$) and $m(\lambda_k)$ (respectively $m(\mu_k)$) their respective multiplicities.

PROOF. To prove the result, we consider the Hodge decomposition (Theorem 1.14)

$$\Omega^1(M) = \mathcal{H}^1(M) \oplus dC^\infty(M) \oplus \delta\Omega^2(M),$$

which allows to write in a unique way any 1-form α as $\alpha = \alpha_0 + \mathrm{d}g + \delta\omega$, where α_0 is a harmonic 1-form, g is a real function and ω is a 2-form on M. The space of harmonic 1-forms, $\mathcal{H}^1(M)$, is the kernel of Δ and its dimension is the first Betti number of M: $\beta_1(M)$.

Hodge's decomposition of 1-forms together with the above decomposition of the Jacobi fields (coming from the special decomposition of $so(2n+2)$) suggest us to split $\mathcal{L} = \mathcal{L}_1 \oplus \mathcal{L}_2$, where

$$\begin{aligned}\mathcal{L}_1 : \mathcal{H}^1(M) \oplus \delta\Omega^2(M) &\to \mathcal{H}^1(M) \oplus \delta\Omega^2(M)\\ \mathcal{L}_1(\alpha) &= \Delta\alpha + 2(n-1)\alpha,\end{aligned}$$

and

$$\begin{aligned}\mathcal{L}_2 : \mathrm{d}C^\infty(M) \oplus C^\infty(M) &\to \mathrm{d}C^\infty(M) \oplus C^\infty(M)\\ \mathcal{L}_2(\mathrm{d}g, f) &= (\Delta \mathrm{d}g + 2(n-1)\mathrm{d}g + 2\mathrm{d}f, \Delta f - 2\Delta g).\end{aligned}$$

So, it is clear that

$$\operatorname{Ind}(M) = \operatorname{Ind}(\mathcal{L}_1) + \operatorname{Ind}(\mathcal{L}_2), \quad \operatorname{Nul}(M) = \operatorname{Nul}(\mathcal{L}_1) + \operatorname{Nul}(\mathcal{L}_2). \tag{5.28}$$

Since $\mathcal{L}_1 = \Delta + 2(n-1)\, Id$, it follows that

$$\operatorname{Ind}(\mathcal{L}_1) = \beta_1(M) + \sum_{0<\mu_k<2(n-1)} m(\mu_k), \quad \operatorname{Nul}(\mathcal{L}_1) = m(2(n-1)), \tag{5.29}$$

where μ_k are the eigenvalues of $\Delta : \delta\Omega^2(M) \to \delta\Omega^2(M)$ and $m(\mu_k)$ are their multiplicities.

Let ρ be a negative eigenvalue of $\mathcal{L}_2$ and $(\mathrm{d}g, f) \in \mathrm{d}C^\infty(M) \oplus C^\infty(M)$ an eigensection corresponding to ρ, i.e. $\mathcal{L}_2((\mathrm{d}g, f)) + \rho(\mathrm{d}g, f) = 0$. Looking at the expression of $\mathcal{L}_2$ we write this equation as

$$\Delta(\mathrm{d}g) + (2(n-1) + \rho)\mathrm{d}g + 2\mathrm{d}f = 0 \quad \text{and} \quad \Delta f - 2\Delta g + \rho f = 0. \tag{5.30}$$

If $g = \sum_{k\geq 0} g_k$ and $f = \sum_{k\geq 0} f_k$ are the decompositions of g and f into eigenfunctions with $\Delta g_k + \lambda_k g_k = 0$, $\Delta f_k + \lambda_k f_k = 0$ and $0 = \lambda_0 < \lambda_1 < \dots$, equations (5.30) become

$$\begin{aligned}(2(n-1) + \rho - \lambda_k)\, \mathrm{d}g_k + 2\,\mathrm{d}f_k &= 0, \quad k \geq 1,\\ 2\lambda_k g_k + (\rho - \lambda_k) f_k &= 0, \quad k \geq 1,\\ \rho f_0 &= 0.\end{aligned} \tag{5.31}$$

Since $\rho < 0$ and $\lambda_k > 0$ for $k \geq 1$, from (5.31) we get that

$$f_0 = 0, \quad f_k = \frac{2\lambda_k}{\lambda_k - \rho} g_k, \quad \left(2(n-1) + \rho - \lambda_k + \frac{4\lambda_k}{\lambda_k - \rho}\right) \mathrm{d}g_k = 0, \tag{5.32}$$

for any $k \geq 1$.

Using (5.30), $\mathrm{d}g = 0$ implies that $f = 0$, and so from (5.32) we get that

$$\rho = h(\lambda) := \lambda - (n-1) - \sqrt{(n-1)^2 + 4\lambda}, \tag{5.33}$$

for certain positive eigenvalues λ of Δ.

From (5.32) it follows that

$$(\mathrm{d}g, f) = \sum_{h(\lambda_k)=\rho} \left(\mathrm{d}g_k, \frac{2\lambda_k}{\lambda_k - \rho} g_k\right),$$

and thus the eigenspace V_ρ corresponding to a negative eigenvalue ρ of $\mathcal{L}_2$ is given by

$$V_\rho = \left\{ \left(\mathrm{d}f, \frac{2\lambda}{\lambda-\rho} f\right) / \Delta f + \lambda f = 0 \quad \text{and} \quad h(\lambda) = \rho \right\}.$$

Taking into account the above considerations, we obtain

$$\mathrm{Ind}(\mathcal{L}_2) = \sum_{0<\lambda_k<2(n+1)} m(\lambda_k),$$

where λ_k are eigenvalues of Δ and $m(\lambda_k)$ are their multiplicities.

To finish the proof, we will compute $\mathrm{Nul}(\mathcal{L}_2)$. If $\rho = 0$, reasoning as above, we get

$$f_k = 2g_k \quad \text{and} \quad (2(n+1) - \lambda_k)\mathrm{d}g_k = 0, \quad \text{for any} \quad k \geq 1.$$

So there are only two possibilities for $(\mathrm{d}g, f)$: either $(\mathrm{d}g, f) = (\mathrm{d}g, 2g)$ with $\Delta g + 2(n+1)g = 0$ or $(\mathrm{d}g, f) = (0, a)$ with $a \in \mathbb{R}$. Hence

$$\mathrm{Nul}(\mathcal{L}_2) = 1 + m(2(n+1)). \qquad \square$$

Suppose now that $\Phi : M^2 \to \mathbb{S}^5$ is a minimal Legendrian immersion of a compact orientable surface M. Then, the star operator $\star$ (see Section 1.3) is an isomorphism from $\delta\Omega^2(M)$ to $\mathrm{d}C^\infty(M)$ commuting with Δ, which implies that the eigenvalues of Δ acting on $\delta\Omega^2(M)$ are the nonzero eigenvalues of Δ acting on $C^\infty(M)$ and with the same multiplicity. So we get the following result.

COROLLARY 5.15 ([**Urb03**]). *Let $\Phi : M^2 \to \mathbb{S}^5$ be a minimal Legendrian immersion of a compact orientable surface M of genus g. Then*

$$\mathrm{Ind}(M) = 2g + 2\sum_{0<\lambda_k<2} m(\lambda_k) + \sum_{2\leq\lambda_k<6} m(\lambda_k)$$
$$\mathrm{Nul}(M) = 1 + m(2) + m(6),$$

where λ_k are the eigenvalues of Δ of M acting on $C^\infty(M)$ and $m(\lambda_k)$ are their respective multiplicities.

We will introduce an important example of a minimal Legendrian submanifold of $\mathbb{S}^{2n+1}$.

EXAMPLE 5.16 (Legendrian n-torus). *We consider*

$$T^n = \left\{ (z_1, \dots, z_{n+1}) \in \mathbb{C}^{n+1} \mid \|z_i\|^2 = \frac{1}{n+1}, 1 \leq i \leq n, \prod_{i=1}^n z_i = \frac{\overline{z_{n+1}}}{(\sqrt{n+1})^{n-1}} \right\}.$$

Then T^n is a minimal Legendrian submanifold of $\mathbb{S}^{2n+1}$ isometric to a flat n-dimensional torus. Moreover,

$$\mathrm{Ind}(T^n) \begin{cases} \geq n(n+2) & \text{for } n \geq 3, \\ = 8 & \text{for } n = 2, \end{cases} \qquad \mathrm{Nul}(T^n) \begin{cases} \geq 2n(n+1)+1 & \text{for } n \geq 3, \\ = 13 & \text{for } n = 2. \end{cases}$$

PROOF. Let V be the hyperplane of $\mathbb{R}^n$ given by

$$V=\{(x_1,\dots,x_n)\in\mathbb{R}^n\mid\sum_{i=1}^n x_i=0\}$$

and

$$\{e_1=(a_{11},\dots,a_{1n}),\dots,e_{n-1}=(a_{n-1,1},\dots,a_{n-1,n})\}$$

a basis of V satisfying $\langle e_i,e_j\rangle=(n+1)\delta_{ij}$, $1\le i,j\le n-1$. We observe that $\{\frac{1}{\sqrt{n+1}}e_1,\dots,\frac{1}{\sqrt{n+1}}e_{n-1},\frac{1}{\sqrt{n}}(1,\dots,1)\}$ is an orthonormal basis of $\mathbb{R}^n$.

Consider the following immersion $\Phi:\mathbb{R}^n\to\mathbb{S}^{2n+1}\subset\mathbb{C}^{n+1}$

$$\Phi(x_1,\dots,x_n)=\frac{1}{\sqrt{n+1}}\left(e^{i\left(\sum_{j=1}^{n-1}a_{j1}x_j+\frac{x_n}{\sqrt{n}}\right)},\dots,e^{i\left(\sum_{j=1}^{n-1}a_{jn}x_j+\frac{x_n}{\sqrt{n}}\right)},e^{-i\sqrt{n}x_n}\right).$$

It is clear that $\Phi(\mathbb{R}^n)=T^n$, that the metric induced by Φ on $\mathbb{R}^n$ is the Euclidean one, and that Φ is a Legendrian immersion. Also, since

$$\frac{\partial^2\Phi}{\partial x_i^2}=\frac{1}{\sqrt{n+1}}\left(-a_{i1}^2e^{i\left(\sum_{j=1}^{n-1}a_{j1}x_j+\frac{x_n}{\sqrt{n}}\right)},\dots,-a_{in}^2e^{i\left(\sum_{j=1}^{n-1}a_{jn}x_j+\frac{x_n}{\sqrt{n}}\right)},0\right),$$

$$\frac{\partial^2\Phi}{\partial x_n^2}=\frac{1}{\sqrt{n+1}}\left(-\frac{1}{n}e^{i\left(\sum_{j=1}^{n-1}a_{j1}x_j+\frac{x_n}{\sqrt{n}}\right)},\dots,-\frac{1}{n}e^{i\left(\sum_{j=1}^{n-1}a_{jn}x_j+\frac{x_n}{\sqrt{n}}\right)},-ne^{-i\sqrt{n}x_n}\right),$$

for $1\le i\le n-1$, we have that the mean curvature vector field $\vec{H}$ of Φ is

$$n\vec{H}=\sum_{i=1}^n\frac{\partial^2\Phi}{\partial x_i^2}-\Big\langle\sum_{i=1}^n\frac{\partial^2\Phi}{\partial x_i^2},\Phi\Big\rangle\Phi=0,$$

which proves that Φ is minimal.

Also, if Γ is the lattice on $\mathbb{R}^n$ generated by $\{v_1,\dots,v_n\}$ with

$$v_i=\frac{2\pi}{n+1}\left(a_{1i},\dots,a_{(n-1)i},\frac{n+1}{\sqrt{n}}\right),\quad 1\le i\le n,$$

then it is easy to check that Φ induces an isometric embedding

$$\widetilde{\Phi}:(\mathbb{R}^n,\langle,\rangle)/\Gamma\to T^n\subset\mathbb{S}^{2n+1}.$$

Now, to have a control of the spectrum of T^n, we make use of the following result:

THEOREM ([**BGM71**]). *Let $(\mathbb{R}^n,\langle,\rangle)/\Gamma$ where $\langle,\rangle$ is the Euclidean metric on $\mathbb{R}^n$ and Γ is a lattice. Then:*

(1) *The spectrum of Δ on the flat torus $\mathbb{R}^n/\Gamma$ is given by*

$$\operatorname{Spec}(\Delta)=\{4\pi^2\|v^*\|^2\mid v^*\in\Gamma^*\},$$

where $\Gamma^*=\{v^*\in\mathbb{R}^n\mid\langle v,v^*\rangle\in\mathbb{Z},\forall v\in\Gamma\}$ is the dual lattice of Γ.

(2) *If λ is an eigenvalue of Δ, its multiplicity is the cardinality of the intersection of Γ^* with the circle centered at the origin of $\mathbb{R}^n$ and having radius $\frac{\sqrt{\lambda}}{2\pi}$.*

First, the dual lattice of Γ is generated by $\{v_1^*,\dots,v_n^*\}$, where

$$v_i^*=\frac{1}{2\pi}\left(a_{1i},\dots,a_{(n-1)i},\frac{1}{\sqrt{n}}\right),\quad 1\le i\le n.$$

Since $4\pi^2\|v_i^*\|^2 = n$ and $4\pi^2\langle v_i^*, v_j^*\rangle = -1$ for $1 \leq i,j \leq n$ and $i \neq j$, we have that the spectrum of the Laplacian of T^n is given by

$$\Big\{n\sum_{i=1}^n m_i^2 - \sum_{i\neq j} m_i m_j \mid (m_1,\dots,m_n) \in \mathbb{Z}^n\Big\}.$$

In particular, n and $2(n-1)$ are the first two nonzero eigenvalues of the Laplacian of T^n, with multiplicities $2n$ and $n(n-1)$ respectively. Using Theorem 5.14 and Corollary 5.15, we get that

$$\operatorname{Ind}(T^n) \geq n + 2n + n(n-1) = n(n+2), \quad n \geq 3, \quad \text{and} \quad \operatorname{Ind}(T^2) = 8.$$

To estimate the nullity, we observe that $2(n+1)$ is an eigenvalue of the Laplacian of T^n acting on $C^\infty(M)$ with multiplicity $n(n+1)$. Also, from (5.27), $2(n-1)$ is an eigenvalue of Δ acting on $\delta\Omega^2(M)$ with multiplicity at least $n(n+1)$, since it is easy to check that the 1-form α associated to any $B \in so^-(2n+2)$ is non-trivial. Considering again the Theorem 5.14 and the Corollary 5.15, we get that

$$\operatorname{Nul}(T^n) \geq 1 + 2n(n+1), \quad n \geq 3 \quad \text{and} \quad \operatorname{Nul}(T^2) = 13. \qquad \square$$

REMARK 5.17. If $n = 2$, the torus $T^2 \equiv \mathbb{R}^2/\Gamma$, where Γ is the lattice on $\mathbb{R}^2$ generated by

$$\Big\{\Big(\tfrac{\sqrt{2}\pi}{\sqrt{3}}, \sqrt{2}\pi\Big), \Big(-\tfrac{\sqrt{2}\pi}{\sqrt{3}}, \sqrt{2}\pi\Big)\Big\}.$$

The conformal structure of T^2 corresponds to the **equilateral torus.**

LEMMA 5.18. *Let $\Phi : M^2 \to \mathbb{S}^5$ be a minimal Legendrian immersion of a compact orientable surface M of genus h.*

(1) *If $h = 0$, then Φ is an embedding and $\Phi(M)$ is the totally geodesic*
$$\mathbb{S}^2 = \{(z_1, z_2, z_3) \in \mathbb{S}^5 \mid \Im(z_i) = 0,\ i = 1,2,3\}.$$
(2) *If M is flat, then $\Phi(M)$ is congruent to the Legendrian torus T^2 (Example 5.16).*

PROOF. We will use the complex structure of M and the holomorphic 3-differential on M associated to the minimal Legendrian immersion Φ. Let $z = x+iy$ be a complex coordinate on M and let $\partial_z = \frac{1}{2}(\frac{\partial}{\partial x} - i\frac{\partial}{\partial y})$ and $\partial_{\bar z} = \frac{1}{2}(\frac{\partial}{\partial x} + i\frac{\partial}{\partial y})$ the Cauchy-Riemann operators. If the induced metric is given by $g = e^{2u}\|dz\|^2$, then we have that

$$\begin{aligned} \Phi_{zz} &= 2u_z\Phi_z + \sigma(\partial_z, \partial_z) \\ \Phi_{z\bar z} &= -\frac{e^{2u}}{2}\Phi, \end{aligned} \tag{5.34}$$

where we have extended to the complexified bundles all the operators associated to the immersion Φ. On M we define a 3-differential by

$$\Theta(z) = \langle \sigma(\partial_z, \partial_z), J\Phi_z\rangle \otimes (dz)^3.$$

From the Codazzi equation (2.2), (5.24) and (5.34), we obtain that

$$\begin{aligned} \partial_{\bar z}\langle\sigma(\partial_z,\partial_z), J\Phi_z\rangle &= \langle(\nabla\sigma)(\partial_z,\partial_z,\partial_{\bar z}), J\Phi_z\rangle - \frac{e^{2u}}{2}\langle\sigma(\partial_z,\partial_z), J\Phi\rangle \\ &= \langle\nabla^\perp_z\sigma(\partial_z,\partial_{\bar z}), J\Phi_z\rangle - 2u_z\langle\sigma(\partial_z,\partial_{\bar z}), J\Phi_z\rangle = 0, \end{aligned}$$

and so, Θ is a holomorphic 3-differential on the Riemann surface M.

If $h = 0$, the Riemann-Roch theorem implies that $\Theta \equiv 0$. Using (5.24), this implies that Φ is totally geodesic, and since Φ is Legendrian, 1 follows.

If M is flat, then M is a torus and so, Θ has not zeros. With a suitable change of the complex parameter, we can take $\Theta = \frac{1}{2\sqrt{2}}(dz)^3$. From (5.34) we get

$$\Phi_{xx} = \frac{1}{\sqrt{2}} J\Phi_x - \Phi, \quad \Phi_{yy} = -\frac{1}{\sqrt{2}} J\Phi_x - \Phi, \quad \text{and} \quad \Phi_{xy} = -\frac{1}{\sqrt{2}} J\Phi_y.$$

From these formulas we can easily deduce that $(\Phi_{yy} + \frac{3}{2}\Phi)_y = 0$ and so,

$$\Phi(x,y) = \cos\left(\tfrac{\sqrt{3}}{\sqrt{2}}y\right) A_1(x) + \sin\left(\tfrac{\sqrt{3}}{\sqrt{2}}y\right) A_2(x) + B(x),$$

for certain curves A_1, A_2, B in $\mathbb{S}^5$, that are related to the initial immersion by

$$A_1(x) = -\tfrac{2}{3}\Phi_{yy}(x,0),\ A_2(x) = \tfrac{\sqrt{2}}{\sqrt{3}}\Phi_y(x,0) \text{ and } B(x) = \Phi(x,0) + \tfrac{2}{3}\Phi_{yy}(x,0).$$

Considering the conformality of Φ, we deduce that

$$\begin{aligned} \|A_i(x)\|^2 &= \frac{2}{3},\ j = 1,2, \quad \text{and} \quad \|B(x)\|^2 = \frac{1}{3}, \\ \langle A_1(x), A_2(x)\rangle &= \langle A_j(x), B(x)\rangle = 0,\ j = 1,2, \\ \langle A_1(x), JA_2(x)\rangle &= \langle A_j(x), JB(x)\rangle = 0,\ j = 1,2. \end{aligned}$$

Also,

$$A_j'(x) = -\frac{1}{\sqrt{2}} JA_j(x),\ j = 1,2, \quad \text{and} \quad B'(x) = \sqrt{2}JB(x).$$

From which we deduce that

$$\begin{aligned} A_j(x) &= \cos\left(\tfrac{1}{\sqrt{2}}x\right) A_j(0) - \sin\left(\tfrac{1}{\sqrt{2}}x\right) JA_j(0),\ j = 1,2, \\ B(x) &= \cos\left(\sqrt{2}x\right) B(0) + \sin\left(\sqrt{2}x\right) JB(0). \end{aligned}$$

Up to an isometry, we can choose $A_1(0) = \frac{1}{\sqrt{3}}(1,1,0)$, $A_2(0) = \frac{1}{\sqrt{3}}(1,-1,0)$ and $B(0) = \frac{1}{\sqrt{3}}(0,0,1)$. Finally, we deduce that

$$\Phi(x,y) = \frac{1}{\sqrt{3}}\left(e^{\frac{i}{\sqrt{2}}(-x+\sqrt{3}y)}, e^{\frac{i}{\sqrt{2}}(-x-\sqrt{3}y)}, e^{i\sqrt{2}x}\right),$$

and hence $\Phi(M) = T$. Note that the obtained immersion differs from the one presented in Example 5.16 by a reparametrization. □

COROLLARY 5.19 ([**Urb03**]). *Let $\Phi : M^2 \to \mathbb{S}^5$ be a minimal (not totally geodesic) Legendrian immersion of a compact orientable surface M. Then:*

(1) $\mathrm{Ind}(M) \geq 8$ *and the equality holds if and only if Φ is an embedding and $\Phi(M)$ is the Legendrian torus T^2 (Example* 5.16*).*
(2) $\mathrm{Nul}(M) \geq 13$ *and the equality holds if and only if M Φ is an embedding and $\Phi(M)$ is the Legendrian torus T^2.*

PROOF. First, from Example 5.16, the equilateral torus T has $\mathrm{Ind}(T) = 8$ and $\mathrm{Nul}(T) = 13$.

Furthermore, it follows from Corollary 5.15 that $\mathrm{Ind}(M) \geq 2 + m(2)$. Since Φ is a full minimal immersion of M into $\mathbb{S}^5$, $m(2) \geq 6$ and thus $\mathrm{Ind}(M) \geq 8$. Moreover, if $\mathrm{Ind}(M) = 8$, then M is a torus and $\lambda_1 = 2$, and using the main result in [**ESI00**] we get that M is flat. Lemma 5.18 implies that $\Phi(M)$ is the Legendrian torus

T^2. So M is a finite covering of T^2 and, since $\mathrm{Ind}(M) = 8$, M is congruent to the Legendrian torus T^2.

Since $m(2) \geq 6$, to prove 2 we need to estimate $m(6)$. To do this, we define a linear map (see (5.26))

$$F : so^+(6) \to V_6 = \{f \,/\, \Delta f + 6f = 0\}$$
$$B \mapsto f_B = \langle B\Phi, J\Phi \rangle,$$

and thus $m(6) \geq \dim \mathrm{img}\, F = 8 - \dim \ker F$. Now, if $B \in \ker F$, then $f_B = 0$ and thus $\nabla f_B = 0$, which means that $B\Phi$ is tangent to M. So the Killing field X_B on $\mathbb{S}^{2n+1}$ is tangent to M, hence X_B is also a Killing vector field on M. This means that we can define a linear map

$$G : \ker F \to \mathcal{K}(M)$$
$$B \mapsto X_B,$$

which is a monomorphism. Since Φ is not totally geodesic (and thus not isometric to a sphere), the dimension of the isometry group of M is not greater than 2, which implies that $\dim \ker F \leq 2$. The above inequality implies that $m(6) \geq 6$. Now Corollary 5.15 together with the estimate of $m(2)$, implies that $\mathrm{Nul}(M) \geq 13$.

If $\mathrm{Nul}(M) = 13$, then in particular $m(6) = 6$ and the dimension of the isometry group of M is 2. This implies that M is either a sphere or a torus. Lemma 5.18 implies that M is a torus with two linear independent Killing vector fields. This implies that the torus M is flat, and the proof of 2 follows as in 1. □

5.3. Cones over minimal submanifolds of the sphere

In this section we study the stability of certain minimal submanifolds of the Euclidean space: the cones over minimal submanifolds of spheres. The solution of Bernstein's problem (see Theorem 4.42 and Remark 4.43) is closely related to the hypercones of Euclidean space. E. De Giorgi proved that if the only area-minimizing hypercones of $\mathbb{R}^n$ are hyperplanes, then Bernstein's problem has a positive answer, i.e., the only entire graphs in $\mathbb{R}^{n+1}$ are the hyperplanes. In Corollary 5.28 we will prove that the only stable hypercones of $\mathbb{R}^n$ with $n \leq 7$ are hyperplanes, which allowed to J. Simons to solve the Bernstein problem for $n \leq 7$. However, in Corollary 5.29, we prove that for $n \geq 8$ there exist stable hypercones of $\mathbb{R}^n$ which are not hyperplanes.

DEFINITION 5.20. Let $\Phi : M^n \to \mathbb{S}^m$ be an immersion of a manifold M. The **cone** over M is the immersion $C(\Phi) : C(M) := \mathbb{R}^+ \times M \to \mathbb{R}^{m+1}$ defined by

$$C(\Phi)(t,p) = t\Phi(p), \quad t \in \mathbb{R}^+,\, p \in M.$$

We denote by $\langle , \rangle$ the Euclidean metric on $\mathbb{R}^{m+1}$, as well as the induced metric on $\mathbb{S}^m$. As in the previous sections, the induced metric on M is denoted by g. Then the induced metric on $C(M)$ by $C(\Phi)$ is the warped metric $\widetilde{g} = \mathrm{d}t^2 + t^2 g$, where t is the parameter on $\mathbb{R}^+$. If M is a totally geodesic sphere $\mathbb{S}^n \subset \mathbb{S}^m$, then its cone $C(\mathbb{S}^n)$ is a linear subspace of $\mathbb{R}^{m+1}$, and thus the cone can be smoothly extended to $t = 0$.

The tangent space of the cone at a point (t,p) is given by $T_{(t,p)}C(M) = \mathbb{R} \times T_pM$. Furthermore, since $d(C(\Phi))_{(t,p)}(1,0) = \Phi$, it follows that the normal space of the immersion $C(\Phi)$ is

$$T^{\perp}_{(t,p)}C(M) = T^{\perp}_p M, \quad \text{for any} \quad (t,p) \in C(M).$$

We will denote by ∂_t the vector field on $C(M)$ given by $(\partial_t)_{(t,p)} = (1,0) \in T_{(t,p)}C(M)$. Also, if X is a vector field on M, we define a vector field $\widetilde{X}$ on $C(M)$ by $\widetilde{X}_{(t,p)} = (0, X_p)$. So if $\{e_1,\dots,e_n\}$ is a local orthonormal reference on M, we have that a local orthonormal reference on $C(M)$ is given by $\{\partial_t, \frac{1}{t}\widetilde{e}_1,\dots,\frac{1}{t}\widetilde{e}_n\}$.

Finally, given a normal section to the cone, $\xi \in \mathfrak{X}^\perp(C(M))$, we have that for each $t\in\mathbb{R}^+$ the vector field ξ_t defined by $(\xi_t)_p = \xi_{(t,p)}$ is a normal section to Φ.

In the next result we relate the fundamental equations of the immersions Φ and $C(\Phi)$.

PROPOSITION 5.21. *Let $\Phi : M^n \to \mathbb{S}^m$ be an immersion of a manifold M and let $C(\Phi) : C(M) \to \mathbb{R}^{m+1}$ be its cone. Then*

(1) *If $X, Y \in \mathfrak{X}(M)$, then*

$$\widetilde{\nabla}_{\partial_t}\partial_t = 0, \quad \widetilde{\nabla}_{\partial_t}\widetilde{X} = \widetilde{X}, \quad \widetilde{\nabla}_{\widetilde{X}}\partial_t = \widetilde{X}, \quad \widetilde{\nabla}_{\widetilde{X}}\widetilde{Y} = t\widetilde{\nabla_X Y} - tg(X,Y)\partial_t,$$

where $\widetilde{\nabla}$ (respectively ∇) is the Levi-Civita connection on $C(M)$ of the metric $\widetilde{g}$ (respectively on M of the metric g).

(2) *If $\widetilde{\sigma}$ is the second fundamental form of $C(\Phi)$ and $X, Y \in \mathfrak{X}(M)$, then*

$$\widetilde{\sigma}(\partial_t, -) = 0, \quad \widetilde{\sigma}(\widetilde{X}, \widetilde{Y}) = t\sigma(X,Y),$$

where σ is the second fundamental form of Φ.

(3) *If $\xi \in \mathfrak{X}^\perp(C(M))$ and $X \in \mathfrak{X}(M)$, then*

$$\widetilde{A}_\xi \widetilde{X} = \frac{1}{t}\widetilde{A_{\xi_t}X}, \quad \widetilde{\nabla}^\perp_{\widetilde{X}}\xi = \nabla^\perp_X \xi_t,$$

where $\widetilde{A}$ (respectively A) is the Weingarten endomorphism of $C(\Phi)$ (respectively Φ) and $\widetilde{\nabla}^\perp$ (respectively $\nabla^\perp$) is the normal connection of $C(\Phi)$ (respectively the normal connection of Φ).

PROOF. Since $C(\Phi)_*(\partial_t) = \Phi$, if $\widehat{\nabla}$ is the Levi-Civita connection on $\mathbb{R}^{m+1}$, then

$$\widehat{\nabla}_{\partial_t} C(\Phi)_*(\partial_t) = \widehat{\nabla}_{\partial_t}\Phi = 0 \quad \text{and} \quad \widehat{\nabla}_{\widetilde{X}} C(\Phi)_*(\partial_t) = \widehat{\nabla}_{\widetilde{X}}\Phi = \widetilde{X},$$

and so, $\widetilde{\nabla}_{\partial_t}\partial_t = 0$, $\widetilde{\nabla}_{\widetilde{X}}\partial_t = \widetilde{X}$ and $\widetilde{\sigma}(\partial_t, -) = 0$.

Also,

$$\widehat{\nabla}_{\partial_t} C(\Phi)_*(\widetilde{X}) = \widehat{\nabla}_{\partial_t} t\widetilde{X} = \widetilde{X} \quad \text{and} \quad \widehat{\nabla}_{\widetilde{X}} C(\Phi)_*(\widetilde{Y}) = \widehat{\nabla}_{\widetilde{X}} t\widetilde{Y} = t\widehat{\nabla}_{\widetilde{X}}\widetilde{Y},$$

hence $\widetilde{\nabla}_{\partial_t}\widetilde{X} = \widetilde{X}$, $\widetilde{\nabla}_{\widetilde{X}}\widetilde{Y} = t\widetilde{\nabla_X Y} - tg(X,Y)\partial_t$ and $\widetilde{\sigma}(\widetilde{X},\widetilde{Y}) = t\sigma(X,Y)$ and we have proved 1 and 2. The proof of 3 follows in the same way. □

COROLLARY 5.22. *Let $\Phi : M^n \to \mathbb{S}^m$ be an immersion of a manifold M and $C(\Phi) : C(M) \to \mathbb{R}^{m+1}$ its cone. Then*

(1) *The mean curvature vectors $\vec{H}_C$ and $\vec{H}$ of $C(\Phi)$ and Φ respectively are related by*

$$(n+1)\vec{H}_C = \frac{n}{t}\vec{H}.$$

Thus, $C(\Phi)$ is a minimal immersion if and only if Φ is a minimal immersion.

(2) *The lengths of the second fundamental forms are related by $\|\widetilde{\sigma}\| = \frac{1}{t}\|\sigma\|$.*

LEMMA 5.23 ([**Sim68**]). *Let $\Phi : M^n \to \mathbb{S}^m$ be a minimal immersion of a manifold M and $C(\Phi) : C(M) \to \mathbb{R}^{m+1}$ its cone. If $\widetilde{\mathcal{Q}}$ is the index form of $C(M)$ and $\xi \in \mathfrak{X}^\perp(C(M))$, then*

$$\widetilde{\mathcal{Q}}(\xi) = -\int_{\mathbb{R}^+\times M} \langle t^2 \widetilde{\nabla}^\perp_{\partial_t}\widetilde{\nabla}^\perp_{\partial_t}\xi + tn\widetilde{\nabla}^\perp_{\partial_t}\xi + \mathcal{L}\xi_t - n\xi_t, t^{n-2}\xi\rangle \, \mathrm{d}t \times \mathrm{d}v,$$

where $\mathcal{L}$ is the Jacobi operator of the immersion Φ and $\mathrm{d}t \times \mathrm{d}v$ is the measure on $\mathbb{R}^+ \times M$ of the product metric $\mathrm{d}t^2 + g$.

PROOF. First, given $(t,p) \in C(M)$, we will compute $\widetilde{\mathcal{L}}(\xi)$ at the point (t,p), where ξ is any normal section of the cone and $\widetilde{\mathcal{L}}$ is the Jacobi operator of the cone. Let $\{e_1, \dots, e_n\}$ be a local orthonormal reference of M. Then, since $\{\partial_t, \frac{\widetilde{e}_1}{t}, \dots, \frac{\widetilde{e}_n}{t}\}$ is a local orthonormal reference on $C(M)$, we have, considering Proposition 5.21, that

$$\begin{aligned}\widetilde{\Delta}^\perp\xi &= \widetilde{\nabla}^\perp_{\partial_t}\widetilde{\nabla}^\perp_{\partial_t}\xi + \frac{1}{t^2}\sum_{i=1}^n \left(\widetilde{\nabla}^\perp_{\widetilde{e}_i}\widetilde{\nabla}^\perp_{\widetilde{e}_i}\xi - \widetilde{\nabla}^\perp_{\widetilde{\nabla}_{\widetilde{e}_i}\widetilde{e}_i}\xi\right) \\ &= \widetilde{\nabla}^\perp_{\partial_t}\widetilde{\nabla}^\perp_{\partial_t}\xi + \frac{n}{t}\widetilde{\nabla}^\perp_{\partial_t}\xi + \frac{1}{t^2}\sum_{i=1}^n \nabla^\perp_{e_i}\nabla^\perp_{e_i}\xi_t - \frac{1}{t}\sum_{i=1}^n \nabla_{\nabla_{e_i}e_i}\xi_t.\end{aligned}$$

Also, from Proposition 5.21, we have that $\widetilde{\mathcal{A}}(\xi) = \frac{1}{t^2}\mathcal{A}(\xi_t)$, and so, using the property of the local reference, we have that

$$\widetilde{\mathcal{L}}\xi_{(t,p)} = \left(\widetilde{\nabla}^\perp_{\partial_t}\widetilde{\nabla}^\perp_{\partial_t}\xi\right)_{(t,p)} + \frac{n}{t}\left(\widetilde{\nabla}^\perp_{\partial_t}\xi\right)_{(t,p)} + \frac{1}{t^2}(\mathcal{L}\xi_t - n\xi_t)_{(t,p)}.$$

Now, if $\mathrm{d}\widetilde{v}$ is the canonical measure of the metric $\widetilde{g}$, then $\mathrm{d}\widetilde{v} = t^n(\mathrm{d}t \times \mathrm{d}v)$, and the lemma follows from the last expression of $\widetilde{\mathcal{L}}\xi$. □

We can consider a local orthonormal reference $\{\eta_1, \dots, \eta_{m-n}\}$ on $T^\perp M$ in such a way that a section $\xi \in \mathfrak{X}^\perp(C(M))$ can be written (locally) as

$$\xi = \sum_{i=1}^{m-n} h_i(t,p)\eta_i(p), \quad h_i \in C^\infty(\mathbb{R}^+ \times M).$$

This fact, together with Lemma 5.23, suggests to consider the Sturm-Liouville operator

$$L : C^\infty(\mathbb{R}^+) \to C^\infty(\mathbb{R}^+), \quad Lf(t) = t^2 f''(t) + ntf'(t).$$

Under these conditions, if ξ is a normal section of the cone given by $\xi(t,p) = f(t)\eta(p)$ with $f \in C^\infty(\mathbb{R}^+)$ and $\eta \in \mathfrak{X}^\perp(M)$, then from Lemma 5.23 we get that

$$(5.35)\quad \widetilde{\mathcal{Q}}(\xi) = -\int_{\mathbb{R}^+\times M} fL(f)t^{n-2}\|\eta\|^2 \mathrm{d}t\times\mathrm{d}v - \int_{\mathbb{R}^+\times M} t^{n-2}f^2\langle(\mathcal{L}-n)\eta,\eta\rangle \mathrm{d}t\times\mathrm{d}v.$$

LEMMA 5.24 ([**Sim68**]). *Let $C^\infty_*[a,b]$, with $0 < a < b$, be the space of smooth functions on $[a,b]$ that vanish at a and b. Then, the Sturm-Liouville eigenvalue problem with weight t^{n-2}*

$$0 = (t^n f')' + \mu t^{n-2} f = t^{n-2}(Lf + \mu f), \quad f \in C^\infty_0[a,b], \quad 0 < a < t < b,$$

has as eigenfuctions f_k with eigenvalues μ_k the following ones:

$$f_k = t^{-\frac{n-1}{2}} \sin\left(\frac{k\pi}{\log(\frac{b}{a})}\log(\frac{t}{a})\right), \quad \mu_k = \left(\frac{n-1}{2}\right)^2 + \left(\frac{k\pi}{\log(\frac{b}{a})}\right)^2,$$

with $\mu_1 < \mu_2 < \cdots < \mu_k < \cdots \to \infty$. *If* $k \neq l$ *then* $\int_a^b t^{n-2} f_k f_l \, \mathrm{d}t = 0$ *and any function* $f \in C^\infty_*[a,b]$ *can be written as* $f = \sum_{k=1}^\infty a_k f_k$ *for unique constants* a_k.

PROOF. Direct computations and standard arguments show that the functions $f_k \in C^\infty_*[a,b]$ satisfy $Lf_k + \mu_k f_k = 0$ and that L has no other eigenvalues. Also, if $f, h \in C^\infty_*[a,b]$, then

$$\int_a^b t^{n-2} hLf \, \mathrm{d}t = \int_a^b (t^n f'h)' \, \mathrm{d}t - \int_a^b t^n f'h' \, \mathrm{d}t = -\int_a^b t^n f'h' \, \mathrm{d}t = \int_a^b t^{n-2} fLh \, \mathrm{d}t,$$

and thus, for $k \neq l$ it follows that $\int_a^b t^{n-2} f_k f_l \, \mathrm{d}t = 0$. The expansion of f is standard. □

LEMMA 5.25 ([**Sim68**]). *Let* $\Phi : M^n \to \mathbb{S}^m$ *be a minimal immersion of a compact manifold* M *and* $C(\Phi) : C(M) \to \mathbb{R}^{m+1}$ *its cone. Then,* $C(M)$ *is unstable if and only if there exist* $0 < a < b$ *such that* $n + \lambda_1 + \mu_1 < 0$, *where* λ_1 *is the first eigenvalue of the Jacobi operator of* M *and* μ_1 *is the first eigenvalue of* L *acting on* $C^\infty_*[a,b]$.

PROOF. Suppose that $n + \lambda_1 + \mu_1 < 0$ for some $0 < a < b$. We define a normal section of the cone $C(M)$ by

$$\xi(t,p) = \begin{cases} f(t)\eta(p) & \text{for } a \le t \le b, \quad p \in M, \\ 0 & \text{for } 0 < t < a, \quad b < t, \quad p \in M, \end{cases}$$

where $Lf + \mu_1 f = 0$ and $\mathcal{L}\eta + \lambda_1\eta = 0$. The normal section ξ belongs to the Sobolev space $H^{1,2}_0(C(M), T^\perp C(M))$ (see Definition 2.19). So, from (5.35) we have

$$\widetilde{\mathcal{Q}}(\xi) = \int_{[a,b]\times M} (n + \lambda_1 + \mu_1) t^{n-2} f^2 \|\eta\|^2 \, \mathrm{d}t \times \mathrm{d}v < 0,$$

which proves that $C(M)$ is unstable.

Conversely, if $C(M)$ is unstable, then there exist $a, b \in \mathbb{R}$ with $0 < a < b$ such that the cone $(a,b) \times M$ is also unstable. Furthermore, if ξ is a normal section of $(a,b) \times M$ with $\xi(a,-) = \xi(b,-) = 0$, then ξ is written as

$$\xi = \sum_{k,m=1}^\infty a_{km} f_k \eta_m, \quad a_{km} \in \mathbb{R}, \quad L\eta_m + \lambda_m \eta_m = 0,$$

where f_k are the functions given in Lemma 5.24. It follows from Lemma 5.23 and (5.35) that

$$\begin{aligned}
\widetilde{\mathcal{Q}}(\xi) &= \sum_{k,m,l,n=1}^\infty a_{km} a_{ln} \int_{[a,b]\times M} \langle (\mu_k + n + \lambda_m) f_k \eta_m, t^{n-2} f_l \eta_n \rangle \, \mathrm{d}t \times \mathrm{d}v \\
&= \sum_{k,m,l,n=1}^\infty a_{km} a_{ln} (\mu_k + n + \lambda_m) \Big(\int_a^b f_k f_l t^{n-2} \, \mathrm{d}t \Big) \Big(\int_M \langle \eta_m, \eta_n \rangle \, \mathrm{d}v \Big) \\
&= \sum_{k,m=1}^\infty a_{km}^2 (\mu_k + n + \lambda_m) \Big(\int_a^b f_k^2 t^{n-2} \, \mathrm{d}t \Big) \Big(\int_M \|\eta_m\|^2 \, \mathrm{d}v \Big).
\end{aligned}$$

Since $(a,b) \times M$ is unstable, then there exists a normal section ξ with $\widetilde{\mathcal{Q}}(\xi) < 0$. Thus, the last formula implies that there exist μ_k and λ_m such that $\mu_k + n + \lambda_m < 0$. Since $\mu_1 + n + \lambda_1 \le \mu_k + n + \lambda_m$, we end the proof. □

COROLLARY 5.26. *Let $\Phi : M^n \to \mathbb{S}^{2n+1}$ be a minimal Legendrian immersion of a compact manifold M. Then, the cone $C(M)$ of M is stable.*

REMARK 5.27. From Proposition 5.21, it is very easy to check that the cone over a minimal Legendrian submanifold of a sphere is a minimal Lagrangian submanifold of $\mathbb{C}^{n+1} \equiv \mathbb{R}^{2n+2}$ (see Definition 6.6). In Corollary 6.10 we will prove that every minimal Lagrangian submanifold of $\mathbb{C}^{n+1}$ is stable. Even more, in Proposition 6.19, we will see that these submanifolds are calibrated for the special Lagrangian calibration.

PROOF. From (5.33) it is easy to check that the first eigenvalue of the Jacobi operator of M satisfies $\lambda_1 \geq -2(n-1)$ for $n \geq 3$, and $\lambda_1 \geq -9/4$ for $n = 2$. So,

$$n + \lambda_1 + \mu_1 \geq \frac{(n-3)^2}{4} + \left(\frac{\pi}{\log(\frac{b}{a})}\right)^2 \geq 0, \quad \text{for} \quad n \geq 3,$$

and $n + \lambda_1 + \mu_1 \geq \left(\frac{\pi}{\log(\frac{b}{a})}\right)^2 \geq 0$, for $n = 2$, for any pair of real numbers $0 < a < b$. Lemma 5.25 tells us that the cone over a compact minimal Legendrian submanifold is stable. □

COROLLARY 5.28 ([**Sim68**]). *Let $\Phi : M^n \to \mathbb{S}^{n+1}$ be a (non-totally geodesic) minimal immersion of an orientable compact manifold M and $C(\Phi) : C(M) \to \mathbb{R}^{n+2}$ its cone. If $n \leq 5$, then $C(M)$ is unstable.*

PROOF. From Theorem 5.2,2, we know that the first eigenvalue λ_1 of the Jacobi operator of M satifies $\lambda_1 \leq -2n$. Thus,

$$\lambda_1 + n + \mu_1 \leq -n + \frac{(n-1)^2}{4} + \left(\frac{\pi}{\log(\frac{b}{a})}\right)^2 = \frac{n^2 - 6n + 1}{4} + \left(\frac{\pi}{\log(\frac{b}{a})}\right)^2.$$

If $n \leq 5$, the number $n^2 - 6n + 1$ is negative and we can choose a, b so that $\lambda_1 + n + \mu_1 < 0$. The result follows from Lemma 5.25. □

Corollary 5.28 is false for $n \geq 6$, as the following result proves.

COROLLARY 5.29 ([**CHS84**]). *Let $\Phi : M^n \to \mathbb{S}^{n+1}$ be a minimal immersion of an orientable compact manifold M with constant $\|A\|^2$. Then its cone $C(\Phi) : C(M) \to \mathbb{R}^{n+2}$ is stable if and only if $\|A\|^2 \leq \frac{(n-1)^2}{4}$. In particular:*

1. *The cone over the Clifford hypersurface $\mathbb{S}^{n_1}(r_1) \times \mathbb{S}^{n_2}(r_2) \subset \mathbb{S}^{n_1+n_2+1}$ (Example 2.50), with $n_1 + n_2 \geq 6$, is stable.*
2. *The cones over the minimal hypersurfaces of the sphere given in Remark 2.51 are stable, except the cones over $SO(3)/\mathbb{Z}_2 \times \mathbb{Z}_2 \subset \mathbb{S}^4$ and $SU(3)/T^2 \subset \mathbb{S}^7$, which are unstable.*

PROOF. Since $\|A\|^2$ is constant, the first eigenvalue of the Jacobi operator $\mathcal{L} = \Delta + \|A\|^2 + n$ of M is $\lambda_1 = -(\|A\|^2 + n)$. So, from Lemma 5.25 it follows that the cone is unstable if and only if $\|A\|^2 > \frac{(n-1)^2}{4}$. This proves the first part of the corollary.

In Example 2.50 we proved that the Clifford hypersurface has $\|A\|^2 = n_1 + n_2$. Moreover, $n_1 + n_2 \leq \frac{(n_1+n_2-1)^2}{4}$ if and only if $n_1 + n_2 \geq 6$. This proves 1. Also, the examples given in Remark 2.51 have $\|A\|^2 = 2n$, where n is the dimension of the hypersurface. So, $\|A\|^2 \leq \frac{(n-1)^2}{4}$ for all examples except for $SO(3)/\mathbb{Z}_2 \times \mathbb{Z}_2$ and $SU(3)/T^2$. This proves 2. □

5.4. Minimal submanifolds of the real projective space

In Proposition 2.55 we gave a criterion to study the index of minimal submanifolds of the real projective space, depending on whether or not these submanifolds admit a lifting to the corresponding sphere. However, as was shown in Section 5.1, in the study of the index of the minimal submanifolds of the sphere it has been fundamental to see the sphere isometrically embedded in Euclidean space.

We begin this section by recalling the best isometric embedding of the real projective space in Euclidean space, which was given by S.-S. Tai in [**Tai68**]. We will use it to prove some of the most important results in this section.

Let $S(m+1) = \{B \in gl(m+1, \mathbb{R}) \mid B = B^t\}$ be the linear space of real symmetric matrices of order $m+1$, endowed with the metric $\langle B, C\rangle = \frac{1}{2} \operatorname{trace} BC$. Then, $(S(m+1), \langle,\rangle)$ is isometric to the Euclidean space $\mathbb{R}^{\frac{(m+1)(m+2)}{2}}$.

The real projective space $\mathbb{RP}^m$ is embedded in $S(m+1)$ by the embedding $\Psi : \mathbb{RP}^m \to S(m+1)$ defined by

$$\Psi(\Pi(x)) = x^t x \in S(m+1), \quad x = (x_1, \ldots, x_{m+1}) \in \mathbb{S}^m \subset \mathbb{R}^{m+1},$$

where $\Pi : \mathbb{S}^m \to \mathbb{RP}^m$ is the projection induced by the antipodal map of $\mathbb{S}^m$. The image of $\mathbb{RP}^m$ by Ψ is given by

$$\Psi(\mathbb{RP}^m) = \{B \in S(m+1) \mid BB = B, \operatorname{trace} B = 1\}.$$

In the next result we collect the most important properties of the above embedding.

LEMMA 5.30.

(1) *The metric induced by $\langle,\rangle$ on $\mathbb{RP}^m$ is the standard metric of constant sectional curvature 1.*
(2) *The second fundamental form $\overline{\sigma}$ of Ψ is parallel, i.e., $\overline{\nabla}\overline{\sigma} = 0$.*
(3) *If $\overline{A}$ is the Weingarten endomorphism of the embedding Ψ, then*

$$\overline{A}_{\overline{\sigma}(v,w)} z = 2\langle v, w\rangle z + \langle w, z\rangle v + \langle v, z\rangle w,$$

for all vectors $v, w, z \in T\mathbb{RP}^m$, where $\langle,\rangle$ also denotes the induced metric on $\mathbb{RP}^m$.

Unlike the case of the sphere, where there are no stable compact minimal submanifolds, there are stable ones in projective space. In the following result we characterize the stable ones and give a lower bound on the nullity, very similar to the one given in Theorem 5.1.

THEOREM 5.31 ([**Ohn86**]). *Let $\Phi : M^n \to \mathbb{RP}^m$ be a minimal immersion of a compact manifold M. Then*

(1) *M is stable if and only if Φ is an embedding and $\Phi(M)$ is the totally geodesic real projective space $\mathbb{RP}^n$ (Example 2.57).*
(2) $\operatorname{Nul}(M) \geq (m-n)(n+1)$*, and the equality holds if and only if either Φ is an embedding and $\Phi(M)$ is the totally geodesic real projective space $\mathbb{RP}^n$ or Φ is the totally geodesic immersion of the sphere $\mathbb{S}^n$ (Example 2.58).*

PROOF. **Proof of 1.** The sufficient condition was proved in Example 2.57.

Now assume that M is stable. We will use Proposition 2.42 when $\mathbb{R}^p$ is $(S(m+1), \langle,\rangle)$. If $\{B_1, \dots, B_p\}$, with $p = \frac{(m+1)(m+2)}{2}$, is an orthonormal basis of $S(m+1)$, then it follows from Proposition 2.42 that

$$\sum_{\mu=1}^{p} \mathcal{Q}(B_\mu^\perp) = \sum_{i=1}^{n} \sum_{r=1}^{m-n} \int_M \{-\overline{R}(\xi_r, e_i, e_i, \xi_r) + \|\overline{\sigma}(e_i, \xi_r)\|^2\}\, dv,$$

where $\mathcal{Q}$ is the index form of Φ, $B_\mu^\perp$ is the normal component in $T^\perp M$ of B_μ, and $\{e_1, \dots, e_n\}$ (respectively $\{\xi_1, \dots, \xi_{m-n}\}$) is a local orthonormal reference on TM (respectively on $T^\perp M$). Now, considering the expression of the curvature $\overline{R}$ and property 3 of Lemma 5.30, we get that

$$\sum_{\mu=1}^{p} \mathcal{Q}(B_\mu^\perp) = -n(m-n) + n(m-n) = 0.$$

Since M is stable we have that $\mathcal{Q}(B_\mu^\perp) \geq 0$, and so, the last formula implies that $\mathcal{Q}(B_\mu^\perp) = 0$, for any $\mu \in \{1, \dots, p\}$. This implies that $\mathcal{L}B_\mu^\perp = 0$, for any $\mu \in \{1, \dots, p\}$, and thus $\mathcal{L}B^\perp = 0$ for any $B \in S(m+1)$.

Now, using (2.31) and Lemma 5.30 we get, for any $B \in S(m+1)$, that

$$\begin{aligned} 0 = \mathcal{L}B^\perp = 2nB^\perp &- \sum_{i,j=1}^{n} \langle \overline{\sigma}(e_i, e_j), B\rangle \sigma(e_i, e_j) \\ &+ \sum_{i=1}^{n} \sum_{r=1}^{m-n} \{\langle \overline{\sigma}(e_i, \nabla_{e_i}^\perp \xi_r), B\rangle \xi_r + \langle \overline{\sigma}(e_i, \xi_r), B\rangle \nabla_{e_i}^\perp \xi_r\}. \end{aligned}$$

Given any point $p \in M$ and any tangent vectors $v, w \in T_pM$, let $B = \overline{\sigma}(v, w)$. Taking the above formula at the point p and using Lemma 5.30, we get that

$$0 = -\sum_{i,j=1}^{n} \langle \overline{\sigma}(e_i, e_j), \overline{\sigma}(v, w)\rangle \sigma(e_i, e_j)$$

$$+ \sum_{i=1}^{n} \sum_{r=1}^{m-n} \{\langle \overline{\sigma}(e_i, \nabla_{e_i}^\perp \xi_r), \overline{\sigma}(v, w)\rangle \xi_r + \langle \overline{\sigma}(e_i, \xi_r), \overline{\sigma}(v, w)\rangle \nabla_{e_i}^\perp \xi_r\} = -2\sigma(v, w),$$

which implies that $\sigma(v, w) = 0$ for any $v, w \in T_pM$ and any $p \in M$. So, Φ is totally geodesic and either Φ is the totally geodesic immersion of $\mathbb{S}^n$, which is unstable, or Φ is the totally geodesic embedding of $\mathbb{RP}^n$. This concludes the proof of 1.

Proof of 2. In Examples 2.57 and 2.58 it was proved that the totally geodesic immersions of $\mathbb{S}^n$ and $\mathbb{RP}^n$ have nullity $(m-n)(n+1)$.

Now, given the immersion Φ and following Proposition 2.55, we have two possibilities. The first one is that Φ admits a lift to $\mathbb{S}^m$ with the same nullity. In this case, 2 follows from Theorem 5.1,2. The second one is that Φ does not admit a lift to $\mathbb{S}^m$. In this case, the covering $\Pi : \mathbb{S}^m \to \mathbb{RP}^m$ induces a two-sheeted covering $\widetilde{\Pi} : \widetilde{M} \to M$ and there exists a minimal immersion $\Psi : \widetilde{M} \to \mathbb{S}^m$ such that $\Pi \circ \Psi = \Phi \circ \widetilde{\Pi}$. It also follows from Proposition 2.55 that $\mathrm{Nul}(M) = \mathrm{Nul}(\mathcal{L}^-)$, where $\mathcal{L}^- = \mathcal{L}_{|\mathfrak{X}_-^\perp(\widetilde{M})}$, where $\mathcal{L}$ is the Jacobi operator of Ψ,

$$\mathfrak{X}_-^\perp(\widetilde{M}) = \{\xi \in \mathfrak{X}^\perp(\widetilde{M}) \mid \xi_{\tau(\tilde{p})} = -\xi_{\tilde{p}}, \forall \tilde{p} \in \widetilde{M}\},$$

and τ is the deck transformation of $\widetilde{M}$ with $\tau^2 = Id$.

It is clear that any Killing vector field X of $\mathbb{S}^m$ satisfies $X_{-p} = -X_p$ for any $p \in \mathbb{S}^m$. Hence, the normal component ξ of its restriction to $\widetilde{M}$ also satisfies $\xi_{\tau(p)} = -\xi_p$ for any $p \in \widetilde{M}$. So $\xi \in \mathfrak{X}^\perp_-(\widetilde{M})$ and, from Theorem 5.1,2, we have that $\mathrm{Nul}(M) = \mathrm{Nul}(\mathcal{L}^-) \geq (m-n)(n+1)$ and the equality holds if and only if $\widetilde{M}$ is the totally geodesic $\mathbb{S}^n$. So, in this second case, $\mathrm{Nul}(M) \geq (m-n)(n+1)$ and the equality holds if and only if M is the totally geodesic $\mathbb{RP}^n$. □

If we consider hypersurfaces of $\mathbb{RP}^m$, we can go deeper about the index, as the following result proves.

THEOREM 5.32 ([**dCRR00, ACS18, Che24**]). *Let $\Phi : M^n \to \mathbb{RP}^{n+1}$ be a minimal hypersurface of a compact manifold M.*

(1) *If Φ is two-sided, then* $\mathrm{Ind}(M) = 1$ *if and only if either Φ is the totally geodesic immersion of $\mathbb{S}^n$ (Example* 2.58*) or Φ is an embedding and $\Phi(M)$ is a Clifford hypersurface (Example* 2.59*).*
(2) *If Φ is an one-sided unstable embedding, then* $\mathrm{Ind}(M) \geq n+2$.
(3) $\mathrm{Ind}(M) \geq \frac{2}{(n+2)(n+1)}\beta_1(M)$, *where $\beta_1(M)$ is the first Betti number of M.*

REMARK 5.33. If, as in Theorem 5.2, we apply Proposition 2.43 to the isometric embedding of $\mathbb{RP}^n$ in $(S(n+1), \langle , \rangle)$ (see Lemma 5.30), we get nothing relevant.

In [**Che24**] S. Chen, besides proving part 2 of the above theorem, calculates the index of the projections on the projective space of isoparametric hypersurfaces with three constant principal curvatures of the sphere studied by B. Solomon. Using Solomon's index calculation (see Remark 2.51), she proved that the four isoparametric hypersurfaces of the projective space have index $n+2$.

In the same article, S. Chen, using the calculation of the index of the Lawson surfaces $\xi_{1,h}$ of $\mathbb{S}^3$ obtained by N. Kapouleas and D. Wiygul (see Remark 5.7), calculates the index of the projected surfaces in $\mathbb{RP}^3$, $\widehat{\xi}_{1,h} = \xi_{1,h}/\{\pm Id\}$ with h odd, proving that their index is h.

PROOF. **Proof of 1.** In Examples 2.58 and 2.59 it was proved that the totally geodesic immersion of $\mathbb{S}^n$ and the Clifford hypersurfaces are two-sided minimal immersions in $\mathbb{RP}^{n+1}$ of index one. According to Proposition 2.55, we have two possibilities for Φ. The first one is that Φ admits a lift Ψ to $\mathbb{S}^{n+1}$ and $\mathrm{Ind}(\Phi) = \mathrm{Ind}(\Psi)$. In this case, if Φ is two-sided and $\mathrm{Ind}(\Phi) = 1$, from Theorem 5.1,2, the lift is the totally geodesic $\mathbb{S}^n$ and hence the immersion Φ is the totally geodesic $\mathbb{S}^n$ in $\mathbb{RP}^{n+1}$.

The second and more complicated situation for Φ is that Φ does not admit a lift to $\mathbb{S}^{n+1}$. In this case, the covering $\Pi : \mathbb{S}^{n+1} \to \mathbb{RP}^{n+1}$ induces a two-sheeted covering $\widetilde{\Pi} : \widetilde{M} \to M$ and there exists a minimal immersion $\Psi : \widetilde{M} \to \mathbb{S}^{n+1}$ such that $\Pi \circ \Psi = \Phi \circ \widetilde{\Pi}$. It also follows from Proposition 2.55 that $1 = \mathrm{Ind}(M) = \mathrm{Ind}(\mathcal{L}^-)$, where $\mathcal{L}^- = \mathcal{L}_{|\mathfrak{X}^\perp_-(\widetilde{M})}$, where $\mathcal{L}$ is the Jacobi operator of Ψ,

$$\mathfrak{X}^\perp_-(\widetilde{M}) = \{\xi \in \mathfrak{X}^\perp(\widetilde{M}) \mid \xi_{\tau(p)} = -\xi_p, \forall p \in \widetilde{M}\},$$

and τ is the deck transformation of $\widetilde{M}$ with $\tau^2 = Id$.

Since Φ is two-sided, then Ψ is also two-sided and so the corresponding unit normal vector field N of Ψ projects into M. Hence $N \circ \tau = -N$. This implies that

$\mathfrak{X}^{\perp}_{-}(\widetilde{M})$ can be described by

$$C^{\infty}_{+}(\widetilde{M}) = \{f \in C^{\infty}(\widetilde{M}) \mid f \circ \tau = f\},$$

and $\mathcal{L}^-$ can be identified with the operator $\mathcal{L}^+ : C^{\infty}_{+}(\widetilde{M}) \to C^{\infty}_{+}(\widetilde{M})$.

Since $\operatorname{Ind}(\mathcal{L}^+) = 1$, let $\phi \in C^{\infty}_{+}(\widetilde{M})$ be an eigenfunction of the only negative eigenvalue of $\mathcal{L}^+$. Then, for any function $f \in C^{\infty}_{+}(\widetilde{M})$ with $\int_{\widetilde{M}} \phi f \, d\tilde{v} = 0$ we have that $\mathcal{Q}(f) \geq 0$ and if the equality holds then $\mathcal{L}^+ f = 0$, where $\mathcal{Q}$ is the index form of the minimal hypersurface $\widetilde{M}$.

Now we will find suitable test functions to apply the above inequality. In fact, for any $a, b \in \mathbb{R}^{n+2}$ we consider the function on $\widetilde{M}$ valuated in $\mathbb{R}^{n+2}$ defined by

$$f_{a,b} = \langle \Psi, a \rangle \Psi + \langle N, a \rangle N + \langle \Psi, b \rangle N,$$

where $\langle , \rangle$ is the standard metric on $\mathbb{R}^{n+2}$.

Since $\Psi \circ \tau = -\Psi$ and $N \circ \tau = -N$, then $f_{a,b} \in C^{\infty}_{+}(\widetilde{M}, \mathbb{R}^{n+2})$.

CLAIM. For any vector $a \in \mathbb{R}^{n+2}$, there exists a vector $b \in \mathbb{R}^{n+2}$ such that $\int_{\widetilde{M}} \phi f_{ab} \, d\tilde{v} = 0$.

To prove the claim, we consider the linear map $F : \mathbb{R}^{n+2} \to \mathbb{R}^{n+2}$ given by

$$F(b) = \int_{\widetilde{M}} \phi \langle \Psi, b \rangle N \, d\tilde{v}.$$

If $b \in \ker F$, as $\langle \Psi, b \rangle N \in C^{\infty}_{+}(\widetilde{M}, \mathbb{R}^{n+2})$, we get

$$0 \leq \mathcal{Q}(\langle \Psi, b \rangle N) = \sum_{i=1}^{n+2} \mathcal{Q}(\langle \Psi, b \rangle \langle N, a_i \rangle),$$

and if the equality holds then $\mathcal{L}(\langle \Psi, b \rangle \langle N, a_i \rangle) = 0$ for any a_i, where $\{a_1, \ldots, a_{n+2}\}$ is an orthonormal basis of $\mathbb{R}^{n+2}$. But

$$\Delta(\langle \Psi, b \rangle \langle N, a_i \rangle) = -(n + \|A\|^2) \langle \Psi, b \rangle \langle N, a_i \rangle - 2 \langle A b^{\top}, a_i \rangle,$$

where $\top$ stands for the tangential component, and so,

$$\mathcal{Q}(\langle \Psi, b \rangle N) = 2 \int_M \sum_{i=1}^{n+2} \langle \Psi, b \rangle \langle A b^{\top}, a_i \rangle \langle N, a_i \rangle \, d\tilde{v} = 0.$$

This implies that $0 = \mathcal{L}(\langle \Psi, b \rangle \langle N, a_i \rangle) = -2 \langle A b^{\top}, a_i \rangle$ for any $1 \leq i \leq n+2$. Thus we get that $A b^{\top} = 0$. This means that the function $\langle N, b \rangle$ is constant, and since $\langle N, b \rangle \circ \tau = -\langle N, b \rangle$, this constant is zero. Now, the fact that $\langle N, b \rangle = 0$ implies that the Hessian of $\langle \Psi, b \rangle$ satisfies the equation $\nabla^2 \langle \Psi, b \rangle + n \langle \Psi, b \rangle = 0$. If $\langle \Psi, b \rangle$ is not equal to zero, then a result of Obata, [**Oba62**], says that $\widetilde{M}$ is a sphere of constant curvature one. This means that Ψ is totally geodesic and hence, Φ is the totally geodesic embedding of $\mathbb{RP}^n$ whose index is zero. Finally we conclude that $\langle \Psi, b \rangle = 0$ and thus b is tangent to M. Derivating $b = b^{\top}$ with respect to any tangent vector $v \in T\widetilde{M}$ we get that $\langle b, v \rangle = 0$, which implies that $b = 0$. Thus, $\ker F = 0$ and F is a linear isomorphism.

Now, given $a \in \mathbb{R}^{n+2}$, we consider $\lambda = -\int_{\widetilde{M}} \phi(\langle \Psi, a \rangle \Psi + \langle N, a \rangle N) \, d\tilde{v} \in \mathbb{R}^{n+2}$. Since F is an isomorphism, there exists $b \in \mathbb{R}^{n+2}$ such that $F(b) = \lambda$ and we prove the claim.

Let $\{a_1, \dots, a_{n+2}\}$be an orthonormal basis of $\mathbb{R}^{n+2}$. Then, from the claim, there exists $\{b_1, \dots, b_{n+2}\} \subset \mathbb{R}^{n+2}$ such that $\int_{\widetilde{M}} \phi f_{a_i b_i} \, d\tilde{v} = 0$ for $1 \leq i \leq n+2$. But it is easy to check that

$$\mathcal{L} f_{a_i b_i} = (\|A\|^2 - n)(\langle \Psi, a_i \rangle \Psi - \langle N, a_i \rangle N) + 2(A^2 a_i^\top + a_i^\top - A b_i^\top), \tag{5.36}$$

and so,

$$0 \leq \mathcal{Q}(f_{a_i b_i}) = \int_{\widetilde{M}} (\|A\|^2 - n)(-\langle \Psi, a_i \rangle^2 + \langle N, a_i \rangle^2 + \langle N, a_i \rangle \langle \Psi, b_i \rangle) \, d\tilde{v}. \tag{5.37}$$

On the other hand,

$$\begin{aligned} \operatorname{div}(\langle N, a_i \rangle \nabla \langle \Psi, b_i \rangle - \langle \Psi, b_i \rangle \nabla \langle N, a_i \rangle) &= \langle N, a_i \rangle \Delta \langle \Psi, b_i \rangle - \langle \Psi, b_i \rangle \Delta \langle N, a_i \rangle \\ &= (\|A\|^2 - n)(\langle \Psi, b_i \rangle \langle N, a_i \rangle. \end{aligned}$$

Using the divergence theorem, from (5.37) we obtain that

$$\sum_{i=1}^{n+2} \mathcal{Q}(f_{a_i b_i}) = \sum_{i=1}^{n+2} \int_{\widetilde{M}} (\|A\|^2 - n)(-\langle \Psi, a_i \rangle^2 + \langle N, a_i \rangle^2) \, d\tilde{v} = 0.$$

This fact together with (5.37) implies that $\mathcal{Q}(f_{a_i b_i}) = 0$ for any $1 \leq i \leq n+2$, and hence $\mathcal{L} f_{a_i b_i} = 0$ for any $1 \leq i \leq n+2$. Now, from (5.36) we get that $\|A\|^2 = n$ and using [**CdCK70**] we get that $\widetilde{M}$ is one of the Clifford hypersurfaces. Thus, part 1 is proved.

Proof of 2. As in part 1, we use Proposition 2.55. If Φ admits a lift Ψ to $\mathbb{S}^{n+1}$, since Ψ is also one-sided, Ψ is not totally geodesic and so, from Theorem 5.1,2, $\operatorname{Ind}(\Phi) = \operatorname{Ind}(\Psi) \geq n+2$.

If Φ does not admit a lift to $\mathbb{S}^{n+1}$, the covering $\Pi : \mathbb{S}^{n+1} \to \mathbb{RP}^{n+1}$ induces a two-sheeted covering $\widetilde{\Pi} : \widetilde{M} \to M$ and there exists a minimal immersion $\Psi : \widetilde{M} \to \mathbb{S}^{n+1}$ such that $\Pi \circ \Psi = \Phi \circ \widetilde{\Pi}$. It also follows from Proposition 2.55 that $\operatorname{Ind}(M) = \operatorname{Ind}(\mathcal{L}^-)$, where $\mathcal{L}^- = \mathcal{L}_{|\mathfrak{X}^\perp_-(\widetilde{M})}$, where $\mathcal{L}$ is the Jacobi operator of Ψ,

$$\mathfrak{X}^\perp_-(\widetilde{M}) = \{\xi \in \mathfrak{X}^\perp(\widetilde{M}) \mid \xi_{\tau(p)} = -\xi_p, \, \forall p \in \widetilde{M}\},$$

and τ is the deck transformation of $\widetilde{M}$ with $\tau^2 = Id$. In this case, since Φ is an embedding, Ψ is also an embedding and so, Ψ is two-sided. If N is a unit normal vector field to Ψ, since Φ is one-sided, $N \circ \tau = N$. This implies that $\mathfrak{X}^\perp_-(\widetilde{M})$ can be identified with

$$C^\infty_-(\widetilde{M}) = \{f \in C^\infty(M) \mid f \circ \tau = -f\},$$

and $\mathcal{L}^-$ with the operator $\mathcal{L}^- : C^\infty_-(\widetilde{M}) \to C^\infty_-(\widetilde{M})$.

Since Φ is unstable and one-sided, from Theorem 5.31, Φ is not totally geodesic, and so, the space $\{\langle \Psi, a \rangle \mid a \in \mathbb{R}^{n+2}\}$ has dimension $n+2$. As $\Psi \circ \tau = -\Psi$, the functions $\langle \Psi, a \rangle \in C^\infty_-(\widetilde{M})$ and

$$\mathcal{Q}(\langle \Psi, a \rangle) = -\int_{\widetilde{M}} \|A\|^2 \langle \Psi, a \rangle^2 d\tilde{v} < 0, \quad \forall a \in \mathbb{R}^{n+2}.$$

This means that $\operatorname{Ind}(\mathcal{L}_-) \geq n+2$ and part 2 is proved.

Proof of 3. Using again Proposition 2.55, if Φ admits a lift Ψ to $\mathbb{S}^{n+1}$, it follows from Theorem 5.2,4 and Remark 5.4 that

$$\operatorname{Ind}(\Phi) = \operatorname{Ind}(\Psi) \geq \frac{2\beta_1(M)}{(n+2)(n+1)}.$$

If Φ does not admit a lift to $\mathbb{S}^{n+1}$, the covering $\Pi : \mathbb{S}^{n+1} \to \mathbb{RP}^{n+1}$ induces a two-sheeted covering $\widetilde{\Pi} : \widetilde{M} \to M$ and there exists a minimal immersion $\Psi : \widetilde{M} \to \mathbb{S}^{n+1}$ such that $\Pi \circ \Psi = \Phi \circ \widetilde{\Pi}$. It also follows from Proposition 2.55 that $\mathrm{Ind}(M) = \mathrm{Ind}(\mathcal{L}^-)$, where $\mathcal{L}^- = \mathcal{L}_{|\mathfrak{X}_-^\perp(\widetilde{M})}$, where $\mathcal{L}$ is the Jacobi operator of Ψ,

$$\mathfrak{X}_-^\perp(\widetilde{M}) = \{\xi \in \mathfrak{X}^\perp(\widetilde{M}) \mid \xi_{\tau(p)} = -\xi_p, \forall p \in \widetilde{M}\},$$

and τ is the deck transformation of $\widetilde{M}$ with $\tau^2 = Id$.

If X is a harmonic vector field on M, then its lift $\widetilde{X}$ to $\widetilde{M}$ is also a harmonic vector field satisfying $\widetilde{X}_{\tau(p)} = d\tau_p(\widetilde{X}_p)$, for any $p \in \widetilde{M}$. Thus, given $a \in \mathbb{R}^{n+2}$, the function $f = \langle \widetilde{X}, a\rangle$ satisfies

$$f(\tilde{\tau}(p)) = \langle \mathrm{d}\Psi_{\tilde{\tau}(p)}(\widetilde{X}_{\tau(p)}), a\rangle = \langle \mathrm{d}(\Psi \circ \tau)_p(\widetilde{X}_p), a\rangle = -\langle \mathrm{d}\Psi_p(\widetilde{X}_p), a\rangle = -f(p),$$

for any $p \in \widetilde{M}$. It is also clear that $a^\perp_{\tau(p)} = a^\perp_p$ for any $p \in \widetilde{M}$.

Thus, for any $a, b \in \mathbb{R}^{n+2}$, the normal section $\xi^{\widetilde{X}}_{ab} = \langle \widetilde{X}, b\rangle a^\perp - \langle \widetilde{X}, a\rangle b^\perp$ satisfies $\big(\xi^{\widetilde{X}}_{ab}\big)_{\tau(p)} = -\big(\xi^{\widetilde{X}}_{ab}\big)_p$, i.e., $\xi^{\widetilde{X}}_{ab} \in \mathfrak{X}_-^\perp(\widetilde{M})$.

Now, following the proof of Theorem 5.2,4 with the space $\{\widetilde{X} \mid X \in \mathcal{H}(M)\}$, whose dimension is $\beta_1(M)$, and the normal sections $\{\xi^{\widetilde{X}}_{\lambda\mu} \in \mathfrak{X}_-^\perp(\widetilde{M}) \mid 1 \le \lambda, \mu \le n+2\}$, we get from Proposition 2.43

$$\sum_{1 \le \lambda < \mu \le n+2} \mathcal{Q}(\xi^{\widetilde{X}}_{\lambda\mu}) = -2(n-1) \int_{\widetilde{M}} \|\widetilde{X}\|^2 \, \mathrm{d}\tilde{v}.$$

If $\mathrm{Ind}(M) = m$, $\lambda_1 \le \lambda_2 \le \cdots \le \lambda_m$ are the negative eigenvalues of $\mathcal{L}^-$ and $\{\eta_1, \ldots, \eta_m\}$ the corresponding eigensections, we define a linear map $F : \mathcal{H}(M) \to \mathbb{R}^{\frac{(n+2)(n+1)m}{2}}$ by

$$F(X) = \left(\int_{\widetilde{M}} \langle \xi^{\widetilde{X}}_{\lambda\mu}, \eta_i\rangle \, \mathrm{d}\tilde{v}\right)_{1 \le \lambda < \mu \le n+2,\, 1 \le i \le m}.$$

If $X \in \ker F$, then every $\xi^{\widetilde{X}}_{\lambda\mu}$, $\lambda < \mu$, is L^2-orthogonal to η_i, $1 \le i \le m$, and so, $\mathcal{Q}(\xi^{\widetilde{X}}_{\lambda\mu}) \ge 0$. It follows that $-2(n-1)\int_{\widetilde{M}} \|\widetilde{X}\|^2 d\tilde{v} \ge 0$, which implies that $X = 0$. Thus, $\ker F = 0$ and $\beta_1(M) = \dim \mathcal{H}(M) \le \frac{1}{2}(n+2)(n+1)m$, which proves the result. □

5.5. Minimal submanifolds of the product of spheres

In this section we will consider as an ambient manifold the product of two spheres and we will study the index of its compact minimal submanifolds, classifying the stable ones. It is worth remembering that in Examples 2.52, 2.53 and Propositions 2.40, 2.41 we studied the index of some examples that are minimal submanifolds of a product manifold.

First of all we show some geometric properties of the product of two spheres $\mathbb{S}^{n_1}(r_1) \times \mathbb{S}^{n_2}(r_2)$ of radii r_1 and r_2, endowed with the product metric $g = g_1 \times g_2$ of the canonical metrics g_i of $\mathbb{S}^{n_i}(r_i)$, $i = 1, 2$.

The **product structure** P on $\mathbb{S}^{n_1}(r_1) \times \mathbb{S}^{n_2}(r_2)$ is defined as the tensor

$$P(v) = P(v_1, v_2) := (v_1, -v_2), \quad v = (v_1, v_2) \in T_{(p,q)}\mathbb{S}^{n_1}(r_1) \times \mathbb{S}^{n_2}(r_2).$$

We note that, $T\mathbb{S}^{n_1}(r_1) = \{v \in T(\mathbb{S}^{n_1}(r_1) \times \mathbb{S}^{n_2}(r_2))) \mid Pv = v\}$ and that $T\mathbb{S}^{n_2}(r_2) = \{v \in T(\mathbb{S}^{n_1}(r_1) \times \mathbb{S}^{n_2}(r_2)) \mid Pv = -v\}$. It is clear that:

(1) P is a linear isometry,
(2) P is parallel, i.e. $\overline{\nabla}P = 0$,
(3) $P^2 = Id$ and trace $P = n_1 - n_2$.

The second fundamental form $\overline{\sigma}$ of the embedding $\mathbb{S}^{n_1}(r_1)\times\mathbb{S}^{n_2}(r_2)\subset\mathbb{R}^{n_1+1}\times\mathbb{R}^{n_2+1}=\mathbb{R}^{n_1+n_2+2}$ is given by

$$\overline{\sigma}(v,w)=\left(-\frac{g_1(v_1,w_1)}{r_1^2}p,-\frac{g_2(v_2,w_2)}{r_2^2}q\right) \tag{5.38}$$

where $v,w\in T_{(p,q)}\mathbb{S}^{n_1}(r_1)\times\mathbb{S}^{n_2}(r_2)$, $v=(v_1,v_2)$ and $w=(w_1,w_2)$. Using the above expression, the curvature $\overline{R}$ of $\mathbb{S}^{n_1}(r_1)\times\mathbb{S}^{n_2}(r_2)$ is given as follows:

$$\overline{R}(v,w)x=\left(\frac{1}{r_1^2}(g_1(w_1,x_1)v_1-g_1(v_1,x_1)w_1),\frac{1}{r_2^2}(g_2(w_2,x_2)v_2-g_2(v_2,x_2)w_2)\right),$$

where $v=(v_1,v_2), w=(w_1,w_2)$ and $x=(x_1,x_2)$. Considering that, if $v=(v_1,v_2)$ is a tangent vector to $\mathbb{S}^{n_1}(r_1)\times\mathbb{S}^{n_2}(r_2)$ then $v_1=\frac{v+Pv}{2}$ and $v_2=\frac{v-Pv}{2}$, it is easy to see that the Ricci tensor $\overline{S}$ of $\mathbb{S}^{n_1}(r_1)\times\mathbb{S}^{n_2}(r_2)$ is given by

$$\overline{S}(v,w)=\frac{1}{2}\left(\frac{n_1-1}{r_1^2}+\frac{n_2-1}{r_2^2}\right)g(v,w)+\frac{1}{2}\left(\frac{n_1-1}{r_1^2}-\frac{n_2-1}{r_2^2}\right)g(Pv,w). \tag{5.39}$$

So, $\mathbb{S}^{n_1}(r_1)\times\mathbb{S}^{n_2}(r_2)$ is an Einstein manifold if and only if $r_2^2(n_1-1)=r_1^2(n_2-1)$.

If $n_1=n_2=2$, the manifold $\mathbb{S}^2(r_1)\times\mathbb{S}^2(r_2)$ has two structures of Kähler manifolds. In fact, if we denote by J the complex structure of $\mathbb{S}^2(r_1)$, as well as the corresponding one in $\mathbb{S}^2(r_2)$, then

$$J_1(v_1,v_2)=(Jv_1,Jv_2)\quad\text{and}\quad J_2(v_1,v_2)=(Jv_1,-Jv_2)$$

define complex structures on $\mathbb{S}^2(r_1)\times\mathbb{S}^2(r_2)$ and the manifolds $(\mathbb{S}^2(r_1)\times\mathbb{S}^2(r_2),g,J_i)$ with $i=1,2$ are Kähler manifolds. In this case, it is interesting to note that the product structure P is given by $P=-J_1J_2=-J_2J_1$.

We start by classifying the stable compact minimal submanifolds of the product of two spheres.

THEOREM 5.34 ([**TU14**]). *Let $\Phi: M^n\to\mathbb{S}^{n_1}(r_1)\times\mathbb{S}^{n_2}(r_2)$ be a minimal immersion of a compact n-manifold M in the product of two spheres of radius r_1 and r_2. Then, Φ is stable if and only if one of the following possibilities holds:*

(1) $M=\mathbb{S}^{n_1}(r_1)$ and $\Phi(M)$ is a slice $\mathbb{S}^{n_1}(r_1)\times\{q\}$ with q a point of $\mathbb{S}^{n_2}(r_2)$.
(2) $M=\mathbb{S}^{n_2}(r_2)$ and $\Phi(M)$ is a slice $\{p\}\times\mathbb{S}^{n_2}(r_2)$ with p a point of $\mathbb{S}^{n_1}(r_1)$.
(3) $n_1=n_2=n=2$, M is orientable and Φ is a complex immersion (see Definition 6.1*) of the Riemann surface M in $\mathbb{S}^2(r_1)\times\mathbb{S}^2(r_2)$ with respect to one of the two complex structures that $\mathbb{S}^2(r_1)\times\mathbb{S}^2(r_2)$ has.*

PROOF. From Proposition 2.40 the examples given in 1 and 2 are stable. Also, from Theorem 6.4 the examples given in 3 are also stable.

Now suppose that Φ is stable. Then we consider the embedding

$$\mathbb{S}^{n_1}(r_1)\times\mathbb{S}^{n_2}(r_2)\subset\mathbb{R}^{n_1+1}\times\mathbb{R}^{n_2+1}=\mathbb{R}^{n_1+n_2+2},$$

and, for any $a\in\mathbb{R}^{n_1+n_2+2}$, the normal component $a^\perp$ of a in the normal bundle of the immersion Φ. Thus, the stability condition says that the index form $\mathcal{Q}$ satisfies $\mathcal{Q}(a^\perp)\geq 0$ for any $a\in\mathbb{R}^{n_1+n_2+2}$.

In contrast to the Proposition 2.42, in this case we will use only vectors $a=(a_1,0)\in\mathbb{R}^{n_1+1}\times\mathbb{R}^{n_2+1}$. It is useful to write $\Phi=(\phi,\psi)$, where ϕ and ψ are the projections of Φ over $\mathbb{S}^{n_1}(r_1)$ and $\mathbb{S}^{n_2}(r_2)$ respectively. Also, without loss of generality, we can assume that $n_1 \geq n_2$. Then, given $a = (a_1, 0)$ and, as $\langle \phi, a_1 \rangle = \langle \Phi, a \rangle$, we can write

$$a = a^\top + a^\perp + \frac{\langle \Phi, a \rangle}{r_1^2}(\phi, 0), \tag{5.40}$$

where $a^\top$ is the tangential component of a and $\langle , \rangle$ is the Euclidean metric in $\mathbb{R}^{n_1+1}$, in $\mathbb{R}^{n_2+1}$, and in $\mathbb{R}^{n_1+n_2+2}$. Derivating (5.40) with respect to a tangent vector $v \in TM$ and using the Gauss and Weingarten formulas (2.1) we get

$$\nabla_v a^\top = A_{a^\perp} v - \frac{\langle \Phi, a \rangle}{2r_1^2}(v + (Pv)^\top), \quad \nabla_v^\perp a^\perp = -\sigma(v, a^\top) - \frac{\langle \Phi, a \rangle}{2r_1^2}(Pv)^\perp.$$

From the above equations we get, taking into account that Φ is minimal and the Codazzi equation (2.2), that

(5.41)

$$\begin{aligned}
\Delta^\perp a^\perp &= \sum_{i=1}^n \left((\nabla\sigma)(e_1, a^\top, e_i) + \sigma(e_i, \nabla_{e_i} a^\top) + \frac{\langle \Phi, a \rangle}{2r_1^2} \nabla_{e_i}(Pe_i)^\perp \right) - \frac{(Pa^\top)^\perp}{2r_1^2} \\
&= \sum_{i=1}^n \Big\{ (\overline{R}(a^\top, e_i)e_i)^\perp - \sigma(e_i, A_{a^\perp} e_i) \\
&\quad + \frac{\langle \Phi, a \rangle}{2r_1^2}\sigma(e_i, (Pe_i)^\top) - \frac{\langle \Phi, a \rangle}{2r_1^2}\nabla_{e_i}(Pe_i)^\perp \Big\} - \frac{(Pa^\top)^\perp}{2r_1^2},
\end{aligned}$$

where $\{e_1, \ldots, e_n\}$ is an orthonormal reference on M. We will denote by B the square n-matrix function $B_{ij} = g((Pe_i)^\top, e_j)$, where g is the induced metric on M. Now, taking normal components in the derivative with respect to e_i of the expression $Pe_i = (Pe_i)^\top + (Pe_i)^\perp$, we get that

$$\sum_{i=1}^n \nabla_{e_i}^\perp (Pe_i)^\perp = \sum_{i=1}^n \sigma(e_i, (Pe_i)^\top),$$

and thus, from (5.41) we have

$$\mathcal{L}a^\perp = \sum_{i=1}^n \left\{ (\overline{R}(a^\top + a^\perp, e_i)e_i)^\perp + \frac{\langle \Phi, a \rangle}{r_1^2} \sum_{j=1}^n B_{ij}\sigma(e_i, e_j) \right\} - \frac{(Pa^T)^\perp}{2r_1^2}. \tag{5.42}$$

From (5.40), $a^\top + a^\perp = \left(a_1 + \frac{\langle \Phi, a \rangle}{r_1^2}\phi, 0\right)$ and so, using the curvature expression $\overline{R}$ we deduce that

$$\begin{aligned}
\sum_{i=1}^n (\overline{R}(a^\top + a^\perp, e_i)e_i)^\perp &= \frac{1}{4r_1^2}\sum_{i=1}^n \{\|e_i + Pe_i\|^2 a^\perp - \langle e_i + Pe_i, a^\top + a^\perp \rangle (Pe_i)^\perp\} \\
&= \frac{1}{2r_1^2}\{(n + \operatorname{trace} B)a^\perp - (Pa^\top)^\perp\}.
\end{aligned}$$

Thus, finally (5.42) becomes in

$$\mathcal{L}a^\perp = \frac{1}{2r_1^2}\{(n + \operatorname{trace} B)a^\perp - 2(Pa^\top)^\perp + 2\langle \Phi, a \rangle \sum_{i,j=1}^n B_{ij}\sigma(e_i, e_j)\}.$$

Let now $\{a_1^1,\dots,a_1^{n_1+1}\}$ be an orthonormal basis of $\mathbb{R}^{n_1+1}$ and $a_i=(a_1^i,0)$, $1\le i\le n_1+1$. Then, from the last formula we have

$$\begin{aligned}\sum_{\alpha=1}^{n_1+1}\mathcal{Q}(a_\alpha^\perp)&=\frac{1}{2r_1^2}\sum_{\alpha=1}^{n_1+1}\int_M\{-(n+\operatorname{trace}B)\|a_\alpha^\perp\|^2+2g((Pa_\alpha^\top)^\perp,a_\alpha^\perp)\}\,dv\\&\quad-\frac{1}{r_1^2}\sum_{\alpha=1}^{n_1+1}\int_M\langle\Phi,a_\alpha\rangle\sum_{i,j=1}^{n}B_{ij}\langle\sigma(e_i,e_j),a_\alpha\rangle\,dv.\end{aligned}\tag{5.43}$$

But from (5.40) it follows that

$$\begin{aligned}\sum_{\alpha=1}^{n_1+1}\|a_\alpha^\perp\|^2&=\sum_{\alpha=1}^{n_1+1}\left(1-\|a_\alpha^\top\|^2-\frac{\langle\Phi,a_\alpha\rangle^2}{r_1^2}\right)=n_1-\sum_{\alpha=1}^{n_1+1}\sum_{i=1}^{n}\langle a_\alpha,e_i\rangle^2\\&=n_1-\frac14\sum_{i=1}^{n}\|e_i+Pe_i\|^2=n_1-\frac{n+\operatorname{trace}B}{2}.\end{aligned}$$

Also,

$$\begin{aligned}\sum_{\alpha=1}^{n_1+1}g((Pa_\alpha^\top))^\perp,a_\alpha^\perp)&=\sum_{\alpha=1}^{n_1+1}\sum_{i=1}^{n}\langle a_\alpha,e_i\rangle\langle Pe_i-(Pe_i)^\top,a_\alpha\rangle\\&=\frac14\sum_{i=1}^{n}\|e_i+Pe_i\|^2-\frac14\sum_{i,j=1}^{n}B_{ij}g(e_i+Pe_i,e_j+Pe_j)\\&=\frac12(n-\operatorname{trace}B^2).\end{aligned}$$

Similarly, we have

$$\sum_{\alpha=1}^{n_1+1}\langle\Phi,a_\alpha\rangle\langle\sigma(e_i,e_j),a_\alpha\rangle=\frac12\langle\sigma(e_i,e_j)+P\sigma(e_i,e_j),\phi\rangle=0.$$

So, using the stability inequality $0\le\sum_{\alpha=1}^{n_1+1}\mathcal{Q}(a_\alpha^\perp)$ and the last three equations in (5.43), we finally get

$$0\le\int_M\{-2\operatorname{trace}B^2+(\operatorname{trace}B)^2-2(n_1-n)\operatorname{trace}B-n(2n_1-n-2)\}\,dv\tag{5.44}$$

Now, since $n_1\ge n_2$ and $n\ge 2$ we have that $n_1\ge 2$. We distinguish two cases:
First case: $n\le n_1$.

The Schwarz inequality implies that $(\operatorname{trace}B)^2\le n\operatorname{trace}B^2$ and the equality holds if and only if $B=f\,Id$ for some function f on M. So (5.44) becomes

$$0\le\int_M(\operatorname{trace}B+n)\left(\frac{n-2}{n}\operatorname{trace}B-2n_1+n+2\right)dv.\tag{5.45}$$

Since $-n\le\operatorname{trace}B\le n$, the integrand in (5.45) is non-positive, and hence the equality occurs in (5.44) and (5.45). This implies that $B=f\,Id$ and one of the following possibilities is true: $\operatorname{trace}B=-n$ or $\operatorname{trace}B=n=n_1>2$ or $n=n_1=2$. So, we get that either $B=-Id$, or $B=Id$ with $n=n_1>2$, or $B=f\,Id$ and $n=n_1=2$, for some function f on the surface M. Now we analyze these three cases.

If $B=-Id$, then $Pv=-v+(Pv)^\perp$ for any $v\in T_pM$ and any $p\in M$. Since P is an isometry, then $(Pv)^\perp=0$ and so, $Pv=-v$ for any tangent vector v. Thus,

$\phi_*(v) = 0$ for any tangent vector v and this implies that $\phi : M \to \mathbb{S}^{n_1}(r_1)$ is a constant function. Hence $\psi : M \to \mathbb{S}^{n_2}(r_2)$ is either a stable minimal immersion (if $n < n_2$) or $n = n_2$ and ψ is a covering map. From Theorem 5.1,2, the first situation is not possible, and since $\mathbb{S}^{n_2+1}$ is simply connected, case 1 in the theorem occurs.

If $B = Id$ and $n = n_1 > 2$, reasoning as above, we get that $\psi : M \to \mathbb{S}^{n_2}(r_2)$ is a constant function and so, ϕ is a covering map. In this case we get 2.

If $B = f\,Id$ and $n = n_1 = 2$, then n_2 is either one or two. But if $n_2 = 1$, then at any point of M there exists a nonzero vector v tangent to M such that $\psi_* v = 0$. This implies that $Pv = v$ and, since $B = f\,Id$, we get that $f = 1$ and $B = Id$. Again we get 2.

We are left with the case $n = n_1 = n_2 = 2$ and $B = f\,Id$. In this case, as we pointed out at the beginning of this section, $\mathbb{S}^2(r_1) \times \mathbb{S}^2(r_2)$ has two structures of Kähler manifold with complex structures J_1 and J_2. If ω_i denote the Kähler 2-forms of the above Kähler structures, and $\{e_1, e_2\}$ is an orthonormal reference on TM, then we have that

$$J_i(e_j) = (\Phi^*\omega_i)(e_j, e_k)e_k + \xi_{ij}, \quad k \neq j, \quad 1 \leq i, j, k \leq 2, \tag{5.46}$$

where ξ_{ij} denotes the normal component of $J_i(e_j)$. Then, since

$$B_{ij} = g((Pe_i)^\top, e_j) = \langle J_1 e_i, J_2 e_j \rangle,$$

it follows that

$$(B_{11} - B_{22})^2 + 4B_{12}^2 = \big(\langle \xi_{11}, \xi_{21}\rangle - \langle \xi_{12}, \xi_{22}\rangle\big)^2 + 4\langle \xi_{11}, \xi_{22}\rangle^2. \tag{5.47}$$

From (5.46), it easily follows that $\|\xi_{ij}\|^2 = 1 - \frac{\|\Phi^*\omega_i\|^2}{2}$, $\langle \xi_{i1}, \xi_{i2}\rangle = 0$ for $1 \leq i, j \leq 2$ and $\langle \xi_{11}, \xi_{22}\rangle = \langle \xi_{21}, \xi_{12}\rangle$. So, (5.47) will become in

$$\begin{aligned}(B_{11} - B_{22})^2 + 4B_{12}^2 &= \left(1 - \frac{\|\Phi^*\omega_1\|^2}{2}\right)\left(1 - \frac{\|\Phi^*\omega_2\|^2}{2}\right)(4\cos^2\alpha + 4\sin^2\alpha) \\ &= \big(2 - \|\Phi^*\omega_1\|^2\big)\big(2 - \|\Phi^*\omega_2\|^2\big),\end{aligned}$$

where α is the angle of the vectors ξ_{11} and ξ_{21}.

Since $B = fId$, it follows from the last equation that $\big(2 - \|\Phi^*\omega_1\|^2\big)\big(2 - \|\Phi^*\omega_2\|^2\big) = 0$. But minimal surfaces of Kähler surfaces are either complex or the set of their complex points are isolated (see [**Wol89**]). So, we get that either $\|\Phi^*\omega_1\|^2 = 2$ or $\|\Phi^*\omega_2\|^2 = 2$. This means that Φ is a complex immersion with respect to either J_1 or J_2.

Second case: $n > n_1$.

In this setting, for any point $x \in M$ we have that $\dim\ker d\phi_x \geq n - n_1$ and so, there exists an $(n - n_1)$-dimensional linear subspace $V_x \subset T_xM$ such that $V_x \subset \ker d\phi_x$, i.e., that $Pv = -v$ for any $v \in V_x$. So, we can decompose $T_xM = V_x \oplus Z_x$, where Z_x is orthogonal to V_x and $\dim Z_x = n - (n - n_1) = n_1$. Thus our matrix B can be written as $B = -Id \oplus \widehat{B}$, where $\widehat{B}_{ij} = g(Pz_i, z_j)$, and $\{z_1, \dots, z_{n_1}\}$ is an orthonormal basis of Z_x. In particular,

$$\operatorname{trace} B = n_1 - n + \operatorname{trace}\widehat{B}, \quad \operatorname{trace}(B^2) = n - n_1 + \operatorname{trace}(\widehat{B}^2).$$

Hence, the stability inequality (5.44) becomes

$$0 \leq \int_M \big(-2\operatorname{trace}(\widehat{B}^2) + (\operatorname{trace}\widehat{B})^2 - n_1(n_1 - 2)\big)\, dv. \tag{5.48}$$

The Schwarz inequality implies that $(\operatorname{trace}\widehat{B})^2 \leq n_1 \operatorname{trace}(\widehat{B}^2)$ and the equality holds if and only if $\widehat{B} = f\,Id$ for some function f on M. So, (5.48) becomes

$$0 \leq \frac{n_1-2}{n_1}\int_M \left((\operatorname{trace}\widehat{B})^2 - n_1^2\right) dv. \tag{5.49}$$

Since $-n_1 \leq \operatorname{trace}\widehat{B} \leq n_1$, the integrand of (5.49) is non-positive and we get the equality in (5.48) and (5.49), which means that either $\operatorname{trace}\widehat{B} = \pm n_1$, $\widehat{B} = f\,Id$ and $n_1 > 2$, or $\widehat{B} = f\,Id$ and $n_1 = 2$. Since $n_2 \leq n_1 = 2 < n$, we finally get that either $\widehat{B} = \pm Id$ with $n_1 > 2$, or $\widehat{B} = f\,Id$, $n_1 = n_2 = 2$ and $n = 3$. So, we get that $n_1 > 2$ and either $B = -Id$, or $B = -Id_{n-n_1} \oplus Id_{n_1}$, or $n_1 = n_2 = 2$, $n = 3$ and $B = -Id_1 \oplus f\,Id_2$.

If $B = -Id$, reasoning as above, we get case 1.

If $n_1 = n_2 = 2$, $n = 3$ and $B = -Id_1 \oplus f\,Id_2$, then for any point $p \in M$ there exists a nonzero vector $v \in T_pM$ such that $d\psi_p(v) = 0$. This means that $Pv = v$ and hence $f(p) = 1$. Since this is true for every point $p \in M$, we get that $f = 1$. So this case is included in the more general case where $B = -Id_{n-n_1} \oplus Id_{n_1}$. In this situation, we have two orthogonal distributions D_1 and D_2 on M of dimensions n_1 and $n - n_1$ respectively, defined by

$$D_1 = \{v \in TM \mid Pv = v\} \quad D_2 = \{v \in TM \mid Pv = -v\},$$

so that $TM = D_1 \oplus D_2$. In fact, $\phi : M \to \mathbb{S}^{n_1}(r_1)$ is a Riemannian submersion for which D_2 defines the vertical subspaces and D_1 defines the horizontal ones. Since P is parallel, these distributions are totally geodesic foliations on M, and thus, $M = \mathbb{S}^{n_1}(r_1) \times M_2$ and $\Phi = Id \times \Phi_2$ where $\Phi_2 : M_2 \to \mathbb{S}^{n_2}(r_2)$ is an immersion. Moreover, since Φ is stable, Proposition 2.41 says that Φ_2 is also stable, which is impossible by Theorem 5.1, 2. □

Now, using again the ideas developed in Theorems 5.2 and 5.32, we will obtain a lower bound on the index of minimal hypersurfaces of the product of two spheres of equal radius in terms of their first Betti numbers.

THEOREM 5.35 ([**ACS18**]). *Let $\Phi : M^n \to \mathbb{S}^{n_1}\times\mathbb{S}^{n_2}$ be a minimal hypersurface ($n_1 + n_2 = n+1$) with $n \geq 3$ and $(n_1, n_2) \neq (2,2)$ of a compact manifold M into the product of two spheres of radius one. Then*

$$\operatorname{Ind}(M) \geq \frac{2}{(n_1+n_2+1)(n_1+n_2)}\beta_1(M),$$

where $\beta_1(M)$ is the first Betti number of M.

PROOF. To prove the theorem, we use Proposition 2.43 with

$$\Psi : \mathbb{S}^{n_1}\times\mathbb{S}^{n_2} \subset \mathbb{R}^{n_1+n_2+2} = \mathbb{R}^{n_1+1}\times\mathbb{R}^{n_2+1}.$$

If X is a harmonic vector field on M, $\{a_1, \dots, a_{n_1+n_2+2}\}$ is an orthonormal basis of $\mathbb{R}^{n_1+n_2+2}$, and

$$\xi^X_{\lambda\mu} = \langle X, a_\mu\rangle a_\lambda^\perp - \langle X, a_\lambda\rangle a_\mu^\perp, \qquad 1 \leq \lambda, \mu \leq n_1+n_2+2,$$

then, from Proposition 2.43 and (5.38), we get

(5.50)
$$\sum_{1\leq\lambda<\mu\leq n_1+n_2+2} \mathcal{Q}(\xi^X_{\lambda\mu}) = -(n-3)\int_M \|X\|^2\, dv - \frac{1}{2}\int_M C(n_1-n_2+C)\|X\|^2\, dv - \int_M \left\{\langle PX,N\rangle^2 + \frac{n_1-n_2-C}{2}\langle PX,X\rangle\right\} dv,$$

where N is a local unit normal vector field on M and C is the well-defined function on M given by $C=\langle PN,N\rangle$.

We will consider three different cases. Without loss of generality, we can assume that $n_1\geq n_2$.

First case: $n_2\geq 3$. In this case, $n\geq 6$. From (5.50) and considering that $-1\leq C\leq 1$ and $-\|X\|^2\leq\langle PX,X\rangle\leq\|X\|^2$, it follows that

(5.51)
$$\begin{aligned}\sum_{1\leq\lambda<\mu\leq n_1+n_2+2} \mathcal{Q}(\xi^X_{\lambda\mu}) &\leq -(n-3)\int_M \|X\|^2\, dv \\ &\quad -\frac{1}{2}\int_M C(n_1-n_2)\|X\|^2 dv - \int_M \frac{n_1-n_2-C}{2}\langle PX,X\rangle dv \\ &\leq \int_M \left\{(3-n+\frac{n_1-n_2}{2})\|X\|^2 + \frac{n_1-n_2+1}{2}\|X\|^2\right\} dv \\ &= \int_M (\frac{9}{2}-2n_2)\|X\|^2\, dv \leq -\frac{3}{2}\int_M \|X\|^2\, dv.\end{aligned}$$

Second case: $n_2=1$. In this case, $n_1=n\geq 3$. The vector fields X and N can be written as $X=(X_1,X_2)$ and $N=(N_1,N_2)$. Hence $PX=(X_1,-X_2)$ and $PN=(N_1,-N_2)$. This implies that

$$\langle PX,X\rangle = \|X_1\|^2-\|X_2\|^2, \quad C=\|N_1\|^2-\|N_2\|^2.$$

Since $1=\|N\|^2=\|N_1\|^2+\|N_2\|^2$, we get that $\|N_1\|^2=\frac{1+C}{2}$ and $\|N_2\|^2=\frac{1-C}{2}$.

On the other hand, $0=\langle X,N\rangle=\langle X_1,N_1\rangle+\langle X_2,N_2\rangle$ and so $\langle PX,N\rangle = \langle X_1,N_1\rangle-\langle X_2,N_2\rangle = 2\langle X_1,N_1\rangle = -2\langle X_2,N_2\rangle$. Since X_2 and N_2 are vector fields tangent to $\mathbb{S}^1$, we get that $\langle X_2,N_2\rangle^2=\|X_2\|^2\|N_2\|^2=\frac{1-C}{2}\|X_2\|^2$. Hence, $\langle PX,N\rangle^2=2(1-C)\|X_2\|^2$. Taking these relations into account, (5.50) becomes

(5.52)
$$\begin{aligned}\sum_{1\leq\lambda<\mu\leq n_1+n_2+2} \mathcal{Q}(\xi^X_{\lambda\mu}) &= -\frac{1}{2}\int_M (C^2+(n-2)C+3n-7)\|X_1\|^2\, dv \\ &\quad -\frac{1}{2}\int_M (C^2+(n-4)C+n-1)\|X_2\|^2\, dv \\ &\leq \int_M (\frac{5}{2}-n)\|X_1\|^2 dv - \frac{1}{2}\int_M \|X_2\|^2 dv \\ &\leq -\frac{1}{2}\int_M \|X\|^2 dv.\end{aligned}$$

Third case: $n_2 = 2$ and $n_1 \geq 3$. In this case, $n = n_1 + 1$. From (5.50) if follows that

$$\sum_{1\leq\lambda<\mu\leq n_1+n_2+2} \mathcal{Q}(\xi^X_{\lambda\mu}) = -\frac{1}{2}\int_M \{C^2 + (n_1-3)C + 3(n_1-2)\}\|X_1\|^2\,dv \tag{5.53}$$
$$-\frac{1}{2}\int_M \{C^2 + (n_1-1)C + (n_1-2)\}\|X_2\|^2\,dv - \int_M \langle PX, N\rangle^2\,dv$$
$$\leq -\frac{1}{2}\int_M (3+C^2)\|X_1\|^2\,dv - \frac{1}{2}\int_M (1+C)^2\|X_2\|^2\,dv - \int_M \langle PX, N\rangle^2\,dv,$$

and the equality occurs if and only if $n_1 = 3$.

Suppose that m is the index of M. If $\lambda_1 \leq \lambda_2 \leq \cdots \leq \lambda_m$ are the negative eigenvalues of the Jacobi operator and $\{\eta_1, \ldots, \eta_m\}$ the corresponding eigensections, we define a map $F : \mathcal{H}(M) \to \mathbb{R}^{\frac{(n_1+n_2+1)(n_1+n_2)m}{2}}$ by

$$F(X) = \left(\int_M g(\xi^X_{\lambda\mu}, \eta_i)\,dv\right)_{1\leq\lambda<\mu\leq n_1+n_2+2,\, 1\leq i\leq m},$$

where $\mathcal{H}(M)$ is the space of harmonic vector fields on M. It is clear that F is a linear map. Moreover, if $X \in \ker F$, then every $\xi^X_{\lambda\mu}$, $\lambda < \mu$, is L^2-orthogonal to η_i, $1 \leq i \leq m$, and so $\mathcal{Q}(\xi^X_{\lambda\mu}) \geq 0$. In the first and second cases, it follows from (5.51) and (5.52), that $X = 0$. So $\ker F = \{0\}$ and the result follows.

In the third case, from (5.53), we have two possibilities: either $X = 0$ or $n_1 = 3$, $X_1 = 0$ and $C = -1$. If $X = 0$, then $\ker F = \{0\}$ and the result follows. If $n_1 = 3$, $X_1 = 0$ and $C = -1$, then Φ is the totally geodesic embedding of $\mathbb{S}^3 \times \mathbb{S}^1$ in $\mathbb{S}^3 \times \mathbb{S}^2$, whose index is one (see Example 2.53). Since in this case $\beta_1(M) = 1$, the result also follows. □

THEOREM 5.36 ([**Che21**]). *Let $\Phi : M^n \to \mathbb{S}^{n_1} \times \mathbb{S}^{n_2}$ be a two-sided minimal hypersurface of a compact manifold M into the product of two spheres with $n_1 \neq n_2$. Then,* Ind$(M) = 1$ *if and only if one of the following possibilities occurs:*

(1) *$n_2 \geq n_1 - 1$, $n_1 \geq 2$, and Φ is the totally geodesic embedding of $M = \mathbb{S}^{n_1-1} \times \mathbb{S}^{n_2}$ in $\mathbb{S}^{n_1} \times \mathbb{S}^{n_2}$ (Example* 2.52*).*
(2) *$n_1 \geq n_2 - 1$, $n_2 \geq 2$, and Φ is the totally geodesic embedding of $M = \mathbb{S}^{n_1} \times \mathbb{S}^{n_2-1}$ in $\mathbb{S}^{n_1} \times \mathbb{S}^{n_2}$ (Example* 2.53*).*

In view of Theorems 5.35 and 5.36, when the ambient manifold is $\mathbb{S}^2 \times \mathbb{S}^2$, the arguments used therein are not applicable. In this case it is reasonable to conjecture:

The only compact orientable minimal hypersurfaces of $\mathbb{S}^2 \times \mathbb{S}^2$ of index one are the totally geodesic ones: $\mathbb{S}^1 \times \mathbb{S}^2$ and $\mathbb{S}^2 \times \mathbb{S}^1$.

REMARK 5.37. Although H. Chen in [**Che21**] considers the case where the spheres have different radii, we only prove the special case where both spheres have the same radius.

PROOF. From Examples 2.52 and 2.53, the hypersurfaces given in 1 and 2 have index one.

Now, if $\Phi : M^n \to \mathbb{S}^{n_1} \times \mathbb{S}^{n_2}$ is a two-sided minimal hypersurface, from (2.17), the Jacobi operator $\mathcal{L} : C^\infty(M) \to C^\infty(M)$ is given by

$$\mathcal{L} = \Delta + \|A\|^2 + \overline{\mathrm{Ricc}}(N),$$

where N is a unit normal vector field to Φ. If $C = \langle PN, N\rangle$, then using (5.39), we get that

$$\mathcal{L} = \Delta + \|A\|^2 + \frac{(n_1-1)(1+C)}{2} + \frac{(n_2-1)(1-C)}{2}. \tag{5.54}$$

We consider the vector valued function $\Phi : M \to \mathbb{R}^{n_1+n_2+2}$ as test function. First we compute its Laplacian. It is clear that $\Delta\Phi = \sum_{i=1}^n \overline{\sigma}(e_i, e_i)$, where $\{e_1, \ldots, e_n\}$ is an orthonormal reference on M and $\overline{\sigma}$ is the second fundamental form of the embedding $\mathbb{S}^{n_1} \times \mathbb{S}^{n_2} \subset \mathbb{R}^{n_1+n_2+2}$. Using (5.38) and writing $\Phi = (\phi, \psi)$, we get that

$$\Delta\Phi = -\left(\frac{n + \operatorname{trace} B}{2}\phi, \frac{n - \operatorname{trace} B}{2}\psi\right),$$

where B is the matrix $B_{ij} = g((Pe_i)^\top, e_j)$.

As $\operatorname{trace} B = n_1 - n_2 - C$, using the above expression of $\Delta\Phi$ in (5.54) we finally get that

$$\mathcal{L}\Phi = \|A\|^2\Phi + \left(\frac{(-n_1+n_2-1)(1-C)}{2}\phi, \frac{(n_1-n_2-1)(1+C)}{2}\psi\right).$$

Now we take an arbitrary vector $a = (a_1, 0) \in \mathbb{R}^{n_1+1} \times \mathbb{R}^{n_2+1}$. Since $\langle \phi, a_1\rangle = \langle \Phi, a\rangle$, from the last equation we get

$$\mathcal{L}\langle \Phi, a\rangle = \left(\|A\|^2 + \frac{(-n_1+n_2-1)(1-C)}{2}\right)\langle \Phi, a\rangle, \quad a = (a_1, 0). \tag{5.55}$$

Similarly, if $a = (0, a_2) \in \mathbb{R}^{n_1+1} \times \mathbb{R}^{n_2+1}$, as $\langle \psi, a_2\rangle = \langle \Phi, a\rangle$, from the last equation we obtain

$$\mathcal{L}\langle \Phi, a\rangle = \left(\|A\|^2 + \frac{(n_1-n_2-1)(1+C)}{2}\right)\langle \Phi, a\rangle, \quad a = (0, a_2). \tag{5.56}$$

We will prove that if M is not totally geodesic, then $\operatorname{Ind}(M) \geq 2$.

First we assume that $n_1 > n_2$. Then $n_1 - n_2 - 1 \geq 0$ and since $1 + C \geq 0$, it follows from (5.56) that

$$\mathcal{Q}(\langle \Phi, a\rangle) \leq -\int_M \|A\|^2 \langle \Phi, a\rangle^2 \, dv \leq 0, \quad \text{for any} \quad a = (0, a_2) \in \mathbb{R}^{n_1+1} \times \mathbb{R}^{n_2+1}.$$

Let V be the linear subspace of $C^\infty(M)$ defined by $V = \{\langle \Phi, a\rangle \mid a = (0, a_2),\ a_2 \in \mathbb{R}^{n_2+1}\}$. If $\dim V \leq 1$, as $n_2 + 1 \geq 2$, there exists a non-null $a_2 \in \mathbb{R}^{n_2+1}$ such that $\langle \Phi, (0, a_2)\rangle = 0$. This implies that $\Phi : M \to \mathbb{S}^{n_1} \times \mathbb{S}^{n_2-1}$ is a local isometry and thus, M is totally geodesic, which is a contradiction. Hence, $\dim V \geq 2$ and $\mathcal{Q}(f) < 0$ for any $f \in V$. This means that $\operatorname{Ind}(M) \geq 2$.

If $n_2 > n_1$, use (5.55) and proceed in a similar way to get that $\operatorname{Ind}(M) \geq 2$.

So if $\operatorname{Ind}(M) = 1$, then M is totally geodesic. Therefore, if $\overline{\nabla}$ is the Levi-Civita connection of $\mathbb{S}^{n_1} \times \mathbb{S}^{n_2}$, then $\overline{\nabla}_v N = 0$ for any $v \in TM$ and hence the function $C = \langle PN, N\rangle$ is constant. We next prove that $C^2 = 1$. In fact, if $C^2 < 1$, then the tangent vector field $X = PN - CN$ has constant length $\sqrt{1 - C^2}$ and $\nabla_v X = 0$ for any $v \in TM$. So, $0 = \operatorname{Ricc}(X) = \overline{\operatorname{Ricc}}(X)$, and from (5.39), we get that

$$0 = (1 - C^2)\big((1 - C)n_1 + (1 + C)n_2\big),$$

which is impossible. This implies that either $C = 1$ or $C = -1$. In the first case, $N = (N, 0)$ and it is easy to see that either $\Phi(M) = \mathbb{S}^{n_1-1} \times \mathbb{S}^{n_2}$ with $n_1 \geq 2$, or $\Phi(M)$ is a slice $\{p\} \times S^{n_2}$ with $n_1 = 1$. In the second case, $N = (0, N)$ and either

$\Phi(M)=\mathbb{S}^{n_1}\times\mathbb{S}^{n_2-1}$ with $n_2\geq 2$ or $\Phi(M)$ is a slice $\mathbb{S}^{n_1}\times\{q\}$ with $n_2=1$. Now the result follows from Theorem 5.34 and Examples 2.52 and 2.53. □

In Theorem 5.35, we obtained a lower bound for the index of a compact minimal hypersurface of the product of two spheres as a function of the first Betti number of the submanifold, when the dimension of the hypersurface is at least three. To conclude this section, we will tackle the previous problem for surfaces.

LEMMA 5.38. *Let $\Phi: M\to\mathbb{S}^2\times\mathbb{S}^1$ be a minimal immersion of an orientable surface M. If ξ is the parallel unit vector field on $\mathbb{S}^2\times\mathbb{S}^1$ given by $\xi_{(p,q)}=(0,iq)$, where we consider $\mathbb{S}^1\subset\mathbb{C}\equiv\mathbb{R}^2$, then:*

(1) *If Φ is not the totally geodesic slice $\Phi(p)=(\phi(p),a)$, $\forall p\in M$, with $\phi: M\to\mathbb{S}^2$ a local isometry, then the tangential component $\xi^\top$ of ξ is a harmonic vector field on M with only a discrete number of zeros.*
(2) *If $(M,\tau)\to M_0$ is the orientable double covering of a nonorientable surface M_0 and Φ is the lift to M of a minimal immersion $\Phi_0: M_0\to\mathbb{S}^2\times\mathbb{S}^1$, then $\xi^\top\circ\tau=\xi^\top$ and $J\xi^\top\circ\tau=-J\xi^\top$, where J is the complex structure of M.*
(3) *If X is a harmonic vector field on M and $\mathcal{Q}$ is the index form of Φ, then*

$$\mathcal{Q}(X)=-2\int_M(1-\|\xi^\top\|^2)\langle X,\xi\rangle^2\,\mathrm{d}v.$$

PROOF. **Proof of 1.** Given that $\xi=\xi^\top+\xi^\perp$ is a parallel vector field, it is clear that for every v,w tangent to M we have that

$$g(\nabla_v\xi^\top,w)=\langle\sigma(v,w),\xi^\perp\rangle.$$

Since σ is a traceless symmetric two-tensor, the above equation implies that $\xi^\top$ is a harmonic vector field on M.

If $z=x+iy$ is a conformal parameter on M with induced metric $e^{2u}\|\mathrm{d}z\|^2$ and $\partial_z=(\partial_x-i\partial_y)/2$, $\partial_{\bar z}=(\partial_x+i\partial_y)/2$ are the corresponding complex operators, then the harmonicity of $\xi^\top$ implies that the 1-differential on M

$$\Theta(z)=\langle\Phi_z,\xi\rangle\otimes\mathrm{d}z$$

is holomorphic. Since $\|\Theta\|^2=e^{2u}\|\xi^\top\|^2/2$, we have that either $\xi^\top=0$ (i.e. $\Theta\equiv 0$) or $\{p\in M\mid \xi^\top(p)=0\}$ is isolated. In the first case, ξ is normal to Φ, and then Φ is a totally geodesic slice, i.e., there exists $a\in\mathbb{S}^1$ such that $\Phi(p)=(\phi(p),a)$. This proves 1.

Proof of 2. If $\Pi: M\to M_0$ is the projection, then $\Phi=\Phi_0\circ\Pi$ and so, $\Phi\circ\tau=\Phi$. This implies that $\xi^\top\circ\tau=\xi^\top$. Also, since $\mathrm{d}\tau\circ J=-J\circ\mathrm{d}\tau$, we have that $J\xi^\top\circ\tau=-J\xi^\top$. This proves 2.

Proof of 3. We consider $\mathbb{S}^2\times\mathbb{S}^1\subset\mathbb{R}^5$. If ∇^0 is the connection on $\mathbb{R}^5$ and $\overline{\sigma}$ is the second fundamental form of $\mathbb{S}^2\times\mathbb{S}^1$ in $\mathbb{R}^5$, then for any vector field V tangent to M we have

$$\nabla^0_V X=\nabla_V X+g(AV,X)N+\overline{\sigma}(V,X),$$

where N is a unit normal vector field to Φ. Hence,

$$\Delta^0 X=(K-\|A\|^2/2)X+2g(A,\nabla X)N$$
$$+\sum_{i=1}^2\{-\overline{A}_{\overline{\sigma}(e_i,X)}e_i+2\overline{\sigma}(e_i,\nabla_{e_i}X+g(AX,e_i)N)\},$$

where $\{e_1, e_2\}$ is an orthonormal reference on M and $\overline{A}$ is the Weingarten endomorphism of $\mathbb{S}^2 \times \mathbb{S}^1$ in $\mathbb{R}^5$. Thus, from (5.54) and the Gauss equation (2.2)

$$\langle \mathcal{L}X, X\rangle = \left(\overline{K} + \frac{1+C}{2}\right)\|X\|^2 - \sum_{i=1}^{2}\|\overline{\sigma}(e_i, X)\|^2, \tag{5.57}$$

where $\overline{K} = \overline{R}(e_1, e_2, e_2, e_1)$.

From the expression of the curvature $\overline{R}$ of $\mathbb{S}^2 \times \mathbb{S}^1$ given at the beginning of this section, it is easy to check that

$$\overline{K} = \frac{1}{2}\left(1 + B_{11}B_{22} - B_{12}^2\right) = \frac{1}{2}(1 - C),$$

where $B_{ij} = g((Pe_i)^\top, e_j)$, $1 \le i, j \le 2$.

It also follows from (5.38) that

$$\sum_{i=1}^{2}\|\overline{\sigma}(X, e_i)\|^2 = \|X\|^2 - 2(1 - \|\xi^\top\|^2)\langle X, \xi\rangle^2.$$

Finally, using the last two equations in (5.57) we prove 3. □

THEOREM 5.39 ([**Urb13**]). *Let $\Phi : M \to \mathbb{S}^2 \times \mathbb{S}^1$ be a minimal immersion of a complete surface M.*

(1) *If M is not compact, then* $\operatorname{Ind}(M) = \infty$.
(2) *If M is compact , then* $\operatorname{Ind}(M) \ge \frac{1}{5}(\beta_1(M) - 1)$,

where $\beta_1(M)$ is the first Betti number of M.

PROOF. **Proof of 1.** The proof will be similar to the proof of Theorem 5.5,1, with only a few differences. Since the scalar curvature of $\mathbb{S}^2 \times \mathbb{S}^1$ is $\rho = 2$, we only need to analyze the case where Φ is one-sided and the orientable double covering $\widetilde{M}$ of M is of infinite genus.

Let τ be the change of sheet in $\widetilde{M}$ and $\Pi : \widetilde{M} \to M$ be the projection. Then, $\widetilde{\Phi} = \Phi \circ \Pi$ is also a minimal immersion which is two-sided. If $\widetilde{\mathcal{L}} : C^\infty(\widetilde{M}) \to C^\infty(\widetilde{M})$ is the Jacobi operator of $\widetilde{\Phi}$, from Definition 2.31, we have that $\operatorname{Ind}(M) = \operatorname{Ind}(\widetilde{\mathcal{L}}^-)$, where $\widetilde{\mathcal{L}}^-$ is the restriction of $\widetilde{\mathcal{L}}$ to the space $C^\infty_-(\widetilde{M}) = \{f \in C^\infty(\widetilde{M}) \mid f \circ \tau = -f\}$.

To prove 1, we assume that $\operatorname{Ind}(\Phi) = \operatorname{Ind}(\widetilde{\mathcal{L}}^-) = k$ and we will find a contradiction.

From Theorem 2.30,2, there are $L^2(\widetilde{M})$-functions $\{v_1, \dots, v_k\}$ with $v_i \circ \tau = -v_i$ and $\widetilde{\mathcal{L}}v_i + \lambda_i v_i = 0$, $\lambda_i < 0$, such that, if $f \in C_0^\infty(\widetilde{M})$ satisfies $f \circ \tau = -f$ and is L^2-orthogonal to v_i, $1 \le i \le k$, then $\widetilde{\mathcal{Q}}(f) \ge 0$.

Since the genus of $\widetilde{M}$ is infinite, the space $\mathcal{H}(\widetilde{M}) \cap L^2(\widetilde{M}, T\widetilde{M})$ of L^2-harmonic vector fields on $\widetilde{M}$ has infinite dimension. Let V be an l-dimensional ($l > 5k$) subspace of $\mathcal{H}^-(\widetilde{M}) = \{X \in H(\widetilde{M}) \cap L^2(\widetilde{M}, T\widetilde{M}) \mid X \circ \tau = -X\}$. We will take V with another restriction. If the harmonic vector field $J\xi^\top$ (J is the complex structure of $\widetilde{M}$) satisfies $\int_{\widetilde{M}} \|J\xi^\top\|^2 d\tilde{v} < \infty$, i.e., $J\xi^\top \in \mathcal{H}^-(\widetilde{M})$, we take V being L^2-orthogonal to the line spanned by $J\xi^\top$. If $\int_{\widetilde{M}} \|J\xi^\top\|^2 d\tilde{v} = \infty$, we will impose no further restrictions on V.

Let $\{B_p(n) \mid n \in \mathbb{N}\}$ be an exhaustion of the complete surface $\widetilde{M}$ with p a point of $\widetilde{M}$ and $\{\varphi_n \mid n \in \mathbb{N}\}$ cut-off functions on $\widetilde{M}$ with $\varphi_n = 1$ on $B_p(n)$, $\operatorname{supp}(\varphi_n) \subset B_p(2n)$, $\|\widetilde{\nabla}\varphi_n\|^2 \le 1$ and $\varphi_n \circ \tau = \varphi_n$.

For each $n\in\mathbb{N}$, let $F_n : V\to\mathbb{R}^{5k}$ be the linear map given by

$$F_n(X)=\left(\int_{\widetilde{M}}\varphi_n v_1 X\,\mathrm{d}\tilde{v},\dots,\int_{\widetilde{M}}\varphi_n v_k X\,\mathrm{d}\tilde{v}\right).$$

Since $l>5k$, let $X_n\in\ker F_n$ with $\int_{\widetilde{M}}\|X_n\|^2\,\mathrm{d}\tilde{v}=1$. Thus, for any n, the function $\varphi_n X_n$ has compact support, is L^2-orthogonal to v_i, $1\le i\le k$ and, since $X_n\in\mathcal{H}^-(\widetilde{M})$, we have that $\varphi_n X_n\circ\tau=-\varphi_n X_n$. Hence,

$$\widetilde{\mathcal{Q}}(\varphi_n X_n)\ge 0,\quad \forall n\in\mathbb{N}.$$

Now,

$$\begin{aligned}\widetilde{\mathcal{Q}}(\varphi_n X_n)=&-\int_{\widetilde{M}}\varphi_n(\widetilde{\Delta}\varphi_n)\|X_n\|^2\,\mathrm{d}\tilde{v}-\frac{1}{2}\int_{\widetilde{M}}g(\widetilde{\nabla}\varphi_n^2,\widetilde{\nabla}\|X_n\|^2)\,\mathrm{d}\tilde{v}\\&-\int_{\widetilde{M}}\varphi_n^2\langle\widetilde{\mathcal{L}}X_n,X_n\rangle\,\mathrm{d}\tilde{v}.\end{aligned}$$

As $\operatorname{div}(\|X_n\|^2\widetilde{\nabla}\varphi_n^2)=g(\widetilde{\nabla}\varphi_n^2,\widetilde{\nabla}\|X_n\|^2)+2(\varphi_n\widetilde{\Delta}\varphi_n+\|\widetilde{\nabla}\varphi_n\|^2)\|X_n\|^2$, we finally obtain that

$$\int_{\widetilde{M}}\varphi_n^2\langle\widetilde{\mathcal{L}}X_n,X_n\rangle\,\mathrm{d}\tilde{v}\le\int_{\widetilde{M}}\|\widetilde{\nabla}\varphi_n\|^2\|X_n\|^2\,\mathrm{d}\tilde{v},\quad\forall n\in\mathbb{N}.$$

Using Lemma 5.38, we get that

$$\int_{\widetilde{M}}\|\widetilde{\nabla}\varphi_n\|^2\|X_n\|^2\,\mathrm{d}\tilde{v}\ge 2\int_{\widetilde{M}}\varphi_n^2(1-\|\xi^\top\|^2)\langle X_n,\xi\rangle^2\,\mathrm{d}\tilde{v}.\tag{5.58}$$

It is clear that the sequence of harmonic vector fields X_n, up to the extraction of a subsequence, converges in V to a harmonic vector field X with $\int_{\widetilde{M}}\|X\|^2\,\mathrm{d}\tilde{v}=1$. Furthermore, if $\{V^1,\dots,V^l\}$ is an L^2-orthonormal basis of V, then the function h on $\widetilde{M}$ defined by $h=k^2\max\{\tilde{g}(V_i,V_j)\,,\,1\le i,j\le l\}$ is integrable and $\|X_n\|^2\le h$ for any $n\in\mathbb{N}$. Thus, using the Theorem 1.7,2, we have that

$$\lim_{n\to\infty}\int_{\widetilde{M}}\|\widetilde{\nabla}\varphi_n\|^2\|X_n\|^2\,\mathrm{d}\tilde{v}=\int_{\widetilde{M}}\lim_{n\to\infty}\|\widetilde{\nabla}\varphi_n\|^2\|X_n\|^2\,\mathrm{d}\tilde{v}=0,$$

because $\|\widetilde{\nabla}\varphi_n\|^2\to 0$.

So, from (5.58) we get that

$$\begin{aligned}0=\lim_{n\to\infty}\int_{\widetilde{M}}\varphi_n^2(1-\|\xi^\top\|^2)\langle X_n,\xi\rangle^2\,\mathrm{d}\tilde{v}&=\int_{\widetilde{M}}(1-\|\xi^\top\|^2)\langle X,\xi\rangle^2\,\mathrm{d}\tilde{v}\\&=\int_{\widetilde{M}}(1-\|\xi^\top\|^2)\langle X,\xi^\top\rangle^2\,\mathrm{d}\tilde{v}.\end{aligned}$$

If $\widetilde{\Phi}$ is a slice, then $\widetilde{M}=\mathbb{S}^2$ which does not have infinite genus. Lemma 5.38 says that $\xi^\top$ has only a discrete number of zeros and the last formula says that either $\|\xi^\top\|^2=1$ or $\langle X,\xi\rangle=0$. In the first case, $\widetilde{M}$ is a covering of the totally geodesic torus $\mathbb{S}^1\times\mathbb{S}^1$, and so, its genus cannot be infinite. Thus, $\langle X,\xi\rangle=0$, and from Lemma 5.38, $X=\lambda J\xi^\top$ for some nonzero real number λ. Hence, $1=\lambda^2\int_{\widetilde{M}}\|J\xi^\top\|^2\,\mathrm{d}\tilde{v}$ and $J\xi^\top\in\mathcal{H}^-(\widetilde{M})$. Since in this case $J\xi^\top$ is orthogonal to V, we again get a contradiction. This concludes the proof of 1.

Proof of 2. First, we assume that Φ is two-sided, i.e., M is orientable. Then, if h is the genus of M, $\beta_1(M)=2h$ and we can assume that $h\ge 2$, because if $h=0$ the inequality in the theorem is irrelevant and if $h=1$ it is known, because from the Theorem 5.34 the only stable compact minimal surfaces of $\mathbb{S}^2 \times \mathbb{S}^1$ have genus zero.

If $m = \mathrm{Ind}(M)$, let $\{\varphi_1, \dots, \varphi_m\}$ be eigenfunctions of $\mathcal{L}$ corresponding to the negative eigenvalues $\{\lambda_1, \dots, \lambda_m\}$. Considering $\mathbb{S}^2 \times \mathbb{S}^1 \subset \mathbb{R}^5$ and $\{a_1, \dots, a_5\}$ an orthonormal basis of $\mathbb{R}^5$, we define the linear map $F : \mathcal{H}(M) \to \mathbb{R}^{5m}$ given by

$$F(X) = \left(\int_M \varphi_1 X \, dv, \dots, \int_M \varphi_m X \, dv \right),$$

where the harmonic vector field X is being considered as a $\mathbb{R}^5$-valued function.

Suppose $X \in \ker F$. Then X is L^2-orthogonal to every φ_i, $1 \leq i \leq m$, and hence $\mathcal{Q}(X) \geq 0$. From Lemma 5.38 we get

$$0 \leq \mathcal{Q}(X) = -\int_M \langle X, \xi \rangle^2 (1 - \|\xi^\top\|^2) \, dv \leq 0.$$

Considering that $h \geq 2$ and arguing as in the proof of 1, we deduce that $X = \lambda J \xi^\top$, with $\lambda \in \mathbb{R}$, that is, $\dim \ker F \leq 1$. Since $2h = \dim \mathcal{H}(M) = \dim \ker F + \dim \mathrm{img}\, F \leq 1 + 5m$, we get the result.

If Φ is one-sided, i.e., M is not orientable, let $\Pi : \widetilde{M} \to M$ be the orientable double covering of M and τ be the change of sheet in $\widetilde{M}$. Using the same argument as in the previous results, the index of M is the index of the Jacobi operator $\widetilde{\mathcal{L}}$ of the minimal immersion $\widetilde{\Phi} = \Phi \circ \Pi$ acting on $C^\infty_-(\widetilde{M}) = \{f \in C^\infty(\widetilde{M}) \mid f \circ \tau = -f\}$. Let $\{\varphi_1, \dots, \varphi_m\}$ be eigenfunctions of $\widetilde{\mathcal{L}}$, with $\varphi_i \circ \tau = -\varphi_i$, corresponding to the negative eigenvalues $\lambda_1, \dots, \lambda_m$, where m is the index of Φ. In this case, we define the linear map $F : \widetilde{\mathcal{H}}^-(\widetilde{M}) = \{X \in \widetilde{\mathcal{H}}(M) \mid X \circ \tau = -X\} \to \mathbb{R}^{5m}$ given by

$$F(X) = \left(\int_{\widetilde{M}} \varphi_1 X \, d\tilde{v}, \dots, \int_{\widetilde{M}} \varphi_m X \, d\tilde{v} \right),$$

where X is considered to be a $\mathbb{R}^5$-valued function. Following the same idea as above, and considering that if h is the genus of $\widetilde{M}$ then $\dim \mathcal{H}^-(\widetilde{M}) = h = \beta_1(M)$, we prove 2. $\square$

5.6. Minimal submanifolds of the elliptic Berger spheres

At the end of Section 2.9 we introduced the elliptic Berger spheres $\mathbb{S}^{2m+1}_\tau$ with $\tau \in (0, 1]$ and proved that, for certain values of τ, $\mathbb{S}^{2m+1}_\tau$ admits stable compact minimal submanifolds (Example 2.63). In this section we prove two results on the stability of minimal submanifolds of $\mathbb{S}^{2m+1}_\tau$ which characterize a larger family of stable examples than the one studied in Example 2.63.

Following the notation of Section 2.9, the Berger metric on $\mathbb{S}^{2m+1}_\tau$ is denoted by $\langle , \rangle$ and the standard metric on $\mathbb{S}^{2m+1}$ is denoted by g. If $\Phi : M^d \to \mathbb{S}^{2m+1}_\tau \subset \mathbb{C}^{n+1}$, $\tau \in (0, 1]$, is a minimal immersion of a compact manifold M^d, the induced metric on M will also be denoted by $\langle , \rangle$. Using Definition 2.10 and (2.35), the Jacobi operator $\mathcal{L}$ of Φ is

(5.59)

$$\mathcal{L}\eta = \Delta^\perp \eta + \mathcal{A}\eta + (d - (1-\tau^2)\|\xi^\top\|^2)\eta - d(1-\tau^2)\langle \eta, \xi \rangle \xi^\perp - 3(1-\tau^2) \left[J(J\eta)^\top \right]^\perp,$$

where $\top$ and $\perp$ denote the tangential and normal components to M, respectively, $\xi \in \mathfrak{X}(\mathbb{S}^{2m+1}_\tau)$ is the vector field $\xi_z = \frac{1}{\tau} J z$ for any $z \in \mathbb{S}^{2m+1}_\tau$ and J is the complex structure of $\mathbb{C}^{n+1}$.

If M is a two-sided hypersurface ($d = 2m$), (2.17) transforms the Jacobi operator into the following operator acting on $C^\infty(M)$:

$$\mathcal{L} = \Delta + \|A\|^2 + 2m - 2(1-\tau^2)\big((m+1)\nu^2 - 1\big),$$

where Δ is the Laplacian of M, N is a unit normal vector field to Φ, and $\nu = \langle \xi, N\rangle$ is the so-called angle function.

The first property about the index of submanifolds of $\mathbb{S}^{2m+1}$ is the following:

PROPOSITION 5.40. *Every two-sided compact minimal hypersurface of $\mathbb{S}^{2m+1}_\tau$ is unstable and the first eigenvalue of $\mathcal{L}$ is $\lambda_1 \leq -2m\tau^2$.*

PROOF. Considering the constant function 1 as a test function, we have

$$\begin{aligned}\mathcal{Q}(1) &= -\int_M \|A\|^2\,\mathrm{d}v - \int_M 2(m+1-\tau^2)\,\mathrm{d}v + \int_M 2(m+1)(1-\tau^2)\nu^2\,\mathrm{d}v\\ &\leq -\int_M 2(m+1-\tau^2)\,\mathrm{d}v + \int_M 2(m+1)(1-\tau^2)\,\mathrm{d}v = -\int_M 2m\tau^2\,\mathrm{d}v,\end{aligned}$$

and the result follows. □

Next, we will define an important family of minimal submanifolds of $\mathbb{S}^{2m+1}_\tau$ which are a generalization of the example given in Example 2.63. Their behavior with respect to the index will be very similar to that of Example 2.63, providing new examples of stable minimal submanifolds of $\mathbb{S}^{2m+1}_\tau$. We observe that the embedding $\Phi : \mathbb{S}^{2n+1}_\tau \to \mathbb{S}^{2m+1}_\tau$ given in Example 2.63 makes the following diagram commutative:

$$\begin{array}{ccc}\mathbb{S}^{2n+1}_\tau & \xrightarrow{\ \Phi\ } & \mathbb{S}^{2m+1}_\tau\\ \big\downarrow{\scriptstyle\widetilde{\Pi}} & & \big\downarrow{\scriptstyle\Pi}\\ \mathbb{CP}^n & \xrightarrow{\ \Psi\ } & \mathbb{CP}^m,\end{array}$$

where Π and $\widetilde{\Pi}$ are the Hopf fibrations and Ψ is the totally geodesic complex embedding $\Psi([z]) = [(z,0)]$ (Example 2.60). Furthermore, $\mathbb{S}^{2n+1}_\tau$ is the induced bundle on $\mathbb{CP}^n$ of the Hopf bundle Π.

Instead of the totally geodesic complex embedding Ψ, we consider any complex immersion $\Psi : N^{2n} \to \mathbb{CP}^m$ of a complex manifold N. We define

$$M^{2n+1} = \{(p,x) \in N \times \mathbb{S}^{2m+1}_\tau \mid \Psi(p) = \Pi(x)\},$$

where $\Pi : \mathbb{S}^{2m+1}_\tau \to \mathbb{CP}^n$ is the Hopf fibration. We define $\widetilde{\Pi} : M \to N$ by $\widetilde{\Pi}(p,x) = p$ and $\Phi : M \to \mathbb{S}^{2m+1}_\tau$ by $\Phi(p,x) = x$. Then, the following diagram is commutative:

$$\begin{array}{ccc}M^{2n+1} & \xrightarrow{\ \Phi\ } & \mathbb{S}^{2m+1}_\tau\\ \big\downarrow{\scriptstyle\widetilde{\Pi}} & & \big\downarrow{\scriptstyle\Pi}\\ N^{2n} & \xrightarrow{\ \Psi\ } & \mathbb{CP}^m\end{array}$$

and M^{2n+1} is a $\mathbb{S}^1$-bundle over N^{2n}, where the action $\mathbb{S}^1 \times M \to M$ is given by $z \cdot (p,x) = (p, zx)$ for any $z \in \mathbb{S}^1$ and $(p,x) \in M$. Now, since ξ is tangent to M, i.e. $\xi^\perp = 0$, and Ψ is a minimal immersion, it is easy to check that Φ is also a minimal immersion, which will be called the **induced bundle by the Hopf fibration over the complex immersion** $\Psi : N^{2n} \to \mathbb{CP}^m$. Note that Φ is an embedding if and only if Ψ is an embedding.

Now we show some interesting properties of this kind of minimal submanifolds of Berger spheres. First, since $\xi^\perp = 0$, we can decompose orthogonally $TM = \mathcal{D} \oplus \langle \xi \rangle$. As Ψ is a complex immersion, $\mathcal{D}$ and the normal bundle $T^\perp M$ are invariant by the complex structure J of $\mathbb{CP}^{n+1}(4(1-\tau^2))$. Thus, using the expression of the Jacobi operator $\mathcal{L}$ given above, we obtain that

$$\mathcal{L}\eta = \Delta^\perp \eta + \mathcal{A}\eta + (2n + \tau^2)\eta. \tag{5.60}$$

Moreover, since $TM = \mathcal{D} \oplus \langle \xi \rangle$, then from (2.37) and the fact that $\mathcal{D}$ is invariant by J, we get

$$\nabla_\xi \xi = 0, \quad \nabla_X \xi = \tau JX, \quad \sigma(\xi, \xi) = 0, \quad \sigma(X, \xi) = 0, \tag{5.61}$$

for any vector field $X \in \Gamma(\mathcal{D})$. If $\overline{D}$ is the Levi-Civita connection of $\mathbb{CP}^{m+1}(4(1-\tau^2))$, using the Gauss and Weingarten formulas (2.1) together with (2.34) and (5.61) in the equation $\overline{D}J = 0$, it is easy to prove that

$$\begin{aligned} \nabla_X JY = J(\nabla_X Y + \tau\langle JX, Y\rangle\xi) - \tau\langle X, Y\rangle\xi, &\qquad \sigma(X, JY) = J\sigma(X, Y) \\ A_{J\eta}X = JA_\eta X = -A_\eta JX, &\qquad \nabla^\perp_X J\eta = J\nabla^\perp_X \eta, \end{aligned} \tag{5.62}$$

for all $X, Y \in \Gamma(\mathcal{D})$ and all $\eta \in \mathfrak{X}^\perp(M)$.

Using (5.62) we easily get that $\Delta^\perp J = J\Delta^\perp$ and $\mathcal{A}J = J\mathcal{A}$, and so, $\mathcal{L}J = J\mathcal{L}$. This implies that the eigenspace associated to every eigenvalue of $\mathcal{L}$ is invariant under J and thus, of even dimension. In particular, the index and nullity of any bundle induced by the Hopf fibration over a complex immersion $\Psi : N^{2n} \to \mathbb{CP}^m$ is even.

THEOREM 5.41 ([**TU22**]). *Let $\Phi : M^{2n+1} \to \mathbb{S}^{2m+1}_\tau$ be an induced bundle by the Hopf fibration over a complex immersion $\Psi : N^{2n} \to \mathbb{CP}^m$, of a compact manifold N. If $0 < \tau^2 \le \frac{1}{2n+2}$, then Φ is stable.*

PROOF. Using the above expression of the Jacobi operator of Φ, the index form $\mathcal{Q}$ is

$$\mathcal{Q}(\eta) = \int_M \{\|\nabla^\perp \eta\|^2 - \langle \mathcal{A}\eta, \eta\rangle - (2n + \tau^2)\|\eta\|^2\}\, dv, \qquad \forall \eta \in \mathfrak{X}^\perp(M). \tag{5.63}$$

We consider, for any section $X \in \Gamma(\mathcal{D})$, the operator $D_X : \mathfrak{X}^\perp(M) \to \mathfrak{X}^\perp(M)$ given by

$$D_X \eta = \nabla^\perp_{JX}\eta - J\nabla^\perp_X \eta.$$

Let $\{e_1, \dots, e_n, Je_1, \dots, Je_n, \xi\}$ be a local orthonormal reference on M. Then,

$$\begin{aligned} \sum_{i=1}^n \|D_{e_i}\eta\|^2 &= \sum_{i=1}^n \{\|\nabla^\perp_{Je_i}\eta\|^2 + \|J\nabla^\perp_{e_i}\eta\|^2 - 2\langle \nabla^\perp_{Je_i}\eta, J\nabla^\perp_{e_i}\eta\rangle\} \\ &= \|\nabla^\perp \eta\|^2 - \|\nabla^\perp_\xi \eta\|^2 - 2\sum_{i=1}^n \langle \nabla^\perp_{Je_i}\eta, J\nabla^\perp_{e_i}\eta\rangle. \end{aligned}$$

We consider, for each $\eta \in \mathfrak{X}^\perp(M)$, the 1-form on M defined by

$$\alpha(u) = \langle \nabla^\perp_{(Ju)^\top}\eta, J\eta\rangle, \qquad \forall u \in TM,$$

where $\top$ is the tangential component to M. Then, considering (5.61) and (5.62), its codifferential is given by

$$\begin{aligned}\delta\alpha &= \sum_{i=1}^n\{e_i(\alpha(e_i)) - \alpha(\nabla_{e_i}e_i) + (Je_i)(\alpha(Je_i)) - \alpha(\nabla_{Je_i}Je_i)\} + \xi(\alpha(\xi)) - \alpha(\nabla_\xi\xi)\\
&= \sum_{i=1}^n\{\langle\nabla^\perp_{e_i}\nabla^\perp_{Je_i}\eta, J\eta\rangle - \langle\nabla^\perp_{Je_i}\nabla^\perp_{e_i}\eta, J\eta\rangle\}\\
&\quad + 2\sum_{i=1}^n\{\langle\nabla^\perp_{Je_i}\eta, J\nabla^\perp_{e_i}\eta\rangle - \alpha(\nabla_{e_i}e_i) - \alpha(\nabla_{Je_i}Je_i)\}\\
&= \sum_{i=1}^n\{R^\perp(e_i, Je_i, \eta, J\eta) + \langle\nabla^\perp_{[e_i,Je_i]}\eta, J\eta\rangle\}\\
&\quad + 2\sum_{i=1}^n\{\langle\nabla^\perp_{Je_i}\eta, J\nabla^\perp_{e_i}\eta\rangle - \alpha(\nabla_{e_i}e_i) - \alpha(\nabla_{Je_i}Je_i)\},\end{aligned}$$

where $R^\perp$ is the normal curvature of the immersion Φ. Now, using again (5.62), we easily get that

$$\sum_{i=1}^n\langle\nabla^\perp_{[e_i,Je_i]}\eta, J\eta\rangle - \alpha(\nabla_{e_i}e_i) - \alpha(\nabla_{Je_i}Je_i) = -2\tau n\langle\nabla^\perp_\xi\eta, J\eta\rangle.$$

Furthermore, from the Ricci equation (2.2) and using (2.35) and (5.62),

$$\begin{aligned}\sum_{i=1}^n R^\perp(e_i, Je_i, \eta, J\eta) &= \sum_{i=1}^n \overline{R}(e_i, Je_i, \eta, J\eta) + \langle[A_\eta, A_{J\eta}]e_i, Je_i\rangle\\
&= -2n(1-\tau^2)\|\eta\|^2 - \langle\mathcal{A}\eta, \eta\rangle.\end{aligned}$$

As a result of all the above calculations, we get that

$$\sum_{i=1}^n\|D_{e_i}\eta\|^2 = \|\nabla^\perp\eta\|^2 - \langle\mathcal{A}\eta,\eta\rangle - \|\nabla^\perp_\xi\eta\|^2 - 2n(1-\tau^2)\|\eta\|^2 - 2\tau n\langle\nabla^\perp_\xi\eta, J\eta\rangle - \delta\alpha.$$

Using the last expression in (5.63), the quadratic form $\mathcal{Q}$ over any normal vector field η can be written as

(5.64)

$$\begin{aligned}\mathcal{Q}(\eta) &= \int_M\left(\sum_{i=1}^n\|D_{e_i}\eta\|^2 + \|\nabla^\perp_\xi\eta\|^2 - \tau^2(2n+1)\|\eta\|^2 + 2\tau n\langle\nabla^\perp_\xi\eta, J\eta\rangle\right)dv\\
&\geq \int_M\left(\|\nabla^\perp_\xi\eta\|^2 - \tau^2(2n+1)\|\eta\|^2 + 2\tau n\langle\nabla^\perp_\xi\eta, J\eta\rangle\right)dv,\end{aligned}$$

where we have used that $\|D_{e_i}\eta\|^2 \geq 0$.

Now, to get a lower bound of (5.64), we change the Berger metric $\langle,\rangle$ in $\mathbb{S}^{2m+1}_\tau$ by the standard one g. It is clear that the volume forms dv of $(M, \Phi^*\langle,\rangle)$ and dv_g of (M, Φ^*g) are related by $dv = \tau dv_g$. Also, thanks to (2.36) and (5.61), we obtain

$$\text{(5.65)}\qquad \nabla^\perp_\xi\eta = \tfrac{1}{\tau}\left({}^g\nabla^\perp_V\eta - (1-\tau^2)J\eta\right), \qquad A^g_\eta V = 0,$$

where A^g is the Weingarten operator of $\Phi: M\to\mathbb{S}^{2m+1}$, ${}^g\nabla^\perp$ its normal connection, and V is the tangent vector field on M defined by $V_p = Jp$, $\forall p\in M$. Thus, since $\xi_z = \frac{1}{\tau} Jz$, by the previous formula the inequality in (5.64) becomes

$$\mathcal{Q}(\eta) \geq \tfrac{1}{\tau} \int_M \left[\| {}^g\nabla^\perp_V \eta\|_g^2 + (\tau^2(2n+2)-2) g({}^g\nabla^\perp_V \eta, J\eta) + (1-\tau^2(2n+2))\|\eta\|_g^2\right] dv_g.$$

CLAIM. The operator $G : \mathfrak{X}^\perp(M) \to \mathfrak{X}^\perp(M)$ given by

$$G\eta = {}^g\nabla^\perp_V {}^g\nabla^\perp_V \eta - {}^g\nabla^\perp_{\nabla_V V}\eta = {}^g\nabla^\perp_V {}^g\nabla^\perp_V \eta,$$

is a self-adjoint operator with spectrum $\{k^2 \mid k \in \mathbb{Z},\, k \geq 0\}$.

We postpone the proof of the claim to prove that Φ is stable. We can write $\eta = \sum_{k\geq 0} \eta_k$, where η_k an eigensection of G associated to the eigenvalue k^2, i.e., $G\eta_k + k^2\eta_k = 0$. In particular η_0 satisfies $\nabla^\perp_V \eta_0 = 0$. So, from the last expression for the quadratic form Q and, as $G({}^g\nabla^\perp_V \eta_k) + k^2\, {}^g\nabla^\perp_V \eta_k = 0$, we get

(5.66)

$$\begin{aligned}
\mathcal{Q}(\eta) &\geq \tfrac{1}{\tau} \int_M (1-\tau^2(2n+2))\|\eta_0\|_g^2\, dv_g \\
&\quad + \tfrac{1}{\tau}\sum_{k\geq 1} \int_M \left[\| {}^g\nabla^\perp_V \eta_k\|_g^2 + (\tau^2(2n+2)-2) g({}^g\nabla^\perp_V \eta_k, J\eta_k)\right. \\
&\quad \left. + (1-\tau^2(2n+2))\|\eta_k\|_g^2\right] dv_g \\
&\geq \tfrac{1}{\tau}\sum_{k\geq 1} \int_M \left[(k^2+1-\tau^2(2n+2))\, \|\eta_k\|_g^2 + (\tau^2(2n+2)-2) g({}^g\nabla^\perp_V \eta_k, J\eta_k)\right] dv_g,
\end{aligned}$$

where we have used that $1-\tau^2(2n+2) \geq 0$ and

$$\int_M \| {}^g\nabla^\perp_V \eta_k\|_g^2\, dv_g = -\int_M g(\eta_k, G\eta_k)\, dv_g = k^2 \int_M \|\eta_k\|^2\, dv_g.$$

Now, for any $k \geq 1$, we have that

$$0 \leq \frac{1}{2}\int_M \| {}^g\nabla^\perp_V \eta_k - kJ\eta_k\|_g^2\, dv_g = k^2 \int_M \|\eta_k\|_g^2\, dv_g - k\int_M g({}^g\nabla^\perp_V \eta_k, J\eta_k)\, dv_g.$$

Since $\tau^2(2n+2)-2 < 0$, using the above inequality in (5.66) we obtain

$$Q(\eta) \geq \tfrac{1}{\tau} \sum_{k\geq 1} (k-1)\left[(k-1) + 2\tau^2(n+1)\right] \int_M \|\eta_k\|_g^2\, dv_g \geq 0.$$

So, Φ is stable.

PROOF OF THE CLAIM. First, since $\operatorname{div}_g V = 0$,

$$\begin{aligned}
0 &= \int_M \operatorname{div}_g(g({}^g\nabla^\perp_V \eta, \zeta)V)\, dv_g = \int_M \left[g({}^g\nabla^\perp_V {}^g\nabla^\perp_V \eta, \zeta) + g({}^g\nabla^\perp_V \eta, {}^g\nabla^\perp_V \zeta)\right] dv_g, \\
&= \int_M g(G\eta, \zeta) + g({}^g\nabla^\perp_V \eta, {}^g\nabla^\perp_V \zeta)\, dv_g,
\end{aligned}$$

and so, $\int_M g(G\eta, \zeta)\, dv_g = \int_M g(\eta, G\zeta)\, dv_g$, which proves that G is self-adjoint.

Moreover, G is the vertical normal Laplacian with respect to the fibration $\widetilde{\Pi} : M^{2n+1} \to N^{2n}$, because if $p \in M$ and F_p is the fiber through p, then

$$(G\eta)_{|F_p} = \Delta^\perp_p(\eta|F_p),$$

where $\Delta^\perp_p$ is the normal Laplacian of the totally geodesic immersion $F_p \subset M^{2n+1} \to \mathbb{S}^{2m+1}$. This immersion is congruent to the immersion : $\nu : \mathbb{S}^1 \to \mathbb{S}^{2m+1}$ given by

$z\mapsto(z,0\dots,0)$, and so, the eigenvalues of the normal Laplacian $\Delta_p^\perp$ will be those of the normal Laplacian of ν. If $\Delta_\nu^\perp$ is the normal Laplacian of the immersion ν, $\{e_1,\dots,e_m,Je_1,\dots,Je_m\}$ is a global orthonormal reference of the normal bundle, and $\varphi=\sum_{i=1}^m\{f_ie_i+g_iJe_i\}$ is a normal section, then

$$\Delta_\nu^\perp\varphi=\sum_{i=1}^m\{(\Delta f_i)e_i+(\Delta g_i)Je_i\}.$$

Thus, the eigenvalues of $\Delta_\nu^\perp$ will be those of the Laplacian of $\mathbb{S}^1$ endowed with the metric induced by ν, i.e., the set $\{k^2\mid k\in\mathbb{Z},k\geq 0\}$. □

Since every compact minimal submanifold of the sphere $(\mathbb{S}^{2m+1},g)$ is unstable, every compact minimal submanifold of $\mathbb{S}^{2m+1}_\tau$ is also unstable for τ close to 1. In the following result we get the first value of τ for which the unstability disappears. This value is $\tau^2=\frac{1}{d+1}$, where d is the dimension of the submanifold and for this τ, we classify the stable compact minimal embedded submanifolds.

THEOREM 5.42 ([**TU22**]). *Let $\Phi:M^d\to\mathbb{S}^{2m+1}_\tau$ be a minimal embedding of a compact d-manifold M in the Berger sphere $\mathbb{S}^{2m+1}_\tau$. If $\frac{1}{d+1}\leq\tau^2<1$, then Φ is stable if and only if $\tau^2=\frac{1}{d+1}$, $d=2n+1$ and M is the induced bundle by the Hopf fibration over a compact complex embedded submanifold $\Psi:N^{2n}\to\mathbb{CP}^m$.*

PROOF. The sufficient condition was proved in Theorem 5.41.

To prove the necessary condition, we begin with the following property:

CLAIM 1. If $\frac{1}{d+1}\leq\tau^2\leq 1$ and Φ is stable then $\tau^2=\frac{1}{d+1}$ and $\xi^\perp=0$.

PROOF OF CLAIM 1. First, we define certain vector fields on $\mathbb{S}^{2m+1}_\tau$, whose normal components will be test sections for the quadratic form $\mathcal{Q}$. To do this, for each $a\in\mathbb{C}^{m+1}$, let $X_a\in\mathfrak{X}(\mathbb{S}^{2m+1})=\mathfrak{X}(\mathbb{S}^{2m+1}_\tau)$ be the vector field given by

$$(X_a)_p=a-g(a,p)p,\quad\text{for all }p\in\mathbb{S}^{2m+1}.$$

If $\widetilde{\nabla}$ is the Levi-Civita connection of the Euclidean metric in $\mathbb{C}^{m+1}$, then for any $u\in T_p\mathbb{S}^{2m+1}$ we get

$$0=\widetilde{\nabla}_ua=\widetilde{\nabla}_u(X_a+g(a,p)p)={}^g\overline{\nabla}_uX_a+g(a,p)u,$$

where we recall that ${}^g\overline{\nabla}$ is the Levi-Civita connection of $(\mathbb{S}^{2m+1},g)$. So, ${}^g\overline{\nabla}_uX_a=-f_au$, where $f_a:\mathbb{S}^{2m+1}\to\mathbb{R}$ is given by $f_a(p)=g(a,p)$.

Now, using the relation between the Levi-Civita connections $\overline{\nabla}$ of $\mathbb{S}^{2m+1}_\tau$ and ${}^g\overline{\nabla}$ given in (2.36), we get the following behavior of the vector field X_a with respect to $\overline{\nabla}$:

$$\langle\overline{\nabla}_uX_a,v\rangle=-f_a\langle u,v\rangle-\tfrac{1-\tau^2}{\tau}\left(\langle X_a,\xi\rangle\langle Ju,v\rangle+\langle u,\xi\rangle\langle JX_a,v\rangle\right).\tag{5.67}$$

for any $u,v\in T_p\mathbb{S}^{2m+1}_\tau$.

In this situation, given the immersion $\Phi:M^d\to\mathbb{S}^{2m+1}_\tau$, we decompose $X_a\in\mathfrak{X}(\mathbb{S}^{2m+1}_\tau)$ into its tangent a normal component of Φ, i.e. $X_a=X_a^\top+X_a^\perp$. Our goal is to compute the Jacobi operator acting on $X_a^\perp$.

From (5.67) we deduce that, for any $u\in T_pM$,

$$\nabla_uX_a^\top=-f_au+A_{X_a^\perp}u-\tfrac{1-\tau^2}{\tau}\left[\langle X_a,\xi\rangle(Ju)^\top+\langle u,\xi\rangle(JX_a)^\top\right],\tag{5.68}$$

$$\nabla_u^\perp X_a^\perp=-\sigma(u,X_a^\top)-\tfrac{1-\tau^2}{\tau}\left[\langle X_a,\xi\rangle(Ju)^\perp+\langle u,\xi\rangle(JX_a)^\perp\right].\tag{5.69}$$

We do the computation of $\mathcal{L}X_a^\perp$ at a point $p \in M$ and we use an orthonormal tangent reference $\{e_1, \ldots, e_d\}$ to M which satisfies $(\nabla_{e_k} e_j)_p = 0$. Thus, taking normal derivatives with respect to e_i in (5.69) and using the Codazzi equation we deduce

$$\begin{aligned}\nabla^\perp_{e_i}\nabla^\perp_{e_i}X_a^\perp &= (\overline{R}(X_a^\top, e_i)e_i)^\perp - \sigma(e_i, \nabla_{e_i}X_a^\top) \\ &\quad - \tfrac{1-\tau^2}{\tau}\nabla^\perp_{e_i}\left(\langle X_a, \xi\rangle (Je_i)^\perp + \langle e_i, \xi\rangle (JX_a)^\perp\right).\end{aligned}$$

But, using (5.68), the minimality assumption and that $\sum_{i=1}^d \sigma(e_i, (Je_i)^\top) = 0$ because of the skew-symmetry of J, we get

$$\sum_{i=1}^d \sigma(e_i, \nabla_{e_i}X_a^\top) = \mathcal{A}X_a^\perp - \tfrac{1-\tau^2}{\tau}\sigma(\xi^\top, (JX_a)^\top).$$

As a result,

$$\begin{aligned}\Delta^\perp X_a^\perp &= \sum_{i=1}^d (\overline{R}(X_a^\top, e_i)e_i)^\perp - \mathcal{A}X_a^\perp + \tfrac{1-\tau^2}{\tau}\sigma(\xi^\top, (JX_a)^\top) \\ &\quad - \tfrac{1-\tau^2}{\tau}\sum_{i=1}^d \nabla^\perp_{e_i}\left(\langle X_a, \xi\rangle (Je_i)^\perp + \langle e_i, \xi\rangle (JX_a)^\perp\right).\end{aligned} \tag{5.70}$$

Now we calculate the final sum. First, using the minimality assumption, (2.37) and (5.67), we get

$$e_i(\langle X_a, \xi\rangle) = -f_a\langle e_i, \xi\rangle - \tau\langle JX_a, e_i\rangle \quad \text{and} \quad \sum_{i=1}^d e_i(\langle e_i, \xi\rangle) = 0. \tag{5.71}$$

As a direct consequence of the last equation,

$$\begin{aligned}\sum_{i=1}^d \nabla^\perp_{e_i}\left(\langle X_a, \xi\rangle (Je_i)^\perp + \langle e_i, \xi\rangle (JX_a)^\perp\right) &= -f_a(J\xi^\top)^\perp - \tau[J(JX_a)^\top]^\perp \\ &\quad + \sum_{i=1}^d \left(\langle X_a, \xi\rangle \nabla^\perp_{e_i}(Je_i)^\perp + \langle e_i, \xi\rangle \nabla^\perp_{e_i}(JX_a)^\perp\right).\end{aligned}$$

Now, to compute the last term of the above formula, we will get a general expression for $\nabla^\perp_u(JY)^\perp$, where Y is any vector field in $\mathbb{S}^{2m+1}_\tau$ and $u \in T_pM$. To do this, we will consider $\mathbb{S}^{2m+1}_\tau$ isometrically embedded in $\mathbb{CP}^{m+1}(4(1-\tau^2))$. On the one hand, decomposing $JY \in \mathfrak{X}(\mathbb{CP}^{m+1}(4(1-\tau^2)))$ as $JY = (JY)^\top + (JY)^\perp + \langle Y, \xi\rangle J\xi$ and using (2.34) and (2.37) we deduce

$$\begin{aligned}(\overline{D}_u JY)^\perp &= (\overline{D}_u (JY)^\top)^\perp + (\overline{D}_u (JY)^\perp)^\perp + \langle Y, \xi\rangle (\overline{D}_u J\xi)^\perp \\ &= \sigma(u, (JY)^\top) + \nabla^\perp_u (JY)^\perp + \tfrac{1-\tau^2}{\tau}\langle Y, \xi\rangle\langle u, \xi\rangle \xi^\perp,\end{aligned}$$

where $\overline{D}$ is the Levi-Civita connection of $\mathbb{CP}^{n+1}(4(1-\tau^2))$.

On the other hand, using (2.34)

$$(\overline{D}_u JY)^\perp = (J\overline{D}_u Y)^\perp = (J\overline{D}_u Y)^\perp - \left(\tau\langle u, Y\rangle - \tfrac{1-\tau^2}{\tau}\langle u, \xi\rangle\langle Y, \xi\rangle\right)\xi^\perp.$$

So from both expressions of $(\overline{D}_u JY)^\perp$, we get

$$\nabla^\perp_u (JY)^\perp = (J\overline{\nabla}_u Y)^\perp - \sigma(u, (JY)^\top) - \tau\langle u, Y\rangle \xi^\perp. \tag{5.72}$$

As a consequence, taking into account the minimality of Φ and (5.72), we get

$$\sum_{i=1}^{d}\nabla^\perp_{e_i}(Je_i)^\perp = -\tau d\xi^\perp, \tag{5.73}$$

and also, by (5.67) and (5.72),

$$\begin{aligned}\sum_{i=1}^{d}\langle e_i,\xi\rangle\nabla^\perp_{e_i}(JX_a)^\perp = &- f_a(J\xi^\top)^\perp - 2\tfrac{1-\tau^2}{\tau}\langle X_a,\xi\rangle\|\xi^\top\|^2\xi^\perp + \tfrac{1-\tau^2}{\tau}\|\xi^\top\|^2X_a^\perp\\ &-\tau\langle X_a,\xi^\top\rangle\xi^\perp - \sigma(\xi^\top,(JX_a)^\top).\end{aligned}$$

Finally, using (2.35) we get

$$\sum_{i=1}^{d}(\overline{R}(X_a^\top,e_i)e_i)^\perp = -3(1-\tau^2)[J(JX_a^\top)^\top]^\perp + (1-\tau^2)(1-d)\langle X_a^\top,\xi\rangle\xi^\perp. \tag{5.74}$$

Plugging the above calculations in (5.70) and considering the expression of the Jacobi operator given in (5.59), we finally get

$$\begin{aligned}\mathcal{L}X_a^\perp = &\left(d - \tfrac{1-\tau^2}{\tau^2}\|\xi^\top\|^2\right)X_a^\perp + 2\tfrac{1-\tau^2}{\tau}\sigma(\xi^\top,(JX_a)^\top) + 2\tfrac{1-\tau^2}{\tau}f_a(J\xi^\top)^\perp\\ &-2(1-\tau^2)[J(JX_a)^\top]^\perp + 2(1-\tau^2)\left(\langle X_a,\xi^\top\rangle + \tfrac{1-\tau^2}{\tau^2}\|\xi^\top\|^2\langle X_a,\xi\rangle\right)\xi^\perp.\end{aligned} \tag{5.75}$$

Now we calculate $\sum_j\langle\mathcal{L}X_{a_j}^\perp, X_{a_j}^\perp\rangle$ for a -orthonormal basis of $\mathbb{C}^{m+1}$ $\{a_j : j = 1,\dots,2m+2\}$. Considering the definition of the Berger metric, we get

$$\sum_{j=1}^{2m+2}\langle X_{a_j},u\rangle\langle X_{a_j},v\rangle = \langle u,v\rangle - (1-\tau^2)\langle u,\xi\rangle\langle v,\xi\rangle. \tag{5.76}$$

As a direct consequence of the previous formula,

$$\sum_{j=1}^{2m+2}\langle X_{a_j}^\perp,X_{a_j}^\perp\rangle = \sum_{\alpha=1}^{q}\sum_{j=1}^{2m+2}\langle X_{a_j},\zeta_\alpha\rangle^2 = q-(1-\tau^2)\|\xi^\perp\|^2,$$

$$\begin{aligned}\sum_{j=1}^{2m+2}f_{a_j}\langle(J\xi^\top)^\perp,X_{a_j}\rangle &= \sum_{j=1}^{2m+2}g(a_j,p)g((J\xi^\top)^\perp,a_j)\\ &\quad-(1-\tau^2)\sum_{j=1}^{2m+2}g((J\xi^\top)^\perp,ip)g(a_j,ip)\\ &= g((J\xi^\top)^\perp,p)-(1-\tau^2)g((J\xi^\top)^\perp,ip)g(p,ip)=0,\end{aligned}$$

$$\sum_{j=1}^{2m+2}\langle J(JX_{a_j})^\top,X_{a_j}^\perp\rangle = -\sum_{i=1}^{d}\sum_{j=1}^{2m+2}\langle X_{a_j},Je_i\rangle\langle X_{a_j},(Je_i)^\perp\rangle = -\sum_{i=1}^{d}\|(Je_i)^\perp\|^2,$$

$$\sum_{j=1}^{2m+2}\langle X_{a_j},\xi^\top\rangle\langle X_{a_j},\xi^\perp\rangle = -(1-\tau^2)\|\xi^\perp\|^2\|\xi^\top\|^2,$$

$$\sum_{j=1}^{2m+2}\langle X_{a_j},\xi\rangle\langle X_{a_j},\xi^\perp\rangle = \tau^2\|\xi^\perp\|^2,$$

where $\{\zeta_\alpha\colon \alpha = 1,\ldots,q\}$ is an orthonormal reference of the normal bundle of Φ. Also, again using (5.76) and considering that $\sigma(\xi^\top,u) = \tau(Ju)^\perp - \nabla^\perp_u\xi^\perp$ from (2.37),

$$\sum_{j=1}^{2m+2}\langle\sigma(\xi^\top,(JX_{a_j})^\top),X^\perp_{a_j}\rangle = -\sum_{i,\alpha}\langle\sigma(\xi^\top,e_i),\zeta_\alpha\rangle\sum_{j=1}^{2m+2}\langle X_{a_j},Je_i\rangle\langle X_{a_j},\zeta_\alpha\rangle$$

$$= -\sum_{i=1}^{d}\langle\sigma(\xi^\top,e_i),Je_i\rangle = -\tau\sum_{i=1}^{d}\|(Je_i)^\perp\|^2 + \sum_{i=1}^{d}\langle\nabla^\perp_{e_i}\xi^\perp,Je_i\rangle =$$

$$= -\tau\sum_{i=1}^{d}\|(Je_i)^\perp\|^2 + \tau d\|\xi^\perp\|^2 + \delta\beta,$$

where β is the 1-form $\beta(X) = \langle\xi^\perp,JX\rangle$, whose codifferential is given by

$$\delta\beta = \sum_{i=1}^{d}\left(\langle\nabla^\perp_{e_i}\xi^\perp,Je_i\rangle + \langle\xi^\perp,\nabla^\perp_{e_i}(Je_i)^\perp\rangle\right) = \sum_{i=1}^{d}\langle\nabla^\perp_{e_i}\xi^\perp,Je_i\rangle - \tau d\|\xi^\perp\|^2,$$

thanks to (5.73).

Finally, after a long but straightforward calculation, we write $\|\xi^\top\|^2 = 1 - \|\xi^\perp\|^2$, and

$$\sum_{j=1}^{2n+2}\langle\mathcal{L}X^\perp_{a_j},X^\perp_{a_j}\rangle = -P(\|\xi^\perp\|^2) + 2\tfrac{1-\tau^2}{\tau}\delta\beta,$$

where P is the polynomial

$$P(x) = \tfrac{(1-\tau^2)^2}{\tau^2}x^2 - \tfrac{1-\tau^2}{\tau^2}\left(1+q+\tau^2(d-1)\right)x - q\left(d+1-\tfrac{1}{\tau^2}\right). \tag{5.77}$$

Now, if M is stable and $\frac{1}{d+1} \leq \tau^2 < 1$ then we have, using the divergence theorem, that

$$0 \leq \sum_{j=1}^{2n+2}\mathcal{Q}(X^\perp_{a_j}) = \int_M P(\|\xi^\perp\|^2)\,dv$$

$$\leq -\int_M\left\{\tfrac{(1-\tau^2)}{\tau^2}(q+\tau^2 d)\|\xi^\perp\|^2 + q\left(d+1-\tfrac{1}{\tau^2}\right)\right\}dv$$

$$\leq -q\left(d+1-\tfrac{1}{\tau^2}\right)\operatorname{vol}(M) \leq 0,$$

where we have used that $\|\xi^\perp\|^4 \leq \|\xi^\perp\|^2$ in the second inequality and that $\|\xi^\perp\|^2 \geq 0$ in the third one. Therefore, $\tau^2 = \frac{1}{d+1}$ and $\xi^\perp = 0$. This finishes the proof of Claim 1.

CLAIM 2. Let $\Phi : M^d \to \mathbb{S}^{2m+1}_\tau$ be a stable minimal immersion of a compact d-dimensional manifold. If $\tau^2 = \frac{1}{d+1}$ and $\xi^\perp = 0$, then the normal bundle $T^\perp M$ and the subbundle $\mathcal{D}$ defined by $TM = \mathcal{D}\oplus\langle\xi\rangle$ are invariant under J.

To prove this claim we will use Proposition 2.42, the isometrically embedding $\mathbb{S}^{2m+1}_\tau$ in Euclidean space by using the Tai embbedding of $\mathbb{CP}^{m+1}(4(1-\tau^2))$ in $\mathbb{R}^k$ with $k = (m+2)^2$ (see Section 6.4) and seeing $\mathbb{S}^{2m+1}_\tau$ as a geodesic sphere of $\mathbb{CP}^{m+1}(4(1-\tau^2))$. In fact, we take in Proposition 2.42 $\overline{M} = \mathbb{S}^{2m+1}_\tau$ and $\Psi : \mathbb{S}^{2m+1}_\tau \to \mathbb{R}^p$ the isometric embedding $\Psi = \widetilde{\Psi}\circ F$ where $F : \mathbb{S}^{2m+1}_\tau \to \mathbb{CP}^{m+1}(4(1-\tau^2))$ is the embedding given at the end of Chapter 2, and $\widetilde{\Psi} : \mathbb{CP}^{m+1}(4(1-\tau^2)) \to H(m+1)$ is the Tai embedding defined at the beginning of Section 6.4, where $H(m+1)$ is the linear space of Hermitian matrices of order $m+1$. If $\{B_1,\dots,B_p\}$ with $p=(m+1)^2$ is an orthonormal basis of $H(m+1)$, $\{e_1,\dots,e_d\}$ is an orthonormal reference on M, and $\{\xi_1,\dots,\xi_{2m+1-p}\}$ is an orthonormal reference of the normal bundle of Φ, then

$$\sum_{\mu=1}^{p}\mathcal{Q}(B_\mu^\perp)=\sum_{i=1}^{d}\sum_{r=1}^{2m+1-d}\int_M\{-\langle\overline{\sigma}(e_i,e_i),\overline{\sigma}(\xi_r,\xi_r)\rangle+2\|\overline{\sigma}(e_i,\xi_r)\|^2\}\,\mathrm{d}v,$$

where $B_\mu^\perp\in\mathfrak{X}^\perp(M)$ is the normal component of B_μ and $\overline{\sigma}$ is the second fundamental form of Ψ. But $\overline{\sigma}=\widehat{\sigma}+\widetilde{\sigma}$, where $\widehat{\sigma}$ is the second fundamental form of F given in (2.34) and $\widetilde{\sigma}$ is the second fundamental form of the Tai embedding $\widetilde{\Psi}$ given in Lemma 6.29. With this and $\tau^2=1/d+1$ we get

$$\sum_{\mu=1}^{p}\mathcal{Q}(B_\mu^\perp)=-4(1-\tau^2)\sum_{i=1}^{d}\int_M\|(Je_i)^\perp\|^2\,\mathrm{d}v,$$

where we have taken into account, when using Lemma 6.29, that the curvature of the complex projective space is $4(1-\tau^2)$. Since M is stable, we get that $(Je_i)^\perp=0$ for $1\le i\le d$, which proves claim 2.

Using claims 1 and 2, $\xi^\perp=0$ and the normal bundle of Φ is invariant under the complex structure J, so the normal bundle is of even dimension and hence, $d=2n+1$ for some integer $n\ge 1$. Also, since $\xi^\perp=0$, the restriction $\widehat{\xi}$ of ξ on M is tangent to M, and so, TM decomposes orthogonally as $TM=\mathcal{D}\oplus\langle\widehat{\xi}\rangle$, where the sub-bundle $\mathcal{D}$ is invariant under J.

If η is the 1-form on M given by $\eta(v)=\langle\widehat{\xi},v\rangle$, then from (2.36) and $\xi^\perp=0$, it follows that the differential of η is given by $\mathrm{d}\eta(X,Y)=2\tau\langle JX,Y\rangle$ for arbitrary vector fields X,Y tangent to M. So, if $\{e_1,\dots,e_n,Je_1,\dots,Je_n\}$ is an orthonormal basis of $\mathcal{D}_p$, $p\in M$, we have that

$$(\eta\wedge(\mathrm{d}\eta)^n)_p(\widehat{\xi}_p,e_1,\dots,e_n,Je_1,\dots,Je_n)\neq 0,$$

and thus, η defines a contact structure on M. Note that M is orientable. Now, from a result of W. Boothby and H. Wang [**BW58**] and A. Morimoto [**Mor64**], we have that M is an $\mathbb{S}^1$-principal bundle over a complex manifold N^{2n}. If $\widehat{\Pi}:M\to N$ is the projection, then we define $\Psi:N\to\mathbb{CP}^m$ by $\Psi(q)=\Pi(\Phi(p))$, where $p\in\widehat{\Pi}^{-1}(q)$. It is clear that Ψ is well defined, i.e., it is independent of the point p in $\widehat{\Pi}^{-1}(q)$, and so Ψ defines a complex immersion of the manifold N in $\mathbb{CP}^m$ satisfying $\Psi\circ\widehat{\Pi}=\Pi\circ\Phi$.

ASSERTION. *If $\Phi_0:M_0^{2n+1}\to\mathbb{S}^{2m+1}_\tau$ is the induced $\mathbb{S}^1$-bundle over N, then there exists an integer $s\ge 1$ and an s-sheeted covering map $\widetilde{\pi}:M\to M_0$ such that $\Pi_0\circ\widetilde{\Pi}=\widehat{\Pi}$ and $\Phi=\Phi_0\circ\widetilde{\Pi}$, where $\Pi_0:M_0\to N$ is the projection. In this case, the following diagrams are commutative:*

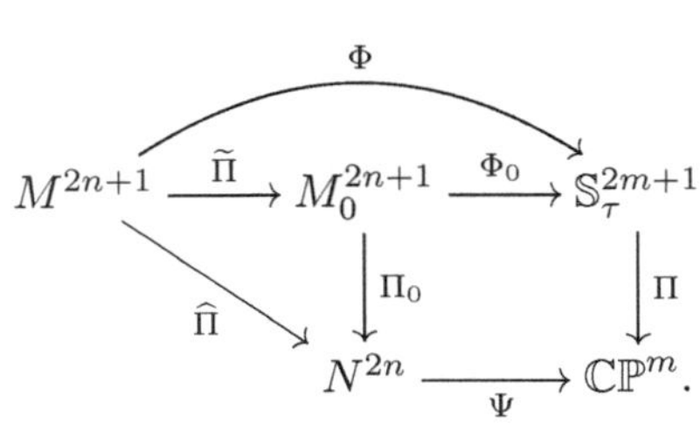

Since Φ is an embedding, $s = 1$ and so, M is the induced bundle over N. Since Φ is an embedding, the complex immersion Ψ is also an embedding.

PROOF OF THE ASSERTION. Let $\varphi : \mathbb{R}\times M \to M$ be the uniparametric group of $\widehat{\xi}$, where $\widehat{\xi}$ is the restriction of ξ to M. Then, $(\Phi\circ\varphi)(t,p) = \zeta(t,\Phi(p))$. If $\lambda \in \mathbb{R}$ is the minimum period of all the curves φ_p (see [**BW58**]), then $\zeta(t+\lambda,\Phi(p)) = \zeta(t,\Phi(p))$ for any $p \in M$, and so, there exists an integer $s \geq 1$ such that $\lambda = 2\pi\tau s$.

The action $\mathbb{S}^1 \times M \to M$ is given by

$$e^{it} \cdot p = \varphi(s\tau t, p), \quad 0 \leq t \leq 2\pi,$$

hence the subgroup $G_s \subset \mathbb{S}^1$ given by $G_s = \{1, e^{\frac{i2\pi}{s}}, \dots, e^{\frac{i2\pi(s-1)}{s}}\}$ acts on M, $\widehat{\Phi} : M/G_s \to \mathbb{S}^{2m+1}_\tau$ given by $\widehat{\Phi}([p]) = \Phi(p)$ is well defined, and it is also an $\mathbb{S}^1$-bundle over N compatible with the Hopf fibration.

Then it is easy to see that

$$\begin{aligned} F : M/G_s &\to M_0 \\ [p] &\mapsto F([p]) = (\widehat{\Pi}(p), \Phi(p)) \end{aligned}$$

s an $\mathbb{S}^1$-bundle isomorphism and so, the assertion follows. □

CHAPTER 6

Index of Minimal Submanifolds of Kähler Manifolds

In this last chapter we will study the index and nullity of submanifolds of Kähler manifolds, with special emphasis on the cases where the ambient manifold is the complex Euclidean space and the complex projective space.

6.1. Complex submanifolds

We begin by considering a very special class of minimal submanifolds of Kähler manifolds, the complex submanifolds. They inherit the structure of the Kähler manifold and, besides being minimal submanifolds, are always stable. In fact, we will prove a stronger property of being calibrated submanifolds for certain calibrations defined in any Kähler manifold.

DEFINITION 6.1. An immersion $\Phi : M \to (\overline{M}^m, g, J)$ of a manifold M into a Kähler manifold $\overline{M}$ of complex dimension m is called **complex** if $J(d\Phi_p(T_pM)) \subset d\Phi_p(T_pM)$, for any $p \in M$, where J is the complex structure of $\overline{M}$. This implies that J induces an almost complex structure on M, which is denoted by J again. In particular, the dimension of M is even and $J(T_p^\perp M) = T_p^\perp M$, for any $p \in M$.

PROPOSITION 6.2. *Let $\Phi : M^n \to (\overline{M}^m, g, J)$ be a complex immersion. Then:*

(1) $\nabla J = 0$, *i.e., (M, g, J) is a Kähler manifold.*
(2) $\sigma(v, Jw) = J\sigma(v, w), \quad \forall\, v, w \in TM.$
(3) $A_{J\xi} = JA_\xi = -A_\xi J, \quad \forall\, \xi \in T^\perp M.$
(4) $\nabla^\perp J = 0.$

PROOF. Since $\overline{M}$ is a Kähler manifold, $(\overline{\nabla}_X J)Y = 0$ for all tangent vector fields X, Y to M. Using (2.1) we get 1 and 2.

Also, as $(\overline{\nabla}_X J)\xi = 0$ for any vector fields X tangent to M and ξ normal to M, from (2.1) and 2 we get 3 and 4. □

COROLLARY 6.3. *Let $\Phi : M^n \to (\overline{M}^m, g, J)$ be a complex immersion. Then:*

(1) *The immersion Φ is minimal.*
(2) *The Jacobi operator $\mathcal{L}$ of Φ satisfies $\mathcal{L}(J\xi) = J(\mathcal{L}\xi)$ for any $\xi \in \mathfrak{X}^\perp(M)$. In particular, the eigenvalues of $\mathcal{L}$ have even multiplicities.*
(3) *The normal bundle $T^\perp M$ has a unique structure of holomorphic bundle on M such that a section $\xi \in \mathfrak{X}^\perp(M)$ is holomorphic if and only if $D_v\xi = 0$ for any $v \in TM$, where*

$$D_v\xi = \nabla^\perp_{Jv}\xi - J\nabla^\perp_v\xi. \tag{6.1}$$

PROOF. From Proposition 6.2, 2, it is clear that $\sigma(Jv, Jw) = -\sigma(v, w)$ for any tangent vectors v, w. Now, since M is Kähler, we can take at any point $p \in M$ an orthonormal basis $\{e_1, \dots, e_n, Je_1, \dots, Je_n\}$ and thus the mean curvature vector $\vec{H}$ of Φ satisfies

$$\vec{H} = \frac{1}{2n}\sum_{i=1}^{n}\{\sigma(e_i, e_i) + \sigma(Je_i, Je_i)\} = 0,$$

which proves 1.

On the other hand, it follows from Proposition 6.2, 4 that $\Delta^{\perp}J\xi = J\Delta^{\perp}\xi$ for any $\xi \in \mathfrak{X}^{\perp}(M)$, and from Proposition 6.2,2,3, it follows that $\mathcal{A}J\xi = J\mathcal{A}\xi$. Also, the standard properties of the curvature of Kähler manifolds given in Section 1.6, show that $\mathcal{R}J\xi = J\mathcal{R}\xi$. So, $\mathcal{L}J\xi = J\mathcal{L}\xi$ for any $\xi \in \mathfrak{X}^{\perp}(M)$, which proves 2.

Finally, as $\nabla^{\perp}J = 0$, statement 3 follows from the Koszul-Malgrange theorem ([**KM58**]). □

In the next result we study the index and nullity of complex submanifolds of Kähler manifolds.

THEOREM 6.4 ([**LS73**]). *Let $\Phi : M^n \to (\overline{M}^m, g, J)$ be a complex immersion. Then, Φ is stable and, if M is compact,* Nul(M) *is twice the complex dimension of the space of holomorphic sections of the normal bundle.*

PROOF. Let $\xi \in \mathfrak{X}^{\perp}(M)$ be an arbitrary normal section on Φ and

$$\{e_i, \dots, e_n, Je_1, \dots, Je_n\}$$

a local orthonormal reference on M. It follows from (6.1) that

$$\begin{aligned}\|D\xi\|^2 &= \sum_{i=1}^{n}\{\|D_{e_i}\xi\|^2 + \|D_{Je_i}\xi\|^2\} = 2\sum_{i=1}^{n}\|D_{e_i}\xi\|^2 \\ &= 2\|\nabla^{\perp}\xi\|^2 - 4\sum_{i=1}^{n} g(\nabla^{\perp}_{Je_i}\xi, J\nabla^{\perp}_{e_i}\xi).\end{aligned} \tag{6.2}$$

On the other hand, from (2.2) and Proposition 6.2, we have

$$\begin{aligned}\sum_{i=1}^{n}\overline{R}(e_i, Je_i, \xi, J\xi) &= \sum_{i=1}^{n}\{g(\nabla^{\perp}_{e_i}\nabla^{\perp}_{Je_i}\xi, J\xi) - g(\nabla^{\perp}_{Je_i}\nabla^{\perp}_{e_i}\xi, J\xi) \\ -g(\nabla^{\perp}_{[e_i,Je_i]}\xi.J\xi) - g([A_\xi, A_{J\xi}]e_i, Je_i)\} &= \sum_{i=1}^{n}\{e_i(g(\nabla^{\perp}_{Je_i}\xi, J\xi)) - Je_i(g(\nabla^{\perp}_{e_i}\xi, J\xi)) \\ &\quad -2g(\nabla^{\perp}_{Je_i}\xi, J\nabla^{\perp}_{e_i}\xi) - g(\nabla^{\perp}_{[e_i,Je_i]}\xi, J\xi) + 2g(A_\xi e_i, A_\xi e_i)\}.\end{aligned} \tag{6.3}$$

Now, if α is the 1-form on M defined by $\alpha(v) = g(\nabla^{\perp}_{Jv}\xi, J\xi)$, it is easy to check that the codifferential of α is given by

$$\begin{aligned}\delta\alpha &= \sum_{i=1}^{n}\{e_i(\alpha(e_i)) + Je_i(\alpha(Je_i)) - \alpha(\nabla_{e_i}e_i - \alpha(\nabla_{Je_i}Je_i)\} \\ &= \sum_{i=1}^{n}\{e_i(g(\nabla^{\perp}_{Je_i}\xi, J\xi)) - Je_i(g(\nabla^{\perp}_{e_i}\xi, J\xi)) - g(\nabla^{\perp}_{[e_i,Je_i]}\xi, J\xi)\}.\end{aligned} \tag{6.4}$$

The Jacobi identity also implies that

$$\begin{aligned}\overline{R}(e_i, Je_i, \xi, J\xi) &= -\overline{R}(\xi, e_i, Je_i, J\xi) - \overline{R}(Je_i, \xi, e_i, J\xi) \\ &= -\overline{R}(\xi, e_i, e_i, \xi) - \overline{R}(\xi, Je_i, Je_i, \xi).\end{aligned} \tag{6.5}$$

From (6.2), (6.3), (6.4) and (6.5), we conclude that

$$\frac{1}{2}\|D\xi\|^2 = \|\nabla^\perp\xi\|^2 - g(\mathcal{A}\xi,\xi) - g(\mathcal{R}\xi,\xi) - \delta\alpha.$$

If ξ has compact support, then α also has compact support, and using Theorem 1.15, we conclude that

$$\mathcal{Q}(\xi) = \frac{1}{2}\int_M \|D\xi\|^2 \,\mathrm{d}v \geq 0, \qquad \forall\, \xi \in \mathfrak{X}_0^\perp(M).$$

The above formula means that M is stable and that, if M is compact, the eigenspace corresponding to the eigenvalue 0 is the space of holomorphic normal sections (see Corollary 6.3). □

To conclude this section, we will prove that the complex submanifolds of Kähler manifolds are calibrated submanifolds for certain calibrations defined in any Kähler manifold.

THEOREM 6.5 ([**HL82**]). *If $(\overline{M}^m, g, J)$ is a Kähler manifold and $\omega \in \Omega^2(\overline{M})$ is its Kähler 2-form, then for any $1 \leq k \leq m-1$, $\omega^k/k! \in \Omega^{2k}(\overline{M})$ is a calibration on $\overline{M}$, whose calibrated submanifolds are the complex k-dimensional submanifolds of $\overline{M}$.*

PROOF. First, since $\mathrm{d}\omega = 0$, it follows that $\mathrm{d}\omega^k = 0$, for $1 \leq k \leq m-1$.

Next we prove the Wirtinger inequality, which states that if $\{e_1, \ldots, e_{2k}\}$ are orthonormal vectors in $T_p\overline{M}$ with $p \in \overline{M}$, then

$$\frac{\omega^k}{k!}(e_1, \ldots, e_{2k}) \leq 1,$$

and the equality holds if and only if $\{e_1, \ldots, e_{2k}\}$ spans a complex subspace of $T_p\overline{M}$.

In fact, if $V \subset T_p\overline{M}$ is the subspace with basis $\{e_1, \ldots, e_{2k}\}$, then ω on V can be written as

$$\omega = \sum_{i=1}^k \omega(v_i, w_i)\, v_i^* \wedge w_i^*,$$

where $\{v_1, w_1, \ldots, v_k, w_k\}$ is certain orthonormal basis of V defining the same orientation as $\{e_1, \ldots, e_{2k}\}$. Furthermore, for any $1 \leq i \leq k$, $\omega(v_i, w_i) = g(Jv_i, w_i) \leq 1$ and the equality holds if and only if $Jv_i = w_i$. From the expression of ω it follows that ω^k on V is given by

$$\omega^k = k!\,\omega(v_1, w_1)\ldots\omega(v_k, w_k)\, v_1^* \wedge w_1^* \wedge \cdots \wedge v_k^* \wedge w_k^*,$$

and so,

$$\omega^k(e_1, \ldots, e_{2k}) = k!\,\omega(v_1, w_1)\ldots\omega(v_k, w_k)\,\det A = k!\,\omega(v_1, w_1)\ldots\omega(v_k, w_k) \leq k!,$$

where $A \in SO(2k)$ is the change of basis matrix. Moreover, the equality in the above inequality holds if and only if $Jv_i = w_i$ for all $1 \leq i \leq k$, that is, V is a complex subspace of $T_p\overline{M}$.

Thus, we have proved that $\omega^k/k!$ is a calibration on $\overline{M}$ and that M^{2k} is a calibrated submanifold of the calibration $\omega^k/k!$ if and only if M is a complex submanifold. □

6.2. Minimal Lagrangian submanifolds

In this section we consider another important family of submanifolds of a Kähler manifold, the Lagrangian submanifolds. Unlike the complex submanifolds, their behavior with respect to the complex structure is completely opposite, since the Kähler form of the ambient manifold vanishes over such submanifolds. If the Ricci curvature of the ambient manifold is non-positive, the minimal Lagrangian submanifolds are always stable. In the special case where the ambient Kähler manifold is Calabi-Yau, these submanifolds are calibrated with respect to the so-called special Lagrangian calibration.

Let $(\overline{M}^n, g, J)$ be a Kähler manifold and ω its Kähler 2-form. If $\Phi : M \to (\overline{M}^n, g, J)$ is an immersion and the induced 2-form $\Phi^*\omega = 0$, then $g(Jv, w) = 0$ for all vectors v, w tangent to M. So, J defines a monomorphism $J : T_pM \to T_p^\perp M$ for any $p \in M$, which implies that $\dim M \leq n$.

DEFINITION 6.6. An immersion $\Phi : M \to (\overline{M}^n, g, J)$ of a manifold M into a Kähler manifold $\overline{M}$ of complex dimension n is called **Lagrangian** if $\Phi^*\omega = 0$ and the dimension of M is n.

PROPOSITION 6.7. *Let $\Phi : M^n \to (\overline{M}^n, g, J)$ be a Lagrangian immersion. Then:*

(1) *$J : TM \to T^\perp M$ defines a fiber bundle isomorphism and thus, an isomorphism between $\mathfrak{X}(M)$ and $\mathfrak{X}^\perp(M)$.*
(2) *$\nabla^\perp J = J\nabla$.*
(3) *$JA_{Jv}w = \sigma(v, w)$, $\forall v, w \in TM$ and the trilinear map $g(\sigma(\_,\_), J\_)$ is totally symmetric.*
(4) *$R(v, w, x, y) = R^\perp(v, w, Jx, Jy)$, $\forall v, w, x, y \in TM$.*
(5) *If $\alpha_{\vec{H}}$ is the 1-form on M defined by $\alpha_{\vec{H}} = \vec{H} \lrcorner \omega$, then its differential is given by $\mathrm{d}\alpha_{\vec{H}} = \frac{1}{n}\Phi^*\eta$, where $\vec{H}$ is the mean curvature vector of Φ and η is the Ricci 2-form on $\overline{M}$. (see Section 1.6).*

PROOF. Assertion 1 is a direct consequence of the definition of Lagrangian immersion.

Since $\overline{M}$ is a Kähler manifold, $(\overline{\nabla}_X J)Y = 0$ for all $X, Y \in \mathfrak{X}(M)$. Using (2.1) we get 2 and 3.

Statement 4 is an simple consequence of 2. Finally, using 2 and the fact that $\alpha_{\vec{H}}(X) = g(J\vec{H}, X)$, we get

$$(\mathrm{d}\alpha_{\vec{H}})(X, Y) = -g(\nabla^\perp_X \vec{H}, JY) + g(\nabla^\perp_Y \vec{H}, JX), \quad \forall X, Y \in \mathfrak{X}(M).$$

If $\{e_1, \ldots, e_n\}$ is a local orthonormal reference around $p \in M$ with $\left(\nabla_{e_j} e_i\right)_p = 0$, $1 \leq i \leq n$, then using 3, (2.2) and general curvature properties we get

$$\begin{aligned}
(\mathrm{d}\alpha_{\vec{H}})(X, Y)(p) &= -\frac{1}{n}\sum_{i=1}^n \{g((\nabla\sigma)(X, e_i, e_i), JY) + g((\nabla\sigma)(Y, e_i, e_i), JX)\}(p) \\
&= -\frac{1}{n}\sum_{i=1}^n \{g((\nabla\sigma)(X, Y, e_i), Je_i) + g((\nabla\sigma)(Y, X, e_i), Je_i)\}(p) \\
&= -\frac{1}{n}\sum_{i=1}^n \overline{R}(X, Y, e_i, Je_i)(p) = \frac{1}{n}\sum_{i=1}^n \{\overline{R}(e_i, X, Y, Je_i) + \overline{R}(Y, e_i, X, Je_i)\}(p)
\end{aligned}$$

$$= \frac{1}{n}\sum_{i=1}^{n}\{\overline{R}(Je_i, JX, Y, Je_i) + \overline{R}(e_i, JX, Y, e_i)\}(p) = \frac{1}{n}\overline{S}(X,Y)(p),$$

which proves 5. □

REMARK 6.8. From Proposition 6.7,5, it follows that if Φ is a minimal immersion, then $\Phi^*\eta = 0$. This means that $\overline{S}(Jv, w) = 0$ for any vector $v, w \in TM$, where $\overline{S}$ is the Ricci tensor of $\overline{M}$. This property of the Ricci tensor of $\overline{M}$ is irrelevant if $\overline{M}$ is an Einstein-Kähler manifold.

THEOREM 6.9 ([**Oh90**]). *Let $\Phi : M^n \to (\overline{M}^n, g, J)$ be a minimal Lagrangian immersion of a manifold M into a Kähler manifold $(\overline{M}^n, g, J)$. Then:*

(1) *There exists an isomorphism $F : \mathfrak{X}^\perp(M) \to \Omega^1(M)$ given by*

$$F(\xi) = \xi \lrcorner \omega, \qquad F(\xi)(X) = g(J\xi, X), \quad \forall X \in \mathfrak{X}(M),$$

where ω is the Kähler 2-form of $\overline{M}$. Furthermore,

$$g(F(\xi), F(\eta)) = g(\xi, \eta), \qquad \forall \xi, \eta \in \mathfrak{X}^\perp(M).$$

(2) *The Jacobi operator $\mathcal{L}$ of Φ satisfies $F \circ \mathcal{L} = (\Delta + P) \circ F$, where Δ is the Hodge Laplacian on M acting on 1-forms and $P : \Omega^1(M) \to \Omega^1(M)$ is the endomorphism given by $(P\alpha)(V) = \overline{S}(\alpha^\sharp, V)$, where $\alpha^\sharp$ is the dual vector field of α, i.e., $g(\alpha^\sharp, V) = \alpha(V)$ for any $V \in \mathfrak{X}(M)$. To study the index and nullity of M, we can replace the Jacobi operator by the operator $\mathcal{L} = \Delta + P : \Omega^1(M) \to \Omega^1(M)$.*

(3) *The index form $\mathcal{Q} : \Omega^1_0(M) \to \mathbb{R}$ is given by*

$$\mathcal{Q}(\alpha) = \int_M \{\|d\alpha\|^2 + \|\delta\alpha\|^2 - \overline{S}(\alpha^\sharp, \alpha^\sharp)\}\, dv.$$

PROOF. The linear map F is clearly an isomorphism with inverse $F^{-1}(\alpha) = -J\alpha^\sharp$. Furthermore, if $\{e_1, \ldots, e_n\}$ is a local orthonormal reference on M, then

$$g(F(\xi), F(\eta)) = \sum_{i=1}^{n} F(\xi)(e_i)F(\eta)(e_i) = \sum_{i=1}^{n} g(J\xi, e_i)g(J\eta, e_i) = g(\xi, \eta),$$

which proves 1.

Now, given $\xi \in \mathfrak{X}^\perp(M)$, we want to compute $F(\mathcal{L}\xi)$. To do this, it follows from Proposition 6.7 that the rough Laplacian $\Delta : \mathfrak{X}(M) \to \mathfrak{X}(M)$, defined in Section 1.3 by $\Delta = \nabla_{e_i}\nabla_{e_i} - \nabla_{\nabla_{e_i}e_i}$, being $\{e_1, \ldots, e_n\}$ a local orthonormal reference on M, and the normal Laplacian $\Delta^\perp$ are related by $\Delta^\perp J = J\Delta$.

Using this property together with Proposition 6.7 and (2.2), we have that for any tangent vector field V to M,

$$F(\mathcal{L}\xi)(V) = g(J\mathcal{L}\xi, V) = g(J\Delta^\perp\xi, V) - \sum_{i=1}^{n}\{g(A_{Je_i}A_\xi e_i, V) + \overline{R}(\xi, e_i, e_i, JV)\}$$

$$= g(\Delta J\xi, V) - \sum_{i=1}^{n}\{\overline{R}(e_i, V, \xi, Je_i) - R^\perp(e_i, V, \xi, Je_i) + \overline{R}(\xi, e_i, e_i, JV)\}$$

$$= g(\Delta J\xi, V) + \overline{S}(J\xi, V) - S(J\xi, V).$$

But the Laplacian acting on 1-forms and vector fields are related by

$$\Delta\alpha = (\Delta\alpha^\sharp)^\flat - S(\alpha^\sharp, -), \quad \forall \alpha \in \Omega^1(M),$$

where $X^\flat$ is the dual 1-form of X, i.e., $X^\flat(V) = g(X,V)$. Using this relation, we finally get that

$$F(\mathcal{L}\xi)(V) = \Delta(F\xi)(V) + P(F\xi)(V), \quad \forall \xi \in \mathfrak{X}^\perp(M), \quad \forall V \in \mathfrak{X}(M),$$

which proves 2.

Finally, if $\alpha \in \Omega_0^1(M)$, as $\Delta = \mathrm{d}\delta + \delta\mathrm{d}$ and d and δ are adjoint operators, we get that

$$\begin{aligned}\mathcal{Q}(\alpha) &= -\int_M g(\alpha,(\Delta+P)\alpha)\,\mathrm{d}v - \int_M \{g(\alpha,(\mathrm{d}\delta+\delta\mathrm{d})\alpha) + g(\alpha,P\alpha)\}\,\mathrm{d}v \\ &= \int_M \{\|\mathrm{d}\alpha\|^2 + \|\delta\alpha\|^2 - \overline{S}(\alpha^\sharp,\alpha^\sharp)\}\,\mathrm{d}v,\end{aligned}$$

which proves 3. □

Corollary 6.10 ([**Oh90, CLN13**]). *Let $\Phi : M \to (\overline{M}^n, g, J)$ be a minimal Lagrangian immersion of a manifold M into a Kähler manifold $(\overline{M}^n, g, J)$.*

(1) *If the Ricci curvature of $\overline{M}$ is non-positive, then M is stable.*
(2) *If the Ricci curvature of $\overline{M}$ is negative, then M is strictly stable.*
(3) *If the Ricci curvature of $\overline{M}$ is positive and M is compact, then* $\mathrm{Ind}(M) \geq \beta_1(M)$, *where $\beta_1(M)$ is the first Betti number of M. In particular, if M is stable, then $H^1(M,\mathbb{R}) = 0$.*
(4) *If $\overline{M}$ is Einstein with Ricci tensor $\overline{S} = \lambda g$, $\lambda \in \mathbb{R}$, and M is compact, then* $\mathrm{Index}(M)$ *is the number of eigenvalues (counted with multiplicities) of $\Delta : \Omega^1(M) \to \Omega^1(M)$ less than λ and* $\mathrm{Nul}(M)$ *is the multiplicity of λ as an eigenvalue of Δ. In particular, if $\overline{M}$ is Ricci flat, then* $\mathrm{Nul}(M) = \beta_1(M)$.

Proof. By Theorem 6.9,3, if the Ricci curvature of $\overline{M}$ is non-positive (or negative), then $\mathcal{Q}(\alpha) \geq 0$ (or $\mathcal{Q}(\alpha) > 0$) for any $\alpha \in \Omega_0^1(M)$. This implies that M is stable (or strictly stable), proving 1 and 2.

If M is compact, and the Ricci curvature of $\overline{M}$ is positive, we have $\mathcal{Q}(\alpha) < 0$ for any nonzero $\alpha \in \mathcal{H}^1(M)$. So, $\mathrm{Ind}(M) \geq \beta_1(M)$. Moreover, if M is stable then $\beta_1(M) = 0$ and thus, the first cohomology group of M with real coefficients is trivial. This proves 3.

If $\overline{M}$ is Einstein with $\overline{S} = \lambda g$, then the operator P of Theorem 6.9 is $P = \lambda\, Id$, and so, the Jacobi operator $\mathcal{L} = \Delta + \lambda$. Then 4 follows easily, again taking into account that the eigenspace of the zero eigenvalue of Δ is the space of harmonic 1-forms on M, whose dimension is the first Betti number $\beta_1(M)$. □

In [**Tak84**] M. Takeuchi, using techniques of symmetric spaces that are beyond the scope of this book, made an important contribution to the study of the stability of Lagrangian submanifolds of Hermitian symmetric spaces, proving the following result:

Theorem 6.11 ([**Tak84**]). *Let $\Phi : M^n \to \overline{M}$ be a totally geodesic Lagrangian embedding of a compact manifold M in a Hermitian symmetric space of compact type $\overline{M}$. Then M is stable if and only if M is simply connected.*

Given a Kähler manifold $(\overline{M}^n, g, J)$, a vector field X on $\overline{M}$ is a **Hamiltonian vector field** if there exists a smooth function F on $\overline{M}$ such that $X = J\overline{\nabla}F$. The diffeomorphisms of the flow $\{\phi_t\}$ of X satisfy that $\phi_t^*\omega = \omega$, and so, they transform Lagrangian submanifolds into Lagrangian submanifolds. If $\Phi : M^n \to \overline{M}^n$ is a

Lagrangian immersion, then the normal component of a Hamiltonian vector field $J\overline{\nabla}F$ is the normal vector field $J\nabla f$, where $f = F \circ \Phi$. This fact justifies the following concept.

DEFINITION 6.12 ([**Oh90**]). A minimal Lagrangian immersion $\Phi : M^n \to \overline{M}^n$ of a manifold M into a Kähler manifold $(\overline{M}^n, g, J)$ is called **Hamiltonian stable** if $\mathcal{Q}(\mathrm{d}f) \geq 0$ for any $f \in C_0^\infty(M)$. This means that the immersion Φ is stable under Hamiltonian deformations.

A particularly interesting case occurs when the ambient space is an Einstein-Kähler manifold with positive Ricci curvature. This includes interesting examples of Kähler manifolds, and for these ambient manifolds the behavior of the Jacobi operator of their minimal Lagrangian submanifolds is nice.

COROLLARY 6.13 ([**Oh90**]). *Let $\Phi : M^n \to (\overline{M}^n, g, J)$ be a minimal Lagrangian immersion of a compact manifold M into a Kähler-Einstein manifold $(\overline{M}^n, g, J)$ with Ricci tensor $\overline{S} = \lambda g$, $\lambda \in \mathbb{R}$. Then:*

(1) *If $\lambda > 0$, we define the **Hamiltonian index** of M, $\mathrm{Ind}_0(M)$, as the number of positive eigenvalues (counted with multiplicities) of $\Delta : C^\infty(M) \to C^\infty(M)$ less than λ. So M is Hamiltonian stable if and only if $\mathrm{Ind}_0(M) = 0$.*

(2) *If $\lambda > 0$, we define $\mathrm{Ind}_1(M)$ as the number of positive eigenvalues (counted with multiplicities) of $\Delta : \Omega^2(M) \to \Omega^2(M)$ less than λ. Then*
$$\mathrm{Ind}(M) = \beta_1(M) + \mathrm{Ind}_0(M) + \mathrm{Ind}_1(M).$$

(3) *If $\lambda > 0$, $n = 2$ and M is an orientable surface of genus h, then $\mathrm{Ind}(M) = 2(h + \mathrm{Ind}_0(M))$.*

(4) *If $\lambda > 0$, $n = 2$, M is not orientable, $\Pi : \widetilde{M} \to M$ is the orientable double covering of M and $\Phi \circ \Pi : \widetilde{M} \to \overline{M}^n$ the corresponding minimal Lagrangian immersion, then $\mathrm{Ind}(M) = \frac{1}{2}\mathrm{Ind}(\widetilde{M})$.*

PROOF. The Hodge-de Rham theorem (Theorem 1.14) states that
$$\Omega^1(M) = \mathcal{H}^1(M) \oplus \mathrm{d}C^\infty(M) \oplus \delta\Omega^2(M).$$
According to T. Nagano, [**Nag61**], any 1-form α is uniquely written as $\alpha = \alpha_0 + df + \delta\beta$, with α_0 harmonic, although the function f and the 2-form β are not unique.

Since $\Delta \circ \mathrm{d} = \mathrm{d} \circ \Delta$ and $\Delta \circ \delta = \delta \circ \Delta$, if $\Delta\alpha + \lambda\alpha = 0$, with $\lambda \neq 0$ and $\alpha \in \Omega^1(M)$, then it is possible to choose f and β such that $\alpha = \mathrm{d}f + \delta\beta$, $\Delta f + \lambda f = 0$ and $\Delta\beta + \lambda\beta = 0$. Thus, the multiplicity of λ as a nonzero eigenvalue of Δ is the sum of the multiplicity of λ as an eigenvalue of $\Delta : C^\infty(M) \to C^\infty(M)$ plus the multiplicity of λ as an eigenvalue of $\Delta : \Omega^2(M) \to \Omega^2(M)$. Hence, taking into account the Corollary 6.10,4, the statements 1 and 2 follow easily.

If M is an orientable compact surface and $\star : C^\infty(M) \to \Omega^2(M)$ is the star operator (see Section 1.3), then it is known that $\Delta \circ \star = \star \circ \Delta$, and thus the non-null eigenvalues of $\Delta : C^\infty(M) \to C^\infty(M)$ are exactly the non-null eigenvalues of $\Delta : \Omega_2(M) \to \Omega_2(M)$. So, $\mathrm{Ind}_0(M) = \mathrm{Ind}_1(M)$ and, since $\beta_1(M) = 2h$, we have proved 3.

To prove 4 we need to relate the spaces of forms of M and $\widetilde{M}$. Let M be a nonorientable Riemannian compact surface and $\Pi : \widetilde{M} \to M$ be the orientable double Riemannian covering. If $\tau : \widetilde{M} \to \widetilde{M}$ is the isometry such that $\{Id, \tau\}$ is the transformation group of the covering, then the spaces of forms on $\widetilde{M}$ can be decomposed as follows:

$$\Omega^i(\widetilde{M}) = \Omega^i_+(\widetilde{M}) \oplus \Omega^i_-(\widetilde{M}), \quad i = 0, 1, 2, \quad \Omega^0(\widetilde{M}) = C^\infty(\widetilde{M}),$$

where

$$\Omega^i_\pm(\widetilde{M}) = \{\alpha \in \Omega^i(\widetilde{M}) \,/\, \tau^*\alpha = \pm\alpha\}.$$

Also, the space of harmonic 1-forms on $\widetilde{M}$ decomposes into two subspaces $\mathcal{H}^1(\widetilde{M}) = \mathcal{H}^1_+(\widetilde{M}) \oplus \mathcal{H}^1_-(\widetilde{M})$, where again $\mathcal{H}^1_\pm(\widetilde{M}) = \{\alpha \in \mathcal{H}^1(\widetilde{M}) \,/\, \tau^*\alpha = \pm\alpha\}$. In this way, we get

$$\Omega^1_\pm(\widetilde{M}) = \mathcal{H}^1_\pm(\widetilde{M}) \oplus d\,\Omega^0_\pm(\widetilde{M}) \oplus \delta\,\Omega^2_\pm(\widetilde{M}).$$

As $\Pi \circ \tau = \Pi$, the map $\alpha \in \Omega^i(M) \mapsto \pi^*\alpha \in \Omega^i(\widetilde{M})$ allows to identify $\mathcal{H}^1(M) \equiv \mathcal{H}^1_+(\widetilde{M})$ and $\Omega^i(M) \equiv \Omega^i_+(\widetilde{M})$, $i = 0, 1, 2$. Since M is not orientable, the star operator $\star$ on $\widetilde{M}$ satisfies $\star\tau^* = -\tau^*\star$, and so, $\star$ identifies $\Omega^0_-(\widetilde{M}) \equiv \Omega^2_+(\widetilde{M})$. This gives us the identification

$$\Omega^2(M) \equiv \Omega^0_-(\widetilde{M})$$
$$\alpha \equiv f,$$

where $\pi^*\alpha = f\omega_0$, where ω_0 is the volume 2-form on $\widetilde{M}$.

Since M is not orientable, $H^2(M, \mathbb{R}) = 0$ and thus M has no harmonic 2-forms. This implies that the eigenvalues of $\Delta : \Omega^2(M) \to \Omega^2(M)$ are positive, and thus, taking into account the above remarks, $\text{Ind}_1(M)$ is the number of eigenvalues (counted with multiplicity) of $\Delta : \Omega^0_-(\widetilde{M}) \to \Omega^0_-(\widetilde{M})$ less than λ. Also, since $\text{Ind}_0(M)$ is the number of positive eigenvalues (counted with multiplicity) of $\Delta : \Omega^0_+(\widetilde{M}) \to \Omega^0_+(\widetilde{M})$ less than λ, we get that

$$\text{Ind}_0(M) + \text{Ind}_1(M) = \text{Ind}_0(\widetilde{M}). \tag{6.6}$$

Finally, if h is the genus of $\widetilde{M}$, then $\beta_1(M) = h$, and so, taking into account 2 and 3 we get that $2\,\text{Ind}(M) = \text{Ind}(\widetilde{M})$. □

To conclude this section, we will introduce an important family of Lagrangian submanifolds which are critical points for the volume functional with a Lagrangian constraint.

A vector field X on a Kähler manifold $(\overline{M}^n, J, \omega)$ is called **symplectic** if the 1-form $X \lrcorner \omega$ is closed. The diffeomorphisms of the flow ϕ_t of X preserve ω, i.e., $\phi_t^*\omega = \omega$ and so, they transform Lagrangian submanifolds into Lagrangian submanifolds. The Hamiltonian vector fields defined above are exactly those symplectic vector fields where $X \lrcorner \omega$ is exact.

From Proposition 6.7,5 it follows that if M^n is a Lagrangian submanifold of an Einstein-Kähler manifold $\overline{M}^n$, then $d\alpha_{\vec{H}} = 0$, where $\alpha_{\vec{H}} = \vec{H} \lrcorner \omega$, and therefore, the mean curvature vector $\vec{H}$ is an infinitesimal symplectic motion. This fact suggests that it is natural to consider variational problems for the volume with a Lagrangian constraint.

DEFINITION 6.14 ([**Oh93**],[**SW01**]). A Lagrangian immersion $\Phi : M^n \to \overline{M}^n$ of a manifold M into a Kähler manifold $(\overline{M}^n, g, J, \omega)$ is called **Lagrangian stationary** if it is a critical point of the volume functional for any compactly supported variation F of Φ through Lagrangian submanifolds. The immersion Φ is called

Hamiltonian stationary or H-minimal if it is a critical point of the volume functional for any compactly supported Hamiltonian variation F of Φ.

PROPOSITION 6.15 ([**Oh93**],[**SW01**]). *Let $\Phi : M^n \to (\overline{M}^n, g, J)$ be a Lagrangian immersion.*

(1) *If $\overline{M}$ is Einstein-Kähler, then M is Lagrangian stationary if and only if M is minimal.*
(2) *M is Hamiltonian stationary if and only if $\delta\alpha_{\vec{H}} = 0$, i.e, the tangent vector field $J\vec{H}$ satisfies $\operatorname{div} J\vec{H} = 0$.*

PROOF. It is trivial that if M is minimal then M is Lagrangian stationary. Conversely, if M is Lagrangian stationary, from Theorem 2.6, we have that if Φ_t is a compactly supported Lagrangian variation of Φ through Lagrangian submanifolds, then

$$\left.\frac{\mathrm{d}\, V_F(t)}{\mathrm{d}t}\right|_{t=0} = -n \int_M g(X, \vec{H})\, \mathrm{d}v = 0,$$

where X is the variational vector field of a variation F. Like $\Phi_t^*\omega = 0$, $\forall t$, then $\frac{\mathrm{d}}{\mathrm{d}t}\Phi_t^*\omega = 0$, which implies that the 1-form on M given by $\alpha_X = X \lrcorner \omega$ is closed. So

$$0 = \int_M g(X, \vec{H})\, \mathrm{d}v = \int_M g(\alpha_X, \alpha_{\vec{H}})\, \mathrm{d}v.$$

Since $\alpha_{\vec{H}}$ is also closed, if M is Lagrangian stationary we get that $\vec{H} = 0$. This proves 1.

To prove 2, if $X = J\overline{\nabla}F$, then $X^\perp = J\nabla f$ with $f = F \circ \Phi$. Hence,

$$\left.\frac{\mathrm{d}\, V_F(t)}{\mathrm{d}t}\right|_{t=0} = -n \int_M g(X, \vec{H})\, \mathrm{d}v = -n \int_M g(J\nabla f, \vec{H})\, \mathrm{d}v = n \int_M f \operatorname{div} J\vec{H}\, \mathrm{d}v.$$

So, M is Hamiltonian stationary if and only if $\operatorname{div} J\vec{H} = 0$, which proves 2. □

H-minimal Lagrangian submanifolds were defined by Y. Oh in [**Oh93**]. In this paper he calculated the second variation of the volume and studied the stability of certain tori. Since then, many authors have studied this type of Lagrangian submanifolds, either by constructing new examples or by studying stability problems in various Einstein-Kähler ambient manifolds (see [**BC12, CU98, HR02, HR05, Lee12**]).

6.3. Special Lagrangian calibrations on complex Euclidean space

By Corollary 6.10, minimal Lagrangian submanifolds of the complex Euclidean space $\mathbb{C}^n$ are stable. In this section we will construct a parametric family of calibrations on $\mathbb{C}^n$ (called special Lagrangian calibrations) and prove that their calibrated submanifolds are exactly the orientable minimal Lagrangian submanifolds of $\mathbb{C}^n$. The entire content of this section can be found in the work of R. Harvey and H. B. Lawson [**HL82**].

On $\mathbb{C}^n = \{(z_1, \ldots, z_n) \mid z_j = x_j + iy_j \in \mathbb{C}, 1 \leq j \leq n\}$, we consider the holomorphic n-form Ω given by

$$\Omega = \mathrm{d}z_1 \wedge \cdots \wedge \mathrm{d}z_n.$$

If ω is the Kähler 2-form on $\mathbb{C}^n$ (see section 1.6), we have that

$$\omega = \sum_{j=1}^{n}\{\mathrm{d}x_j \wedge \mathrm{d}y_j\} = \frac{i}{2}\sum_{j=1}^{n}\{\mathrm{d}z_j \wedge \mathrm{d}\bar{z}_j\}.$$

Now, it is easy to to check that

$$\omega^n = \frac{(-1)^{\frac{n(n-1)}{2}} i^n n!}{2^n}\Omega \wedge \overline{\Omega},$$

where $i = \sqrt{-1}$.

PROPOSITION 6.16 ([**HL82**]). *For every $\theta \in [0, 2\pi)$, we define $\Omega_\theta = e^{-i\theta}\Omega$. If $\Re$ and $\Im$ denote the real and imaginary parts, then the n-form of $\mathbb{C}^n$*

$$\Re(\Omega_\theta) = (\cos\theta)\,\Re(\Omega) + (\sin\theta)\,\Im(\Omega)$$

is a calibration in the complex Euclidean space $\mathbb{C}^n$, called the ***special Lagrangian calibration with phase*** θ .

PROOF. Since $\Omega_\theta = \Re(\Omega_\theta) + i\Im(\Omega_\theta)$ and $\mathrm{d}\Omega_\theta = e^{-i\theta}\mathrm{d}\Omega = 0$, we have that $\mathrm{d}\Re(\Omega_\theta) = 0$, that is, $\Re(\Omega_\theta)$ is a closed n-form on $\mathbb{C}^n$.

Let V be an n-dimensional real subspace of $\mathbb{C}^n \equiv \mathbb{R}^{2n}$ and $\{v_1, \ldots, v_n\}$ an orthonormal reference of V. Then

$$\begin{aligned}\big(\Re(\Omega_\theta)(v_1,\ldots,v_n)\big)^2 &\le \big(\Re(\Omega_\theta)(v_1,\ldots,v_n)\big)^2 + \big(\Im(\Omega_\theta)(v_1,\ldots,v_n)\big)^2\\ &= \|\Omega_\theta(v_1,\ldots,v_n)\|^2 = \|\Omega(v_1,\ldots,v_n)\|^2.\end{aligned}$$

If $\{e_i = (0,\ldots,1,\ldots,0) \mid 1 \le i \le n\}$ is the standard basis of $\mathbb{C}^n$ and A is the complex matrix such that $Ae_i = v_i$ for all $1 \le i \le n$, then

$$\Omega(v_1,\ldots,v_n) = \big(\det A\big)\Omega(e_1,\ldots,e_n) = \det A.$$

So the above inequality becomes in

$$\big(\Re(\Omega_\theta)(v_1,\ldots,v_n)\big)^2 \le \|\det A\|^2 = \det B,$$

where B is the real representation of A, i.e.

$$B = \begin{pmatrix}\Re(A) & -\Im(A)\\ \Im(A) & \Re(A)\end{pmatrix}.$$

Using the Hadamard inequality, we have

$$\det B \le \|v_1\|\ldots\|v_n\|\|Jv_1\|\ldots\|Jv_n\| = 1,$$

where J is the complex structure of $\mathbb{C}^n$. So we prove that

$$\big(\Re(\Omega_\theta)(v_1,\ldots,v_n)\big)^2 \le 1,$$

and thus, $\Re(\Omega_\theta)$ is a calibration on $\mathbb{C}^n$. Moreover, the equality in the above inequality holds if and only if $\Im(\Omega_\theta)(v_1,\ldots,v_n) = 0$ and $\{v_1,\ldots,v_n,Jv_1,\ldots,Jv_n\}$ is an orthonormal basis of $\mathbb{R}^{2n}$ (i.e., V is a Lagrangian subspace of $\mathbb{C}^n$). □

DEFINITION 6.17 ([**HL82**]). An immersion $\Phi : M^n \to \mathbb{C}^n$ of an oriented manifold M is called a **special Lagrangian with phase** θ if Φ is a calibrated immersion for the special Lagrangian calibration $\Re(\Omega_\theta)$, i.e., $\Phi^*\Re(\Omega_\theta)$ is the volume form of M.

PROPOSITION 6.18 ([**HL82**]). *Let $\Phi : M^n \to \mathbb{C}^n$ be an immersion of an oriented manifold M. Then:*

(1) *Φ is a special Lagrangian immersion with phase θ if and only if Φ is a Lagrangian immersion, i.e. $\Phi^*\omega = 0$, and $\Phi^*\Im(\Omega_\theta) = 0$.*
(2) *If Φ is a special Lagrangian immersion with phase θ and F is the holomorphic isometry of $\mathbb{C}^n$ given by $F(p) = Ap + a$ with $A \in U(n)$ and $a \in \mathbb{C}^n$, then $\Psi = F \circ \Phi$ is a special Lagrangian immersion with phase $\theta + \alpha$, where* $\det A = e^{i\alpha}$.

PROOF. **Proof of 1**. If Φ is a special Lagrangian immersion with phase θ, then, for any $p \in M$ and any oriented orthonormal basis $\{v_1, \ldots, v_n\}$ of T_pM, we have

$$1 = \Phi^*\Re(\Omega_\theta)(v_1, \ldots, v_n) = \Re(\Omega_\theta)(\mathrm{d}\Phi(v_1), \ldots, \mathrm{d}\Phi(v_n)).$$

So, using the last part of Proposition 6.16, we get that $\mathrm{d}\Phi(T_pM)$ is a Lagrangian subspace of $\mathbb{C}^n$ and $\Im(\Omega_\theta)(\mathrm{d}\Phi(v_1), \ldots, \mathrm{d}\Phi(v_n)) = 0$ for any $p \in M$. This means that Φ is a Lagrangian immersion and $\Phi^*\Im(\Omega_\theta) = 0$.

Conversely, if Φ is a Lagrangian immersion and $\Phi^*\Im(\Omega_\theta) = 0$, then clearly $\Phi^*\Re(\Omega_\theta)$ is a volume form on M, and so, Φ is a special Lagrangian immersion.

Proof of 2. Since Φ is a special Lagrangian immersion, $\Phi^*\Re(\Omega_\theta)$ is the volume form of M. Hence

$$\begin{aligned}\Psi^*\Re(\Omega_{\theta+\alpha}) &= (\Phi^* \circ F^*)\Re(\Omega_{\theta+\alpha}) = \Phi^*\Re(e^{-i(\theta+\alpha)}F^*\Omega) \\ &= \Phi^*\Re(e^{-i(\theta+\alpha)}(\det A)\,\Omega) = \Phi^*\Re(\Omega_\theta),\end{aligned}$$

and 2 follows. □

PROPOSITION 6.19 ([**HL82**]). *Let $\Phi : M^n \to \mathbb{C}^n$ be a Lagrangian immersion of an oriented manifold M. Then, Φ is minimal if and only if Φ is a special Lagrangian immersion of phase θ for some $\theta \in [0, 2\pi)$.*

PROOF. The space of oriented Lagrangian n-planes in $\mathbb{C}^n$ is identified with the symmetric space $U(n)/SO(n)$. We denote the determinant map by $\det : U(n)/SO(n) \to \mathbb{S}^1$. If $\nu : M \to U(n)/SO(n)$ is the Gauss map of the Lagrangian immersion Φ, then $\det \circ \nu = e^{i\beta}$, for a function $\beta : M \to \mathbb{R}/2\pi\mathbb{Z}$, called the Lagrangian angle map of Φ. This means that $e^{i\beta(p)} = \Omega(\mathrm{d}\Phi_p(T_pM))$, for any $p \in M$.

Let $\{E_1, \ldots, E_n\}$ be an oriented orthonormal frame on an open subset $O \subset M$, $p \in O$ such that $\nabla_v E_i = 0$, $\forall v \in T_pM$, $i = 1, \ldots, n$. We define $A : O \to U(n)$ by $A = \{\Phi_*(E_1), \ldots, \Phi_*(E_n)\}$. Then, on O, $e^{i\beta} = \det A$. Hence,

$$i\langle \nabla\beta, v\rangle e^{i\beta} = \det A \text{ trace } \left(v(A)\overline{A}^t\right),$$

where $\overline{A}^t$ denotes the transpose conjugate matrix of A. As

$$v(A) = \{\sigma(v, E_1(p)), \ldots, \sigma(v, E_n(p))\},$$

from Proposition 6.7 we get that

$$i\langle \nabla\beta, v\rangle = i\sum_{j=1}^{n}\langle \sigma(v, E_j(p)), JE_j(p)\rangle = in\langle \vec{H}(p), Jv\rangle,$$

which implies that $n\vec{H} = J\nabla\beta$, where $\vec{H}$ is the mean curvature vector of Φ.

Thus, the Lagrangian immersion Φ is minimal if and only if the Lagrangian angle map β is constant. Since

$$e^{i\beta} = \Omega(\Phi_*(E_1), \ldots, \Phi_*(E_n)) = (\Phi^*\Omega)(E_1, \ldots, E_n),$$

we get that β is a constant θ if and only if $\Re\Phi^*(\Omega_\theta)$ is a volume form on M. This proves the proposition. □

Next we will prove that, in the special case $n = 2$, special Lagrangian calibrations in $\mathbb{C}^2$ are complex calibrations for appropriate complex structures in $\mathbb{R}^4$, and so, special Lagrangian surfaces of $\mathbb{C}^2$ are complex surfaces of $\mathbb{R}^4$ endowed with appropriate complex structures.

PROPOSITION 6.20 ([**HL82**]). *In $\mathbb{C}^2 = \{(z_1, z_2) \mid z_j = x_j + iy_j,\ j = 1, 2\}$, the special Lagrangian calibration $\Re(\Omega_\theta)$, $\theta \in [0, 2\pi)$, is given by $\Re(\Omega_\theta) = \omega_\theta$, where ω_θ is the Kähler 2-form of the complex structure on $\mathbb{R}^4$ given by*

$$J_\theta(x_1, y_1, x_2, y_2) = \cos\theta\,(-x_2, y_2, x_1, -y_1) + \sin\theta\,(-y_2, -x_2, y_1, x_1).$$

So, a special Lagrangian surface of $\mathbb{C}^2$ with phase θ is a complex surface of $(\mathbb{R}^4, J_\theta)$.

PROOF. It is easy to check that

$$\Re(\Omega) = dx_1 \wedge dx_2 - dy_1 \wedge dy_2, \quad \Im(\Omega) = dx_1 \wedge dy_2 + dy_1 \wedge dx_2,$$
$$\omega_\theta = \cos\theta\,\big(dx_1 \wedge dx_2 - dy_1 \wedge dy_2\big) + \sin\theta\,\big(dx_1 \wedge dy_2 + dy_1 \wedge dx_2\big),$$

and so, the proof follows. □

We end this section by giving important examples of special Lagrangian submanifolds of $\mathbb{C}^n$, and thus examples of stable minimal orientable submanifolds of $\mathbb{C}^n$.

EXAMPLE 6.21 (Lagrangian graphs [**HL82**]). *Given a smooth function $f : O \subset \mathbb{R}^n \to \mathbb{R}$ on an open set O, the graph of ∇f*

$$\begin{aligned} \Phi : O &\to \mathbb{C}^n = \mathbb{R}^n \oplus i\mathbb{R}^n \\ x &\mapsto \Phi(x) = (x, (\nabla f)_x) \end{aligned}$$

is a special Lagrangian submanifold with phase 0 of $\mathbb{C}^n$ if and only if $\Im(\det(I + i\nabla^2 f)) = 0$. In the case $n = 2$, the above equation of f translates into $\Delta f = 0$, i.e., f is a harmonic function. In the case $n = 3$, the above equation of f translates into the Monge-Ampère equation $\Delta f = \det \nabla^2 f$.

PROOF. First, we prove that Φ is a Lagrangian immersion. Indeed,

$$\mathrm{d}\Phi_p(v) = (v, \nabla_v \nabla f) \equiv v + i\nabla_v \nabla f, \quad \forall v \in T_pM, \quad p \in O.$$

Hence,

$$(\Phi^*\omega)(v, w) = \langle J\mathrm{d}\Phi_p(v), \mathrm{d}\Phi_p(w)\rangle = -\langle \nabla_v \nabla f, w\rangle + \langle \nabla_w \nabla f, v\rangle = 0,$$

which means that Φ is a Lagrangian immersion.

Now, if we prove that $\Phi^*\Im(\Omega) = 0$ if and only if $\Im\big(\det(I + i\nabla^2 f)\big) = 0$, the result follows from Proposition 6.18.

In fact, the metric induced by Φ on O is given by

$$g(v, w) = \langle v, w\rangle + \langle \nabla_v \nabla f, \nabla_w \nabla f\rangle, \quad \forall\, v, w \in T_pO,\, p \in O.$$

Now, let $\{v_1, \ldots, v_n\}$ be an oriented orthonormal basis of $(T_pO, \langle,\rangle)$ which diagonalizes to the Hessian $\nabla^2 f$, i.e., $\nabla_{v_i} \nabla f = \lambda_i v_i$, $1 \le i \le n$. Then, $g(v_i, v_j) =$

$(1+\lambda_i\lambda_j)\delta_{ij}$ and so, $\left\{w_1 = \frac{v_1}{\sqrt{1+\lambda_1^2}}, \ldots, w_n = \frac{v_n}{\sqrt{1+\lambda_n^2}}\right\}$ is an oriented orthonormal basis of (T_pM, g). Hence,

$$\begin{aligned}\Phi^*\Im(\Omega)(w_1,\ldots,w_n) &= \Im(\Omega)(\mathrm{d}\Phi_p(w_1),\ldots,\mathrm{d}\Phi_p(w_n))\\ &= \frac{1}{\prod_{j=1}^n\sqrt{1+\lambda_j^2}}\Im(\Omega)\big((1+i\lambda_1)v_1,\ldots,(1+i\lambda_n)v_n\big)\\ &= \frac{\Im\left(\prod_{j=1}^n(1+i\lambda_j)\right)}{\prod_{j=1}^n\sqrt{1+\lambda_j^2}} = \frac{\Im\big(\det(I+\nabla^2 f)(p)\big)}{\prod_{j=1}^n\sqrt{1+\lambda_j^2}}.\end{aligned}$$

It follows from this equality that $\Phi^*\Im(\Omega) = 0$ if and only if $\Im\big(\det(I+i\nabla^2 f)\big) = 0$. □

DEFINITION 6.22. An immersion $\Phi : M^m \to \mathbb{R}^n$ is called **austere** if for any normal vector $\nu \in T_p^\perp M$ and any $p \in M$,

$$s_{2k-1}(A_\nu) = 0, \quad \text{for} \quad k = 1,\ldots,\lfloor \tfrac{m+1}{2}\rfloor,$$

where s_{2k-1} is the elementary symmetric function of the eigenvalues of A_ν and $\lfloor\cdot\rfloor$ stands for the integer part.

Austere submanifolds were introduced by R. Harvey and B. Lawson, [**HL82**], in the context of special Lagrangian calibrations. Since $s_1(A_\nu) = 0$, such submanifolds are minimal, and for $m = 2$, the austere surfaces are exactly minimal surfaces of Euclidean space. The property of being austere can be interpreted algebraically as the nonzero eigenvalues of the second fundamental form in any normal direction at any point occur in oppositely signed pairs. Thus, every complex submanifold of $\mathbb{C}^m$ is an austere submanifold of $\mathbb{R}^{2m}$. Some authors have studied low dimensional austere submanifolds ([**Bry91, DF01**]).

The next result shows the relation between the austere submanifolds of the Euclidean space and the special Lagrangian submanifolds of the complex Euclidean space.

EXAMPLE 6.23 ([**HL82**]). *Let $\Psi : M^m \to \mathbb{R}^n$ be an austere immersion. Then, $\Phi : T^\perp M \to \mathbb{C}^n = \mathbb{R}^n \oplus i\mathbb{R}^n$ given by*

$$\Phi(\nu) = (\Psi(p), \nu), \quad \nu \in T_p^\perp M, \quad p \in M,$$

is a special Lagrangian immersion with phase $\arg(i^{n-m})$.

PROOF. First, we prove that Φ is a Lagrangian immersion. If $(p,\nu) \in T^\perp M$, then

$$T_{(p,\nu)}T^\perp M = \{(v,\eta) \mid v \in T_pM,\ \eta \in T_p^\perp M\}.$$

Given $(v,0) \in T_{(p,\nu)}T^\perp M$, the curve $\beta(t) = (\alpha(t), \nu_{\alpha(t)}) \subset T^\perp M$ with $\beta(0) = (p,\nu)$, $\alpha'(0) = v$ and $\nu_{\alpha(t)}$ a parallel vector field along $\alpha(t)$ with respect to the normal connection of Ψ, satisfies $\beta'(0) = (v, 0)$. Thus,

$$\mathrm{d}\Phi_{(p,\nu)}((v,0)) = \big(\mathrm{d}\Psi_p(v), -\mathrm{d}\Psi_p(A_\nu v)\big),$$

where A_ν is the Weingarten endomorphism of Ψ.

Also, given $(0,\eta) \in T_{(p,\nu)}T^\perp M$, we have

$$\mathrm{d}\Phi_{(p,\nu)}((0,\eta)) = (0,\eta).$$

Therefore,

$$\begin{aligned}(\Phi^*\omega)((v,\eta),(w,\xi)) &= \langle J\mathrm{d}\Phi_{(p,\nu)}(v,\eta), \mathrm{d}\Phi_{(p,\nu)}(w,\xi)\rangle \\ &= \langle (\mathrm{d}\Psi_p(A_\nu v) - \eta, \mathrm{d}\Psi_p(v)), (\mathrm{d}\Psi_p(w), -\mathrm{d}\Psi_p(A_\nu w) + \xi)\rangle \\ &= \langle \sigma(v,w),\nu\rangle - \langle \sigma(w,v),\nu\rangle = 0,\end{aligned}$$

where σ is the second fundamental form of Ψ. This means that Φ is a Lagrangian immersion.

Now, if we prove that $\Phi^*\Im(i^{m-n}\Omega) = 0$ if and only if Ψ is an austere immersion, the result follows from Proposition 6.18.

In fact, the metric induced by Φ on $T^\perp M$ is given by

$$g\big((v,\eta),(w,\xi)\big) = \langle v,w\rangle + \langle A_\nu v, A_\nu w\rangle + \langle \eta,\xi\rangle, \quad \forall\,(v,\eta),(w,\xi) \in T_{(p,\nu)}T^\perp M.$$

Let $\{v_1,\dots,v_m\}$ be an orthonormal basis of $(T_pM,\langle,\rangle)$ which diagonalizes to A_ν, i.e., $A_\nu v_j = \lambda_j v_j$, $1 \le j \le m$, and $\{\nu_1,\dots,\nu_{n-m}\}$ an orthonormal basis of $T_p^\perp M$ such that the basis $\{v_1,\dots,v_m,\nu_1,\dots,\nu_{n-m}\}$ is an oriented basis of $T_{(p,\nu)}T^\perp M$. Then, $g((v_i,\nu_k),(v_j,\nu_l)) = (1+\lambda_i\lambda_j)\delta_{ij} + \delta_{kl}$ and

$$\left\{ w_1 = \left(\frac{v_1}{\sqrt{1+\lambda_1^2}}, 0\right), \dots, w_m = \left(\frac{v_m}{\sqrt{1+\lambda_m^2}}, 0\right), (0,\nu_1),\dots,(0,\nu_{n-m}) \right\}$$

is an oriented orthonormal basis of $(T_{(p,\nu)}T^\perp M, g)$. Hence,

$$\begin{aligned}&\Phi^*\Im(i^{m-n}\Omega)(w_1,\dots,w_m,\nu_1,\dots,\nu_{n-m}) \\ &= \Im(i^{m-n}\Omega)\left(\frac{(1-i\lambda_1)\mathrm{d}\Psi_p(v_1)}{\sqrt{1+\lambda_1^2}},\dots,\frac{(1-i\lambda_m)\mathrm{d}\Psi_p(v_m)}{\sqrt{1+\lambda_m^2}}, i\nu_1,\dots,i\nu_{n-m}\right) \\ &= \frac{\Im\left(\prod_{j=1}^m (1-i\lambda_j)\right)}{\prod_{j=1}^m \sqrt{1+\lambda_j^2}}.\end{aligned}$$

From this equality follows that $\Phi^*\Im(i^{n-m}\Omega) = 0$ if and only if $\Im(\prod_{j=1}^m(1-i\lambda_j)) = 0$. But $\Im(\prod_{j=1}^m(1-i\lambda_j)) = 0$ means that Ψ is austere. □

In Corollary 5.26 it was proved that the cone over a minimal Legendrian submanifold of the sphere is stable. Also, in Remark 5.27, it was mentioned that such a cone is a special Lagrangian submanifold of the complex Euclidean space. In the following example, we will not only prove the content of Remark 5.27, but also generalize it. This will provide important examples of special Lagrangian submanifolds.

EXAMPLE 6.24 ([**CU99**]). *Let $\Psi : N^{n-1} \to \mathbb{S}^{2n-1} \subset \mathbb{C}^n$ be a minimal Legendrian immersion of an orientable manifold N and let $\alpha : I \to \mathbb{C}^*$ be a regular curve such that the curve α^n has curvature zero, where I is an open interval of $\mathbb{R}$. Then, the map $\Phi : I \times N \to \mathbb{C}^n$ defined by*

$$\Phi(t,p) = \alpha(t)\Psi(p), \quad \forall\, t \in I,\ p \in N,$$

is a special Lagrangian immersion with phase θ for some $\theta \in [0,2\pi)$.

REMARK 6.25. Since α^n has zero curvature, it is a straight line on $\mathbb{C}$. If this straight line passes through $0 \in \mathbb{C}$, then α is also contains in a straight line in $\mathbb{C}$ passing through $0 \in \mathbb{C}$, and so, up to rotations, α can be taken as $\alpha(t) = t \in \mathbb{C}$.

In this case, the examples of special Lagrangian immersions given in Example 6.24 are the cones given in Corollary 5.26.

If the straight line α^n does not contain $0 \in \mathbb{C}$, and since rotations of α of angle θ produce rotations of α^n of angle $n\theta$, up to rotations, we can take α^n as the straight line $\Im(z) = c > 0$. Also, up to homotheties of center 0, we can take $c = 1$ and parametrize α^n by $\alpha^n(t) = \sinh(nt) + i$. So, the curve α is the embedded curve given by

$$\alpha(t) = \cosh^{1/n}(nt)e^{i\beta(t)}, \qquad \beta(t) = \frac{\pi}{2n} - \frac{2}{n}\arctan(\tanh\frac{nt}{2}).$$

It is clear that rotations and homotheties performed on the α curve in the example give rise to rotations and homotheties of the corresponding special Lagrangian immersions.

PROOF. First we prove that Φ is a Lagrangian immersion. Indeed,

$$\Phi_t = \alpha'(t)\Psi, \quad \mathrm{d}\Phi_{(t,p)}(0,v) = \alpha(t)\mathrm{d}\Psi_p(v),$$

which implies that Φ is an immersion. Now, using that Ψ is a Legendrian immersion (Definition 5.13), we have

$$\begin{aligned}(\Phi^*\omega)\big((\partial_t,0),(0,v)\big) &= \langle J\Phi_t, \alpha(t)\mathrm{d}\Psi_p(v)\rangle = \Im\left(\overline{\alpha}'\alpha\sum_{i=1}^n \overline{\Psi_i}(\mathrm{d}\Psi_p(v)_i\right)\\ &= \Im\left(\overline{\alpha}'\alpha\{\langle\Psi,\mathrm{d}\Psi_p(v)\rangle + i\langle J\Psi,\mathrm{d}\Psi_p(v)\rangle\}\right) = 0.\end{aligned}$$

Also,

$$(\Phi^*\omega)\big((0,v),(0,w)\big) = \|\alpha\|^2\langle J\mathrm{d}\Psi_p(v),\mathrm{d}\Psi_p(w)\rangle = 0,$$

and so, the immersion Φ is Lagrangian.

Now, if $\langle,\rangle$ also denotes the induced metric on N by Ψ, then the induced metric on $I\times N$ by Φ is given by $\|\alpha'\|^2\mathrm{d}t^2 + \|\alpha\|^2\langle,\rangle$. So, if $\{v_2,\dots,v_n\}$ is an oriented orthonormal basis of T_pN, then

$$\left\{w_1 = \left(\frac{\partial_t}{\|\alpha'\|},0\right), w_2 = \left(0,\frac{v_2}{\|\alpha\|}\right),\dots, w_n = \left(0,\frac{v_n}{\|\alpha\|}\right)\right\}$$

is an oriented orthonormal basis of $T_{(t,p)}I\times N$. Hence,

$$\begin{aligned}(\Phi^*\Omega)(w_1,\dots,w_n) &= \frac{1}{\|\alpha'(t)\|\|\alpha(t)\|^{n-1}}\Omega\big(\Phi_t,\mathrm{d}\Phi_{(t,p)}(0,v_2),\dots,\mathrm{d}\Phi_{(t,p)}(0,v_n)\big)\\ &= \frac{\alpha'(t)\alpha(t)^{n-1}}{\|\alpha'(t)\|\|\alpha(t)\|^{n-1}}\Omega\big(\Psi(p),\mathrm{d}\Psi_p(v_2),\dots,\mathrm{d}\Psi_p(v_n)\big)\\ &= \frac{\alpha'(t)\alpha(t)^{n-1}}{\|\alpha'(t)\|\|\alpha(t)\|^{n-1}}(\det B)(p),\end{aligned}$$

where B is the matrix $\{\Psi(p),\mathrm{d}\Psi_p(v_2),\dots,\mathrm{d}\Psi_p(v_n)\}$.

Calling $\beta(t) = \alpha(t)^n$, we have

$$\frac{\beta'(t)}{\|\beta'(t)\|} = \frac{\alpha'(t)\alpha(t)^{n-1}}{\|\alpha'(t)\|\|\alpha(t)\|^{n-1}},$$

and, as β has zero curvature, $\frac{\beta'(t)}{\|\beta'(t)\|}$ is a constant e^{ia} for some $a\in[0,2\pi)$. Thus, $\frac{\alpha'(t)\alpha(t)^{n-1}}{\|\alpha'(t)\|\|\alpha(t)\|^{n-1}} = e^{ia}$, for all $t\in I$.

On the other hand, if $\{V_2, \ldots, V_n\}$ is an orientable local orthonormal reference around a point $p \in N$ such that $(V_i)_p = v_i$, $\nabla_v V_i = 0$ for any $v \in T_pN$, then

$$B = \{\Psi, \Psi_*(V_2), \ldots, \Psi_*(V_n)\}.$$

Since Ψ is Legendrian, B is a unitary matrix, and so, for any $v \in T_pN$, we have

$$v(\det B) = (\det B)(p) \operatorname{trace}\big(v(B)\overline{B}^t(p)\big).$$

But

$$v(B) = \big\{\mathrm{d}\Psi_p(v), \sigma(v, V_2(p)) - \langle v, V_2(p)\rangle\Psi(p), \ldots, \sigma(v, V_n(p)) - \langle v, V_n(p)\rangle\Psi(p)\big\},$$

and so, taking into account that Ψ is a minimal Legendrian immersion, we deduce

$$\begin{aligned}\operatorname{trace}\big(v(B)\overline{B}^t(p)\big) &= \langle \mathrm{d}\Psi_p(v), \Psi(p)\rangle + i\langle \mathrm{d}\Psi_p(v), J\Psi(p)\rangle \\ &\quad + \sum_{j=2}^n \{\langle \sigma(v, V_j(p)) - \langle v, V_j(p)\rangle\Psi(p), \mathrm{d}\Psi_p(V_j(p))\rangle\} \\ &\quad + i\sum_{j=2}^n \{\langle \sigma(v, V_j(p)) - \langle v, V_j(p)\rangle\Psi(p), J\mathrm{d}\Psi_p(V_j(p))\rangle\} \\ &= i\sum_{j=2}^n \{\langle \sigma(V_j(p), V_j(p)), \mathrm{d}\Psi_p(v)\rangle\} = (n-1)i\langle \vec{H}(p), \mathrm{d}\Psi_p(v)\rangle = 0.\end{aligned}$$

Thus, $\det B$ is a constant function on N, i.e., $\det B = e^{ib}$ for some $b \in [0, 2\pi)$.

Now it is clear that $\Phi^*(e^{-i(a+b)}\Omega)(w_1, \ldots, w_n) = 1$ and so, Φ is a special Lagrangian immersion with phase $a + b$. □

Next, we are going to show some special cases of the construction made in Example 6.24.

EXAMPLE 6.26 (Lagrangian catenoid). *We consider the totally geodesic Legendrian embedding*

$$\Psi : \mathbb{S}^{n-1} \subset \mathbb{R}^n \to \mathbb{S}^{2n-1} \subset \mathbb{C}^n, \qquad \Psi(x_1, \ldots, x_n) = (x_1, \ldots, x_n).$$

Then, $\Phi : \mathbb{R} \times \mathbb{S}^{n-1} \to \mathbb{C}^n$, *given by* $\Phi(t, x) = \cosh^{1/n}(nt)e^{i\beta(t)}x$, *where* $\beta(t) = \frac{\pi}{2n} - \frac{2}{n}\arctan(\tanh\frac{nt}{2})$, *is a special Lagrangian embedding with phase* 0.

Denoting $z = x + iy \in \mathbb{C}^n = \mathbb{R}^n + i\mathbb{R}^n$, it is easy to see that $M = \Phi(\mathbb{R} \times S^{n-1})$ is given by

$$M = \Big\{x + iy \in \mathbb{C}^n \mid \|y\|x = \|x\|y,\, \Im((\|x\| + i\|y\|)^n) = 1,\, \|y\| < \|x\|\tan\frac{\pi}{n}\Big\}.$$

This example was studied by R. Harvey and H. Lawson in [**HL82**], Theorem 3.5, and it is invariant under the natural action of $SO(n)$ on $\mathbb{C}^n$. Other characterizations of this example have been given in [**CU99**], and the similarity of its properties to the catenoid of Euclidean space, makes it known as the **Lagrangian catenoid**. As we indicated in Proposition 6.20, when n is two, the Lagrangian catenoid corresponds the complex surface of $\mathbb{C}^2$ given by $\{(z_1, z_2) \in \mathbb{C}^2 \mid z_1z_2 = 1\}$.

EXAMPLE 6.27. *We consider the minimal Legendrian embedding* $\Psi : T^{n-1} \to \mathbb{S}^{2n-1}$ *given in Example* 5.16. *Then,* $\Phi : \mathbb{R} \times T^{n-1} \to \mathbb{C}^n$, *given by* $\Phi(t, z) = \cosh^{1/n}(nt)e^{i\beta(t)}\Psi(z)$, *where* $\beta(t) = \frac{\pi}{2n} - \frac{2}{n}\arctan(\tanh\frac{nt}{2})$, *is a special Lagrangian embedding.*

This example is invariant under the action of the subgroup

$$\{\mathrm{diag}(e^{i\theta_1}, \dots, e^{i\theta_n}) \mid \theta_1 + \dots + \theta_n = 0\} \subset SU(n)$$

and it was studied by R. Harvey and B. Lawson in [**HL82**], Theorem 3.1.

Special Lagrangian calibrations on the complex Euclidean space $\mathbb{C}^n$ can also be defined on a broader family of Kähler manifolds, the Calabi-Yau manifolds (see Definition 1.29). In fact, it follows from Proposition 1.30 that if $(\overline{M}^n, g, J)$ is a Calabi-Yau manifold of complex dimension n, then $\overline{M}$ has a non-trivial holomorphic n-form Ω such that

$$\omega^n = \frac{(-1)^{\frac{n(n-1)}{2}} i^n n!}{2^n} \Omega \wedge \overline{\Omega},$$

where ω is the Kähler 2-form on $\overline{M}$. The following result can be proved in a similar way to the case where the ambient manifold is the complex Euclidean space.

PROPOSITION 6.28. *Let $(\overline{M}^n, g, J, \Omega)$ be a Calabi-Yau manifold of complex dimension n.*

(1) *For $\theta \in [0, 2\pi)$, we define $\Omega_\theta = e^{-i\theta}\Omega$. Then, the n-form*
$$\Re(\Omega_\theta) = (\cos\theta)\, \Re(\Omega) + (\sin\theta)\, \Im(\Omega)$$
is a calibration in $\overline{M}$, called the ***special Lagrangian calibration with phase θ****.*

(2) *$\Phi : M^n \to \overline{M}^n$ is a calibrated immersion for Ω_θ if and only if Φ is a Lagrangian immersion, i.e. $\Phi^*\omega = 0$ and $\Phi^*\Im(\Omega_\theta) = 0$.*

(3) *Let $\Phi : M^n \to \overline{M}^n$ be a Lagrangian immersion of an oriented manifold M. Then, Φ is minimal if and only if Φ is a calibrated immersion of Ω_θ for some $\theta \in [0, 2\pi)$.*

For a detailed study of Calabi-Yau manifolds and their calibrated submanifolds we recommend the book by M. Gross, D. Huybrechts and D. Joyce, [**GHJ03**].

6.4. Compact minimal submanifolds of the complex projective space

Let $\mathbb{CP}^m$ be the complex projective space endowed with the Fubini-Study metric g of constant holomorphic sectional curvaturte 4. We know (see Section 1.6) that $\mathbb{CP}^m$ is an Einstein-Kähler manifold with Ricci tensor $\overline{S} = 2(m+1)\, g$.

Similar to the case of real projective space (see Section 5.4), let us recall the best isometric embedding of $\mathbb{CP}^m$ in Euclidean space given by S.-S. Tai in [**Tai68**]. As in the previous cases, such an embedding will be a good tool to study the index of the minimal submanifolds of $\mathbb{CP}^m$.

Let $H(m+1) = \{B \in gl(m+1, \mathbb{C}) \mid \overline{B} = B^t\}$ be the space of Hermitian matrices of order $m+1$, endowed with the metric $\langle B, C\rangle = \frac{1}{2}\,\mathrm{trace}\, BC$. Then, $(H(m+1), \langle , \rangle)$ is isometric to the Euclidean space $\mathbb{R}^{(m+1)^2}$. The complex projective space $(\mathbb{CP}^m, g)$ is embedded in $H(m+1)$ by the embedding $\Psi : \mathbb{CP}^m \to H(m+1)$ defined by

$$\Psi(\Pi(z)) = z^t\overline{z} \in H(m+1), \quad z = (z_1, \dots, z_{m+1}) \in \mathbb{S}^{2m+1} \subset \mathbb{C}^{m+1},$$

where $\Pi : \mathbb{S}^{2m+1} \to \mathbb{CP}^m$ is the Hopf fibration. The image of $\mathbb{CP}^m$ in $H(m+1)$ is

$$\Psi(\mathbb{CP}^m) = \{B \in H(m+1) \mid BB = B,\ \mathrm{trace}\, B = 1\}.$$

In the next result we collect the most important properties of the embedding Ψ.

LEMMA 6.29.

(1) *The metric induced by* $\langle , \rangle$ *on* $\mathbb{CP}^m$ *is the Fubiny-Study metric* g *of constant holomorphic sectional curvature 4.*
(2) *The second fundamental form* $\overline{\sigma}$ *of* Ψ *is parallel, i.e.,* $\overline{\nabla}\overline{\sigma} = 0$.
(3) $\overline{\sigma}(v, w) = \overline{\sigma}(Jv, Jw)$*, for all* $v, w \in T\mathbb{CP}^m$.
(4) *The identity matrix* I *satisfies* $I \in \mathfrak{X}^\perp(\mathbb{CP}^m)$ *and* $\langle \overline{\sigma}(v, w), I\rangle = 0$*, for all* $v, w \in T\mathbb{CP}^m$.
(5) *If* $\overline{A}$ *is the Weingarten endomorphism of* Ψ*, then*

$$\overline{A}_{\overline{\sigma}(v,w)}z = 2g(v, w)z + g(w, z)v + g(v, z)w + g(Jw, z)Jv + g(Jv, z)Jw,$$

for all $v, w, z \in T\mathbb{CP}^m$.
(6) *If* $A \in \mathbb{CP}^m$ *and* $\{e_1, \dots, e_m, e_{1*} = Je_1, \dots, e_{m*} = Je_m\}$ *is a basis of* $T_A\mathbb{CP}^m$*, then* $\mathcal{B} = \{\overline{\sigma}_{ij}, \overline{\sigma}_{ij*}, 1 \le i < j \le m\} \cup \{\overline{\sigma}_{ii}, 1 \le i \le m\}$ *is an orthogonal basis of the subspace* V *of* $H(m + 1)$ *defined by* $V = \{B \in T_A^\perp\mathbb{CP}^m \mid \langle B, I\rangle = 0\}$*, whose dimension is* m^2.
(7) *If* $\mathcal{K}(\mathbb{CP}^m)$ *denotes the space of Killing vector fields on* $\mathbb{CP}^m$*, then*

$$\mathcal{K}(\mathbb{CP}^m) = \{JX_B \mid B \in H(m + 1), \langle B, I\rangle = 0\},$$

where X_B *is the tangential component in* $\mathbb{CP}^m$ *of* B.

PROOF. If p is a point of $\mathbb{CP}^m$ and $A = \Psi(p)$, then it is not difficult to check that

$$\Psi_*(T_p\mathbb{CP}^m) = \{B \in H(m + 1) \mid AB + BA = B\},$$
$$T_p^\perp\mathbb{CP}^m = \{B \in H(m + 1) \mid BA = AB\}.$$

From here it is easy to prove 1 and that the complex structure J and the second fundamental form $\overline{\sigma}$ are given by

$$JB = i(I - 2A)B, \quad \overline{\sigma}(B, C) = (BC + CB)(I - 2A), \quad \forall B, C \in \Psi_*(T_p\mathbb{CP}^m).$$

Using the above information, it is not difficult to check 2 and it is straightforward to prove 3, 4 and 5. Now 6 is a simple consequence of 3, 4 and 5.

Finally, in order to prove 7, we recall that the isometry group of the complex projective space $\mathbb{CP}^m$ has dimension $(m + 1)^2 - 1$ and its Lie algebra is $\mathcal{K}(\mathbb{CP}^m)$.

Given $B \in H(m + 1)$ with $\langle B, I\rangle = 0$, we write $B = X_B + B^N$, with B^N the normal component of B. Then, we have that $\overline{\nabla}_V X_B = \overline{A}_{B^N}V$, for any tangent vector field V to $\mathbb{CP}^m$. So, if W is another vector field tangent to $\mathbb{CP}^m$, we have that

$$g(\overline{\nabla}_V JX_B, W) = -\langle \overline{\sigma}(V, JW), B\rangle.$$

Now, property 3 of the lemma says that JX_B is a Killing vector field of $\mathbb{CP}^m$. Also, $X_B = 0$ if and only if $\langle \Psi, B\rangle =$ constant. Given suitable matrices $C's$ in $\Psi(\mathbb{CP}^m) \subset H(m + 1)$, the above condition says that $B = \lambda I$ for some real number λ. Since $\langle B, I\rangle = 0$, we get $\lambda = 0$ and so, $B = 0$. Hence the kernel of the linear map

$$\{B \in H(m + 1) \mid \langle B, I\rangle = 0\} \to \mathcal{K}(\mathbb{CP}^m), \quad B \mapsto JX_B$$

is zero, and since both spaces have dimension $(m + 1)^2 - 1$, point 7 follows. □

Using the above embedding, we will classify the stable compact minimal submanifolds of the complex projective space.

THEOREM 6.30 ([**LS73**]). *Let* $\Phi : M^n \to (\mathbb{CP}^m, g)$ *be a minimal immersion of a compact manifold* M*. Then,* M *is stable if and only if* Φ *is a complex immersion.*

PROOF. The sufficient condition was proved in Theorem 6.4.

Suppose that M is stable. We will use Proposition 2.42 if $(\mathbb{R}^p, \langle,\rangle)$ is $(H(m+1), \langle,\rangle)$. If $\{B_1, \dots, B_{(m+1)^2}\}$ is an orthonormal basis of $H(m+1)$, then it follows from Proposition 2.42 that

$$\sum_{\mu=1}^{(m+1)^2} \mathcal{Q}(B_\mu^\perp) = \sum_{i=1}^{n}\sum_{r=1}^{2m-n} \int_M \{-\overline{R}(\xi_r, e_i, e_i, \xi_r) + \|\overline{\sigma}(e_i, \xi_r)\|^2\}\, dv,$$

where $\mathcal{Q}$ is the index form of Φ, $B_\mu^\perp$ is the normal component in $T^\perp M$ of B_μ, and $\{e_1, \dots, e_n\}$ (respectively $\{\xi_1, \dots, \xi_{2m-n}\}$) is a local orthonormal reference on TM (respectively on $T^\perp M$). Now, considering the expression of the curvature $\overline{R}$ of $\mathbb{CP}^m$ given in (1.7) and property 4 of Lemma 6.29, we get that

$$\sum_{i=1}^{n}\sum_{r=1}^{2m-n} \overline{R}(e_i, \xi_r, \xi_r, e_i) = n(2m-n) + 3\sum_{i=1}^{n} \|(Je_i)^\perp\|^2,$$

$$\sum_{i=1}^{n}\sum_{r=1}^{2m-n} \|\overline{\sigma}(e_i, \xi_r)\|^2 = n(2m-n),$$

and thus,

$$\sum_{\mu=1}^{(m+1)^2} \mathcal{Q}(B_\mu^\perp) = -3\sum_{i=1}^{n} \int_M \|(Je_i)^\perp\|^2\, dv,$$

where $(Je_i)^\perp$ is the normal component of Je_i. Since M is stable, we have that $\mathcal{Q}(B_\mu^\perp) \geq 0$ for any $\mu \in \{1, \dots, (m+1)^2\}$, and so, the last formula says that $(Je_i)^\perp = 0$ for any $i \in \{1, \dots, n\}$. This means that Φ is a complex immersion. □

Using again the Tai embedding and arguing similarly as in Theorem 5.1,3, we obtain a lower bound of the nullity of any compact complex submanifold of $\mathbb{CP}^m$, characterizing the totally geodesic embedding of the Example 2.60 as the only one that achieves this bound.

THEOREM 6.31 ([**Kim77**]). *Let $\Phi : M^n \to \mathbb{CP}^m$ be a complex immersion of a compact manifold M of complex dimension n. Then,* $\mathrm{Nul}(M) \geq 2(n+1)(m-n)$, *and the equality holds if and only if Φ is the totally geodesic embedding of $\mathbb{CP}^n$ (Example* 2.60*).*

PROOF. First, from Example 2.60, the nullity of the totally geodesic embedding is $2(n+1)(m-n)$.

Following Lemma 6.29, let $\mathcal{K}(\mathbb{CP}^m)$ be the space of Killing vector fields on $\mathbb{CP}^m$, whose dimension is $m(m+2)$ and let

$$\mathcal{K}_M(\mathbb{CP}^m) = \{(JX_B)_{|M} \mid JX_B \in \mathcal{K}(\mathbb{CP}^m)\}$$

be the space of restrictions tu M of the Killing vector fields on $\mathbb{CP}^m$. To have some control over its dimension, we define the linear map

$$F : \mathcal{K}(\mathbb{CP}^m) \to \mathcal{K}_M(\mathbb{CP}^m), \qquad F(JX_B) = (JX_B)_{|M}.$$

Then $\dim \mathcal{K}_M(\mathbb{CP}^m) = m(m+2) - \dim\ker F$. If $JX_B \in \ker F$, then over M we have that $B = B^N$, where B^N is the normal component of B to the Tai embedding Ψ. If p is a point of M, $A = \Phi(p)$, and $\{e_1, \dots, e_m, e_{1*} = Je_1, \dots, e_{m*} = Je_m\}$ is an orthonormal basis of $T_A\mathbb{CP}^m$ with $\{e_1,\dots,e_n,e_{1*},\dots,e_{n*}\}$ a basis of $\mathrm{d}\Phi(T_pM)$, then, using Lemma 6.29, we have that $\overline{A}_B v = \overline{A}_{(B^N)_p} v = 0$, for any $v\in T_pM$ and

$$B=(B^N)_p=\sum_{1\le i<j\le m}\{\lambda_{ij}\overline{\sigma}_{ij}+\lambda_{ij^*}\overline{\sigma}_{ij^*}\}+\sum_{i=1}^m \mu_i\overline{\sigma}_{ii},\qquad \lambda_{ij},\lambda_{ij*},\mu_i\in\mathbb{R}.$$

Hence, using Lemma 6.29,5, and that $\overline{A}_B v = 0$ for any $v\in T_pM$, it is easy to get that

$$B=\sum_{n+1\le i<j\le m}\{\lambda_{ij}\overline{\sigma}_{ij}+\lambda_{ij^*}\overline{\sigma}_{ij^*}\}+\sum_{i=n+1}^m \mu_i\overline{\sigma}_{ii}.$$

Thus, B belongs to a subspace of dimension $(m-n)^2$ of $T_A^\perp\mathbb{CP}^m$ and so, $\dim\ker F\le (m-n)^2$. This implies that $\dim\mathcal{K}_M(\mathbb{CP}^m)\ge 2m+n(2m-n)$.

Now, we consider the linear map $G:\mathcal{K}_M(\mathbb{CP}^m)\to\mathcal{J}(M)$ given by

$$G((JX_B)_{|M})=X_B^\perp,$$

discussed in Proposition 2.12, where $\mathcal{J}(M)$ is the space of Jacobi fields on M (Definition 2.11). Then,

$$\mathrm{Nul}(M)\ge\dim\,\mathcal{J}(M)\ge\dim\,\mathcal{K}_M(\mathbb{S}^m)-\dim\,\ker(G)\ge 2m+n(2m-n)-\dim\,\ker(G).$$

If $X_B\in\ker(G)$, then the tangential component of X_B in M is also a Killing vector field on M, and so, using [**Lic58**], $\dim\,\ker(G)\le\dim\mathrm{Iso}(M)\le n(n+2)$. Finally we get that $\mathrm{Nul}(M)\ge 2(n+1)(m-n)$ and if the equality holds, then Φ is the totally geodesic embedding of $\mathbb{CP}^n$. This concludes the proof. □

In Example 2.61 we saw that the Clifford hypersurfaces of $\mathbb{CP}^n$ have index one. The following open problem is interesting:

Classify the two-sided compact minimal hipersurfaces of $\mathbb{CP}^n$ with index one. Are they the Clifford hypersurfaces?

Next we will study **minimal Lagrangian immersions of compact manifolds in $\mathbb{CP}^n$**. In this case (see Corollary 6.13), the Jacobi operator $\mathcal{L}:\Omega^1(M)\to\Omega^1(M)$ is given by $\mathcal{L}=\Delta+2(n+1)$, where Δ is the Hodge Laplacian on the compact submanifold M. From Theorem 6.30, M is unstable, and so, it is interesting to study properties of its index and nullity. We begin by giving some important examples.

EXAMPLE 6.32. *Let us consider the totally geodesic Legendrian embedding of the unit sphere $\mathbb{S}^n$ in $\mathbb{S}^{2n+1}$:*

$$(x_1,\dots,x_{n+1})\in\mathbb{S}^n\subset\mathbb{R}^{n+1}\overset{i}{\mapsto}(x_1,\dots,x_{n+1})\in\mathbb{S}^{2n+1}\subset\mathbb{C}^{n+1}.$$

If $\Pi:\mathbb{S}^{2n+1}\to\mathbb{CP}^n$ is the Hopf fibration, then $\Phi=\Pi\circ i:\mathbb{S}^n\to\mathbb{CP}^n$ is a totally geodesic Lagrangian immersion. Furthermore, $\mathrm{Ind}_0(\Phi)=n+1$, $\mathrm{Ind}_1(\Phi)=\frac{1}{2}n(n+1)$, $\mathrm{Ind}(\Phi)=\frac{1}{2}(n+1)(n+2)$, *and* $\mathrm{Nul}(\Phi)=\frac{1}{2}n(n+3)$.

PROOF. Since i is a totally geodesic Legendrian embedding, then Φ is a totally geodesic Lagrangian immersion. Moreover, from [**GM75**], the three first eigenvalues of $\Delta:\Omega^1(\mathbb{S}^n)\to\Omega^1(\mathbb{S}^n)$ are: n, with multiplicity $n+1$; $2(n-1)$, with multiplicity $\frac{n(n+1)}{2}$ and $2(n+1)$, with multiplicity $\frac{n(n+3)}{2}$. Also n is an eigenvalue of Δ acting on $\mathrm{d}(C^\infty(\mathbb{S}^n))\subset\Omega^1(\mathbb{S}^n)$ and $2(n-1)$ is an eigenvalue of Δ acting on $\delta(\Omega^2(\mathbb{S}^n))\subset\Omega^1(\mathbb{S}^n)$. So, the result follows from Corollary 6.13. □

EXAMPLE 6.33. *The embedding* $i : \mathbb{S}^n \to \mathbb{S}^{2n+1}$ *of Example* 6.32 *projects into the totally geodesic Lagrangian embedding* $\Phi : \mathbb{RP}^n \to \mathbb{CP}^n$, *making the following diagram commutative:*

$$\begin{array}{ccc} \mathbb{S}^n & \xrightarrow{i} & \mathbb{S}^{2n+1} \\ \downarrow \Pi' & & \downarrow \Pi \\ \mathbb{RP}^n & \xrightarrow{\Phi} & \mathbb{CP}^n. \end{array}$$

Moreover, $\text{Ind}_0(\Phi) = 0$, $\text{Ind}_1(\Phi) = \text{Ind}(\Phi) = \frac{1}{2}n(n+1)$ *and* $\text{Nul}(\Phi) = \frac{1}{2}n(n+3)$.

PROOF. The action of $\mathbb{S}^1$ on $\mathbb{S}^{2n+1}$ induces the action of the antipodal map on $\mathbb{S}^n$, and hence, Φ is an embedding. It is also clear that Φ is totally geodesic and Lagrangian. Moreover, the eigenvalues of $\Delta : \Omega^1(\mathbb{RP}^n) \to \Omega^1(\mathbb{RP}^n)$ are the eigenvalues of $\mathbb{S}^n$ whose corresponding eigenforms project into $\mathbb{RP}^n$. It is clear that the eigenforms of $\mathbb{S}^n$ of the eigenvalue n are not invariant under the antipodal map and so, they do not project to $\mathbb{RP}^n$. However, the eigenforms corresponding to the eigenvalues $2(n-1)$ and $2(n+1)$ of the sphere are invariant under the antipodal map, and so, the first two nonzero eigenvalues of $\mathbb{RP}^n$ are $2(n-1)$ and $2(n+1)$ with the same multiplicity as in $\mathbb{S}^n$. From here the result follows, again using Corollary 6.13. □

EXAMPLE 6.34. *Let* $\Psi : T^n \to \mathbb{S}^{2n+1}$ *be the Legendrian submanifold given in Example* 5.16. *Then, it induces a minimal Lagrangian embedding* $\Phi : \mathbb{T}^n = T^n/\mathbb{Z}_{n+1} \to \mathbb{CP}^n$ *of a flat torus* $\mathbb{T}^n$, *making the following diagram commutative:*

$$\begin{array}{ccc} T^n & \xrightarrow{\Psi} & \mathbb{S}^{2n+1} \\ \downarrow \Pi' & & \downarrow \Pi \\ \mathbb{T}^n & \xrightarrow{\Phi} & \mathbb{CP}^n. \end{array}$$

Moreover, $\text{Ind}_0(\Phi) = \text{Ind}_1(\Phi) = 0$, $\text{Ind}(\Phi) = n$ *and* $\text{Nul}(\Phi) = n^2(n+1)$. *When* $n = 2$, $\mathbb{T}^2$ *will be called the* **Lagrangian Clifford torus**.

REMARK 6.35. We call $\mathbb{T}^2$ the Lagrangian Clifford torus because its geometric properties are similar to those of the Clifford torus of $\mathbb{S}^3$. Both are characterized as the unique embedded flat minimal tori of $\mathbb{S}^3$ and $\mathbb{CP}^2$ respectively. However, they are intrinsically very different, since the conformal structure of the Clifford torus of $\mathbb{S}^3$ is the square one, while the conformal structure of the Lagrangian Clifford torus is the hexagonal one.

S. Brendle, in [**Bre13**], proved the Lawson conjecture by showing that the only embedded minimal torus in $\mathbb{S}^3$ is the Clifford one. Y. Oh, in [**Oh94**], conjectured that the Lagrangian Clifford torus is the only embedded Lagrangian minimal torus in $\mathbb{CP}^2$. As far as I know, this conjecture remains unsolved. Considering that (see [**EGT85**]) when the genus of a compact orientable minimal Lagrangian surface of $\mathbb{CP}^2$ is greater than or equal to two, such a surface cannot be embedded, Y. Oh actually conjectured that:

> *The Lagrangian Clifford torus and* $\mathbb{RP}^2$ *are the only embedded minimal Lagrangian compact surfaces in* $\mathbb{CP}^2$.

PROOF. It is clear that the action of $\mathbb{S}^1$ on $\mathbb{S}^{2n+1}$ of the Hopf fibration $\Pi : \mathbb{S}^{2n+1} \to \mathbb{CP}^n$ induces an action of

$$\mathbb{Z}_{n+1} = \left\{ e^{i\theta} \mid \theta = \frac{2\pi k}{n+1}, \, 0 \le k \le n \right\} \subset \mathbb{S}^1$$

on T^n, so that $\Pi \circ \Psi = \Phi \circ \Pi'$. Now it is easy to check that Φ is a minimal Lagrangian embedding and that the induced metric on $\mathbb{T}^n$ is flat.

On the other hand, from Lemma 1.16 and Theorem 1.7,6, every harmonic 1-form on the flat torus $\mathbb{T}^n$ is parallel, and so we can choose harmonic 1-forms $\{\omega_1, \dots, \omega_n\}$, which generate $H^1(\mathbb{T}^n, \mathbb{R})$, and which are an orthonormal reference on T_p^*M for every point $p \in M$. Thus, any 1-form α on $\mathbb{T}^n$ is written as $\alpha = \sum_{i=1}^n f_i\omega_i$, with $f_i \in C^\infty(\mathbb{T}^n)$. Since ω_i are parallel, we get that $\Delta\alpha = \sum_{i=1}^n(\Delta f_i)\omega_i$. Hence,

$$\Delta\alpha + \lambda\alpha = 0 \quad \text{if and only if} \quad \Delta f_i + \lambda f_i = 0,\ 1 \le i \le n,$$

which means that the eigenvalues of $\Delta : \Omega^1(\mathbb{T}^n) \to \Omega^1(\mathbb{T}^n)$ are the eigenvalues of $\Delta : C^\infty(\mathbb{T}^n) \to C^\infty(\mathbb{T}^n)$ and their multiplicities are n times the multiplicities of the corresponding ones for functions.

Proposition B.I.2 of [**BGM71**] describes the eigenfunctions of the flat torus $T^n \equiv \mathbb{R}^n/\Gamma$ corresponding to the eigenvalue $4\pi^2\|v^*\|^2$ with $v^* \in \Gamma^*$ (see Example 5.16). Following this result, it is easy to prove that the eigenfunctions of T^n corresponding to the eigenvalues less than $2(n+1)$ are not invariant under the action of $\mathbb{Z}_{n+1}$, but the eigenfunctions corresponding to the eigenvalue $2(n+1)$ are invariant. Thus, the first non-null eigenvalue of the laplacian of $\mathbb{T}^n$ is $2(n+1)$, which implies, taking into account Corollary 6.13, that $\mathrm{Ind}_0(\mathbb{T}^n) = \mathrm{Ind}_1(\mathbb{T}^n) = 0$ and thus, $\mathrm{Ind}(\mathbb{T}^n) = n$. Since the multiplicity of $2(n+1)$, as an eigenvalue of the laplacian of $\mathbb{T}^n$, is $n(n+1)$ (see Example 5.16), the above reasoning implies that the nullity of $\mathbb{T}^n$ is $n^2(n+1)$. □

EXAMPLE 6.36 ([**Nai81b**]). *For $n \ge 3$, consider the product $\mathbb{S}^1 \times \mathbb{S}^{n-1} \subset \mathbb{C} \times \mathbb{R}^n \subset \mathbb{C} \times \mathbb{C}^n$. If $\Pi : \mathbb{S}^{2n+1} \to \mathbb{CP}^n$ is the Hopf fibration, then the map $\Phi : \mathbb{S}^1 \times \mathbb{S}^{n-1} \to \mathbb{CP}^n$ defined by*

$$\Phi(z,x) = \Pi\left(\frac{1}{\sqrt{n+1}}\left(\sqrt{n}\frac{x}{\sqrt[n+1]{z}}, \sqrt[n+1]{z^n}\right)\right)$$

is a minimal Lagrangian immersion. If $\mathbb{Z}_2$ denotes the group generated by the diffeomorphism $\tau : \mathbb{S}^1 \times \mathbb{S}^{n-1} \to \mathbb{S}^1 \times \mathbb{S}^{n-1}$ given by $\tau(z,x) = (-z,-x)$, then Φ induces a minimal Lagrangian embedding $\widetilde{\Phi} : (\mathbb{S}^1 \times \mathbb{S}^{n-1})/\mathbb{Z}_2 \to \mathbb{CP}^n$, which makes the following diagram commutative:

$$\begin{array}{ccc} \mathbb{S}^1 \times \mathbb{S}^{n-1} & & \\ \downarrow \widetilde{\Pi} & \searrow^{\Phi} & \\ (\mathbb{S}^1 \times \mathbb{S}^{n-1})/\mathbb{Z}_2 & \xrightarrow{\widetilde{\Phi}} & \mathbb{CP}^n. \end{array}$$

Moreover, $\mathrm{Ind}_0(\widetilde{\Phi}) = 0$, $\mathrm{Ind}_1(\widetilde{\Phi}) = \frac{1}{2}n(n-1)$ *and* $\mathrm{Ind}(\widetilde{\Phi}) = 1 + \frac{1}{2}n(n-1)$.

REMARK 6.37. It is interesting to note that the manifold $(\mathbb{S}^1 \times \mathbb{S}^{n-1})/\mathbb{Z}_2$ is orientable if and only if n is even.

PROOF. First, it is clear that Φ is a well-defined map. Also, it is not difficult to check that Φ is a minimal Lagrangian immersion and that the induced metric on $\mathbb{S}^1 \times \mathbb{S}^{n-1}$ by the immersion Φ is $\frac{n}{n+1}(\frac{1}{n+1}g_1 \times g_2)$, where g_1 (respectively g_2) are the standard metrics on $\mathbb{S}^1$ (respectively on $\mathbb{S}^{n-1}$).

Now, using that the first two nonzero eigenvalues of Δ acting on $C^\infty(\mathbb{S}^n)$, ($\mathbb{S}^n$ endowed with its standard metric of constant curvature 1), are n and $2(n+1)$, it follows that the first two nonzero eigenvalues of Δ of $(\mathbb{S}^1, \frac{n}{(n+1)^2}g_1)$ (respectively

$(\mathbb{S}^{n-1}, \frac{n}{n+1}g_2))$ are $\frac{(n+1)^2}{n}$ and $\frac{4(n+1)^2}{n}$ (respectively $\frac{n^2-1}{n}$ and $2(n+1)$). Thus, from Theorem 1.18, $\frac{n^2-1}{n}$, $\frac{(n+1)^2}{n}$ and $2(n+1)$ are the first three nonzero eigenvalues of Δ of $\left(\mathbb{S}^1 \times \mathbb{S}^{n-1}, \frac{n}{n+1}(\frac{1}{n+1}g_1 \times g_2)\right)$. Since the eigenfunctions corresponding to the first two eigenvalues are odd with respect to τ, it follows that there are no nonzero eigenvalues of Δ of $\mathbb{S}^1 \times \mathbb{S}^{n-1}/\mathbb{Z}_2$ less than $2(n+1)$. This means that $\widetilde{\Phi}$ is Hamiltonian stable.

On the other hand, the only eigenvalue of Δ acting on $\mathcal{H}^1(\mathbb{S}^n) \oplus \delta\Omega^2(\mathbb{S}^n)$ ($\mathbb{S}^n$ endowed with its standard metric of constant curvature 1) less that $2(n+1)$ is $2(n-1)$ with multiplicity $\frac{n(n+1)}{2}$. By the above reasoning, the eigenvalues of Δ of $\left(\mathbb{S}^1 \times \mathbb{S}^{n-1}, \frac{n}{n+1}(\frac{1}{n+1}g_1 \times g_2)\right)$ less than $2(n+1)$, are 0 (with multiplicity 1) and $\frac{2(n-2)(n+1)}{n}$ (with multiplicity $\frac{n(n-1)}{2}$). Since in this case the corresponding eigenforms are invariant under τ, we obtain that $\beta_1((\mathbb{S}^1 \times \mathbb{S}^{n-1})/\mathbb{Z}_2) = 1$ and $\mathrm{Ind}_1(\widetilde{\Phi}) = \frac{n(n-1)}{2}$. This completes the proof. □

In the next result we will study the nullity of compact minimal Lagrangian submanifolds of $\mathbb{CP}^n$.

THEOREM 6.38 ([**Oh91**]). *Let $\Phi : M^n \to (\mathbb{CP}^n, g)$ be a minimal Lagrangian immersion of a compact manifold M. Then:*

(1) $2(n+1)$ *is an eigenvalue of* $\Delta : C^\infty(M) \to C^\infty(M)$ *and its multiplicity* m *satisfies* $m \geq (n+1)^2 - 1 - \dim \mathrm{Iso}(M, g)$.
(2) $\mathrm{Nul}(M) \geq \frac{1}{2}n(n+3)$, *and the equality holds if and only if Φ is either the totally geodesic immersion of the unit sphere $\mathbb{S}^n$ (Example* 6.32*) or the totally geodesic embedding of the real projective space $\mathbb{RP}^n$ (Example* 6.33*).*

PROOF. **Proof of 1.** From Lemma 6.29,7 we know that the space of Killing vector fields of $\mathbb{CP}^n$ is $\mathcal{K}(\mathbb{CP}^n) = \{JX_B \mid B \in H(n+1), \langle B, I\rangle = 0\}$, where X_B is the tangential component in $\mathbb{CP}^n$ of B. We consider the linear map

$$G : \mathcal{K}(\mathbb{CP}^n) \to \mathcal{K}_M(\mathbb{CP}^n) = \{X_{|M} \mid X \in \mathcal{K}(\mathbb{CP}^n)\}$$

defined by $G(JX_B) = (JX_B)_{|M}$. We will prove that $\ker G = 0$. In fact, if $(JX_B)_{|M} = 0$, then $B_{|M}$ is normal to $\mathbb{CP}^n$, and so, derivating with respect to a vector v tangent to M, we have that $0 = \overline{A}_B v$. Thus, $0 = \langle \overline{A}_B v, w\rangle = \langle \overline{\sigma}(v, w), B\rangle = 0$, for any vectors $v \in TM$ and $w \in TM \oplus T^\perp M$. Now, if $\{e_1, \ldots, e_n\}$ is an orthonormal basis of T_pM, for some point $p \in M$, then $\{Je_1, \ldots, Je_n\}$ is an orthonormal basis of $T_p^\perp M$. Hence from Lemma 6.29, 6 it follows that $B = 0$, and thus, our claim is proved.

We consider the linear map $\widehat{G} : \mathcal{K}_M(\mathbb{CP}^n) \to \mathcal{J}(M)$ from $\mathcal{K}_M(\mathbb{CP}^n)$ into the space of Jacobi vector fields of M given by (see Proposition 2.12)

$$\widehat{G}((JX_B)_{|M}) = ((JX_B)_{|M})^\perp, \quad B \in H(n+1), \quad \langle B, I\rangle = 0,$$

where $\perp$ stands for normal component to Φ. Since $((JX_B)_{|M})^\perp = J(X_B^T)$, the isomorphism F given in Theorem 6.9 applies $F(((JX_B)_{|M})^\perp) = -\alpha_B$, where α_B is the 1-form on M defined by $\alpha_B(v) = \langle v, B\rangle$, $\forall v \in TM$. Since $(JX_B)^\perp$ is a Jacobi vector field, then $\Delta\alpha_B + 2(n+1)\,\alpha_B = 0$, for any $B \in H(n+1)$ with $\langle B, I\rangle = 0$. We note that $\alpha_B = \mathrm{d}f_B$ where f_B is the function on M given by $f_B(p) = \langle(\Psi \circ \Phi)(p), B\rangle$. So, we have a linear map

$$H : \mathcal{K}_M(\mathbb{CP}^n) \to \{\alpha \in \mathrm{d}C^\infty(M) \mid \Delta\alpha + 2(n+1)\,\alpha = 0\}, \quad H((JX_B)_{|M}) = \alpha_B.$$

Since $\alpha_B = \mathrm{d}f_B$, we have that $2(n+1)$ is also an eigenvalue of $\Delta : C^\infty(M) \to C^\infty(M)$ whose multiplicity satisfies $m \geq \dim \operatorname{img}(H) = (n+1)^2 - 1 - \dim \ker(H)$.

But if $X_B \in \ker(H)$ with $X_B \neq 0$, then $\alpha_B = 0$ and thus, $(JX_B)^T$ is also a non-null Killing vector field on M. Hence, $\ker(H) \subset \mathcal{K}(M)$ and $\dim \ker(H) \leq \dim \operatorname{Iso}(M, g)$. This proves 1.

Proof of 2. Since $\operatorname{Nul}(M) \geq m$ and $\dim \operatorname{Iso}(M, g) \leq \frac{n(n+1)}{2}$, [**Kob72**], the inequality in 2 follows from 1. If the equality holds, then $\dim \operatorname{Iso}(M, g) = \frac{n(n+1)}{2}$ and hence (M, g) is isometric to either $\mathbb{S}^n$ or $\mathbb{RP}^n$ endowed with metrics of constant curvature (see [**Kob72**]). But it is known, [**CO74**], that the only compact minimal Lagrangian submanifolds of $\mathbb{CP}^n$ with positive constant curvature are the totally geodesic immersions of $\mathbb{S}^n$ and $\mathbb{RP}^n$ given in Examples 6.32 and 6.33, respectively. □

To give a general result about the index of compact minimal Lagrangian submanifolds of $\mathbb{CP}^n$, we need to introduce the concept of level as defined by P. Seidel (see [**Sei00**] for completeness).

If $\Pi : \mathbb{S}^{2n+1} \to \mathbb{CP}^n$ is the Hopf fibration of the unit sphere on the complex projective space, then for any integer $p \geq 1$ we consider the projection $\Pi_p : \mathbb{S}^{2n+1} \to \mathbb{L}_p^{2n+1} = \mathbb{S}^{2n+1}/\mathbb{Z}_p$ onto the Lens space $\mathbb{L}_p^{2n+1}$. It is clear that Π factorizes over $\mathbb{L}_p^{2n+1}$, by a projection $\widetilde{\Pi}_p$, which makes the following diagram commutative:

$$\begin{array}{ccc} \mathbb{S}^{2n+1} & \xrightarrow{\Pi_p} & \mathbb{L}_p^{2n+1} \\ & \searrow^{\Pi} & \downarrow{\scriptstyle \widetilde{\Pi}_p} \\ & & \mathbb{CP}^n. \end{array}$$

It is well known, [**GH78**], that the canonical bundle $K_{\mathbb{CP}^n}$ of the complex projective space, which is the holomorphic line bundle over $\mathbb{CP}^n$ of holomorphic n-forms $\Omega^{n,0}(\mathbb{CP}^n)$, can be identified with the bundle $(\mathbb{L}_{n+1}^{2n+1}, \widetilde{\Pi}_{n+1})$.

Now, if $\Phi : M^n \to \mathbb{CP}^n$ is a minimal Lagrangian immersion of an orientable manifold M, the bundle induced in M by $K_{\mathbb{CP}^n}$ is the trivial bundle $\Lambda^n T^*M \otimes \mathbb{C}$, and so, Φ is factorized through $\mathbb{L}_{n+1}^{2n+1}$ by a horizontal minimal immersion $\widetilde{\Phi}$:

$$\begin{array}{ccc} & & \mathbb{L}_{n+1}^{2n+1} \\ & \nearrow^{\widetilde{\Phi}} & \downarrow{\scriptstyle \widetilde{\Pi}_{n+1}} \\ M & \xrightarrow{\Phi} & \mathbb{CP}^n \end{array} \qquad \text{satisfying} \quad \widetilde{\Pi}_{n+1} \circ \widetilde{\Phi} = \Phi.$$

If M is not orientable, we can repeat the above argument with the holomorphic line bundle $K_{\mathbb{CP}^n} \otimes K_{\mathbb{CP}^n}$, and, in this case, Φ admits a horizontal lift $\widetilde{\Phi} : M \to \mathbb{L}_{2(n+1)}^{2n+1} \equiv K_{\mathbb{CP}^n} \otimes K_{\mathbb{CP}^n}$.

On the other hand, given two positive integers p and q, $\mathbb{L}_p^{2n+1}$ is a covering of $\mathbb{L}_q^{2n+1}$ if and only if $p|q$. So the following definition is natural.

DEFINITION 6.39. Let $\Phi : M^n \to \mathbb{CP}^n$ be a minimal Lagrangian immersion. The level of Φ is the smallest positive integer $p|2(n+1)$ such that Φ admits a minimal horizontal lift to $\mathbb{L}_p^{2n+1}$. In particular, $1 \leq \operatorname{level}(\Phi) \leq n+1$ if M is orientable and $1 \leq \operatorname{level}(\Phi) \leq 2(n+1)$ if M is not orientable. This means that a covering of M with level(Φ) sheets admits a Legendrian minimal immersion in $\mathbb{S}^{2n+1}$.

In the sequel we compute the level of some examples of minimal Lagrangian submanifolds of $\mathbb{CP}^n$ given above.

EXAMPLE 6.40 ([**HMU06**]).

(1) *It is clear that the totally geodesic Lagrangian immersion $\Phi : \mathbb{S}^n \to \mathbb{CP}^n$ (Example 6.32) has level one.*

(2) *Now consider the totally geodesic Lagrangian embedding $\Phi : \mathbb{RP}^n \to \mathbb{CP}^n$ given in Example 6.33. It is clear that Φ admits a horizontal lift to $\mathbb{L}_2^{2n+1} = \mathbb{RP}^{2n+1}$. So, the level of Φ is either two or one. But if* level $(\Phi) =$ *1, then Φ admits a Legendrian lift $\widetilde{\Phi} : \mathbb{RP}^n \to \mathbb{S}^{2n+1}$, and thus, the non-trivial homomorphism $\Phi^* : H^2(\mathbb{CP}^n, \mathbb{Z}) \equiv \mathbb{Z} \to H^2(\mathbb{RP}^n, \mathbb{Z}) \equiv \mathbb{Z}_2$ factorizes through the trivial group $H^2(\mathbb{S}^{2n+1}, \mathbb{Z}) = 0$, which is impossible. Hence, the level of Φ is two.*

(3) *Consider the embedding of the flat torus $\Phi : \mathbb{T}^n = T^n/\mathbb{Z}_{n+1} \to \mathbb{CP}^n$ given in Example 6.34. Since $\mathbb{T}^n$ is orientable, Φ admits a horizontal lift to $\mathbb{L}_{n+1}^{2n+1}$ which is given by*

$$\Pi'(z_1, \ldots, z_{n+1}) \overset{\widetilde{\Phi}}{\mapsto} [(z_1, \ldots, z_{n+1})] \in \mathbb{L}_{n+1}^{2n+1} = \mathbb{S}^{2n+1}/\mathbb{Z}_{n+1},$$

for all $(z_1, \ldots, z_{n+1}) \in T^n$, where Π' is the projection of T^n onto $\mathbb{T}^n$.

It is not difficult to check that the homomorphism induced on the fundamental group:

$$\widetilde{\Phi}_* : \Pi_1(\mathbb{T}^n) \equiv \mathbb{Z} \oplus \overset{n}{\cdots} \oplus \mathbb{Z} \to \Pi_1(\mathbb{S}^{2n+1}/\mathbb{Z}_{n+1}) \equiv \mathbb{Z}_{n+1}$$

is surjective. If the level of Φ is $p < n+1$, then Φ admits a horizontal lift to $\mathbb{L}_p^{2n+1}$ with $p|n+1$, and so the above homomorphism is factorized through $\Pi_1(\mathbb{L}_p^{2n+1}) \equiv \mathbb{Z}_p$. But there are no epimorphisms from $\mathbb{Z}_p$ to $\mathbb{Z}_{n+1}$ with $p|n+1$. So, the level of Φ is $n+1$.

(4) *Finally, we consider the minimal Lagrangian immersion $\Phi : \mathbb{S}^1 \times \mathbb{S}^{n-1} \to \mathbb{CP}^n$ given in Example 6.36. It is clear that $\Psi : \mathbb{S}^1 \times \mathbb{S}^{n-1} \to \mathbb{L}_{n+1}^{2n+1} = \mathbb{S}^{2n+1}/\mathbb{Z}_{n+1}$ given by*

$$\Psi(z, x) = \left[\left(\frac{1}{\sqrt{n+1}} (\sqrt{n} \frac{x}{\sqrt[n+1]{z}}, \sqrt[n+1]{z^n}) \right) \right],$$

is a horizontal lift to $\mathbb{L}_{n+1}^{2n+1}$ of Φ, where square brackets represent equivalent classes in the quotient of $\mathbb{S}^{2n+1}$. Similar to (3) above, we can easily see that the homomorphism induced on the fundamental group

$$\Psi_* : \Pi_1(\mathbb{S}^1 \times \mathbb{S}^{n-1}) \equiv \mathbb{Z} \to \Pi_1(\mathbb{L}_{n+1}^{2n+1}) \equiv \mathbb{Z}_{n+1}$$

is surjective. By reasoning similarly to (3) above, we conclude that the level of Φ is $n+1$.

If n is even, the horizontal lift Ψ of Φ can be induced in the quotient $(\mathbb{S}^1 \times \mathbb{S}^{n-1})/\mathbb{Z}_2$ (see Example 6.36), making the next diagram commutative

$$\begin{array}{ccc} \mathbb{S}^1 \times \mathbb{S}^{n-1} & & \\ \downarrow \widetilde{\Pi} & \searrow^{\Psi} & \\ (\mathbb{S}^1 \times \mathbb{S}^{n-1})/\mathbb{Z}_2 & \xrightarrow{\widetilde{\Psi}} & \mathbb{L}_{n+1}^{2n+1}. \end{array}$$

This means that the level of $\widetilde{\Phi} : \mathbb{S}^1 \times \mathbb{S}^{n-1}/\mathbb{Z}_2 \to \mathbb{CP}^n$ is $n+1$.

If n is odd, the above reasoning does not work, and in this case it is not difficult to check that the level of $\widetilde{\Phi}$ is $2n+2$.

THEOREM 6.41 ([**HMU06**]). *Let $\Phi : M^n \to (\mathbb{CP}^n, g)$ be a minimal Lagrangian immersion of a compact manifold M.*

(1) *If the level of Φ is one, then $\text{Ind}_0(M) \geq n+1$, and the equality holds if and only if Φ is the totally geodesic Lagrangian immersion of $\mathbb{S}^n$ (Example 6.32). Moreover, if Φ is not totally geodesic, then $\text{Ind}_0(M) \geq 2(n+1)$.*
(2) *If the level of Φ is one, then $\text{Ind}_1(M) \geq \frac{n(n+1)}{2}$, and the equality holds if and only if Φ is the totally geodesic Lagrangian immersion of $\mathbb{S}^n$.*
(3) *If the level of Φ is one, then $\text{Ind}(M) \geq \beta_1(M) + \frac{(n+1)(n+2)}{2}$, and the equality holds if and only if Φ is the totally geodesic Lagrangian immersion of $\mathbb{S}^n$.*
(4) *If the level of Φ is two, then $\text{Ind}(M) \geq \beta_1(M) + \text{Ind}_1(M) \geq \beta_1(M) + \frac{n(n+1)}{2}$, and the equality holds if and only if Φ is the totally geodesic Lagrangian embedding of $\mathbb{RP}^n$ (Example 6.33).*

REMARK 6.42. In [**Sei00**], P. Seidel proved that if $\Phi : M^n \to \mathbb{CP}^n$ is a minimal Lagrangian immersion of a compact manifold and $H^1(M, \mathbb{Z}_{2(n+1)}) = 0$, then the level of Φ is one. From Theorem 6.41 it follows that if $\text{Ind}_0(\Phi) < n+1$ then $H^1(M, \mathbb{Z}_{2(n+1)}) \neq 0$, and hence $H^1(M, \mathbb{Z}) \neq 0$. In particular, there are no simply connected Hamiltonian stable compact minimal Lagrangian submanifolds of $\mathbb{CP}^n$.

PROOF. **Proof of 1**. Suppose the level of Φ is one. Then, Φ admits a minimal Legendrian lift $\widetilde{\Phi} : M \to \mathbb{S}^{2n+1}$. The induced metrics on M by Φ and $\widetilde{\Phi}$ are equal. Since $\widetilde{\Phi}$ is minimal, for any $a \in \mathbb{R}^{2n+2}$ we have that $\Delta\langle\widetilde{\Phi}, a\rangle + n\langle\widetilde{\Phi}, a\rangle = 0$, and so, n is an eigenvalue of Δ, where $\langle , \rangle$ denotes the Euclidean metric on $\mathbb{R}^{2n+2}$. We will estimate its multiplicity. We define the linear map

$$F : \mathbb{R}^{2n+2} \to C^\infty(M), \quad F(a) = \langle\widetilde{\Phi}, a\rangle.$$

If $a \in \ker F$ with $a \neq 0$, then $\langle\widetilde{\Phi}, a\rangle = 0$, which implies that $\langle v, a\rangle = 0$ for any $v \in TM$. This implies that $\langle Ja, J\widetilde{\Phi}\rangle = \langle Ja, Jv\rangle = 0$, for any $v \in TM$. Since $\widetilde{\Phi}$ is a Legendrian immersion, we have that Ja has no normal component to $\widetilde{\Phi}$. Thus,

$$\nabla^2\langle\widetilde{\Phi}, Ja\rangle(v,w) = \langle\widetilde{\sigma}(v,w), Ja\rangle - g(v,w)\langle\widetilde{\Phi}, Ja\rangle = -g(v,w)\langle\widetilde{\Phi}, Ja\rangle, \forall\, v, w \in TM,$$

where $\widetilde{\sigma}$ is the second fundamental form of $\widetilde{\Phi}$. Hence $\nabla^2\langle\widetilde{\Phi}, Ja\rangle + \langle\widetilde{\Phi}, Ja\rangle g = 0$, and using the well-known Obata theorem, [**Oba62**], we get that either $\langle\widetilde{\Phi}, Ja\rangle = 0$ or M has constant curvature 1. If the first case is true, then $a = 0$ because $\langle a, \widetilde{\Phi}\rangle = \langle a, J\widetilde{\Phi}\rangle = 0$ and so $\langle a, v\rangle = \langle a, Jv\rangle = 0$, $\forall v \in TM$. Hence M has constant curvature 1 and so, Φ is the totally geodesic Legendrian embedding of $\mathbb{S}^n$. In this case, $\dim \ker F = n+1$. Therefore, we get that either $\dim \ker F = 0$ or $\dim \ker F = n+1$ and M is the totally geodesic $\mathbb{S}^n$. Hence, the multiplicity μ of n as an eigenvalue of Δ satisfies

$$\mu \geq \dim \text{img } F = 2n+2 - \dim \ker F \geq n+1,$$

and the equality holds if and only if Φ is the totally geodesic embedding of $\mathbb{S}^n$. Note that if M is not totally geodesic, then $\dim \ker F = 0$ and so $\mu \geq 2(n+1)$. This proves 1.

Proof of 2. To do this we need some 1-form tests coming from the Killing vector fields of $\mathbb{S}^{2n+1} \subset \mathbb{C}^{n+1}$. In the sequel we use the notation of Section 5.2.

The vector space of Killing vector fields on $\mathbb{S}^{2n+1}$ can be identified with the space of skew-symmetric matrices $\mathrm{so}(2n+2)$ in such a way that the Killing field corresponding to a matrix $B \in \mathrm{so}(2n+2)$ is given by $X_B(p) = Bp$ for any $p \in \mathbb{S}^{2n+1}$. We consider the subspace $\mathrm{so}^-(2n+2)$ of $\mathrm{so}(2n+2)$ defined by

$$\mathrm{so}^-(2n+2) = \{B \in \mathrm{so}(2n+2) \,/\, BJ = -JB\},$$

where the skew-symmetric matrix J is the complex structure of $\mathbb{C}^{n+1}$. It is clear that $\dim \mathrm{so}^-(2n+2) = n(n+1)$.

If $\widetilde{\Phi} : M \to \mathbb{S}^{2n+1}$ is the minimal Legendrian lift of Φ, for each $B \in \mathrm{so}^-(2n+2)$, we define a 2-form γ_B in M by

$$\gamma_B(V, W) = \langle BV, JW\rangle, \quad \forall V, W \in \mathfrak{X}(M).$$

We will prove that $\Delta\gamma_B + 2(n-1)\gamma_B = 0$. In fact, it is easy to check that $\mathrm{d}\gamma_B = 0$, $\forall B \in \mathrm{so}^-(2n+2)$. On the other hand, if $\{e_1, \ldots, e_n\}$ is a local orthonormal reference on TM, its codifferential is given by

$$\begin{aligned}\delta\gamma_B(V) &= \sum_{i=1}^{n}\{e_i(\gamma_B(e_i, V) - \gamma_B(\nabla_{e_i}e_i, V) - \gamma_B(e_i, \nabla_{e_i}V)\}\\ &= -(n-1)\langle B\widetilde{\Phi}, JV\rangle + \sum_{i=1}^{n}\langle Be_i, J\widetilde{\sigma}(e_i, V)\rangle = -(n-1)\langle B\widetilde{\Phi}, JV\rangle,\end{aligned}$$

because, as $\widetilde{\Phi}$ is a Legendrian immersion,

$$\begin{aligned}\sum_{i=1}^{n}\langle Be_i, J\widetilde{\sigma}(e_i, V)\rangle &= -\sum_{i=1}^{n}\langle Be_i, \widetilde{A}_{JV}e_i\rangle = -\sum_{i,j=1}^{n}\langle Be_i, e_j\rangle\langle \widetilde{A}_{JV}e_i, e_j\rangle\\ &= \sum_{i,j=1}^{n}\langle Be_j, e_i\rangle\langle \widetilde{A}_{JV}e_j, e_i\rangle = -\sum_{j=1}^{n}\langle Be_j, J\widetilde{\sigma}(e_j, V)\rangle,\end{aligned}$$

where $\widetilde{A}_{JV}$ is the Weingarten endomorphism associated with JV. Thus,

$$\begin{aligned}\Delta\gamma_B(V, W) = \mathrm{d}\delta\gamma_B(V, W) &= -(n-1)V\langle B\widetilde{\Phi}, JW\rangle + (n-1)W\langle B\widetilde{\Phi}, JV\rangle\\ +(n-1)\langle B\widetilde{\Phi}, J[V, W]\rangle &= -(n-1)\langle BV, JW\rangle + (n-1)\langle BW, JV\rangle\\ &= -2(n-1)\gamma_B(V, W).\end{aligned}$$

Again, we will estimate its multiplicity. We define the linear map

$$G : \mathrm{so}^-(2n+2) \to \{\gamma \in \Omega^2(M) \mid \Delta\gamma + 2(n-1)\gamma\}, \qquad G(B) = \gamma_B.$$

If $B \in \ker G$ with $B \neq 0$, then $\delta\gamma_B = 0$ and so $\langle B\widetilde{\Phi}, Jv\rangle = 0$, $\forall v \in TM$. Since $\langle B\widetilde{\Phi}, J\widetilde{\Phi}\rangle = \langle B\widetilde{\Phi}, \widetilde{\Phi}\rangle = 0$, we get that $B\widetilde{\Phi}$ is a tangent vector field to M. Furthermore, $B\widetilde{\Phi}$ is a nonzero Killing vector field on M since

$$\langle \nabla_v(B\widetilde{\Phi}), v\rangle = \langle Bv, v\rangle = 0, \quad \forall v \in TM.$$

Hence, $\dim \ker G \leq \dim \mathrm{Iso}(M, g) \leq \frac{n(n+1)}{2}$ (see [**Kob72**]), which is the largest dimension of the isometry group of M. If the equality occurs, from [**Kob72**] M has constant curvature and so, again using [**CO74**], M is the totally geodesic $\mathbb{S}^n$. Thus, the multiplicity μ of $2(n-1)$ satisfies

$$\mu \geq \dim \mathrm{img}\, G = n(n+1) - \dim \ker G \geq \frac{n(n+1)}{2},$$

and the equality holds if and only if M is the totally geodesic $\mathbb{S}^n$. Now, taking into account the Corollary 6.10, we prove 2.

Proof of 3. Statement 3 is a consequence of 1 and 2 together with Corollary 6.13,3.

Proof of 4. If the level of Φ is two, then Φ admits a minimal horizontal lift $\widetilde{\Phi}$ to $\mathbb{RP}^{2n+1}$. This means that a 2-fold covering of M admits a Legendrian minimal immersion $\widehat{\Phi}$ in $\mathbb{S}^{2n+1}$, and the next diagram is commutative:

$$\begin{array}{ccc} & & \mathbb{S}^{2n+1} \\ & \overset{\widehat{\Phi}}{\nearrow} & \downarrow \Pi_2 \\ \widetilde{M} & & \mathbb{RP}^{2n+1} \\ \downarrow \widehat{\Pi} & \overset{\widetilde{\Phi}}{\nearrow} & \downarrow \widetilde{\Pi}_2 \\ M & \xrightarrow{\ \Phi\ } & \mathbb{CP}^n. \end{array}$$

If τ is the involutive isometry of $\widetilde{M}$, then $\widehat{\Phi} \circ \tau = -\widehat{\Phi}$ and then it follows that $\tau^*\gamma_B = \gamma_B$ for any $B \in \mathrm{so}^-(2n+2)$. So, γ_B projects into a 2-form $\widehat{\gamma}_B$ on M, i.e., $\widehat{\Pi}^*\widehat{\gamma}_B = \gamma_B$. It is now clear that $\Delta\widehat{\gamma}_B + 2(n-1)\widehat{\gamma}_B = 0$. Using the same reasoning as in the proof of 2, we can prove 4. □

In the theorem above we saw that for compact minimal Lagrangian submanifolds M^n of $\mathbb{CP}^n$ with level one or two, $2(n-1)$ is an eigenvalue of $\Delta : \Omega^1(M) \to \Omega^1(M)$. Now, without any assumptions about the level, we will see that $2(n-1)$ is an upper bound for the first eigenvalue of Δ acting on the subspace $\mathcal{H}^1(M) \oplus \delta\Omega^2(M) \subset \Omega^1(M)$. To do this, we again use the first standard embedding $\Psi : \mathbb{CP}^n \to (H(n+1), \langle , \rangle)$ given at the beginning of this section.

LEMMA 6.43. *Let $\Phi : M^n \to \mathbb{CP}^n$ be a minimal Lagrangian immersion of a compact manifold M. We define a 1-form β on M valuated in $H(n+1)$ by*

$$\beta(v) = Jv, \quad \forall v \in TM, \tag{6.7}$$

i.e., if $\{B_1, \dots, B_{(n+1)^2}\}$ is a basis of $H(n+1)$, then $\beta = (\beta_1, \dots, \beta_{(n+1)^2})$ where β_i is the 1-form on M given by $\beta_i(v) = \langle Jv, B_i\rangle$. Then:

(1) $\delta\beta = 0$. *In particular, $\beta_i \in \mathcal{H}^1(M) \oplus \delta\Omega^2(M)$ for any $i \in \{1, \dots, (n+1)^2\}$.*
(2) $\Delta\beta + 2(n-1)\beta = \gamma$, *where γ is the 1-form on M valuated in $H(n+1)$ defined by*

$$\gamma(v) = -2\sum_{i=1}^{n} \overline{\sigma}(A_{Jv}e_i, e_i), \quad \forall v \in TM,$$

$\{e_i, \dots, e_n\}$ is a local orthonormal reference on M and $\overline{\sigma}$ is the second fundamental form of $\Psi : \mathbb{CP}^n \to (H(n+1), \langle , \rangle)$.

PROOF. If $\widetilde{\nabla}$ is the Levi-Civita connection on $H(n+1)$, then the codifferential of β is given by

$$\delta\beta = \sum_{i=1}^{n}\{\widetilde{\nabla}_{e_i}\beta(e_i) - \beta(\nabla_{e_i}e_i)\} = \sum_{i=1}^{n}\overline{\sigma}(e_i, Je_i) = 0,$$

where we have used the minimality of M and Lemma 6.29,3.

Also, using Lemma 6.29,3 again, we have

$$d\beta(V,W) = \widetilde{\nabla}_V\beta(W) - \widetilde{\nabla}_W\beta(V) - \beta([V,W]) = \overline{\sigma}(V,JW) - \overline{\sigma}(W,JV) = 2\overline{\sigma}(V,JW),$$

for arbitrary tangent vector fields V, W, and so, using that Φ is minimal and Lemma 6.29,2,4, we get

$$\Delta\beta(V) = \delta\mathrm{d}\beta(V) = \sum_{i=1}^{n}\{\widetilde{\nabla}_{e_i}\mathrm{d}\beta(e_i, V) - \mathrm{d}\beta(\nabla_{e_i}e_i, V) - \mathrm{d}\beta(e_i, \nabla_{e_i}V)\}$$

$$= -2\sum_{i=1}^{n}\{\overline{A}_{\overline{\sigma}(e_i,JV)}e_i + \overline{\sigma}(e_i, A_{JV}e_i)\} = -2(n-1)\beta(V) + \gamma(V). \qquad \square$$

Using β as a test 1-form to estimate the first eigenvalue of Δ, we prove the following result.

THEOREM 6.44 ([**Urb93**]). *Let $\Phi : M^n \to \mathbb{CP}^n$ be a minimal Lagrangian immersion of a compact manifold M. If λ_1 is the first eigenvalue of $\Delta : \mathcal{H}^1(M) \oplus \delta\Omega^2(M) \to \mathcal{H}^1(M) \oplus \delta\Omega^2$, then $\lambda_1 \leq 2(n-1)$ and the equality holds if and only if Φ is either the totally geodesic Lagrangian immersion of $\mathbb{S}^n$ (Example* 6.32*) or the totally geodesic Lagrangian embedding of $\mathbb{RP}^n$ (Example* 6.33*). In particular, $\beta_1(M) + \mathrm{Ind}_1(M) \geq 1$ and thus $\mathrm{Ind}(M) \geq 1 + \mathrm{Ind}_0(M)$.*

PROOF. If β is the 1-form defined in (6.7), we have that

$$\mathcal{Q}(\beta) \geq \lambda_1 \int_M \|\beta\|^2\, \mathrm{d}v,$$

and the equality holds if and only if $\Delta\beta + \lambda_1\beta = 0$. By Lemma 6.43, we get that

$$\mathcal{Q}(\beta) = -\int_M \langle\beta, \Delta\beta\rangle\, \mathrm{d}v = 2(n-1)\int_M \|\beta\|^2\, \mathrm{d}v - \int_M \langle\beta, \gamma\rangle\, \mathrm{d}v,$$

and hence, since $\|\beta\|^2 = n$ and $\langle\beta, \gamma\rangle = 0$, we get $n(2(n-1) - \lambda_1)\,\mathrm{vol}(M) \geq 0$, which implies that $\lambda_1 \leq 2(n-1)$.

If the equality holds, then $\Delta\beta + 2(n-1)\beta = 0$, and thus, from Lemma 6.43 we get that $\gamma = 0$. From the definition of γ it follows that

$$\sum_{i=1}^{n}\overline{\sigma}(A_{Jv}e_i, e_i) = 0, \quad \forall\, v \in TM,$$

where $\{e_1, \ldots, e_n\}$ is a local orthonormal reference on M. Thus, for any vectors x, y tangent to M and using Lemma 6.29, we have

$$0 = \sum_{i=1}^{n}\langle\overline{\sigma}(A_{Jv}e_i, e_i), \overline{\sigma}(x, y)\rangle = g(\sigma(x, y), Jv), \quad \forall\, v, x, y \in TM.$$

This means that Φ is a totally geodesic immersion and hence, Φ is either the totally geodesic immersion of the sphere $\mathbb{S}^n$ or the totally geodesic embedding of the real projective space $\mathbb{RP}^n$. Finally, it follows from [**GM75**] that $2(n-1)$ is the first eigenvalue of Δ acting on $\mathcal{H}^1(M) \oplus \delta\Omega^2(M)$ when M is either $\mathbb{S}^n$ or $\mathbb{RP}^n$.

The last statement follows from Corollary 6.13. $\square$

Now we will use the spectral decomposition of the 1-form β to get an inequality involving the first two eigenvalues of Δ. We do this only for minimal Lagrangian surfaces.

THEOREM 6.45 ([**Urb93**]). *Let $\Phi : M^2 \to \mathbb{CP}^2$ be a minimal Lagrangian immersion of a compact surface M. If λ_1 and λ_2 are the first two eigenvalues of $\Delta : \mathcal{H}^1(M) \oplus \delta\Omega^2(M) \to \mathcal{H}^1(M) \oplus \delta\Omega^2(M)$, then*

$$(12 - 2(\lambda_1 + \lambda_2) + \lambda_1\lambda_2)\mathrm{Area}(M) \geq 16\pi\chi(M),$$

where $\chi(M)$ is the Euler characteristic of M. Moreover, the equality holds if and only if Φ is the totally geodesic immersion of $\mathbb{S}^2$ (Example 6.32), the totally geodesic embedding of $\mathbb{RP}^2$ (Example 6.33) or the Lagrangian Clifford torus $\mathbb{T}^2$ (Example 6.34).

PROOF. Let $\{\lambda_1 < \lambda_2 < \cdots < \lambda_k < \dots\}$ be the spectrum of Δ acting on $\mathcal{H}^1(M) \oplus \delta\Omega^2(M)$. If V_{λ_k} is the eigenspace corresponding to λ_k, then $\oplus_{k=1}^{\infty} V_{\lambda_k}$ is dense in $\mathcal{H}^1 \oplus \delta\Omega^2(M)$ and thus, the 1-form β defined in (6.7) can be written as $\beta = \sum_{k=1}^{\infty} \beta_k$, with $\Delta\beta_k + \lambda_k\beta_k = 0$. Thus, it is easy to get that

$$\begin{aligned}\int_M \|\Delta\beta\|^2\,\mathrm{d}v + (\lambda_1+\lambda_2)\int_M \langle\Delta\beta,\beta\rangle\,\mathrm{d}v + \lambda_1\lambda_2\int_M \|\beta\|^2\,\mathrm{d}v \\ = \sum_{k=3}^{\infty}(\lambda_k-\lambda_1)(\lambda_k-\lambda_2)\int_M \|\beta_k\|^2\,\mathrm{d}v \geq 0.\end{aligned} \tag{6.8}$$

Also, the equality in (6.8) is true if and only if $\beta = \beta_1 + \beta_2$. Now we will compute the terms of (6.8).

Using Lemma 6.43 and Lemma 6.29,4 we get that $\|\Delta\beta\|^2 = 4\|\beta\|^2 + \|\gamma\|^2 = 8(1 + \|\sigma\|^2)$, $\langle\Delta\beta,\beta\rangle = -4$, and so, using the Gauss equation, (6.8) becomes

$$(4 - 2(\lambda_1+\lambda_2) + \lambda_1\lambda_2)\mathrm{Area}(M) \geq -4\int_M \|\sigma\|^2\,\mathrm{d}v = -8\mathrm{Area}(M) + 8\int_M K\,\mathrm{d}v,$$

which together with the Gauss-Bonnet theorem proves the inequality.

If the equality holds, then $\beta = \beta_1 + \beta_2$ with $\Delta\beta_i + \lambda_i\beta_i = 0$, $i = 1, 2$. So, we get that $\Delta^2\beta + (\lambda_1+\lambda_2)\Delta\beta + \lambda_1\lambda_2\beta = 0$. Using Lemma 6.43, the above equation becomes

$$\Delta\gamma = -(2-\lambda_1)(2-\lambda_2)\beta + (2-(\lambda_1+\lambda_2))\gamma. \tag{6.9}$$

Since β is a 1-form with values in the normal bundle $T^\perp M$, we only compute the component of $\Delta\gamma$ in $T^\perp M$. Since $\Delta\beta = -2\beta + \gamma$ (see Lemma 6.43), and $\delta\beta = 0$, we get that $\delta\gamma = \delta\Delta\beta = \Delta\delta\beta = 0$. So, $\Delta\gamma = \delta\mathrm{d}\gamma$.

If V, W are vector fields tangent to M, then using Proposition 6.7, Lemmas 6.29 and 6.43 and the Codazzi equation (2.2), we get that

$$\begin{aligned}\mathrm{d}\gamma(V,W) &= \widetilde{\nabla}_V\gamma(W) - \widetilde{\nabla}_W\gamma(V) - \gamma([V,W]) \\ &= -4(A_{JW}V - A_{JV}W) + 2\sum_{i=1}^{2}\{\overline{\sigma}((\nabla\sigma)(V,W,e_i),Je_i) - \overline{\sigma}((\nabla\sigma)(W,V,e_i),Je_i)\} \\ &\quad -4\sum_{i=1}^{2}\{\overline{\sigma}(A_{JW}e_i,\sigma(V,e_i)) - \overline{\sigma}(A_{JV}e_i,\sigma(W,e_i))\} \\ &= 2\sum_{i=1}^{2}\overline{R}(V,W,e_i,Je_i) + 8\sum_{i=1}^{2}\overline{\sigma}(A_{JV}e_i,\sigma(W,e_i)) = 8\sum_{i=1}^{2}\overline{\sigma}(A_{JV}e_i,\sigma(W,e_i)),\end{aligned}$$

where $\{e_1, e_2\}$ is a local orthonormal reference on TM and we have used the expressions of γ and the curvature $\overline{R}$ of $\mathbb{CP}^2$. Hence,

$$(\Delta\gamma(V))^\perp = (\delta \mathrm{d}\gamma(V))^\perp = (\sum_{i=1}^{2}\{\widetilde{\nabla}_{e_i}\mathrm{d}\gamma(e_i, V) - \mathrm{d}\gamma(\nabla_{e_i}e_i, V) - \mathrm{d}\gamma(e_i, \nabla_{e_i}V)\})^\perp$$

$$= -8\Big(\sum_{i,j=1}^{2}\overline{A}_{\overline{\sigma}(A_{Je_i}e_j, \sigma(V,e_j))}e_i\Big)^\perp = 8\sum_{i=1}^{2} JA_{\sigma(V,e_i)}e_i,$$

where we have used again Lemma 6.43. Now, since $\gamma^\perp = 0$, we get from (6.9) that

$$8\sum_{i=1}^{2} JA_{\sigma(v,e_i)}e_i = -(2-\lambda_1)(2-\lambda_2)Jv, \quad \forall v \in TM,$$

which, by the Gauss equation of Φ, implies that

$$-(2-\lambda_1)(2-\lambda_2)g(v,w) = 8\sum_{i=1}^{2} g(\sigma(v,e_i), \sigma(w,e_i)) = 8(1-K)g(v,w), \forall v, w \in TM.$$

This means that the surface M has constant curvature $K = \frac{1}{8}(12-2(\lambda_1+\lambda_2)+\lambda_1\lambda_2)$ and therefore, Φ is either the totally geodesic immersion of $\mathbb{S}^2$ ($K=1$), the totally geodesic embedding of $\mathbb{RP}^2$ ($K=1$), or the generalized Clifford torus ($K=0$) (see [**Yau74b**]).

Conversely, from Examples 6.32, 6.33 and 6.34 we see that these three examples satisfy the equality of the theorem. □

COROLLARY 6.46 ([**Urb93, Urb07**]). *Let $\Phi : M^2 \to \mathbb{CP}^2$ be a minimal Lagrangian immersion of a compact surface M.*

(1) *If M is orientable and Φ is Hamiltonian stable, then either Φ is an embedding and $\Phi(M)$ is the Lagrangian Clifford torus (Example* 6.34*), or the genus h of M is $h \geq 5$.*
(2) *If M is not orientable and Φ is Hamiltonian stable, then either Φ is an embedding and $\Phi(M)$ is the totally geodesic $\mathbb{RP}^2$ (Example* 6.33*), or the Euler characteristic is $\chi(M) \leq -1$.*
(3) $\mathrm{Ind}(M) \geq 2$, *and* $\mathrm{Ind}(M) = 2$ *if and only if Φ is the Lagrangian Clifford torus $\mathbb{T}^2$.*

PROOF. **Proof of 1.** First, we assume that M is orientable with genus h. Since M is Hamiltonian stable, using Corollary 6.13 and Theorem 6.38, the first positive eigenvalue of $\Delta : C^\infty(M) \to C^\infty(M)$ is 6. Now we use a familiar argument. From Brill-Noether theory ([**GH78**]) we can obtain a non-constant holomorphic map $\phi : M \to \mathbb{S}^2$ of degree $d \leq 1 + \lfloor \frac{h+1}{2} \rfloor$, where $\lfloor\cdot\rfloor$ stands for the integer part. By Proposition 1.8, there exists a Möbius transformation $\psi : \mathbb{S}^2 \to \mathbb{S}^2$ such that $\int_M (\psi\circ\phi)\,\mathrm{d}v = 0$, and since 6 is the first nonzero eigenvalue of Δ, we have that

$$\int_M \|\nabla(\psi\circ\phi)\|^2\,\mathrm{d}v \geq 6\int_M \|\psi\circ\phi\|^2\,\mathrm{d}v = 6\mathrm{Area}(M).$$

But $\int_M \|\nabla(\psi\circ\phi)\|^2\,\mathrm{d}v = 8\pi\deg(\psi\circ\phi) = 8\pi\deg(\phi) \leq 8\pi(1+\lfloor\frac{h+1}{2}\rfloor)$. Thus, we get $3\mathrm{Area}(M) \leq 4\pi(1+\lfloor\frac{h+1}{2}\rfloor)$.

Also, S. Montiel and F. Urbano proved in [**MU02**] that $\text{Area}(M) \geq 2\pi\mu$ (where μ is the maximum multiplicity of the immersion Φ), and that the equality holds if and only if the surface is the totally geodesic Lagrangian $\mathbb{S}^2$. So, we get that

$$3\mu \leq 2\big(1 + \lfloor \tfrac{h+1}{2} \rfloor\big),$$

and the equality implies that M is the totally geodesic $\mathbb{S}^2$.

Using [**EGT85**], we know that Φ is not an embedding (i.e., $\mu \geq 2$) if $h \geq 2$. Thus, in this case $2 \leq \lfloor \frac{h+1}{2} \rfloor$, which is a contradiction if $h = 2, 3, 4$. If $h = 0$, from [**Yau74b**], our surface is the totally geodesic Lagrangian $\mathbb{S}^2$, which is Hamiltonian unstable. Hence our surface must be a torus ($h = 1$), and using Theorem 6.45, we conclude that M is the Lagrangian Clifford torus, taking into account that, since M is orientable, the eigenvalues of Δ acting on $\mathrm{d}C^\infty(M)$ are also the eigenvalues of Δ acting on $\delta\Omega^2(M)$. So 1 is proved.

Proof of 2. Now suppose M is not orientable and $\chi(M) \geq 0$. In this case, either $\chi(M) = 1$ and from [**Yau74b**] M is the totally geodesic Lagrangian $\mathbb{RP}^2$ which is Hamiltonian stable, or $\chi(M) = 0$ and M is a Klein bottle. If M is a Hamiltonian stable Klein bottle, then the first nonzero eigenvalue of $\Delta : C^\infty(M) \to C^\infty(M)$ is 6. Using a result of N. Nadirashvili ([**Nad87**]) we get that its multiplicity m is $m \leq 7$, and then Theorem 6.38,1 says that $\dim \text{Iso}(M, g) \geq 1$. This means that our Klein bottle admits a nonzero Killing vector field coming from a one-parameter group of isometries of $\mathbb{CP}^2$. In [**Urb07**] this family of minimal Lagrangian Klein bottles of $\mathbb{CP}^2$ was classified and their Hamiltonian index was computed, proving that their Hamiltonian index is greater than or equal to 6. This is a contradiction, and we have proved 2.

Proof of 3. First we assume that M is orientable. By Corollary 6.13 we have $\text{Ind}(M) = 2h + 2\,\text{Ind}_0(M)$, where h is the genus of M. If $h = 0$, then we know from [**Yau74b**] that Φ is the totally geodesic immersion of $\mathbb{S}^2$. In this case, $\text{Ind}_0(M) = 3$ and so, $\text{Ind}(M) = 6$. If $h \geq 1$, then $\text{Ind}(M) \geq 2$ and the equality holds if and only if $h = 1$ and $\text{Ind}_0(M) = 0$. Thus, 1 implies that Φ is an embedding and $\Phi(M)$ is the Clifford torus.

Now suppose M is not orientable. If $\chi(M) = 1$, then from [**Yau74b**] we have that Φ is the totally geodesic embedding of $\mathbb{RP}^2$ whose index is 3. If $\chi(M) = 0$, then M is a Klein bottle and from 2 we have that $\text{Ind}_0(M) \geq 1$. Also using Theorem 6.45 and considering that the first eigenvalue of Δ acting on $\mathcal{H}^1(M) \oplus \delta\Omega^2(M)$ is 0, we have that the second one is $\lambda_2 \leq 6$. Since equality cannot be achieved by a Klein bottle, then $\lambda_2 < 6$ and this means that $\text{Ind}_1(M) \geq 1$. So, in this case $\text{Ind}(M) = 1 + \text{Ind}_0(M) + \text{Ind}_1(M) \geq 3$.

If $\chi(M) = -1$, then the orientable double covering $\widetilde{M}$ of M is an orientable surface of genus 2. By 1, $\widetilde{M}$ is Hamiltonian unstable, and thus, $\text{Ind}(\widetilde{M}) \geq 6$. Now the Corollary 6.13,5 says that $\text{Ind}(M) \geq 3$.

Finally, if $\chi(M) \leq -2$, then $\text{Ind}(M) = 1 - \chi(M) + \text{Ind}_0(M) + \text{Ind}_1(M) \geq 3$. This completes the proof. □

REMARK 6.47. By the last result, it is reasonable to conjecture that

> *the only Hamiltonian stable compact minimal Lagrangian surfaces of* $\mathbb{CP}^2$ *are* $\mathbb{RP}^2$ *and* $\mathbb{T}^2$.

This conjecture, together with the one given in Remark 6.35, suggests considering a similar problem for higher dimensions.

In Theorem 6.38 it was proved that $2(n+1)$ is an eigenvalue of the Laplacian, acting on functions, of any compact minimal Lagrangian submanifold of $\mathbb{CP}^n$. It is interesting to know when $2(n+1)$ is the first nonzero eigenvalue of the Laplacian. In this sense, it seems reasonable to pose the following problem, similar to the Yau's conjecture mentioned in Remark 5.3:

> *Is $2(n+1)$ the first eigenvalue of the Laplacian of any compact minimal Lagrangian submanifold* ***embedded*** *in $\mathbb{CP}^n$?*

Taking into account Corollary 6.13, this problem is equivalent to the following:

> *Is every compact minimal Lagrangian submanifold* ***embedded*** *in $\mathbb{CP}^n$ Hamiltonian stable?*

H. Naitoh and M. Takeuchi classified in [**NT82**] the Lagrangian submanifolds with parallel second fundamental form of $\mathbb{CP}^n$. In their classification, in addition to Examples 6.32, 6.33, 6.34 and 6.36, important examples of symmetric spaces appear as minimal Lagrangian submanifolds in $\mathbb{CP}^n$.

A. Amarzaya and Y. Ohnita proved in [**AO03**] that certain quotients of these examples are Hamiltonian stables. To prove this result, the authors study the first nonzero eigenvalue of the Laplacian acting on functions, by using techniques of symmetric spaces (Casimir operator, etc). To conclude this section, we study the above examples and include their results in this chapter without proof.

EXAMPLE 6.48 ([**Nai81a**]). *We consider the embedding $i: SU(n) \to \mathbb{S}^{2n^2-1} \subset gl(n,\mathbb{C})$ given by $A \mapsto \frac{1}{\sqrt{n}}A$. It is not difficult to check that i is minimal and Legendrian, and so, i induces a minimal Lagrangian immersion*

$$\Phi = \Pi \circ i : SU(n) \to \mathbb{CP}^{n^2-1},$$

where $\Pi : \mathbb{S}^{2n^2-1} \to \mathbb{CP}^{n^2-1}$ is the Hopf fibration. It is clear that the level of Φ is one.

The immersion Φ is not an embedding because given $A, B \in SU(n)$, $\Phi(A) = \Phi(B)$ if and only if $B = zA$ with $z \in \mathbb{Z}_n \subset \mathbb{S}^1$. Thus, Φ induces an embedding

$$\widetilde{\Phi} : [A] \in SU(n)/\mathbb{Z}_n \mapsto \Pi\big(\tfrac{1}{\sqrt{n}}A\big) \in \mathbb{CP}^{n^2-1},$$

which is minimal and Lagrangian. It is clear that

$$\tilde{i} : [A] \in SU(n)/\mathbb{Z}_n \mapsto \{\tfrac{1}{\sqrt{n}}A\} \in \mathbb{S}^{2n^2-1}/\mathbb{Z}_n$$

is a Legendrian lift of this embedding $\widetilde{\Phi}$, which induces an isomorphism

$$\tilde{i}_* : \Pi_1\big(SU(n)/\mathbb{Z}_n\big) \equiv \mathbb{Z}_n \to \Pi_1(\mathbb{S}^{2n^2-1}/\mathbb{Z}_n) \equiv \mathbb{Z}_n$$

between its fundamental groups. This shows that the level of $\widetilde{\Phi}$ is n.

EXAMPLE 6.49 ([**Nai81a**]). *The symmetric space $SU(n)/SO(n)$ can be embedded in the unit sphere of the vector space $Sym(m,\mathbb{C})$ of symmetric complex matrices in the following minimal Legendrian way:*

$$i : A \cdot SO(n) \in SU(n)/SO(n) \mapsto \frac{1}{\sqrt{n}}AA^t \in \mathbb{S}^{n^2+n-1} \subset Sym(m,\mathbb{C}).$$

Thus, this i yields a minimal Lagrangian immersion in the corresponding complex projective space $\Phi = \Pi \circ i : SU(n)/SO(n) \to \mathbb{CP}^{\frac{1}{2}(n^2+n-2)}$. Moreover, the level of Φ is one.

As in the case above, Φ itself is not an injective map, but it induces an embedding

$$\widetilde{\Phi} : [A \cdot SO(n)] \in SU(n)/SO(n) \cdot \mathbb{Z}_n \mapsto \Pi\left(\frac{1}{\sqrt{n}}AA^t\right) \in \mathbb{CP}^{\frac{1}{2}(n^2+n-2)}$$

which has $\tilde{i}$ as a Legendrian lift to $\mathbb{S}^{n^2+n-1}/\mathbb{Z}_n$, given by

$$\tilde{i} : [A \cdot SO(n)] \in SU(n)/SO(n) \cdot \mathbb{Z}_n \mapsto \left\{\frac{1}{\sqrt{n}}AA^t\right\} \in \mathbb{S}^{n^2+n-1}/\mathbb{Z}_n.$$

EXAMPLE 6.50 ([**Nai81a**]). *Let $so(2n,\mathbb{C})$ be the Lie algebra of skew-symmetric complex $2n$-matrices and consider the minimal Legendrian embedding of the symmetric space $SU(2m)/Sp(m)$ into its unit sphere given by*

$$i : A \cdot Sp(n) \in SU(2n)/Sp(n) \to \frac{1}{\sqrt{2n}}AJA^t \in \mathbb{S}^{4n^2-1} \subset so(2n,\mathbb{C}),$$

Projecting into the complex projective space we get a minimal Lagrangian immersion

$$\Phi = \Pi \circ i : SU(2n)/Sp(n) \to \mathbb{CP}^{2n^2-1}$$

whose Legendrian lift is the above i. So, the level of Φ is one.

Studying again the equivalence relation associated to the map Φ, we can construct a minimal Lagrangian embedding by dividing by a suitable discrete subgroup, namely

$$\widetilde{\Phi} : [A \cdot Sp(n)] \in SU(2n)/Sp(n) \cdot \mathbb{Z}_{2n} \to \Pi\left(\frac{1}{\sqrt{2n}}AJA^t\right) \in \mathbb{CP}^{2n^2-1}.$$

EXAMPLE 6.51 ([**Nai81a**]). *Let C be the Cayley algebra over $\mathbb{R}$ endowed with the canonical conjugation and let*

$$H(3,C) = \{A \in gl(3,C) \mid \overline{A}^t = A\}$$

which is a 27-dimensional real vector space. On this matrix space one can define a C-valued determinant det *and a C-inner product $((,))$. Now consider the complexification $V = H(3,C)\otimes\mathbb{C}$ and extend* det *and $((,))$ $\mathbb{C}$-linearly on V. Then, if τ denotes the complex conjugation, $(,) = ((\tau,))$ is a positive definite Hermitian product on V. In this setting, the Lie group E_6 can be viewed as the group of C-linear $(,)$-isometries of V preserving* det, *and the Lie subgroup $F_4 \subset E_6$ as the isotropy subgroup of the identity matrix $I \in V$. Naitoh considered the following minimal Legendrian embedding*

$$i : g \cdot F_4 \in E_6/F_4 \mapsto \frac{1}{\sqrt{3}}g(I) \in \mathbb{S}^{53} \subset V$$

which induces a minimal Lagrangian immersion $\Phi = \Pi \circ i : E_6/F_4 \to \mathbb{CP}^{26}$ with level one.

It is clear that the group of cubic roots of the unit $\mathbb{Z}_3 \subset \mathbb{S}^1$ acts on E_6 as follows:

$$(zg)(A) = zg(A) \quad \forall z \in \mathbb{Z}_3, \quad g \in E_6, \quad A \in V,$$

and that $\Phi(g \cdot F_4) = \Phi(h \cdot F_4)$ if and only if $h = zg$ for some $z \in \mathbb{Z}_3$. Thus, we get a minimal Lagrangian embedding

$$\widetilde{\Phi} : [g \cdot F_4] \in E_6/F_4 \cdot \mathbb{Z}_3 \mapsto \Pi\left(\frac{1}{\sqrt{3}}g(I)\right) \in \mathbb{CP}^{26}.$$

REMARK 6.52. After a calculation similar to the one done in Example 6.48 to compute the level of the embedding $\widetilde{\Phi}$, it can be seen that the levels of the $\widetilde{\Phi}$'s embeddings in Examples 6.49, 6.50, and 6.51 are respectively

$$\begin{cases} n & n \text{ is odd} \\ \frac{n}{2} & n \text{ is even} \end{cases}, \quad \begin{cases} 2n & n \text{ is odd} \\ n & n \text{ is even} \end{cases}, \quad 3.$$

THEOREM 6.53 ([**AO03**]). *The compact minimal Lagrangian submanifolds embedded in $\mathbb{CP}^m$ of the following list:*

(1) $SU(n)/\mathbb{Z}_n, m = n^2 - 1$ *(Example* 6.48*),*
(2) $SU(n)/SO(n) \cdot \mathbb{Z}_n, m = (n-1)(n+2)/2$ *(Example* 6.49*),*
(3) $SU(2n)/Sp(n) \cdot \mathbb{Z}_{2n}, m = (n-1)(2n+1)$ *(Example* 6.50*),*
(4) $E_6/F_4 \cdot \mathbb{Z}_3, m = 26$ *(Example* 6.51*),*

are Hamiltonian stable.

6.5. Minimal Lagrangian surfaces of the product of two 2-spheres

In this last section of the book we will study the second variation of compact minimal Lagrangian surfaces of $\mathbb{S}^2 \times \mathbb{S}^2$. As already shown in Section 5.5, the 4-manifold $\mathbb{S}^2 \times \mathbb{S}^2$ admits two Kähler surface structures and it is an Einstein manifold. In fact, together with the complex projective plane, it is the only Hermitian symmetric space of compact type of complex dimension two. Before studying the index of such surfaces, we will give a brief introduction to them, exposing the most important examples that will be characterized by the behavior of their indices (see [**CU07**] for details).

Let $\mathbb{S}^2$ be the 2-dimensional sphere, $\langle\, , \rangle$ its standard metric, J its complex structure and ω its Kähler 2-form. We endow $\mathbb{S}^2 \times \mathbb{S}^2$ with the product metric (also denoted by $\langle\, , \rangle$) and the complex structures

$$J_1 = (J, J) \quad \text{and} \quad J_2 = (J, -J),$$

which define two structures of Kähler surface on $\mathbb{S}^2 \times \mathbb{S}^2$. It is clear that if $\mathrm{Id} : \mathbb{S}^2 \to \mathbb{S}^2$ is the identity map and $F : \mathbb{S}^2 \to \mathbb{S}^2$ is any antiholomorphic isometry, then $(\mathrm{Id}, F) : \mathbb{S}^2 \times \mathbb{S}^2 \to \mathbb{S}^2 \times \mathbb{S}^2$ is a holomorphic isometry of $(\mathbb{S}^2 \times \mathbb{S}^2, \langle,\rangle, J_1)$ to $(\mathbb{S}^2 \times \mathbb{S}^2, \langle,\rangle, J_2)$.

The Kähler 2-forms ω_j of J_j, $j = 1, 2$, are given by $\omega_1 = \pi_1^*\omega + \pi_2^*\omega$ and $\omega_2 = \pi_1^*\omega - \pi_2^*\omega$ and so,

$$\omega_1 \wedge \omega_1 = -\omega_2 \wedge \omega_2 = 2(\pi_1^*\omega \wedge \pi_2^*\omega),$$

where π_j, $j = 1, 2$, are the projections onto the factors.

An immersion $\Phi : M^2 \to \mathbb{S}^2 \times \mathbb{S}^2$ is Lagrangian if Φ is Lagrangian with respect to one of the two complex structures of $\mathbb{S}^2 \times \mathbb{S}^2$, that is, either $\Phi^*\omega_1 = 0$ or $\Phi^*\omega_2 = 0$. But since both complex structures are holomorphically isometric, the Lagrangian surfaces with respect to J_1 and J_2 are congruent, and thus, it will suffice to consider only the Lagrangian surfaces with respect to J_1.

EXAMPLE 6.54. *Let $M_0 = \{(x, -x) \in \mathbb{S}^2 \times \mathbb{S}^2 \mid x \in \mathbb{S}^2\}$ be the graph of the antipodal map of $\mathbb{S}^2$. If $i : M_0 \subset \mathbb{S}^2 \times \mathbb{S}^2$ is the inclusion, then*

$$i^*\omega_1 = (\pi_1 \circ i)^*\omega + (\pi_2 \circ i)^*\omega = \omega - \omega = 0,$$

which says that M_0 is a Lagrangian surface. It is also easy to check that M_0 is totally geodesic and the induced metric has constant curvature 1/2*. It is also interesting to* *observe that M_0 is a complex surface with respect to J_2 and so, from Theorem 6.4, M_0 is a stable minimal surface.*

EXAMPLE 6.55. *Let $T = \{(x,y) \in \mathbb{S}^2 \times \mathbb{S}^2 \mid x_1 = y_1 = 0\}$ be the product $\mathbb{S}^1 \times \mathbb{S}^1$ embedded in $\mathbb{S}^2 \times \mathbb{S}^2$ as the product of two equators. If $i : T \subset \mathbb{S}^2 \times \mathbb{S}^2$ is the inclusion, then*

$$\begin{aligned}\big((\pi_j \circ i)^*\omega\big)_{(a_1,a_2)}((v_1,v_2),(w_1,w_2)) &= \omega_{(0,a_j)}((0,v_j),(0,w_j)) \\ &= \det\{(0,a_j),(0,v_j),(0,w_j)\} = 0,\end{aligned}$$

for all $(v_1,v_2),(w_1,w_2) \in T_{(a_1,a_2)}T$ and $(a_1,a_2) \in T$. This implies that T is Lagrangian with respect to both complex structures J_1 and J_2. It is clear that T is a totally geodesic surface and that the induced metric is flat.

EXAMPLE 6.56 ([**ESGJ06**],[**JNP06**]). *We highlight an interesting example of a Klein bottle studied by A. El Soufi, H. Giacomini and M. Jazar ([**ESGJ06**]) and D. Jacobson, N. Nadirashvili and I. Polterovich ([**JNP06**]). Considering $\mathbb{S}^2 = \{(x,z) \in \mathbb{R} \times \mathbb{C} \mid x^2 + \|z\|^2 = 1\}$, we define*

$$B = \left\{((x,z),(y,w)) \in \mathbb{S}^2 \times \mathbb{S}^2 \mid 2x = y, \Re(\sqrt{z}w) = \Im(\sqrt{z}w)\right\}, \tag{6.10}$$

where $\sqrt{\ }$ is the main branch of the square root. Note that $\|z\|^2 = 1 - y^2/4 \geq 3/4$. A conformal parametrization of the universal covering of B by elementary Jacobi elliptic functions (for background see [**BF71**]*) is given by*

$$\Psi = (\Psi_1, \Psi_2) : \mathbb{R}^2 \to \mathbb{S}^2 \times \mathbb{S}^2$$

with

$$\begin{aligned}\Psi_1(t,s) &= \frac{1}{3\,\mathrm{dn}(\sqrt{3}t)}\left(-\mathrm{sn}(2\sqrt{3}t), i\big(2\,\mathrm{sn}^2(\sqrt{3}t) - 3\big)e^{\frac{4is}{\sqrt{3}}}\right)\\ \Psi_2(t,s) &= \frac{1}{3\,\mathrm{dn}(\sqrt{3}t)}\left(-2\,\mathrm{sn}(2\sqrt{3}t), -i\big(4\,\mathrm{sn}^2(\sqrt{3}t) - 3\big)e^{\frac{-2is}{\sqrt{3}}}\right),\end{aligned}$$

where sn, cn *and* dn *are the sine, cosine and delta amplitudes, respectively, with modulus* $2\sqrt{2}/3$.

The corresponding group of transformations in $\mathbb{R}^2$ defining the Klein bottle B is given by

$$(t,s) \mapsto \left(t + \frac{2\sqrt{3}K}{3}, s\right), \qquad (t,s) \mapsto \left(\frac{\sqrt{3}K}{3} - t, s + \frac{\sqrt{3}\pi}{2}\right),$$

where K is the complete elliptic integral of the first kind with modulus $2\sqrt{2}/3$.

We note that Ψ is a minimal Lagrangian immersion and that $\Psi(\mathbb{R}^2) \subset \mathbb{S}^4(\sqrt{2})$ (note that B lies in the hyperplane $2x = y$ of $\mathbb{R}^6$). Moreover, B is a minimal surface of $\mathbb{S}^4(\sqrt{2})$ and the orientable double covering of B is an $\mathbb{S}^1$-equivariant minimal torus in $\mathbb{S}^4(\sqrt{2})$.

In [**JNP06**]*, it was proved (see Theorems 1.3.1, 1.4.1) that the first positive eigenvalue λ_1 of the Laplacian Δ (acting on functions) of the Klein bottle B is $\lambda_1(B) = 1$.*

Now we show some properties of the minimal Lagrangian surfaces of $\mathbb{S}^2 \times \mathbb{S}^2$ proved in [**CU07**].

PROPOSITION 6.57 ([**CU07**]). *Let $\Phi : M^2 \to \mathbb{S}^2 \times \mathbb{S}^2$ be a minimal Lagrangian immersion of a surface M. Then:*

(1) *If M is compact and orientable and Φ is an embedding, then the genus of M is $h \leq 1$.*
(2) *If M is a sphere, then $\Phi(M)$ is congruent to M_0 (Example 6.54). In particular, a real projective plane cannot be immersed in $\mathbb{S}^2 \times \mathbb{S}^2$ as a minimal Lagrangian surface.*
(3) *If the Gauss curvature K of M is constant, then $\Phi(M)$ is congruent to some open subset of the totally geodesic Lagrangian surfaces M_0 or T (Example 6.55).*

THEOREM 6.58 ([**CU07**]). *Let $\Phi : M^2 \to \mathbb{S}^2 \times \mathbb{S}^2$ be a minimal Lagrangian immersion of a compact surface M. Then:*

(1) *If M is stable, then $\Phi(M)$ is the totally geodesic Lagrangian sphere M_0 (Example 6.54).*
(2) *If M is Hamiltonian stable and M is orientable with genus $h \leq 2$, then Φ is an embedding and $\Phi(M)$ is either the totally geodesic sphere M_0 or the totally geodesic torus T (Example 6.55).*
(3) *If M is a Hamiltonian stable Klein bottle, then Φ is an embedding and $\Phi(M)$ is the Klein bottle B described in Example 6.56.*
(4) *If M is unstable, then* $\operatorname{Ind}(M) \geq 2$ *and the equality holds if and only if Φ is an embedding and $\Phi(M)$ is the totally geodesic torus T.*

REMARK 6.59. Part 1 shows that the result proved in Theorem 3.46,6 is the best possible. Furthermore, since the totally geodesic sphere M_0 is a complex surface with respect to the complex structure $J = (J, -J)$ on $\mathbb{S}^2 \times \mathbb{S}^2$, we have that M_0 is area minimizing in its homology class (see Theorem 6.5). In addition, in [**IOS03**] it was proved that the totally geodesic torus T is area minimizing under Hamiltonian deformations of $\mathbb{S}^2 \times \mathbb{S}^2$.

PROOF. First, it is important to show that since $\mathbb{S}^2 \times \mathbb{S}^2 \subset \mathbb{S}^5(\sqrt{2})$, the immersion $\Phi : M \to \mathbb{S}^5(\sqrt{2})$ is also minimal. In fact, if $\overline{\sigma}$ is the second fundamental form of M in $\mathbb{R}^6$, then the mean curvature vector field $\vec{H}$ of M in $\mathbb{S}^5(\sqrt{2})$ is given by

$$2\vec{H} = \Phi + \sum_{i=1}^{2} \overline{\sigma}(e_i, e_i),$$

where $\{e_1, e_2\}$ is a local orthonormal reference on TM. But, from (5.38) and the fact that $\{e_1, e_2, J_1e_1, J_1e_2\}$ is a local orthonormal reference of $T(\mathbb{S}^2 \times \mathbb{S}^2)$ on M, it is easy to see thatt $\sum_{i=1}^{2} \overline{\sigma}(e_i, e_i) = -\Phi$, which implies that $\vec{H} = 0$. This proves the above statement.

Second, we prove the following property:

CLAIM. Under the conditions of the theorem, if μ is the maximum multiplicity of Φ, then $\operatorname{Area}(M) \geq 8\pi\mu$ and the equality holds if and only if Φ is an embedding and $\Phi(M)$ is congruent to M_0, whose area is 8π.

To prove the claim, we start by recalling a result of L. Simon, [**Sim93**]. Let $\Psi : M^2 \to \mathbb{R}^n$ be an immersion of a compact surface M with mean curvature vector $\vec{H}$ and maximum multiplicity μ. Then

$$\int_M \|\vec{H}\|^2 \, dv \geq 4\pi\mu,$$

and the equality holds if and only if $\vec{H}$ is given on $\widetilde{M} = M \setminus \{p_1, \dots, p_\mu\}$ by $\vec{H} = \frac{-2(\Psi-a)^\perp}{\|\Psi-a\|^2}$, where $\{p_1, \dots, p_\mu\} = \Psi^{-1}(a)$. This condition on the mean curvature $\vec{H}$ implies that $\frac{\Psi-a}{\|\Psi-a\|^2} : \widetilde{M} \to \mathbb{R}^n$ is a minimal immersion.

In this setting, the minimal Lagrangian immersion Φ produces an immersion $\Psi : M \to \mathbb{R}^6$ which is minimal in $\mathbb{S}^5(\sqrt{2})$. In this case, $\vec{H} = -\Phi/2$ and we get that $\text{Area}(M) \geq 8\pi\mu$, and the equality holds if and only if $\Phi = 4\frac{(\Phi-a)^\perp}{\|\Phi-a\|^2}$ on $\widetilde{M}$, where μ is the maximum multiplicity of Φ. Since $a \in \mathbb{S}^5(\sqrt{2})$ and Φ is normal to the surface, the last equation becomes in $\langle \Phi, a\rangle\Phi = 2a^\perp$. From this it is not difficult to conclude that Φ is the totally geodesic sphere M_0. This proves the claim.

Proof of 1. Since $\mathbb{S}^2 \times \mathbb{S}^2$ is Einstein-Kähler with Ricci tensor $\overline{S} = g$, the Jacobi operator of M is $\mathcal{L} = \Delta + 1 : \Omega^1(M) \to \Omega^1(M)$. Then, if M is stable, then $\beta_1(M) = 0$ by Corollary 6.13, which implies, taking into account Proposition 6.57, that $\Phi(M)$ is congruent to M_0. This proves 1.

Proof of 2. We only need to consider the cases $h = 1, 2$. If the genus of M is 1, i.e., M is a Hamiltonian stable torus, then the first eigenvalue of the Laplacian acting on functions is 1. So, we have a minimal immersion $\Phi : M \to \mathbb{S}^5(\sqrt{2})$ of a torus M where 1 is the first eigenvalue of the Laplacian. A result of A. El Soufi and S. Ilias, [**ESI00**], tells us that M is flat, and then Proposition 6.57 shows that $\Phi(M)$ is the totally geodesic torus T. If the genus of M is 2, we use a known argument. From Brill-Noether theory we get a non-constant meromorphic map $\phi : M \to \mathbb{S}^2$ of degree $d \leq 2$. By Proposition 1.8 there exists a conformal transformation $\psi : \mathbb{S}^2 \to \mathbb{S}^2$ such that $\int_M (\psi \circ \phi)\, dv = 0$, and using that the first positive eigenvalue of Δ is 1, we have

$$\int_M \|\nabla(\psi \circ \phi)\|^2\, dv \geq \int_M \|\psi \circ \phi\|^2\, dv = \text{Area}(M).$$

But $\int_M \|\nabla(\psi\circ\phi)\|^2\, dv = 8\pi \deg(\psi\circ\phi) = 8\pi \deg(\phi) \leq 16\pi$. Thus, we get $\text{Area}(M) \leq 16\pi$. By the claim we finally get that $\mu \leq 2$. Proposition 6.57 implies that $\mu \geq 2$, which gives the equality in the last inequality. This gives a contradiction, because the equality can only be obtained by the totally geodesic M_0. This concludes the proof of 2.

Proof of 3. Now suppose M is a Hamiltonian stable Klein bottle. Thus we have a minimal immersion $\Phi : M \to \mathbb{S}^5(\sqrt{2})$ of a Klein bottle M such that 1 is the first positive eigenvalue of Δ. From Theorem 1.2 in [**ESGJ06**] we deduce that our immersion is an embedding and the surface is the Klein bottle B. This proves 3.

Proof of 4. If M is unstable and orientable, using part 1, the genus h of M satisfies that $h \geq 1$ and, from Corollary 6.13, $\text{Ind}(M) \geq 2$ and the equality holds if and only if M is a Hamiltonian stable torus, which implies, using part 2, that it is the totally geodesic torus T.

If M is not orientable, then $\Pi : \widetilde{M} \to M$ is the orientable double covering and $\Phi \circ \Pi : \widetilde{M} \to \mathbb{S}^2 \times \mathbb{S}^2$ is the corresponding minimal Lagrangian immersion, and Corollary 6.13 implies that $\text{Ind}(M) = \frac{1}{2}\text{Ind}(\widetilde{M})$ and $\text{Ind}_0(M) + \text{Ind}_1(M) = \text{Ind}_0(\widetilde{M})$.

According to Proposition 6.57, there are no minimal Lagrangian immersions of real projective planes in $\mathbb{S}^2 \times \mathbb{S}^2$, and so, we have $\chi(M) \leq 0$. If $\chi(M) \leq -2$, then $\beta_1(M) \geq 3$, and Corollary 6.13 implies that $\text{Ind}(M) \geq 3$. If $\chi(M) = -1$ then

$\beta_1(M) = 2$ and again using Corollary 6.13 we get that

$$\operatorname{Ind}(M) = 2 + \operatorname{Ind}_0(M) + \operatorname{Ind}_1(M) = 2 + \operatorname{Ind}_0(\widetilde{M}).$$

But $\widetilde{M}$ is an oriented compact surface of genus 2, and part 2 leads to $\operatorname{Ind}_0(\widetilde{M}) \geq 1$. This implies that $\operatorname{Ind}(M) \geq 3$.

Finally, if M is a Klein bottle, then $\beta_1(M) = 1$. If $\operatorname{Ind}_1(M) = 0$, using the description of $\operatorname{Ind}_1(M)$ given in the proof of Corollary 6.13, the first eigenvalue $\Delta : \Omega^0_-(\widetilde{M}) \to \Omega^0_-(\widetilde{M})$ satisfies $\lambda_1 \geq 1$. Hence,

$$\int_{\widetilde{M}} \|\nabla f\|^2 \, d\tilde{v} \geq \int_{\widetilde{M}} f^2 \, d\tilde{v}, \quad \forall f \in C^\infty(\widetilde{M}) \quad \text{such that} \quad f \circ \tau = -f.$$

From Theorem 1 in [**RS94**], we can get a non-constant meromorphic map $\phi : \widetilde{M} \to \mathbb{S}^2$ satisfying $\phi \circ \tau = -\phi$ of degree $d \leq 2$. So, we get

$$\int_{\widetilde{M}} \|\nabla \phi\|^2 \, d\tilde{v} \geq \int_{\widetilde{M}} \|\phi\|^2 \, d\tilde{v} = \operatorname{Area}(\widetilde{M}).$$

But $\int_{\widetilde{M}} \|\nabla \phi\|^2 \, d\tilde{v} = 8\pi \deg(\phi) \leq 16\pi$. Thus we deduce that $\operatorname{Area}(\widetilde{M}) \leq 16\pi$. Since $\operatorname{Area}(\widetilde{M}) = 2 \operatorname{Area}(M)$, we have from the claim that $\mu \leq 1$. So, $\mu = 1$ and the equality holds in the claim, which is impossible because $\widetilde{M}$ would be the totally geodesic sphere M_0. Hence, $\operatorname{Ind}_1(M) \geq 1$ for any minimal Lagrangian Klein bottle M of $\mathbb{S}^2 \times \mathbb{S}^2$. Part 3 says that $\operatorname{Ind}_0(M) \geq 1$ for any minimal Lagrangian Klein bottle of $\mathbb{S}^2 \times \mathbb{S}^2$ except for B. Thus, we obtain that every minimal Lagrangian Klein bottle M not congruent to B satisfies $\operatorname{Ind}(M) \geq 3$. To finish the proof, we must check that $\operatorname{Ind}_1(B) \geq 2$. In fact, following Example 6.56, we consider the functions $f, g : \widetilde{B} \to \mathbb{R}$ defined by $f(\Pi(t,s)) = \cos(2s/\sqrt{3})$, $g(\Pi(t,s)) = \sin(2s/\sqrt{3})$, where $\Pi : \mathbb{R}^2 \to \widetilde{B}$ is the projection and $\widetilde{B}$ is the orientable doble covering of B. As the involution $\tau : \widetilde{B} \to \widetilde{B}$ is induced by $(t,s) \mapsto (\sqrt{3}K/3 - t, s + \sqrt{3}\pi/2)$, it is clear that $f \circ \tau = -f$ and $g \circ \tau = -g$. The induced metric on $\mathbb{R}^2$ is given by $e^{2u(t)} = 4 \cosh \log(\sqrt{3} \operatorname{dn}(\sqrt{3}t))$ and so, $e^{2u(t)} \geq 4$. Then, for any real numbers a and b we have that

$$\begin{aligned} \Delta(af + bg)(\Pi(t,s)) = e^{-2u(t)} \frac{d^2}{ds^2} \left(a \cos \frac{2s}{\sqrt{3}} + b \sin \frac{2s}{\sqrt{3}} \right) \\ - \frac{4}{3} e^{-2u(t)} (af + bg)(\Pi(t,s)). \end{aligned}$$

Since $e^{-2u(t)} \leq 1/4$, it follows that

$$-(af + bg)\Delta(af + bg) \leq \frac{1}{3}(af + bg)^2,$$

which implies that

$$\widetilde{\mathcal{Q}}(af + bg) \leq -\frac{2}{3} \int_{\widetilde{B}} (af + bg)^2 \, d\tilde{v}.$$

In conclusion, we have shown that there exists a 2-dimensional subspace of $\Omega^0_-(\widetilde{B})$ where the quadratic form associated to the Jacobi operator $\Delta + 1$ of $\widetilde{B}$ is negative definite. So, $\operatorname{Ind}_1(B) \geq 2$ and this concludes the proof. $\square$

Glossary

$B(R)$: Geodesic ball of center 0 and radius R in $\mathbb{R}^n$.
$B_p(R)$: Geodesic ball of center p and radius R.
$\mathbb{C}^n$: Complex Euclidean space of complex dimension n.
$\mathbb{CP}^n$: Complex projective space of complex dimension n with constant holomorphic sectional curvature 4.
$C^\infty(M)$: Space of smooth functions of M.
$C_0^\infty(M)$: Space of compact support smooth functions of M.
Δ: Laplacian.
div: Divergence.
dv: Measure associated to a Riemannian metric.
d: Differential of a form.
δ: Codifferential of a form.
exp: Exponential map.
$\mathcal{H}(M)$: Harmonic vector fields of M.
$\mathcal{H}^k(M)$: Harmonic k-forms of M.
$\mathbb{H}^n$: Hyperbolic space of dimension n with constant curvature -1.
$\mathrm{Iso}(M, g)$: Isometry group of (M, g).
Ind: Index.
J: Complex structure.
$\mathcal{J}(M)$: Jacobi vector fields of M.
K: Gauss curvature.
$\mathcal{K}(M)$: Killing vector fields of M.
$\mathcal{L}$: Jacobi operator of a minimal immersion.
$L^p(M)$: Space of measurable functions on M with $|f|^p$ integrable.
$L^p(M, E)$: Space of measurable sections ξ of a vector bundle E on M with $\|\xi\|^p$ integrable.
$\mathrm{O}(n)$: Orthogonal group.
Nul: Nullity.
∇: Gradient, Levi-Civita connection.
$\nabla^\perp$: Normal connection.
∇^2: Hessian.
σ: Second fundamental form.
$\nabla\sigma$: Covariant derivative of the second fundamental form.
A_ξ: Weingarten endomorphism.
A: Weingarten endomorphism of a two-sided hypersurface.
∇A: Covariant derivative of the Weingarten endomorphism of a two-sided hypersurface.
R: Curvature operator.
$R^\perp$: Normal curvature operator.

$\mathcal{Q}$: Index form of a minimal immersion.
Ricc: Ricci curvature.
ρ: Scalar curvature.
$\mathbb{R}^n$: Euclidean space of dimension n.
$\mathbb{RP}^n$: Real projective space of dimension n and constant curvature 1.
$\mathbb{RP}^n(r)$: Real projective space of dimension n and constant curvature $1/r^2$.
$\mathbb{S}^n$: Sphere of dimension n with constant curvature 1.
$\mathbb{S}^n(r)$: Sphere of dimension n with constant curvature $1/r^2$.
S: Ricci tensor.
Spin(n): Spinorial group.
Sp(n): Symplectic group.
SU(n): Special unitary group.
U(n): Unitary group.
$H^{1,2}(M), H_0^{1,2}(M)$: Sobolev spaces of M.
$\mathfrak{X}(M)$: Space of vector fields of M.
$\mathfrak{X}_0(M)$: Space of compact support vector fields of M.
$\mathfrak{X}^\perp(M)$: Space of normal vector fields to an immersion of M in N.
$\mathfrak{X}_0^\perp(M)$: Space of compact support normal vector fields to an immersion of M in N.
$\Omega^p(M)$: Space of p-forms on M.

Bibliography

[ACS18] Lucas Ambrozio, Alessandro Carlotto, and Ben Sharp, *Comparing the Morse index and the first Betti number of minimal hypersurfaces*, J. Differential Geom. **108** (2018), no. 3, 379–410, DOI 10.4310/jdg/1519959621. MR3770846

[AHS78] M. F. Atiyah, N. J. Hitchin, and I. M. Singer, *Self-duality in four-dimensional Riemannian geometry*, Proc. Roy. Soc. London Ser. A **362** (1978), no. 1711, 425–461, DOI 10.1098/rspa.1978.0143. MR506229

[Alm66] F. J. Almgren Jr., *Some interior regularity theorems for minimal surfaces and an extension of Bernstein's theorem*, Ann. of Math. (2) **84** (1966), 277–292, DOI 10.2307/1970520. MR200816

[AMP02] Claudio Arezzo, Mario J. Micallef, and Gian Pietro Pirola, *Stable minimal surfaces of finite total curvature*, Comm. Anal. Geom. **10** (2002), no. 1, 11–22, DOI 10.4310/CAG.2002.v10.n1.a2. MR1894139

[And86] Michael Anderson, *The compactification of a minimal submanifold in euclidean space by the gauss map*, Preprint. Dept. of Math., California Institute of Technology, Pasadena, CA 91125., 1986.

[AO03] Amartuvshin Amarzaya and Yoshihiro Ohnita, *Hamiltonian stability of certain minimal Lagrangian submanifolds in complex projective spaces*, Tohoku Math. J. (2) **55** (2003), no. 4, 583–610. MR2017227

[Aub82] Thierry Aubin, *Nonlinear analysis on manifolds. Monge-Ampère equations*, Grundlehren der mathematischen Wissenschaften [Fundamental Principles of Mathematical Sciences], vol. 252, Springer-Verlag, New York, 1982, DOI 10.1007/978-1-4612-5734-5739. MR681859

[Bar75] João Lucas Marquês Barbosa, *On minimal immersions of S^2 into S^{2m}*, Trans. Amer. Math. Soc. **210** (1975), 75–106, DOI 10.2307/1997123. MR375166

[BC12] Adrian Butscher and Justin Corvino, *Hamiltonian stationary tori in Kähler manifolds*, Calc. Var. Partial Differential Equations **45** (2012), no. 1-2, 63–100, DOI 10.1007/s00526-011-0451-6. MR2957651

[BDGG69] E. Bombieri, E. De Giorgi, and E. Giusti, *Minimal cones and the Bernstein problem*, Invent. Math. **7** (1969), 243–268, DOI 10.1007/BF01404309. MR250205

[Bel25] Costante Bellettini, *Extensions of Schoen-Simon-Yau and Schoen-Simon theorems via iteration à la De Giorgi*, Invent. Math. **240** (2025), no. 1, 1–34, DOI 10.1007/s00222-025-01317-0. MR4871955

[Ber27] Serge Bernstein, *Über ein geometrisches Theorem und seine Anwendung auf die partiellen Differentialgleichungen vom elliptischen Typus* (German), Math. Z. **26** (1927), no. 1, 551–558, DOI 10.1007/BF01475472. MR1544873

[BF71] Paul F. Byrd and Morris D. Friedman, *Handbook of elliptic integrals for engineers and scientists: Second edition, revised*, Die Grundlehren der mathematischen Wissenschaften, Band 67, Springer-Verlag, New York-Heidelberg, 1971. MR277773

[BGM71] Marcel Berger, Paul Gauduchon, and Edmond Mazet, *Le spectre d'une variété riemannienne* (French), Lecture Notes in Mathematics, Vol. 194, Springer-Verlag, Berlin-New York, 1971. MR282313

[Bre13] Simon Brendle, *Embedded minimal tori in S^3 and the Lawson conjecture*, Acta Math. **211** (2013), no. 2, 177–190, DOI 10.1007/s11511-013-0101-2. MR3143888

[Bry82] Robert L. Bryant, *Conformal and minimal immersions of compact surfaces into the 4-sphere*, J. Differential Geometry **17** (1982), no. 3, 455–473. MR679067

[Bry84] Robert L. Bryant, *A duality theorem for Willmore surfaces*, J. Differential Geom. **20** (1984), no. 1, 23–53. MR772125

[Bry91] Robert L. Bryant, *Some remarks on the geometry of austere manifolds*, Bol. Soc. Brasil. Mat. (N.S.) **21** (1991), no. 2, 133–157, DOI 10.1007/BF01237361. MR1139562

[BW58] W. M. Boothby and H. C. Wang, *On contact manifolds*, Ann. of Math. (2) **68** (1958), 721–734, DOI 10.2307/1970165. MR112160

[Bé91] Pierre Bérard, *Remarques sur l'équation de J. Simons* (French, with English summary), Differential geometry, Pitman Monogr. Surveys Pure Appl. Math., vol. 52, Longman Sci. Tech., Harlow, 1991, pp. 47–57. MR1173032

[Cal67] Eugenio Calabi, *Minimal immersions of surfaces in Euclidean spheres*, J. Differential Geometry **1** (1967), 111–125. MR233294

[CdCK70] S. S. Chern, M. do Carmo, and S. Kobayashi, *Minimal submanifolds of a sphere with second fundamental form of constant length*, Functional Analysis and Related Fields (Proc. Conf. for M. Stone, Univ. Chicago, Chicago, Ill., 1968), Springer, New York-Berlin, 1970, pp. 59–75. MR273546

[CG72] Jeff Cheeger and Detlef Gromoll, *On the structure of complete manifolds of nonnegative curvature*, Ann. of Math. (2) **96** (1972), 413–443, DOI 10.2307/1970819. MR309010

[CG82] Chi Cheng Chen and Fritz Gackstatter, *Elliptische und hyperelliptische Funktionen und vollständige Minimalflächen vom Enneperschen Typ* (German), Math. Ann. **259** (1982), no. 3, 359–369, DOI 10.1007/BF01456948. MR661204

[CG86] Jeff Cheeger and Mikhael Gromov, *Collapsing Riemannian manifolds while keeping their curvature bounded. I*, J. Differential Geom. **23** (1986), no. 3, 309–346. MR852159

[Cha06] Isaac Chavel, *Riemannian geometry: A modern introduction*, 2nd ed., Cambridge Studies in Advanced Mathematics, vol. 98, Cambridge University Press, Cambridge, 2006, DOI 10.1017/CBO9780511616822. MR2229062

[Che70a] Jeff Cheeger, *A lower bound for the smallest eigenvalue of the Laplacian*, Problems in analysis (Sympos. in honor of Salomon Bochner, Princeton Univ., Princeton, N.J., 1969), Princeton Univ. Press, Princeton, NJ, 1970, pp. 195–199. MR402831

[Che70b] Shiing Shen Chern, *On the minimal immersions of the two-sphere in a space of constant curvature*, Problems in analysis (Sympos. in honor of Salomon Bochner, Princeton Univ., Princeton, N.J., 1969), Princeton Univ. Press, Princeton, NJ, 1970, pp. 27–40. MR362151

[Che21] Hang Chen, *Minimal hypersurfaces in the product of two spheres with index one*, Calc. Var. Partial Differential Equations **60** (2021), no. 4, Paper No. 134, 13, DOI 10.1007/s00526-021-02007-4. MR4279393

[Che24] Shuli Chen, *Index of minimal hypersurfaces in real projective spaces*, Int. Math. Res. Not. IMRN **10** (2024), 8592–8608, DOI 10.1093/imrn/rnad312. MR4749175

[Cho90] Jaigyoung Choe, *Index, vision number and stability of complete minimal surfaces*, Arch. Rational Mech. Anal. **109** (1990), no. 3, 195–212, DOI 10.1007/BF00375088. MR1025170

[CHS84] Luis Caffarelli, Robert Hardt, and Leon Simon, *Minimal surfaces with isolated singularities*, Manuscripta Math. **48** (1984), no. 1-3, 1–18, DOI 10.1007/BF01168999. MR753722

[CL23] Otis Chodosh and Chao Li, *Stable anisotropic minimal hypersurfaces in* $\mathbf{R}^4$, Forum Math. Pi **11** (2023), Paper No. e3, 22, DOI 10.1017/fmp.2023.1. MR4546104

[CL24] Otis Chodosh and Chao Li, *Stable minimal hypersurfaces in* $\mathbf{R}^4$, Acta Math. **233** (2024), no. 1, 1–31, DOI 10.4310/acta.2024.v233.n1.a1. MR4816633

[CLMS24] Otis Chodosh, Chao Li, Paul Minter, and Douglas Stryker, *Stable minimal hypersurfaces in* $\mathbf{R}^5$, `arXiv:2401.01492` [math.DG].

[CLN13] Bang-yen Chen and Tadashi Nagano, *Totally geodesic submanifolds of symmetric spaces. II*, Duke Math. J. **45** (1978), no. 2, 405–425. MR487902

[CLS22] Otis Chodosh, Chao Li, and Douglas Stryker, *Complete stable minimal hypersurfaces in positively curved 4-manifolds*, J. Eur. Math. Soc. (2024), published online first.

[CM11] Tobias Holck Colding and William P. Minicozzi II, *A course in minimal surfaces*, Graduate Studies in Mathematics, vol. 121, American Mathematical Society, Providence, RI, 2011, DOI 10.1090/gsm/121. MR2780140

[CM16] Otis Chodosh and Davi Maximo, *On the topology and index of minimal surfaces*, J. Differential Geom. **104** (2016), no. 3, 399–418. MR3568626

[CM23] Otis Chodosh and Davi Maximo, *On the topology and index of minimal surfaces II*, J. Differential Geom. **123** (2023), no. 3, 431–459, DOI 10.4310/jdg/1683307005. MR4584858

[CMR24] Giovanni Catino, Paolo Mastrolia, and Alberto Roncoroni, *Two rigidity results for stable minimal hypersurfaces*, Geom. Funct. Anal. **34** (2024), no. 1, 1–18, DOI 10.1007/s00039-024-00662-1. MR4706440

[CO67] Shiing-shen Chern and Robert Osserman, *Complete minimal surfaces in euclidean n-space*, J. Analyse Math. **19** (1967), 15–34, DOI 10.1007/BF02788707. MR226514

[CO74] Bang-yen Chen and Koichi Ogiue, *On totally real submanifolds*, Trans. Amer. Math. Soc. **193** (1974), 257–266, DOI 10.2307/1996914. MR346708

[Cos84] Celso J. Costa, *Example of a complete minimal immersion in $\mathbf{R}^3$ of genus one and three embedded ends*, Bol. Soc. Brasil. Mat. **15** (1984), no. 1-2, 47–54, DOI 10.1007/BF02584707. MR794728

[Cos91] C. J. Costa, *Classification of complete minimal surfaces in $\mathbf{R}^3$ with total curvature 12π*, Invent. Math. **105** (1991), no. 2, 273–303, DOI 10.1007/BF01232268. MR1115544

[CSZ97] Huai-Dong Cao, Ying Shen, and Shunhui Zhu, *The structure of stable minimal hypersurfaces in $\mathbf{R}^{n+1}$*, Math. Res. Lett. **4** (1997), no. 5, 637–644, DOI 10.4310/MRL.1997.v4.n5.a2. MR1484695

[CU98] Ildefonso Castro and Francisco Urbano, *Examples of unstable Hamiltonian-minimal Lagrangian tori in $\mathbf{C}^2$*, Compositio Math. **111** (1998), no. 1, 1–14, DOI 10.1023/A:1000332524827. MR1611051

[CU99] Ildefonso Castro and Francisco Urbano, *On a minimal Lagrangian submanifold of $\mathbf{C}^n$ foliated by spheres*, Michigan Math. J. **46** (1999), no. 1, 71–82, DOI 10.1307/mmj/1030132359. MR1682888

[CU07] Ildefonso Castro and Francisco Urbano, *Minimal Lagrangian surfaces in $\mathbb{S}^2 \times \mathbb{S}^2$*, Comm. Anal. Geom. **15** (2007), no. 2, 217–248, DOI 10.4310/cag.2007.v15.n2.a1. MR2344322

[CV35] Stefan Cohn-Vossen, *Kürzeste Wege und Totalkrümmung auf Flächen* (German), Compositio Math. **2** (1935), 69–133. MR1556908

[CW83] Hyeong In Choi and Ai Nung Wang, *A first eigenvalue estimate for minimal hypersurfaces*, J. Differential Geom. **18** (1983), no. 3, 559–562. MR723817

[CY75] S. Y. Cheng and S. T. Yau, *Differential equations on Riemannian manifolds and their geometric applications*, Comm. Pure Appl. Math. **28** (1975), no. 3, 333–354, DOI 10.1002/cpa.3160280303. MR385749

[dC81] Manfredo P. do Carmo, *Stability of minimal submanifolds*, Global differential geometry and global analysis (Berlin, 1979), Lecture Notes in Math., vol. 838, Springer, Berlin, 1981, pp. 129–139. MR636273

[dCD83] M. do Carmo and M. Dajczer, *Rotation hypersurfaces in spaces of constant curvature*, Trans. Amer. Math. Soc. **277** (1983), no. 2, 685–709, DOI 10.2307/1999231. MR694383

[dCP79] M. do Carmo and C. K. Peng, *Stable complete minimal surfaces in $\mathbf{R}^3$ are planes*, Bull. Amer. Math. Soc. (N.S.) **1** (1979), no. 6, 903–906, DOI 10.1090/S0273-0979-1979-14689-5. MR546314

[dCP82] M. do Carmo and C. K. Peng, *Stable complete minimal hypersurfaces*, Proceedings of the 1980 Beijing Symposium on Differential Geometry and Differential Equations, Vol. 1, 2, 3 (Beijing, 1980), Sci. Press Beijing, Beijing, 1982, pp. 1349–1358. MR714373

[dCRR00] Manfredo do Carmo, Manuel Ritoré, and Antonio Ros, *Compact minimal hypersurfaces with index one in the real projective space*, Comment. Math. Helv. **75** (2000), no. 2, 247–254, DOI 10.1007/PL00000373. MR1774704

[DF01] Marcos Dajczer and Luis A. Florit, *A class of austere submanifolds*, Illinois J. Math. **45** (2001), no. 3, 735–755. MR1879232

[DG61] Ennio De Giorgi, *Frontiere orientate di misura minima* (Italian), Editrice Tecnico Scientifica, Pisa, 1961. Seminario di Matematica della Scuola Normale Superiore di Pisa, 1960-61. MR179651

[EGT85] Jost-Hinrich Eschenburg, Irwen Válle Guadalupe, and Renato de Azevedo Tribuzy, *The fundamental equations of minimal surfaces in $\mathbf{C}P^2$*, Math. Ann. **270** (1985), no. 4, 571–598, DOI 10.1007/BF01455305. MR776173

[Eji83] Norio Ejiri, *The index of minimal immersions of S^2 into S^{2n}*, Math. Z. **184** (1983), no. 1, 127–132, DOI 10.1007/BF01162011. MR711733

[EK93] Norio Ejiri and Motoko Kotani, *Index and flat ends of minimal surfaces*, Tokyo J. Math. **16** (1993), no. 1, 37–48, DOI 10.3836/tjm/1270128981. MR1223287

[ESGJ06] Ahmad El Soufi, Hector Giacomini, and Mustapha Jazar, *A unique extremal metric for the least eigenvalue of the Laplacian on the Klein bottle*, Duke Math. J. **135** (2006), no. 1, 181–202, DOI 10.1215/S0012-7094-06-13514-7. MR2259925

[ESI00] Ahmad El Soufi and Saïd Ilias, *Riemannian manifolds admitting isometric immersions by their first eigenfunctions*, Pacific J. Math. **195** (2000), no. 1, 91–99, DOI 10.2140/pjm.2000.195.91. MR1781616

[FC85] D. Fischer-Colbrie, *On complete minimal surfaces with finite Morse index in three-manifolds*, Invent. Math. **82** (1985), no. 1, 121–132, DOI 10.1007/BF01394782. MR808112

[FCS80] Doris Fischer-Colbrie and Richard Schoen, *The structure of complete stable minimal surfaces in 3-manifolds of nonnegative scalar curvature*, Comm. Pure Appl. Math. **33** (1980), no. 2, 199–211, DOI 10.1002/cpa.3160330206. MR562550

[FK92] H. M. Farkas and I. Kra, *Riemann surfaces*, 2nd ed., Graduate Texts in Mathematics, vol. 71, Springer-Verlag, New York, 1992, DOI 10.1007/978-1-4612-2034-3. MR1139765

[Fri84] Thomas Friedrich, *On surfaces in four-spaces*, Ann. Global Anal. Geom. **2** (1984), no. 3, 257–287, DOI 10.1007/BF01876417. MR777909

[FS23] Ailana Fraser and Richard Schoen, *The first Steklov eigenvalue, conformal geometry, and minimal surfaces*, Adv. Math. **226** (2011), no. 5, 4011–4030, DOI 10.1016/j.aim.2010.11.007. MR2770439

[Gau87] P. Gauduchon, *Les immersions super-minimales d'une surface compacte dans une variété riemannienne orientée de dimension* 4 (French, with English summary), Astérisque **154-155** (1987), 9, 151–180, 351 (1988). Théorie des variétés minimales et applications (Palaiseau, 1983–1984). MR0955064

[GH78] Phillip Griffiths and Joseph Harris, *Principles of algebraic geometry*, Pure and Applied Mathematics, Wiley-Interscience [John Wiley & Sons], New York, 1978. MR507725

[GHJ03] M. Gross, D. Huybrechts, and D. Joyce, *Calabi-Yau manifolds and related geometries*, Universitext, Springer-Verlag, Berlin, 2003. Lectures from the Summer School held in Nordfjordeid, June 2001, DOI 10.1007/978-3-642-19004-9. MR1963559

[GM75] S. Gallot and D. Meyer, *Opérateur de courbure et laplacien des formes différentielles d'une variété riemannienne* (French), J. Math. Pures Appl. (9) **54** (1975), no. 3, 259–284. MR454884

[Gro96] M. Gromov, *Positive curvature, macroscopic dimension, spectral gaps and higher signatures*, Functional analysis on the eve of the 21st century, Vol. II (New Brunswick, NJ, 1993), Progr. Math., vol. 132, Birkhäuser Boston, Boston, MA, 1996, pp. 1–213, DOI 10.1007/s10107-010-0354-x. MR1389019

[GT01] David Gilbarg and Neil S. Trudinger, *Elliptic partial differential equations of second order*, Classics in Mathematics, Springer-Verlag, Berlin, 2001. Reprint of the 1998 edition. MR1814364

[HK97] David Hoffman and Hermann Karcher, *Complete embedded minimal surfaces of finite total curvature*, Geometry, V, Encyclopaedia Math. Sci., vol. 90, Springer, Berlin, 1997, pp. 5–93, DOI 10.1007/978-3-662-03484-2_2. MR1490038

[HL82] Reese Harvey and H. Blaine Lawson Jr., *Calibrated geometries*, Acta Math. **148** (1982), 47–157, DOI 10.1007/BF02392726. MR666108

[HM90] David Hoffman and William H. Meeks III, *Embedded minimal surfaces of finite topology*, Ann. of Math. (2) **131** (1990), no. 1, 1–34, DOI 10.2307/1971506. MR1038356

[HMU06] O. Hijazi, S. Montiel, and F. Urbano, Spin^c *geometry of Kähler manifolds and the Hodge Laplacian on minimal Lagrangian submanifolds*, Math. Z. **253** (2006), no. 4, 821–853, DOI 10.1007/s00209-006-0936-8. MR2221101

[HR02] Frédéric Hélein and Pascal Romon, *Hamiltonian stationary Lagrangian surfaces in* $\mathbb{C}^2$, Comm. Anal. Geom. **10** (2002), no. 1, 79–126, DOI 10.4310/CAG.2002.v10.n1.a5. MR1894142

[HR05] Frédéric Hélein and Pascal Romon, *Hamiltonian stationary tori in the complex projective plane*, Proc. London Math. Soc. (3) **90** (2005), no. 2, 472–496, DOI 10.1112/S002461150401500X. MR2142135

[Hub57] Alfred Huber, *On subharmonic functions and differential geometry in the large*, Comment. Math. Helv. **32** (1957), 13–72, DOI 10.1007/BF02564570. MR94452

[Igu54] Jun-ichi Igusa, *On the structure of a certain class of Kaehler varieties*, Amer. J. Math. **76** (1954), 669–678, DOI 10.2307/2372709. MR63740

[IOS03] Hiroshi Iriyeh, Hajime Ono, and Takashi Sakai, *Integral geometry and Hamiltonian volume minimizing property of a totally geodesic Lagrangian torus in $S^2 \times S^2$*, Proc. Japan Acad. Ser. A Math. Sci. **79** (2003), no. 10, 167–170. MR2028342

[JM83] Luquésio P. Jorge and William H. Meeks III, *The topology of complete minimal surfaces of finite total Gaussian curvature*, Topology **22** (1983), no. 2, 203–221, DOI 10.1016/0040-9383(83)90032-0. MR683761

[JNP06] Dmitry Jakobson, Nikolai Nadirashvili, and Iosif Polterovich, *Extremal metric for the first eigenvalue on a Klein bottle*, Canad. J. Math. **58** (2006), no. 2, 381–400, DOI 10.4153/CJM-2006-016-0. MR2209284

[Jos06] Jürgen Jost, *Compact Riemann surfaces: An introduction to contemporary mathematics*, 3rd ed., Universitext, Springer-Verlag, Berlin, 2006, DOI 10.1007/978-3-540-33067-7. MR2247485

[Kim77] Yoshio Kimura, *The nullity of compact Kähler submanifolds in a complex projective space*, J. Math. Soc. Japan **29** (1977), no. 3, 561–580, DOI 10.2969/jmsj/02930561. MR482605

[KM58] J.-L. Koszul and B. Malgrange, *Sur certaines structures fibrées complexes* (French), Arch. Math. (Basel) **9** (1958), 102–109, DOI 10.1007/BF02287068. MR131882

[KN96] Shoshichi Kobayashi and Katsumi Nomizu, *Foundations of differential geometry. Vol. II*, Reprint of the 1969 original. Wiley Classics Library, A Wiley-Interscience Publication. John Wiley & Sons, Inc., New York, 1996. MR1393941

[Kob72] Shoshichi Kobayashi, *Transformation groups in differential geometry*, Ergebnisse der Mathematik und ihrer Grenzgebiete [Results in Mathematics and Related Areas], Band 70, Springer-Verlag, New York-Heidelberg, 1972. MR355886

[Kod49] Kunihiko Kodaira, *Harmonic fields in Riemannian manifolds (generalized potential theory)*, Ann. of Math. (2) **50** (1949), 587–665, DOI 10.2307/1969552. MR31148

[Kus87] Rob Kusner, *Conformal geometry and complete minimal surfaces*, Bull. Amer. Math. Soc. (N.S.) **17** (1987), no. 2, 291–295, DOI 10.1090/S0273-0979-1987-15564-9. MR903735

[KW20] Nikolaos Kapouleas and David Wiygul, *The index and nullity of the Lawson surfaces $\xi_{g,1}$*, Camb. J. Math. **8** (2020), no. 2, 363–405, DOI 10.4310/CJM.2020.v8.n2.a3. MR4091028

[KW24] Rob Kusner and Peng Wang, *On the index of minimal 2-tori in the 4-sphere*, J. Reine Angew. Math. **806** (2024), 9–22, DOI 10.1515/crelle-2023-0066. MR4685083

[Law70a] H. Blaine Lawson Jr., *Complete minimal surfaces in S^3*, Ann. of Math. (2) **92** (1970), 335–374, DOI 10.2307/1970625. MR270280

[Law70b] H. Blaine Lawson Jr., *Rigidity theorems in rank-*1 *symmetric spaces*, J. Differential Geometry **4** (1970), 349–357. MR267492

[Lee12] Yng-Ing Lee, *The existence of Hamiltonian stationary Lagrangian tori in Kähler manifolds of any dimension*, Calc. Var. Partial Differential Equations **45** (2012), no. 1-2, 231–251, DOI 10.1007/s00526-011-0457-0. MR2957657

[Li80] Peter Li, *On the Sobolev constant and the p-spectrum of a compact Riemannian manifold*, Ann. Sci. École Norm. Sup. (4) **13** (1980), no. 4, 451–468. MR608289

[Li93] Peter Li, *Lecture notes on geometric analysis*, Lecture Notes Series, vol. 6, Seoul National University, Research Institute of Mathematics, Global Analysis Research Center, Seoul, 1993. MR1320504

[Li04] Peter Li, *Lectures on harmonic functions*, Lectures at University of California, Irvine, 2004.

[Li17] Chao Li, *Index and topology of minimal hypersurfaces in $\mathbb{R}^n$*, Calc. Var. Partial Differential Equations **56** (2017), no. 6, Paper No. 180, 18, DOI 10.1007/s00526-017-1272-z. MR3722074

[Lic58] André Lichnerowicz, *Géométrie des groupes de transformations*, Travaux et Recherches Mathématiques, III, Dunod, Paris, 1958.

[LM99] Francisco J. López and Francisco Martín, *Complete minimal surfaces in* $\mathbf{R}^3$, Publ. Mat. **43** (1999), no. 2, 341–449, DOI 10.5565/PUBLMAT_43299_01. MR1744617

[Ló92] Francisco J. López, *The classification of complete minimal surfaces with total curvature greater than* -12π, Trans. Amer. Math. Soc. **334** (1992), no. 1, 49–74, DOI 10.2307/2153972. MR1058433

[LR89] Francisco J. López and Antonio Ros, *Complete minimal surfaces with index one and stable constant mean curvature surfaces*, Comment. Math. Helv. **64** (1989), no. 1, 34–43, DOI 10.1007/BF02564662. MR982560

[LR91] Francisco J. López and Antonio Ros, *On embedded complete minimal surfaces of genus zero*, J. Differential Geom. **33** (1991), no. 1, 293–300. MR1085145

[LS73] H. Blaine Lawson Jr. and James Simons, *On stable currents and their application to global problems in real and complex geometry*, Ann. of Math. (2) **98** (1973), 427–450, DOI 10.2307/1970913. MR324529

[LT92] Peter Li and Luen-Fai Tam, *Harmonic functions and the structure of complete manifolds*, J. Differential Geom. **35** (1992), no. 2, 359–383. MR1158340

[LW02] Peter Li and Jiaping Wang, *Minimal hypersurfaces with finite index*, Math. Res. Lett. **9** (2002), no. 1, 95–103, DOI 10.4310/MRL.2002.v9.n1.a7. MR1892316

[LW04] Peter Li and Jiaping Wang, *Stable minimal hypersurfaces in a nonnegatively curved manifold*, J. Reine Angew. Math. **566** (2004), 215–230, DOI 10.1515/crll.2004.005. MR2039328

[LY82] Peter Li and Shing Tung Yau, *A new conformal invariant and its applications to the Willmore conjecture and the first eigenvalue of compact surfaces*, Invent. Math. **69** (1982), no. 2, 269–291, DOI 10.1007/BF01399507. MR674407

[LY83] Peter Li and Shing Tung Yau, *On the Schrödinger equation and the eigenvalue problem*, Comm. Math. Phys. **88** (1983), no. 3, 309–318. MR701919

[Maz24] Laurent Mazet, *Stable minimal hypersurfaces in* $\mathbb{R}^6$, `arXiv:2405.14676` [math.DG].

[Mee81] William H. Meeks III, *The classification of complete minimal surfaces in* $\mathbf{R}^3$ *with total curvature greater than* -8π, Duke Math. J. **48** (1981), no. 3, 523–535. MR630583

[Mic84] Mario J. Micallef, *Stable minimal surfaces in Euclidean space*, J. Differential Geom. **19** (1984), no. 1, 57–84. MR739782

[MM88] Mario J. Micallef and John Douglas Moore, *Minimal two-spheres and the topology of manifolds with positive curvature on totally isotropic two-planes*, Ann. of Math. (2) **127** (1988), no. 1, 199–227, DOI 10.2307/1971420. MR924677

[MN14] Fernando C. Marques and André Neves, *Min-max theory and the Willmore conjecture*, Ann. of Math. (2) **179** (2014), no. 2, 683–782, DOI 10.4007/annals.2014.179.2.6. MR3152944

[MN16] Fernando C. Marques and André Neves, *Morse index and multiplicity of min-max minimal hypersurfaces*, Camb. J. Math. **4** (2016), no. 4, 463–511, DOI 10.4310/CJM.2016.v4.n4.a2. MR3572636

[MN21] Fernando C. Marques and André Neves, *Morse index of multiplicity one min-max minimal hypersurfaces*, Adv. Math. **378** (2021), Paper No. 107527, 58, DOI 10.1016/j.aim.2020.107527. MR4191255

[Mor64] Akihiko Morimoto, *On normal almost contact structures with a regularity*, Tohoku Math. J. (2) **16** (1964), 90–104, DOI 10.2748/tmj/1178243735. MR163246

[Mor81] Hiroshi Mori, *Minimal surfaces of revolution in* H^3 *and their global stability*, Indiana Univ. Math. J. **30** (1981), no. 5, 787–794, DOI 10.1512/iumj.1981.30.30057. MR625602

[Mor09] Filippo Morabito, *Index and nullity of the Gauss map of the Costa-Hoffman-Meeks surfaces*, Indiana Univ. Math. J. **58** (2009), no. 2, 677–707, DOI 10.1512/iumj.2009.58.3476. MR2514384

[MP12] William H. Meeks III and Joaquín Pérez, *A survey on classical minimal surface theory*, University Lecture Series, vol. 60, American Mathematical Society, Providence, RI, 2012, DOI 10.1090/ulect/060. MR3012474

[MR91] Sebastián Montiel and Antonio Ros, *Schrödinger operators associated to a holomorphic map*, Global differential geometry and global analysis (Berlin, 1990), Lecture Notes in Math., vol. 1481, Springer, Berlin, 1991, pp. 147–174, DOI 10.1007/BFb0083639. MR1178529

[MS39] S. B. Myers and N. E. Steenrod, *The group of isometries of a Riemannian manifold*, Ann. of Math. (2) **40** (1939), no. 2, 400–416, DOI 10.2307/1968928. MR1503467

[MS73] J. H. Michael and L. M. Simon, *Sobolev and mean-value inequalities on generalized submanifolds of R^n*, Comm. Pure Appl. Math. **26** (1973), 361–379, DOI 10.1002/cpa.3160260305. MR344978

[MU97] Sebastián Montiel and Francisco Urbano, *Second variation of superminimal surfaces into self-dual Einstein four-manifolds*, Trans. Amer. Math. Soc. **349** (1997), no. 6, 2253–2269, DOI 10.1090/S0002-9947-97-01933-8. MR1422905

[MU02] Sebastián Montiel and Francisco Urbano, *A Willmore functional for compact surfaces in the complex projective plane*, J. Reine Angew. Math. **546** (2002), 139–154, DOI 10.1515/crll.2002.039. MR1900995

[MW93] Mario J. Micallef and Jon G. Wolfson, *The second variation of area of minimal surfaces in four-manifolds*, Math. Ann. **295** (1993), no. 2, 245–267, DOI 10.1007/BF01444887. MR1202392

[Nad87] N. S. Nadirashvili, *Multiple eigenvalues of the Laplace operator* (Russian), Mat. Sb. (N.S.) **133(175)** (1987), no. 2, 223–237, 272, DOI 10.1070/SM1988v061n01ABEH003204; English transl., Math. USSR-Sb. **61** (1988), no. 1, 225–238. MR905007

[Nag61] Tadashi Nagano, *On the minimum eigenvalues of the Laplacians in Riemannian manifolds*, Sci. Papers College Gen. Ed. Univ. Tokyo **11** (1961), 177–182. MR144283

[Nai81a] Hiroo Naitoh, *Isotropic submanifolds with parallel second fundamental form in $P^m(c)$*, Osaka Math. J. **18** (1981), no. 2, 427–464. MR628843

[Nai81b] Hiroo Naitoh, *Totally real parallel submanifolds in $P^n(c)$*, Tokyo J. Math. **4** (1981), no. 2, 279–306, DOI 10.3836/tjm/1270215155. MR646040

[Nay92] Shin Nayatani, *Morse index of complete minimal surfaces*, The problem of Plateau, World Sci. Publ., River Edge, NJ, 1992, pp. 181–189. MR1209216

[Nay93] Shin Nayatani, *Morse index and Gauss maps of complete minimal surfaces in Euclidean 3-space*, Comment. Math. Helv. **68** (1993), no. 4, 511–537, DOI 10.1007/BF02565834. MR1241471

[NT82] Hiroo Naitoh and Masaru Takeuchi, *Totally real submanifolds and symmetric bounded domains*, Osaka Math. J. **19** (1982), no. 4, 717–731. MR687769

[Oba62] Morio Obata, *Certain conditions for a Riemannian manifold to be isometric with a sphere*, J. Math. Soc. Japan **14** (1962), 333–340, DOI 10.2969/jmsj/01430333. MR142086

[Oh90] Yong-Geun Oh, *Second variation and stabilities of minimal Lagrangian submanifolds in Kähler manifolds*, Invent. Math. **101** (1990), no. 2, 501–519, DOI 10.1007/BF01231513. MR1062973

[Oh91] Yong-Geun Oh, *Tight Lagrangian submanifolds in* $\mathbf{C}P^n$, Math. Z. **207** (1991), no. 3, 409–416, DOI 10.1007/BF02571398. MR1115173

[Oh93] Yong-Geun Oh, *Volume minimization of Lagrangian submanifolds under Hamiltonian deformations*, Math. Z. **212** (1993), no. 2, 175–192, DOI 10.1007/BF02571651. MR1202805

[Oh94] Yong-Geun Oh, *Mean curvature vector and symplectic topology of Lagrangian submanifolds in Einstein-Kähler manifolds*, Math. Z. **216** (1994), no. 3, 471–482, DOI 10.1007/BF02572335. MR1283083

[Ohn86] Yoshihiro Ohnita, *Stable minimal submanifolds in compact rank one symmetric spaces*, Tohoku Math. J. (2) **38** (1986), no. 2, 199–217, DOI 10.2748/tmj/1178228488. MR843807

[Oss69] Robert Osserman, *A survey of minimal surfaces*, Van Nostrand Reinhold Co., New York-London-Melbourne, 1969. MR256278

[Oss79] Robert Osserman, *Bonnesen-style isoperimetric inequalities*, Amer. Math. Monthly **86** (1979), no. 1, 1–29, DOI 10.2307/2320297. MR519520

[Pal91] Bennett Palmer, *Stability of minimal hypersurfaces*, Comment. Math. Helv. **66** (1991), no. 2, 185–188, DOI 10.1007/BF02566644. MR1107838

[Per01] Oscar Perdomo, *Low index minimal hypersurfaces of spheres*, Asian J. Math. **5** (2001), no. 4, 741–749, DOI 10.4310/AJM.2001.v5.n4.a8. MR1913819

[Per02] Oscar Perdomo, *First stability eigenvalue characterization of Clifford hypersurfaces*, Proc. Amer. Math. Soc. **130** (2002), no. 11, 3379–3384, DOI 10.1090/S0002-9939-02-06451-1. MR1913017

[Pog81] A. V. Pogorelov, *On the stability of minimal surfaces* (Russian), Dokl. Akad. Nauk SSSR **260** (1981), no. 2, 293–295. MR630142

[Pro60] M. H. Protter, *Unique continuation for elliptic equations*, Trans. Amer. Math. Soc. **95** (1960), 81–91, DOI 10.2307/1993331. MR113030

[Ros06] Antonio Ros, *One-sided complete stable minimal surfaces*, J. Differential Geom. **74** (2006), no. 1, 69–92. MR2260928

[Ros25] Antonio Ros, *Index one minimal surfaces in positively curved 3-manifolds*, J. Differential Geom. **129** (2025), no. 2, 523–540, DOI 10.4310/jdg/1738164150. MR4859582

[RS94] Marty Ross and Chad Schoen, *Stable quotients of periodic minimal surfaces*, Comm. Anal. Geom. **2** (1994), no. 3, 451–459, DOI 10.4310/CAG.1994.v2.n3.a4. MR1305713

[Sav10] Alessandro Savo, *Index bounds for minimal hypersurfaces of the sphere*, Indiana Univ. Math. J. **59** (2010), no. 3, 823–837, DOI 10.1512/iumj.2010.59.4013. MR2779062

[Sch83] Richard M. Schoen, *Uniqueness, symmetry, and embeddedness of minimal surfaces*, J. Differential Geom. **18** (1983), no. 4, 791–809 (1984). MR730928

[Sei00] Paul Seidel, *Graded Lagrangian submanifolds* (English, with English and French summaries), Bull. Soc. Math. France **128** (2000), no. 1, 103–149. MR1765826

[Ser51] Jean-Pierre Serre, *Homologie singulière des espaces fibrés. Applications* (French), Ann. of Math. (2) **54** (1951), 425–505, DOI 10.2307/1969485. MR45386

[Sim68] James Simons, *Minimal varieties in riemannian manifolds*, Ann. of Math. (2) **88** (1968), 62–105, DOI 10.2307/1970556. MR233295

[Sim81] Barry Simon, *Spectrum and continuum eigenfunctions of Schrödinger operators*, J. Functional Analysis **42** (1981), no. 3, 347–355, DOI 10.1016/0022-1236(81)90094-X. MR626449

[Sim93] Leon Simon, *Existence of surfaces minimizing the Willmore functional*, Comm. Anal. Geom. **1** (1993), no. 2, 281–326, DOI 10.4310/CAG.1993.v1.n2.a4. MR1243525

[Sol90a] Bruce Solomon, *The harmonic analysis of cubic isoparametric minimal hypersurfaces. I. Dimensions* 3 *and* 6, Amer. J. Math. **112** (1990), no. 2, 157–203, DOI 10.2307/2374713. MR1047297

[Sol90b] Bruce Solomon, *The harmonic analysis of cubic isoparametric minimal hypersurfaces. II. Dimensions* 12 *and* 24, Amer. J. Math. **112** (1990), no. 2, 205–241, DOI 10.2307/2374714. MR1047298

[Spi79] Michael Spivak, *A comprehensive introduction to differential geometry. Vol. III*, 2nd ed., Publish or Perish, Inc., Wilmington, DE, 1979. MR532832

[SSY75] R. Schoen, L. Simon, and S. T. Yau, *Curvature estimates for minimal hypersurfaces*, Acta Math. **134** (1975), no. 3-4, 275–288, DOI 10.1007/BF02392104. MR423263

[ST69] I. M. Singer and J. A. Thorpe, *The curvature of* 4*-dimensional Einstein spaces*, Global Analysis (Papers in Honor of K. Kodaira), Univ. Tokyo Press, Tokyo, 1969, pp. 355–365. MR256303

[SW01] R. Schoen and J. Wolfson, *Minimizing area among Lagrangian surfaces: the mapping problem*, J. Differential Geom. **58** (2001), no. 1, 1–86. MR1895348

[SY76] Richard Schoen and Shing Tung Yau, *Harmonic maps and the topology of stable hypersurfaces and manifolds with non-negative Ricci curvature*, Comment. Math. Helv. **51** (1976), no. 3, 333–341, DOI 10.1007/BF02568161. MR438388

[SZ98] Yi-Bing Shen and Xiao-Hua Zhu, *On stable complete minimal hypersurfaces in* $\mathbf{R}^{n+1}$, Amer. J. Math. **120** (1998), no. 1, 103–116. MR1600268

[Tai68] Shin-sheng Tai, *Minimum imbeddings of compact symmetric spaces of rank one*, J. Differential Geometry **2** (1968), 55–66. MR231395

[Tak75] Ryoichi Takagi, *Real hypersurfaces in a complex projective space with constant principal curvatures*, J. Math. Soc. Japan **27** (1975), 43–53, DOI 10.2969/jmsj/02710043. MR355906

[Tak84] Masaru Takeuchi, *Stability of certain minimal submanifolds of compact Hermitian symmetric spaces*, Tohoku Math. J. (2) **36** (1984), no. 2, 293–314, DOI 10.2748/tmj/1178228853. MR742600

[Tan79] Shûkichi Tanno, *The first eigenvalue of the Laplacian on spheres*, Tohoku Math. J. (2) **31** (1979), no. 2, 179–185, DOI 10.2748/tmj/1178229837. MR538918

[TU14] Francisco Torralbo and Francisco Urbano, *On stable compact minimal submanifolds*, Proc. Amer. Math. Soc. **142** (2014), no. 2, 651–658, DOI 10.1090/S0002-9939-2013-11810-1. MR3134005

[TU22] Francisco Torralbo and Francisco Urbano, *Index of compact minimal submanifolds of the Berger spheres*, Calc. Var. Partial Differential Equations **61** (2022), no. 3, Paper No. 104, 35.

[Tuz91] A. A. Tuzhilin, *Morse-type indices for two-dimensional minimal surfaces in* $\mathbf{R}^3$ *and* $\mathbf{H}^3$ (Russian), Izv. Akad. Nauk SSSR Ser. Mat. **55** (1991), no. 3, 581–607; English transl., Math. USSR-Izv. **38** (1992), no. 3, 575–598. DOI 10.1070/IM1992v038n03ABEH002215. MR1129827

[Tys89] Johan Tysk, *Finiteness of index and total scalar curvature for minimal hypersurfaces*, Proc. Amer. Math. Soc. **105** (1989), no. 2, 429–435, DOI 10.2307/2046961. MR946639

[TZ09] Luen-Fai Tam and Detang Zhou, *Stability properties for the higher dimensional catenoid in* $\mathbb{R}^{n+1}$, Proc. Amer. Math. Soc. **137** (2009), no. 10, 3451–3461, DOI 10.1090/S0002-9939-09-09962-6. MR2515414

[Urb90] Francisco Urbano, *Minimal surfaces with low index in the three-dimensional sphere*, Proc. Amer. Math. Soc. **108** (1990), no. 4, 989–992, DOI 10.2307/2047957. MR1007516

[Urb93] Francisco Urbano, *Index of Lagrangian submanifolds of* $\mathbf{C}\mathrm{P}^n$ *and the Laplacian of* 1*-forms*, Geom. Dedicata **48** (1993), no. 3, 309–318, DOI 10.1007/BF01264074. MR1254100

[Urb03] Francisco Urbano, *Second variation of compact minimal Legendrian submanifolds of the sphere*, Michigan Math. J. **51** (2003), no. 2, 437–447, DOI 10.1307/mmj/1060013206. MR1992956

[Urb07] Francisco Urbano, *Hamiltonian stability and index of minimal Lagrangian surfaces of the complex projective plane*, Indiana Univ. Math. J. **56** (2007), no. 2, 931–946, DOI 10.1512/iumj.2007.56.2818. MR2317551

[Urb13] Francisco Urbano, *Second variation of one-sided complete minimal surfaces*, Rev. Mat. Iberoam. **29** (2013), no. 2, 479–494, DOI 10.4171/RMI/727. MR3047425

[War83] Frank W. Warner, *Foundations of differentiable manifolds and Lie groups: Corrected reprint of the 1971 edition*, Graduate Texts in Mathematics, vol. 94, Springer-Verlag, New York-Berlin, 1983. MR722297

[Wol89] Jon G. Wolfson, *Minimal surfaces in Kähler surfaces and Ricci curvature*, J. Differential Geom. **29** (1989), no. 2, 281–294. MR982175

[Wol11] Joseph A. Wolf, *Spaces of constant curvature*, 6th ed., AMS Chelsea Publishing, Providence, RI, 2011, DOI 10.1090/chel/372. MR2742530

[WW09] Guofang Wei and Will Wylie, *Comparison geometry for the Bakry-Emery Ricci tensor*, J. Differential Geom. **83** (2009), no. 2, 377–405, DOI 10.4310/jdg/1261495336. MR2577473

[Yau74a] Shing Tung Yau, *Non-existence of continuous convex functions on certain Riemannian manifolds*, Math. Ann. **207** (1974), 269–270, DOI 10.1007/BF01351342. MR339008

[Yau74b] Shing Tung Yau, *Submanifolds with constant mean curvature. I, II*, Amer. J. Math. **96** (1974), 346–366; ibid. 97 (1975), 76–100, DOI 10.2307/2373638. MR370443

[Yau75] Shing Tung Yau, *Isoperimetric constants and the first eigenvalue of a compact Riemannian manifold*, Ann. Sci. École Norm. Sup. (4) **8** (1975), no. 4, 487–507. MR397619

[Yau82] Shing Tung Yau, *Problem section*, Seminar on Differential Geometry, Ann. of Math. Stud., No. 102, Princeton Univ. Press, Princeton, NJ, 1982, pp. 669–706. MR645762

Index

Index of authors

Selected Published Titles in This Series

292 **Francisco Urbano,** Morse Index of Minimal Submanifolds, 2025

287 **Jacques Sauloy,** Basic Modern Theory of Linear Complex Analytic q-Difference Equations, 2024

286 **Zalman Balanov, Wieslaw Krawcewicz, Dmitrii Rachinskii, Hao-Pin Wu, and Jianshe Yu,** Degree Theory and Symmetric Equations Assisted by GAP System, 2025

285 **Niles Johnson and Donald Yau,** Bimonoidal Categories, E_n-Monoidal Categories, and Algebraic K-Theory, 2024

284 **Donald Yau,** Bimonoidal Categories, E_n-Monoidal Categories, and Algebraic K-Theory, 2024

283 **Donald Yau,** Bimonoidal Categories, E_n-Monoidal Categories, and Algebraic K-Theory, 2024

282 **Michael Kapovich and Pranab Sardar,** Trees of Hyperbolic Spaces, 2024

281 **Horst R. Thieme,** Discrete-Time Dynamics of Structured Populations and Homogeneous Order-Preserving Operators, 2024

280 **Tadashi Ochiai,** Iwasawa Theory and Its Perspective, Volume 2, 2024

279 **Ellen Elizabeth Eischen, Wee Teck Gan, Aaron Pollack, and Zhiwei Yun,** Automorphic Forms Beyond GL_2, 2024

278 **Delaram Kahrobaei, Ramón Flores, Marialaura Noce, Maggie E. Habeeb, and Christopher Battarbee,** Applications of Group Theory in Cryptography, 2024

277 **George Metcalfe, Francesco Paoli, and Constantine Tsinakis,** Residuated Structures in Algebra and Logic, 2023

276 **Balázs Bárány, Károly Simon, and Boris Solomyak,** Self-similar and Self-affine Sets and Measures, 2023

275 **Alekos Vidras and Alain Yger,** Multidimensional Residue Theory and Applications, 2023

274 **Willi Freeden and M. Zuhair Nashed,** Recovery Methodologies: Regularization and Sampling, 2023

273 **Gennadiy Feldman,** Characterization of Probability Distributions on Locally Compact Abelian Groups, 2023

272 **Tadashi Ochiai,** Iwasawa Theory and Its Perspective, Volume 1, 2023

271 **Hideto Asashiba,** Categories and Representation Theory, 2022

270 **Leslie Hogben, Jephian C.-H. Lin, and Bryan L. Shader,** Inverse Problems and Zero Forcing for Graphs, 2022

269 **Ralph S. Freese, Ralph N. McKenzie, George F. McNulty, and Walter F. Taylor,** Algebras, Lattices, Varieties, 2022

268 **Ralph S. Freese, Ralph N. McKenzie, George F. McNulty, and Walter F. Taylor,** Algebras, Lattices, Varieties, 2022

267 **Dragana S. Cvetković Ilić,** Completion Problems on Operator Matrices, 2022

266 **Kate Juschenko,** Amenability of Discrete Groups by Examples, 2022

265 **Nabil H. Mustafa,** Sampling in Combinatorial and Geometric Set Systems, 2022

264 **J. Scott Carter and Seiichi Kamada,** Diagrammatic Algebra, 2021

263 **Vugar E. Ismailov,** Ridge Functions and Applications in Neural Networks, 2021

262 **Ragnar-Olaf Buchweitz,** Maximal Cohen–Macaulay Modules and Tate Cohomology, 2021

261 **Shiri Artstein-Avidan, Apostolos Giannopoulos, and Vitali D. Milman,** Asymptotic Geometric Analysis, Part II, 2021

For a complete list of titles in this series, visit the
AMS Bookstore at **www.ams.org/bookstore/survseries/**.

VOLCANIC ASH

Grant Heiken
Kenneth Wohletz

LOS ALAMOS NATIONAL LABORATORY

UNIVERSITY OF CALIFORNIA PRESS
Berkeley · Los Angeles · London

UNIVERSITY OF CALIFORNIA PRESS
Berkeley • Los Angeles • London

University of California Press
Berkeley and Los Angeles, California

University of California Press
London, England

ISBN 0520-05241-2
Printed in the United States of America

1 2 3 4 5 6 7 8 9

CONTENTS

PREFACE

The characterization of volcanic ash and its relationship to eruption phenomena have been of interest to scientists since the time of Aristotle. Interest in this subject was renewed during the mid-1800s when European naturalists began studying volcanic fields in the Mediterranean, South America, and Indonesia. With the invention of the scanning electron microscope (SEM), a full range of pyroclast sizes could be investigated and chemical changes on grain surfaces measured. Used in conjunction with optical microscopy and the electron microprobe, the SEM has allowed us to begin systematic characterization of volcanic ash.

The purpose of this book is to provide a reference and guide to the characteristics of volcanic ash. We began assembling data for the book after the eruption of Mount St. Helens, when we received many inquiries regarding the physical and chemical properties of volcanic ash. Questions came not only from geologists but from people in the fields of environmental science, public health, engineering, aeronautics, agriculture, and the atmospheric sciences. The book is designed to be used by that broad spectrum of scientists and engineers.

We began the book on our own time but eventually were supported by research funds from Los Alamos National Laboratory. For their encouragement throughout the project, we thank John Whetten, Wes Myers, John Birely, David Sharp, and L. M. Simmons of Los Alamos. Technical support in the laboratory was provided by Roland Hagen and David Mann.

We collected most of the samples ourselves, but we also received many samples and much help from the volcanological community. Those who contributed include: William Rose, Jr., Michigan Technological University; Stephen Self, University of Texas/Arlington; Michael Sheridan, Arizona State University; Richard Fisher, University of California/Santa Barbara; Richard Hay, University of Illinois; Johan Varekamp, Wesleyan University; Colin Wilson, University of Auckland; Steve Carey, University of Rhode Island; Rodey Batiza, Washington University; J. V. Wright, British Petroleum Co.; Arturo Alcarez, Philippine Commission on Volcanology; Gudrun Larson, Nordic Volcanological Institute; Barry Dawson, University of Sheffield; R. Funiciello, University of Rome; David McKay (who was one of the very early collaborators on this project), Johnson Space Center; Leopold Gélinas, École Polytechnique, Montreal; John Stix, University of Toronto; the Portuguese Geological Survey; The Smithsonian Institution/ Natural History Museum; the late Sigurdur Thorarinsson; and the late Roald Fryxell.

Many people participated in construction of the book. Typing of the various drafts was done by Barbara Hahn and Marcia Jones. Line illustrations were drawn by John Tubb and Janet Repa. The many photos were reduced in the photo lab under supervision of Dan Morse. Book design was by Lisa Mirski of the University of California Press. Technical reviews were provided by Richard Fisher and Stephen Self. Editing this book was a difficult job that required strict attention to conformity among the various sections and steadfast patience; this job was undertaken by Jody Heiken. Her persistence in demanding grammatical, technical, and visual consistency greatly improved this book. E. Katherine Valdez took the challenging responsibility of composing, layout, and continual design revisions required for the complex presentation of this book. She not only devoted every spare moment but also worked into evenings and over weekends to satisfy the time constraints for completion. We greatly appreciate Kathy's effort and professional job.

We especially acknowledge Richard Fisher and Michael Sheridan, who have continually offered scientific support and enthusiasm. And finally: without the understanding and good sense of our wives, Jody Heiken and Ann Wohletz, this book would not have been possible.

This book is a compilation of a large amount of volcanic ash data that is, we hope, presented in a usable form. Along with the data, we have offered both conservative and liberal interpretations in an attempt to fit observations into both classical and contemporary volcanological thought.

Ash particles are like snowflakes in that no two are alike. They present an infinite variety of morphological, chemical, and physical characteristics. To the novice, systematic classification may seem an insurmountable task, but with persistence this practice yields increasing organization and understanding of this new aspect of volcanology. Although this work is in its infancy, it is becoming increasingly apparent that, with careful progress and the incorporation of techniques and logic from other related sciences, the unifying potential of ash analysis to many aspects of the geologic, planetary, and atmospheric sciences can be great. In one sense, each ash grain is like a field area to be systematically mapped. On a statistical basis, the distinguishing characteristics of ash sample sets, consisting of thousands of grains, define a general relationship that is readily incorporated into the macroscopic world. Because each ash grain yields a multitude of data, much of which has not yet been adequately described or measured, ash deposits are a new world to be studied, utilized, and interwoven into geological practice—sufficient work for generations of scientists. This book represents our present view of what should prove to be only the initial stages in the study of volcanic ash.

Houses buried in volcanic ash—Sakurazima, Japan, 1914.

INTRODUCTION

Explosive volcanic activity has always captured the public's attention. Indirectly, there is interest in such events as the last days of Pompeii and Herculaneum, the eruption of Krakatau, and the spectacular eruptive activity on Io, one of Jupiter's moons, as seen by the Voyager Spacecraft. More recently, villages and towns in many parts of the Earth have been affected directly. The 1980 eruptions of Mount St. Helens, Washington, paralyzed much of the Pacific Northwest with products of explosive activity—volcanic ash. Volcanic ash was also responsible for many deaths and the evacuation of thousands after the eruptions of El Chichón, in the State of Chiapas, Mexico. The eruption plume from El Chichón has also been responsible for spectacular twilight effects seen over much of the Earth during 1982 and may yet have an effect on climate. Around the house, volcanic ash is used in the form of mechanic's soap, tooth powder, and household cleansers, and as aggregates for cement blocks and insulation.

Volcanic ash deposits are found throughout the entire geologic record. Pyroclastic rocks are present as layers in metamorphic rocks over a billion years old, interbedded with sedimentary rocks deposited throughout Paleozoic and Mesozoic time, and in recent deposits only a few years old. Many of these deposits have served as the source or host of a variety of mineral resources such as uranium, lithium, copper, and diamonds. Reserves of zeolite minerals (molecular sieves) are most often found in altered volcanic ash deposits. Volcanic ash is also an important constituent of soil in many areas of the world; because it often determines the fertility of soils, it is critical for development of agriculture. Blocks of lithified volcanic ash are cut and used for construction of houses and public buildings throughout all parts of the world where there are volcanoes.

This book is intended to serve as a reference for those interested in the identification of volcanic ash and its physical properties. For the geologist and volcanologist, we have attempted to provide examples of ash from all eruption types and from a variety of eruption sequences. These examples may be used in interpretation of explosive eruption phenomena and physical characteristics of the magmas. Ash characteristics, along with chemical characterization, may be used in tephrochronology, the study and correlation of ash layers in the geologic record. Tephrochronologists determine the ages of volcanic eruptions and locate source vents for ash layers; these data are especially important for the appraisal of volcanic hazards (Westgate and Gold, 1974). We have provided a number of examples of relict volcanic ash textures in soils and meta-tuffs to help identify ancient eruption sequences.

Examples of volcanic ash collected in eruption plumes may be of interest to those in the atmospheric and environmental sciences. There is increasing research on the atmospheric effects of volcanic ash since the eruption of Mount St. Helens. Characterization of ashes from ash fall deposits and eruption plumes could also be of interest to engineers in the aerospace industry; in 1982, three jet aircraft were forced down because they had incurred engine damage after flying into eruption plumes.

This book is devoted to descriptions of volcanic ash from different eruption types, sequences of ash layers, and weathered and metamorphosed ash. Divisions of eruption and, therefore, ash types are based upon the mechanism of ash formation (discussed later in this chapter) and the chemical composition of the ash. The samples upon which this book is based were collected by the authors or thoughtfully contributed by members of the volcanological community; in our opinion, the samples typify each major type of explosive eruption and its ash deposits. Each example is presented independently; plates and text are on facing pages so that this book can be used as a reference text for laboratory application. Appropriate references follow each section. Because the terminology used here may be unfamiliar to many readers, an illustrated glossary is included. If used as a text, this book should be used in conjunction with *Pyroclastic Rocks* by Fisher and Schmincke (1984). The Fisher-Schmincke book is a thorough study of the genesis and field occurrence of volcanic ash deposits.

REFERENCES: Fisher, R. and Schmincke, H. (1984). *Pyroclastic rocks*, Springer-Verlag, Berlin, 472 pp.

Westgate, J. A. and Gold, C. M., 1974. *World bibliography and index of Quaternary tephrochronology*. Univ. of Alberta, Edmonton, 528 pp.

Glossary

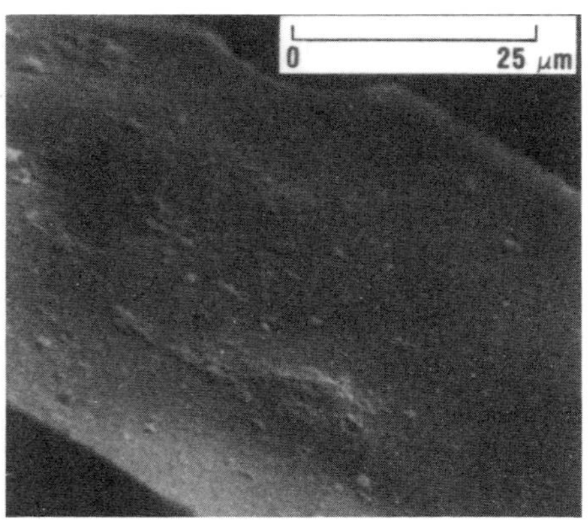

Adhering dust

Aggregate

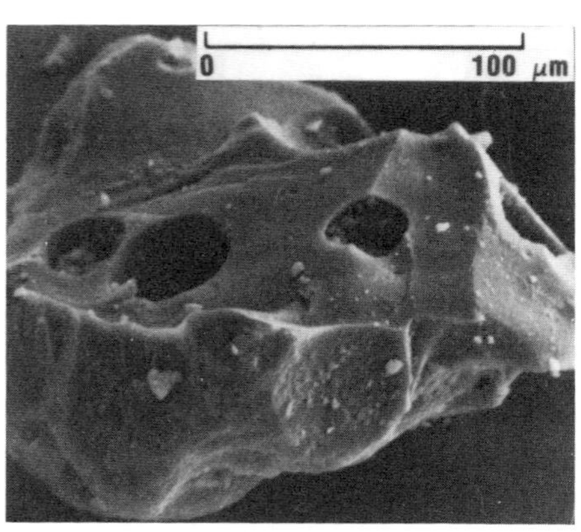

Blocky morphology (shape)

Achnelith
Term used by Walker and Croasdale (1972) to describe droplets from Strombolian lava fountaining. It is from the Greek *achne*—spray.

Accretionary lapilli
More or less spherical masses of cemented ash ranging in diameter from a few millimeters to several centimeters. Sometimes referred to as "volcanic hailstones," "pisolites," "volcanic mud-balls," or "chalazoidites" (in New Zealand).

Adhering dust
Fine-grained volcanic dust sticking to the surfaces of coarser ash grains.

Aerosol
A suspension of colloidal particles in a gas. The particles may be fine ash, sublimates, ice and water, or acid droplets present in a dilute volcanic plume.

Agglutinates
Vitric pyroclasts consisting of a cluster of smaller pyroclasts that are annealed or "welded" to one another.

Aggregate
A mass of pyroclasts bonded by electrostatic forces, sublimates, or alteration products such as carbonate, zeolites, or clay.

Aphyric (non-porphyritic)
An igneous texture in which phenocrysts are absent in a fine-grained or glassy matrix.

Ash
Pyroclasts 2 mm or less in diameter. (Fisher, 1961; Schmid, 1981).

Ash fall
A rain of airborne volcanic ash falling from an eruption cloud and the resulting deposit. An ash fall deposit is generally characterized by mantle bedding of well-to-moderately sorted ash. Syn: "ash fallout."

Ash flow
Synonym for "pyroclastic flow" but referring to pyroclastic flows with pyroclasts under 2 mm in diameter.

An
Abbreviation for anorthite ($CaAl_2Si_2O_8$). Plagioclase compositions are expressed as the percentage of anorthite in solid solution with albite ($NaAlSi_3O_8$).

Base surge
A ring-shaped cloud that moves horizontally outward from the base of a vertical explosion column, initially at hurricane velocities, as a turbulent density current. Base surge deposits are thinly bedded and consist of dune-like beds, inversely graded flat-lying beds, and massive beds showing little or no internal grading. Base surges were first described in association with nuclear weapons tests in the Pacific and then in the deposits of maar volcanoes.

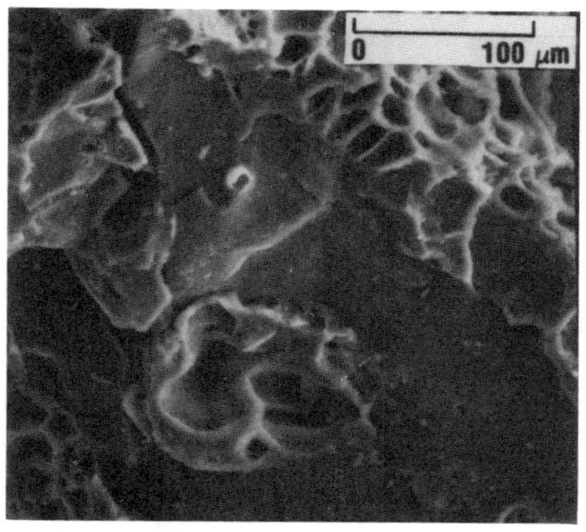

Bubble-wall texture

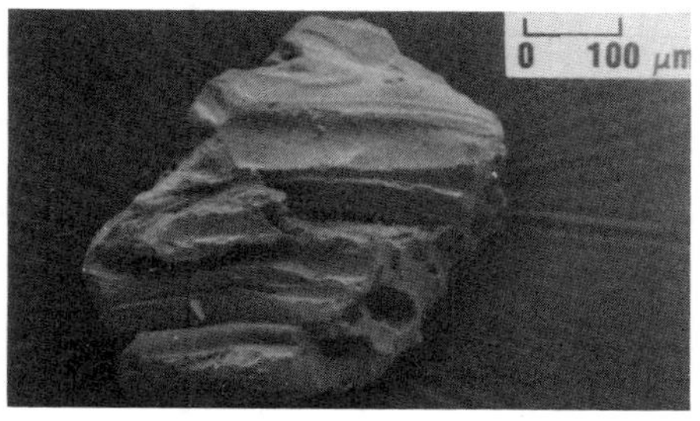

Chipped edges

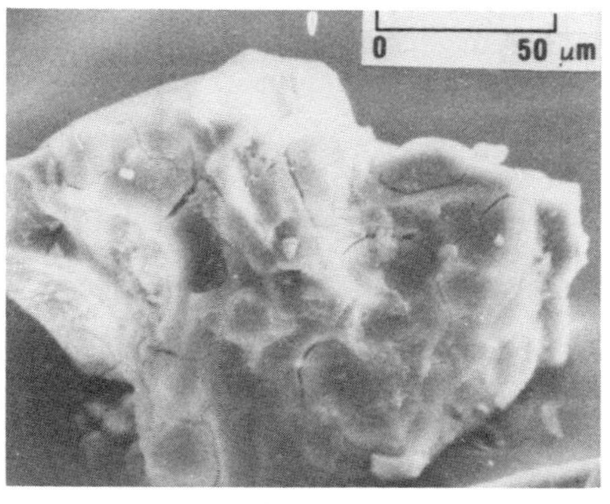

Cracks

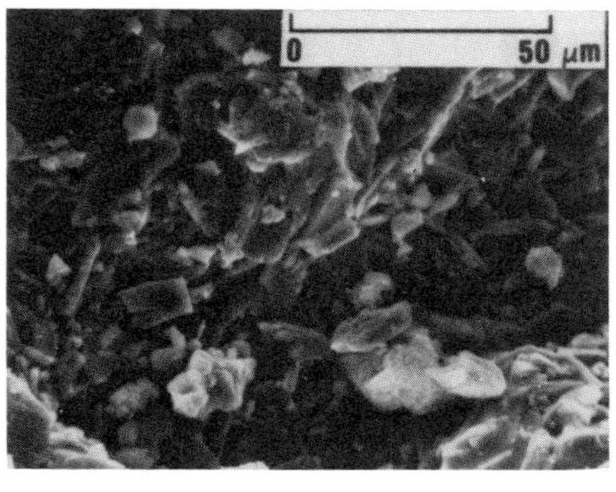

Diktytaxitic

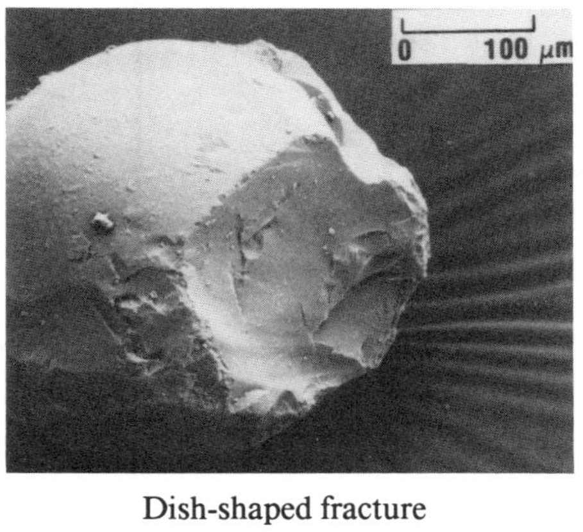

Dish-shaped fracture

Block
: A pyroclast larger than 64 mm that was solid when ejected (AGI). It may have been juvenile or torn from older rocks in the vent or crater walls.

Blocky morphology (shape)
: Overall grain shape characterized by planar or curviplanar surfaces that intersect nearly at right angles. Equant, blocky grains have a nearly isometric form that may grade into a tabular or prismatic form. Typical of phreatomagmatic ash grains.

Bomb
: A pyroclast larger than 64 mm that was viscous (partly or completely fluid) when ejected (AGI).

Botyryoidal
: Having the form of a bunch of grapes (AGI). Clusters of glass or minerals consisting of spherical and/or ovoid shapes.

Breccia
: A fragmental rock made of highly angular coarse fragments with or without a finer grained matrix.

Bubble-wall texture
: Texture shown by phenocrysts with a thin coating of vesicular glass (Fisher, 1963).

Chipped edges
: A textural feature of pyroclasts in which small, concave fractures (dish-shaped, v-shaped, or uneven) are concentrated along thin, angular edges of a pyroclast. This feature is present when angular grains are rounded because of transport abrasion.

Cinder
: Synonym of "scoria." Used in ancient times when volcanoes were thought to contain fire from combustibles within the Earth.

Cracks
: Cracks in volcanic ash grains form by mechanical (fragmentation and percussion) and chemical (alteration) processes. The latter, illustrated here, are tens of micrometers long and several micrometers wide, project through a glass skin several micrometers thick, and result from volume changes during hydration.

Diatreme
: A general term for a volcanic vent or pipe drilled upward through enclosing rocks (usually flat-lying sedimentary rocks) by the explosive energy of gas-charged magmas. The surface expression of a diatreme is a maar crater and, after extensive erosion, is a tuff or tuff-breccia neck.

Diktytaxitic
: A textural term for an igneous rock with irregular, angular cavities between groundmass minerals. In many instances, the groundmass minerals have a thin coating of glass.

Dish-shaped fracture
: Fracture surfaces characterized by circular, or nearly circular, concave areas that are the result of conchoidal fracture of glassy material. Such fractures are typical of grain modification during transport abrasion.

EDS
: Energy dispersive spectral analysis. A method of chemical analysis based upon the energy spectrum of x-rays emitted from a sample illuminated by an electron beam. This method is readily adaptable for use with a scanning electron microscope.

Eruption plume
: A convoluted, rolling mass of partly condensed water vapor, dust, and ash that often rises to great heights. Syn: "eruption cloud."

Explosion pit
: A bowl-shaped crater surrounded by a low rim of ejecta; generally formed by phreatic or phreatomagmatic explosions in which little or no juvenile material is ejected.

Fluidal texture
: Smooth, round surfaces of pyroclasts formed by surface tension within a fluid droplet before it is chilled.

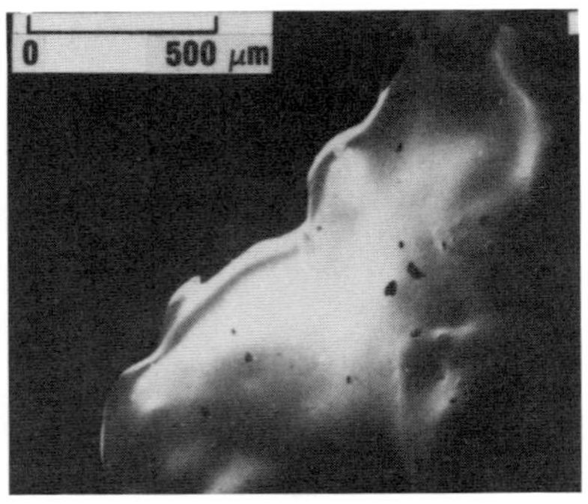

Fluidal texture

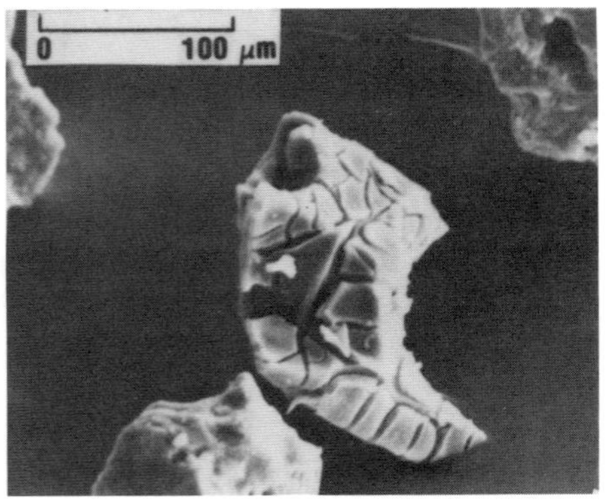

Hydrated surface

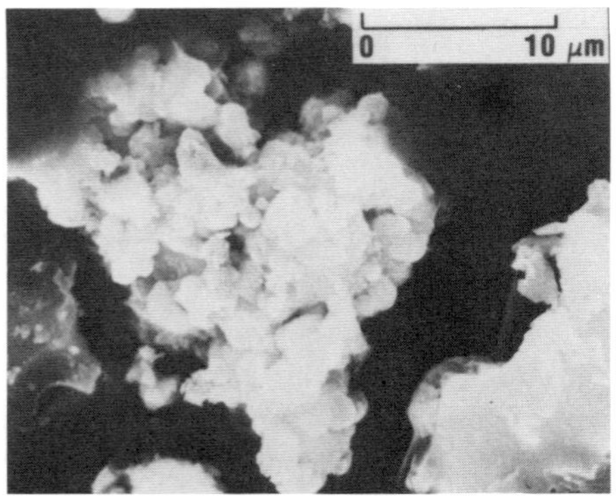

Moss-like morphology

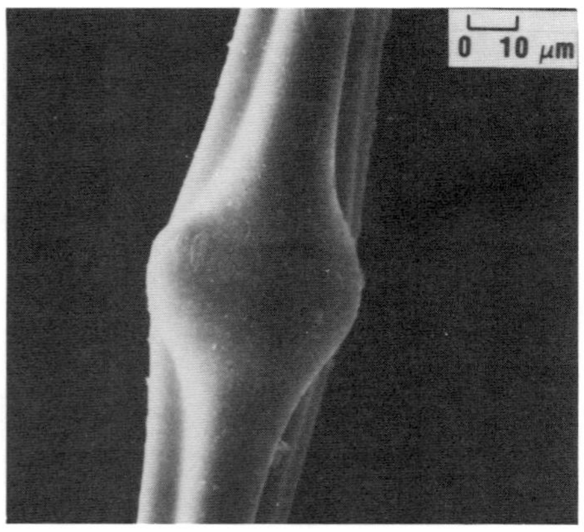

Pele's hair

Fo
: Abbreviation for forsterite (Mg_2SiO_3). Olivine compositions are expressed as the percentage of forsterite in solid solution with fayalite (Fa—Fe_2SiO_3).

Glomeroporphyritic
: Igneous texture denoting clustering of phenocrysts in a finely crystalline groundmass.

Hackly
: Showing jagged points on a fracture surface (AGI).

Hawaiian eruption
: Eruption of low-viscosity gas-charged magmas as lava fountains and thin, rapidly moving lava flows.

Hyaloclastite
: Volcaniclastic rocks generated by non-explosive to mildly explosive granulation of volcanic glass when basaltic magmas are quenched on contact with water. These are common products of deep-sea volcanism (Honnorez and Kirst, 1975) and subglacial eruptions.

Hyalocrystalline
: A textural term applied to igneous porphyritic rocks composed of equal or nearly equal amounts of crystals (phenocrysts) and glassy groundmass.

Hyalotuff
: Pyroclastic rocks generated by phreatomagmatic explosions when volcanoes erupt in shallow water (Honnorez and Kirst, 1975). The term is applied mainly to the tuffs of nearshore maar volcanoes.

Hydrated surface
: A lighter colored, vitreous skin covering all or part of a vitric pyroclast; it is formed by hydration of the outer part of a pyroclast. The surface is often cracked by desiccation of the skin.

Hydromagmatic
: A general term to include all processes, subsurface or surface, involving interaction of magma or magmatic heat with meteoric or connate water in the Earth. Syn: "phreatomagmatic" when describing volcanic eruptions.

Hydrovolcanic
: A term encompassing all volcanic activity that results from the interaction of lava, magmatic heat, or gases with meteoric or connate water at or near the surface of the Earth. This volcanic activity is generally associated with vapor explosions. Syn: "phreatomagmatic."

Ignimbrite
: A rock-unit term applied to tuffs deposited by pyroclastic flows (AGI).

Juvenile pyroclast
: Pyroclast derived directly from magma reaching the surface (AGI).

Lahar
: A debris flow or mudflow of pyroclastic material that is mainly derived from pyroclastic deposits.

Lapilli (s.: lapillus)
: Pyroclasts between 2 and 64 mm (Fisher, 1961).

Lava fountain
: A jet of low-viscosity lava shot into the air by release of volatiles from the melt (AGI).

Lithic pyroclasts
: Pyroclasts consisting of previously formed rocks; fragments of igneous, metamorphic, or sedimentary rock or pieces of earlier lavas in the vent area that are included in a pyroclastic deposit.

Littoral cone
: A tuff cone formed on or adjacent to a lava flow when it enters a body of water. Such cones are the result of steam explosions (AGI). Syn: "rootless cone."

Maar volcano
: A small volcano characterized by a crater that is wide (several hundred meters to several kilometers diameter) relative to its height and having a crater floor that commonly lies below the general level of surrounding topography. Maars form during phreatomagmatic eruptions.

Magma
: Molten rock within a terrestrial planet.

Pele's hair has intrigued geologists and historians for centuries. Typically, it consists of long strands that are circular in cross-section, with diameters from 1 to 300 μm. The thinner varieties may or may not be vesicular and are near-perfect cylinders. Elongate vesicles within the hairs may be as thin as several micrometers in diameter. Larger (>100-μm-diameter) varieties may be smooth or ribbed cylinders; the ribs reflect elongate vesicles near and parallel to the hair surface. Lumps in the hairs are present over phenocrysts or ovoid vesicles. In some cases, elongate glass threads are bent or twisted into knots. Some are attached to ovoid droplets, with shapes akin to a "tail" (Duffield et al., 1977).

Near-surface acceleration of lava in fountains, caused by expansion of gases, drives jets of low-viscosity basaltic lava (10^2 to 10^3 poise), which form Pele's hair. Unstable jets that break up into a spray produce melt droplets, formed by surface tension, which are then chilled to glass while in flight. Many of the streamers of chilled glass (Pele's hair) and smaller droplets are carried off by wind, although some fall back into the fountain to form composite droplets after multiple collisions with rising melt droplets. Many Pele's hairs are broken up by air turbulence and the hair and droplets are frequently broken upon hitting the ground or colliding with larger lapilli and bombs. No one has measured the length of a Pele's hair because they are so delicate. Pele's hair also forms at lava cascades and lava falls, where clots of low-viscosity lava are "spun" by strong air currents passing over the turbulent lava.

Reticulite [the thread lace scoria of Wentworth (1938) or "Limu" of the native Hawaiians] is a remarkable open network of lattice-shaped, triangular basaltic glass rods of extremely low bulk density. Nearly all of the vesicles are coalesced, although some thin glass membranes remain as vesicle walls. Reticulite is extremely delicate and generally is broken into lumps of several centimeters or less in diameter. Reticulite is formed by the extreme attenuation of a low-viscosity liquid during vesiculation. Deposits of reticulite are rarely preserved because they are so easily broken down mechanically and chemically.

Other pyroclast forms include ragged "cinders" or spinose, vesiculated droplets (Walker and Croasdale, 1972). Although the range of shapes is great, most pyroclasts have the smooth skin formed by surface tension within droplets of melt ejected from lava fountains.

Ultramafic Tephra Formed in Lunar Lava Fountains. Fine-grained, homogeneous glass spheres and composite spheres (those with glass mounds adhering to their surfaces) of mafic to ultramafic composition have been found in clastic horizons deposited on lunar lava plains at three Apollo landing sites. It has been proposed that these glass spheres were produced in fountains of ultramafic lava (Heiken et al., 1974) because they are so similar to pyroclasts produced in Hawaiian eruptions. Some differences, including the remarkable sphericity, may be the result of very low viscosities of the erupting lavas (calculated viscosities of 3 to 10 poise at 1250°C). For comparison, Hawaiian lava fountains have viscosities of $\sim 10^2$ to $\sim 10^3$ poise. It was first believed that these glass spheres might have been of impact-melt origin; none, however, contain shock damaged crystals or exhibit the heterogeneity so common to glass particles of impact origin.

Strombolian Eruptions. Strombolian eruptions of slightly more viscous lava ($\sim 10^3$ to $\sim 6 \times 10^3$ poise) produce distinctive tephra that are considerably different than those formed in Hawaiian eruptions. They are characterized by the ejection of partly molten bombs and ash that forms cinder cones and their associated tephra blankets. Ash produced during Strombolian activity consists of a variety of pyroclasts, from irregular sideromelane droplets to blocky tachylite (black to dark brown microcrystalline pyroclasts).

Sideromelane pyroclasts are smooth-skinned, vesicular droplets that form an irregular, partly open framework of vesicle walls. Where smooth "skin" is preserved, it consists of lumpy, botryoidal surfaces. Many grain shapes reflect interior vesicle shapes, which range from spherical to deformed ovoids. Blocky sideromelane fragments are mostly droplet fragments. The gradation from clear sideromelane droplets into microcrystalline tachylite grains reflects the degree of chilling and crystallization of the melt droplet.

Angular, blocky tachylite pyroclasts range in shape from those with irregular, open-vesicle networks to nonvesicular blocky grains. Vesicularities (with vesicle volumes of 10 to 20%) tend to be lower than in the sideromelane pyroclasts. Grain surfaces and vesicle walls are not smooth, but rough; they exhibit both fracture surfaces and diktytaxitic textures.

In both sideromelane and tachylite pyroclasts, melt viscosities are not low enough to allow formation of the smooth-skinned spheres and ovoids that are typical of tephra produced by Hawaiian eruptions. Most of these pyroclasts are rapidly quenched after eruption; few appear to have been plastically deformed upon hitting the ground or have particles welded to grain surfaces (agglutinates).

The variety of pyroclasts indicate the variety of conditions that may exist in a Strombolian eruption. Sideromelane pyroclasts exhibit the most pronounced deformation by surface tension and, thus, the lowest viscosities. They are produced by a higher temperature magma, which may be the case if lava flux from the conduit is high or if the vent is open and some jetting is possible (similar to characteristic Hawaiian activity). If the vent is clogged or lava flux is low, occasional bursts of cooler, less vesicular melt, which pools for a while, and "accidental" pyroclasts derived from slumps into the crater will produce an abundance of the blocky, tachylite pyroclasts.

Ash of Peralkaline Silicic Compositions:

Explosive eruptions of rhyolitic-trachytic magmas produce both pyroclastic flows and ash fall tephra, although the volume of these units, associated with stratovolcanoes and calderas, is small when compared with that of some rhyolitic volcanoes (Schmincke, 1974). These magmas have relatively low viscosities when erupted (e.g., $10^{2.8}$ to 10^6 poise at temperatures of $>950°C$ and $>2\%$ water).

Peralkaline pyroclast shapes vary in samples from basal to upper portions of thin pyroclastic flow units. Tricuspate (Y-shaped) or blade-like shards and pumice with elongate (tubular) or spherical vesicles are most common near the base. Within the upper portion of these pyroclastic flows, vitric pyroclasts have spherical shapes; some of the larger ones (generally 0.2 to 3 mm) are hollow. Some spheres up to 5 cm in diameter, described at Mt. Suswa, Kenya, have highly vesicular crystalline cores and thin glass rims. In some deposits, glass spheres form the matrix material of coarser grained pumice and obsidian fragments (Hay and Hildreth, 1976; Schmincke, 1974). Ash fall pyroclastic units of glass spheres of melilitic composition from Kenya have also been described (R. L. Hay, 1976, personal communication).

There is considerable disagreement about the genesis of these spherical pyroclasts. In his comprehensive review of peralkaline silicic welded tuffs, Schmincke (1974) suggests that at least two processes are responsible for their genesis: (1) "pulling together" of hot plastic shards within the pyroclastic cloud before final deposition and (2) expansion of volatiles within shards after deposition. Hay and Hildreth (1976) believe that peralkaline glass spheres are formed when jets of low-viscosity magmas break up into sprays, which is similar to the process of droplet formation in Hawaiian eruptions. Supporting their hypothesis is the presence of spherical pyroclasts of melilitic composition in some Kenyan ash fall units.

All of the above processes may account, to some degree, for the spherical pyroclasts present in tuff deposits of peralkaline silicic composition.

Ash of Andesitic to Rhyolitic Composition:

Petrologic and morphologic characteristics of andesitic to rhyolitic pyroclasts are related to higher viscosities (10^3 to 10^6 poise) and the higher volatile contents of their source magmas. Viscosities of above 10^3 poise do not allow the formation of droplets by surface tension; pyroclast shapes are dependent upon the geometry of vesicles developed within the highly viscous magmas. The resulting tephra consist mostly of varieties of pumice or their broken products and are characteristic of nearly all large-volume pyroclastic deposits. They make up the extensive ash fall and pyroclastic flow deposits produced during highly explosive eruptions and are associated with composite cones, some ash and pumice cones, and large calderas. These pyroclast types represent some of the products of Vulcanian and most products of Peléean and Plinian eruptions.

Andesitic Volcanic Ash. Andesitic ashes consist of complex mixtures of vitric, crystal, and lithic components. Vitric components consist of brownish to colorless, homogeneous to heterogeneous glasses, the latter having been derived from mixed magmas. In equant to elongate pumice fragments the particle shape depends on vesicle density and geometry. Many have parallel, highly elongate "pipe-like" vesicles that are parallel to the long axis of the pyroclast. There are no smooth particle surfaces; exposed broken vesicle walls provide sharp, irregular surfaces. Only the vesicle walls are smooth. Smaller particles, which are bits of vesicle wall, are Y-shaped curved plates or simple needle-like forms.

Phenocrysts broken out during explosive eruptions result in a crystal component in these ashes. Broken, angular mineral fragments are common; the particle shapes are governed by the fracture pattern of the mineral. Most lithic fragments (a wide variety of rock types from vent and crater walls) are equant to elongate and angular to rounded. Surface textures are entirely dependent on the physical properties of the rock type.

Dacitic and Rhyolitic Volcanic Ash. Most dacitic and rhyolitic pyroclasts are pumices or their broken products. Pumice has been the object of scholarly studies for a few thousand years; in the 4th century B.C., Theophrastos called it "worm-eaten wood" and thought it to be derived from sea-foam and fire. A thorough historical background of pumice studies may be found in an article by Perlaki (1966).

Nearly all pumices are angular and highly vesicular. Narrow, elongate pumices usually contain highly elongate vesicles (with aspect ratios of >20:1) that are circular to lens-like in cross section and 5 to 20 µm in diameter. Pumice pyroclasts containing phenocrysts are generally heterogeneous. Some of the highly elongate varieties contain large, ovoid cavities around which thinner, elongate vesicles are bent. Larger, less elongate pumices contain 100- to 500-µm-long ovoid vesicles with aspect ratios of less than 20:1. Another pumice variety is massive with remarkably small vesicles (5 to 15 µm in diameter) and relatively thick (10- to 100-µm) vesicle walls.

Most finer grained silicic tephras consist of comminution products of pumice (shards), individual phenocrysts (crystal fraction), and some lithic fragments (xenoliths). Merrill (1885) described curved or Y-shaped glass plates from a vitric-crystal ash from Nebraska; he assumed that the shards had been formed by the pulverization of pumice. Judd (1888) came to a similar conclusion in his studies of silicic volcanic ash.

Many phenocrysts are coated with vesicular glass rinds; vesicles are flattened ovoids that are parallel or perpendicular to phenocryst surfaces and range from 80 to 300 µm long. This is the "bubble-wall texture" of Fisher (1963) and Meyer (1972). Meyer proposed that the volume of glass clinging to a phenocryst was inversely correlated to the "explosivity" or kinetic energy of an eruption.

A rare and unusual variety of rhyolitic ash was produced during a phase of the eruption at Novarupta (in the Valley of 10,000 Smokes), Alaska in 1912. Perret (1950) describes minute hollow bubbles, thickened on one side and often girdled by a raised ridge or septum. Some are divided by a median wall. The mechanism by which this pyroclast type originates is not known.

Many dacitic to rhyolitic domes and flows may represent the last phases of volatile-depleted activity. There are, however, minor ash deposits associated with these domes. Explosive release of volatiles (gas eruption) often occurs within the dome vents, producing lithic and vitric ashes (Rose, 1972). The coarser tephra include pumiceous bombs and massive blocks. Dacitic ashes from this eruption type consist of equant, hyalocrystalline to holocrystalline, slightly vesicular lithic ash fragments. Most grain surfaces are hackly to conchoidal, depending on the glass component of the dome lavas.

Original pyroclast shapes characteristic of silicic tephra may be drastically modified by heating, plastic deformation, and welding within ash flow tuff units. For a complete synthesis of welding and post eruption modification of pyroclasts, see Ross and Smith (1961).

Formation of Rhyolitic and Dacitic Volcanic Ash. In a early analysis of volcanic ash formation, Verhoogen (1951) addressed the problem of nucleation, bubble growth, and rise of these bubbles within a magma. In that analysis, he assumed that volcanic ash forms when the rising, expanding bubbles coalesce, which causes the lava to lose its cohesion. Later work by McBirney (1963) and others shows that the rate of bubble rise within a magma is far too slow to be responsible for the formation of pyroclasts except in basaltic volcanoes with magmas of relatively low viscosities. On the basis of observations that silicic pyroclastic deposits nearly all contain an abundance of pumice pyroclasts that are composed of multiple, intact vesicles and are not composed entirely of bubble walls (shards), it is most unlikely that ash is formed by bubble coalescence alone. The lack of glass droplets within silicic pyroclastic deposits rules out the jetting of melt and formation of a spray that are characteristic of Hawaiian eruptions.

McBirney and Murase (1970) determined that if vesicularity within a melt is low, ash formation takes place when the gas pressure within the bubbles exceeds the tensile strength of the liquid. When the melt is highly vesicular, the loss of coherence of the melt occurs only when gas pressure exceeds the surface tension of the bubble walls. McBirney and Murase, however, also assume bubble coalescence is a requirement for disruption of the magma.

Variations within the form and petrography of erupted tephra may provide clues to variations of temperature, viscosity, and volatile content within a magma body before its eruption. To illustrate this hypothesis, the following is an example using data from Murase and McBirney (1973) for a rhyolite magma that contains 2 to 3% H_2O. Gas pressure should exceed lithostatic pressures at depths of 1 to 2 km. The magma body (at least the upper portion) begins to vesiculate in place. This process continues until rock units overlying the magma chamber are fractured and an eruption begins. Expansion of gases in the vesiculating magma is concurrent with a temperature drop from 1000 to 750°C and a subsequent increase in viscosity from $\sim 10^5$ to 10^8 poise. The vesiculated mass becomes too viscous to flow rapidly through the vent; therefore, another mechanism is called upon to disrupt the froth. This disruption may be accomplished by the large pressure discontinuity at the magma/atmosphere interface. The pressure change creates a shock wave that moves ahead of the ash out of the vent and an expansion wave that passes down into the magma chamber and allows further vapor expansion and disruption of the froth. This phenomenon is described by Rittmann (1936) and Bennett (1974). A high-pressure, condensed fluid (steam), suddenly released, will undergo an unsteady, simple wave expansion of the type already known in conventional shock-tube flow. In a shock tube, one-dimensional expansion of a compressed driver gas creates and drives a shock wave ahead of the contact surface between the driver and ambient gas. At the same time, an expansion wave moves into the driver gas and allows its decompression. The expansion wave moving into a magma chamber disrupts nearly solid foam particles, which are subsequently accelerated out of the vent (Sandford et al., 1975; Heiken, 1978). Because of the high angularity of pumice fragments and their comminution products, this mechanism for the disruption of semi-molten or nearly solid, highly viscous froth by explosive expansion is an acceptable mechanism for formation of dacitic or rhyolitic volcanic pumice and shards.

The theory proposed by Bennett (1974) is compatible with observed tephra, except in one respect. Bennett proposes that ash is formed by condensation of magma after eruption (expansion of melt). Pyroclasts produced by condensation should be mainly droplets or melt blobs (spheres, etc.). This is not the case for tephra from highly explosive eruptions of silicic magmas.

Whatever hypothesis is used to explain the formation of volcanic ash, as the eruption proceeds, hotter, less-viscous melt moves into the void left by the melt previously erupted, vesiculates, and continues the eruption. If most volatile components are concentrated in the top of a magma chamber, a highly vesicular froth is eventually depleted as the eruption proceeds. In many large-volume pyroclastic deposits this is indeed the case, and less vesicular, higher density pumice is ejected during waning phases of the eruption.

The degree of vesicle elongation within pumice is most likely due to movement of the vesiculating magma and not due to elongation toward the magma/atmosphere interface, as proposed by Perlaki (1966). Viscosities are too high for a bubble to grow in that manner.

Rhyolitic pumices from many eruptions are texturally heterogeneous; they consist of "pockets" of large, ovoid vesicles and thin, highly elongate vesicles. The "pockets" consist of 1- to 2-mm-long, ovoid vesicles in clusters that surround or are adjacent to phenocrysts. These pockets, in turn, are enclosed in vesicular glass that is characterized by highly elongate vesicles having much smaller volumes (Heiken, 1983). Differences in vesicle types can be explained by heterogeneous vesicle development. It is possible for vesicle growth to begin at depths of 2 or 3 km in rhyodacitic or rhyolitic melts with a water content of 3% or more (Sparks, 1978). Early vesicle growth may begin at phenocryst surfaces, when those surfaces act as bubble nuclei. Vesicle pockets that are radial to phenocryst surfaces have been described by Heiken and Eichelberger (1980). The larger pockets with thin internal vesicle walls have a higher bulk viscosity than does the surrounding melt and may resist deformation by flow. As magma rises, vesiculation begins around the pockets. Closer to the surface, later stage vesicles are sheared into highly elongate, tube-like forms as they flow around the competent, stronger framework that is created by pockets of ovoid vesicles. At or near the surface, this heterogeneous, vesicular mass could be fragmented, perhaps by a pressure wave passing down into the vent.

Finer grained tephra (shards) are the products of pumice comminution; it is not yet clear to what degree the finer shards are products of explosive disintegration of the magma froth or the mechanical interaction (grinding) of pumices against each other as they are erupted. Both processes appear to be important.

Ash of Carbonatite Composition: Unique pyroclasts containing a substantial portion of sodium carbonate are erupted from the strato-volcano Oldoinyo Lengai, Tanzania (Dawson et al., 1968). The activity varies from the eruption of minor ash plumes to major Plinian eruptions.

The ash consists of sodium carbonate (trona) with crystals of nepheline, pyroxene, wollastonite, apatite, melanite, and pyrite. It is a crystal ash mostly composed of perfect euhedral to anhedral nepheline crystals. Ash particles are coated with clumps of blade-shaped trona crystals. Ash fall tephra of carbonatite composition may contain vitric spheres in addition to other components (R. L. Hay, personal communication).

VOLCANIC ASH FROM PHREATOMAGMATIC (HYDROVOLCANIC) ERUPTIONS

Phreatomagmatic (sometimes called *hydrovolcanic*) eruptions are caused by the explosive generation of steam when rising magma contacts bodies of water (such as the sea, lakes, and marshes), ground water, snow, or ice. Although the dominant factor in phreatomagmatic eruptions is the process of steam explosion, magma composition and physical properties are also important factors determining the mechanism of fragmentation during interaction of magma with external water.

In contrast with magmatic fragmentation, which has been studied from both theoretical approaches and experimental evidence, phreatomagmatic fragmentation has been investigated largely through field observation. *Hyaloclastite* is the term suggested by Honnorez and Kirst (1975) for glass found with pillow basalts that were produced by nonexplosive quenching and fracturing of basaltic glass; *hyalotuff* is used to identify rocks formed by explosive fragmentation of glass during phreatomagmatic eruptions in shallow water. Hyaloclastite may be formed at water depths of over 500 m, where hydrostatic pressure is high enough to inhibit vesiculation in basaltic magma (McBirney, 1963). Consideration of the explosive mechanism and the presence of blocky, equant glass shapes has resulted in the general conclusion that phreatomagmatic ash is formed by thermal contraction and shattering of glass (von Waltershausen, 1853; Peacock, 1926; Fisher and Waters, 1969; Bonatti, 1970; Heiken, 1971). This conclusion is supported by Carlisle (1963), who observed nonvesicular teardrop- and spindle-shaped hyaloclasts in hyaloclastite deposits and in experimentally produced curved splinters and pointed chips of sideromelane. Also, Honnorez and Kirst (1975) observed blocky grain shapes in blast furnace slags that were quenched by water. Recent experimental studies, however, indicate a much more complex fragmentation mechanism for phreatomagmatic ash formation. These experiments have studied the explosive interaction of molten materials with water: a phenomenon commonly termed *fuel-coolant interaction* (FCI-Buchannan, 1974; Peckover et al., 1973; Nelson et al., 1981).

The explosive contact of water with melt begins with the formation and collapse of steam films on the melt surface. This process is cyclic on a micro- or millisecond time scale; it results in the generation of fluid instabilities at the contact, water jet penetration of the melt, and stress waves propagating into the melt. These mechanisms fragment the melt, thereby increasing surface area and heat-transfer rates. The effects of superheating and detonation waves are critical to the vaporization of water. Superheating is a process involving nonequilibrium heat transfer and homogeneous nucleation that causes instantaneous vaporization. It is still poorly understood, but it causes explosive efficiencies several orders of magnitude higher than in normal boiling processes. Detonation waves are shock waves that propagate through the melt to cause fragmentation and mixing with water by fluid instabilities and vapor film collapse. These waves are sensitive to the system size and permit interaction of large volumes of melt and water within a short time.

Ash morphology has been used to elucidate the mechanism of phreatomagmatic eruption; Walker and Croasdale (1972), Heiken (1972; 1974), and Honnorez and Kirst (1975) used optical microscopy and scanning electron microscopy (SEM) to characterize hydrovolcanic ash produced in Surtseyan eruptions and submarine extrusion of pillow basalts. Equant, blocky shapes and curviplanar surfaces with a paucity of vesicles resulted when fragmentation was due to contact of magma with meteoric water. Grain surfaces are mostly planar, quite often meeting at right angles; these flat fracture surfaces are characteristic of hyaloclastic pyroclasts. Vesicularities range from low (mostly less than 30% of volume) to nonvesicular; vesicles are usually spherical and widely spaced.

Particle size as well as shape has also been studied in assessing the formation of volcanic ash. Numerous published size analyses of pyroclasts have shown the relationship of particle size to mechanism of transport and deposition. Walker (1971) demonstrated the size characteristics of flow and fallout tephra by plotting the sorting coefficient versus median diameter, and later, Walker (1973) showed that phreatomagmatic pyroclasts have a much higher degree of fragmentation than do magmatic pyroclasts. These studies and others suggest the important relation of pyroclast size to eruptive mechanism. Size distributions of phreatomagmatic pyroclasts are strongly controlled by eruption energy that, in turn, determines dispersal mechanisms and the resulting deposit types.

Comparison of natural samples and ash debris from experiments modeling hydrovolcanic explosions allowed Wohletz (1983) to develop several theoretical models of hydrovolcanic ash formation. Although there are physical limitations when comparing man-made metallic melts with magma, experiments suggest that the fragmentation mechanisms are basically the same for a wide range of melt compositions. Experimentally produced ash particle shapes are blocky and equant, spherical and drop-like mossy aggregates and the plate-like shapes of broken bubble walls. These shapes, as well as their sizes, show strong resemblance to those of ash produced by hydrovolcanic eruptions. Because rapid vaporization of water is the driving mechanism of hydrovolcanic explosions, studying the pyroclastic material generated by these explosions provides information on the efficiency of the heat-transfer process. Heat transfer is dominantly conductive and requires large surface areas to reach explosive rates. Experimental debris show decreasing grain size with increasing explosive efficiency. The maximum size of experimental explosion debris is 2 to 3 mm; nonexplosive debris are coarser grained. The finest debris are <1 μm. Similarly, explosive hydrovolcanic debris have median grain sizes less than 1 or 2 mm, which is also the median size of many magmatic tephra deposits.

Fragment shapes produced during melt/water interactions reflect fragmentation by brittle failure during passage of stress waves and viscous melt/water mixing by fluid instabilities. The most efficient explosive interactions appear to result in the latter of these two fragmentation mechanisms. Hence, the production of fine-grained debris with large surface areas and fluid shapes is predicted for highly explosive eruptions of fluid basalt. However, melt viscosity and strength strongly encourages development of instabilities. Therefore, for melts of intermediate and silicic composition, brittle fragmentation dominates.

An important consideration is that both magmatic and hydrovolcanic fragmentation mechanisms may operate during an eruption. This situation is illustrated by Self and Sparks (1978) for phreatoplinian silicic eruptions in which magma is initially disrupted by exsolution and expansion of magmatic volatiles, producing relatively coarse-grained pyroclasts; further fragmentation (fine-grained pyroclasts) is caused by explosive interaction with water (Tazieff, 1968). Theoretical consideration of experimental fragmentation mechanisms suggests that stress waves produced by high-pressure vaporization of water at the magma/water interface may, in part, induce vesiculation in the melt.

VOLCANIC ASH FROM LITTORAL CONES

Volcanic ash similar to that of maar volcanoes is characteristic of littoral cones and is formed by steam explosions that are generated when lava flows into a body of water (Fisher, 1968). Evidence at Puu Hou, Hawaii, suggests that explosions occurred when steam was trapped beneath crusted lava flow surfaces. The varied mixture of pyroclasts includes: (1) crystalline basalt fragments and olivine phenocrysts, (2) scalloped sideromelane fragments with abundant vesicles, (3) stretched sideromelane glass with small vesicles, elongate shapes, and fluidal edges (frozen droplets), and (4) irregular or equant sideromelane fragments with few or no vesicles (Fisher, 1968). This mixture is easily characterized by the variety of pyroclast types: blocky and pyramidal sideromelane fragments, rough-surfaced fragments of tachylite, and sideromelane flow crust.

Well-developed vesicularity near lava flow surfaces accounts for the oddly shaped hyaloclastite components of these tephra. When the lava was chilled and the resulting quench was shattered by thermal stresses, irregular pyramidal- and crescent-shaped fragments were formed. The steam explosion also carries out droplets of unquenched lava and bits of solid flow crust. Blocky hyaloclastites also occur as matrix material in pillow lavas and in lava deltas when lava flows into rivers, lakes, and glaciers (Fuller, 1931).

VOLCANIC ASH FROM PHREATIC ERUPTIONS

Active hydrothermal systems often erupt considerable volumes of tephra at stratovolcanoes, domes, or hot springs; for example, the 5×10^6 m^3 of "mud" erupted from Lake City Hot Springs, California in 1951 (White, 1955). Among those who study such eruptions, there is often considerable controversy over the presence or absence of pyroclasts derived from juvenile magma; therefore, the identification of juvenile pyroclasts is important.

Tephra from active hydrothermal (steam) eruptions consist mainly of hydrothermally altered lithic and mineral fragments, commonly in a clay matrix. Included are subangular lithic ash (or lava) fragments in various states of alteration; they range from fragments with a trace of hematite stain to those that are mostly a mixture of clays, zeolites, and pyrite. Only relict textures remain to identify pyroclast types. Particle surfaces are often covered with aggregates of zeolite crystals or clay. Basaltic glass fragments may be rimmed with palagonite, a mixture of smectite clays and iron oxides. Ash particles are similar to the rock units exposed in vent walls and are broken from them by escaping steam; they are erupted to form a local low-volume pyroclastic deposit.

REFERENCES: Bennett, F. D., 1974. On volcanic ash formation. Amer. Jour. Sci., **274**: 648-661.

Bonatti, E., 1970. Deep sea volcanism. Naturwissenschaften, **57**: 379-384.

Buchanan, D. J., 1974. A model for fuel-coolant interactions. J. Phys. D: Appl. Phys., **7**: 1441-1457.

Carlisle, D., 1963. Pillow breccias and their aquagene tuffs in Quadra Island, British Columbia. Jour. Geol., **71**: 48-71.

Dawson, J. B., Bowden, P., and Clark, G. D., 1968. Activity of the carbonatite volcano, Oldoinyo Lengai, 1966. Geol. Rundschau, **57**: 865-879.

Duffield, W. A., Gibson, Jr., E. K., and Heiken, G. H., 1977. Some characteristics of Pele's hair. U.S. Geol. Surv. Jour. Res., **5**: 93-101.

Fisher, R. V., 1963. Bubble-wall texture and its significance. Jour. Sed. Pet., **33**: 224-235.

Fisher, R. V., 1968. Puu Hou littoral cones Hawaii. Geol. Rundschau, **57**: 837-864.

Fisher, R. V. and Waters, A. C., 1969. Bed forms in base surge deposits: Lunar implications. Sci., **165**: 1349-1352.

Fuller, R. E., 1931. The aqueous chilling of basaltic lava on the Columbia River Plateau. Amer. Jour. Sci., **21**: 281-300.

Hay, R. L. and Hildreth, W., 1976. Globule ignimbrites in Kenya. Geol. Soc. Amer. Abs. with Prog., **8**: 589.

Heiken, G. H., 1971. Tuff rings: examples from the Fort Rock-Christmas Lake Valley, south-central Oregon. J. Geophys. Res., **76**: 5615-5626.

Heiken, G., 1972. Morphology and petrography of volcanic ashes. Geol. Soc. Amer. Bull., **83**: 1961-1988.

Heiken, G., 1974. *An atlas of volcanic ash.* Smithsonian Contr. Earth Sci., **12**: 101 pp.

Heiken, G., 1978. Plinian-type eruptions in the Medicine Lake Highland, California and the nature of the underlying magma. J. Volcanol. Geotherm. Res., **4**: 375-402.

Heiken, G., 1983. Textural analysis of tephra from a rhyodacitic eruption sequence, Thira (Santorini), Greece. *In*: Proc., Workshop on characterization and quantification of surface features on clastic and pyroclastic particles (in press).

Heiken, G. and Eichelberger, J., 1980. Eruptions at Chaos Crags, Lassen Volcanic National Park, California. J. Volcanol. Geotherm. Res., **7**: 443-481.

Heiken, G., McKay, D. S., and Brown, R. W., 1974. Lunar deposits of possible pyroclastic origin. Geochim. Cosmochim. Acta, **38**: 1703-1718.

Honnorez, J. and Kirst, P., 1975. Submarine basaltic volcanism: morphometric parameters for discriminating hyaloclastites from hyalotuffs. Bull. Volcanol., **32**: 441-465.

Judd, J. W., 1888. On the volcanic phenomenon of the eruption, and on the nature and distribution of the ejected materials. *In*: G. J. Symons (Ed.), *The Eruption of Krakatoa and Subsequent Phenomena*, London, Trubner and Co., 1-56.

McBirney, A. R., 1963. Deep-sea volcanism. Bull. Volcanol., **26**: 453-469.

McBirney, A. R. and Murase, T., 1970. Factors governing the formation of pyroclastic rocks. Bull. Volcanol., **34**: 372-384.

Merrill, G. P., 1885. On deposits of volcanic dust and sand in southwestern Nebraska. U.S. Nat'l. Mus., Proc., **8**: 99-100.

Meyer, J. D., 1972. Glass crust on intratelluric phenocrysts in volcanic ash as a measure of eruptive violence. Bull. Volcanol., **35**: 358-368.

Murase, T. and McBirney, A. R., 1973. Properties of some common igneous rocks and their melts at high temperatures. Geol. Soc. Amer. Bull., **84**: 3563-3592.

Nelson, L. S., Buxton, L. D., and Planner, H. N., 1980. Steam explosion triggering phenomena, part 2: Corium-A and Corium-E and oxides of iron and cobolt studied with a floodable arc-melting apparatus. Sandia National Laboratories, SAND77-0998, NUREG/CR-0122, 138 pp.

Peacock, M. A., 1926. The basic tuffs. *In*: G. W. Tyrell and M. A. Peacock (Eds.), The petrology of Iceland, Royal Soc. Edinburgh Trans., **45**: 51-76.

Peckover, R. S., Buchanan, D. J., and Ashby, D. E. T. F., 1973. Fuel-coolant interactions in submarine volcanism. Nature, **245**: 307-308.

Perlaki, E., 1966. Pumice and scoria: their notion, criteria, structure and genesis. Acta Geol. Hung., **10**: 13-29.

Perret, F., 1950. Volcanic ejectamenta. *In*: *Volcanic Observations* (F. Perrett, Ed.), Carnegie Inst. Pub. **549**: 115-125.

Rittmann, A., 1936. Vulkane und ihre Tätigkeit. Ferdinand Enke Verlag, Stuttgart, 188 pp.

Rose, Jr., W. I., 1972. Santiaguito volcanic dome, Guatemala. Geol. Soc. Amer. Bull., **83**: 1413-1434.

Ross, C. S. and Smith, R. L., 1961. Ash-flow tuffs: their origin, geologic relations and identification. U.S. Geol. Surv. Prof. Paper **366**: 81 pp.

Sandford, II, M. T., Jones, E. M., and McGetchin, T. R., 1975. Hydrodynamics of caldera-forming eruptions. Geol. Soc. Amer. Abs. with Prog., **7**: 1257-1258.

Schmincke, H. U., 1974. Volcanological aspects of peralkaline silicic welded ash-flow tuffs. Bull. Volcanol., **38**: 594-636.

Self, S. and Sparks, R. S. J., 1978. Characteristics of widespread pyroclastic deposits formed by the interaction of silicic magma and water. Bull. Volcanol., **41-3**: 196-212.

Sparks, R. S. J., 1978. The dynamics of bubble formation and growth in magmas: a review and analysis. Jour. Volcanol. Geotherm. Res., **3**: 1-37.

Tazieff, H., 1968. Sur le mécanisme des eruptions basaltiques sous-marines a faibles profundeurs et la genese d'hyaloclastites associeés. Geol. Rund., **57**: 955-966.

Verhoogen, J., 1951. Mechanisms of ash formation. Amer. Jour. Sci., **249**: 29-739.

von Waltershausen, W. S., 1853. Über die vulkanischen gesteine in Sizilien und Island und ihre submarine umbildung. Gottingen.

Walker, G. P. L., 1971. Grain-size characteristics of pyroclastic deposits. J. Geol., **79**: 696-714.

Walker, G. P. L., 1973. Explosive volcanic eruptions—a new classification scheme. Sond. Geol. Rund., **62**: 431-446.

Walker, G. P. L. and Croasdale, R., 1972. Characteristics of some basaltic pyroclastics. Bull. Volcanol., **35**: 303-317.

Wentworth, C. K., 1938. Ash formations of the Island of Hawaii. Haw. Volc. Obs. 3rd Spec. Rpt., 183 pp.

Wentworth, C. K. and MacDonald, G. A., 1953. Structures and forms of basaltic rocks in Hawaii. U.S. Geol. Surv. Bull., **994**: 98 p.

White, D., 1955. Violent mud-volcano eruption of Lake City Hot Springs, northeastern California. Geol. Soc. Amer. Bull., **66**: 1109-1130.

Wohletz, K. H., 1983. Mechanisms of hydrovolcanic pyroclast formation: size, scanning electron microscopy, and experimental results. Jour. Volcanol. Geotherm. Res., **17**: 31-63.

Classification of Pyroclastic Rocks

Classification is a difficult task, especially because the categories are not precise and usually grade one into another. Classification schemes of volcanic ash are based upon grain size, composition of pyroclasts and magma, genesis of the pyroclasts, and mechanism of deposition. The most commonly used of these schemes is grain size and is similar to that used in sedimentology to classify clastic sediments and sedimentary rocks. Wentworth and Williams (1932) developed the first version using a φ scale (see Glossary). However, the most commonly used version of this classification is the scheme of Fisher (1961) shown in Table 1. A slightly modified version of this classification was accepted by the Subcommission on the Systematics of Igneous Rocks of the International Union of Geological Sciences (IUGS) (Schmid, 1981).

TABLE 1. GRAIN-SIZE LIMITS AND TERMS FOR PYROCLASTIC ROCKS.

		Pyroclastic Deposit	
Clast Size (mm)	**Pyroclast**	**Mainly unconsolidated: tephra**	**Mainly consolidated: pyroclastic rock**
	Bomb, block	Agglomerate	Agglomerate, pyroclastic breccia
64 mm			
	Lapillus	Layer, bed of lapilli (lapilli tephra)	Lapilli tuff (Lapillistone)
2 mm			
	Coarse ash	Coarse ash	Coarse (ash) tuff
1/16 mm			
	Fine ash	Fine ash	Fine (ash) tuff

There are many classifications of tephra on the basis of the constituent composition (pyroclast compositions). For communication purposes, Cook's classification (1965) is very useful (Fig. 1). Within other petrographic classifications, including that of the IUGS, most tuffs are classified as "vitric" and are not easily separated; this problem can be avoided by using Cook's classification scheme.

Walker (1971) has separated pyroclastic flow from pyroclastic fall deposits on the basis of median grain size and sorting. This approach is very useful in unraveling the genesis of stratigraphic sequences of pyroclastic rocks. Further subdivision of pyroclastic flow, surge, and fall deposits has been made by Wright et al. (1980).

The genetic classification used in this book is similar to that used by Heiken (1972; 1974). The classification is based mainly on two categories: magmatic and phreatomagmatic; i.e., explosive eruptions caused primarily by release of magmatic gases and those caused primarily by interaction with surface or ground water. As in most classifications, there are overlaps; both processes may be responsible for the explosive activity of a single volcanic eruption. Within these two categories there are also many variations, depending upon the physical and chemical properties of the magma.

For each example in this book the grain-size classification (median grain size and sorting) is given. Sorting characteristics are described below:

$$\sigma_\varphi = <1\varphi \; - \text{ well sorted}$$

$$\sigma_\varphi = 1 \text{ to } 2\varphi \; - \text{ moderately sorted}$$

$$\sigma_\varphi = >2\varphi \; - \text{ poorly sorted}$$

$$\text{where: } \sigma_\varphi = \frac{\varphi 84 - \varphi 16}{2}$$

Further classification involves the identification of the magmatic composition. All of the above classifications may be combined easily into one phrase to describe a single sample: e.g., "Well-sorted rhyolitic vitric-lithic lapilli-bearing ash from a pyroclastic fall deposit of phreatomagmatic origin."

REFERENCES: Cook, E., 1965. Stratigraphy of Tertiary volcanic rocks in eastern Nevada. Nev. Bur. Mines, Rept. **11**: 66 pp.

Fisher, R., 1961. Proposed classification of volcaniclastic sediments and rocks. Geol. Soc. Amer. Bull., **72**: 1409-1414.

Heiken, G., 1972. Morphology and petrography of volcanic ashes. Geol. Soc. Amer. Bull., **83**: 1961-1988.

Heiken, G., 1974. An atlas of volcanic ash. Smithsonian Contr. Earth Sci. **12**: 101 pp.

Schmid, R., 1981. Descriptive nomenclature and classification of pyroclastic deposits and fragments: Recommendations of the I.U.G.S. Subcommission on the Systematics of Igneous Rocks. Geol., **9**: 41-43.

Walker, G., 1971. Grain-size characteristics of pyroclastic deposits. Jour. Geol., **79**: 696-714.

Wentworth, C. and Williams, H., 1932. The classification and terminology of the pyroclastic rocks. Nat. Acad. Sci. Nat. Res. Council, **89**: 19-53.

Wright, J., Smith, A., and Self, S., 1980. A working terminology of pyroclastic deposits. Jour. Volcanol. Geotherm. Res.: **8**: 315-336.

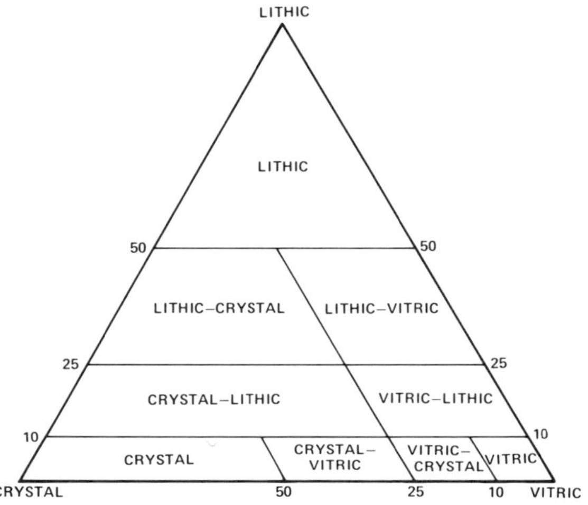

Figure 1. Pyroclastic classification of Cook (1965).

Summary of Physical Methods for Analysis of Tephra and Pyroclastic Rocks

For information about size parameters of unconsolidated tephra and pyroclastic rocks, see *Pyroclastic Rocks* by Fisher and Schmincke. References for detailed physical analysis of tephra are those designed for clastic sediments (Krumbein and Pettijohn, 1938; Pettijohn, Potter, and Siever, 1973; Mueller, 1967).

Grain-size analysis of tephra samples requires considerable judgement on the part of the investigator because of the delicacy of the pyroclasts. If the ash is composed mostly of lithic fragments, for example, a conventional stack of sieves and a ro-tap sieve shaker may be used as one would to analyze river sands.

However, vitric fragments in most ashes are usually very delicate and easily abraded. Vigorous use of a ro-tap can easily turn a coarse vesicular ash into a fine ash by particle abrasion. Wet-sieving and gentle manual shaking of the sieve stack may be necessary to avoid destroying the sample. For fine-grained ashes, a sonic sifter, or other size analysis method (such as a Coulter Counter) may be used.

If size analyses are done, splits of each size fraction should be taken for binocular and thin-section optical inspection as well as for SEM mounts. If samples are not sieved, conventional splitting techniques are adequate to provide samples for SEM, thin sections, and bulk chemical analyses.

Before SEM inspection, the samples should be cleaned to improve image quality. This technique is especially important when samples are partly coated with loosely adhering fine material or a cement such as calcite. By removing this material, one is able to see the original pyroclast surfaces. A "dirty" sample may also make it difficult to obtain the uniform conductive coat that is necessary to counteract sample charging in an electron beam. The most common cleaning technique consists of washing the sample in dilute HCl at low to moderate heat for 3 to 5 minutes, followed by a rinse in distilled H_2O and a drying period that may be accelerated by a second rinse with acetone. This method is not suitable, however, if the dilute acid reacts with cementing agents or alteration products on the pyroclast surfaces. If acid cleansing does not produce good results, washing the sample in acetone or distilled H_2O with ultrasound can loosen unwanted material. Ultrasound cleansing should be done with great care because it may damage the pyroclast surface, especially if it is applied for longer than 30-second intervals.

Mounting samples for SEM examination is simple: (1) coat a sample stub with a thin smear of cement (dilute duco cement is effective) or double-sided adhesive tape, and (2) drop the sample split onto the stub and shake off the excess ash, or place individual grains on the stub with tweezers or a probe. After the sample is mounted, apply a thin carbon coat in a vacuum evaporator. If energy-dispersive x-ray analyses are to be made of grain surfaces, apply a second carbon coat. If the SEM analysis is to produce images only, a gold or gold-palladium coating should be applied over the carbon coat.

A very good polished thin section or grain mount of each sample is absolutely necessary for tephra studies. Either may be used for petrographic descriptions, grain counts (for classification of pyroclasts), and analysis of individual pyroclasts by electron microprobe. If pyroclast textures are finely crystalline or hyalocrystalline, chemical- or ion-etching techniques will enhance the contrast between individual phases. A thin section may then be studied by SEM, reflected-light microscopy, or phase-contrast microscopy.

REFERENCES:

Fisher, R. V. and Schmincke, H.-U., 1984. *Pyroclastic Rocks.* Springer-Verlag, Heidelberg, 472 pp.

Krumbein, W. C. and Pettijohn, F. J., 1938. *Manual of Sedimentary Petrography.* Appleton-Century-Crofts, New York, 549 pp.

Mueller, G., 1967. *Methods in Sedimentary Petrology.* Hafner Pub. Co., New York, 283 pp.

Pettijohn, F., Potter, P., and Siever, R., 1972. *Sand and Sandstone.* Springer-Verlag, Heidelberg, 618 pp.

Lava fountain—1959 eruption of Kilauea Iki, Hawaii.

MAGMATIC ERUPTIONS

Magmatic eruptions produce ashes derived directly from a vesiculating magma that has not interacted with meteoric water. Their characteristics are related to the magma strength and viscosity and the thermodynamic behavior of gas phases within the magma.

Hawaiian Eruptions (Lava Fountaining)

Eruption of low-viscosity basaltic magma often occurs as lava fountains. This behavior is characteristic of eruptions on the island of Hawaii where relatively mild lava fountaining produces tephra. Ash particles from lava fountains generally have fluidal shapes, including droplets and strands.

ALOI-ALAE, HAWAIIAN VOLCANOES NATIONAL PARK

Introduction: The sample, collected 1 km west of Alae Crater, was deposited during lava fountaining from fissures between Aloi and Alae Craters, which are located along the east rift zone of Kilauea Volcano.

The lava fountains of this eruption were similar to many others (e.g., Richter et al., 1970): vigorous, gas-driven fountains, a few meters to hundreds of meters high, of low-viscosity basaltic lava. Pyroclasts ranging from droplets only a few micrometers in diameter to meter-sized bombs were erupted in the fountains. Wind swept away the finer grained fractions described here and deposited them over areas of a few tens of square kilometers. The area of ash fall is considerably larger if multiple fountains erupt along fissures, as is often the case in Hawaii.

Grain Size: Because there is a remarkable variety of pyroclast shapes (especially Pele's hair that may only be 200 μm in diameter but several meters long), grain-size analyses of Hawaiian tephra are nearly impossible. The remarkably delicate nature of these pyroclasts also prevents accurate grain-size measurement. Particle sizes range from fragments of basaltic glass only a few micrometers in diameter to bombs over a meter in diameter that fall around the vents.

Tephra Description: Nearly all the tephra from the Aloi-Alae eruption consist of basaltic glass (sideromelane) with no microlites and only rare olivine phenocrysts. Most pyroclasts have some smooth, curved surfaces from surface tension effects on melt particles from the fountains.

Glass spheres in these tephra deposits are typical of eruptions of basaltic melts with low viscosities. The spheres range in size from a few micrometers to several hundred micrometers. Each sphere has a low bulk density, is highly vesicular, and has numerous (40 to 70% by volume) spherical vesicles.

Many of the angular glass pyroclasts are broken glass spheres or ovoids; therefore, particles from the grain edges of pyroclasts have one smooth, curved surface, whereas interior fragments are remarkably angular. Vesicles are well developed and vesicle walls are thick.

Glass ovoids with pointed tails and "dumbbells" formed by the spinning of melt droplets are transitional to the fibers of Pele's hair. In some cases, these strands are curved back on themselves, forming closed loops.

Pele's hair, in this case, appear to be mostly simple cylindrical strands, a few micrometers to 50 μm in diameter. Some of the coarser hairs have grooved surfaces with furrows located along the strands over parallel vesicles. Vesicles within these strands are highly elongate.

Comments on Pyroclast Formation: All the pyroclasts from this sample of the Aloi-Alae eruption were formed as a spray of basaltic liquid in lava fountains driven by expansion of gases. Viscosities are low, usually less than 10^3 poise. Unbroken particles all have smooth, rounded surfaces formed by surface tension on individual melt droplets. These particles are quenched during cooling by thermal radiation and flow of air over their surfaces before they hit the ground surface. Larger bombs, which land near the vent, are deformed and form spatter ramparts. Finer grained tephra are swept away by wind or ballistically thrown beyond these ramparts.

Some melt particles are "spun" into long, thin, hair-like strands by expanding gases and thermally induced turbulence over the fountain (Pele's hair).

All the basaltic glass particles are delicate and most break upon hitting the ground. Particles with smooth surfaces formed by surface tension are here simply called droplets. A new term *achnelith* (GR., achne = spray) was used by Walker and Croasdale (1972) for this type of pyroclast.

REFERENCES:

Heiken, G., 1974. An atlas of volcanic ash. Smithsonian Contr. Earth Sci., **12**: 101 pp.

Heiken, G. and Lofgren, G., 1971. Terrestrial glass spheres. Geol. Soc. Am. Bull., **82**: 1045-1050.

Richter, D., Eaton, J., Murata, K., Ault, W., and Krivoy, H., 1970. Chronological narrative of the 1959-60 eruption of Kilauea Volcano, Hawaii, U.S. Geol. Surv. Prof. Paper, **537-E**: 73 pp.

Walker, G. P. L. and Croasdale, R., 1972. Characteristics of some basaltic pyroclastics. Bull. Volcanol., **35**: 303-312.

FIGURE 1a. SEM of a 140-μm-diameter spherical pyroclast. The basaltic glass sphere has a smooth surface. Small particles adhering to the surface are partly basaltic glass and partly soluble sublimates from gases driving the eruption.

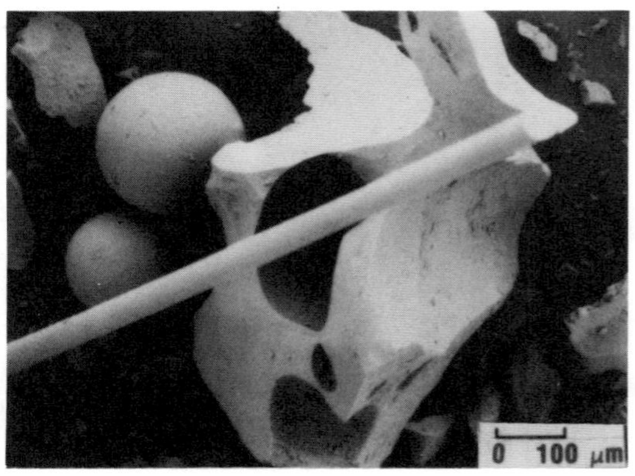

FIGURE 1b. Broken fragments of vesicular droplets, small glass spheres, and a single strand of Pele's hair.

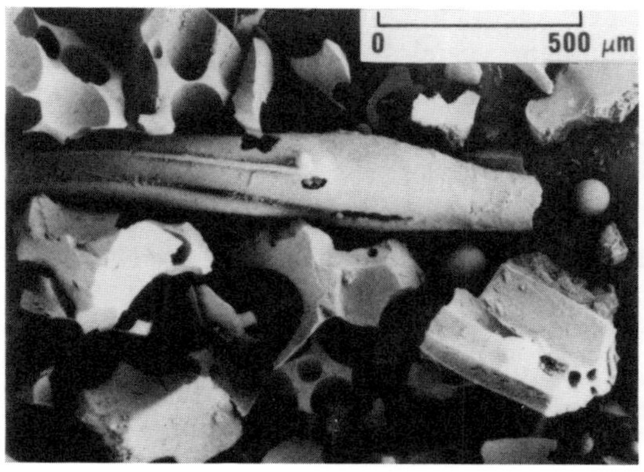

FIGURE 2a. SEM of a typical variety of pyroclasts from this eruption, including spherical droplets, Pele's hair, and broken, angular pyroclasts. The Pele's hair (*center*) has a groove in the surface between two highly elongate vesicles. Angular pyroclasts are mostly interior fragments from broken droplets with large, well-developed vesicles and very thick walls between vesicles.

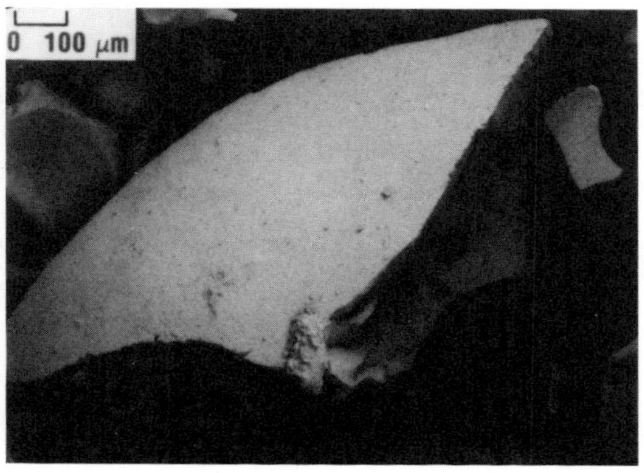

FIGURE 2b. Angular pyroclast; broken droplet fragment with one smooth, curved surface.

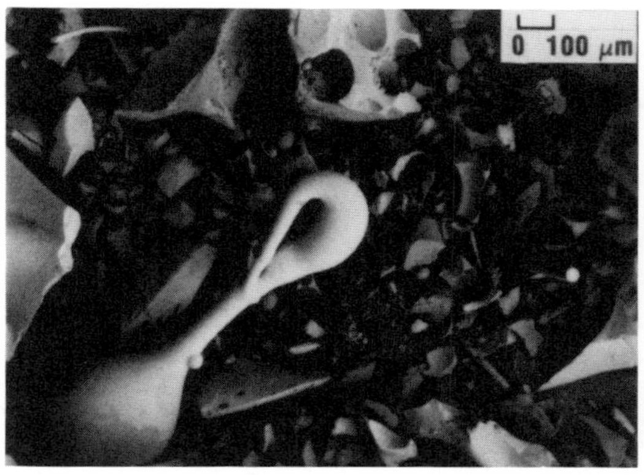

FIGURE 3a. SEM of a representative sample of tephra from Aloi-Alae. Dominating the image is a droplet with an elongate, pointed tail that was bent into a loop and annealed to itself while still molten. A smaller sphere is also welded to this pyroclast; it probably collided with the larger droplet while still partly molten. This particle may represent a transition from spheres and ovoids to strands of Pele's hair. Other particles in the image include vesicular droplet fragments, simple, cylindrical Pele's hair, and spheres.

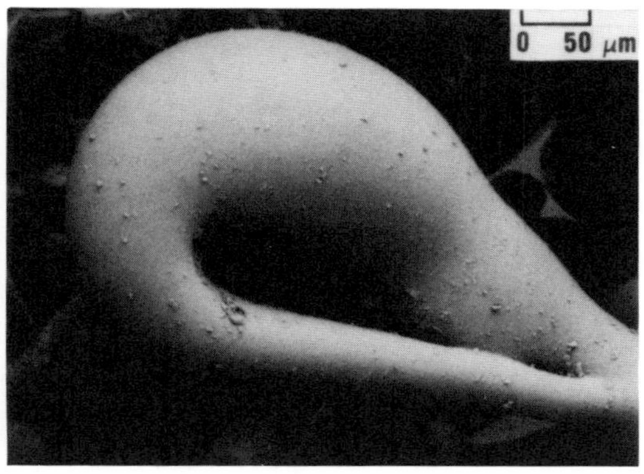

FIGURE 3b. Close-up of the droplet "tail" in *3a*. Bulges in the surface may be over vesicles that continued to expand after the pyroclast's formation.

PELE'S HAIR

Introduction: One of the more intriguing and unusual pyroclastic forms produced during vigorous fountaining of basaltic lava is Pele's hair. Extremely thin (tens to hundreds of micrometers wide) but long (up to several meters) strands are never found complete—they are delicate, breaking up in the wind or when they hit the ground surface. Usually, they are found along a wide zone downwind from the source in deposits consisting of tangled bundles of pale brown fibers.

The only paper dedicated solely to the genesis and characterization of Pele's hair is by Duffield et al. (1977), from which much of the text in this section is condensed.

As discussed in previous sections on basaltic tephra, Pele's hairs are but one pyroclast type formed by the mildly explosive eruption of low-viscosity ($<10^3$ poise) basaltic melt. They are chilled streamers of melt. Pele's hair is, however, also formed in lava cascades and falls (one of our examples is from a 120-m-high lava fall into the Makaopuhi Pit Crater). Strong air currents are induced over hot lava and sweep bits of melt off and away from cascades and falls. Masses of Pele's hair trap small particles and are often coated with sublimates from the fume clouds associated with eruptions.

Tephra Description: Most Pele's hairs are simple cylindrical forms, 1 to 500 μm in diameter. Smaller fibers have no vesicles, but most pyroclasts over 50 μm in diameter are vesicular, with vesicles of high aspect ratio (L:W) parallel to hair axes. Whenever the tips of these hairs are visible, they are tapered. Some forms are twisted into loops and welded to themselves.

Thicker Pele's hairs have multiple, elongate, parallel vesicles; V-shaped, elongate depressions in pyroclast surfaces represent spaces between multiple vesicles. Highly elongate vesicles, circular in cross section, indicate that a melt droplet stretched into a fiber during vesicle growth. Chilling of hair surfaces to form a thin (5- to 10-μm thick) flexible "skin" while the pyroclast interior is still molten may produce Pele's hairs with rumpled surfaces (Heiken, 1972).

Chemistry: Pele's hair is generally associated with low-viscosity, Fe- and Mg-rich melts of basaltic composition (Table 1). The example here, from Duffield et al. (1977), is from the 1969-1971 Mauna Ulu eruption.

TABLE 1. CHEMICAL COMPOSITION OF PELE'S HAIR.*

	(Wt. %)
SiO_2	50.6
Al_2O_3	13.7
Fe_2O_3	0.60
FeO	10.0
MgO	7.6
CaO	11.3
Na_2O	2.2
K_2O	0.51
H_2O^+	0.27
H_2O^-	0.06
TiO_2	2.3
P_2O_5	0.30
MnO	0.18
CO_2	0.02
TOTAL	99.60

*Duffield et al. (1977).

Comments on Pyroclast Formation: Streamers of melt are chilled by thermal radiation from clear, glassy pyroclasts and by air passing over the melt particles; cooling and increased viscosity accompanying the cooling ensures preservation of the thin filaments. Were they not cooled, the streamers of melt would most likely break up into droplets (Isard, 1971).

REFERENCES: Duffield, W. A., Gibson, Everett K., Jr., and Heiken, G., 1977. Some characteristics of Pele's hair. J. Res. U.S. Geol. Survey, **5**: 93-101.

Heiken, G. H., 1972. Morphology and petrography of volcanic ashes. Geol. Soc. Am. Bull., **83**: 1961-1988.

Isard, J. O., 1971. The formation of spherical glass particles on the lunar surface. Proc. Lunar Sci. Conf., 2nd. Geochim. Cosmochim. Acta, Suppl. 3, **3**: 2003-2008.

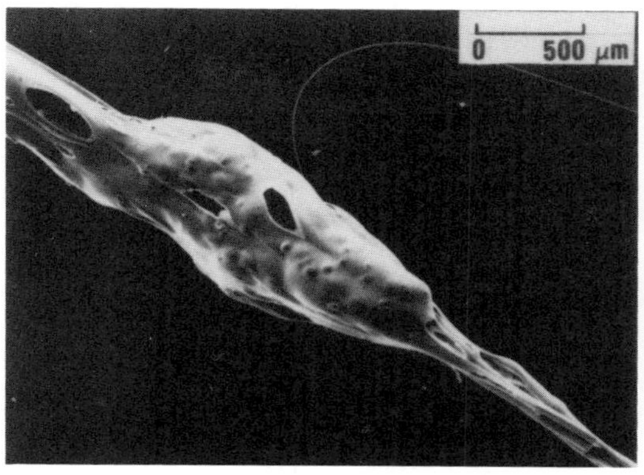

FIGURE 1a. SEM of a Pele's hair from the July 1975 eruption of Mauna Loa, Hawaii. The highly vesicular (50 to 60%), light brown basaltic glass droplet has a smooth surface. Broad, low bulges are over elongate vesicles; some vesicles have ruptured, leaving 100- to 400-μm-long ovoid exterior cavities. Small bulges (30- to 50-μm diameter) are over small olivine and plagioclase phenocrysts, which make up 10% of the glassy pyroclasts.

FIGURE 1b. Close up of (*a*); broken tip of the drawn-out, tapered "hair." Vesicles are 3 to 15 μm in diameter; long axes are much greater but not known. Walls between vesicles are thick: usually around 10 μm.

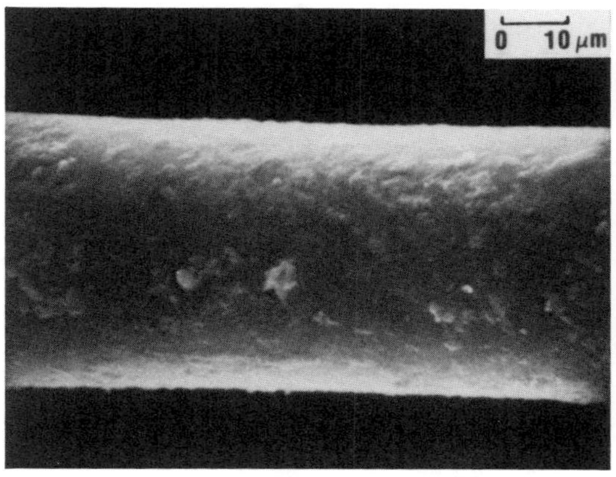

FIGURE 2. SEM of a Pele's hair from Mauna Ulu, Hawaii, October 1, 1970 (collected by W. Duffield). The hair was generated by turbulence over a cascading lava stream. It is a simple, cylindrical Pele's hair coated with patches of sublimates (rough, irregular patches 1 to 2 μm thick). Volatile constituents released upon heating included S, carbonates, sulfates, CO_2, and H_2O (Duffield et al., 1977).

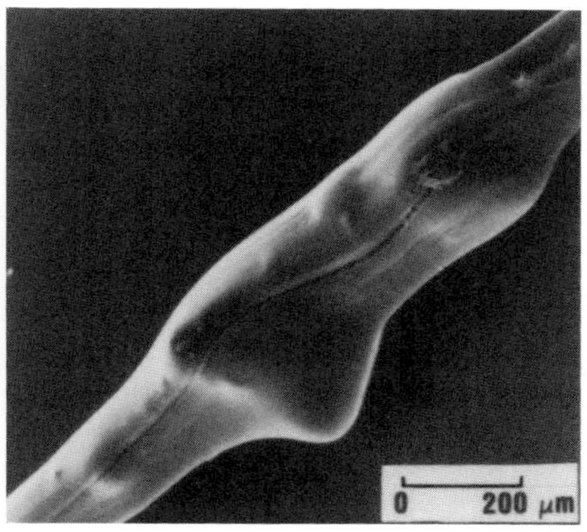

FIGURE 3a. SEM of a Pele's hair from the August 14, 1971 eruption, Kilauea Caldera, Hawaii. Grooves parallel to the pyroclast surface with sharp edges are collapsed vesicles near the pyroclast surface. Note the enormous aspect ratio (L:W) of these vesicles that were stretched during "hair" formation. Large lumps are over phenocrysts (compare with Figs. *4* and *5b*).

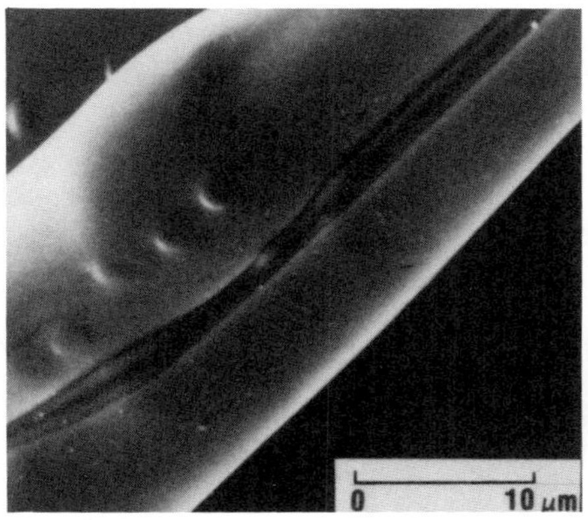

FIGURE 3b. Detail of Pele's hair in Fig. 3a.

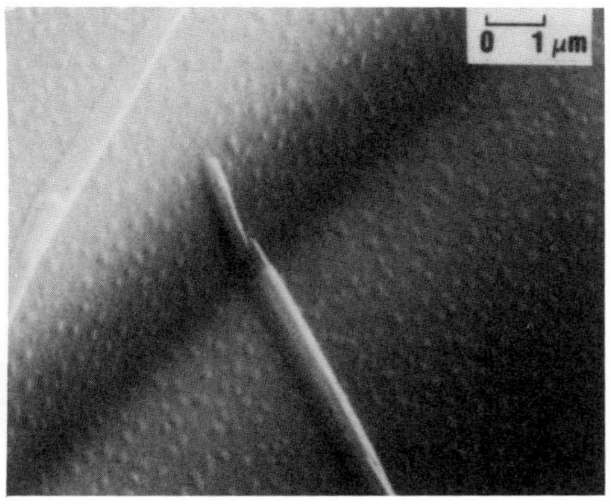

FIGURE 3c. Enlargement of the surface of pyroclast in (*a*), showing 0.5-μm-wide Pele's hair bonded to a larger hair by sublimates. Sublimates are visible as 0.1- to 0.4-μm-diameter mounds.

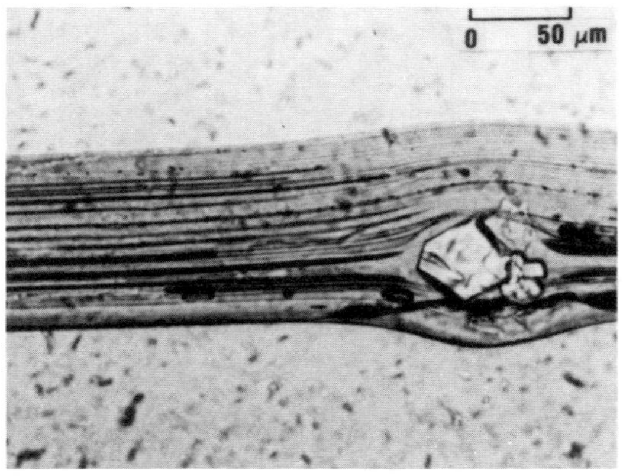

FIGURE 4. Photomicrograph (transmitted light) of a 70-μm-diameter Pele's hair spun from a lava fall by whirlwinds over Makaopuhi Crater, June 11, 1972. The extremely long, very narrow vesicles are similar to those generated in lava fountains and are not similar to most hairs from lava falls. Bulge on the surface is over an olivine phenocryst.

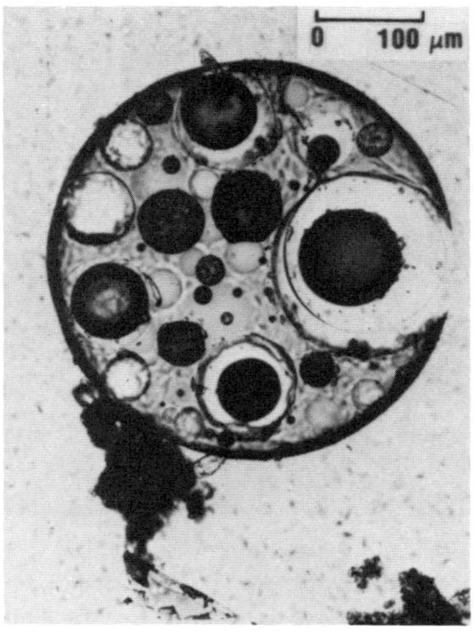

FIGURE 5a. Photomicrograph (transmitted light) of a cross section of Pele's hair from the 1959 eruption of Kilauea Iki, Hawaii. The hair is 460 μm in diameter. Vesicles are circular in cross section only. Note the thick walls of clean basaltic glass (sideromelane) between the vesicle and the thin rim of dark, sheared and jointed glass at surface. Dark patches in the vesicles are grinding compound.

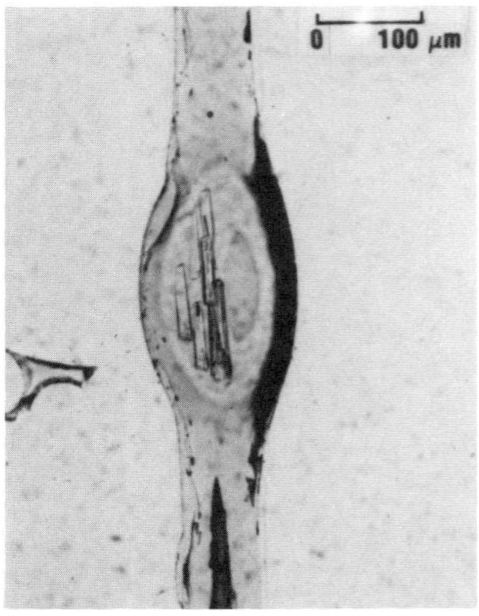

FIGURE 5b. Photomicrograph (transmitted light) of a 50-μm-diameter Pele's hair. Two-thirds of the fiber visible in this photo is vesicle-free; the bulge is over a cluster of small plagioclase phenocrysts.

RETICULITE

Introduction: A very unusual type of pyroclast is reticulite (also known as thread-lace scoria or limu). It is associated with gas-rich basaltic eruptions of lavas that have very low viscosities; there are rare examples of rhyolitic reticulite (S. Self, personal communication). Reticulite is characterized by extreme vesicularity (perhaps as high as 98%; the specific gravity is 0.3). It is formed in lava fountains (pyroclastic) and as thin crusts on shelly pahoehoe flows.

This sample was collected from the "reticulite" bed located at the base of a 2-m-thick sequence of pyroclastic rocks that overlies lava flows on the west rim of Kilauea caldera. Numerous sideromelane droplets (spatter) are also present in the reticulite bed.

Grain Size: Because of the delicate nature of the samples, no attempt was made to sieve them.

Tephra Description: This vitric ash is composed of fragments of sideromelane froth. The fragments are part of a lattice work made up of triangular glass rods. Some thin (<1-μm) glass membranes remain between vesicles, but most of them are broken. These thin, flexible sheets are also present as fragments adhering to the glass rods. However, there is generally a 100-μm to several-millimeter-wide hole between vesicles. These "holes" are rimmed by very sharp, delicate ridges where the thinly stretched membranes had burst or were broken.

Chemical Composition: A chemical analysis of this basaltic reticulite is shown in Table 1.

TABLE 1. CHEMICAL COMPOSITION OF BASALTIC RETICULITE.*

	(Wt. %)
SiO_2	49.29
TiO_2	2.51
Al_2O_3	14.12
Fe_2O_3**	12.79
MnO	0.23
MgO	7.33
CaO	10.44
K_2O	0.35
P_2O_5	0.23
S	0.09
Total	97.38

*XRF analysis by M. Rhodes.
**Fe as Fe_2O_3.

Comments on Pyroclast Formation: Eruption of gas-rich basaltic magma with low viscosity allows maximum development of vesicles. This produces curved, triangular rods at triple junctions (that indicate hexagonal closest packing of vesicles in the melt) and thin membranes between adjacent vesicles. The delicate membranes are mostly broken; they are either burst during the eruption or during weathering or abrasion after deposition.

REFERENCE: Wentworth, C. K., 1938. *Ash Formations of Hawaii*, 3rd Spec. Rpt. of the Hawaiian Volcano Obs.: 183 pp.

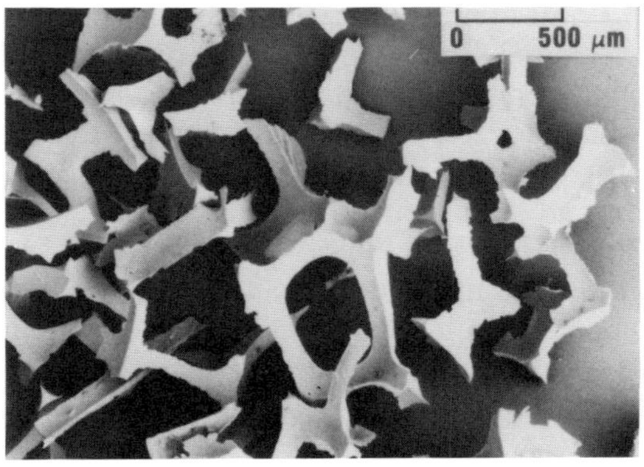

FIGURE 1. Reticulite and reticulite fragments. Triangular rods that make up the lattice are sideromelane and are formed at mutual intersections between three vesicles. Vesicle walls are thin glass membranes (mostly broken after cooling; they now have ragged, sharp edges).

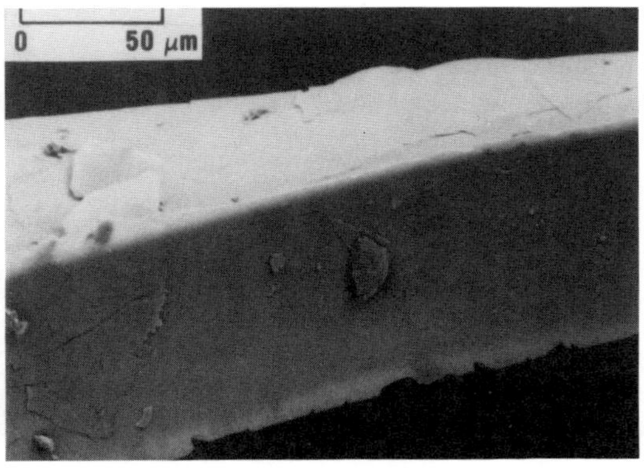

FIGURE 2. Close-up of one of the rods that make up the reticulite lattices. Thin, angular sheets adhering to the rods are pieces of glass membranes that once separated adjacent vesicles.

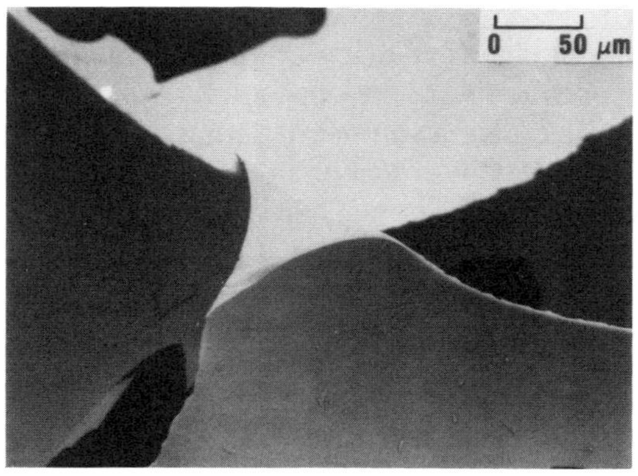

FIGURE 3. Broken rod, showing the typical triangular, curved cross section.

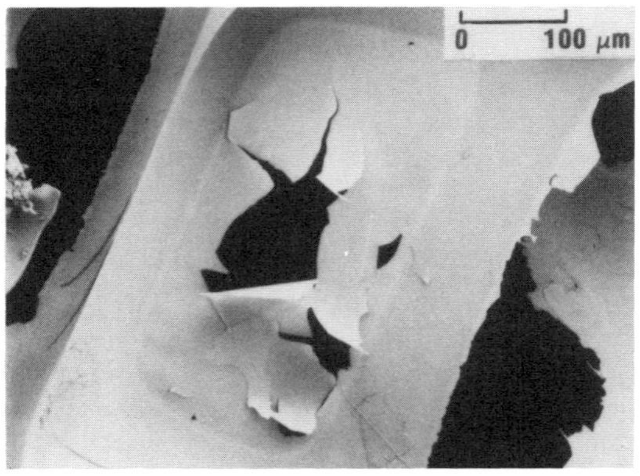

FIGURE 4. Broken glass membrane between adjacent vesicles. Membrane is less than 1 μm thick, quite flexible, but easily torn.

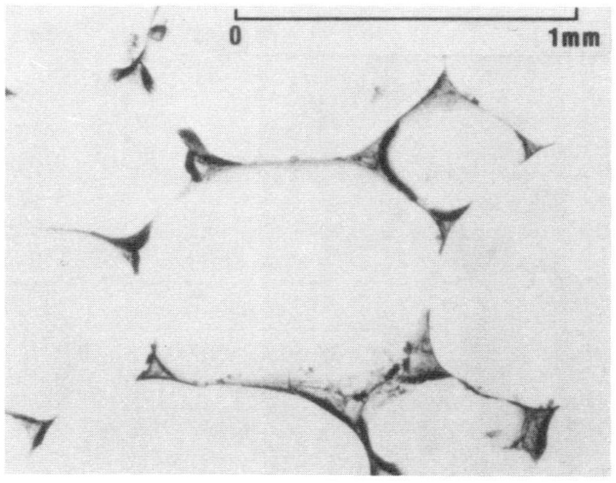

FIGURE 5a. Transmitted-light photomicrographs of reticulite. This thin section demonstrates the remarkable porosity of this tephra. Note the extremely thin vesicle walls.

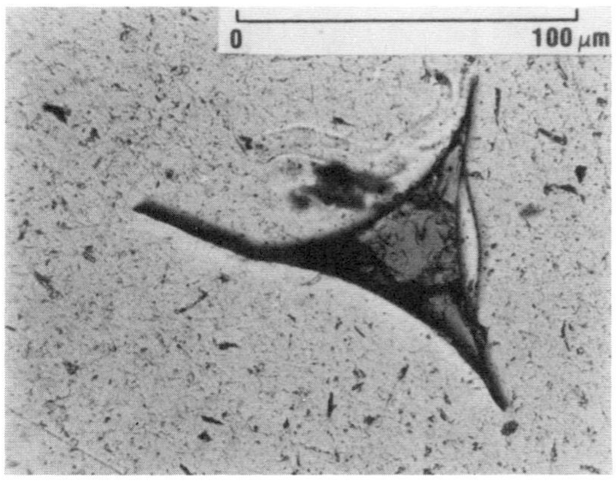

FIGURE 5b. Cross section of a curved, triangular rod. These pass into the extremely thin (<1-μm) basaltic glass membranes that separate adjacent vesicles.

LUNAR PYROCLASTIC ROCKS; "DARK MANTLE DEPOSIT," APOLLO 17 LANDING SITE

Introduction: The "dark mantle deposits" of the Moon have been interpreted as being of pyroclastic origin (e.g., McGetchin and Head, 1973). These deposits are located along the rims of several large mare basins and were sampled during the Apollo 17 mission. In the Valley of Taurus-Littrow (the Apollo 17 site), very dark and orange-tinted regolith samples were evident. Within the rim of Shorty Crater, a 120-m-diameter impact crater, sections of the orange and black clastic deposits were sampled. These are fine-grained clastic rocks consisting of small orange glass droplets and their partly crystalline (quench crystal) equivalent; the partly crystalline droplets are black. These were unlike any regolith soil samples from previous Apollo missions. Although fresh and unaltered, these deposits have an age of about 3.6 billion years and were contemporaneous with interbedded basalt flows at the same site. The samples were collected by Apollo 17 astronauts H. H. Schmitt and E. Cernan.

Grain Size: Mean grain size ranges from 40 to 43 μm (4.64 to 4.53 φ); grains are moderately sorted ($\sigma_\varphi = 1.69$ to 1.58 φ).

Tephra Description: Pyroclasts in these samples consist of spheres, ovoids, and compound droplets. Spheres and ovoids consist of orange glass and partly crystalline orange glass; they have smooth surfaces covered only with sublimates (<1-μm micromounds). Compound forms consist of a central droplet with smaller droplets on the surface that had hit the droplet while still molten. Throughout the stratigraphic sequence sampled by coring, 35 to 74% of the droplets are spheres, 28 to 38% are ovoids, and 4 to 41% have compound shapes. Between 60 and 78% of the droplets are broken

(Heiken and McKay, 1978). Up to 6% of the droplets contain single vesicles; the others are vesicle-free.

The percentage of partly crystalline droplets decreases from 93% near the base to 24% near the top of the 68-cm-long core. Only 0.3 to 2.7% of the droplets contain olivine phenocrysts (composition = Fo_{81}).

Partly crystalline droplets exhibit quench textures. Variations in distinctive olivine quench textures have been used to determine cooling rates and degrees of supercooling of lavas (Donaldson, 1976). Within these droplets there is a progression from the most rapidly to most slowly cooled olivines as follows: (1) branching, dendritic phases, (2) thin, tabular, parallel phases, (3) acicular skeletons and, (4) subequant skeletons.

Chemical Composition: An average chemical analysis of orange glass is shown in Table 1.

TABLE 1. CHEMICAL COMPOSITION OF ORANGE GLASS.*

	Average (Wt. %)	(Variation)
SiO_2	38.63	(0.33)
TiO_2	8.96	(0.47)
Al_2O_3	5.87	(0.36)
Cr_2O_3	0.55	(0.3)
FeO	22.00	(0.32)
MgO	14.79	(1.15)
CaO	7.31	(0.40)
Na_2O	0.37	(0.13)
K_2O	0.08	(0.02)

No. of samples analyzed = 80.
*Microprobe analyses from Reid et al. (1973).

Comments on Pyroclast Formation: Most lunar-sample investigators agree that the ultramafic clastic rocks of the Apollo 17 site are part of a pyroclastic deposit and were formed by lava fountaining. Arguments for and against this hypothesis are summarized by Heiken et al. (1974). Glass droplets from this site are chemically and texturally homogeneous; *none* contain shock-damaged crystals and heterogeneous glass that are typical of glass clasts formed by meteorite impact. Sulfur-rich coatings on pyroclast surfaces appear to have been deposited as sublimates from the volcanic gases that drove the explosive eruptions (Butler and Meyer, 1976).

It is probable that the magma involved in lava fountaining had a very low viscosity; a calculated viscosity of this magma, at a temperature of 1250°C, is 3.5 poise (using the method of Shaw, 1972), which is lower than any determined for Hawaiian eruptions. Such low viscosities would account for the abundance of fine-grained spherical droplets and a dearth of vesicles. Lack of vesicles may be the result of volatiles escaping from a low-viscosity fluid into the lunar vacuum.

Variations in droplet morphology can be explained by variations in the lava fountaining. Fountains with a narrow dispersion angle will contain many spray droplets that fall back into the fountain and subsequently form compound droplets. These droplets remain hot enough for quench crystals to grow. In a broadly dispersed fountain, individual droplets are thrown clear, cool rapidly, and hit the lunar surface as simple glass spheres.

REFERENCES:

Butler Jr., P. and Meyer Jr., C., 1976. Sulfur prevails in coatings on glass droplets: Apollo 15 green and brown glasses and Apollo 17 orange and black (devitrified) glasses. Proc. Lunar Sci. Conf., **7th**: 1561-1581.

Donaldson, C., 1976. An experimental investigation of olivine morphology. Contrib. Mineral. Petrol., **57**: 187-213.

Heiken, G., McKay, D., and Brown, R., 1974. Lunar deposits of possible pyroclastic origin. Geochim. Cosmochim. Acta, **38**: 1703-1718.

Heiken, G. and McKay, D., 1977. A model for eruption behavior of a volcanic vent in eastern Mare Serenitatis. Proc. Lunar Sci. Conf., **8th**: 3243-3253.

Heiken, G. and McKay, D., 1978. Petrology of a sequence of pyroclastic rocks from the Valley of Taurus-Littrow (Apollo 17 landing site). Proc. Lunar Sci. Conf., **9th**: 1933-1943.

McGetchin, T. R. and Head, J. W., 1973. Lunar cinder cones. Science, **180**: 68-70.

Reid, A., Ridley, W., Donaldson, C., and Brown, R., 1973. Glass compositions in the orange and gray soils from Shorty Crater, Apollo 17. Trans. Am. Geophys. Union, **54**: 607-609.

Shaw, H., 1972. Viscosities of magmatic silicate liquids: an empirical method of prediction. Am. J. Sci., **272**: 870-893.

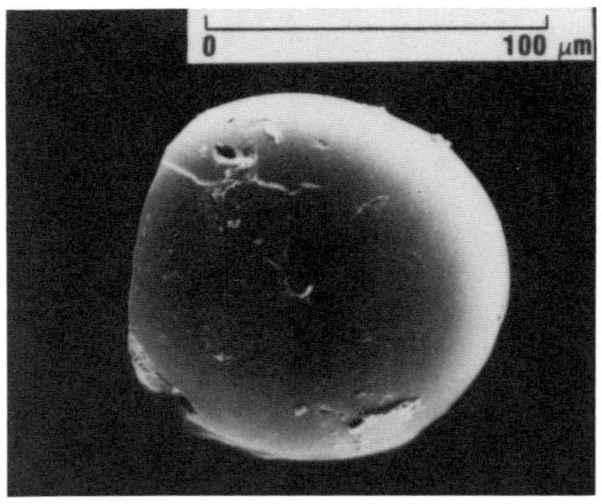

FIGURE 1. SEM of an orange glass sphere. The smooth surface is marred only by spalls and chips caused by low-velocity impacts, perhaps during collision with other particles or when hitting the lunar surface. It is flattened on one side, perhaps upon collision with another particle or the lunar surface.

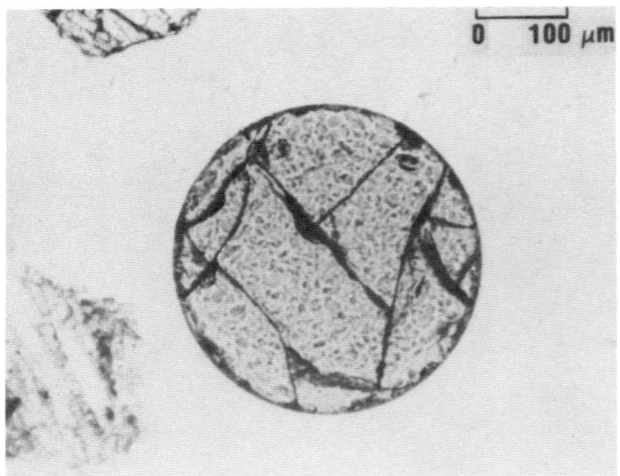

FIGURE 2. Transmitted-light photomicrograph of a 300-μm-diameter orange glass sphere. The glass is homogeneous; fracturing occurred when the thin-section was made.

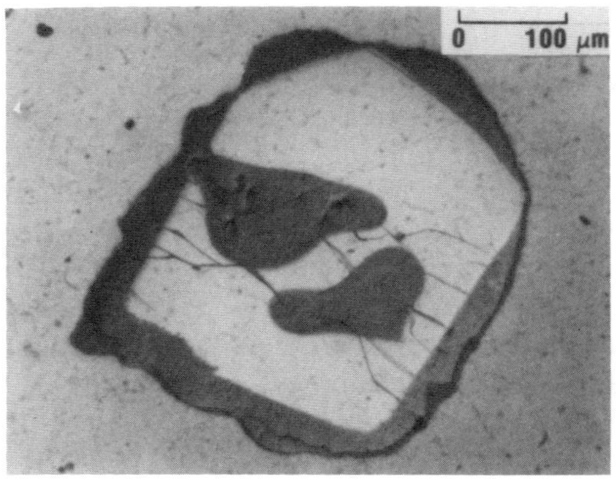

FIGURE 3. Reflected-light photomicrograph of a partly crystalline compound droplet containing an olivine phenocryst (composition = $Fo_{80\text{-}81}$). There are 0.3 to 2.7% olivine phenocrysts in these samples. Note radial growth of dendritic quench-crystal olivine and ilmenite phases from the phenocryst surface.

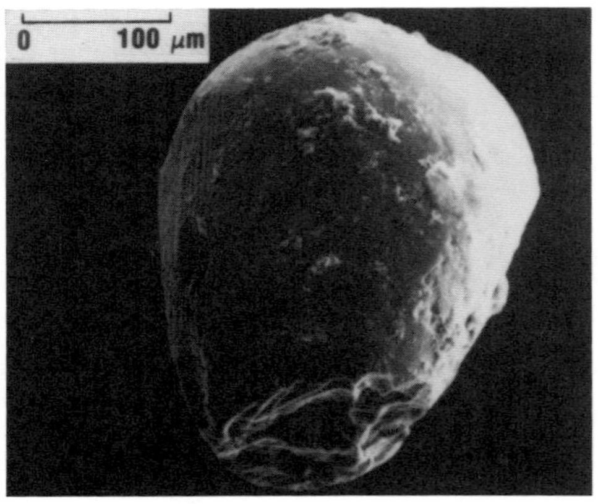

FIGURE 4. SEM of a partly crystalline broken droplet that consists of parallel growths of ilmenite and olivine in orange glass. Parallel lines near the bottom of the droplet are ilmenite phases intersecting the pyroclast surface. Grainy coatings consist of a mixture of sublimates (vapor condensates) and flattened glass particles of the same composition. From Heiken et al. (1974).

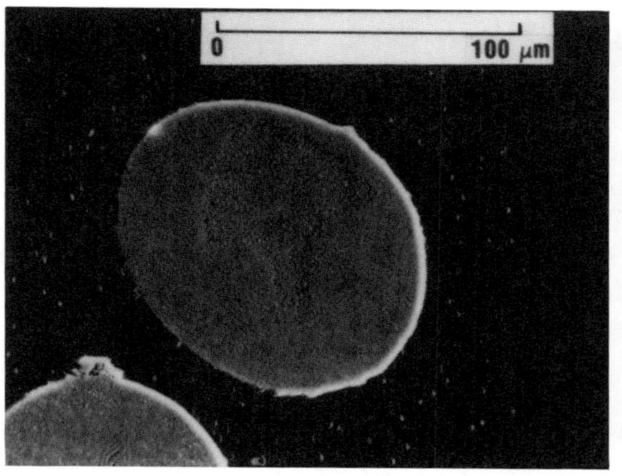

FIGURE 5a. SEM of a polished, etched thin section of a partly crystalline orange glass droplet. Clumps of finely crystalline dendrites (most phases are less than 0.5 μm long) consist of intergrown, feathery olivine and ilmenite phases in an orange glass matrix. The central area of this droplet is somewhat less crowded with dendrites.

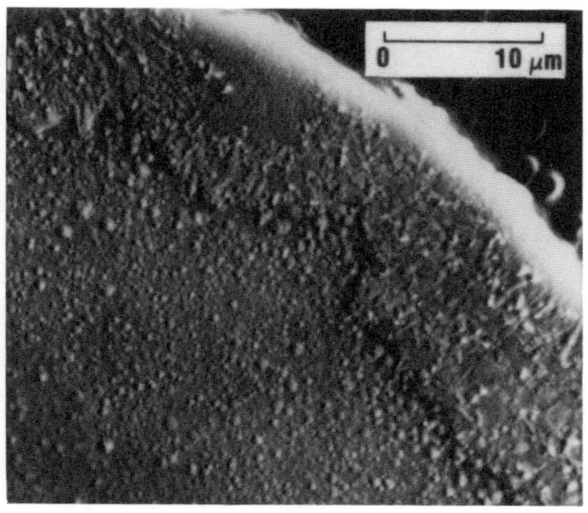

FIGURE 5b. SEM of a polished, etched thin section of a partly crystalline orange glass droplet. Feathery dendrites have grown toward the center of the sphere from the droplet surface to form an irregular, 6- to 15-μm-thick "front." Darker gray phases are olivines and the brighter phases are ilmenite. Mounds in the center are 0.2- to 0.5-μm-diameter dendrites separated by orange glass (low relief).

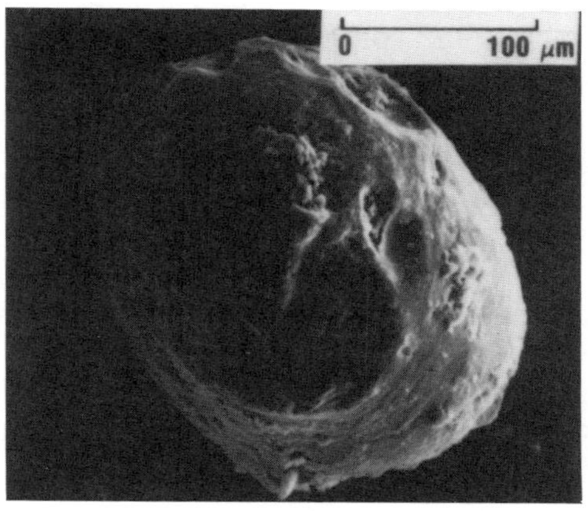

FIGURE 6. SEM of a compound droplet that consists of a partly crystalline, ovoid core with multiple droplets adhering to it. Most of the smaller droplets were flattened when they collided with the larger droplet. From Heiken and McKay (1977).

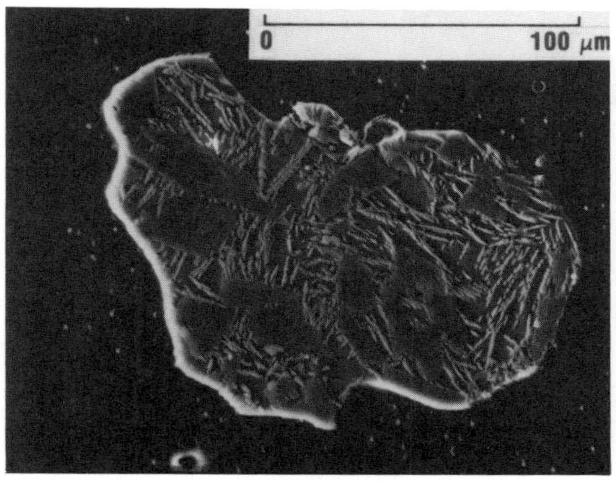

FIGURE 7a. SEM of polished, etched thin sections of compound, partly crystalline droplets. Irregular, broken droplet attached to a sphere (lower). The high-relief, low-contrast crystals are subequant, skeletal olivines (Fo_{76-78}). Ilmenite is the bright, feathery phase and Cr-spinel is the rare, bright, rectangular phase. Areas of low relief are orange glass. From Heiken and McKay (1977).

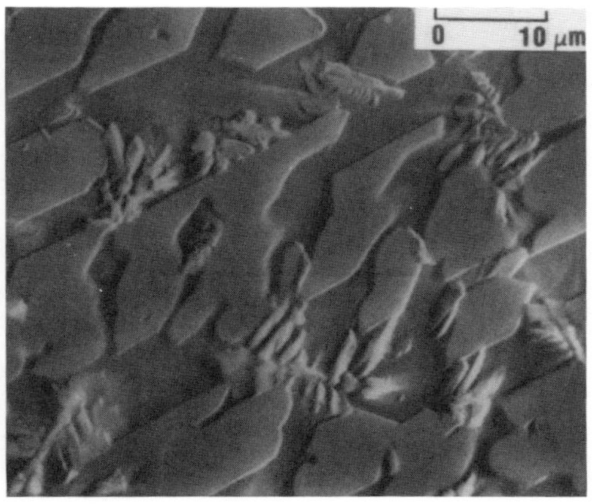

FIGURE 7b. Detail of an ovoid droplet. The ladder-like phases are acicular skeletal olivines (Fo_{70-78}). Areas of low relief are orange glass and the bright, feathery phases are ilmenite. From Heiken and McKay (1977).

Strombolian Eruptions

Stromboli Volcano, called the "lighthouse of the Mediterranean," is characterized by nearly continuous activity. Typical Strombolian activity consists of explosive bursts of solidified and partly solidified ash and scoria. These pyroclasts fall around the crater, building a cinder (scoria) cone, the most common pyroclastic volcanic landform on Earth.

CINDER CONE—LASSEN VOLCANIC NATIONAL PARK, CALIFORNIA

Introduction: Strombolian activity is responsible for construction of cinder cones (also known as scoria cones). The cone-building process is one mainly of ballistic deposition of coarse scoria and ash followed by slumping; some of the finer grained tephra is carried away by the wind (Fries, 1953; McGetchin et al., 1974; and Self et al., 1974). The example used here, Cinder Cone (the geographic name), is located in the northeast corner of Lassen Volcanic National Park, California, on the flank of Prospect Peak. Several periods of activity at Cinder Cone occurred during the last 400 years, and the latest eruption was in 1851 A.D. Eruptions of basalt produced several block lava flows, a cinder cone with two concentric craters (the cone makes up 54% of the volume of tephra produced during these eruptions), and a tephra blanket (scoria fall deposit). Total volume of Cinder Cone tephra is 70×10^6 m^3. Most of the samples studied from this eruption sequence were collected from the tephra deposit surrounding Cinder Cone (Heiken, 1978).

The major tephra units associated with cone-building activity are described in Figures 1 and 2. The lowest unit (Unit 1) consists of poorly bedded, black to dark brown, medium- to coarse-grained vitric ash. Unit 2 is coarsely bedded and composed of highly vesicular, light brown sideromelane bombs and lapilli, which are interbedded with thin layers of dark gray coarse ash. Unit 3 is the most voluminous deposit and makes up 71% of the tephra blanket. It occupies an 18- by 10-km, NNE-trending oval. It consists of graded beds of gray and brown coarse ash. Nearly all beds in Unit 3 are normally graded from vesicular sideromelane lapilli upward into slightly finer grained gray ash.

Grain Size: All samples used for size analysis were collected 1 km SSE of Cinder Cone. Mean grain size ranges from 0.3 φ (0.8 mm = coarse ash) to -1.53 φ (3.1 mm = lapilli) and are moderately to poorly sorted. In plots of mean grain size versus sorting, all of the samples are within the Strombolian "scoria fall" category of Self et al. (1974).

Tephra Description: Although they grade into one another, three categories of pyroclast types occur in this deposit (Table 1).

TABLE 1. PYROCLAST TYPES FROM CINDER CONE.

Cinder Cone Pyroclast Types (%)	Unit 1	Unit 2	Unit 3 (lower)	Unit 3 (upper)
Light brown sideromelane	28.2	47.5	37.0	25.5
Medium brown sideromelane	38.6	28.0	45.0	46.3
Tachylite	32.2	22.0	14.0	25.9
Phenocrysts				
Olivine	0.5	2.5	2.5	2.3
Plagioclase	---	---	0.5	---
Quartz xenocrysts	---	---	1.0	---
Andesite fragments	0.5	---	---	---

Light Brown Sideromelane. These smooth-skinned, light-brown, vesicular sideromelane droplets have vesicularities ranging from 20 to 70%. There are numerous coalesced vesicles, which form irregular, partly open frameworks. Most vesicles are deformed ovoids 20 to 500 μm long. Sideromelane pyroclasts range from highly vesicular, delicate small bombs with low bulk densities to fine-grained ash consisting of broken droplets. There are 2 to 5% olivine phenocrysts and traces of plagioclase laths and quartz xenocrysts within the coarser pyroclasts.

Medium Brown Sideromelane Pyroclasts. These pyroclasts are generally less vesicular (30 to 40%) and exhibit a greater degree of crystallization than do the light brown pyroclasts. There are fewer rounded, glassy surfaces and more rough surfaces caused by fragmentation of partly crystalline vesicular pyroclasts. The groundmass ranges from glass with a few 4- to 20-μm-long feldspar laths to dense networks of finely crystalline phases with <50% glass.

Tachylite Pyroclasts. Completely crystalline, these pyroclasts exhibit a diktytaxitic texture that consists of a network of finely crystalline, glass-coated feldspar and pyroxene crystals separated by irregular vesicles. Pyroclast surfaces are very rough, reflecting the crystalline matrix.

Relative portions of these phenocryst types vary throughout the stratigraphic section. The significance of these variations will be discussed in the following section.

Chemical Composition: Chemical compositions shown in Table 2 represent two types of material: (1) composition of glass between quench crystals in tachylite, and (2) composition of glass in a sideromelane pyroclast.

TABLE 2. CHEMICAL COMPOSITION OF GLASS FROM CINDER CONE.*

	1** (Wt. %)	2*** (Wt. %)
SiO_2	52.95	53.19
TiO_2	1.15	1.00
Al_2O_3	15.95	15.40
FeO	8.37	7.63
MgO	3.95	4.12
CaO	6.72	7.99
Na_2O	4.28	3.91
K_2O	1.13	0.77
P_2O_5	0.16	0.20
Total	94.67	94.20

*Analyses are by Cameca electron microprobe.
**Between quench crystals in a tachylite pyroclast.
***A sideromelane pyroclast.

Comments on Pyroclast Formation: The tephra sequence at Cinder Cone is similar to those described by McGetchin et al. (1974) for Strombolian eruptions at Northeast Crater, Etna, Italy, and at Heimaey, Iceland (Self et al., 1974). MacDonald (1972) describes Strombolian activity as "weak-to-violent ejection of partly fluid blobs." According to McGetchin et al. (1974), clots of partly fluid lava and ash are produced when jets of melt, accelerated by gas release, break up into clusters of pyroclasts.

The type of pyroclast dominating any particular tephra layer appears to have been controlled by clogging of the vent and the lava flux. If the vent were clear and the lava flux were high, an abundant supply of lava at high temperature and lower viscosity could produce highly vesicular, light brown sideromelane pyroclasts typical of Unit 2. If the vent were blocked, either by pooling of lava or slumping of talus into the vent, explosive activity would be suppressed. During periods when the vent was blocked, Strombolian activity could have occured with intermittent, small tephra eruptions, producing tephra similar to those in Units 1 and 3; lava within the crater could have cooled and lost some of its volatile phases. If the crater were blocked by cooling lava, intermittent gas release could disrupt it, breaking the partly molten mass into poorly vesicular tachylite pyroclasts mixed with sideromelane droplets.

REFERENCES: Fries, Jr., C., 1953. Volumes and weights of pyroclastic material, lava, and water erupted by Paricutin Volcano, Michoácan, Mexico. Trans. Amer. Geophys. Union., **34**: 603-616.

Heiken, G., 1978. Characteristics of tephra from Cinder Cone, Lassen Volcanic National Park, California. Bull. Volcanol. **41**: 119-130.

MacDonald, G. A., 1972. Volcanoes. Prentice-Hall, Englewood Cliffs, N. J.: 510 pp.

McGetchin, T. R., Settle, M., and Chouet, B. A., 1974. Cinder cone growth modeled after Northeast Crater, Mount Etna, Sicily. J. Geophys. Res., **79**: 3257-3272.

Self, S., Sparks, R. S. J., Booth, B., and Walker, G. P. L., 1974. The 1973 Heimaey Strombolian scoria deposit, Iceland. Geol. Mag., **111**: 534-548.

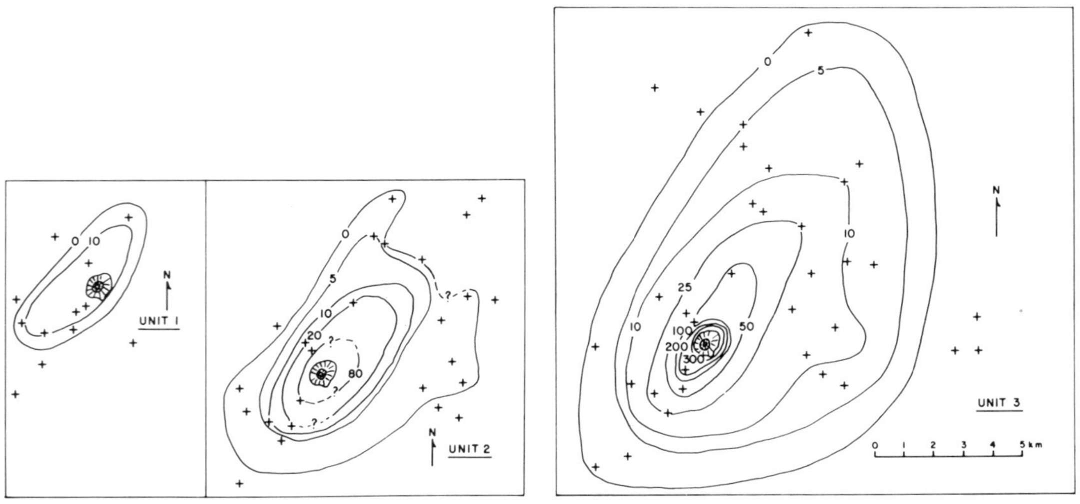

FIGURE 1. Isopach maps of Cinder Cone scoria fall Units 1, 2, and 3. Thicknesses in centimeters.

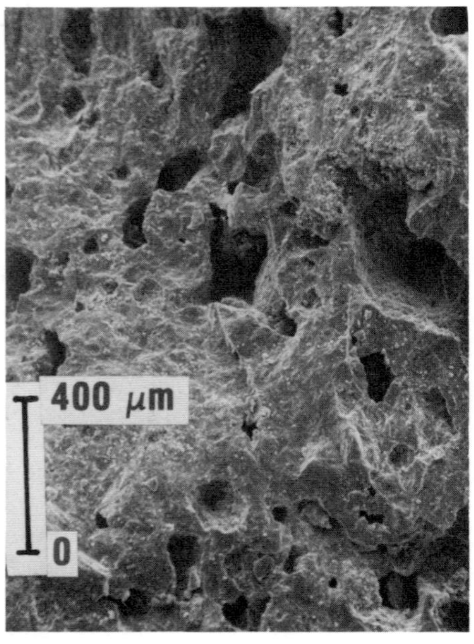

FIGURE 2a. Tachylite pyroclasts, some of the most common particle types in this unit, are irregular, blocky, and vesicle-poor. Pyroclast surface: coalesced, irregular vesicles are common. Diktytaxitic textures result in rough grain surfaces.

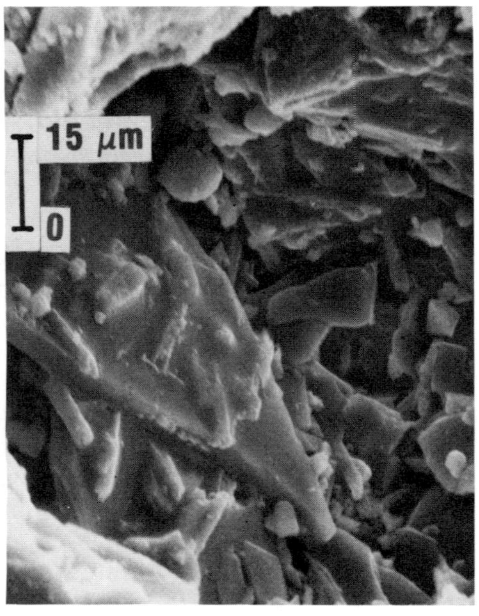

FIGURE 2b. Detail of (*a*). Framework of 10- to 80-μm plagioclase and pyroxene laths surrounded by glass.

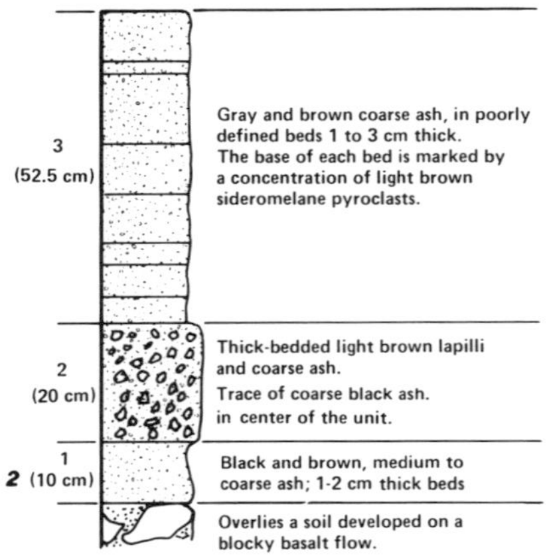

FIG. 2 Tephra sequence at Cinder Cone, 0.5 km N of the cone.

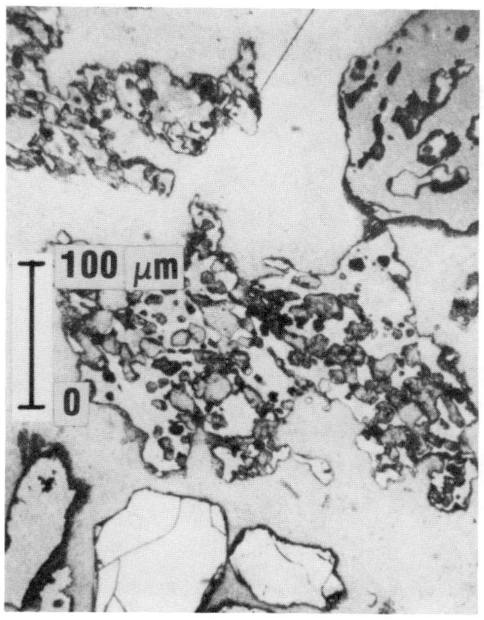

FIGURE 3a. Unit 2 is characterized by mostly light brown, highly vesicular sideromelane pyroclasts. Reflected-light photomicrograph of a representative ash sample from Unit 2. *Center* - 0.32-mm-long, highly vesicular sideromelane pyroclast. Note the irregular, delicate, glassy framework and smooth vesicle and grain surfaces. Other pyroclasts include similar sideromelane droplets, tachylite, and olivine phenocrysts.

FIGURE 3b. SEM of ash from Unit 2. *Center* - subequant sideromelane pyroclast with smooth, glassy surfaces. Highly vesicular, with 20- to 150-μm-diameter vesicles. Other pyroclasts are mostly tachylite with low vesicularities.

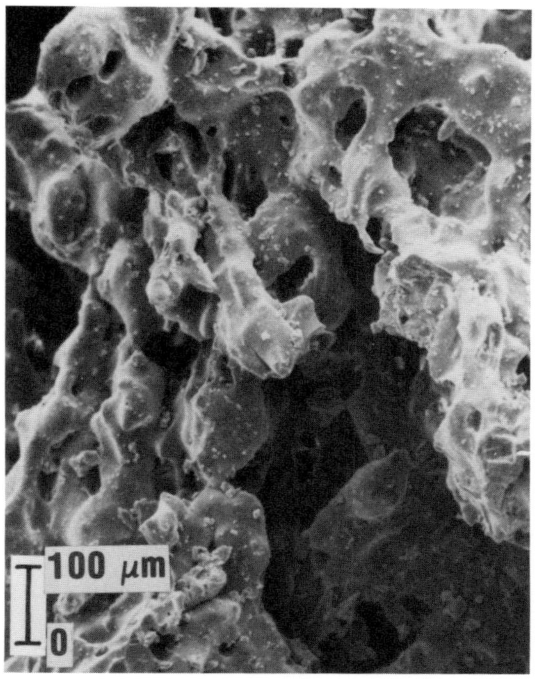

FIGURE 3c. SEM of surface of a light brown sideromelane pyroclast. Traces of small, angular particles are welded to the pyroclast surface.

FIG. 3

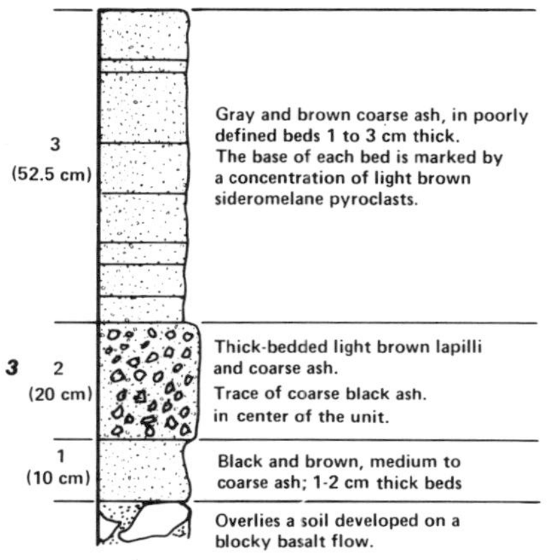

Tephra sequence at Cinder Cone, 0.5 km N of the cone.

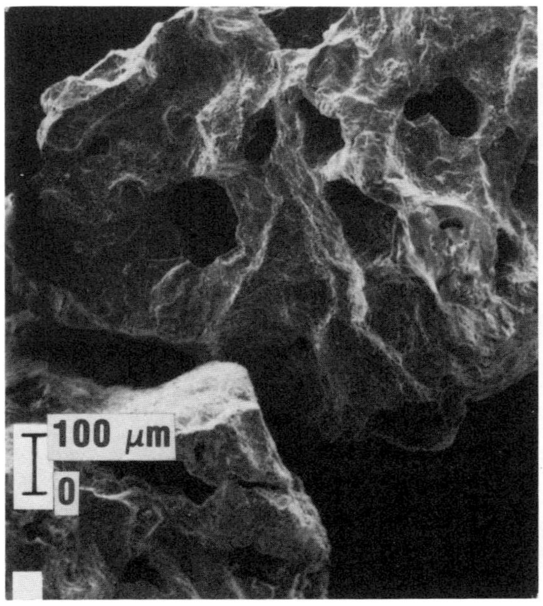

FIGURE 4a. Medium brown sideromelane and tachylite pyroclasts. SEM of a pyroclast with partly crystalline groundmass. It has a low vesicularity but larger vesicles than those in the light brown sideromelane pyroclasts.

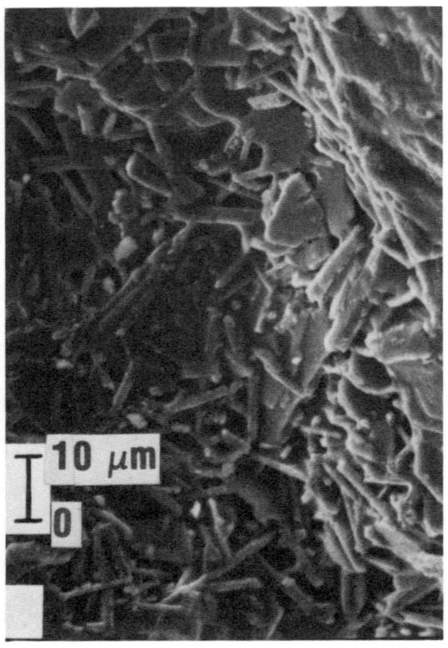

FIGURE 4b. Detail of grain surface in (a). The surface is hyalocrystalline and consists of a network of 5- to 15-μm-long plagioclase and pyroxene laths in a glassy matrix.

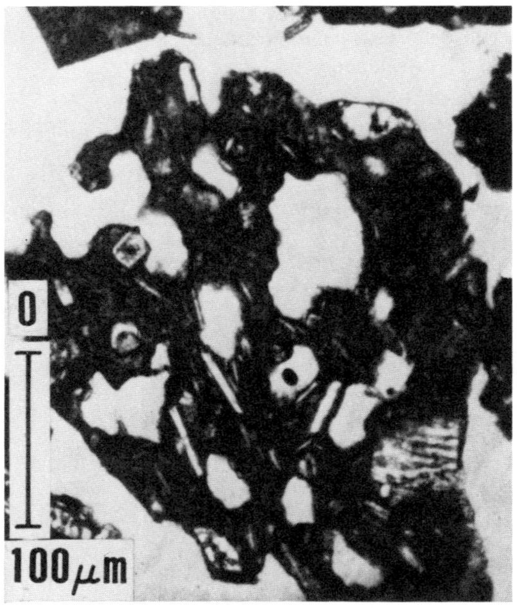

FIGURE 4c. Transmitted-light photomicrograph of a pyroclast similar to the one in (a) and (b), showing the well-developed, coalesced vesicles.

FIG. 4

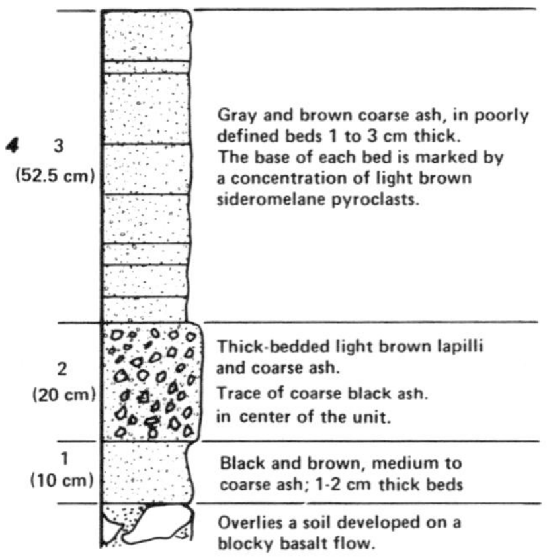

Tephra sequence at Cinder Cone, 0.5 km N of the cone.

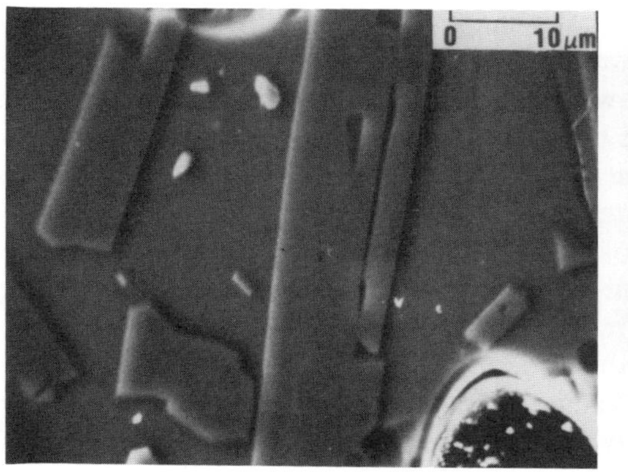

FIGURE 5a. Comparison of sideromelane and tachylite textures. SEM images of ion-etched polished thin sections across pyroclasts. Sideromelane—plagioclase and pyroxene laths, exhibiting quench textures (skeletal forms) are the high-relief phases. The smooth, low-relief phase is basaltic glass.

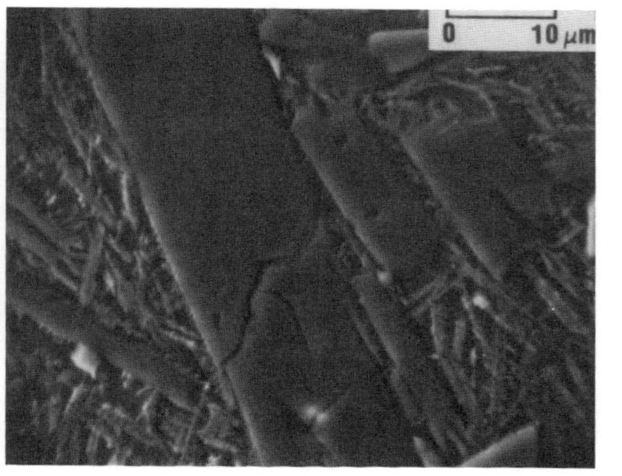

FIGURE 5b. Tachylite—well-developed pyroxene and plagioclase laths surrounded by dendritic growths of 0.2- to 2-µm-wide pyroxene and Fe-Ti oxides. There is a trace of glass (low-relief phase) between dendritic phases.

CERRO NEGRO, NICARAGUA; 1968 ERUPTION

Introduction: Cerro Negro is one of four clustered cinder cones located along a N-S fault line in the central Marabios Range, Nicaragua. Strombolian eruptions are characteristic of this cone, which has been active intermittently since 1850; the most recent eruptions were in 1968, 1969, and 1971. The 1968 eruption ejected abundant blocks and ash, and an ash cloud rose to several thousand meters above the central crater. At the same time, there was lava fountaining on the flank of the cone. Lavas and ash from this eruption are all olivine basalts. This sample was provided by W. I. Rose, Jr.

Grain Size: A weighted average of grain-size estimates was made by Rose et al. (1973). The ash has a mean grain size of 180 µm and is very poorly sorted.

Tephra Description: The vitric ash consists of mostly broken sideromelane droplets and tachylite pyroclasts that contain abundant augite, olivine, and plagioclase phenocrysts. Sideromelane droplets have smooth, botryoidal or lumpy surfaces that are broken by depressions of outermost vesicles. There are cavities (broken vesicles) with both smooth and sharp, irregular rims. Tachylite pyroclasts are less vesicular and have irregular grain surfaces over the outermost groundmass minerals.

Chemical Composition: Chemical analyses of Cerro Negro basalts are given in Table 1.

TABLE 1. CHEMICAL ANALYSES OF OLIVINE BASALT FROM CERRO NEGRO, 1968 ERUPTION.

	Ash* (Wt. %)	Lava** (Wt. %)
SiO_2	49.20	49.50
TiO_2	1.28	0.80
Al_2O_3	19.38	19.50
Fe_2O_3	4.42	4.60
FeO	5.55	4.20
MnO	0.14	n.d.
MgO	5.41	7.25
CaO	11.89	11.50
Na_2O	1.76	2.35
K_2O	0.85	0.30
P_2O_5	0.05	0.19
TOTAL	99.93	100.1

*Analyzed by M. Rhodes.
**Analyzed by J. G. Viramonte (Viramonte and DiScala, 1970).

Comments on Pyroclast Formation: Pyroclast shapes are related to vesicle growth within droplets that are formed as a spray in lava fountains. Viscosities are higher than those of lavas erupted in Hawaiian lava fountains. There are no vesicular spheres or ovoids; instead, there are highly irregular droplets with smooth, glassy surfaces. There are some surface tension effects on outer pyroclast surfaces (sideromelane scoria); some of the outermost depressions with smooth, curved rims were formed when vesicles broke while the pyroclast was still molten. Broken droplets with some angular, sharp edges were formed when chilled solid pyroclasts hit the ground or other pyroclasts. Some may have been broken by rupture of overpressured vesicles within chilled, brittle, vesicular droplets.

Tachylite fragments typically have hyalocrystalline groundmass textures. They may have formed by the disruption of crusts on lava lakes that are present between explosive bursts, or they may have been introduced as accidental debris from the slumping of crater walls.

REFERENCES: Rose, W. I., Jr., Bonis, S., Stoiber, R. E., Keller, M., and Bickford, T., 1973. Studies of volcanic ash from two recent Central American eruptions. Bull. Volcanol., **37**: 338-364.

Viramonte, J. G. and DiScala, L., 1970. Summary of the 1968 eruption of Cerro Negro, Nicaragua. Bull. Volcanol., **34**: 347-351.

FIGURE 1a. Sideromelane scoria lapilli. The unbroken 800- by 700-μm scoria fragment is heavily coated with finer grained ash particles. This scoria pyroclast may be heterogeneous, consisting partly of basaltic glass (sideromelane) and partly of hyalocrystalline groundmass (tachylite).

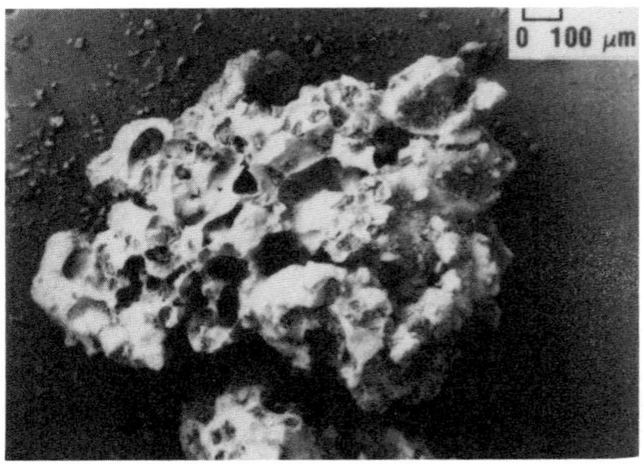

FIGURE 1b. A 1.1- by 0.8-mm sideromelane scoria pyroclast (medium to dark brown). As can be seen in the broken face of this pyroclast, all irregular projections are associated with vesicles or clusters of vesicles. Unbroken surfaces are smooth glass (where not coated with finer grained pyroclast fragments). Vesicles range in length from 10 to 230 μm; most are irregular ovoids deformed at contacts with adjacent vesicles. Highly variable vesicle-wall thicknesses also show the heterogeneity of vesicle size and shape.

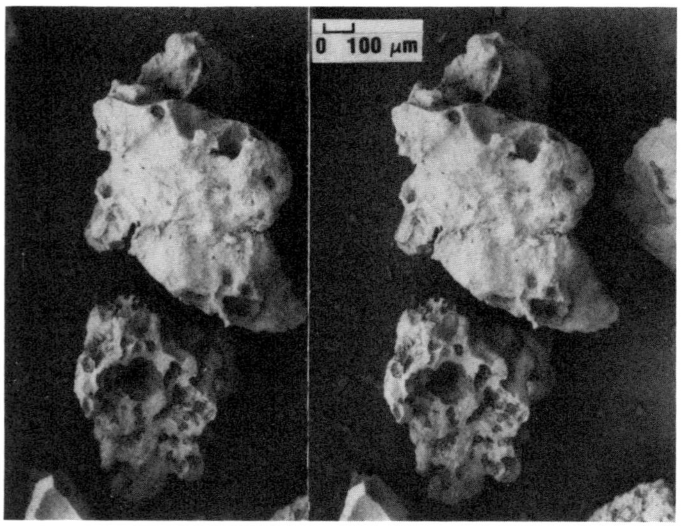

FIGURE 2. Stereophotomicrograph of two broken scoria lapilli. Note the heterogeneity of vesicle size and shape and the lack of deformation by flow (no highly elongate vesicles). The upper lapillus is tachylitic and the lower is composed of mostly sideromelane.

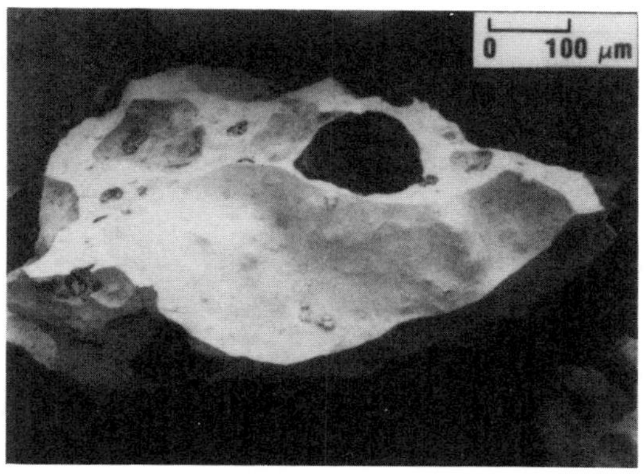

FIGURE 3. Large shard broken from a scoria lapillus; all depressions are vesicle remnants.

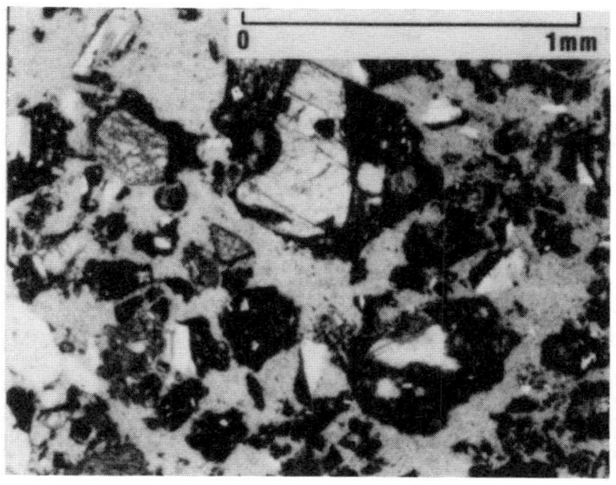

FIGURE 4a. Photomicrograph of tephra from Cerro Negro. Partly plane-polarized transmitted light. Irregular, blocky, phenocryst-rich tachylite pyroclasts make up most of the grains visible here (black, opaque pyroclasts). Some hyalocrystalline sideromelane pyroclasts are present in the lower left corner of the photograph.

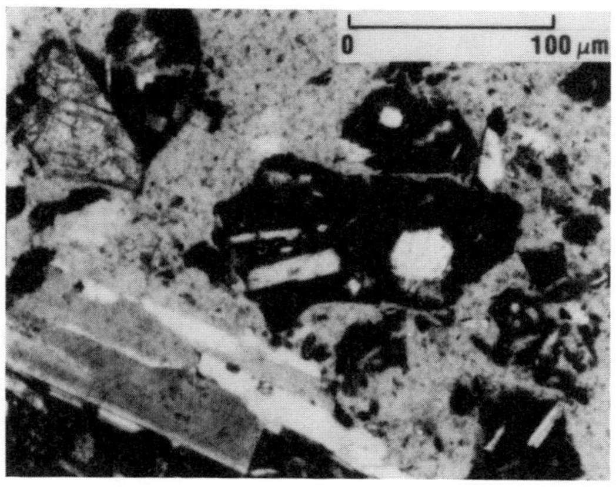

FIGURE 4b. All pyroclasts visible here are irregular, blocky tachylite grains. All are phenocryst-rich, mostly plagioclase and clinopyroxene.

PACAYA, GUATEMALA; 1965-1969 ERUPTION

Introduction: Pacaya is a composite cinder cone located 40 km SSW of Guatemala City. It lies within the crater of an older composite cone (Cerro Chino), which is, in turn, located on the southern rim of Amatitlan caldera (Mooser et al., 1958). Pacaya was formed during recent intermittent Strombolian activity. This particular eruptive phase was most active from 1965 to 1969 and has continued intermittently.

Grain Size: These samples, collected at a distance of 1 km west of the summit vent, have a median grain size of 340 μm (1.57 φ) and are moderately sorted ($\sigma_\varphi = 1.3$ φ).

Tephra Description: In this deposit, 97% of the pyroclasts consist of sideromelane (basaltic glass) droplets or droplet fragments. The pyroclasts are irregularly shaped, yet smooth skinned; the skin is broken by depressions over the outermost vesicles. Vesicles are large (30- to >100-μm) ovoids and spheres, and they have thin walls. Some pyroclasts contain phenocrysts of plagioclase (An_{70}), olivine, and clinopyroxene. Phenocrysts partly coated with basaltic glass are present as individual pyroclasts.

Chemical Composition: The chemical composition of Pacaya lava is shown in Table 1.

TABLE 1. CHEMICAL COMPOSITION OF HIGH-ALUMINA BASALT FROM PACAYA.*

	(Wt. %)
SiO_2	51.80
TiO_2	1.15
Al_2O_3	17.73
Fe_2O_3	2.96
FeO	6.49
MnO	0.22
MgO	4.03
CaO	9.52
Na_2O	4.05
P_2O_5	0.26
S	0.06
TOTAL	99.1

*M. Rhodes, analyst (from Heiken, 1972).

Comments on Pyroclast Formation: Intermittent Strombolian activity produced spray bursts consisting of small molten droplets and bombs. Each droplet vesiculated rapidly before quenching and is irregular because of heterogeneous vesicle growth. Viscosity before quenching was low enough to allow smoothing of pyroclast surfaces, but high enough to prevent formation of vesicular glass spheres. Most of the pyroclasts formed in this way are very delicate and break upon impact with one another or the ground. Some may have broken during rupture of pressurized vesicles, but this is difficult to prove. Slightly curved to curved thin-glass shards are vesicle-wall fragments from broken basaltic lapilli and coarse ash-size droplets.

REFERENCES:

Heiken, G., 1972. Morphology and petrography of volcanic ashes. Geol. Soc. Am. Bull., **83**: 1961-1988.

Mooser, F., Meyer-Abich, H., and McBirney, A., 1958. *Catalogue of the active volcanoes and solfatara fields of Central America. Part VI of the Catalogue of Active Volcanoes of the World.* Int'l. Volcanol. Assoc., Napoli: 146 pp.

FIGURE 1. Basaltic tephra from Strombolian eruptions at Pacaya, Guatemala. Nearly all the pyroclasts visible here are light brown (in thin section) sideromelane droplets. Megascopically, these tephra are dark brown. Most are characterized by high vesicularity, smooth surfaces (except where there are lumps caused by underlying vesicles and phenocrysts), and some breakage.

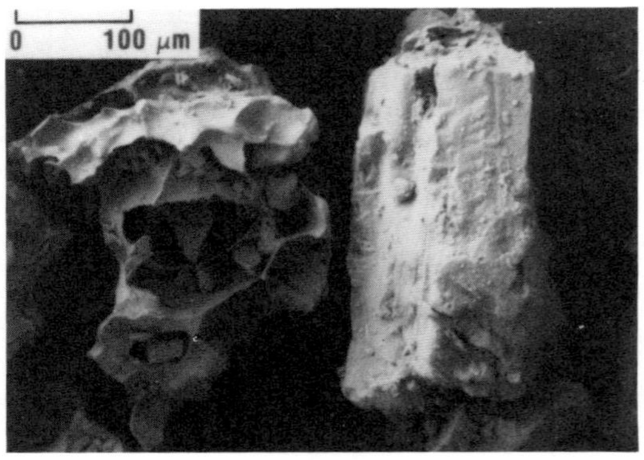

FIGURE 2. Right - Plagioclase phenocryst partly coated with basaltic glass; some twin planes are visible near its center. *Left* - Broken sideromelane droplet with high vesicularity. Vesicles are large (30- to >100-μm) and thin walled (generally ≤ 10 μm).

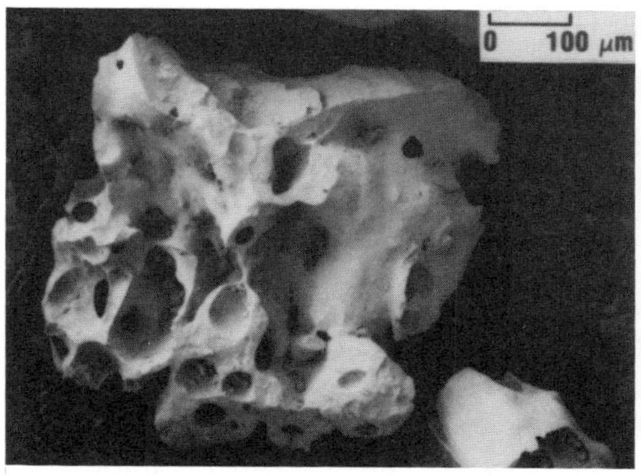

FIGURE 3. Sideromelane lapillus. The outermost vesicles collapsed while the droplet was still molten, leaving smooth-rimmed depressions and smooth glass pillars between vesicles (*center*). Other surfaces are broken, exposing the highly vesicular interior.

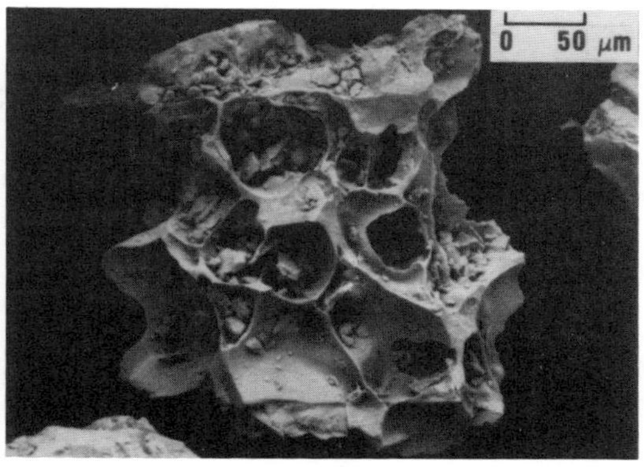

FIGURE 4. Broken sideromelane lapillus. Many vesicles are now polygonal as a result of mutual interference of vesicle walls during vesicle growth. Small clasts adhering to the lapillus surface are angular, broken vesicle-wall fragments (shards).

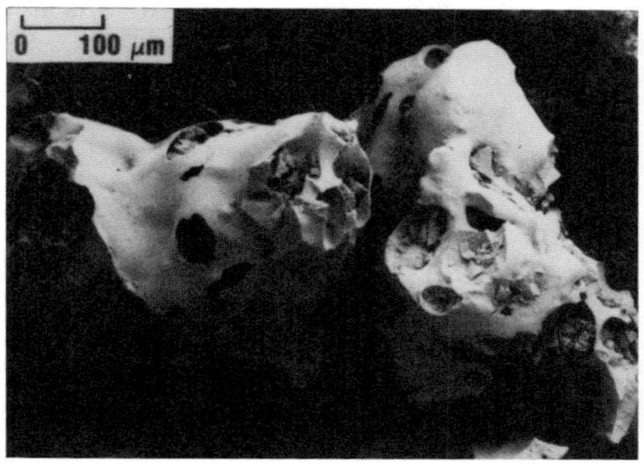

FIGURE 5. Two sideromelane droplets. These pyroclasts have irregular, twisted forms, yet have smooth surfaces. These formed by surface tension within still-molten, vesiculating droplets that were a spray in lava fountains.

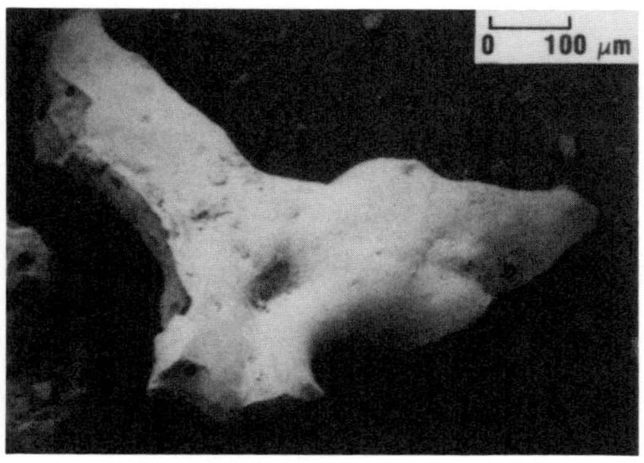

FIGURE 6. Large shard that was broken from a sideromelane droplet.

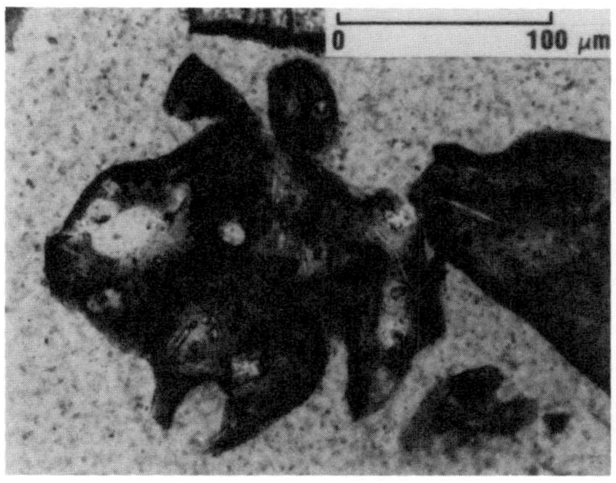

FIGURE 7a. Transmitted-light photomicrographs of Pacaya tephra. Irregular sideromelane droplets. There are small plagioclase phenocrysts (>25 µm) and abundant quench-texture plagioclase and pyroxene phases (generally less than 20 µm long). A rind, only a few micrometers thick, is darker brown than the light brown glass of the interior.

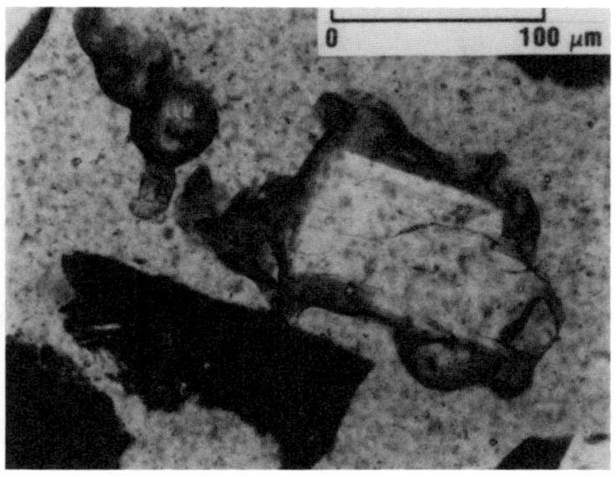

FIGURE 7b. Sideromelane-coated plagioclase phenocryst. Black grains are tachylite pyroclasts.

MT. ETNA, ITALY; 1971 ERUPTION

Introduction: Mt. Etna is a 60- by 40-km, 3300-m-high composite volcano (a shield of lava flows capped with interbedded lava flows and minor pyroclastic deposits) located on the eastern coast of Sicily, Italy. Frequent historical eruptions have produced lava flows and numerous chains of cinder cones at the summit and on the flanks. The April-May 1971 eruption was from SE and SSE radial fissures at about the 3000-m level (Tanguy, 1971; Rittmann, 1971). Continuous explosive activity built new, 30- to 50-m-high cinder cones along the two fissures. Rittmann (1971) noted that there may have been some phreatomagmatic interaction because the snow was up to 4 m deep at the time. Near the end of this eruption, a new explosion crater was formed (May 4 to May 26) several hundred meters NE of the fissures; Booth and Walker (1973) noted that activity at this crater was both Strombolian and phreatomagmatic.

This sample was collected in April 1971 at the 2400-m level (first cable car station) by R. Funiciello, University of Rome.

Grain Size: There is no grain-size data available for this sample. However, McGetchin et al. (1974) note that ash samples from the 1971 vents have median sizes of 100 μm to 1 mm. Ash falls from the explosion crater of 1971, collected 850 m NE of the crater, have median grain sizes of 110 μm (3.2 φ) and 125 μm (3.0 φ) and are moderately sorted ($\sigma_\varphi = 1.25\ \varphi$) (Booth and Walker, 1973).

Tephra Description: The sample is a vitric-lithic ash that consists of mostly angular, poorly vesicular, blocky sideromelane fragments. Also present in this ash are blocky tachylite pyroclasts that exhibit both aphanitic and diktytaxitic textures (the latter texture consists of interwoven, finely crystalline feldspar and pyroxene grains (1- to 2-μm long and 0.2- to 0.3-μm wide) and irregular, broken, vesicular sideromelane droplets. Void spaces between crystalline phases in the tachylite pyroclasts occupy 30 to 50% of the rock by volume. There are also smaller amounts of broken droplets, glassy agglutinates, and clinopyroxene and plagioclase pyroclasts.

Chemical Composition: The chemical composition of a 1971 Etna lava is shown in Table 1.

TABLE 1. CHEMICAL COMPOSITION OF ETNA LAVA.*

	(Wt. %)
SiO_2	47.01
Al_2O_3	17.21
Fe_2O_3	6.83
FeO	4.44
MgO	5.54
CaO	10.77
Na_2O	3.40
K_2O	1.72
TiO_2	1.66
H_2O	0.11
H_2O^-	0.18
TOTAL	98.87

*Lava flow from upper east vent, May 4, 1971 (from Tanguy, 1971); Analyst, A. Lenoble, University of Paris.

Comments on Pyroclast Formation: Published observations concerning the 1971 eruption indicate that activity along the fissures was Strombolian and that chains of cinder cones were formed. Most of the pyroclasts in this sample indeed appear to be of Strombolian origin—irregular, broken sideromelane droplets and tachylite pyroclasts. There are, however, many pyroclasts that appear to be of phreatomagmatic origin—blocky, sideromelane clasts with low vesicularities. This suggests that minor phreatomagmatic explosions were mixed with Strombolian activity and confirms Rittmann's (1971) observation that there may have been lava/water interactions caused by deep snow cover on the mountain.

REFERENCES: Booth, B. and Walker, G. P. L., 1973. Ash deposits from the new explosion crater, Etna, 1971. Phil. Trans. R. Soc. Lond., **A. 274**: 147-151.

McGetchin, T. R., Settle, M., and Chouet, B. A., 1974. Cinder cone growth modeled after Northeast Crater, Mount Etna, Sicily. J. Geophys. Res., **79**: 3257-3272.

Rittmann, A., 1971. The Mt. Etna volcanic eruption of 1971. Smithsonian Inst. Center for Short-Lived Phenomena, Cambridge, Massachusetts, 9 July, 1971: 23 pp.

Tanguy, J. C., 1971. Mt. Etna volcanic eruption, 5 April—12 June 1971. Smithsonian Inst. Center for Short-Lived Phenomena, Cambridge, Mass., 15 December, 1971: 5 pp.

FIGURE 1. Vitric-lithic basaltic ash from the 1971 eruption of Mt. Etna. Most pyroclasts are angular, blocky, and subequant with low vesicularities. Pyroclasts with relatively smooth fracture surfaces are sideromelane; scalloped surfaces are broken vesicles (30- to 250-μm-long vesicles). Particles with rough surfaces are mostly tachylite fragments. The lumpy particle in the lower left corner of the photo is an agglutinate made up of welded sideromelane droplets.

FIGURE 2. There are mostly angular sideromelane pyroclasts in this photograph. Both smooth and conchoidal fracture surfaces are visible. Outermost vesicles are quite broad and flattened (disc shaped). The shard in the lower left corner of the photo appears to have been broken from a highly vesicular foam with large (>500-μm), well-developed vesicles. Distribution and the size of vesicles in these pyroclasts is heterogeneous.

FIGURE 3. Subangular sideromelane pyroclast. This particle has a low vesicularity, but a few flattened vesicles are visible at the surface. In the center of the pyroclast is a small phenocryst that has deformed the wall of a vesicle.

FIGURE 4. A mixture of sideromelane and tachylite pyroclasts. *Left bottom* - Broken sideromelane droplet; smooth, curved surface and few vesicles. *Left top* - Broken, highly vesicular sideromelane fragment with large vesicles. *Right top and bottom* - Subangular to subrounded tachylite pyroclasts, with diktytaxitic texture (*top*) and aphanitic texture and hackly fracture surfaces (*bottom*).

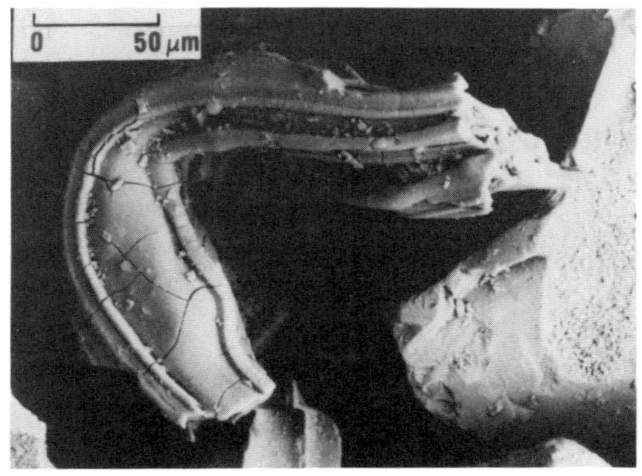

FIGURE 5. Twisted sideromelane droplet. The ribs are over elongate vesicles. Note the vesicles on broken ends of the droplet. This could be classified as Pele's hair.

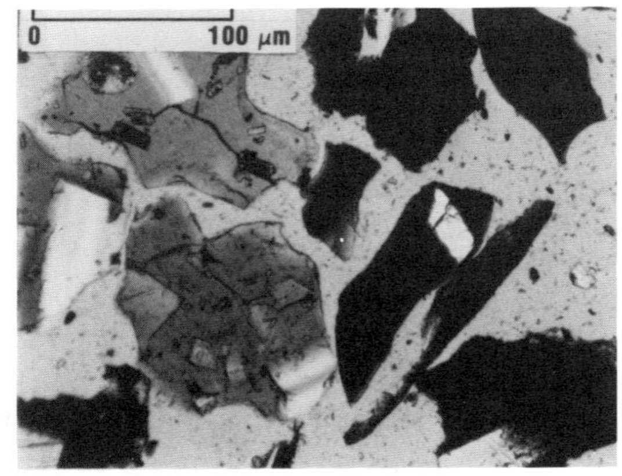

FIGURE 6. Photomicrograph of Etna tephra. On the *right* are mostly broken tachylite droplets and angular, blocky clasts. On the *left* are hyalocrystalline, phenocryst-bearing, blocky, sideromelane phenocrysts.

Vulcanian Eruptions

This activity is typified by the historic, intermittent explosions of ash from Vulcano, Aeolian Islands, Italy. The distinguishing characteristics of this type of ash are not well defined and reflect a mixture of magmatic and phreatomagmatic eruption mechanisms. Typically, Vulcanian ash is composed of varying amounts of juvenile and nonjuvenile fragments. The nonjuvenile glass pyroclasts are generally very angular, reflecting their origin by explosive pulverization of older lava flows or plugs within the vent. Juvenile glass particles may also be vesicular or blocky, depending upon the amount of magmatic and meteoric water involved in the eruption. Vulcanian eruptions can occur at silicic domes where viscous lava is extruded. During explosions, both magmatic and phreatomagmatic, much of the preexisting vent plug as well as new lava may be shattered by shock waves generated by rapidly expanding gases. Vulcanian ash deposits often resemble those of phreatomagmatic eruptions and the reader is refered to that section in this book for further discussion of depositional textures.

VOLCAN FUEGO, GUATEMALA; 1966 ERUPTION

Introduction: Volcan Fuego is a large, composite cone in Guatemala; its activity has been characterized over the last 450 years by frequent, short-lived eruptions. Much of the activity during the last several decades has consisted of extremely energetic Vulcanian eruptions of sulfur-rich, high-alumina basalt (Anderson, 1975; Rose, 1977). The short but intense eruptions of the last 20 years have frequently sent eruption plumes into the stratosphere; these have often been accompanied by small pyroclastic flows that travel down topographic lows on the volcano's flanks.

Activity at Fuego generally fits descriptions of Vulcanian activity: "moderate to violent ejection of solid or very viscous hot fragments of new lava" (Macdonald, 1972). Abundant lithic clasts are also ejected by these eruptions. The extremely energetic eruptions may be related to a high volatile content (Harris, 1979; Rose, 1977).

Grain Size: The sample collected 4 km N of Fuego's active vent, from the 1971 eruption, has a median grain size of 2.5 mm (-0.3 φ) and is moderately sorted ($\sigma_\varphi = 1.3\ \varphi$). Downwind, median grain size decreases from >0.5 mm (within 40 km) to <0.1 mm (beyond 120 km) (Rose et al., 1973).

Tephra Description: The ash mostly consists of angular, fractured, dark brown sideromelane fragments with 30% tachylite and phenocryst fragments. None of these pyroclasts has the smooth grain surfaces typical of tephra from Hawaiian and Strombolian eruptions.

Most vitric pyroclasts are equant to slightly elongate with hackly or conchoidal grain surfaces. A few irregular patches of smooth glass are preserved on some fragments. Tachylite pyroclasts have very irregular, hackly surfaces. All pyroclasts have few vesicles; most have fewer than 30% vesicles, and these are deformed ovoids ranging from 10 to 40 μm.

Chemical Composition: Microprobe analyses of glass inclusions in crystalline phases of tephra from the 1974 eruption of Fuego indicate a sulfur content of 2500 ppm (Harris, 1979; Rose, 1977). The bulk chemical composition of some basaltic ash from Fuego is shown in Table 1.

TABLE 1. CHEMICAL COMPOSITION OF HIGH-ALUMINA BASALTIC ASH FROM THE APRIL 22, 1966 ERUPTION OF FUEGO.*

	(Wt. %)
SiO_2	49.42
Al_2O_3	19.09
Fe_2O_3**	10.61
MgO	6.73
CaO	10.07
Na_2O	3.09
K_2O	0.60
TiO_2	0.87
P_2O_3	0.41
TOTAL	100.83

*A. K. Baird, analyst (from Stoiber and Rose, 1970).
**Total Fe expressed as Fe_2O_3.

Comments on Pyroclast Formation: Little is known about the formation of tephra in Vulcanian eruptions of basaltic magma. Viscosities at the time of eruption are fairly high, as is indicated by the blocky, fragmented nature of the tephra. The large percentage of pyroclasts consisting of basaltic glass shows that the tephra were rapidly quenched. Although the original volatile content of the magma was high, as indicated by analyses of glass inclusions (Harris, 1979), vesicularities of the vitric pyroclasts are quite low. A high, pre-eruption magma viscosity suppressed vesicle growth. It is possible that total pressure within those vesicles that developed was very high, resulting in the short but energetic eruptions. Pyroclasts may have formed during passage of a very pronounced pressure gradient down from the vent surface into a somewhat vesicular melt.

REFERENCES: Anderson, A., 1975. Some basaltic and andesitic gases. Rev. Geophys. Space Phys., **13**: 37-53.

Harris, D., 1979. Pre-eruption variations of H_2O, S, and Cl in a subduction zone basalt. Trans. Am. Geophys. Union, **60**: 968.

Macdonald, G., 1972. *Volcanoes.* Prentice-Hall, Inc., Englewood Cliffs, N.J.: 510 pp.

Rose, W. Jr., 1977. Scavenging of volcanic aerosol by ash: atmospheric and volcanologic implications. Geology, **5**: 621-624.

Rose, W. Jr., Bonis, S., Stoiber, R., Keller, M., and Bickford, T., 1973. Studies of volcanic ash from two recent Central American eruptions. Bull. Volcanol., **37**: 338-364.

Stoiber, R. and Rose, W. Jr., 1970. The geochemistry of Central American volcanic gas condensates. Geol. Soc. Am. Bull., **81**: 2891-2912.

FIGURE 1a. Sideromelane (basaltic glass) pyroclasts. Subequant, angular pyroclast with low vesicularity. There are a few patches of smooth outer skin (*left side*, *right center*), but most surfaces are conchoidal or hackly.

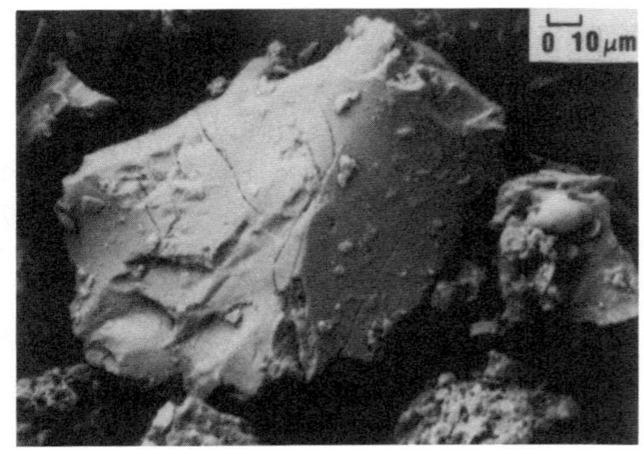

FIGURE 1b. Nonvesicular sideromelane pyroclast with a smooth surface that is broken in places by cracks and conchoidal spalls.

FIGURE 2a. Vitric and crystal pyroclasts. Angular vitric pyroclast with a variety of small, angular, vitric and crystal clasts adhering to the surface. Note the twin planes in plagioclase pyroclast (*upper left*).

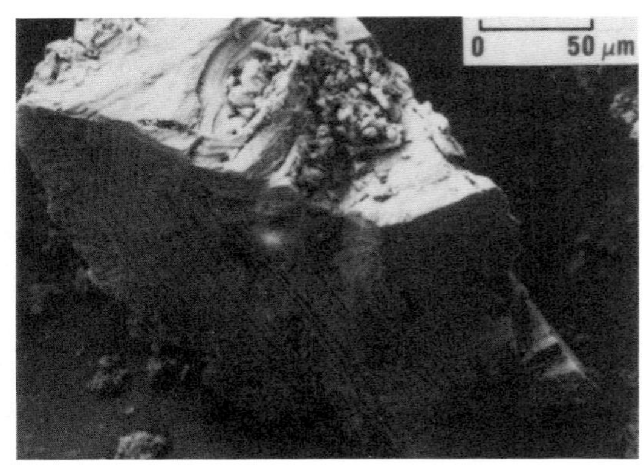

FIGURE 2b. Fractured plagioclase phenocryst. This pyroclast is free of glass (with the exception of aggregates of fine-grained pyroclasts loosely adhering to the surface). Note the complexity of oscillatory zoning that is visible on the face of this phenocryst.

FIGURE 3. Aggregate of the fine-grained fraction of 1966 Fuego tephra. Although mostly very fine grained (<2-µm) angular clasts, there are several rounded glass droplets visible: a rod and small (2-µm) droplet on the right and some delicate, irregular droplets throughout the matrix.

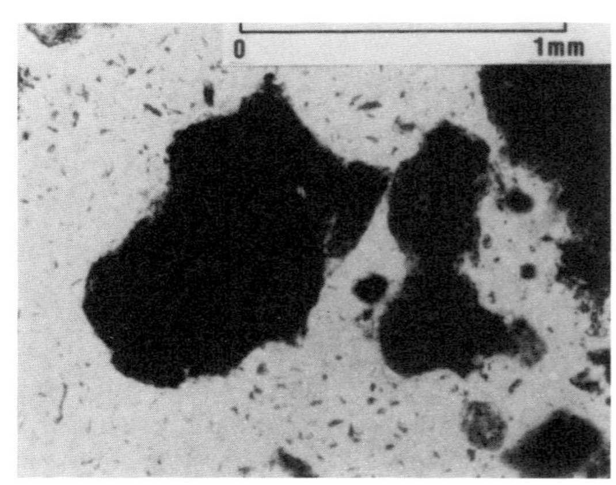

FIGURE 4. Transmitted-light photomicrograph of Fuego tephra. On the *left* is an angular, nonvesicular tachylite pyroclast, and on the *right* is a pyroclast with a tachylite core and sideromelane at the surface.

VULCANO, ITALY; 1888-1890 ERUPTION

Introduction: Vulcano is the southernmost active volcano in the Aeolian archipelago; it is located north of Sicily in the Tyrrhenian Sea (Keller, 1980). Numerous explosive eruptions have occurred at Vulcano in historic times; the most recent occuring from 1888 to 1890 during which Mercalli and Silvestri (1891) made classical observations of Vulcanian activity. Characteristic activity consisted of staccato explosions followed by quiet periods that lasted several minutes to several days. Initial explosions emitted clasts of solidified older lava derived from the vent; later explosions ejected juvenile blocks, bombs, and ashes.

Characteristically, Vulcanian explosions produce a vertically directed eruption cloud that may reach elevations of up to several kilometers. Base surges may emanate from this eruption cloud as directed blasts accompanied by ballistic fall of material onto the slopes of the cone. These eruptions produce coarse layers of ash fall that are interbedded with layers of fine surge ash on the cone flanks. Lahars develop near the base of the cone, usually as the result of heavy rains just after an eruption. The collapse of the eruptive column may occasionally produce pyroclastic flows. Sheridan et al. (1981) and Frazzetta et al. (1983) have described and modeled other eruptive phenomena. Samples described in this report were taken from the 1888 to 1890 eruptive phase (Sheridan and Marshall, 1981).

Grain Size: Size analyses (Table 1) of the most recent eruption products give the average median diameter and sorting for principal bed forms on the flanks of the volcano. These bed forms are: (1) sandwave- and massive-surge beds; (2) fall, coarse and fine layers (sets of fall and surge ash in one depositional unit); and (3) wet-surge deposits that have bed forms transitional between surges and lahars. Most samples consist of two populations, with dominant modes at 0.5 to 1.5 φ and 2.5 to 3.5 φ. Ash falls, however, commonly show an additional mode in the range of -2.0 to -1.0 φ, and wet-surge deposits have a mode finer than 4.5 φ.

TABLE 1. GRAIN-SIZE ANALYSES.

Bed Form*	No.**	Md_φ (mm)	σ_φ
S	35	1.96 (0.26)	1.08
M	10	1.63 (0.32)	1.03
WS	10	2.92 (0.13)	1.26
F	4	1.26 (0.42)	1.40
C + F	6	1.62 (0.32)	1.12

*Bed forms are: S - sandwave surge, M - massive surge, WS - wet surge, F - fall, and C + F - coarse and fine unit.
**Number of samples analyzed.

Tephra Description: The 1888-1890 eruption products form a gray to dark gray mantle of pyroclastic deposits resting on the lighter colored slopes of the Fossa cone. These deposits consist of fresh glass and crystal contituents that show little secondary alteration. The lithic contributuion to the 1888-1890 deposits is small and is not discussed here.

Glass. Glass particles are poorly vesicular to almost nonvesicular. Although the grains are blocky in appearance, they have sharp, acute edges that have suffered little abrasive action. Solution and precipitation effects on some grains, however, give an uneven appearance to curviplanar surfaces. For the most part, grains have broad, planar faces and dish-shape concavities that are typical of fragmentation of the viscous, silicic magma; smooth, ropy surfaces of the more fluid, basic magma are rare.

Crystals. Phenocrysts of clinopyroxene, calcic plagioclase, and olivine are entirely free of glass coatings. Crystal and cleavage faces are modified by stepped fractures and some shallow, dish-shaped concavities. These surface modifications appear to be mostly cleavage controlled; the grains are angular, and their sharp corners show little effect from transport rounding. A small amount of sugary coating appears on some crystal surfaces and may be related to solution and preciptation.

Chemical Composition: Frazzetta et al. (1983) suggest that two magmas of different compositions mixed before and during each of Vulcano's eruptive periods. One is an evolved aphyric rhyolite that was derived from a sanidine trachyte; the other appears to be a tephrite, as inferred from mafic melt inclusions with reaction rims in the rhyolite. This mafic component is also represented by phenocrysts of clinopyroxene, calcic plagioclase, olivine, and magnetite and by the fluidal-shaped pyroclasts mentioned earlier.

Comments on Pyroclast Formation: The blocky, nonvesicular nature of most of the 1888-1890 pyroclasts and their mixed magma glass and phenocryst compositions has led to a complex eruption model proposed by Frazzetta et al. (1983). In that model, eruptive cycles at Vulcano are initiated by tectonic movement, which promotes mixing of the evolved rhyolite and tephrite magmas. The mixing causes viscosity changes and volatile separation that allow the mixed magma to move to the surface. When magma reaches the water-saturated zone beneath the volcano, hydrofracture occurs and promotes the contact of magma and water, which leads to hydrovolcanic surge-producing eruptions. As these eruptions proceed, water access to the vent decreases and vesiculation of the magma begins because of the reduced confining pressure above the magma column. Degassing of the magma by vesiculation causes eruption of pumice falls. After the magma column has degassed, a viscous, silicic, and often trachytic lava flow plugs the vent.

Repetition of this cycle begins with the initial phreatic eruptions that fracture and expel the vent-blocking obsidian flow—"ultra-Vulcanian activity." The stratigraphic column in Fig. 1 illustrates the Vulcanian eruption cycles for recent activity at Vulcano. Characterisitic "Vulcanian" activity produces a complex assemblage of pyroclast types that range from lithic materials of fragmented old lavas to hydrovolcanic materials, which consist of mixed magmas and pumiceous debris of vesiculated lava. This significant component of nonjuvenile material is characteristic of Vulcanian deposits in general. The 1888-1890 materials represent the middle-stage activity of the eruptive cycle at Vulcano.

REFERENCES:

Frazzetta, G., LaVolpe, L., and Sheridan, M. F., 1983. Evolution of the Fossa vent, Vulcano. In: Explosive Volcanism, M. F. Sheridan and F. Barberi (Eds.), J. Volcanol. Geotherm. Res., **17**: 329-360.

Keller, J., 1980. The island of Vulcano. Rend. Soc. Ital. Min. Petrol., **36(1)**: 369-414.

Mercalli, G. and Silvestri, O., 1891. Le eruzioni dell'isola di Vulcano, incominciate il 3 Agosto 1888 e terminate il 22 Marzo 1890. Relazione Scientifica 1891, Ann. Uff. Cent. Meteor. Geodin., **10(4)**: 213 pp.

Sheridan, M. F. and Marshall, J. R., 1981. SEM surface-textural analysis of volcanic ejecta: an interpretation of base-surge behavior on Vulcano. Int. Assoc. Sed., Proc. 2nd European Reg. Meet., Bologna, Italy: pp. 185-188.

Sheridan, M. F., Moyer, T. C., and Wohletz, K. H., 1981. Preliminary report on the pyroclastic products of Vulcano. Mem. Soc. Astron. Ital., **52**: 523-527.

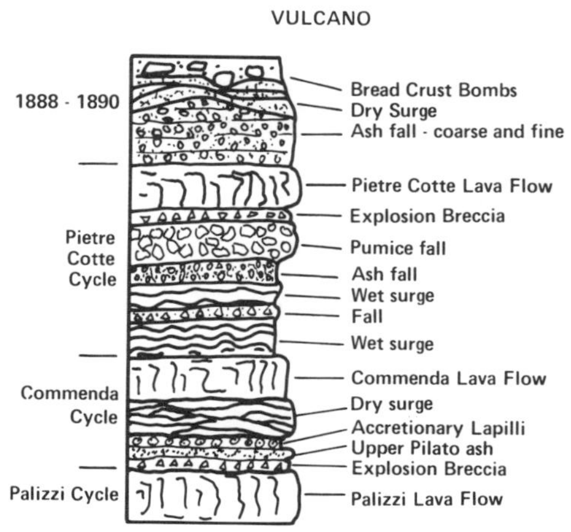

FIGURE 1. Stratigraphy of recent eruption cycles at Vulcano, adapted from Frazzetta et al. (1983) and Sheridan et al. (1981). Each cycle begins with pyroclastic eruptions that trend from hydrovolcanic to magmatic and end with lava-flow emplacement. Samples described are taken from the 1888-1890 deposits.

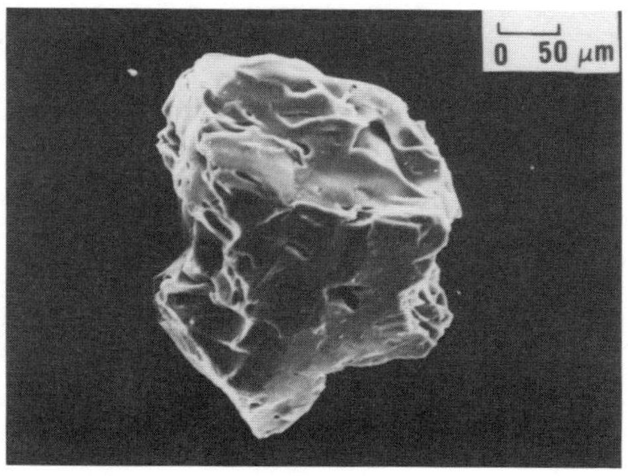

FIGURE 2. This SEM of a nonvesicular, blocky pyroclast shows nearly planar surfaces that have irregularities caused by poorly developed vesicles. The elongate, curving cracks of the hollow areas are likely caused by flow foliations in this obsidian, which developed when small vesicles stretched and bent during flow.

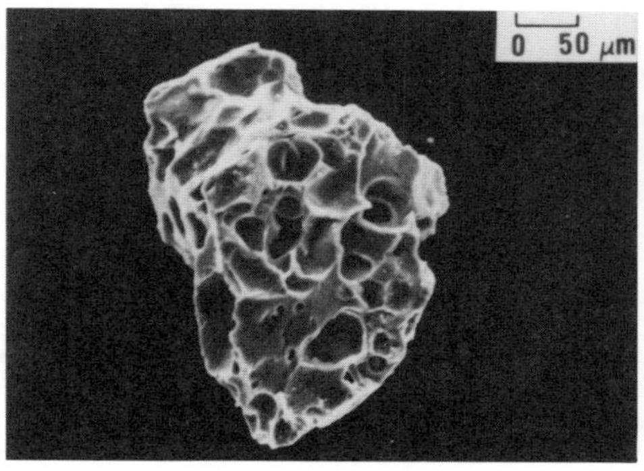

FIGURE 3. SEM of a vesicular pyroclast from magma of more mafic composition. Many of the vesicles are cut by curviplanar fractures as a result of hydrovolcanic fragmentation of the melt.

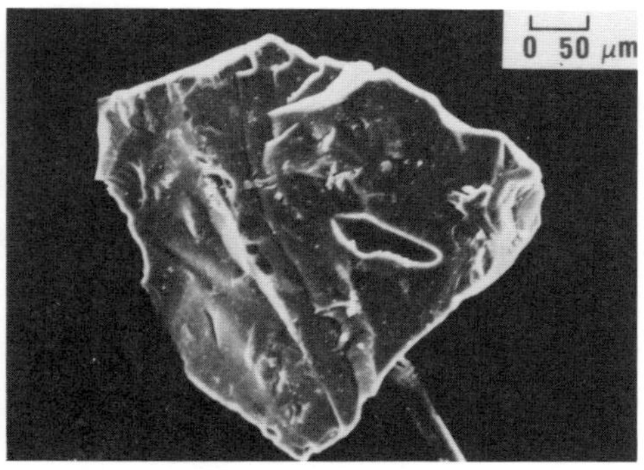

FIGURE 4. SEM of an obsidian pyroclast. The sharp, angular edges show that few chips and scratches developed during transport. Small sets of nearly parallel fractures are caused by conchoidal-like fracture behavior of the viscous glass.

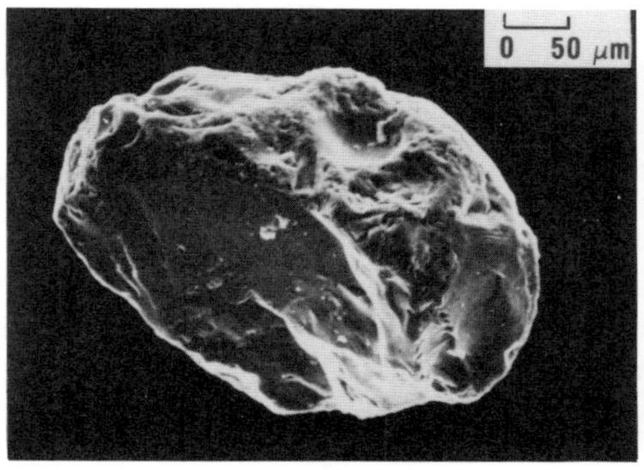

FIGURE 5. This SEM of a pyroclast shows two large, dish-shaped, curviplanar fractures and a frosted surface formed by solution and precipitation.

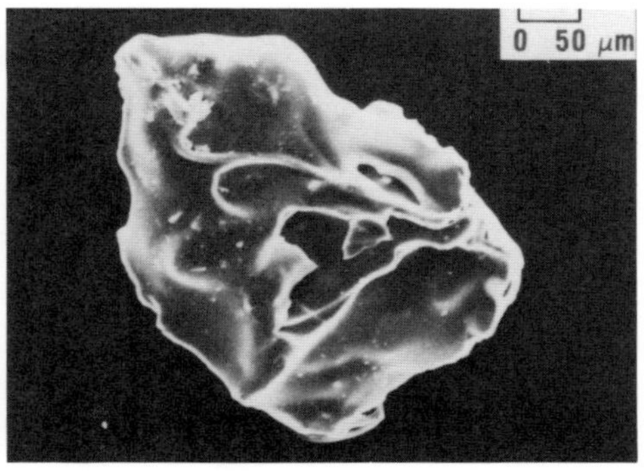

FIGURE 6. SEM of a fluidal-surfaced pyroclast of the mafic magma. Note the smooth, curved, and ropy skin texture, deformed vesicle embayments, and sharp edges of this grain.

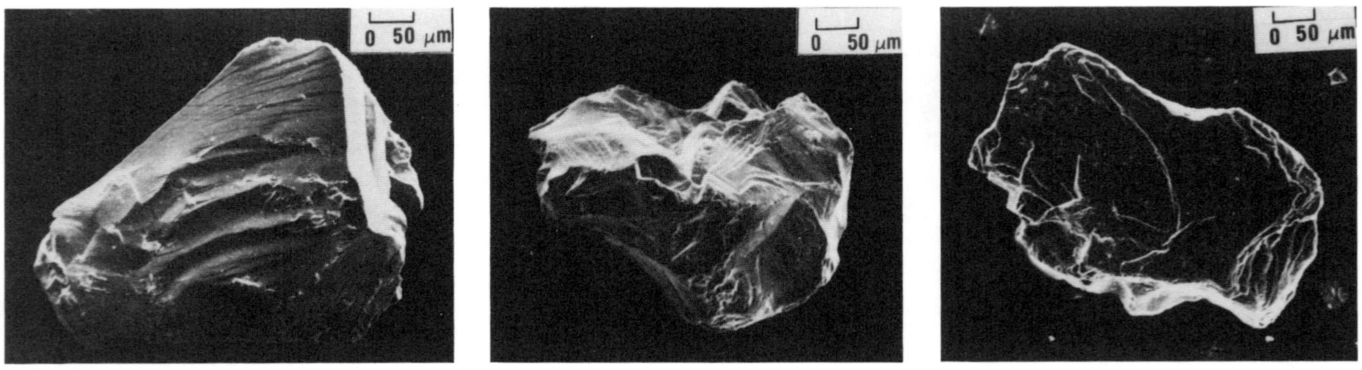

FIGURE 7. SEMs of crystals showing stepped and dish-shaped fracture faces dominantly related to cleavage planes. Note the lack of a glassy coating and the sharpness of fracture surfaces.

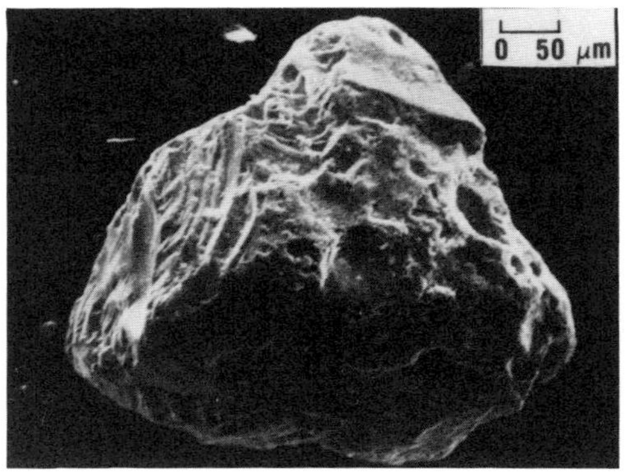

FIGURE 8. SEM of a partly frosted crystal; this grain shows solution pitting and precipitation coating on one side. The uneven surface has a sugary appearance that may be the result of contact with a grain of different composition.

MAYON VOLCANO, PHILIPPINES; 1968 ERUPTION

Introduction: Mayon is a symmetrical composite cone with an elevation of 2462 m. It has had 45 eruptions since 1616—all explosive and many with pyroclastic flows (Simkin et al., 1981). The 1968 eruptions were described by Moore and Melson (1969).

All eruptions were from the summit crater. Incandescent blocks and ash were ejected to a height of 600 m above the crater. Collapse of the eruption column was followed by pyroclastic flows in ravines on all sides of the volcano. Buoyant ash clouds rose to heights of 3 to 10 km (Moore and Melson, 1969). Ash falls, in measurable quantities, covered several hundred square kilometers; at distance, the ash is fine grained. Accretionary lapilli are present in the thicker ash fall deposits (Moore and Melson, 1969).

The sample examined here was supplied by A. Alcarez of the Philippine Commission on Volcanology.

Grain Size: The median grain size of six samples from pyroclastic flows ranges from 0.4 to 1.1 mm (1.3 to -0.1 φ); all are poorly sorted (Moore and Melson, 1969). The sample examined by the authors (distal ash fall) has a median grain size of 90 μm and is poorly sorted.

Tephra Description: The crystal-lithic ash consists of broken augite, hypersthene and plagioclase phenocrysts, and equant to slightly elongate angular glass shards. Scoria pyroclasts make up a few per cent; they have low vesicularity and thin, brown, oxidized rims.

Most pyroclasts are angular and blocky with few or no vesicles and are the products of brittle fracture. A few per cent of the pyroclasts are droplets. They have smooth, curved surfaces formed by surface tension within melt droplets. Some are agglutinated and some have wrinkled surfaces.

Chemical Composition: The chemical composition of Mayon andesitic pyroclasts is shown in Table 1.

TABLE 1. CHEMICAL COMPOSITION OF MAYON ANDESITE.

	Ash*	Core of Breadcrust Bomb**
SiO_2	56.2	54.90
TiO_2	0.68	0.64
Al_2O_3	19.13	19.04
Fe_2O_3	7.81	2.96
FeO	---	4.81
MnO	0.20	0.15
MgO	3.71	4.05
CaO	8.54	8.62
Na_2O	4.15	3.40
K_2O	1.16	1.13
P_2O_5	0.28	0.26
S	0.12	n.d.
TOTAL	102.10	100.00

*M. Rhodes, analyst.
**E. Jarosewich, analyst (from Moore and Nelson, 1969).

Comments on Pyroclast Formation: The formation of this tephra is not well understood. Observations of the eruption by Moore and Melson (1969) indicate that explosions produced vertical ejections of incandescent material, much of which fell back into the slopes forming pyroclastic flows. These deposits were welded within 1 km of the vent and consisted of crystal-rich ash, lapilli, and scoriaceous blocks (with some breadcrust bombs). From the observations, all indications are that this eruption was similar to energetic lava fountaining of intermediate viscosity melts; thus the ash contains droplets similar to those produced by some Strombolian and Vulcanian eruptions. However, droplets form only a few per cent of the ash fall tephra. The vesicularity of pyroclasts is very low, which indicates that disruption of expanding gases within a gas-charged magma probably did not occur (the interpretation by Heiken, 1972, is probably not correct).

Perhaps heterogeneities within the magma are responsible for disruption and fragmentation of the magma. A rarefaction wave (associated with an atmospheric shock wave) could have passed through the rising magma, fracturing more viscous portions to form the fine ash and less viscous portions to form large blobs of molten material. Much of the finer grained material could have been winnowed out or carried higher in the eruption column and dispersed as ash falls. The molten blobs would have fallen back to form most of the agglutinated pyroclastic flow deposits. This hypothesis can account for the extremes in pyroclast types present in these deposits. Another possible cause is similar to that described for Vulcanian eruptions at Ngauruhoe, New Zealand (Nairn and Self, 1978): comminution of brittle material, mixed with molten ejecta occurs during the pressure build-up beneath a partly cooled, solid plug in the vent of a composite core. The heterogeneities are vertical in this case and could account for characteristics of the ash from Mayon.

REFERENCES:

Heiken, G., 1972. Morphology and petrography of volcanic ashes. Geol. Soc. Am. Bull., **83**: 1961-1988.

Moore, J. G. and Melson, W. G., 1969. Nuées ardentes of the 1968 eruption of Mayon Volcano, Philippines. Bull. Volcanol., **33**: 600-620.

Nairn, I. A. and Self, S., 1978. Explosive eruptions and pyroclastic avalanches from Ngauruhoe in February 1975. J. Volcanol. and Geotherm. Res., **3**: 39-60.

FIGURE 1. Equant to slightly elongate angular shards. A few grain surfaces are smooth and curved; they are broken across vesicles. Most are irregular, with numerous conchoidal fracture surfaces. Some of the smaller shards are wedge or needle shaped.

FIGURE 2. The whole spectrum of pyroclast types is visible in this SEM. The largest equant fragments (*center*) are crystal pyroclasts (plagioclase or pyroxene). Most grains are angular equant to subequant vitric pyroclasts with flat or conchoidal surfaces. A few droplets are present. Pyroclasts are nonvesicular or have rare vesicles although, in thin section, some of the small scoria clasts are vesicular.

FIGURE 3. A closer view of angular shards. One (*lower center*) is broken from the intersection of several vesicles. All others are nonvesicular. A broken droplet surface is visible (*lower right*). There are numerous equant, angular particles in the <5-μm range.

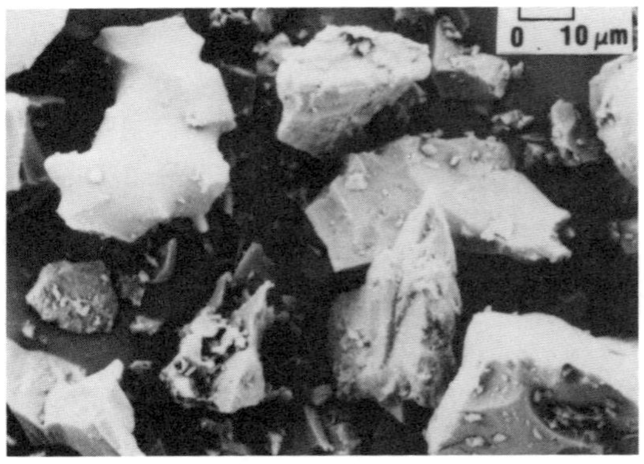

FIGURE 4. All except one pyroclast are subequant, angular shards. There is a broken droplet (*upper left corner*) whose smooth, curved surfaces are caused by flow after comminution. Most clasts are nonvesicular.

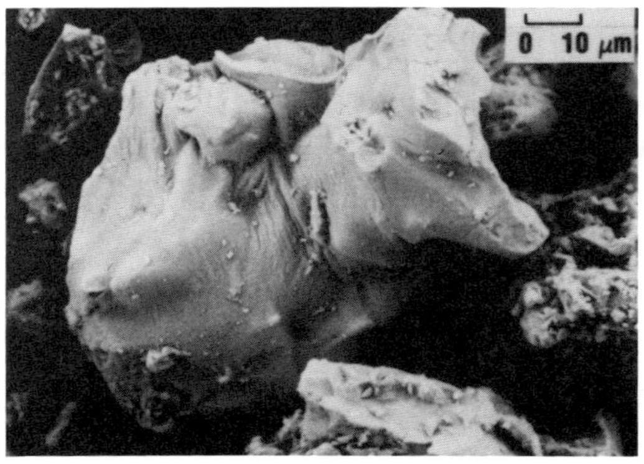

FIGURE 5. Broken droplet with a wrinkled "skin;" this occurs when the pyroclast surface has cooled while the interior was still molten. Another droplet was annealed to the surface (*top-center*) while still molten.

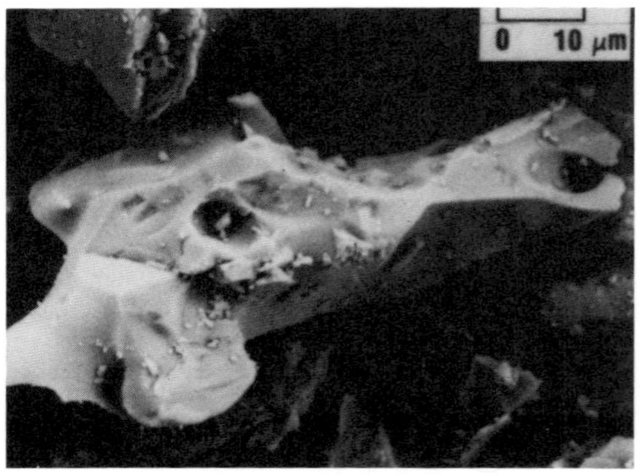

FIGURE 6. Angular, vitric fragment with only two well-developed vesicles visible in this image. It is typical of the vitric pyroclasts with very low vesicularities in this sample.

NGAURUHOE, NEW ZEALAND; 1974 ERUPTION

Introduction: Ngauruhoe Volcano is located in the central North Island of New Zealand. It is part of the Taupo volcanic zone, which extends 250 km SW-NE from Ruapehu to White Island and marks the continuation of the Tonga-Kermadec arc. Recent documented eruptive activity at Ngauruhoe from 1972 through 1975 was typical for that andesitic stratovolcano and consisted mostly of periodic, small Vulcanian explosions. The sample characterized here was donated by S. Self, who witnessed the explosive activity in 1974 (Self, 1975). Although this eruption was relatively small and produced only a small, thin fallout, Nairn (1976) and Nairn and Self (1978) documented highly explosive, cannon-like eruptions in 1975 that produced pyroclastic avalanche deposits.

This 1974 Ngauruhoe ash sample is described by Self (1975) as part of a 0.74×10^6 m^3 fallout (0.5×10^{-3} km^3 dense rock volume) deposited from a plume of fine ash that erupted to 3600 m above the vent and was dispersed by winds to the NW. Most of this deposit was less than several millimeters thick. By comparison, the 1975 eruption produced 2.4×10^6 m^3 of fallout and 1.0 to 1.5×10^6 m^3 of avalanche deposits from ash columns reaching nearly 11,000 m above the vent (Nairn and Self, 1978).

Grain Size: Table 1 lists grain-size characteristics of Ngauruhoe ash collected from the 1974 and 1975 eruptions. Ash particles from the 1974 eruption are generally less than 1 mm in diameter and consist dominantly of moderately sorted, fine ash. The more explosive eruptions of 1975 produced deposits that are much coarser grained; the fallout is well sorted and the avalanche deposits are very poorly sorted. It is likely that much of the fine material from the 1975 eruption was separated from the coarser ash in the high eruptive column before emplacement.

TABLE 1. SIZE CHARACTERISTICS OF NGAURUHOE 1974 AND 1975 DEPOSITS.*

Sample	Md_φ	(mm)	σ_φ
1974 (Fallout)	4 - 5	(0.03-0.06)	1.5
1975 (Fallout)	1.1	(0.46)	0.4
1975 (Avalanche)	−4 - 0	(16.0-1.0)	3.5

*Data taken from Self (1975) and Nairn and Self (1978).

Tephra Description: Mostly nonjuvenile, glassy pyroclasts are represented in the sample illustrated. The nonvesicular, blocky particles may be distinguished from vesicular juvenile ones by their vesicularity and a greater degree of surface alteration, which is manifested as rough surfaces. The rough surface texture is a thin coating of 1- to 5-µm particles and crusty material that gives grains a frosted appearance. Gross surface structure is marked by uneven or hackly fractures. Some surfaces are smooth where phenocrysts have been plucked during fragmentation. Other smooth surfaces are slightly curved and have stepped fractures that appear as faint striations; these features are characteristic of brittle breakage. Many of these nonjuvenile clasts show rounded, smooth edges that appear to be remelted. The v-shaped surfaces marking the edges of many of these grains resulted from subsequent abrasion of grains within the vent after fragmentation.

Juvenile glassy particles show mostly delicate prong-like edges of vesicle walls. The tips of the prongs also show rounding caused by abrasion. Vesicle surfaces are smooth and relatively free of alteration materials and adhering dust grains. Some solution pitting is evident with high magnification of vesicle surfaces, and a frosted appearance of the vesicle prongs and blades is also evident.

Small (10- to 25-µm) grains appear to be vesicular but also blocky; surfaces are planar, cutting across vesicle-wall projections. Fragmentation of these small grains likely involved vesicle bursts followed by brittle fracture. Grains less than 5 µm in diameter are rare.

Chemical Composition: As mentioned above, the 1974 fallout deposit is composed of both essential (juvenile) and accessory (nonjuvenile) components. Data from Self (1975) show that 3 to 4% of the ash is dark, glassy, slightly vesicular juvenile andesite. The nonjuvenile component is mainly from the pulverized 1954 andesite lava plug. This accessory ash is a nonvesicular labradorite-pyroxene andesite. Although bulk chemical analyses are not available for the 1974 material, in Table 2 chemical analyses of selected grain surfaces from that eruption are compared to those of the 1954 andesite lava and the White Island andesite. The variations in oxide abundances for 1974 ash, especially apparent for Al_2O_3, are likely the result of grain-surface alteration.

TABLE 2. CHEMICAL COMPOSITION OF NGAURUHOE ANDESITES.*

	1	2	3	4	5
SiO_2	56.23	58.87	59.87	55.90	62.16
TiO_2	0.30	0.60	0.67	0.76	0.71
Al_2O_3	22.78	17.65	14.96	16.90	14.32
$FeO + Fe_2O_3$	5.07	8.20	10.22	8.40	5.97
MnO	---	0.01	---	0.15	0.15
MgO	3.59	4.23	5.20	5.20	3.97
CaO	7.58	5.29	3.66	8.40	5.89
Na_2O	3.64	3.59	3.40	2.60	2.49
K_2O	0.79	1.53	2.03	1.00	2.37
Total	99.98	99.97	100.01	99.31	98.01

*Analyses 1, 2, and 3 are standardless energy dispersive spectral values of 1974 ash grain surfaces; analysis 4 is of a 1954 andesite lava from Ngauruhoe (Steiner, 1958); and analysis 5 is a hypersthene andesite from White Island (Black, 1970).

Comments on Pyroclast Formation: A high contribution of nonjuvenile ash is typical of Vulcanian eruption products. Self (1975) explained that the 1974 ash was the result when new magma degassed in the volcano's vent conduit. Pulverization of the older lava plug contributed to the ash volume. Nairn and Self (1978) showed that the crater produced in the 1975 eruption required an excavation energy equivalent to about 5 kt of high explosive. At 1000°C and 2 kbar (7.5-km burial depth), water has nearly the same energy density as high explosive. For a phreatomagmatic mechanism of ash formation and ejection, at least 5 kt of water (probably several times more) must have been vaporized to produce the observed crater; this figure is well within annual precipitation limits for the Ngauruhoe cone.

A more recent evaluation of Vulcanian eruption at Ngauruhoe by Self et al. (1979) explains the characteristic volumes and velocities of ash expelled during discrete cannon-like explosions (5×10^5 m^3 dense rock at 400 to 600 m s^{-1}). From geometric considerations, 4 to 6 wt. % water was necessary for explosions; this suggests that both magmatic volatiles and small amounts of vaporized ground water were involved; however, the high pressures (3 to 5 kbar) characteristic of some magma/water interactions were not required for the explosions. Vapor pressures of 100 to 200 bars, which correspond to the tensile strength of an andesite plug several tens of meters thick, satisfy the volume constraints; this indicates that ground water heated in pores and fractures constituted the small phreatic component.

Because most of the Vulcanian sample illustrated here represents fragmented nonjuvenile lava, it is difficult to assess the eruptive contribution of ground water. Both vesicular and nonvesicular clasts are present. Those that appear to be the youngest and freshest are vesicular; however, there is clear evidence that curviplanar fractures, characteristic of ground water mixing with magma, do form surfaces on some vesicular grains.

REFERENCES:

Black, P. M., 1970. Observations of White Island volcano, New Zealand. Bull. Volcanol., **34**: 158-167.

Nairn, I. A. 1976. Atmospheric shock waves and condensation clouds from Ngauruhoe explosive eruptions. Nature, **277**: 440-443.

Nairn, I. A. and Self, S., 1978. Explosive eruptions and pyroclastic avalanches from Ngauruhoe in February 1975. J. Volcanol. Geotherm. Res., **3**: 39-60.

Self, S., 1975. Explosive activity of Ngauruhoe, 27-30 March 1974 (note). N. Z. J. Geol. Geophys., **18(1)**: 189-195.

Self, S., Wilson, L., and Nairn, I. A., 1979. Vulcanian eruption mechanisms. Nature, **277**: 440-443.

Steiner, A., 1958. Petrologic implications of the 1954 Ngauruhoe lava and its xenoliths. N. Z. J. Geol. Geophys., **1**: 325-363.

FIGURE 1. SEM of glassy nonjuvenile pyroclasts. These particles are mostly nonvesicular and blocky, with many angular surfaces. These particles may have been formed when an older lava plug was pulverized within the vent during eruption.

FIGURE 2. SEM of nonjuvenile pyroclasts that show frosted surface textures caused by a thin alteration coating of hydrated glass. Some curviplanar fracture surfaces have tiny adhering particles, whereas others are relatively clean.

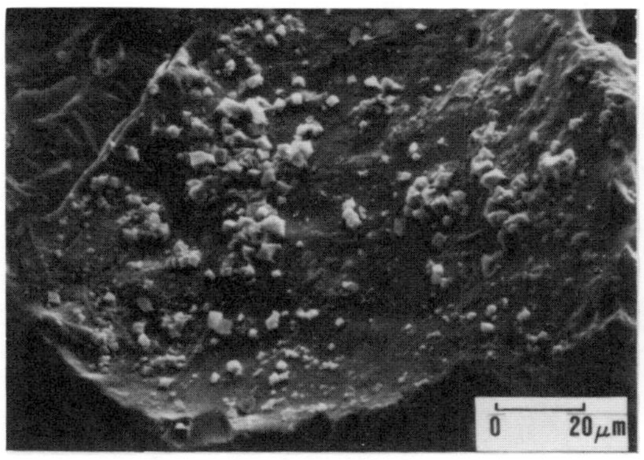

FIGURE 3. SEM closeup of a grain surface from Fig. 2. Small particles on the grain surface are probably crystalline alteration materials of a zeolite structure.

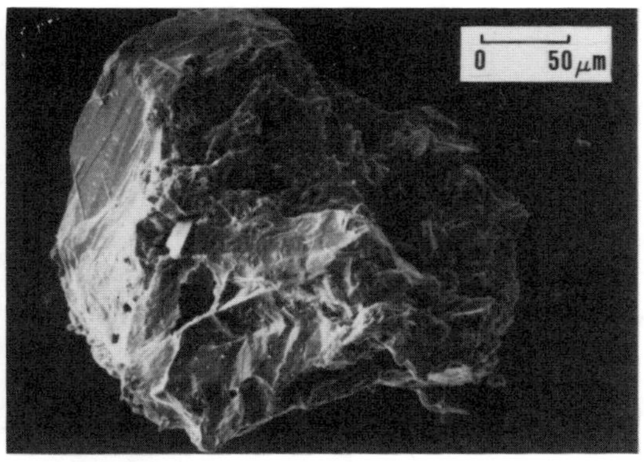

FIGURE 4. SEM of a nonjuvenile particle shows a large crystal mold near a smaller crystal still embedded in the glass.

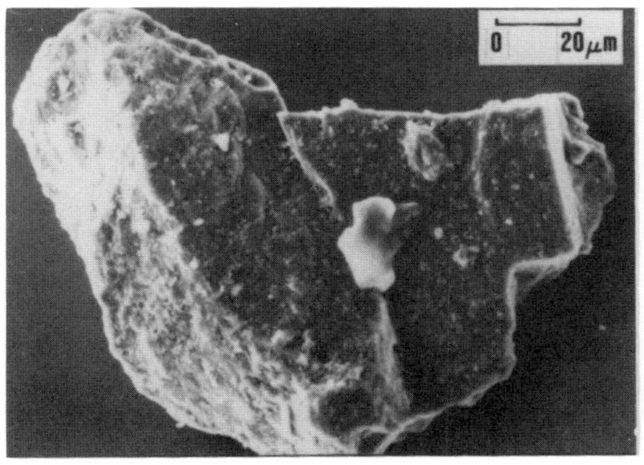

FIGURE 5. SEM showing frosted surface of nonjuvenile, twinned, clinopyroxene crystals. Note the rounded, abraded edges, rough surfaces, and fused piece of glass stuck to the crystal surface.

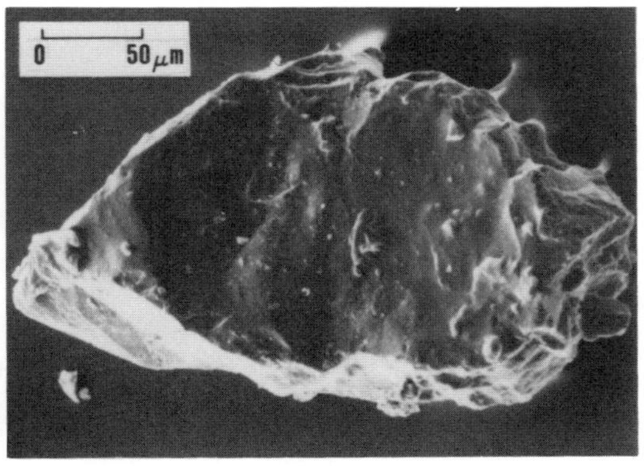

FIGURE 6. SEM of a glassy nonjuvenile particle that shows smooth surfaces indicative of remelting. Angular edge fractures have also been subdued by remelting.

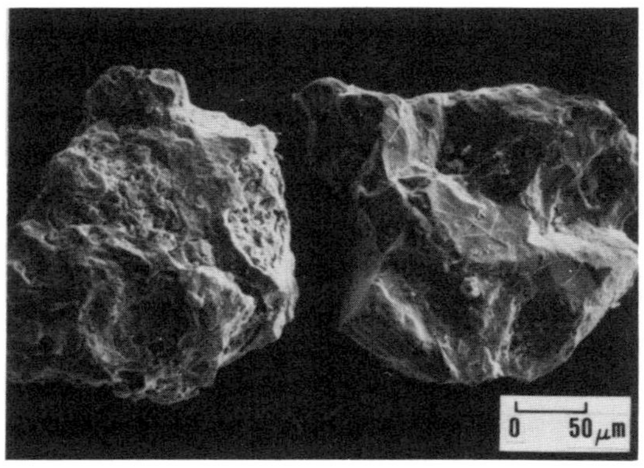

FIGURE 7. SEM of two glassy, nonjuvenile particles. Particle on the left shows extensive alteration and hydration cracks, whereas the other grain shows less alteration. Small secondary crystals form needle-like structures on some surfaces of the less altered grain.

FIGURE 8. SEM field of view showing a few vesicular juvenile particles with other nonjuvenile ones. Juvenile pyroclasts constitute only a few per cent of the total population of this sample.

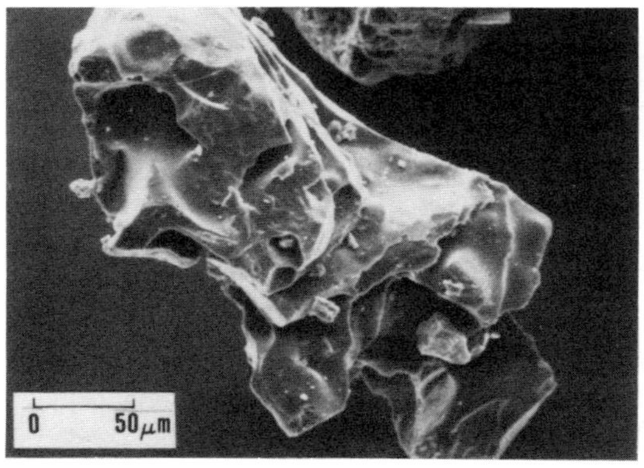

FIGURE 9. SEM of a glassy, juvenile pyroclast with characteristic fresh and embayed surfaces and delicate prong- and blade-like edges.

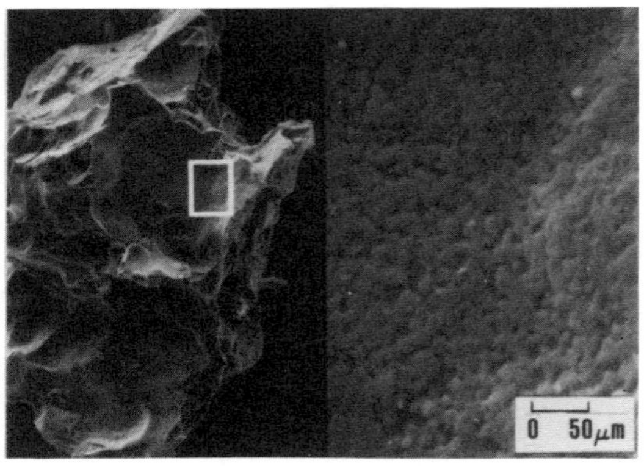

FIGURE 10. SEM of a glassy juvenile particle with a 10X enlargement of the vesicle surface. The enlargement details surface pitting caused by solution and precipitation of amorphous material.

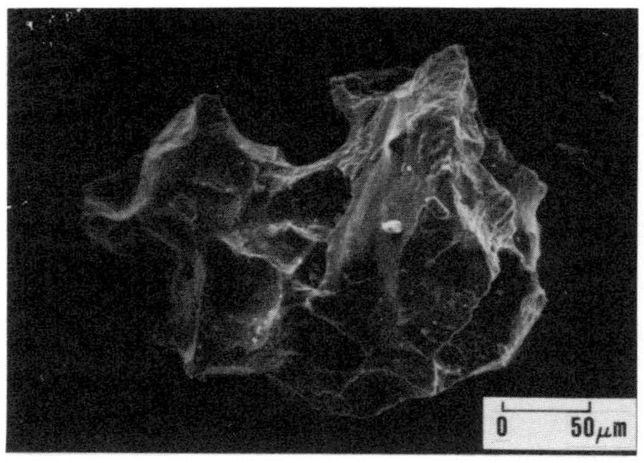

FIGURE 11. SEM of a juvenile particle that shows rounding and chipping of vesicle edges from abrasion.

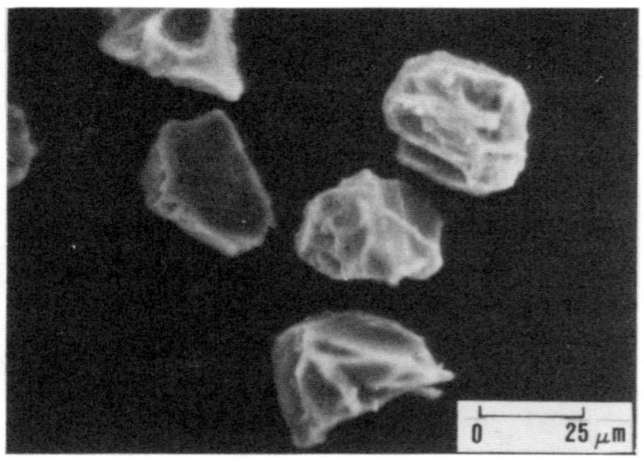

FIGURE 12. SEM of fine-fraction, glassy juvenile particles. These partly vesicular grains are blocky, with surfaces that cut across vesicle embayments; this indicates that brittle fracture occurred after an initial fragmentation caused by vesicle burst.

SANTIAGUITO, GUATEMALA

Introduction: Santiaguito is a dome complex in southwestern Guatemala that consists of dacite domes and flows. Its eruption, which began in 1922, followed the catastrophic eruption of Volcán Santa Maria in 1902 by 20 years and has been the site of constant volcanic activity since that time (Rose, 1972; Rose et al., 1976).

Accompanying the slow growth of this dome complex are sporadic small explosions from several craters located near or over vent areas (Caliente and El Brujo vents). Small ash clouds from these vents rarely exceed 2 km above the vent but occasionally may reach an altitude of 5 km above sea level (Rose et al., 1980). In 1973, a small nuée ardente issued from both the El Brujo and Caliente vents (Rose et al., 1976).

This sample was a fresh ash fall from the Caliente vent and was collected July 28, 1970. Rose et al. (1980) described Santiaguito tephra that was sampled by flying through eruption clouds.

Grain Size: This tephra has a mean grain size of 0.5 mm (1φ) and is very poorly sorted. The grain size is similar to that of nuée ardente deposits from Santiaguito mentioned by Rose et al. (1976). The ash clouds, sampled by aircraft, have a bimodal distribution, with peaks at 2 to 10 μm and 0.2 to 0.8 μm (Rose et al., 1980).

Tephra Descriptions: The ash consists of mostly hyalocrystalline, equant, blocky dacite fragments. Most grain surfaces have hackly to conchoidal surfaces and have low vesicularities or are nonvesicular. Some are glassy enough to be classified as vitric pyroclasts, but most have abundant phenocrysts and microlites in a glassy matrix. There are 25% crystal pyroclasts that are mostly plagioclase with lesser amounts of quartz. Individual phenocrysts are often fractured and in some cases broken along twinning planes. The more glassy pyroclasts are vesicular with sparse, but large, irregular elongate vesicles.

Rose et al. (1976), in their examination of tephra from Santiaguito nuée ardente, found a similar variety of pyroclasts. However, there was a larger fraction of pumiceous pyroclasts in the nuée ardente deposits than in ash fall from routine small explosions of the dome complex. Angular, nonvesicular particles are predominant in eruption cloud samples.

Chemical Composition: Table 1 is a chemical comparison of the slightly more silica-rich Caliente Dome lava with that of the ash fall.

TABLE 1. CHEMICAL COMPOSITION OF SANTIAGUITO ASH AND LAVA.

	Ash Fall (7/28/70)*	Lava (Caliente Dome)**
SiO_2	62.44	64.3
TiO_2	0.68	0.36
Al_2O_3	16.69	16.8
Fe_2O_3	2.98	1.6
FeO	2.08	2.4
MnO	0.20	n.d.
MgO	2.48	1.48
CaO	5.02	4.5
Na_2O	4.56	4.9
K_2O	1.54	1.70
P_2O_5	0.20	0.17
S	0.07	n.d.
Total	98.75	98.63

*M. Rhodes, analyst.
**From Rose (1972).

Comments on Pyroclast Formation: Most of the pyroclasts were formed by vesiculation and disintegration of highly viscous magma near vent areas. The same process occurs occasionally near the ends of blocky dacite flows, when it generates nuées ardentes (Rose et al., 1976). Vesiculation and fragmentation by pressure release may be the main cause of tephra formation. However, as the region is characterized by high rainfall, it is possible that there may be some magma/water interaction, which results in phreatomagmatic eruptions.

REFERENCES: Rose, W. I., Jr., 1972. Santiaguito volcanic dome, Guatemala. Geol. Soc. Am. Bull., **83**: 1413-1434.

Rose, W. I., Jr., Pearson, T., and Bonis, S., 1976. Nuée ardente eruption from the foot of a dacite flow, Santiaguito Volcano, Guatemala. Bull. Volcanol., **40**: 1-16.

Rose, W. I., Jr., Chuan, R., Cadle, R., and Woods, D., 1980. Small particles in volcanic eruption clouds. Am. J. Sci., **280**: 671-696.

FIGURE 1. Blocky, angular, hyalocrystalline dacite pyroclasts typical of tephra from Santiaguito. Vesicularity is usually less than 5%. Conchoidal or hackly fracture surfaces are common. The pyroclast with smooth, planar fracture surface (*lower left*) is a plagioclase phenocryst.

FIGURE 2. Lithic pyroclast with smooth, hackly, and conchoidal fracture surfaces.

FIGURE 3. Hyalocrystalline dacite pyroclast. Most grain surfaces are blocky and angular, but one surface (facing) is smooth and curved; perhaps it bounded a broad, irregular vesicle before comminution. All grain surfaces are coated with sublimates.

FIGURE 4. Lithic pyroclast similar to the one described in Fig. *5*. Grain surfaces include angular, hackly fractures and smooth, curved vesicle walls.

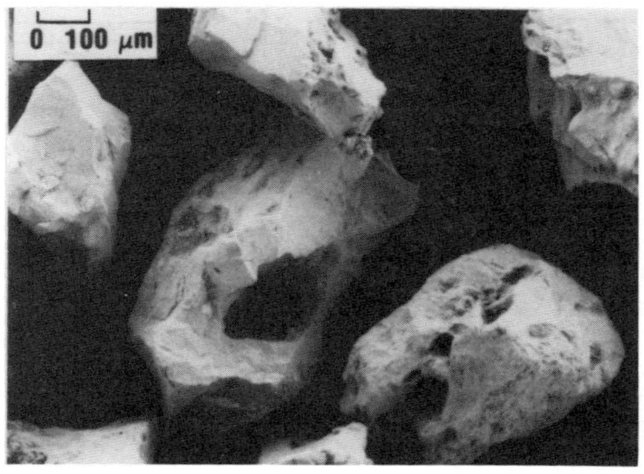

FIGURE 5. Hyalocrystalline lithic pyroclasts with large, irregular vesicles.

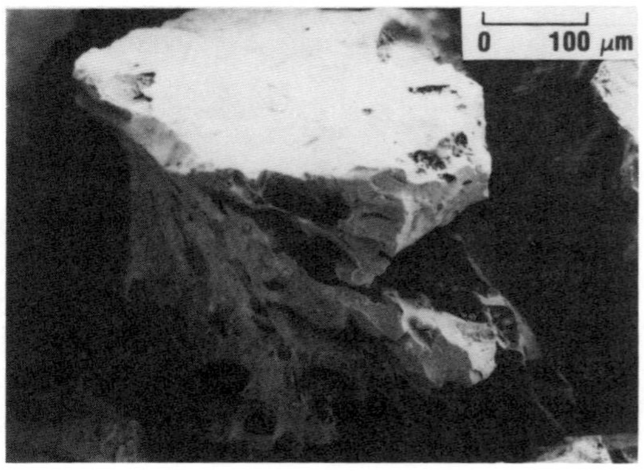

FIGURE 6. Blocky, hyalocrystalline lithic pyroclast with well-developed, lenticular vesicles. The large vesicles have curved yet lumpy walls. Comminution of a highly viscous melt, such as one from the Santiaguito dacite flows with heterogeneous patches of well-developed vesicles, could account for the mixture of blocky, angular pyroclasts and pyroclasts with one or two curved yet lumpy surfaces; the latter are particles broken from vesicle walls.

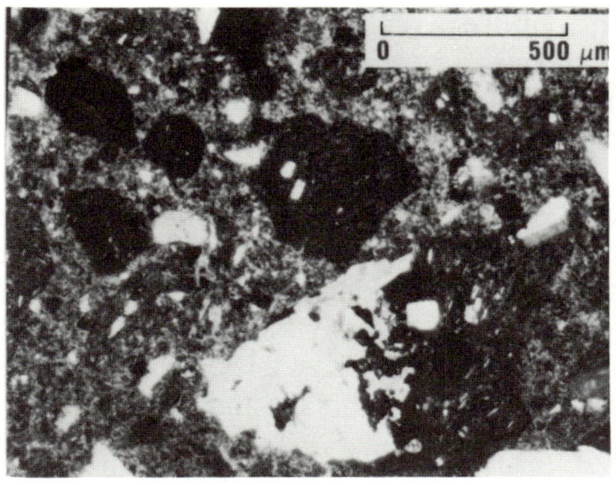

FIGURE 7. Photomicrograph (partly polarized light) of tephra from Santiaguito. There are mostly hyalocrystalline dacite pyroclasts: subequant, blocky forms with abundant, zoned plagioclase phenocrysts. This sample is bimodal, with 0.2- to 1-mm-long dacite pyroclasts in a matrix of <100-μm glass, crystal, and lithic clasts of the same composition.

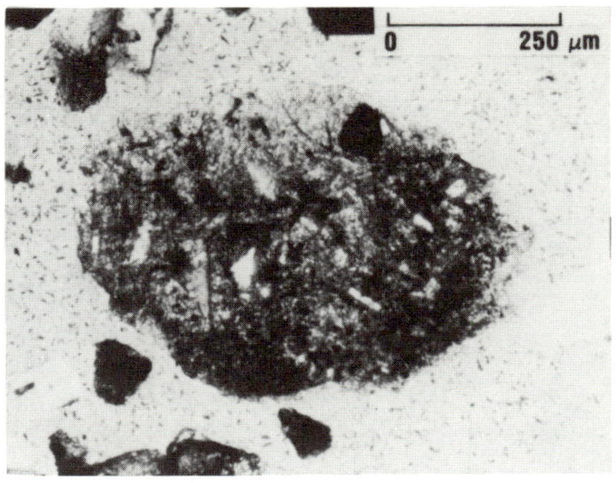

FIGURE 8. Photomicrograph of a hyalocrystalline dacite pyroclast; the groundmass is glassier along the rims (*top*).

Plinian and Peléean Eruptions

Plinian pumice falls and associated pyroclastic flows range from small-volume eruptions associated with domes to some of the truly great volcanic eruptions that produce widespread blankets of tephra and are associated with the collapse of large calderas. They are most often associated with rhyolitic or dacitic magmas that are charged with volatiles. The explosive decompression of the gas-charged magma forms highly vesicular ash that rises to great heights before settling to the Earth's surface. Peléean eruptions are generally associated with the explosive disintegration of a lava dome.

MEDICINE LAKE HIGHLAND, CALIFORNIA

Introduction: The Medicine Lake Highlands, California, is a 25-km-diameter shield volcano located 55 km E of Mount Shasta. Holocene volcanic activity there has been bimodal, producing basaltic eruptions along the flanks of the shield and rhyolitic eruptions near the rim of an 8- by 6-km shallow caldera. Eruptions during the last 1,100 years have generally consisted of highly explosive Plinian activity that mantled the countryside with pumice and was followed by extrusion of thick lava domes and flows. Nearly all of this pumice came from two sources, Little Glass Mountain and Glass Mountain. There is no evidence for development of pyroclastic flows during these eruptions.

Measurable tephra layers from Little Glass Mountain cover about 200 km^2 and have a volume of 0.046 km^3. Within a distance of 3 km from the vent, there are five tephra units that consist mainly of reversely graded, pale-gray pumice blocks and lapilli. These units range in thickness from 20 to 50 cm, and each contains 5 to 20% lithic fragments. At greater distances only one or two beds are visible. The tephra are overlain by the lava flows of Little Glass Mountain, which have a volume of 0.3 km^3

The nearly contemporaneous Glass Mountain eruptions covered a larger area (320 km^2) and have a volume of 0.09 km^3. There are 11 pumice falls with thicknesses of 5 to 102 cm; most beds are reversely graded.

Grain Size: All samples consist of very poorly sorted lapilli (near the vent) to medium to coarse ash (distal). All samples are within the field of tephra fall deposits discussed by Walker (1971).

Tephra Descriptions: Aphyric pumice pyroclasts make up 78 to 100% of each sample. All are remarkably homogeneous in composition and morphology; tephra from Little Glass Mountain and Glass Mountain are nearly identical (Table 1). All are equant to slightly elongate, blocky fragments with average L/W ratios of 1.4 to 1.6. All have sharp, angular grain surfaces either parallel to or normal to long axes of vesicles. Pyroclast surfaces do not exhibit the postfragmentation deformation or glazing that is characteristic of larger blocks or bombs in near-vent pumice cones. Vesicles are small and highly elongate; they range from 5 to 90 μm wide and 20 to 700 μm long. Some are twisted into "vortex" patterns that are interlayered with the highly elongate forms. Nearly all vesicles consist of several coalesced vesicles; vesicle cavities are crossed by threads and ribs of glass that remained after coalescence.

TABLE 1. TEPHRA DESCRIPTIONS OF GLASS MOUNTAIN TEPHRA, MEDICINE LAKE HIGHLANDS.

Composition and Texture	Unit				
	7	8	9	10	11 (Top)
TEXTURE					
Pyroclast shape	equant to slightly elongate	equant to platy	equant to highly elongate	equant to platy	equant to slightly elongate
Length/width ratio	1.4	1.4	1.5	1.2	1.5
Vesicle shapes	ovoid to highly elongate	ovoid to highly elongate	spherical to highly elongate	spherical to elongate ovoid	spherical to elongate
Coalesced vesicle shapes	spherical to highly elongate	ovoid to highly elongate	ovoid to highly elongate	spherical to elongate ovoid	spherical to highly elongate
Lithic fragments	yes	yes	no	no	yes
Vesicularity (%)	40-50	40-50	40-50	40-50	55
Mode	all glass	all glass	n.d.	all glass	all glass
Pumice surface	A (angular)	A	A	A	A
COMPOSITION (%)					
Highly vesicular pumice	77	78	n.d.	80	20
Basalt fragments	---	---	n.d.	---	40
Dacite fragments	33	---	n.d.	---	---
Banded, nonvesicular glass fragments	---	22	n.d.	20	40
Unit thickness (cm)	102	26	33	20	60

Chemical Composition: All samples were analyzed using an electron microprobe (Table 2). There are no mineral phases in Glass Mountain tephra and only traces of plagioclase and pyroxene phenocrysts in Little Glass Mountain tephra. All glasses analyzed were of rhyolitic composition. The compositions are very similar to rhyolite flows erupted following the Plinian activity.

TABLE 2. GLASS ANALYSES OF GLASS MOUNTAIN TEPHRA, MEDICINE LAKE HIGHLANDS.*

Unit	1	2	3	4
SiO_2	73.61	73.17	73.19	73.30
TiO_2	0.28	0.27	0.32	0.27
Al_2O_3	13.73	13.91	14.31	13.80
FeO	1.53	1.55	1.82	1.62
MgO	0.27	0.26	0.36	0.25
CaO	1.22	1.17	1.42	1.13
Na_2O	2.53	3.19	3.02	3.30
K_2O	4.20	4.23	3.99	4.34
F	0.11	0.18	0.12	0.08
Total	97.48	97.93	98.55	98.11
Number of Analyses	4	5	4	5

*Pumice Pit section (2.8 km east). Analyses by Cameca electron microprobe, Model Camebax.

Comments on Pyroclast Formation: The high vesicularity and lack of phenocrysts indicate rapid rise of magma and eruption without much volatile loss. All the pumice pyroclasts contain parallel vesicles with high aspect ratios. It is likely that initially the highly elongate vesicles grew in the magma as spheres and then were stretched into their present shapes by laminar flow of the highly viscous rhyolite magma through dikes. Most exhibit some degree of vesicle coalescence in their ruptured mutual vesicle walls; thin rounded threads and ridges within these vesicles are remnants of burst vesicle walls.

Most vesiculation appears to have occurred before fragmentation of the magma. The majority of pumice pyroclasts here have blocky, angular forms with surfaces parallel to and normal to elongate vesicles. If, indeed, elongation of vesicles were caused by flow within the vent, then fracture surfaces perpendicular to this flow may have formed when an expansion wave moved downward from the surface, much like pyroclast formation models developed by Rittmann (1936), Szekely and Reitan (1971), and Bennett (1974). The angularity of the pyroclasts indicates derivation from a highly viscous, brittle foam. After fragmentation and eruption of pyroclasts, none of the pumices were deformed or "inflated," which would have been the case if expansion of gases had continued after their eruption. After the volatile-rich top of the magma body had erupted, rhyolitic lavas with low (<20%) vesicularities were erupted.

REFERENCES:

Anderson, C. A., 1941. Volcanoes of the Medicine Lake Highland, California. Univ. Calif. Berkeley Publ., Geol. Sci., **23**: 215-276.

Bennett, F. C., 1974. On volcanic ash formation. Am. J. Sci., **274**: 648-661.

Chesterman, C. W., 1956. Pumice, pumicite and volcanic cinders in California. Calif. Div. Mines Bull., **174**: 119 pp.

Heiken, Grant, 1978. Plinian-type eruptions in the Medicine Lake Highland, California and the nature of the underlying magma. J. Volcanol. Geotherm. Res., **4**: 375-402.

Rittmann. A., 1936. Vulkane und ihre Tätigkeit. Ferdinand Enke, Stuttgart, 188 pp.

Szekely, J. and Reitan, P. H., 1971. Dike filling by magma intrusion and by explosive entrainment of fragments. J. Geophys. Res., **76**: 2602-2609.

Walker, G. P. L., 1971. Grain-size characteristics of pyroclastic deposits. J. Geol., **79**: 696-714.

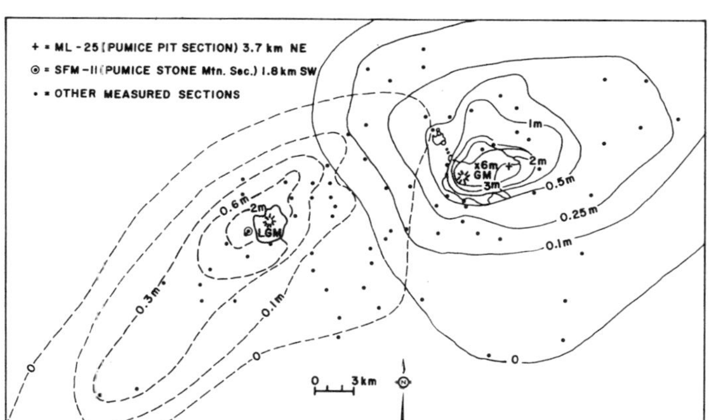

FIGURE 1. Isopach map of Little Glass Mountain (LGM - dashed lines) and Glass Mountain (GM - solid lines) tephra. Stippled areas are flows at the two vent areas. Dots are the locations of measured stratigraphic sections. The cross near Glass Mountain is the location of the thickest section measured.

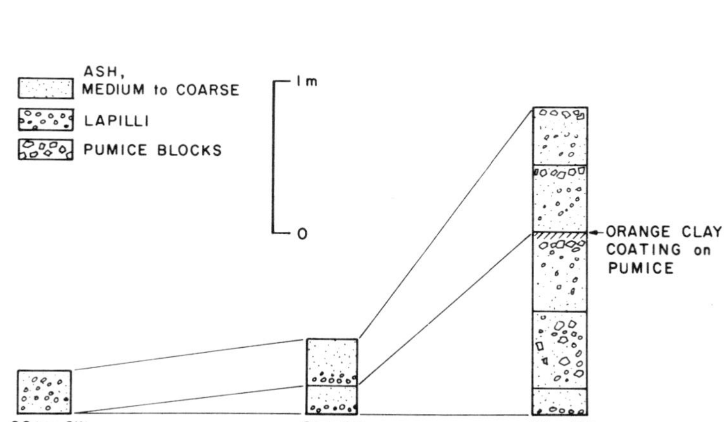

FIGURE 2. Tentative correlation of tephra units along a line SW from Little Glass Mountain. There may have been two episodes of explosive activity at LGM, since near-source units are separated by a thin layer of orange clay (wind blown, infiltrated dust?).

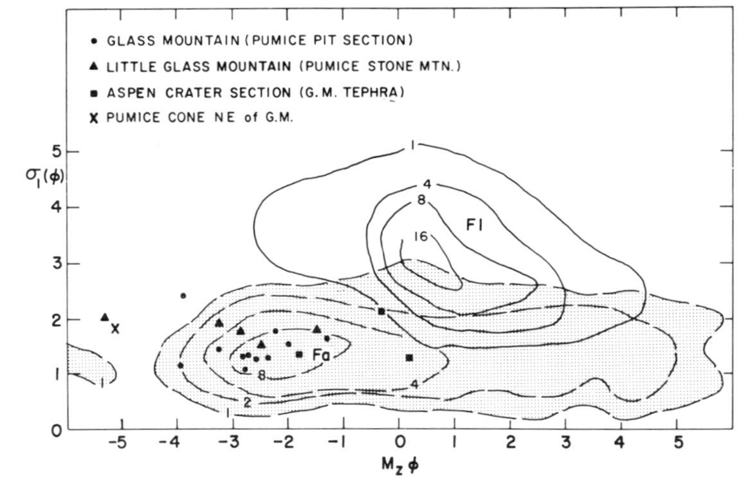

FIGURE 3. Mean grain size Mz_φ plotted against sorting σ_φ of tephra from Little Glass Mountain and Glass Mountain. Outlines of tephra fall (Fa) and tephra flow (FL) deposits are from Walker (1971). Contours represent percentages. Nearly all the samples are within the tephra fall category.

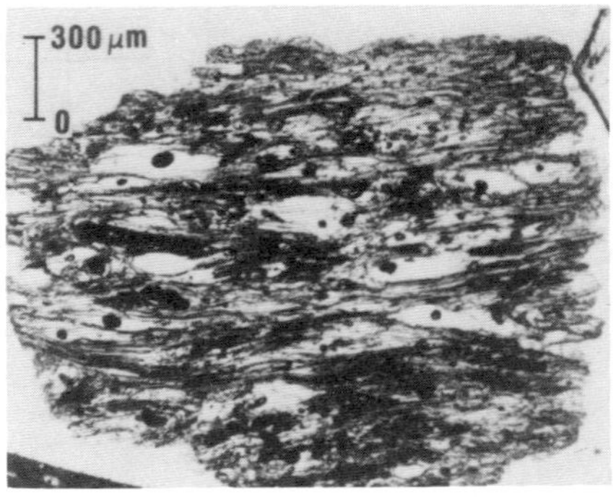

FIGURE 4. Photomicrograph of a typical pumice pyroclast with highly elongate, parallel vesicles. Note the fracture surfaces oriented normal to the long axes of vesicles.

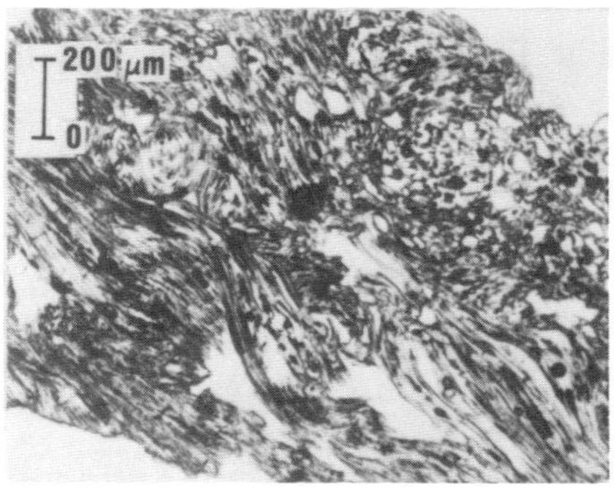

FIGURE 5. Photomicrograph of a pumice pyroclast; it shows deformed elongate vesicles and "vortices."

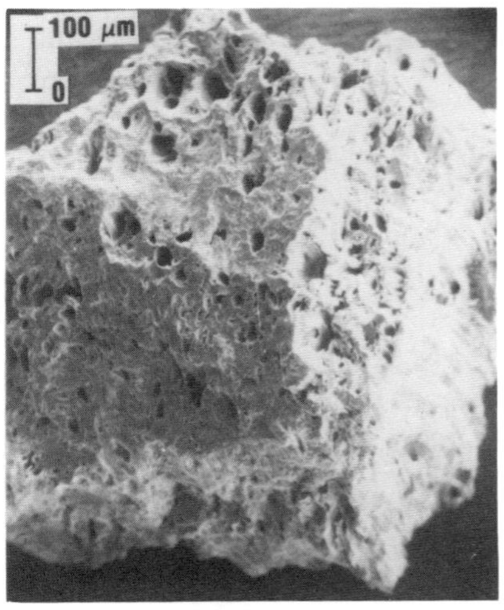

FIGURE 6. End view of a pumice pyroclast that is normal to elongate vesicles.

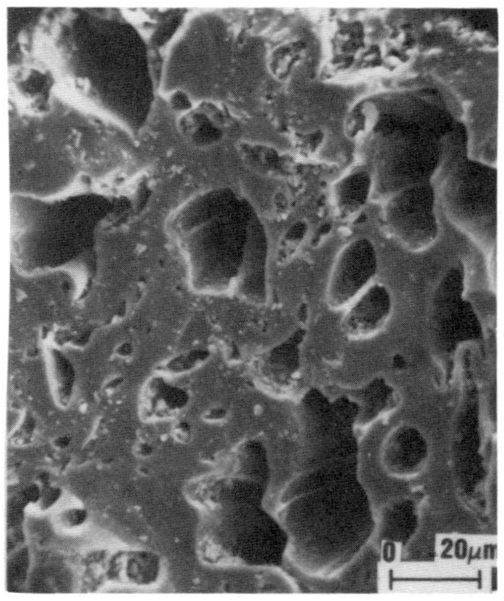

FIGURE 7. Closer view of the pyroclast in Fig. 6, showing cross sections of single and coalesced vesicles. Septae that remain from collapsed walls between coalesced vesicles are visible within the vesicles.

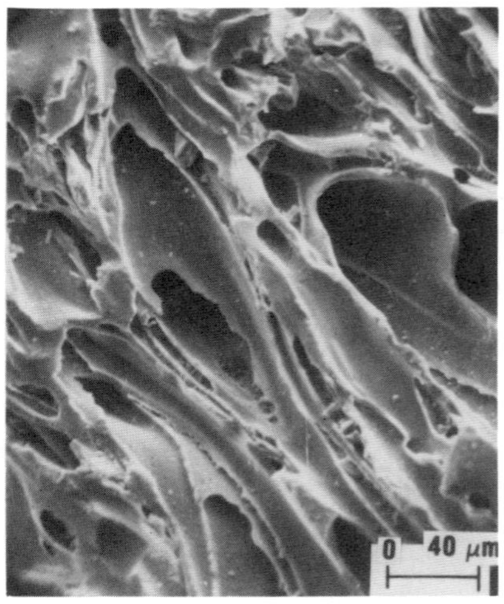

FIGURE 8. Surface of a pumice pyroclast with thin vesicle walls. Tapered ends of highly elongate vesicles are visible.

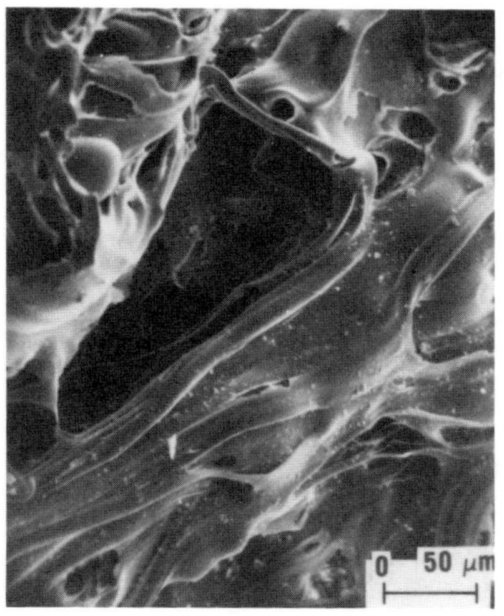

FIGURE 9. A few of the pumice pyroclasts in this deposit exhibit features that are indicative of continued vesicle growth and plastic deformation after comminution and eruption. In this example, the outer surface is deformed and outermost vesicles are intact. Some vesicle walls have collapsed but were drawn into thin glass threads. This texture is similar to that of many coarser pumice bombs found near the vents.

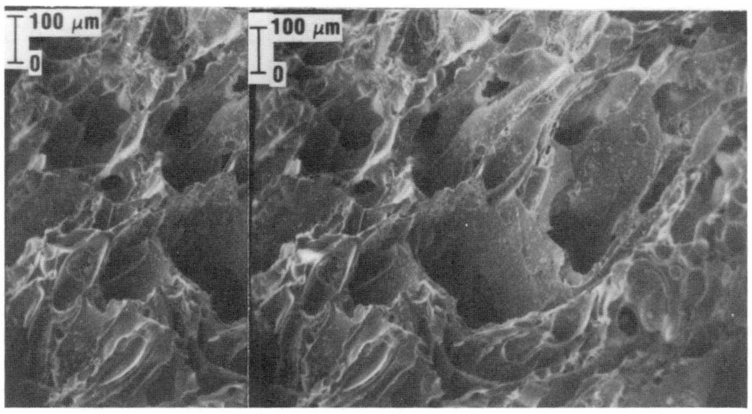

FIGURE 10. Stereo-pair image of a pumice pyroclast surface similar to that in Fig. *8*. Many of the vesicles are not only elongate but flattened. Note the extremely thin, delicate walls between vesicles. All the larger vesicles are compound; within them are septae or wall remnants of coalesced vesicles.

CHAOS CRAGS, LASSEN VOLCANIC NATIONAL PARK, CALIFORNIA

Introduction: The Chaos Crags are a group of silicic lava domes and associated tephra deposits located on the edge of the central plateau of Lassen Volcanic National Park, California. About 0.15 km^3 of rhyodacitic pyroclastic flows were erupted 1,100 years ago (Crandell et al., 1974) from vents beneath the Crags and moved down the valleys of Lost and Manzanita Creeks. Three episodes of pyroclastic flows were defined at stratigraphic sections in Lost Creek Valley, 3 km ENE of the vent.

Nearly all the pyroclastic-flow units consist of massive, nonwelded coarse ash that sometimes contains large pumice blocks and lapilli. The larger blocks have breadcrusted surfaces. Examples of the ash shown here are from a stratigraphic section in the Lost Creek drainage. Units A, B, and C-D are separated by thin silt layers that are possibly from windblown dust deposited between pyroclastic flows.

A section of ash fall tephra that is associated with the pyroclastic flows extends northeast from the Crags. Tephra deposits are overlain by a short, stubby flow and a three-lobed dome (the Chaos Crags). These domes were extruded from or near the vents that produced the tephra and consist of 1.25 km^3 of dacite lava.

Tephra Description: The pyroclastic flows consist of pumice block bearing medium to coarse, vitric-crystal or crystal-vitric ash (Tables 1 and 2). They contain pumice, shards, plagioclase, hornblende, Fe-Ti oxide, and quartz pyroclasts; there are also andesitic and dacitic lithic clasts. There is little variation in pyroclast types from Unit A to Unit D. Plagioclase content decreases from 42% in Unit A to 25% in Unit B and stays at that level throughout. Plagioclase pyroclasts (average composition of An_{30}) generally have a rind of rhyolitic glass.

Pumice pyroclasts are angular and slightly elongate; most pyroclasts have sharp, angular surfaces, but a few have smooth surface "skins." The smooth outer surfaces are only a few μm thick. These pumices have vesicularities of less than 30%. Vesicle shapes range from spheres to highly elongate "tubes," but most are ovoids 10 to 200 μm long. In many pumice pyroclasts, elongate vesicles are radial to phenocryst surfaces. This texture is associated only with larger phenocrysts or glomeroporphyritic clots.

Shards, consisting of vesicle walls broken from pumice pyroclasts, are present in some of the pyroclastic flows. The most common crystal pyroclast is plagioclase; some have bubble-wall textures and traces of vesicular glass adhering to grain surfaces. Lithic pyroclasts make up 2 to 4% of the tephra and are mostly angular to subrounded andesite fragments.

TABLE 1. PYROCLAST TYPES, PYROCLASTIC FLOWS OF LOST CREEK, CHAOS CRAGS.*

							Xenoliths (Lithic Fragments)		
Unit	Pumice	Shards	Plag	Hb	Bio	Qtz	Basalt	Andesite	Dacite
D	53.0	27.5	7.0	tr	1.5	1.0	---	5.5	1.5
C	100.0	---	---	---	---	---	---	---	---
B (upper)	72.7	---	15.0	0.7	2.7	1.7	tr	6.7	0.7
B (lower)	69.1	6.3	10.2	1.6	1.3	2.6	---	3.0	5.6
A**	62.7	3.7	20.6	2.3	3.0	1.7	2.7	1.3	0.3

*Volume per cent determined by grain counts.
**Also contains 0.7% Fe-Ti oxides, 0.3% resorbed plagioclase, and 0.6% pyroxene.

TABLE 2. PETROGRAPHIC CHARACTERISTICS OF TEPHRA, PYROCLASTIC FLOWS OF LOST CREEK, CHAOS CRAGS.

		Mode*											
Unit	Thickness 3 km E of vent (m)	Glass	Plag	Hb	Bio	Qtz	Fe-Ti Oxide	Glomerocrysts	Resorbed Plag	Vesicularity (%)	Vesicle Shape	Vesicle Size (μm)	Ash Classification**
D	0.82	67.5	19.9	2.4	5.1	3.4	0.9	0.4	0.9	15.5	spherical to ovoid	40-650	pumice block (near vent)
D	0.82	64.6	27.9	4.3	3.2	---	---	---	---	46.0	irregular ovoid	5-2100	pumice block (at 3 km)
D	0.82	58.5	27.1	6.2	2.2	3.0	0.9	---	---	24.0	ovoid to highly elongate	5-100	vitric-xl
C	0.6	63.4	21.1	8.9	3.7	0.4	1.9	---	---	30.0	ovoid	10-600	vitric
B	5.7	52.3	31.4	3.8	2.1	8.4	1.3	---	0.4	24.0	ovoid	10-1000	vitric-xl
B	5.7	62.5	24.6	4.7	2.4	4.0	2.0	---	1.2	16.0	spherical to ovoid	5-30	vitric-xl
A	1.2	42.2	42.7	4.6	7.4	1.6	1.6	---	---	19.0	ovoid	20-600	xl-vitric

*Percentages determined from thin section.
**See introduction for classification used here.

Chemical Composition: All glass fractions are rhyolitic; they were analysed by a Cameca electron microprobe (Table 3).

TABLE 3. CHEMICAL COMPOSITION OF TEPHRA, CHAOS CRAGS.*

	Unit A	Unit B	Unit B lower	Unit C upper	Unit D
SiO_2	77.07	77.67	78.72	77.19	75.77
Al_2O_3	12.60	12.53	12.34	12.51	12.17
FeO	0.73	0.74	0.78	0.74	0.71
CaO	0.81	0.83	0.74	0.82	0.82
Na_2O	2.53	1.98	1.58	3.03	2.82
K_2O	2.53	1.78	1.86	3.98	4.02
Total	96.27	95.53	96.02	98.27	96.31
Number of Analyses	7	6	10	8	5

*Glass fraction of tephra for Lost Creek pyroclastic flow.

Comments on Pyroclast Formation: Elongate vesicles, oriented perpendicular to surfaces of large phenocrysts or glomeroporphyritic clots, are common within pyroclasts of this tephra sequence. This texture is anomalous within pumice pyroclasts that are dominated by small, ovoid vesicles. It is present only on phenocrysts larger than 0.7 mm. This texture is associated with plagioclase, hornblende, biotite, quartz, and magnetite phenocrysts. The number of phenocrysts with radial vesicular texture decreases from the bottom to the top of the section (e.g., 45 to 60% of the phenocrysts in Unit C are surrounded by radial vesicles, whereas within Unit D, 0 to 19% of the phenocrysts are associated with this texture).

This texture may have formed in several ways. The most commonly accepted is that vesicle growth was initiated at phenocryst surfaces, and subsequent vesicle growth was normal to those surfaces. Another possible explanation is that of pool boiling. Pool boiling may occur where a hot surface is submerged in a cooler liquid; closely spaced bubbles nucleate at the surface, then move away from it (Tong, 1972). Before vesiculation, phenocrysts and surrounding liquid are in thermal equilibrium, and when vesiculation begins, the liquid is cooled by adiabatic expansion. If the vesiculation process is rapid enough, the temperature gradient between phenocrysts and liquid may be great enough for vapor exsolution near the crystal/melt interface. The process is fast: a 1-mm-diameter phenocryst reaches thermal equilibrium within a vesiculating melt in about 1 second.

Radial vesicle patterns in pumice must form as the melt is being comminuted and accelerated out of the vent; the pattern is not deformed by flow, as would be the case if this process occurred at depth in a melt that was rising to the surface before eruption.

REFERENCES:

Crandell, D. R., Mullineaux, D. R., Sigafoos, R. S., and Rubin, M., 1974. Chaos Crags eruptions and rockfall-avalanches, Lassen Volcanic National Park, California. J. Res. U.S. Geol. Surv., **2**: 49-59.

Heiken, G. and Eichelberger, J., 1980. Eruptions at Chaos Crags, Lassen Volcanic National Park, California. J. Volcanol. Geotherm. Res., **7**: 443-481.

Tong, L. S., 1972. Boiling crisis and critical heat flux. U.S. At. Energy Comm. Crit. Rev. Ser., TD-25897: 82 pp.

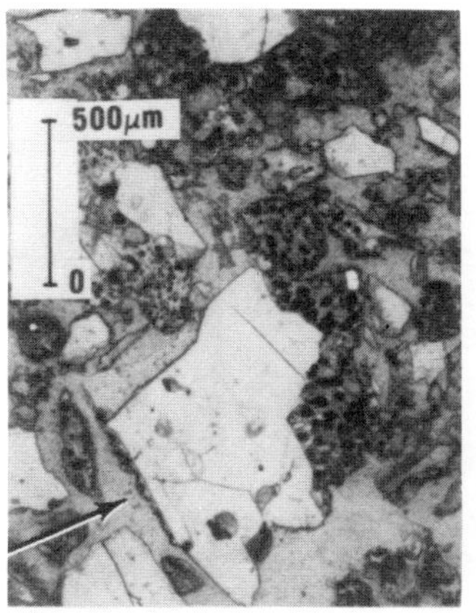

FIGURE 1a. (Unit A) Photomicrograph of crystal-vitric ash, consisting of mostly small, angular pumice pyroclasts. Abundant phenocrysts (mostly plagioclase) are present in the pumice and as separate pyroclasts. Partly crossed polarizer.

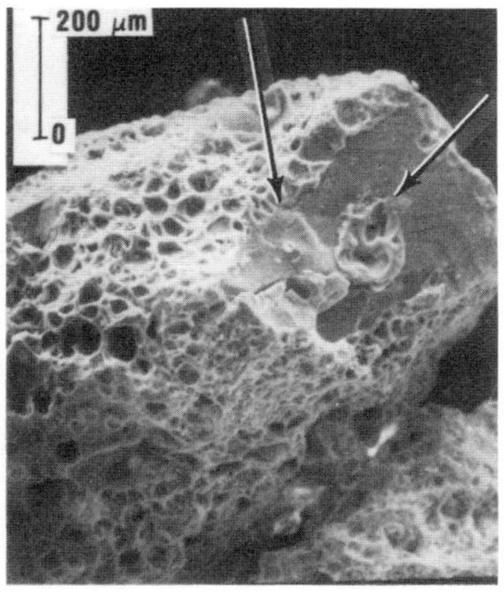

FIGURE 1b. SEM of a pumice pyroclast with small (20- to 600-µm) ovoid pyroclasts that are typical of the tephra in this unit. Within the core is a phenocryst surrounded by vesicular glass, and on one end, a "bubble-wall" texture similar to the texture visible on the left-lower edge of the largest phenocryst visible in *1a* (arrow).

FIG. 1 —

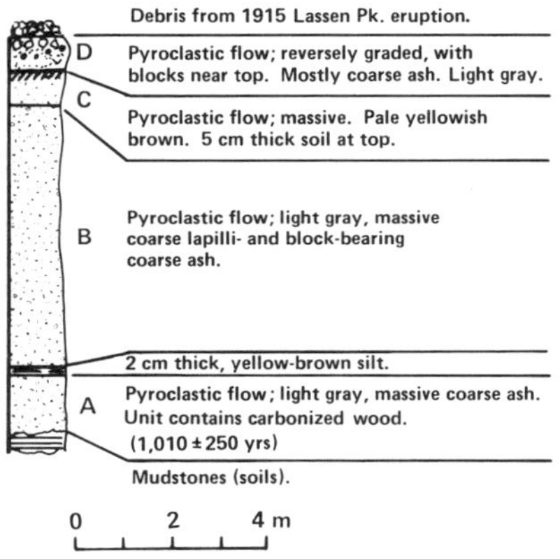

Composite stratigraphic section from the floor of Lost Creek Valley, 3.0 km E and 550 m below the vent. Carbon-14 data from Crandell et al. (1974).

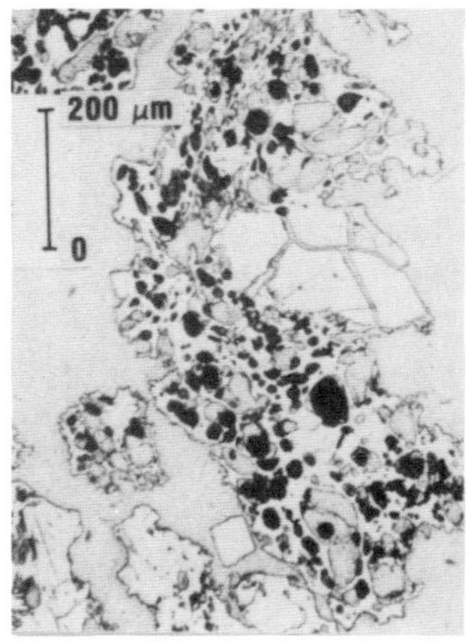

FIGURE 2a. (Unit B) Reflected-light photomicrograph of pumice pyroclast with a relatively high vesicularity. Although it is generally angular, there is some degree of rounding.

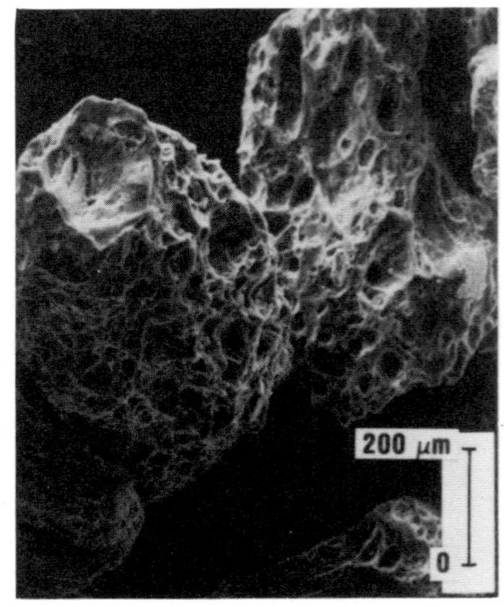

FIGURE 2b. SEM of pumice pyroclasts similar to those in Fig. *2a*. It has mostly ovoid vesicles. Although most pumice surfaces are irregular and angular, some are rounded by viscous deformation of vesicle walls after comminution (arrow).

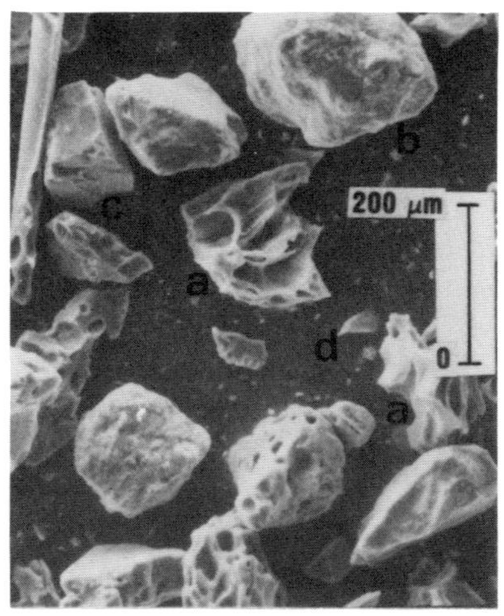

FIGURE 3. (Unit C) SEM of fine-grained tephra that consist of small pumice (*a*), lithic (*b*) and crystal (*c*) pyroclasts, and shards (*d*).

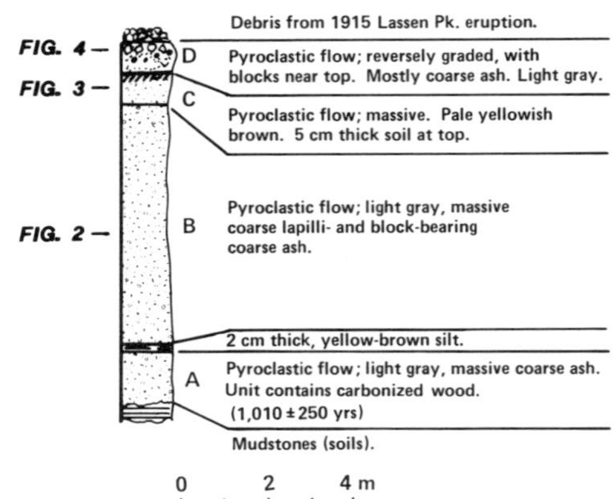

Composite stratigraphic section from the floor of Lost Creek Valley, 3.0 km E and 550 m below the vent. Carbon-14 data from Crandell et al. (1974).

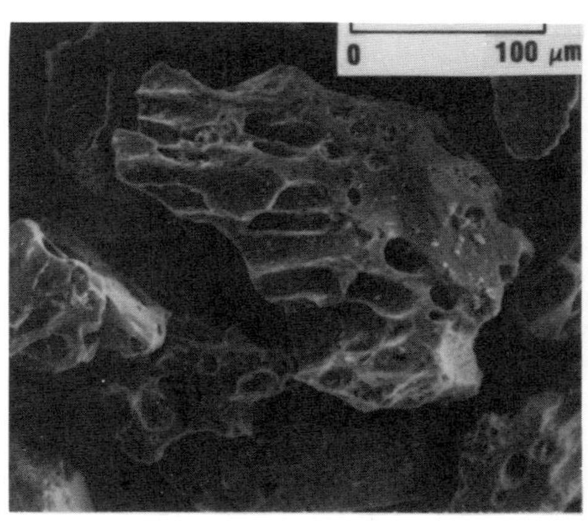

FIGURE 4a. (Unit D) SEM of pumice pyroclasts that show low vesicularity (20 to 40%) with slightly elongate vesicles. Most vesicles are separated by thick (10- to 200-μm) vesicle walls. These pyroclasts have sharp, angular surfaces.

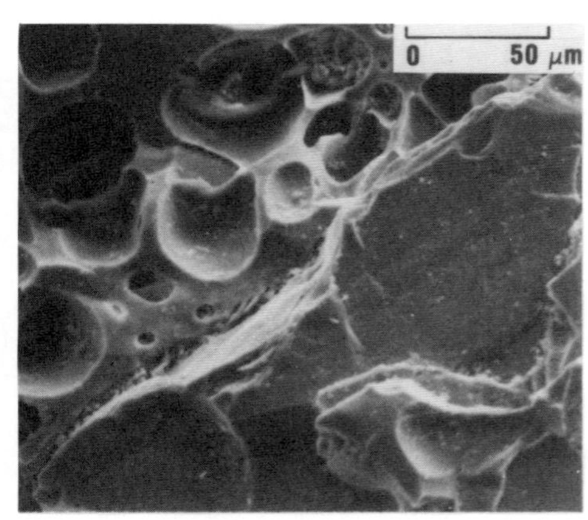

FIGURE 4b. SEM of a phenocryst/glass boundary. In this particular case there appears to be no relation of phenocryst surfaces and vesicle shapes, as discussed in the text and in Fig. *5*. There has been some flattening and deformation of very small vesicles (1 to 2 μm by 10 to 13 μm) that are located at the phenocryst surface. Spherical to ovoid vesicles that are typical of pumices from this eruption sequence are visible 10 μm beyond the phenocryst surface.

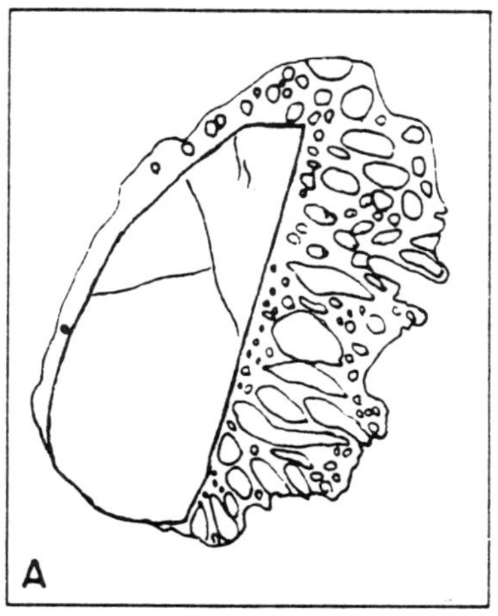

FIGURE 5a. Elongate vesicles oriented radially around phenocrysts. Pumice pyroclast from a pyroclastic flow in Manzanita Creek. Long axes of vesicles are up to 220 μm long and oriented normal to the surface of a plagioclase phenocryst. Base of the sketch is 0.8 mm wide.

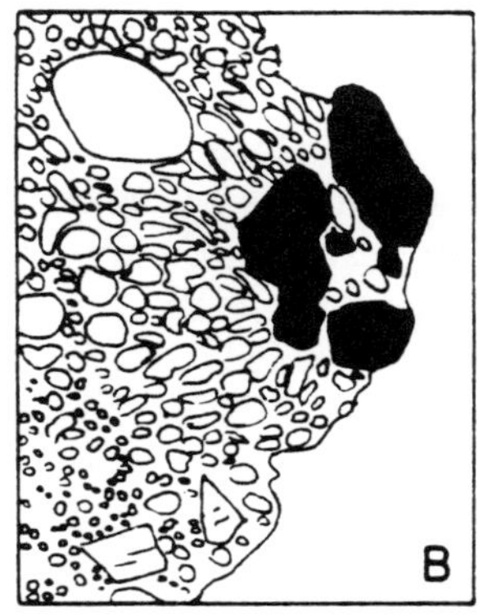

FIGURE 5b. Long axes of vesicles are radially oriented around a cluster of magnetite phenocrysts in this pumice pyroclast. Base of the sketch is 1.2 mm wide.

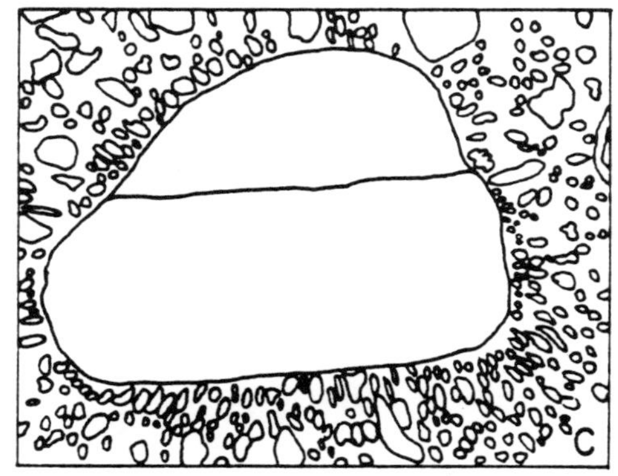

FIGURE 5c. Vesicles are clustered and oriented radially around a plagioclase phenocryst. Base of the sketch is 2.1 mm wide.

TAMBORA (LESSER SUNDA ISLANDS), INDONESIA; APRIL 1815 ERUPTION

Introduction: Tambora is located on the Sangaar Peninsula of the Island of Sumbawa, Lesser Sunda Islands, Indonesia. It is well known for its disastrous eruption of 1815, which resulted in great loss of life and a vast plume of ash and gases in the stratosphere. The overall effect on the Earth's atmosphere in 1816 was large; it caused the "year without a summer" (Rampino and Self, 1982).

The eruption began April 5, 1815, with earthquakes, eruption clouds, and rumbling (van Padang, 1951). Climactic activity began April 10, 1815, with a Plinian eruption that produced large, dense, ash clouds and pumice falls and lasted for 24 to 36 hours. The Island of Madura 500 km away was in total darkness for 3 days (van Padang, 1951). Plinian ash falls were followed by pyroclastic flows and surges. Petroeschevsky (1949) estimated that the ejecta had a volume of 100 km^3, but Self (personal communication) has calculated a bulk volume of 150 to 175 km^3. A 6-km-diameter, 1-km-deep caldera was formed during the eruption.

Grain Size: Samples from the 1815 deposits were provided by S. Self. Grain-size data given in Table 1 show both a decrease in median diameter and sorting from the Plinian phase products to the ignimbrites and surges.

TABLE 1. GRAIN-SIZE DATA FOR TAMBORA 1815 ERUPTION PRODUCTS.

Plinian Phase							
Md_φ	−.90	−.80	−.23				
σ_φ	1.65	2.0	2.14				
Ignimbrite							
Md_φ	−.50	−.80	+2.6	−.44	+1.0	+1.3	−.80
σ_φ	3.05	2.00	1.75	2.7	2.05	2.45	2.55
Surges							
Md_φ	+1.0	−.60					
σ_φ	4.1	3.4					

Tephra Descriptions: Tephra characteristics indicate eruption of a physically heterogeneous, volatile-rich magma. Within the Plinian pumice fall deposits there are at least three kinds of pumice. Two pumice types developed in a very vesicular, viscous (brittle) melt: one has large, ovoid vesicles and thin vesicle walls and the other has parallel, thin, highly elongate vesicles. Both pyroclast types are very angular and have sharp, thin broken vesicle walls that form grain surfaces. Most vesicles are composite.

Coalescence and better development of vesicles around phenocrysts form lens-shaped "pockets" within pumices. In contrast to these highly vesicular pumices, another type of pumice has fewer vesicles and is characterized by thick vesicle walls and rounding of all surfaces by viscous flow after vesiculation.

Tephra from the ignimbrite of 1815 is characterized by pyroclasts with larger, better developed vesicles than those of pumices from the Plinian phase. Overall vesicularity is lower and vesicle walls are thicker in pyroclasts from the ignimbrite. Shards are thick, curved plates or equant forms. All pyroclasts in this deposit are coated with small (2- to 16-μm), equant shards; these adhere to grain surfaces but are not welded to them.

Cross-bedded pyroclastic flow deposits overlying the ignimbrite contain a variety of pyroclasts similar to a mixture of those described for both the Plinian pumice fall and the ignimbrite. Small pyroclasts cling to grain surfaces but are not as pervasive as in the ignimbrite.

Chemical Composition: The chemical composition listed in Table 2 is from S. Self et al. (1984). These two analyses, representative of the compositional range of the 1815 eruption products, are normalized to 100% and are volatile-free.

TABLE 2. CHEMICAL COMPOSITION, TAMBORA.*

Element	1	2
SiO_2	56.00	57.88
TiO_2	0.63	0.58
Al_2O_3	19.78	19.82
Fe_2O_3	5.78	5.36
MgO	1.79	1.34
MnO	0.18	0.21
CaO	4.55	3.94
Na_2O	5.01	5.03
K_2O	5.83	6.01
P_2O_5	0.45	0.33

*Note: Total Fe as Fe_2O_3. From Self et al. (1984).

Comments on Pyroclast Formation: The eruption sequence is a classic example of decreasing volatile content as the eruption proceeded from a Plinian phase to mainly pyroclastic flows and surges. Vesiculation and comminution appear to have developed within a heterogeneous melt; angular, thin-walled pumices were derived from a more viscous melt, and less vesicular pyroclasts with fluidal surfaces came from a less viscous portion. Pyroclasts are clean; the finer grained fraction (if it existed) was winnowed out and carried farther away.

Pyroclastic flow and surge deposits contain a wider range of grain sizes; small particles cling to surfaces of larger pyroclasts of all types and sizes. Pumices and shards were derived from a melt with larger vesicle sizes but lower overall vesicularity. There is less elongation caused by flow of vesicles within these pyroclasts than within those from the Plinian deposits.

There is little evidence of abrasion of pyroclasts within any of the deposits; only minor abrasion of thin, delicate shard edges within the surge deposit can be seen.

REFERENCES:

Rampino, M. and Self, S., 1982. Historic eruptions of Tambora (1815), Krakatau (1883) and Agung (1963); their stratospheric aerosols and climatic impact. Quat. Res., **18**: 127-143.

Self, S., Rampino, M., Newton, M., and Wolff, J., 1984. A volcanological study of the great Tambora eruption of 1815. Submitted to Geology.

van Padang, M. N., 1951. Catalogue of the active volcanoes of Indonesia. Part I of *Catalogue of the Active Volcanoes of the World.* Int'l. Assoc. Volcanol.,Napoli: 271 pp.

Petroeschevsky, W. A., 1949. A contribution to the knowledge of the Gunung Tambora (Sumbawa). Tidschr. Kan. Ned. Aardr. Gen., **66**: 688-703.

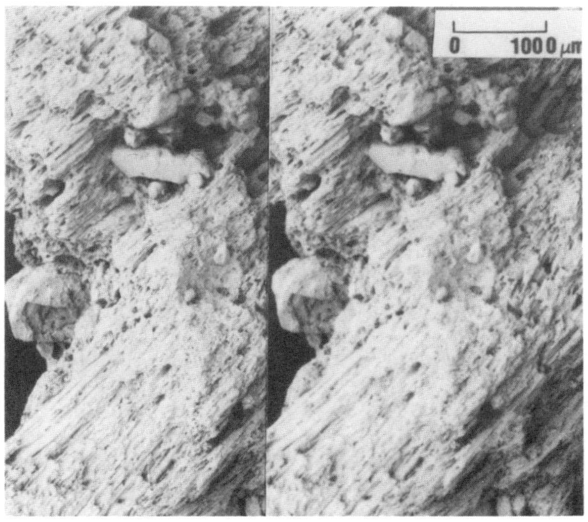

FIGURE 1a. Pumice pyroclast from a fall unit with trachyandesitic pumice from the base. Stereo pair showing the slightly elongate, subrounded pumice. Most vesicles are parallel and have large L:W ratios. There has been considerable coalescence of larger vesicles around phenocrysts, forming irregular "pockets." Elongate vesicles are bent around the "pockets." Pumice surfaces are parallel to and normal to elongate vesicles.

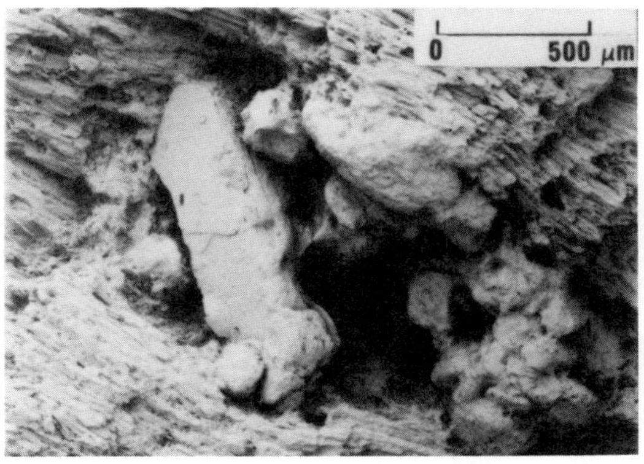

FIGURE 1b. Close-up of Fig. *1a*, showing the "pocket" of coalesced vesicles that developed around a cluster of phenocrysts, including biotite (*center-left*).

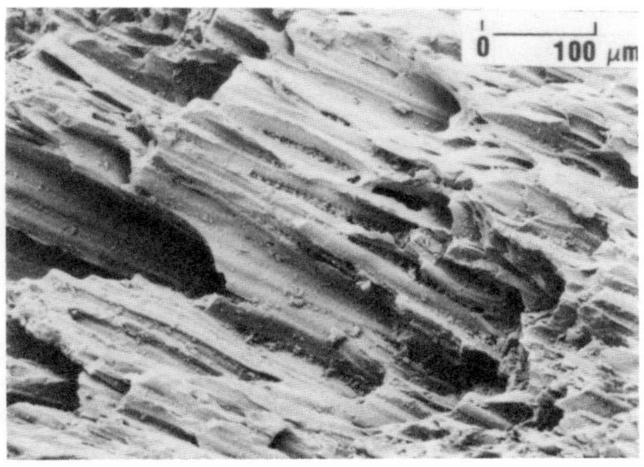

FIGURE 1c. Detail of a pumice surface, showing elongate, parallel vesicles with fairly thick (>3-μm) vesicle walls. Many of these elongate vesicles show considerable coalescence. Note the rounded parallel ridges that are collapsed vesicle walls.

FIGURE 1d. Detail of a pumice surface, showing remarkable smoothness and thickness of vesicle walls. Ridge-like septae are indicative of vesicle coalescence.

FIGURE 2a. Plinian pumice fall sample from near the base. Pumice pyroclast with large, well-developed ovoid vesicles. Most of the ovoid vesicles show collapse of mutual vesicle walls and coalescence to form larger composite vesicles. This is a very angular pyroclast with sharp edges along thin, broken vesicle walls.

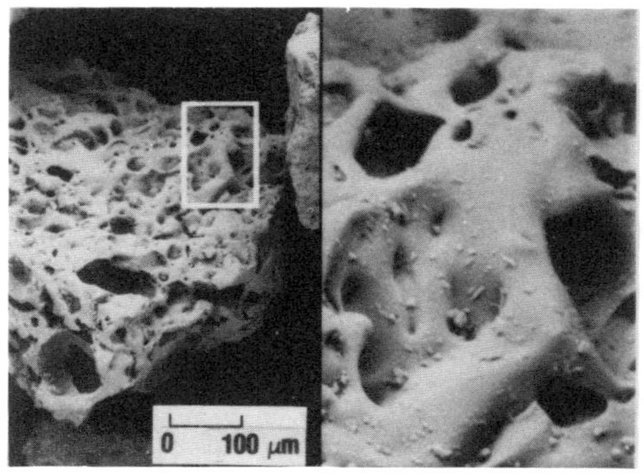

FIGURE 2b. In contrast with the sharp, angular, brittle pumice pyroclast of Fig. *2a*, this pumiceous pyroclast shows evidence of viscous deformation by the smoothing and rounding of all surfaces that must have occurred after vesiculation and separation of individual pyroclasts. Irregular, ovoid vesicles. *Inset* - view of surface, showing smooth, curved surfaces. Note that even margins of vesicles open to the grain surface have been smoothed, perhaps by movement during flow.

FIGURE 2c. Angular pumice pyroclast with closely spaced, parallel, elongate vesicles and thin, angular vesicle walls. *Inset* - close-up of surface; the composite vesicles with rounded and angular parallel ridges within vesicles exceed the volume of single vesicles.

FIGURE 2d. Stereomicrograph of a pumice fall sample. Note the diversity of pumice types, which illustrates magma heterogeneity. Some rounded or subrounded lithic fragments are present.

FIGURE 3a. The ignimbrite of 1815. Fine-grained size fraction of the ignimbrite. In contrast with the Plinian phase pumices, these have large, better developed vesicles but lower overall vesicularity. Vesicle walls are thick (>10 µm). Shards are thick, curved plates or equant forms that were broken from a pumiceous melt with ovoid or spherical vesicles and low overall vesicularity. Many small, angular particles cling to pyroclast surfaces.

*FIGURE 3b.*Slightly closer view of a sample similar to that in Fig. *3a*. Note the 2- to 10-µm particles that cling to shard surfaces.

FIGURE 3c. Rounded pumice pyroclasts from coarser size fraction. Vesicularity is low, vesicles are mostly ovoids, and grain surfaces are coated with fine-grained particles.

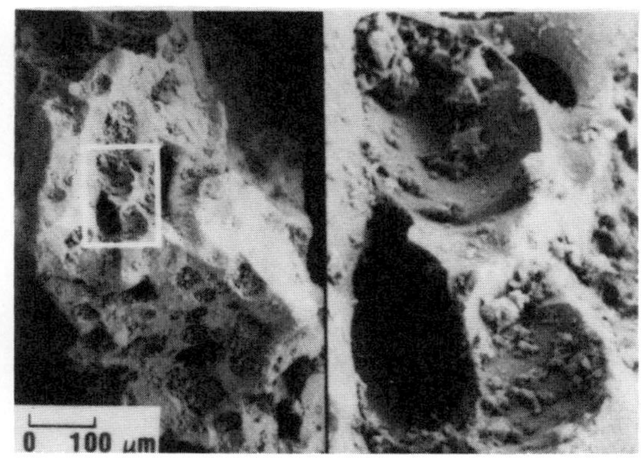

FIGURE 3d. Subangular pyroclast showing relatively low overall vesicularity of the pumices in this deposit. *Inset* - view of ovoid vesicles at and near the grain surface. Small (2- to 16-μm), equant particles adhere to grain surfaces.

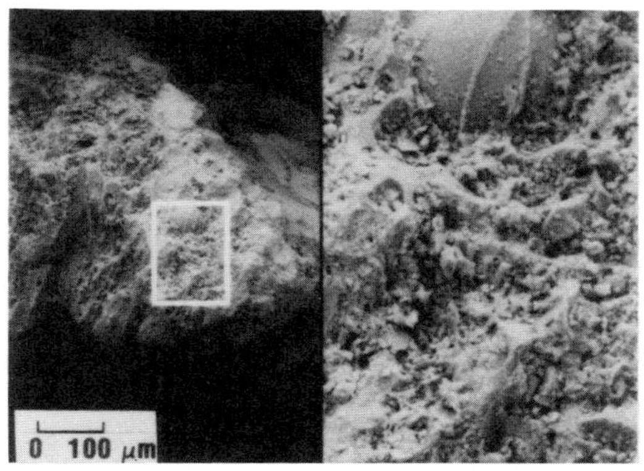

FIGURE 3e. Pumice pyroclast partly coated with fine-grained particles.

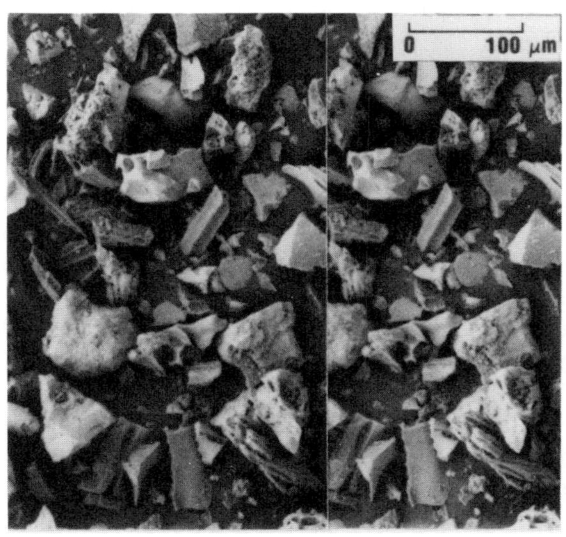

FIGURE 4a. Pyroclasts from a cross-bedded pyroclastic flow deposit *above* the 1815 ignimbrite. Stereomicrograph of the fine-grained size fraction. As in the Plinian fall deposit, this deposit contains a variety of pyroclast types including: 1. Shards derived from pumice containing highly elongate, composite vesicles and thin vesicle walls. 2. Equant or subequant angular blocky pyroclasts with low vesicularity, large ovoid vesicles, and thick vesicle walls. 3. Smooth, rounded droplets. 4. Aggregates of fine-grained (<10-μm), equant shards.

FIGURE 4b. Angular, subequant pumice pyroclast. *Inset* - surface of pumice. Irregular, composite vesicles bounded by both thin and thick vesicle walls. Some small, equant particles cling to pyroclast surfaces.

FIGURE 4c. Coarser grain-size fraction of the surge deposit. The variety of pyroclast types is the same as that described in Fig. *4a*. *Lower center* - a rounded lithic fragment.

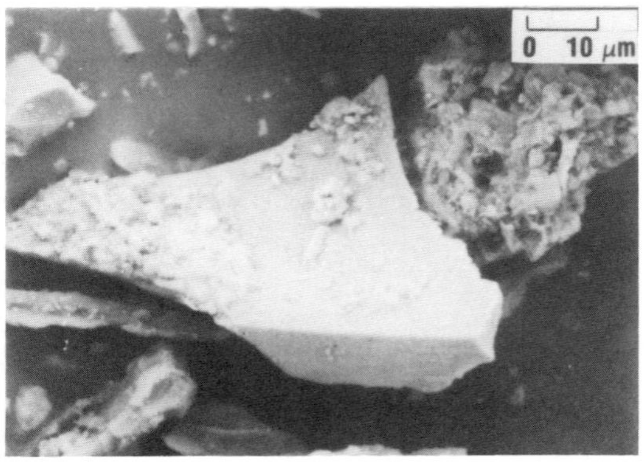

FIGURE 4d. Close-up of fine grain-size fraction. *Center* - angular, curved shard with fine-grained particles adhering to the grain surface. A few of the more delicate grain edges have been abraded. *Upper right* - aggregate of fine-grained shards.

HEKLA, ICELAND; H_4 TEPHRA

Introduction: The volcano Hekla, located along SW-NE-trending fissures in south-central Iceland, is that country's most famous volcano. Five major eruptions over the last 7000 years and 15 eruptions since 1104 A.D. (the first recorded activity) have spread tephra over much of Iceland and occasionally over northern Europe. Nearly all of Hekla's eruptions have begun with explosive eruptions of rhyolitic or rhyodacitic pumice. As these eruptions proceed, the magma becomes more mafic and less explosive and eruptions end after extrusion of basaltic or basaltic-andesite lava flows. Thorarinsson (1981) estimates that Hekla has erupted between 6 and 7 km^3 of silicic tephra over the last 7000 years. The example of Hekla tephra shown here is from the H_4 tephra layer, which is about 4000 ^{14}C years old. It is one of the most widespread Hekla tephra layers, covering 78,000 km^2 (on land), along a zone trending NNE of Hekla (Larsen and Thorarinsson, 1977). The samples for this example came from the Thjórsádalur Valley and from near Hraunfoss Dam, west of Hekla.

Tephra Descriptions: The fine-grained base of H_4 consists of finely comminuted rhyolitic vitric ash. The ash consists of small (25- to 50-μm-long) pumice pyroclasts and shards (1 to 15 μm long). Pumice pyroclasts contain: (1) narrow, elongate and (2) flattened ovoid vesicles. Most shards are slightly curved, but many are flat plates.

Pumices from the base of the Plinian pumice fall (collected in the Thjórsádalur Valley) are subrounded to subangular and equant to subequant. Vesiculation was heterogeneous; "pockets" of coalesced vesicles, 60 to 350 μm long, are surrounded by zones of slightly elongate vesicles that are 1 to 25 μm wide.

Pumice pyroclasts from the top of the pumice-fall are similar in shape to those from the base. They too are characterized by a heterogeneous vesicle distribution. In contrast with the base, however, vesiculation is more advanced; there are more pockets of coalesced vesicles. The pockets are better developed and consist of irregular, 100- to 500-μm-diameter lenticular tubes lined by layers of collapsed vesicle walls.

Chemical Composition: This is one of the most silicic tephras erupted from Hekla; near the base, SiO_2 = 74%, but it gradually decreases toward the top to 57%. Hekla is a classical case of a zoned magma chamber; each eruption begins as a gas-rich rhyolite and ends as a basaltic andesite. These variations are reflected in the nature of the tephra erupted.

Comments on Pyroclast Formation: A thin layer at the base of the H_4 unit appears to be an excellent example of phreatoplinian tephra. It is a highly vesicular rhyolitic magma that mixed with surface water (or ice and snow) and has been broken into extremely fine-grained pyroclasts.

Pumice pyroclasts in the pumice bed show some variation from the base to the top: not in the size or appearance of the coarse-grained pumice pyroclasts but in the size and shape of vesicles. The less rhyolitic pumices in the upper part of the fall unit have better developed vesicles and show more vesicle coalescence. This variation may be related to the lower silica content (and lower viscosity) of the last pumices to be erupted.

REFERENCES:

Larsen, G. and Thorarinsson, S., 1977. H_4 and other acid Hekla tephra layers. Jökull, **27**: 28-46.

Thorarinsson, S., 1981. The application of tephrochronology in Iceland. In *Tephra Studies* (S. Self and S. Sparks, editors), Reidel Publ. Co., Dordrecht: 109-134.

FIGURE 1a. Fine-grained base of H_4 (collected near Hraunfoss Dam). Angular, colorless glass shards. These range from 1-μm-wide slivers to 15-μm irregular forms.

FIGURE 1b. Shards and small pumice pyroclasts. Elongate, thin vesicles and ovoid, flattened vesicles within the pumices provide some idea of the heterogeneous vesicularity that existed in the magma. As would be expected, these fragments are quite similar (in the nature of their vesicles) to pumices from the base of the pumice fall (Fig. *2a, b*).

FIGURE 2a. Pumice from the base of the H_4 pumice fall (from the Thjórsádalur Valley). Subequant, subrounded pumice pyroclast.

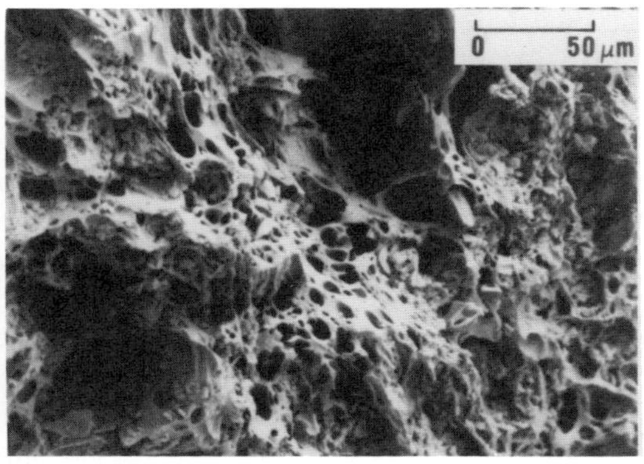

FIGURE 2b. Close-up of the pumice in Fig. *2a*. The heterogeneous nature of the vesicles is obvious: there are large pockets composed of coalesced vesicles and small, elongate vesicles surrounding the pockets; these are separated by relatively thick vesicle walls. Remnant vesicle walls in the "pockets" are thin and tapered.

FIGURE 3a. Pumice pyroclast from the top of the H_4 Plinian pumice-fall. Surface of a subangular pumice pyroclast. Pockets of coalesced vesicles form long, tapered tubes. These are surrounded by well-developed, ovoid, flattened vesicles. This pyroclast is partly covered with fine-grained shards.

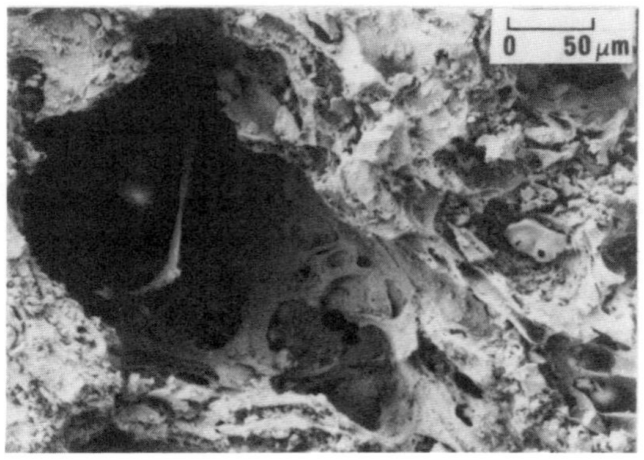

FIGURE 3b. Close-up of one of the tapered, tube-like vesicle pockets. It was formed by inflation and coalescence of many vesicles. Remnant vesicle walls are visible as flattened membranes and twisted fibers.

Peralkaline and Carbonatite Eruptions

Unusual volcanic rocks include the *peralkaline* rocks in which the abundance of Na_2O and K_2O is greater than that of Al_2O_3 and the *carbonatites,* which are composed of mostly carbonate minerals. Typically, the relatively low abundance of Al_2O_3 in peralkaline lavas causes magma viscosities that are much lower than their calcalkaline counterparts with the same silica content; hence pyroclasts from peralkaline eruptions have fluid textures. The abundance of CO_2 in carbonatites causes these magmas to be volatile-rich and eruptions to often be explosive.

PANTELLERIA, ITALY

Introduction: The island of Pantelleria is located in the Channel of Sicily, and its total area of 85 km^2 is underlain by mostly persodic volcanic rocks. Pantelleria is well known for its peralkaline silicic rocks (pantellerites), which were described by Washington (1914) and Villari (1974). Welded tuffs (Borsi et al., 1963) make up nearly half of the volcanic rocks of which the youngest, the Green Tuff, is best studied (Villari, 1969; Wright, 1980; Wolff and Wright, 1981a,b). Schmincke (1974) attributed the Green Tuff to a pyroclastic flow origin; Barberi et al. (1969) and Villari (1974) suggest, based on radiometric dates, that it was erupted within the last 0.12 m.y. (50,000 y. b. p.; Cornette et al., 1983). The widespread distribution of this well-sorted pumice, its mantle-bedding on steep slopes, and alternations of welded and nonwelded layers, are evidence that Sparks and Wright (1979), Wright (1980), and Wolff and Wright (1981a) have given for an air fall (tephra fall) emplacement origin. In fact, both modes of eruption and deposition could account for different parts of the Green Tuff.

The Monastero caldera, which is near the central part of the island, formed over the vent area of the Green Tuff. This vent area, a possible central vent, is older (Orsi, 1982 personal communication) than the Cuiddia di Mida pumice cone vent cited by Mahood and Hildreth (1980, 1983). The samples described here were obtained from G. Orsi and M. Sheridan: they were collected near the north rim of the Monastero caldera on the north side of Bagno dell'Acqua. The samples were from a pumice fall and welded ash flow just below and in contact with the Green Tuff (M. Sheridan, 1983 personal communication); they are pantellerites and may be part of the Green Tuff. In addition to the following descriptions, Schmincke (1974) describes ash from the Green Tuff in reference to overall characteristics of tephra from peralkaline, welded ash flow tuffs.

Tephra Description: The tephra consists of glass, crystal, and lithic fragments; the glass and crystals described here range in size from 0.1 to 0.5 mm in maximum dimension.

Glass. Because of the lower viscosity of peralkaline silicic magmas (as much as 3 orders of magnitude less than that of calcalkaline rhyolite), glass particles show both fluid-formed textures and those of brittle breakage. Typically, the glass is vesicular as a result of its magmatic fragmentation and shows partly ruptured bubbles (see Schmincke, 1974, p. 609). Some vesicles are elongated and half of the studied samples are hydrated; glass surfaces are covered by clay minerals as well as abundant feldspar laths. The posteruptive fluidity of the magma is evident in deformed vesicle surfaces and globular shards (Johnson, 1968; Hay et al., 1979) that resemble Pele's tears formed in Hawaiian lava fountains. Vesicle-wall shards and chips are common in finer size fractions and several generations of vesiculation are evident. Stretched and deformed vesicles developed during postemplacement (rheomorphic) movement of still-fluid pyroclasts are a characteristic of pantelleritic tuffs. Vesicular tricuspate (Y-shaped) shards that resemble calcalkaline Plinian ash are also evident. However, extreme aspect ratios of vesicles and fluidal surfaces along vesicle walls betray their fluid nature.

Crystals. Euhedral to subhedral crystals of anorthoclase, hedenbergitic clinopyroxene, aenigmatite, fayalite, apatite, and magnetite/ilmenite are abundant (>10 wt. %). Clinopyroxenes are distinct because of the glass encrustations that form bubble-wall textures. Feldspars, however, have clean surfaces with no adhering glass.

Chemical Composition: Energy-dispersive spectral (EDS) analyses of samples studied here are compared to previous analyses of fresh samples of the Green Tuff (Table 1). Hydration has affected the studied samples to some degree so that elemental abundances are somewhat altered. However, these analyses also suggest that the tephra at the base of the Green Tuff may be derived from the silica-enriched top of the magma chamber that erupted the Green Tuff.

TABLE 1. CHEMICAL COMPOSITION OF GLASS FROM THE GREEN TUFF.*

	1	2	3	4	5	6
SiO_2	70.0	69.2	74.74	68.30	74.67	72.24
TiO_2	0.5	0.44	0.61	1.04	0.47	0.73
Al_2O_3	7.4	7.8	8.62	14.68	8.75	10.99
$FeO + Fe_2O_3$	8.1	8.4	7.30	4.68	7.00	6.20
MnO	---	0.35	0.04	0.29	0.15	0.17
MgO	---	0.2	0.36	0.66	0.43	0.49
CaO	0.3	0.34	0.46	1.37	0.55	0.88
Na_2O	6.5	6.4	1.64	2.04	1.76	1.83
K_2O	3.6	4.5	6.23	6.94	6.21	6.45

*Analysis of sample 1 is taken from Wolfe and Wright (1981) and that of sample 2 is from Wright (1980), based upon data from Korringa and Noble. Normalized standardless EDS analyses of samples 3, 4, and 5 are those of partly altered pyroclasts of this report, and 6 is an average analysis of all samples studied in this report.

Comments on Tephra Formation: Schmincke (1974) discussed the physical and chemical aspects of peralkaline magmas and the formation of tephra from them. General inferences made from these physio-chemical aspects are that the magmas are hotter, lower in H_20, and of lower viscosity than calcalkaline silicic magmas.

The formation of spherical shards with hollow centers and subspherical globular particles may be the result of several processes: (1) Surface tension effects minimize surface free energy by spherical geometry for fluid fragments—a mechanism occurring during dispersal before deposition. (2) Posteruptive expansion of gas inside volatile-rich shards is especially possible for fluid melt fragments and is manifested by spherical vesicles that occur in collapsed pumice fragments; post-depositional vesicles transect earlier formed vesicles. (3) Finally, postdepositional movement of ash deposits may distort the original spherical shape of vesicles and blade-like shapes of shards (Wolff and Wright, 1981a,b).

All the above processes that contribute to the distinctive shape of peralkaline ash particles are strongly functions of viscosity. Abundant chlorine and fluorine, as well as the small amounts of alumina and the hot eruption temperatures, are responsible for the fluidity of these pyroclasts. That the initial fragmentation of these particles in the vent was entirely magmatic is not yet clear. The existence of both brittle, fractured shards and fluid-deformed particles, the inferred low-volatile content of the peralkaline magma, and the abundance of hydrated glass sugggest that the eruption may have been phreatoplinian or dominantly phreatomagmatic. This hypothesis also explains the uncommonly wide dispersal of the peralkaline deposits by both ash flow and ash fall.

REFERENCES:

Barberi, F., Borsi, S., Ferrara, G., and Innocenti, F., 1969. Strontium isotopic composition of some recent basic volcanites of the southern Tyrrhenian Sea and Sicily Channel. Contr. Mineral. Petrol., **23**: 157-172.

Borsi, B., Marnelli, G., Mazzoncini, F., Mittempergher, M., and Tedesco, C., 1963. Reconnaissance of some ignimbrites of Pantelleria and Eolian Islands. Bull. Volcanol., **25**: 359-367.

Cornette, Y., Crisci, G. M., Gillot, P. Y., and Orsi, G., 1983. Recent volcanic history of Pantelleria: a new interpretation. J. Volcanol. Geotherm. Res., **17**: 361-373.

Hay, R. L., Hildreth, W., and Lambe, R. N., 1979. Globule ignimbrite of Mount Suswa, Kenya. Geol. Am. Spec. Paper **180**: 43-75.

Johnson, R. W., 1968. Volcanic globule rock from Mt. Suswa, Kenya. Roy. Soc. London Trans., **265**: 383-412.

Mahood, G., and Hildreth, W., 1980. Pantelleria, a new interpretation. Trans. Am. Geophys. Union, **61**: 1141.

Mahood, G., and Hildreth, W., 1983. Nested calderas and trapdoor uplift at Pantelleria, Strait of Sicily. Geology **11(2)**: 722-726.

Schmincke, H.-U., 1974. Volcanological aspects of peralkaline silicic welded ash-flow tuffs. Bull. Volcanol., **38**: 594-636.

Sparks, R. S. J. and Wright, J. V., 1979. Welded air-fall tuffs. Geol. Soc. Am. Spec. Paper, **180**: 155-166.

Villari, L., 1969. On particular ignimbrites of the island of Pantelleria (Channel of Sicily). Bull. Volcanol., **38**: 689-724.

Villari, L., 1974. The Island of Pantelleria. Bull. Volcanol., **38**: 680-724.

Washington, H. S., 1914. The volcanoes and rocks of Pantelleria. J. Geol., **22**: 653-670 and 683-713.

Wolff, J. A. and Wright, J. V., 1981a. Formation of the Green Tuff, Pantelleria. Bull. Volcanol., **44**: 681-690.

Wolff, J. A. and Wright, J. V., 1981b. Rheomorphism of welded tuffs. J. Volcanol. Geotherm. Res., **10**: 13-34.

Wright, J. V., 1980. Stratigraphy and geology of the welded air-fall tuffs of Pantelleria, Italy. Geol. Rundsch., **69**: 263-291.

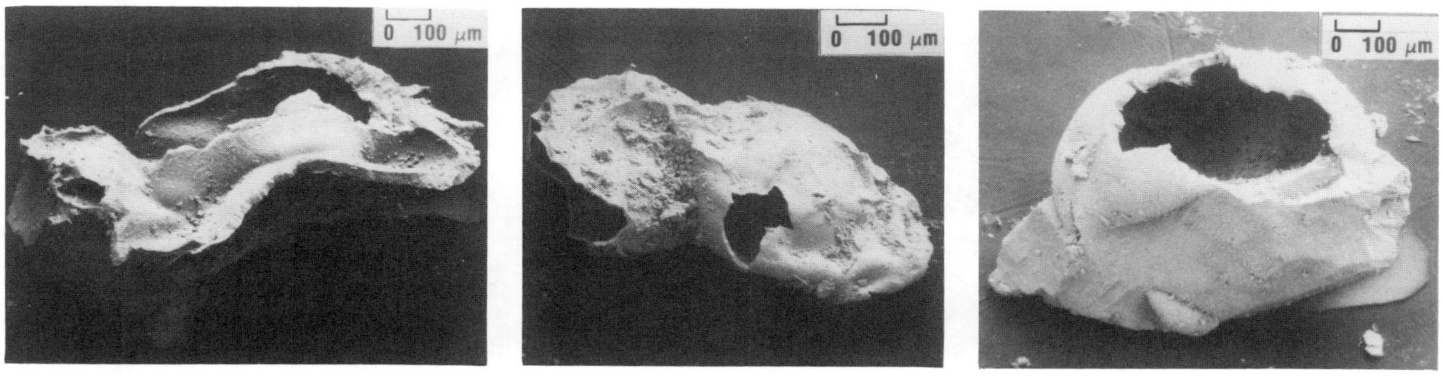

FIGURE 1. SEMs of three vesicular pyroclasts showing partly ruptured bubbles. Note the thin bubble walls; these vesicles may have grown after initial eruptive fragmentation. Spherical or subspherical shards from peralkaline eruptions are often hollow and contain evident secondary vapor-phase minerals (*center*) (Schmincke, 1974).

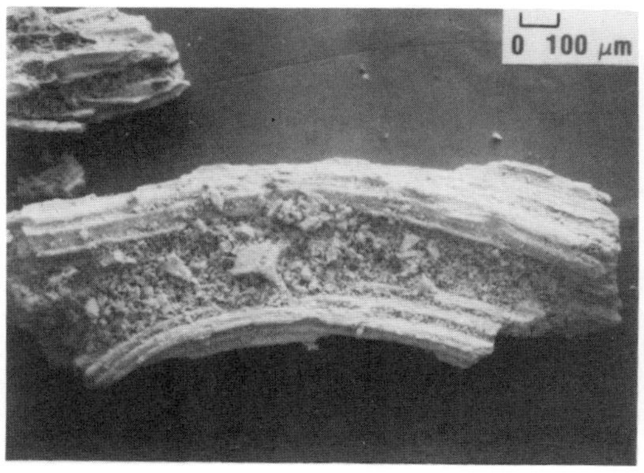

FIGURE 2. SEM of elongate pumice shard with tubular vesicles that have either been stretched during extrusion or flattened during postdepositional compaction. The vesicles are filled with adhering material, most of which is composed of the alteration products, clays and zeolites, that were formed during hydration of the glass. Some of the material is glassy volcanic dust scavenged by shards from the eruptive cloud during dispersal.

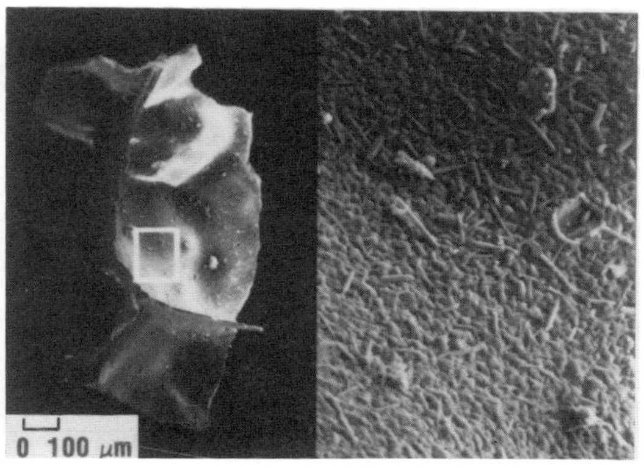

FIGURE 3. Split-image SEM, a 10X enlargement of the surface of a vesicular shard. Note the intricate crust of interlocking feldspar laths, many of which result from vapor-phase crystallization on the glass.

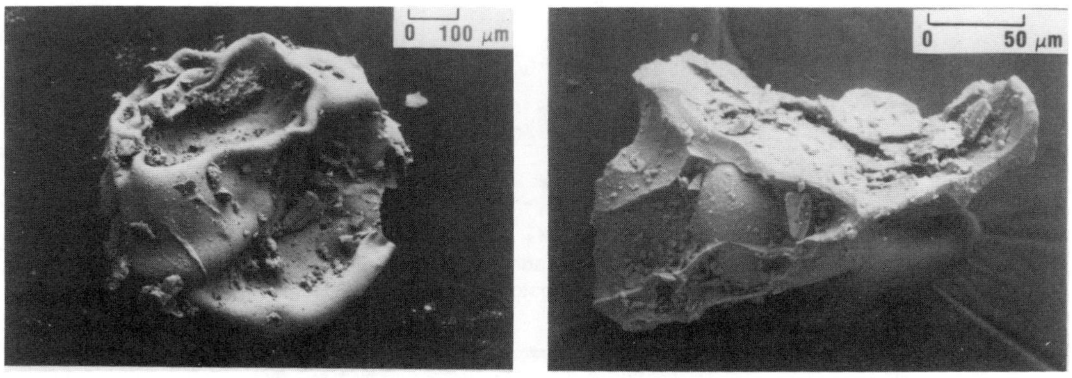

FIGURE 4. SEMs of globular shards. The distorted vesicle surfaces of the shard (*left*) and its ropy surface texture indicate the fluidity of the glass and its rheomorphic behavior after eruption. *Right*—a shard consisting of a subspherical globule surrounded by prong- and blade-like shards. Prongs of the vesicular shard appear to have broken in a brittle fashion, which is unlike the more ductile behavior of the globules.

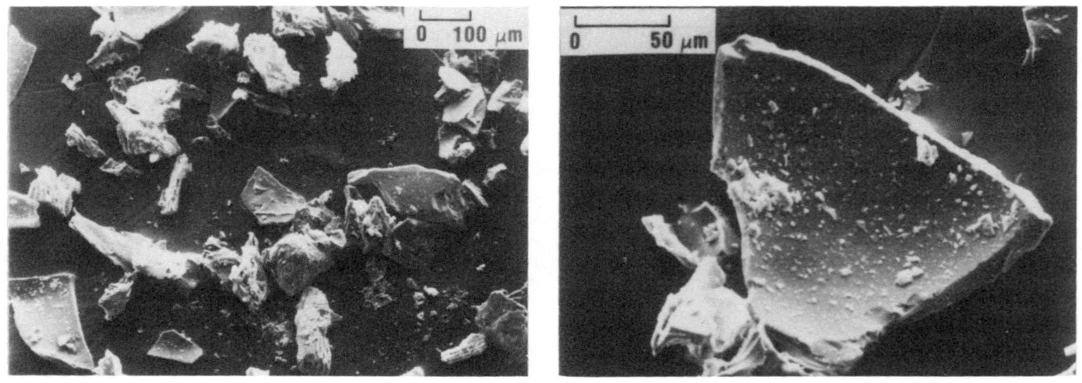

FIGURE 5. SEMs of fractured bubble walls that form curved splinters and chips and constitute much of the fine ash fraction. Globules are absent in the fine debris, and the material appears to have brittly fractured. Fine, adhering dust particles cover the surface of plates and splinters. Hydration cracks are evident on some shards.

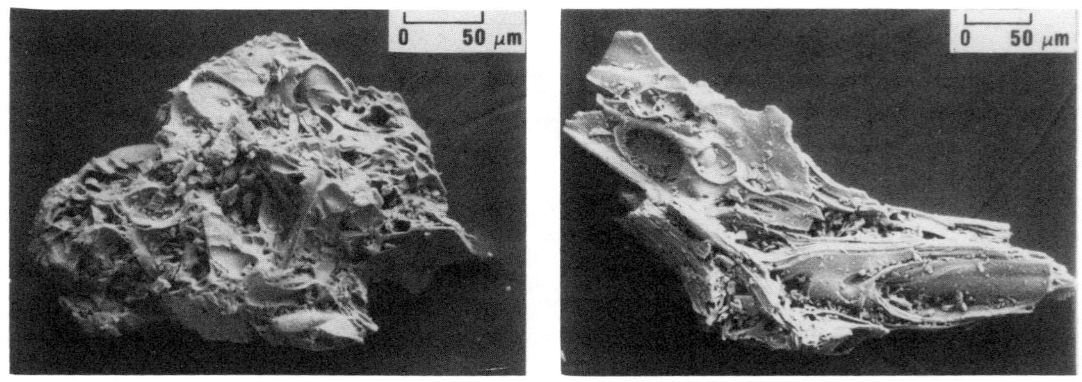

FIGURE 6. SEMs showing multiple vesiculation and deformed vesicles. Secondary vesicle formation is evidenced by bubble walls transecting earlier formed bubbles (*left*) and tubular, stretched vesicles that may be bent as a result of rheomorphic behavior of ash deposits after emplacement.

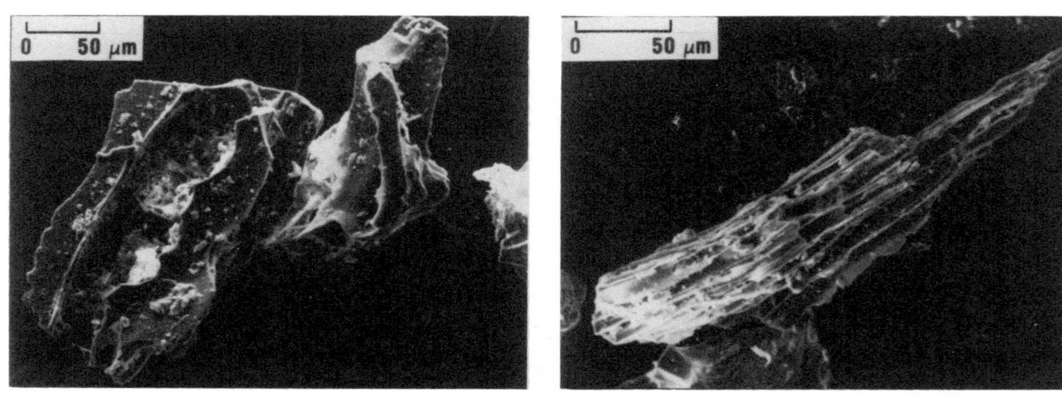

FIGURE 7. SEMs of vesicular shards that resemble typical calcalkaline silicic ash from Plinian eruptions. Abundant, fine, adhering dust is evident on the surface of the vesicle walls. The pumice pyroclast (*right*) shows the typical elongate vesicles.

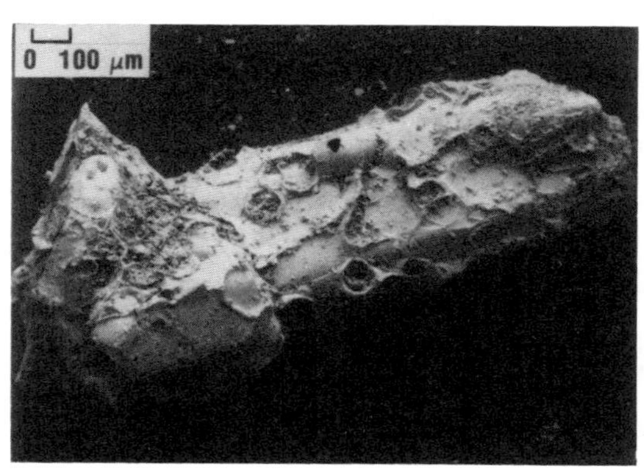

FIGURE 8. SEM of a clinopyroxene phenocryst surrounded by a glassy rind with bubble-wall texture. Note the burst vesicle that formed in the glass coat; it is evidence that the glass was still fluid after the fragment separated from the magma.

FIGURE 9. Sem of a feldspar phenocryst with a clean surface that shows no glass encrustation. Cleavage traces are apparent on this backscattered electron image.

OLDOINYO LENGAI, TANZANIA; 1966 ERUPTION

Introduction: Oldoinyo Lengai is part of the Neogene volcanic province of northern Tanzania and is located 19 km south of Lake Natron. The dominantly pyroclastic cone, with an elevation of 1,920 m, has also produced lavas in which the general trend is phonolite-nephelinite-melanephelinite. Volcanic ash and lavas from the 1966 explosive eruption consist of Ca-Na carbonatites. The eruptions varied from minor emissions of ash to major Plinian and Vulcanian-type behavior. An ash sample from the 1966 eruption was provided by J. B. Dawson.

Grain Size: None available.

Tephra Description: Samples received from the Institute of Geological Sciences, London (Dawson et al., 1968) consist of gray volcanic ash that contains individual mineral pyroclasts and accretionary aggregates of mineral grains cemented by trona. In addition to sodium carbonate (trona) grains, there are phenocrysts of nepheline, pyroxene, wollastonite, apatite, melanite, and pyrrhotite. The ash was originally black but turned white after deposition because of the growth of trona crystals on grain surfaces. The ash consists of mineral fragments (phenocrysts) and aggregates of irregular and tabular trona crystals.

Chemical Composition: The unusual chemical composition of this volcanic ash is shown in Table 1.

TABLE 1. CHEMICAL COMPOSITION OF ASH FROM OLDOINYO LENGAI.*

	(Wt. %)
SiO_2	25.23
TiO_2	0.46
Al_2O_3	5.72
Total FeO	8.71
MnO	0.38
MgO	1.62
CaO	14.10
Na_2O	18.65
K_2O	4.89
P_2O_5	1.49
BaO	1.00
SrO	0.77
SO_3	1.23
S	0.31
Cl	1.13
F	0.62
H_2O^+	1.15
H_2O^-	4.18
CO_2	8.75
Total	100.39

*Analyzed at the Institute of Geological Sciences, London. Analyst was P. Bowden (from Dawson et al., 1968).

Comments on Pyroclast Formation: It is obvious from the Dawson et al. (1968) descriptions of the Oldoinyo Lengai eruption that carbonatite magmas are volatile-rich and may be very energetic; ash analyses support the observation that these magmas are rich in volatile phases. Irregular aggregates of sodium carbonate and phenocrysts were erupted. Carbonatite deposits of the Laetolil Beds of Tanzania near Ngorongoro Caldera are believed to have been ash falls from a volatile-rich eruption of very fluid magma (Hay, 1978). Hay observed numerous carbonatite spheres in these deposits, which implies formation in fountains of low-viscosity magma.

Carbonatite ashes are rapidly altered and might be difficult to recognize. Most of the trona crystals in the Oldoinyo Lengai ash are secondary, having grown on grain surfaces after they were saturated with rain water. Hay (1979) proposes that carbonatite ashes are first cemented by soluble salts such as trona. Diagenetic alteration in the vadose zone forms rocks consisting mainly of calcite, montmorillonite, and phillipsite. Sparry calcite was precipitated as a replacement cement and a pore filling. Carbonatite globules (droplets) are replaced by calcite pseudomorphs. Much of the diagenetic process must be rapid, as footprints, delicate fossils, and raindrop imprints are preserved in these tuff beds. Finer grained carbonatite ash beds are indistinguishable from calcretes of non-carbonatite origin (Hay and Reeder, 1978).

REFERENCES: Dawson, J., Bowden, P., and Clark, G., 1968. Activity of the carbonatite volcano Oldoinyo Lengai, 1966. Geol. Rundschau, **57**: 865-879.

Hay, R., 1978. Melilitite-carbonatite tuffs in the Laetolil beds of Tanzania. Contrib. Mineral. Petrol., **67**: 357-367.

Hay, R. and Reeder, R., 1978. Calcretes of Olduvai Gorge and the Ndolanya Beds of Northern Tanzania. Sedimentology, **25**: 649-673.

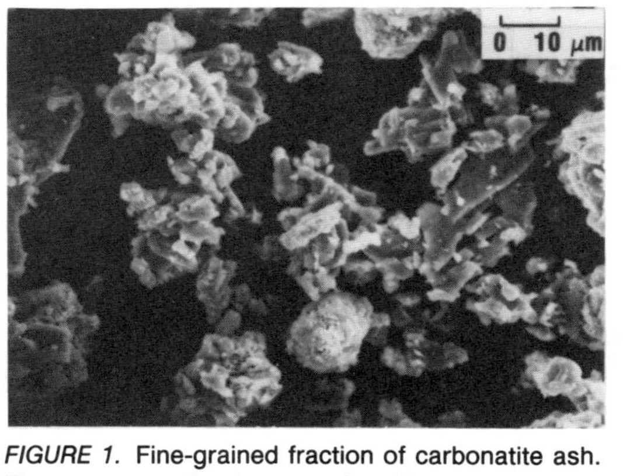

FIGURE 1. Fine-grained fraction of carbonatite ash. There are mostly clusters of sodium and calcium carbonate; these secondary overgrowths have obscured the phenocrysts.

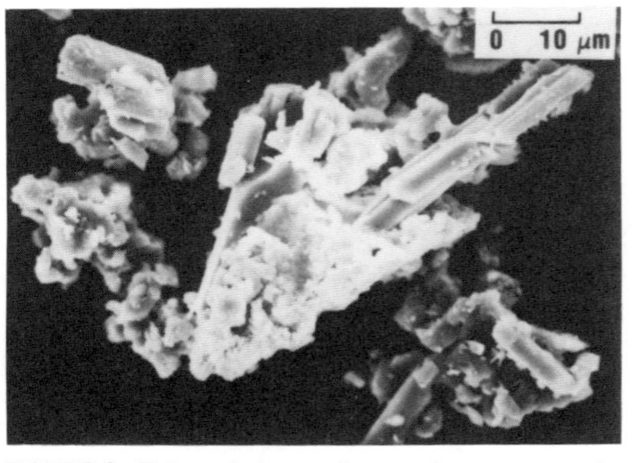

FIGURE 2. Thin, tabular sodium carbonate crystals bonded by an irregular aggregate of the same composition.

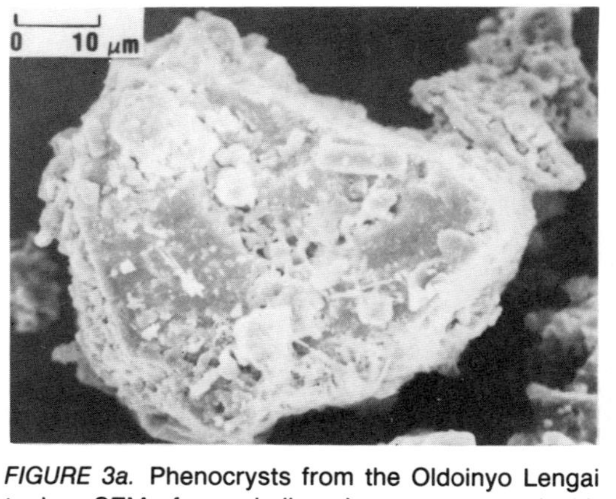

FIGURE 3a. Phenocrysts from the Oldoinyo Lengai tephra. SEM of a nepheline phenocryst coated with both irregular aggregates and well-developed tabular crystals of sodium carbonate.

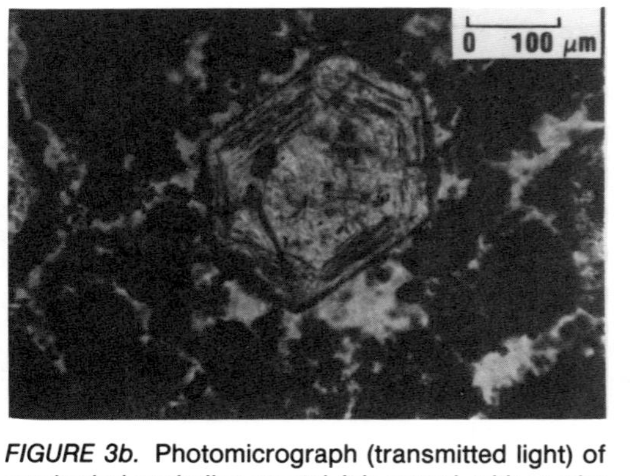

FIGURE 3b. Photomicrograph (transmitted light) of a euhedral nepheline crystal. It is zoned, with apatite crystals growing parallel to the grain rim. Dark clusters around the phenocryst are carbonate aggregates similar to those of Fig. *1*.

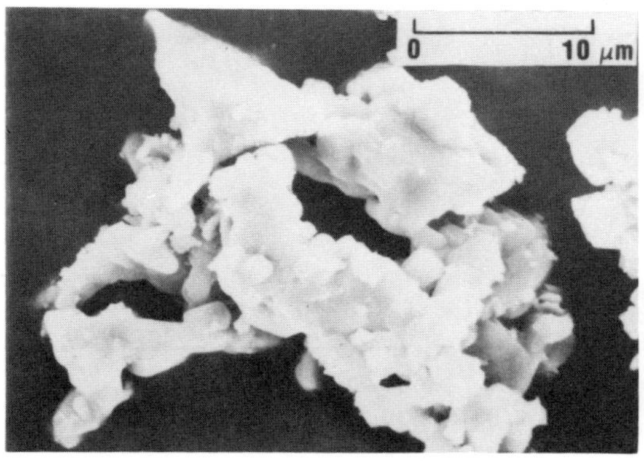

FIGURE 4a. Sodium carbonate (trona) and calcium carbonate phases that grew after ash deposition. Irregular clusters typical of the ash today.

FIGURE 4b. Well-developed trona crystals.

Eruption of Capelinhos, Azores, 1958.

PHREATOMAGMATIC ERUPTIONS

Explosive, ash-producing eruptions may occur when rising magma encounters ground or surface water. Volcanic ash deposits produced during these eruptions are unique and their characteristics may be used to interpret eruptive mechanisms. Generally, two types of vent landforms result from phreatomagmatic eruptions of monogenetic volcanoes. Steep-sloped, cone-shaped vents with wide craters are called *tuff cones* and consist mostly of highly palagonitized and altered tephra that is thickly bedded. *Tuff rings* are vents with a low-profile apron of tephra, which surrounds a wide crater that is often below the general topographic level of the area (termed a *maar* crater); they consist of mostly nonpalagonitized, thinly bedded tephra that are often referred to as a *base surge* or *pyroclastic surge* deposit. Tuff ring and tuff cone deposits may also be associated with polygenetic volcanoes, such as composite cones, and are commonly interbedded with lavas or other pyroclastic debris. Recently, an additional set of terms has been adopted for phreatomagmatic deposits; these terms refer to whether the eruption involved abundant water or not. When phreatomagmatic eruptions occur in lakes or other places where water is abundant, vast amounts of condensing or wet steam are usually erupted with the tephra. Accordingly, these deposits are characterized by bedding and textures referred to as *wet*, such as the *wet-surge* deposits found in tuff cones. Conversely, if little water is involved during eruption, dry or superheated steam is erupted with tephra; hence, *dry-surge* deposits that are usually found in tuff rings are characterized by textures produced by dryer conditions than those of wet surges.

Tuff Cones and Tuff Rings (Mafic Magmas)

Because mafic magmas are generally of low viscosity, their contact with external water results in formation of tephra that are characterized by fluidal textures as well as by the typical blocky textures resulting from brittle behavior. Grain-size distributions of phreatomagmatic ash reflect the energy of the eruption; the strong phreatomagmatic eruptive component produces fine ash (<64 μm in diameter).

SURTSEY, VESTMANN ISLANDS, ICELAND

Introduction: Surtsey is a volcano that erupted in the ocean south of the Icelandic coast from 1963 through 1965. Eventually, the two tuff cones Surtur I and II formed a new island. In his book *Surtsey, The New Island in the North Atlantic*, Thorarinsson (1966) gave day-by-day accounts of the growth of the tuff cone. The main cone, Surtur II, is about 500 m in diameter and has a crater about 60 m below the rim and about 90 m above sea level.

The initial phreatomagmatic eruption of Surtsey excellently characterized this type of activity. The earlier phases of the eruption were typified by rapid cycles of sea water pouring into the vent and alternating with steam explosions as the water came into contact with the rising basaltic magma. This phreatomagmatic activity is now termed *Surtseyan* at other volcanoes (Walker and Croasdale, 1972). Deep explosions produced vertical eruption columns, whereas shallow explosions, caused by interaction of sea water and magma near the surface, produced jets of material that expanded into hemispherically shaped clouds (Thorarinsson, 1966; Thorarinsson et al., 1964). Base surges were associated mainly with the latter events. When the growing cone completely surrounded the vent and shut out sea water, Hawaiian and Strombolian activity began, forming a spatter cone and a small apron of lava flows.

The cone of Surtur II is composed of a number of beds, many of ash fall origin and others that are massive and planar pyroclastic surge deposits. The dips on the beds are variable—nearly horizontal at the base and becoming steeper near the top. Jakobsson (1978) has studied the consolidation and palagonitization of the Surtsey tuffs. He found that shortly after eruption the tuffs were largely unconsolidated and 82 to 88% (by volume) unaltered sideromelane. Consolidation and palagonitization increased over a period of several years.

Recent drilling at Surtsey has revealed a stratigraphy similar to that found at Heimaey (Tomasson, 1967). Pillow lavas occur at depth; they are overlain by hyaloclastites and then by tuffs and breccias. M. F. Sheridan provided these samples of surge ashes from an earlier study (Sheridan, 1971), and S. Thorarinsson provided a fallout sample collected from the deck of a Coast Guard vessel 3 km downwind from Surtsey on December 1, 1963.

Grain Size: Surtsey ash is moderately to poorly sorted in both fall and surge samples. The fall sample (Thorarinsson) has a median size of 0.09 mm (3.5φ), whereas that collected on Surtsey (sample 9) is coarser (1.16 mm; −0.2φ). All surge samples have an abundant fine-ash fraction of 14 wt.% or greater. Figure 2 shows ternary size distributions of the ash from data given in Table 1.

TABLE 1. GRAIN-SIZE DESCRIPTION OF SURTSEY TEPHRA.

Sample	Bed Form*	Md_φ (mm)	σ_φ	Fine Ash (Wt. %)
1	M	1.62 (0.33)	2.48	17.42
2	P	0.57 (0.67)	3.06	14.00
3	M	1.16 (0.45)	2.22	12.46
4	P	0.25 (0.84)	3.15	13.74
5	M	1.48 (0.36)	3.03	26.09
6	M	1.60 (0.33)	2.55	20.33
7	M	1.15 (0.45)	3.01	18.28
8	M	1.26 (0.42)	2.76	16.30
9	F	−0.22 (1.16)	2.17	6.62

*Bed forms are S = sandwave surge, M = massive surge, P = planar surge, and F = fall.

Tephra Description: Vitric pyroclasts can be divided into three types of relatively equal abundance in the samples studied. (1) Vesicular ash dominates in ash fall deposits. Vesicles are spherical and irregularly shaped, and they often determine the overall particle shape. Some vesicular particles, however, show planar breakage surfaces crossing vesicle walls. (2) The second ash type dominates the phreatomagmatic surge deposits. These grains are very angular and poorly vesicular, and they have shapes dominated by curviplanar breakage surfaces. Small grains have some adhering dust, especially in vesicle hollows. Edge modification is developed and manifested as chips, dish-shaped fractures, and hackly or rough areas, especially along grain edges where two planar surfaces intersect. These features are best developed on larger grains that had significant transport abrasion. (3) Finally, a third ash type shows smooth fusiform surfaces between vesicles. These are mostly drop-like pyroclasts.

All samples show varying degrees of alteration to palagonite, which changes the overall appearance of grains. A hydrated skin forms first in vesicles and then over the entire grain surface as palagonitization increases. Note that these samples were collected before extensive palagonitization. Hydrated surfaces tend to crack after extensive alteration. Cracked walls reveal a 1- to 5-µm-thick skin that curls away from the grain along edges in much the same way that desiccation cracks appear in mud.

The sample collected on shipboard has fragments with smooth, planar fracture surfaces that are broken by curved depressions where fractures intersect vesicle surfaces. Vesicularity is low in this sample, suggesting that vesicles must have been widely spaced and poorly developed when fragmentation occurred.

Chemical Composition: A representative analysis of Surtsey basalt from Thorarinsson (1967) is given in Table 2. Table 3 lists a petrographic description of samples 1 through 9; samples from massive deposits show greater amounts of palagonitization than do those from planar beds.

TABLE 2. CHEMICAL COMPOSITION OF SURTSEY TEPHRA.*

	(Wt. %)
SiO_2	46.5
TiO_2	2.3
Al_2O_3	16.8
Fe_2O_3	1.7
FeO	10.8
MnO	0.2
MgO	7.6
CaO	9.4
Na_2O	3.3
P_2O_5	0.3
Total	98.90

*From Thorarinsson, 1967.

TABLE 3. PETROGRAPHIC DESCRIPTION OF SURTSEY TEPHRA.

Sample	500 μm	250 μm	88 μm
1	75% Sideromelane 24% Palagonite	75% Sideromelane 24% Palagonite	97% Sideromelane 3% Crystals* tr. Lithics
2	90% Sideromelane 10% Palagonite tr. Crystal	60% Sideromelane 40% Palagonite	78% Sideromelane 20% Palagonite 2% Crystal and lithic
3	59% Sideromelane 40% Palagonite 1% Crystal	99% Sideromelane 1% Crystal	99% Sideromelane 1% Crystal
4	89% Sideromelane 10% Palagonite 1% Crystal	99% Sideromelane 1% Crystal	99% Sideromelane 1% Crystal
5	79% Sideromelane 20% Palagonite tr. Crystal	59% Sideromelane 40% Palagonite 1% Crystal	58% Sideromelane 40% Palagonite 2% Crystal
6	70% Sideromelane 30% Palagonite	99% Sideromelane 20% Palagonite 1% Crystal	89% Sideromelane 10% Palagonite
7	60% Sideromelane 40% Palagonite	98% Sideromelane 1% Palagonite 1% Crystal	95% Sideromelane 5% Palagonite
8	15% Sideromelane 85% Palagonite tr. Crystal	30% Sideromelane 70% Palagonite tr. Crystal	40% Sideromelane 60% Palagonite
9	24% Sideromelane 75% Palagonite 1% Crystal	38% Sideromelane 60% Palagonite 1% Crystal 1% Lithics	90% Sideromelane 10% Crystal

*Crystals: plagioclase > olivine > clinopyroxene.

Comments on Pyroclast Formation: Surtsey went through stages of phreatomagmatic eruption early in its development; they were followed by Hawaiian and Strombolian ejections and lava flows. Blocky, vitric pyroclasts, showing varying degrees of palagonitization, resulted from sea water interacting with magma as it erupted.

The degree of vesicle development depends in part upon the depth of the magma/water interaction and volatile content of the basalt. Most of the vesicle growth in basaltic magmas apparently occurs within 1 km of the surface (Heiken, 1971). Magma/water interactions deep within the vent can produce nearly vesicle-free tephra, whereas those formed at or near the ground surface may produce scoria. The amount of water pouring into the vent helps determine the energy of vapor explosion and the particle morphology. Where mixing of water is minimal, the magma is fragmented by thermal and acoustic stress waves propagating into the vent. Fusiform particles have drop-like shapes and fluid-formed surfaces between breakage surfaces; these resulted from abundant water physically mixing with magma before quenching and thereby turbulently fragmenting the fluid melt. Some eruptions involved so much water that slurries of ash and water were weakly ejected and flowed from the vent like lahars (Thorarinsson, 1966). Vesicular pyroclasts were ejected during Strombolian phases when access of water to the vent was cut off. This activity dominated the ending phases of explosions when the vent was above sea level and volatiles in expanding vesicles tore apart the lava as it moved out of the vent.

REFERENCES:

Heiken, G., 1971. Tuff rings: examples from the Fort Rock-Christmas Lake Valley, south-central Oregon. J. Geophys. Res., **76**: 5615-5626.

Jakobsson, S. P., 1978. Environmental factors controlling the palagonitization of Surtsey tephra, Iceland. Geol. Soc. Denmark Bull., **27**: 91-105.

Sheridan, M. F., 1971. Textural analysis of Surtsey tephra—a preliminary report. Surtsey Research Progress Report, **6**: 1-4.

Thorarinsson, S., 1967. *Surtsey—The New Island in the North Atlantic.* New York, Viking Press, 47 pp.

Thorarinsson, S., Einarsson, T., Sigvaldasson, G. E., and Ellison, G., 1964. The submarine eruption off the Vestmann Islands, 1963-64. Bull. Volcanol., **27**: 1-11.

Tomasson, J., 1967. On the origin of sedimentary water beneath Vestmann Islands. Jökull, **17**: 300-311.

Walker, G. P. L. and Croasdale, R., 1972. Characteristics of some basaltic pyroclastics. Bull. Volcanol., **35**: 303-317.

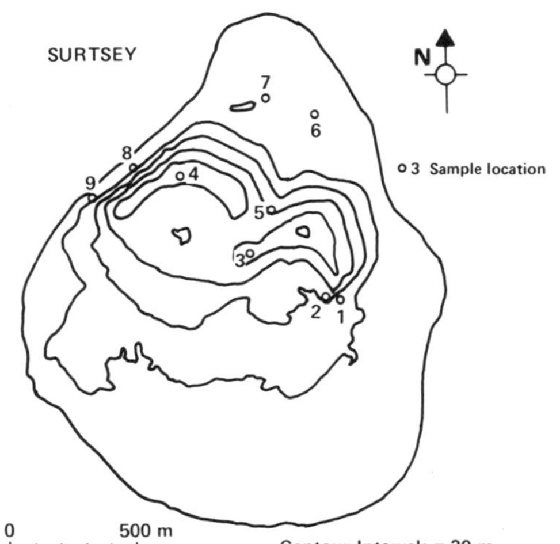

FIGURE 1. Topographic map of Surtsey showing sample locations around the tuff cones.

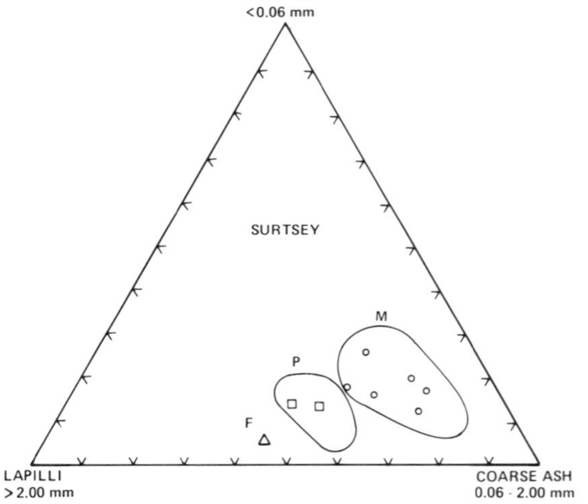

FIGURE 2. Ternary diagram of tephra-size analyses showing fields of planar (P) and massive (M) pyroclastic surge beds.

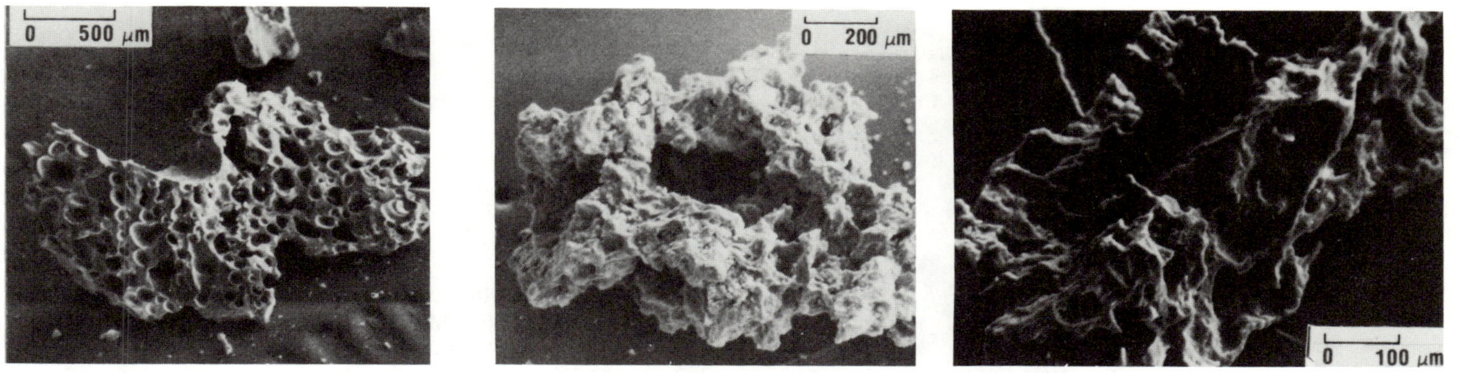

FIGURE 3. SEMs of vesicular particles produced by Strombolian eruptions during which water did not have access to the vent. Note the abundant vesicle surfaces, their irregular shapes, and the jagged outline of particles.

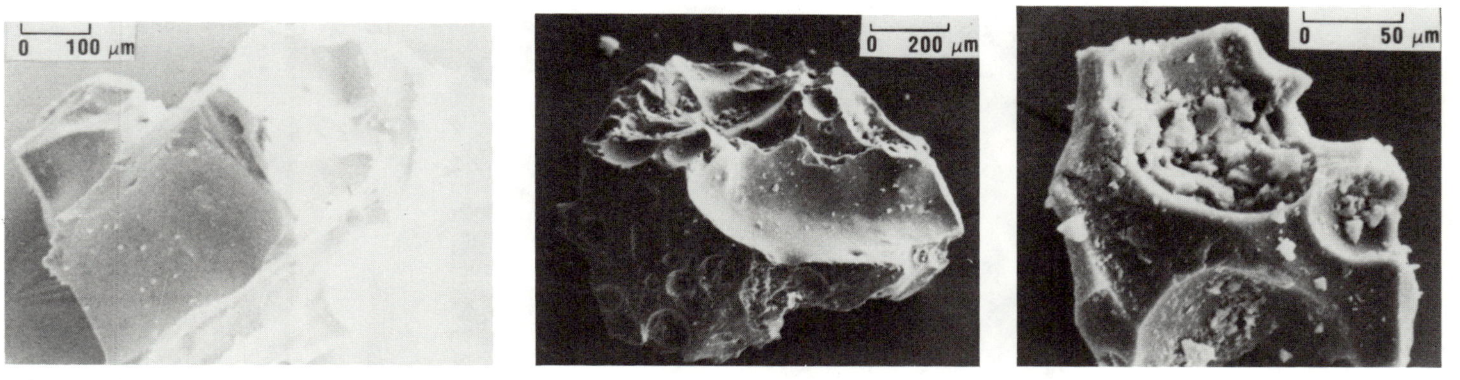

FIGURE 4. SEMs of blocky pyroclasts that have curviplanar fracture surfaces cutting vesicles. Many of the corners where planar surfaces meet have been chipped by transport abrasion. The fine-grained particle *right* shows planar fracture surfaces as well as adhering dust.

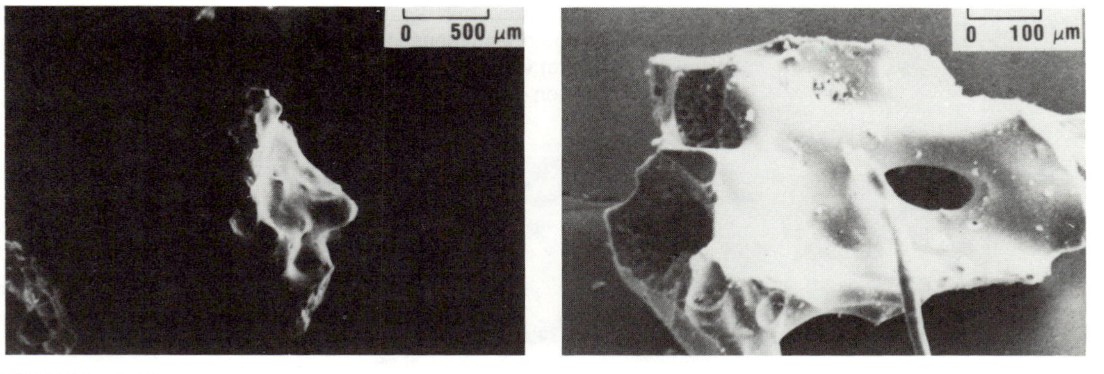

FIGURE 5. SEMs of a drop-like particle *left* and one with fused surfaces between vesicles, which obscure original fracture surfaces. These particles were fragmented by stress waves before they solidified. The smooth, irregular surfaces may also indicate actual physical mixing of particles with water during fragmentation.

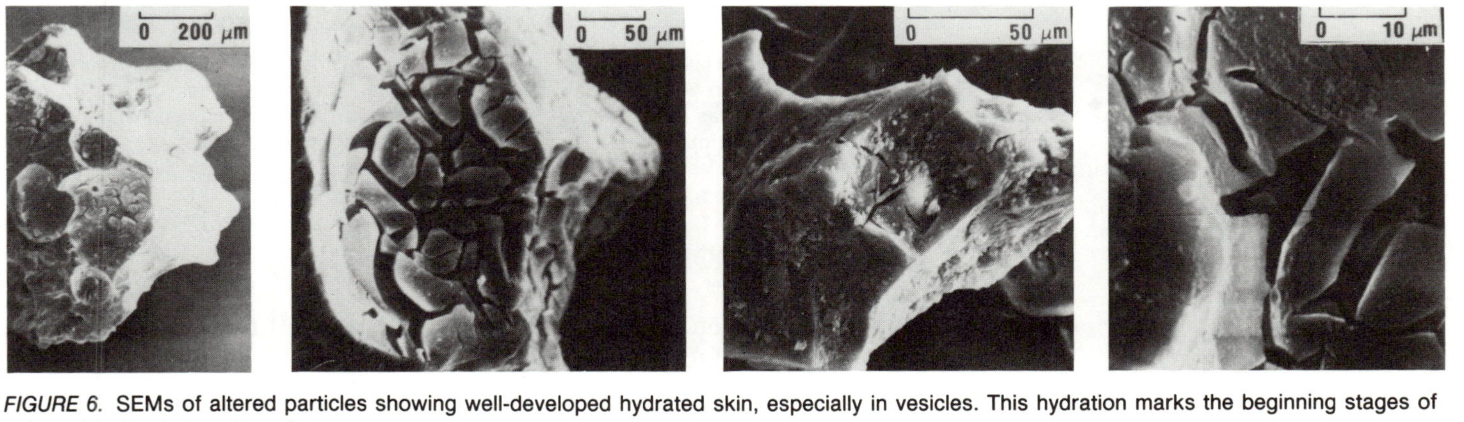

FIGURE 6. SEMs of altered particles showing well-developed hydrated skin, especially in vesicles. This hydration marks the beginning stages of palagonitization in basaltic ash.

FIGURE 7. General SEM view of the Surtsey fallout tephra. Most of the pyroclasts are angular sideromelane (basaltic glass) fragments. Curved surfaces are former vesicle-wall fragments.

FIGURE 8. SEM stereomicrograph of Surtsey fallout tephra. Note both smooth and conchoidal fracture surfaces. Fine-grained (<63-μm) vitric pyroclasts clinging to larger grain surfaces are common in phreatomagmatic tephra.

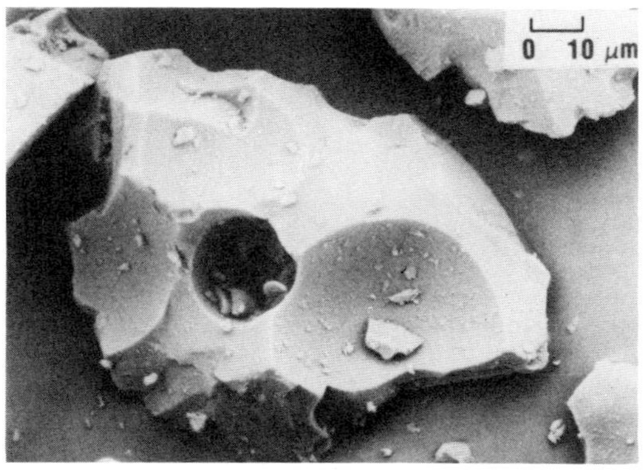

FIGURE 9. SEM closeup of pyroclasts similar to those of Figs. *7* and *8*. There were vesicles in the magma from which this fragment was derived; however, vesicle walls were thick (>20 μm).

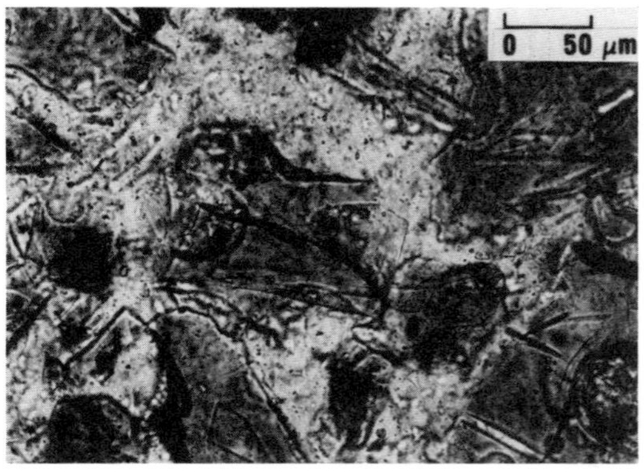

FIGURE 10. Transmitted-light photomicrograph of Surtsey tephra. These are blocky, angular, vitric pyroclasts (light brown basaltic glass) with traces of feldspar laths. Most have remnants of vesicles (curved surfaces).

CAPELINHOS, AZORES; 1957-1958 ERUPTION

Introduction: A submarine eruption began on September 27, 1957, off the west coast of the island of Faial, Azores. All early phases of the activity consisted of phreatomagmatic explosions that constructed a tuff ring. Ash deposition was by base surge and ash fall. Phreatomagmatic activity continued until April 1958, when access of sea water to the crater was cut off. Lava fountaining and lava flows occurred until October 25, 1958 (Servico Geologicos de Portugal, 1959; 1962). The sample described here is from the phreatomagmatic activity and was provided by the Portuguese Geological Survey.

Grain Size: The sample has a median grain size of 132 μm and is poorly sorted. A large sample would most likely include bombs and blocks.

Tephra Description: Much of the tephra consists of small, equant or subequant, basaltic glass fragments. All are angular and bounded by flat, smooth, or conchoidal surfaces or by smooth, curved surfaces where large ovoid vesicles were broken during fragmentation. There are some wedge-shaped and pointed vitric pyroclasts. Most of the sample consists of sideromelane (basaltic glass) pyroclasts with small amounts of tachylite and feldspar, olivine, and clinopyroxene phenocrysts. Vesicularity is very low; only rarely are individual pyroclasts vesicular. Pyroclast surfaces are partly coated with small (0.5- to 4-μm-long), angular vitric shards.

Chemical Composition: Table 1 shows chemical analyses of basaltic pyroclasts erupted from Capelinhos.

TABLE 1. CHEMICAL COMPOSITION OF BASALTIC PYROCLASTS FROM CAPELINHOS, AZORES.*

	Ash (Wt. %)	Bomb (Wt. %)
SiO_2	47.59	46.66
TiO_2	2.65	2.06
Al_2O_3	17.05	19.02
Fe_2O_3	2.97	3.29
FeO	7.44	7.47
MnO	0.20	0.06
MgO	5.47	5.86
CaO	9.09	10.26
Na_2O	4.57	3.28
K_2O	1.85	1.52
P_2O_5	0.50	0.68
Total	99.33	100.16

*Guimares and Viera, analysts, Museu e Laboratorio Mineralogico e Geoligico da Faculdade de Ciencias de Lisboa.

Comments on Pyroclast Formation: Fine-grained ash, consisting of blocky, angular shards with low vesicularity, is formed during interaction of magma and water. Rising basaltic melt with a few well-developed vesicles is mixed with sea water, chilled, fragmented, and erupted as part of a water-steam-tephra mixture. Pyroclast surfaces are mostly fracture planes; uniformity of grain size and particle shape indicates that the process is thorough, with uniform comminution of the glass produced during the magma/water interaction.

REFERENCES:

Servico Geologicos de Portugal, 1959. Le volcanisme de l'isle de Faial et l'eruption de volcan de Capelinhos. Mem. 4 (Nova Serie), 100 pp.

____________________, 1962, Le volcanism de l'isle de Faial et l'eruption de volcan de Capelinhos. Mem. 9 (Nova Serie), 54 pp.

(Note: both monographs listed above provide excellent documentation of the year-long eruption of Capelinhos).

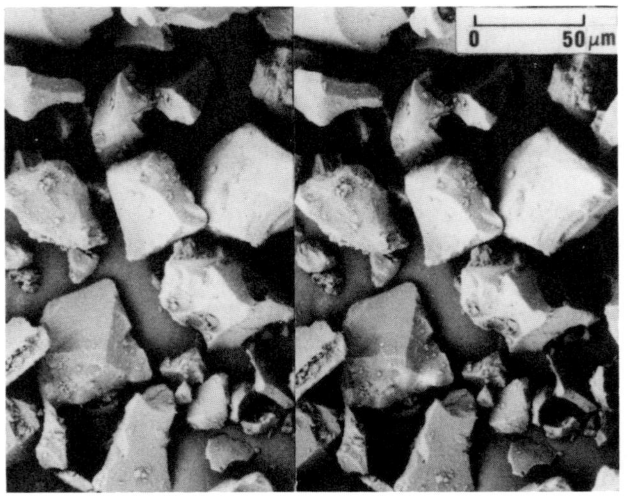

FIGURE 1. Stereomicrograph of ash typical of the phreatomagmatic activity of Capelinhos. All blocky pyroclasts with smooth or conchoidal surfaces are composed of basaltic glass. Smooth, curved surfaces are vesicle-wall remnants.

FIGURE 2. Detailed SEM view of vitric pyroclasts that illustrates surface textures, mainly smooth and conchoidal surfaces. Surfaces are coated, in part, by 0.5- to 5-μm-long, angular, vitric fragments.

FIGURE 3. SEM of vitric (basaltic glass) pyroclasts. Many of these have at least one surface that is part of a vesicle wall. All surfaces are smooth, curved, and partly filled with small, angular pyroclasts similar to those described in Fig. *2*.

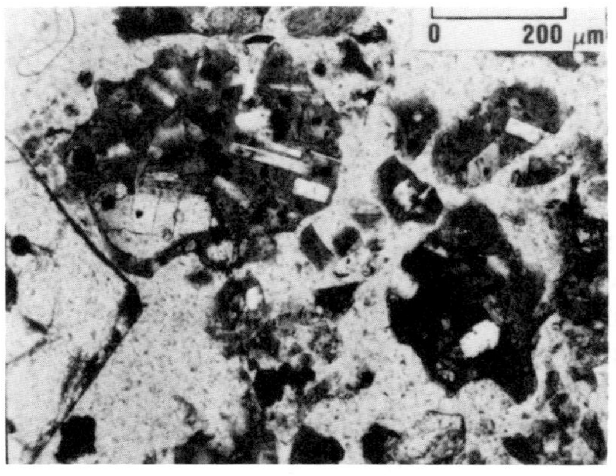

FIGURE 4. Phenocryst-rich, vitric pyroclasts. Photomicrograph in transmitted light. Most pyroclasts are subequant, angular, vesicle-poor sideromelane (basaltic glass). Phenocrysts are feldspar, olivine, and clinopyroxene.

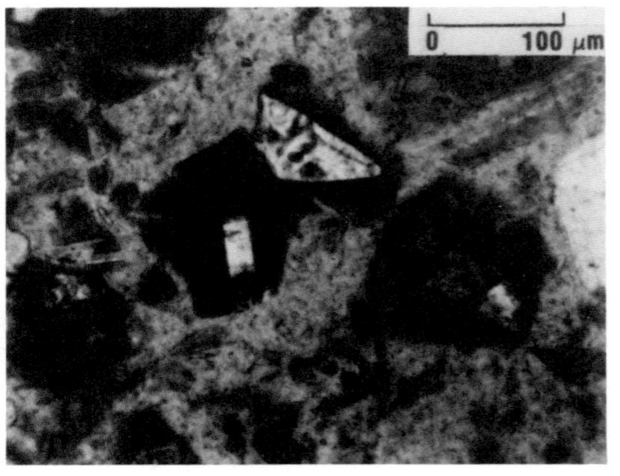

FIGURE 5. Closer view of the ash shown in Fig. *4*. Vitric pyroclasts in this size range are free of vesicles.

KOKO CRATER, HAWAII

Introduction: Koko Crater is a tuff cone that was formed during Quaternary time (Stearns, 1961). It belongs to the Honolulu group, a series of cones formed during at least 30 eruptions of silica-undersaturated basalt. These eruptions are part of the late-stage, posterosional activity that is typical of the Hawaiian Island chain. Other well-known vents in this group are tuff rings and cones such as Diamond Head, Salt Lake Craters, Punchbowl, Manana Island, and Pali. Almost all these tuffs are, to some degree, palagonitized. They have been discussed by Hay and Iijima (1968), Wentworth (1926), Wentworth and Winchell (1947), and Winchell (1947). Fisher

(1977) studied stratigraphic relationships among surge deposits from Koko Crater and other craters in the vicinity.

Koko Crater is 1000 m in diameter and 330 m in elevation from the crater base to the topographic rim. The stratigraphy is complicated by onlapping deposits from Koko Head and Kahauloa Craters, which are located about 1 km to the SE. Beds near the base of the cone dip gently away from the crater at 5 to 10°, whereas beds on the cone dip 20 to 30°. Koko Crater tuffs can be divided into three units: (1) a basal breccia and lapilli fall, (2) a tuff ring sequence of thinly bedded surge deposits, and (3) the cone-building unit composed of highly altered, poorly stratified massive beds.

Grain Size: Table 1 shows the results for the 18 samples analysed. These samples were taken from the dominant bed-form types, which include planar and massive surge beds and ash fall lapilli beds that resulted from short Strombolian bursts during the predominantly phreatomagmatic activity. Massive beds show the finest median grain sizes (0.50 mm = 1φ), and as much as 11 wt.% of the sample is fine ash (<0.06 mm). These beds also show the poorest sorting ($\sigma_\varphi \simeq 2.00\varphi$). Ash fall lapilli beds are better sorted ($\sigma_\varphi \simeq 1.75\varphi$) than surge deposits and have median diameters of 1.00 mm (0.0φ).

TABLE 1. GRAIN-SIZE ANALYSES OF KOKO CRATER TEPHRA.

Sample	Bed Form*	Md_φ (mm)	σ_φ	Fine ash (Wt. %)
1	P	−0.54 (1.45)	1.28	1.89
2	P	−0.11 (1.08)	1.47	2.53
3	M	1.06 (0.48)	2.04	8.43
4	M	1.07 (0.47)	2.09	8.73
5	M	0.76 (0.59)	1.73	4.28
6	M	0.77 (0.58)	1.68	5.14
7	AF	0.31 (0.81)	1.44	2.88
8	AF	−0.21 (1.16)	1.68	2.86
9	M	1.12 (0.46)	2.34	11.03
10	M	1.22 (0.43)	2.27	11.78
11	AF	0.29 (0.82)	1.85	4.93
12	AF	0.14 (0.91)	1.93	5.03
13	AF	−0.27 (1.21)	2.02	2.77
14	AF	−0.35 (1.27)	1.77	2.08
15	M	1.02 (0.49)	1.61	2.65
16	M	0.97 (0.51)	1.61	2.79
17	M	0.96 (0.52)	1.82	6.37
18	M	1.11 (0.46)	1.80	6.41

*Bed forms are AF = ash fall, M = massive surge, and P = planar surge.

Tephra Description: Overall, the tephra sequence at Koko Crater illustrates a typical tuff cone, beginning with alternating phreatomagmatic and Strombolian explosions that produce an explosion breccia and lapilli fall (Unit 1) of medium to dark brown color. The tephra fall beds contain relatively fresh and unpalagonitized fragments with jagged outlines formed by bubble walls, which is typical of Strombolian eruption phases. These lapilli fall beds are 20 to 60 cm thick and show poorly developed normal grading.

Unit 2 resulted from phreatomagmatic eruptions that deposited thinly bedded, massive and planar surge deposits with a few interbedded sandwave beds (dry surge). These beds are a medium gray color where unaltered and unconsolidated or orange-brown where palagonitization has lithified the tuff. Coral fragments are abundant in the tuff of Unit 2 because coral reef and beach material was pierced by the eruption and fragmented within the vent. Typical pyroclasts of this unit are blocky with rounded edges. Some particles have a highly altered, rounded surface that is covered with a cracked, hydrated skin about 1 μm thick.

Unit 3 is phreatomagmatic (wet surge); fragments are totally altered except for lapilli fall beds that contain tephra with vesicle surfaces. Planar bed tephra show less surface alteration than do massive bed tephra; both have few vesicle surfaces and a broken, blocky morphology.

Koko Crater tuffs are composed of sideromelane, tachylite, and palagonitic glass shards; lithic fragments consisting of coral and minor amounts of quartz beach sand; and plagioclase, olivine, and minor pyroxene crystals. Basaltic glass is partly to totally palagonitized and makes up about 90% of the samples; 2 to 4% is crystals, and the rest is lithic materials. Samples taken from near the base of Unit 2 consist of up to 50 wt.% lithic materials from country rock that was comminuted during early vent-clearing eruptions (Table 2).

TABLE 2. PETROGRAPHIC DESCRIPTION OF KOKO CRATER TEPHRA.

Sample	354 µm	177 µm	63 µm
1	95% Sideromelane, highly palagonitized 5% Carbonate grains	90% Sideromelane, partly palagonitized 10% Carbonate grains	80% Sideromelane, partly palagonitized 20% Carbonate grains
3	---	50% Sideromelane, partly palagonitized 50% Quartz and coral beach sand	---
7	85% Sideromelane, highly palagonitized	90% Sideromelane, partly palagonitized	80% Sideromelane highly palagonitized
9	95% Sideromelane 4% Crystals 1% Carbonate	85% Sideromelane 5% Crystals 5% Carbonate	80% Sideromelane 10% Crystals 5% Carbonate
11	100% Sideromelane, totally palagonitized	98% Palagonitized sideromelane 2% Coral fragments	25% Sideromelane 5% Carbonate
15	98% Palagonite 1% Carbonate	---	---
17	---	70% Sideromelane 30% Palagonite and Carbonate	---

Comments on Pyroclast Formation: Koko Crater tuffs illustrate a sequence in which the contribution of hydromagmatic fragmentation increases throughout the eruption. Fresh ash deposited from Strombolian explosions in Unit 1 is magmatic; there is, however, a phreatomagmatic component. Upward in the section, pyroclasts are almost entirely phreatomagmatic and show evidence of brief phases of ash falls produced by Strombolian activity during times when water did not have access to the vent.

REFERENCES:

Fisher, R. V., 1977. Erosion by volcanic base-surge density currents: u-shaped channels. Geol. Soc. Am. Bull., **88**: 1287-1297.

Hay, L. and Iijima, A., 1968. Nature and origin of palagonite tuffs of the Honolulu Group on Oahu, Hawaii. Geol. Soc. Am. Mem., **116**: 331-376.

Stearns, H. T., 1961. Eustatic shorelines on Pacific islands. Zeitschr. Geomorph., **3**: 3-16.

Wentworth, C. K., 1926. Pyroclastic geology of Oahu. Bernice P. Bishop Museum Bull., **30**: 121.

Wenthworth, C. K. and Winchell, H., 1947. Koolau basalt series, Oahu, Hawaii. Geol. Soc. Am. Bull., **58**: 49-78.

Winchell, H., 1947. Honolulu series, Oahu, Hawaii. Geol. Soc. Am. Bull., **58**: 1-48.

FIGURE 1. Photograph of Koko Crater, which rises about 300 m above the surrounding plain and is 1000 m in diameter.

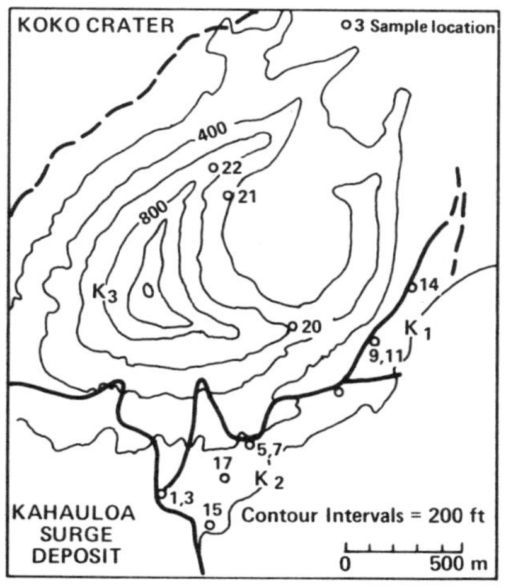

FIGURE 2. Map of Koko Crater that shows distribution of Units 1 (K_1) and 2 (K_2) (early tuff ring), Unit 3 (K_3) (later tuff cone), and sample locations.

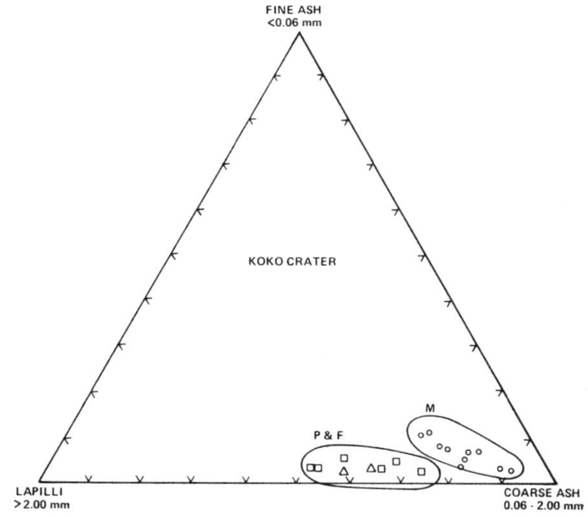

FIGURE 3. Triangular plot of tephra size analyses that shows fields related to ash fall and planar surge beds (F + P) and massive surge beds (M).

FIGURE 4. Explosion breccia and lapilli fall exposed near sample location 10. The base of this unit is not exposed. It is dominantly composed of 5- to 15-cm blocks of basalt with subordinate ash. Note geologic hammer for scale.

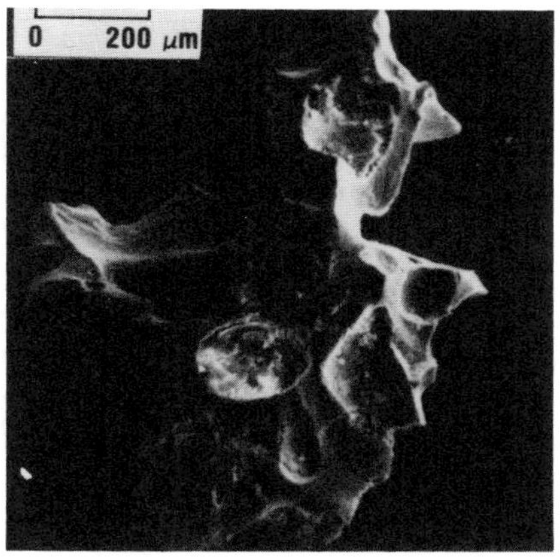

FIGURE 5. SEM of vesicular, fresh sideromelane pyroclast from fall deposit near the top of Unit 1. Note the smooth, curved surfaces and the jagged edges of bubble walls. Alteration products from minor palagonitization fill some vesicles.

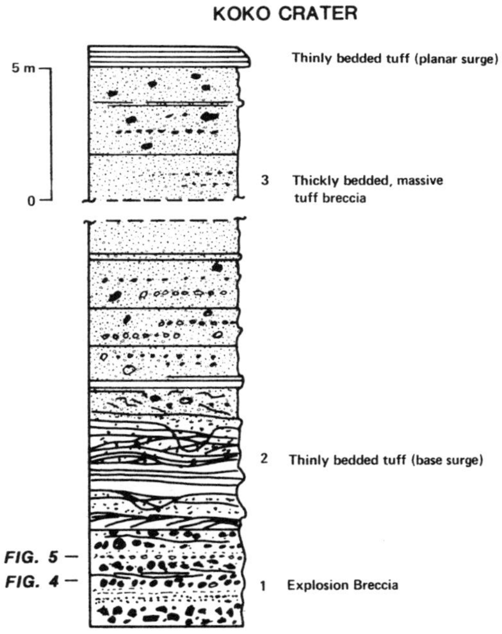

Koko Crater rim stratigraphy measured on the eastern slopes. Parts of Unit 2 and the base of Unit 3 are included in the section described by Fisher (1977).

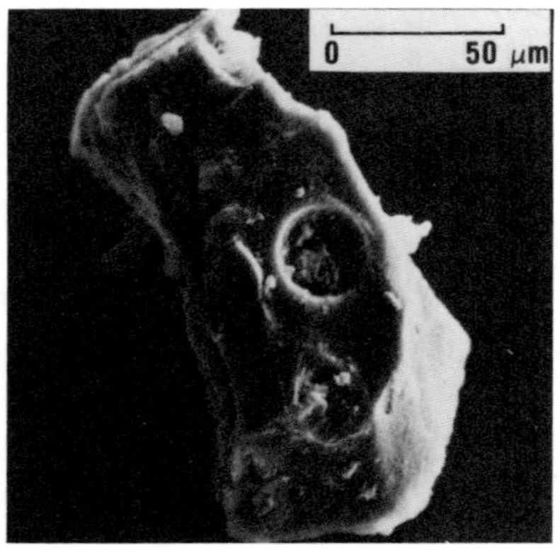

FIGURE 6. SEM of a blocky pyroclast from the opening phases of phreatomagmatic activity. Few vesicle surfaces characterize this grain, which shows curviplanar surfaces that have been chipped and abraded during transport.

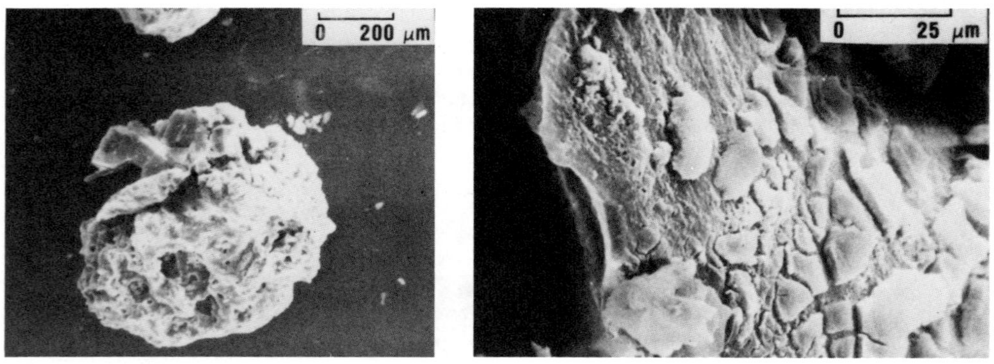

FIGURE 7. SEMs of a phreatomagmatic pyroclast rounded by abrasion during transport in surge (*left*) and covered by hydrated, palagonitic skin, which is shown at higher magnification (*right*).

FIGURE 8. SEM of a sample taken from lapilli fall bed; it was deposited from a Strombolian magmatic burst during a period of little water/magma interaction. Note vesicle surfaces that show only slight palagonitization.

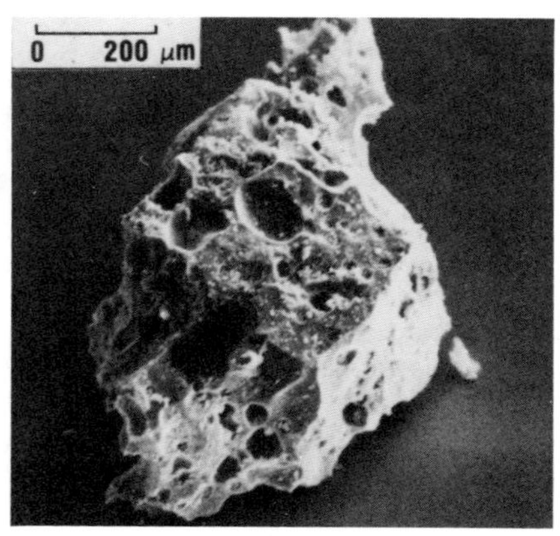

FIGURE 9. SEM of a blocky, partly palagonitized pyroclast from a planar surge bed.

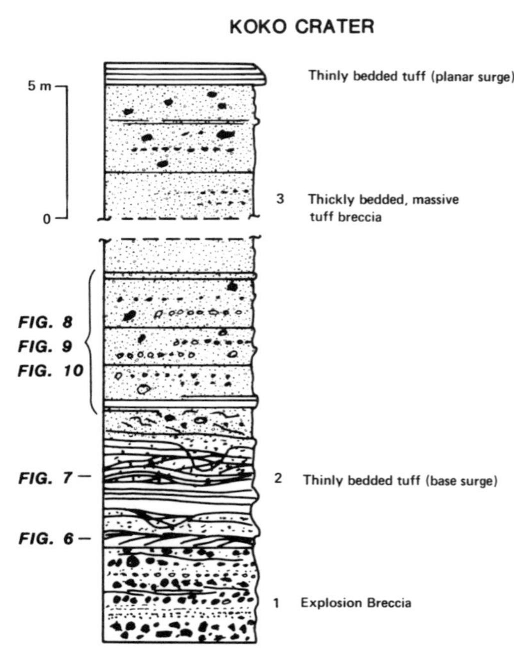

Koko Crater rim stratigraphy measured on the eastern slopes. Parts of Unit 2 and the base of Unit 3 are included in the section described by Fisher (1977).

FIGURE 10. SEM of a rounded, highly palagonitized pyroclast from a massive surge bed. Contrast the surface alteration with that shown in Fig. *9*. This contrast results from the massive beds having been emplaced as much wetter deposits than those of planar beds.

ZUNI SALT LAKE, NEW MEXICO

Introduction: Zuni Salt Lake maar is located in west-central New Mexico. It is the northernmost of a number of vents, many of which are tuff rings, aligned along a SW-trending fault line that is over 15 km long. Early studies of this maar include those of Herrick (1900) and Darton (1905, 1916). Later studies include a comparison with lunar craters (Shoemaker, 1962), a dissertation on the origin of the salt lake (Bradbury, 1967), and a geologic map (Cummings, 1968). Fisher and Waters (1970) and Wohletz and Sheridan (1983) give descriptions of its phreatomagmatic origin and surge deposits.

The crater is 1.6 km in diameter and about 100 m deep. It is underlain by Precambrian crystalline rocks; the Permian Abo, Yeso, Glorieta, and San Andreas Formations; and the Cretaceous Dakota Sandstone, Mancos Shale, and Mesa Verde Group. Basaltic eruptions pierced these strata, and the resulting tephra deposits contain a significant proportion of lithic sandstone fragments.

Water that interacted with basaltic magma to produce phreatomagmatic eruptions appears to have been derived from spring water structurally trapped along the fault or from an underlying sedimentary aquifer. Eruptions produced a tuff ring with rim beds as much as 30 m thick. The tuff is mostly thinly bedded surge with a subordinate amount of ash fall deposits. The ring is mantled by an ash and lapilli fall that is mappable to 8 km from the crater rim. Maar floor collapse during and after eruption produced an arcuate, concentric fault system that now exposes the tuff sequence on the inner walls of the ring.

Grain Size: Table 1 presents size analyses of 17 samples of Zuni Salt Lake tephra. The tuff is a lapilli- and block-bearing ash to a lapilli- and block-bearing fine ash that is moderately to poorly sorted and light gray to gray. Fall ashes are coarser (Md $\simeq$ 1.2 mm) and better sorted ($\sigma_\varphi \simeq 1.45\varphi$) than are surge ashes (Md $\simeq$ 0.2-0.8 mm, $\sigma_\varphi \simeq 1.90\varphi$). However, the surge deposits contain significantly more fine ash than do the fall deposits.

TABLE 1. GRAIN-SIZE CHARACTERISTICS OF ZUNI SALT LAKE TEPHRA.

Sample	Bed Form*	Md_φ (mm)	σ_φ	Fine Ash (Wt. %)
1	P	−0.16 (1.12)	2.39	2.25
2	P	−1.50 (2.83)	2.24	1.65
3	S	1.84 (0.28)	1.65	7.97
4	S	1.80 (0.29)	1.70	12.42
5	S	2.20 (0.22)	1.98	14.14
6	S	2.25 (0.21)	1.99	14.40
7	S	2.38 (0.19)	1.65	9.37
8	S	2.16 (0.22)	2.04	10.94
9	M	0.26 (0.84)	1.83	4.40
10	M	0.31 (0.81)	1.87	4.22
11	F	−0.32 (1.25)	1.45	0.50
12	M	0.83 (0.56)	1.90	5.00
13	M	0.99 (0.50)	1.88	10.73
14	P	0.27 (0.83)	1.43	1.69
15	P	0.35 (0.78)	1.48	1.88
16	F	−0.32 (1.25)	1.38	0.44
17	F	−0.40 (1.32)	1.46	0.70

*Bed forms are P = planar surge, S = sandwave surge, M = massive surge, and F = fall.

Tephra Description: The stratigraphic section here illustrates the upper third of the total section at Zuni Salt Lake. This interval represents the gradual decrease of magma/water interaction as the eruption progressed. During this interval, phreatomagmatic activity, which produced surges, gave way to dominantly Strombolian activity, which mainly deposited ash and lapilli falls. Samples of the sandwave-, massive-, and planar-surge sequence show a gradual upward increase in vesicularity, although particle shapes are dominated by angular, blocky fractures. These samples may be contrasted with fall-deposit samples whose vesicular particles are shaped by bubble walls. Also, there is greater surface alteration and abrasion on the surge samples, which all show some development of altered surfaces: the least of which is typical of particles from sandwave beds, and the greatest is seen on particles from massive beds, especially those of the fine fraction. The surge ashes have angular edges that are smoothed and rounded by laminar and turbulent transport where grain collisions are abundant. The fall clasts, however, retain delicate shapes.

Table 2 shows results of grain counts for size fractions >63 μm. Palagonite appears in greater abundance in surge samples than in fall samples, and there is also a marked increase within the finer grained fractions. The crystal content of these samples ranges from 5 to 10% (distribution not shown in Table 2) and is concentrated in the finer size fractions. Lithic material consists mostly of quartz grains from comminuted sandstone.

TABLE 2. PETROGRAPHIC DESCRIPTION OF ZUNI SALT LAKE TEPHRA.

Sample	354 μm	177 μm	64 μm
1	85% Sideromelane 10% Lithic	60% Sideromelane 30% Lithic	30% Sideromelane 60% Lithic
3	1% Sideromelane 85% Palagonite 9% Lithic	10% Sideromelane 75% Palagonite 50% Lithic	15% Sideromelane 70% Palagonite 5% Altered glass
5	90% Sideromelane 5% Palagonite 3% Lithic	80% Sideromelane 5% Palagonite 5% Lithic	70% Sideromelane 10% Palagonite 10% Lithic
7	45% Sideromelane 40% Palagonite 10% Lithic	40% Sideromelane 40% Palagonite 10% Lithic	---
10	85% Sideromelane 5% Palagonite 5% Lithic	50% Sideromelane 40% Palagonite 5% Lithic	20% Sideromelane 60% Palagonite 10% Lithic
12	5% Sideromelane 5% Palagonite 80% Lithic	2% Sideromelane 90% Lithic	1% Sideromelane 90% Lithic
15	90% Sideromelane 5% Lithic	60% Sideromelane 39% Lithic	40% Sideromelane 55% Lithic
16 & 17	80% Sideromelane 10% Palagonite 5% Lithic	70% Sideromelane 20% Lithic	40% Sideromelane 50% Lithic

Comments on Pyroclast Formation: The eruption sequence illustrated here in part reflects the overall eruptive activity at Zuni Salt Lake. Early explosive eruptions were dominantly phreatomagmatic, producing blocky and angular tephra particles that were then deposited in base surges. Later explosive eruptions became increasingly Strombolian as water access to the vent was impeded. These eruptions produced vesicular ash, which was deposited as fall layers. The last stages of eruption show little explosive water interaction and are marked by the building of cinder cones in the maar. Intrusion of a ring dike, exposed on the inner walls of the maar, may have been responsible for the cessation of water flow into the vent late in the eruptive history.

REFERENCES:

Bradbury, J. P., 1967. Origin, paleolimnology, and limnology of Zuni Salt Lake maar, west central New Mexico. Ph.D. dissertation, University of New Mexico, 246 pp.

Cummings, D., 1968. Geologic map of the Zuni Salt Lake volcanic crater, Catron County, New Mexico. U.S. Geol. Sur. Map. I-544.

Darton, N. H., 1905. The Zuni Salt Lake. J. Geol., **13**: 185-193.

Darton, N. H., 1916. Explosion Craters. Sci. Mon., **3**: 417-430.

Fisher, R. V. and Waters, A. C., 1970. Base surge bed forms in maar volcanoes. Am. J. Sci., **268**: 157-180.

Herrick, D. L., 1900. Report of a geologic reconnaissance in western Socorro and Valencia Counties, New Mexico. Am. Geol., **25**: 331-346.

Shoemaker, E. M. 1962. Primary structures of maar rims and their bearing on the origin of Kilbourne Hole and Zuni Salt Lake, New Mexico (abs.). Geol. Soc. Am., **68**: 1846.

Wohletz, K. H. and Sheridan, M. F., 1983. Hydrovolcanic explosions II. Evolution of basaltic tuff rings and tuff cones. Am. J. Sci., **283**: 385-413.

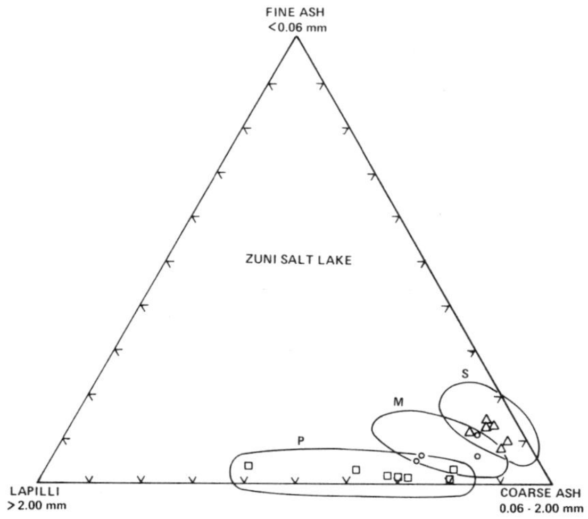

FIGURE 1. Triangular plot of tephra size analyses that shows fields of pyroclastic surge bed forms: S = sandwave; M = massive; P = planar; and F = fall beds.

ZUNI SALT LAKE

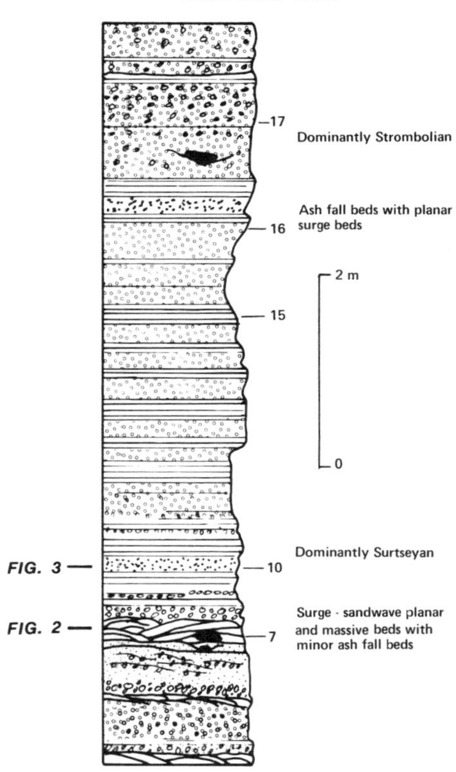

Rim stratigraphic section of Zuni Salt Lake maar. The section shows the upper one-third of the total pyroclastic section. The lower part of this section is dominantly surge beds with intercalated lapilli fall beds. The upper part is dominantly ash and lapilli fall beds with intervals of planar surge deposits.

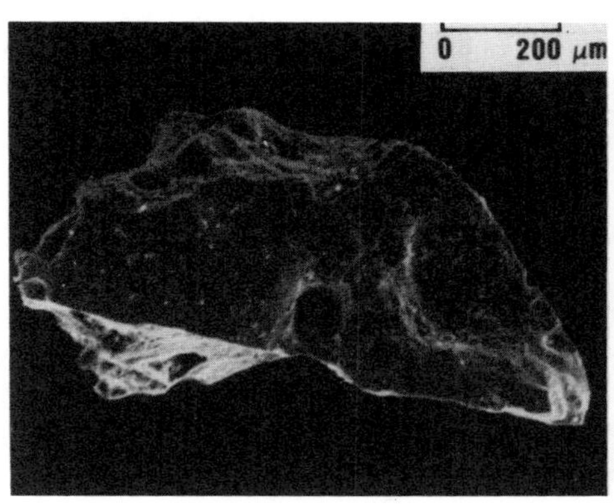

FIGURE 2a. SEMs of pyroclasts from sample 7, from a sandwave surge bed. An angular ash grain with fresh, planar surfaces, chipped edges, and very few vesicles.

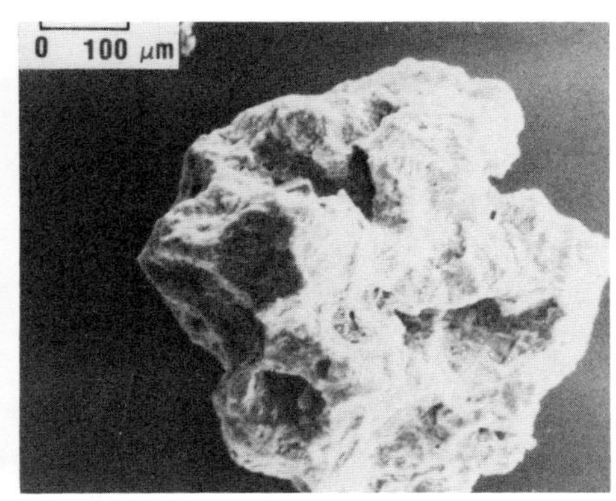

FIGURE 2b. Grain with abrasion pits and chips, which have rounded the original flat surfaces.

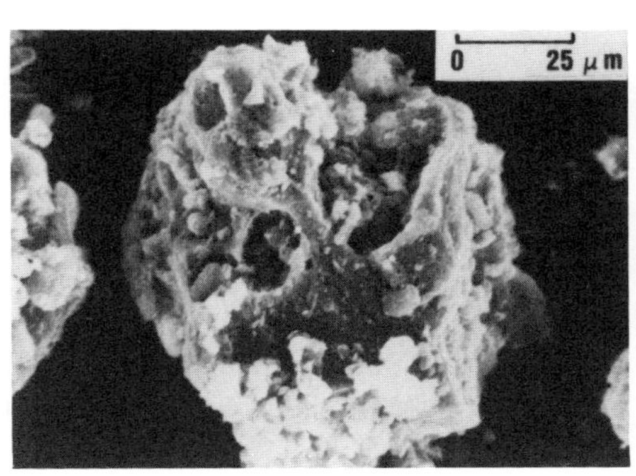

FIGURE 3a. SEMs of pyroclasts from sample 10, from a massive surge bed. Rounded grain covered with adhering alteration material both on flat surfaces and in vesicles.

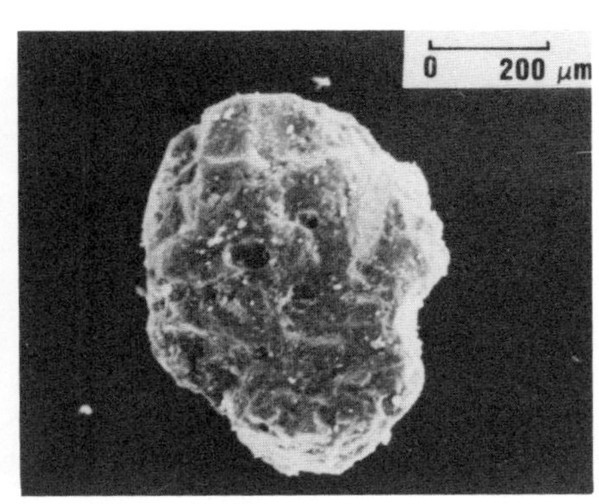

FIGURE 3b. Rounded grain with hydrated surfaces and few vesicles.

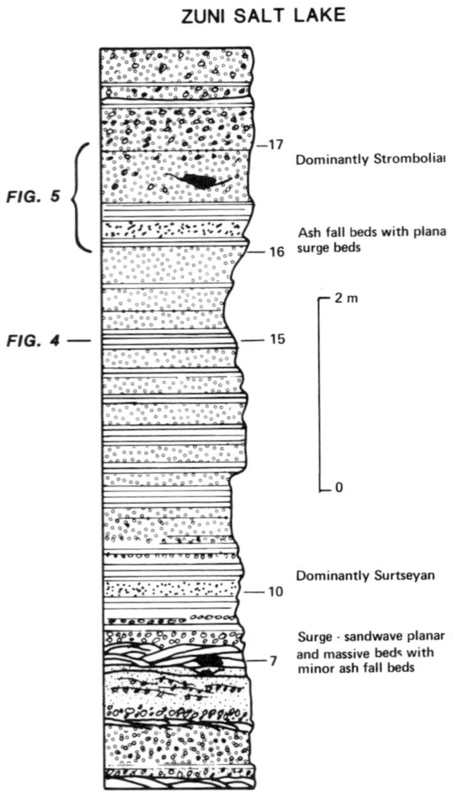

Rim stratigraphic section of Zuni Salt Lake maar. The section shows the upper one-third of the total pyroclastic section. The lower part of this section is dominantly surge beds with intercalated lapilli fall beds. The upper part is dominantly ash and lapilli fall beds with intervals of planar surge deposits.

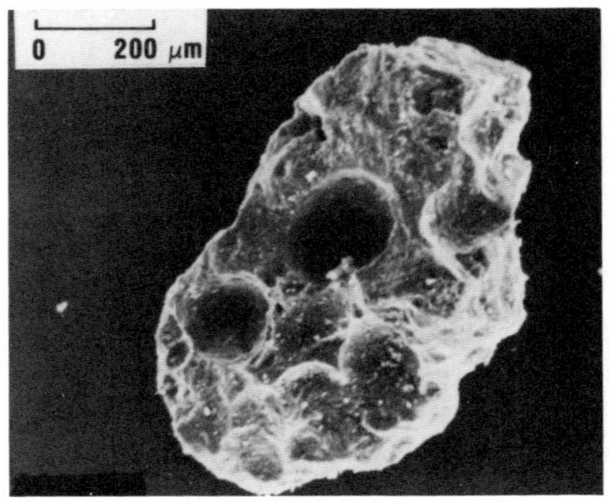

FIGURE 4a. SEMs of sample 15, from a planar surge bed. Blocky, vesicular grain with smoothed edges and pitted surfaces.

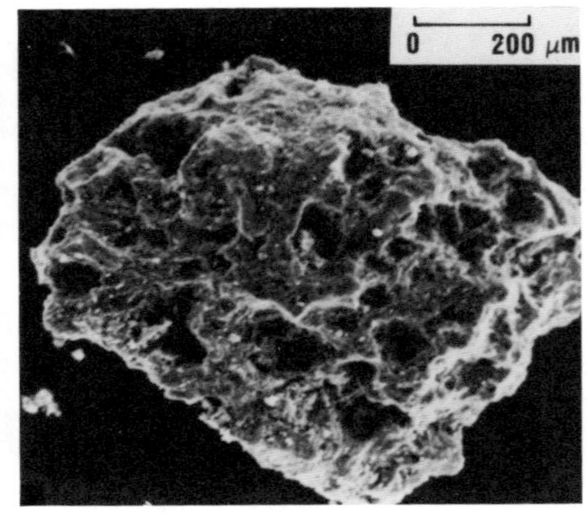

FIGURE 4b. Subangular grain with blocky, vesicular shape and pitted surfaces. Note adhering fine dust.

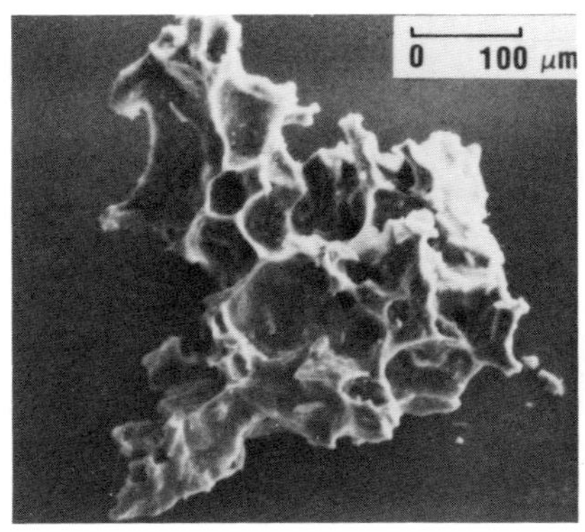

FIGURE 5a. SEMs of pyroclasts from samples 16 and 17, from fall beds. Spongy, highly vesicular grain with a delicate shape. Note the fresh-appearing surfaces and abundance of vesicles.

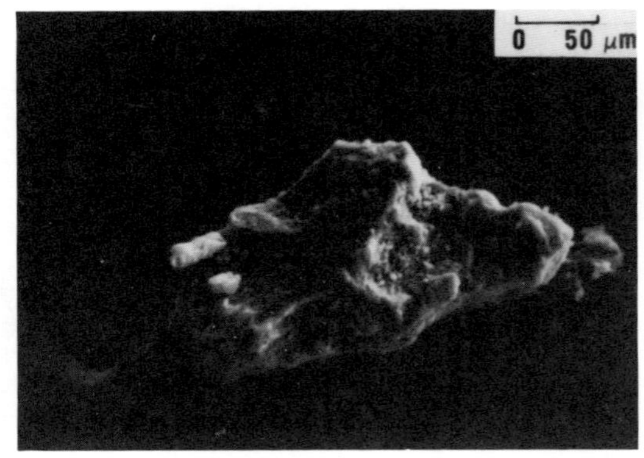

FIGURE 5b. Vesicular grain from the fine fraction. This particle is likely a broken piece of a larger, vesicular grain. Some alteration is apparent in a vesicle hollow on this fine-fraction sample.

KILBOURNE HOLE, NEW MEXICO

Introduction: Kilbourne Hole is one of three maars identified within the Potrillo volcanic field of south-central New Mexico (Hoffer, 1976); the other two are Hunts Hole and Potrillo Maar (Lee, 1907; Darton, 1916; Reiche, 1940; Shoemaker, 1957; DeHon, 1965; Reeves and DeHon, 1965). The maar erupted during Pleistocene time, along the NS-trending Fitzgerald fault system—hence its oblong dimensions of 3.2 by 2.0 km. The maar is about 130 m deep and has a surrounding tuff ring deposit, 30 to 40 m thick, that overlies sediments of the Santa Fe Group and a 5- to 7-m-thick olivine basalt flow. Hoffer (1976) divided the tuff ring into three units: a basal ash fall unit, a middle pyroclastic surge unit, and an upper massive tuff unit that DeHon (1965) considered a lahar. The basal unit is an explosion breccia, approximately 10 to 15 m thick, that crops out on the north and east rims of the maar crater. It consists of angular basalt blocks supported in dark gray, poorly bedded ash and lapilli fall deposits. The middle unit consists of thinly bedded pyroclastic surge beds that are nearly 30 m thick on the east rim. These dry surge beds are a light tan, lithic tuff near the base and grade upward to light gray color as the juvenile constituent increases in abundance. The surge deposits are dominantly sandwave beds near the crater rim; they show repeated gradations from beds of coarse ash and lapilli fall—to sandwave beds—to accretionary lapilli-bearing massive tuff beds. The upper unit, nearly 10 m thick, is composed of dominantly massive beds 3 to 4 m thick and is moderately palagonitized.

Grain Size: The tuff consists of lapilli-bearing ash to lapilli-ash because of the aggregation of fine ash in accretionary lapilli (Table 1). The tuff is moderately sorted and contains very little fine ash.

TABLE 1. GRAIN-SIZE CHARACTERISTICS OF KILBOURNE HOLE TEPHRA.

Sample	Bed Form*	Md_φ (mm)	σ_φ	Fine Ash (Wt. %)
1	S	2.20 (0.22)	1.10	4.85
2	S	2.16 (0.22)	1.11	6.52
3	F	−0.11 (1.08)	1.98	0.87
4	F	−0.16 (1.12)	1.99	0.87
5	F	−0.74 (1.67)	1.60	0.48
6	F	−0.89 (1.85)	1.58	0.42
7	S	2.17 (0.22)	1.03	3.90
8	S	2.10 (0.23)	1.14	3.67
9	M	1.30 (0.41)	1.64	2.64
10	M	1.26 (0.42)	1.69	2.39
11	M	1.67 (0.31)	1.72	3.80
12	M	1.72 (0.30)	1.70	3.57
13	S	2.19 (0.22)	0.96	5.40
14	S	2.19 (0.22)	0.94	6.76
15	S	1.98 (0.25)	1.12	4.12
16	S	2.00 (0.25)	1.12	3.93

*Bed forms are S = sandwave surge, M = massive surge, and F = fall.

Tephra Description: Kilbourne tuffs are lithic-rich (Table 2). The juvenile component within the lower part of the surge unit (beginning of strong phreatomagmatic eruptions) is only 10%. Lithic material decreases near the top of the surge unit where sideromelane is 90% of the erupted materials. Pyroclasts from sandwave beds near the crater rim are poorly vesicular, subangular to subrounded particles that show well-developed frosting. This frosted coating is the result of extensive abrasion pits and scratches that have rounded the angular, planar surfaces. Tiny grains and irregularities on surfaces, which form by silica precipitation, also contribute to the frosted appearance. The silica is apparently derived from quartz-sand lithic material that surrounds sideromelane grains. The latest phreatomagmatic eruptions produced pyroclasts with somewhat smoother, fluidal surfaces despite the presence of abundant lithic material in the beds. Pyroclasts from massive surge beds are strongly abraded, however, and are distinguishable from those of sandwave beds by surface accumulations of small dust particles and hydrated

TABLE 2. PETROGRAPHIC DESCRIPTION OF KILBOURNE HOLE TEPHRA.

Sample	>354 μm	>177 μm	>64 μm
1	10% Sideromelane 89% Lithic 1% Sideromelane	5% Sideromelane 94% Lithic 1% Sideromelane	5% Sideromelane 95% Lithic
3	10% Palagonite 89% Lithic 1% Crystals	5% Sideromelane 94% Lithic 1% Crystals	2% Sideromelane 98% Lithic
5	35% Sideromelane 62% Lithic 3% Crystals	20% Sideromelane 77% Lithic 3% Crystals	10% Sideromelane 89% Lithic 1% Crystals
7	5% Sideromelane 95% Lithic	2% Sideromelane 98% Lithic	1% Sideromelane 99% Lithic
9	89% Sideromelane 10% Crystals 1% Lithic	40% Sideromelane 20% Palagonite 35% Lithic 5% Crystals	5% Sideromelane 15% Palagonite 75% Lithic 5% Crystals
11	35% Sideromelane 62% Lithic 3% Crystals	8% Sideromelane 2% Palagonite 89% Lithic 1% Crystals	5% Sideromelane 95% Lithic
13	9% Sideromelane 1% Palagonite 89% Lithic 1% Crystals	10% Sideromelane 89% Lithic 1% Crystals	5% Sideromelane 95% Lithic
15	5% Sideromelane 5% Palagonite 90% Lithic	2% Sideromelane 98% Lithic	1% Sideromelane 99% Lithic

glass particles. Surge pyroclasts contrast sharply with those of fall beds. Surge ashes are strongly vesicular, angular-to-subangular particles with relatively unabraded surfaces.

Table 2 shows the persistence of a strong lithic contribution in most samples. The lithic particles are derived from quartzo-feldspathic sands and silts of the underlying Santa Fe Group, which are poorly consolidated. Palagonite, where present, is not strongly developed on glass surfaces, and sideromelane dominates juvenile material.

Comments on Pyroclast Formation: The phreatomagmatic eruptions at Kilbourne Hole fragmented a poorly crystalline basalt magma, which, when ejected with expanding steam, excavated a large crater in strata composed of sand and silt. The sand and silt is generously mixed with basalt pyroclasts to form the lithic ash deposit of the tuff ring. It is likely that sediments of the Santa Fe Group were water saturated, which contributed to the steam explosions. The lack of fine-fraction pyroclasts supports a conclusion that magma fragmentation was never very strong and that much of the activity borders on pure phreatic explosions caused by vaporization of water in sedimentary strata.

REFERENCES:

Darton, N. H., 1916. Explosion craters. Sci. Mon., **3**: 417-430.

De Hon, R. A., 1965. Maare of La Mesa. New Mexico Geol. Soc. Guidebook, 16th Field Conf.: 204-209.

Hoffer, J. M., 1976. The Potrillo basalt field, southcentral New Mexico. New Mexico Geol. Soc. Spec. Pub., **5**: 89-92.

Lee, W. T., 1907. Afton Craters. Geol. Soc. Am. Bull., **18**: 214-215.

Reeves, Jr., C. C. and De Hon, R. A., 1965. Geology of the Potrillo Maar, New Mexico and northern Chihuahua, Mexico. Am. J. Sci., **63**: 401-409.

Reiche, P., 1940. The origin of Kilbourne Hole, New Mexico. Am. J. Sci., **238**: 212-225.

Shoemaker, E. M., 1957. Primary structures of maar rims and their bearing on the origin of Kilbourne Hole and Zuni Salt Lake, New Mexico (abs.). Geol. Soc. Am. Bull., **68**: 1846.

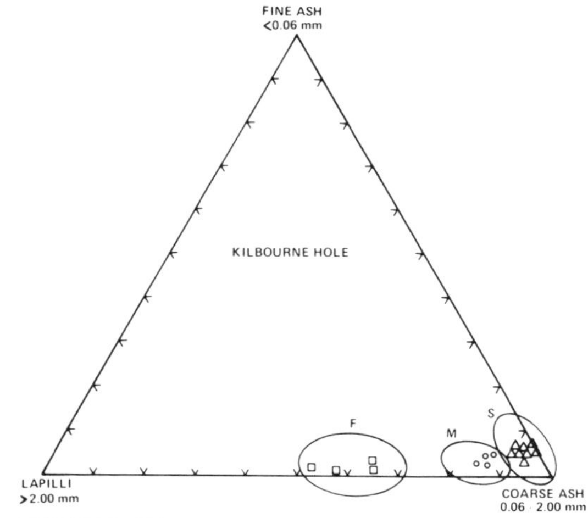

FIGURE 1. Triangular plot of tephra size analyses that shows groups related to bed form: F = fall, M = massive surge, and S = sandwave surge.

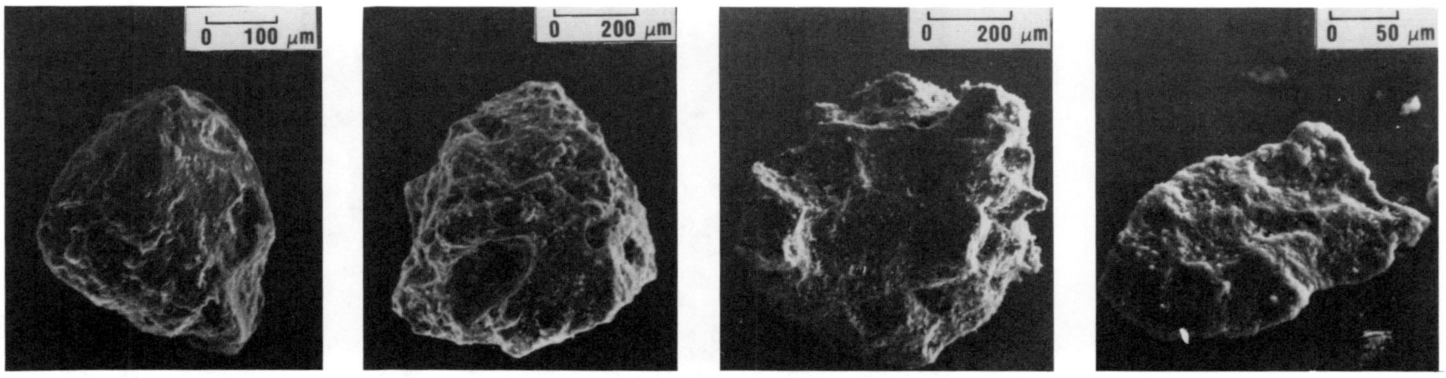

FIGURE 2. SEMs of pyroclasts from sample 1, from a sandwave surge deposit. Note the planar surfaces that have been rounded and the low vesicularity. Chipping and pitting has given the grain and vesicle edges a smooth appearance.

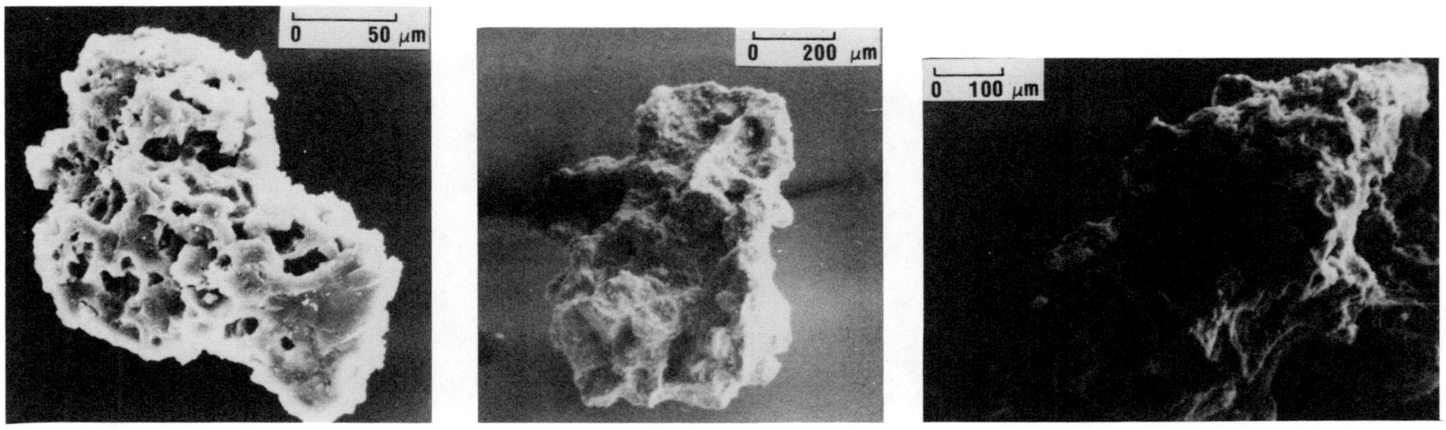

FIGURE 3. SEMs of pyroclasts from samples 3 and 5, from ash-fall deposits. These grains are vesicular and show angular shapes with sharp edges. Silica precipitation on surfaces of the two grains *center* and *right* has produced a somewhat frosted appearance.

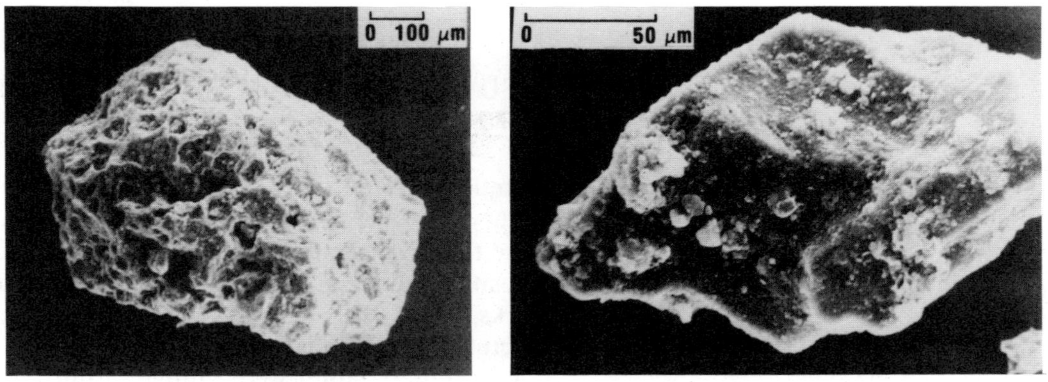

FIGURE 4. SEMs of pyroclasts from sample 7, from a sandwave surge deposit, in which curviplanar surfaces dominate the shapes. Note the edge abrasion on the bottom of the grain on the *right*.

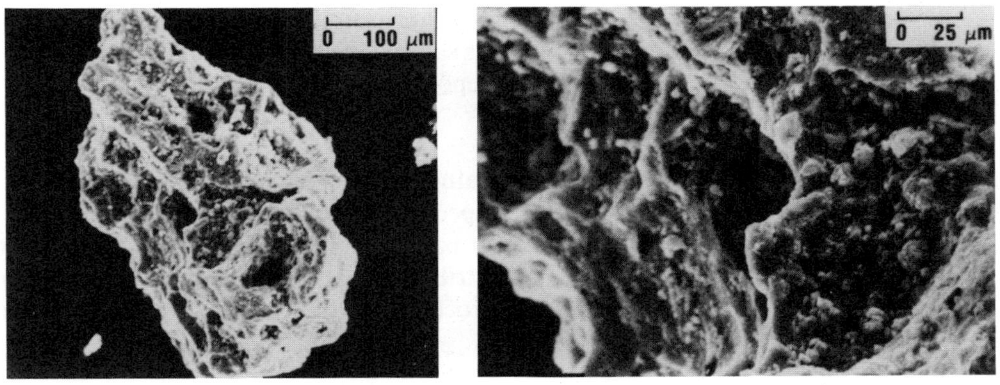

FIGURE 5. SEMs of a pyroclast from sample 9, from a massive surge deposit. *Right* - an enlargement of the central portion of the grain shown on the *left*. Abundant adhering dust particles of fine ash and alteration products exist in vesicle hollows.

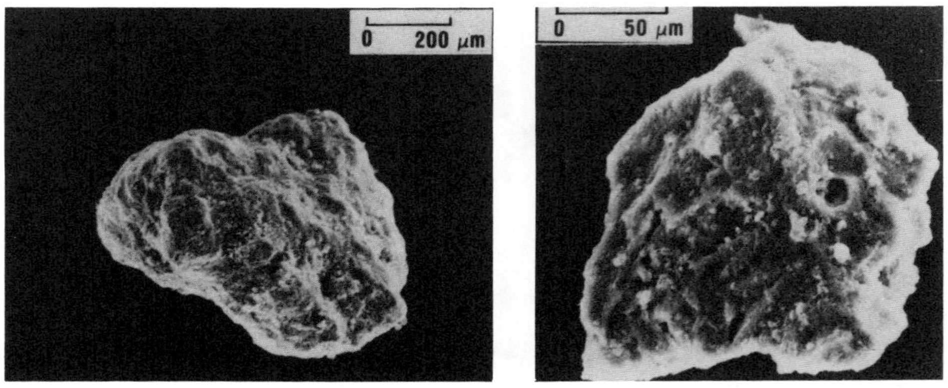

FIGURE 6. SEMs of pyroclasts from sample 11, from a massive surge deposit. The *right* view is of a fine-fraction grain covered with abundant adhering dust particles. The *left* view shows a larger grain with a strongly hydrated surface.

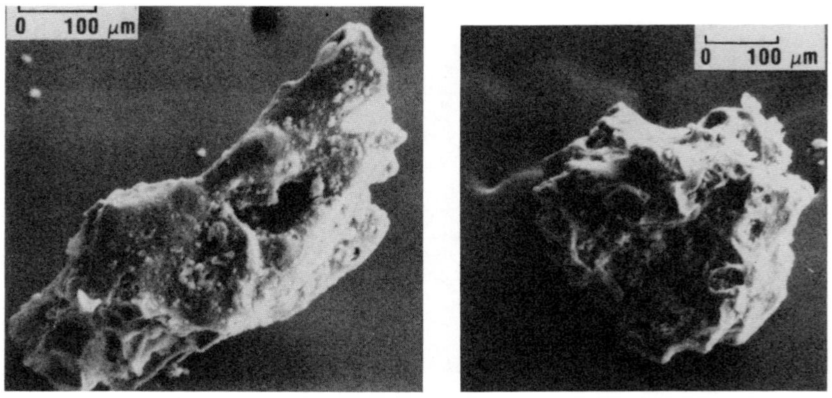

FIGURE 7. SEMs of pyroclasts from sample 15, from a sandwave bed deposited near the end of eruptive activity. Grains show fusiform surfaces between vesicle hollows. These surfaces were likely formed by fluid deformation of particle surfaces after magma fragmentation. *Left* - small vesicles are revealed in the interior of the grain.

VESUVIUS, ITALY; ACCRETIONARY LAPILLI FROM THE ERUPTION OF 79 A.D.

Introduction: Accretionary lapilli have been described in the literature since the turn of the century (LaCroix, 1904; Scrope, 1925; Stearns, 1925). More recent descriptions relate their occurrence to phreatomagmatic eruption of ash falls and base surge deposits (Moore and Peck, 1962; Fisher and Waters, 1970; Heiken, 1971; Self and Sparks, 1978). Sheridan and Wohletz (1983) described in detail the formation and structure of accretionary lapilli from Vesuvius.

Accretionary lapilli were sampled from the Pompeii (A.D. 79) and Avellino (3,500 y.b.p.) Plinian deposits of Vesuvius. The lapilli occur within the fine-ash deposits that rest above thick (>2-m) Plinian pumice fall beds (Fig. 1). Sheridan et al. (1981) proposed that these fine-ash beds are pyroclastic surge deposits erupted from catastrophic hydromagmatic phases of activity shortly after the pumice fall. The samples were studied by optical microscopy and SEM as well as EDS analyses of epoxy-impregnated thin sections.

Grain Size: Grain-size analyses of constituent ash particles were made from SEM images by inspecting approximately 1000 grains from each sample. The median grain size of ash particles ranges from 77 to 24 μm; the common mode is near 45 μm. The distributions are well sorted ($\sigma_\varphi = 1.00\varphi$). These results are characteristic of pyroclasts produced by hydroexplosions (Self and Sparks, 1978; Wohletz, 1983).

Lapilli Description: The lapilli are centimeter-size spheroids consisting of <1-mm fragments of glass, crystals, and lithic clasts. Large clasts (greater than 50-μm diameter) are present in one or two concentric bands or are randomly distributed throughout the lapilli. A matrix of fine-ash particles (less than 1 μm in diameter) has an irregular distribution, but in some lapilli the matrix occurs in distinct layers around the core.

The lapilli have three distinct textural zones. (1) A massive (structureless) core of ash (3- to 6-mm diameter) is surrounded by (2) an intermediate zone with bands of bubbles and larger clasts and (3) a fine-grained rim (0.3 to 1.8 mm thick). Patches of bubbles (50 to 300 μm in diameter) occur in the cores of some lapilli as well as in the bands and patches between cores and rims.

Most clasts in the lapilli have blocky shapes with delicate projections. A small fraction of the clasts have vesicles—2% in Pompeii lapilli and 5% in Avellino samples.

Chemical Composition: Di Girolamo (1968) made chemical analyses of samples from intervals throughout the entire section at Pompeii. These analyses (Table 1) show a significant difference between the composition of samples in the pumice fall and those from the fine-ash horizon. This distinction reflects a general increase in the carbonate lithic constituent of the ash during the eruption. The increase is abrupt from the pumice fall to the fine-ash surge beds. Table 1 also gives pumice analyses of the Avellino and Pompeii deposits (from Barberi et al., 1981).

Table 2 lists representative EDS analyses of samples from selected areas along traverses from the core to the rim of individual lapilli. These analyses also show the general increase in carbonate content.

TABLE 1. CHEMICAL ANALYSES OF POMPEII AND AVELLINO SAMPLES.*

	1	2	3	4	5
SiO_2	53.60	49.44	44.40	52.85	42.88
TiO_2	0.27	0.70	0.56	0.49	0.56
Al_2O_3	21.57	17.33	14.80	18.91	18.84
$Fe_2O_3 + FeO$	2.76	5.66	5.29	4.23	4.51
MnO	0.12	0.13	0.09	0.11	0.12
MgO	0.31	4.51	3.51	2.55	2.12
$CaO + CaCO_3$	2.75	7.65	13.37	6.44	5.68
K_2O	9.71	6.34	5.19	6.56	8.29
Na_2O	5.09	3.28	2.80	5.22	4.42

*Sample descriptions and sources of analyses are as follows:
1 = White pumice from bottom of Pompeii section (see Fig. 1) (di Girolamo, 1968).
2 = Fine-ash sample from directly above pumice fall at Pompeii (di Girolamo, 1968).
3 = Pisolitic (accretionary lapilli) ash sample from 30 cm above pumice fall at Pompeii section (di Girolamo, 1968).
4 = Avellino pumice (Barberi et al., 1981).
5 = Pompeii pumice (Barberi et al., 1981).

TABLE 2. CHEMICAL ANALYSES OF THE CORES AND RIMS OF ACCRETIONARY LAPILLI.*

	Sample 6		Sample 7		Sample 8	
	Core	Rim	Core	Rim	Core	Rim
SiO_2	51.19	50.85	47.57	46.84	44.32	39.49
TiO_2	0.58	0.74	0.59	0.53	0.73	0.72
Al_2O_3	21.70	21.41	19.22	18.31	17.63	17.19
$Fe_2O_3 + FeO$	3.80	4.42	3.38	2.67	3.82	3.51
MgO	2.06	2.59	2.57	2.31	4.60	5.09
$CaO + CaCO_3$	7.84	9.82	17.92	20.15	18.60	24.76
K_2O	5.22	4.14	4.09	4.63	5.49	4.38
Na_2O	6.73	5.18	3.95	3.87	3.96	3.71
Total	99.12	99.15	99.29	99.31	99.15	98.85

*Standardized EDS analyses of samples taken from Pompeii fine-ash deposits at increasing levels above the pumice fall deposit (see Fig. 1).

Comments on Lapilli Formation: The interpretation of lapilli formation is based upon the general model of Plinian eruptions of Vesuvius described by Barberi et al. (1981) and Sheridan et al. (1981). As eruptions proceed, the limestone aquifer walls of the magma chamber disintegrate and allow water to enter the chamber—thus creating hydroexplosions. This phase of activity is reflected by a change from pumice fall deposits to phreatomagmatic surge deposition of fine ash.

The EDS analyses of the lapilli plot on the same trend of compositions (carbonate versus SiO_2) shown in di Girolamo's (1968) ash analyses; this notable increase in carbonate from core to rim for each lapillus indicates that these lapilli accreted in progressively lithic-rich zones of the eruption plume. The chemical, size, and shape analyses of the lapilli suggest the following interpretation of their origin. Cores of the lapilli formed as aggregates in the hot central zone of the eruption plume. As the lapilli moved outward into turbulent margins of the plume, they began to tumble and accrete layers of fine ash that were enriched in lithics from the magma-chamber walls. The turbulent zone of the plume likely contained more water vapor, which promoted the accretion and formed water bubbles in the layered zone of the lapilli. Little accretion occurred after the lapilli moved out of the turbulent plume, and only a thin veneer of carbonate-rich dust was deposited as a rim before the lapilli fell down into advancing surge clouds.

REFERENCES:

Barberi, F., Bizouard, H., Clocchiatti, R., Metrich, N., Santacroce, R., and Sbrana, A., 1981. The Somma-Vesuvius magma chamber: a petrologic and volcanologic approach. Bull. Volcanol., **44**: 295-316.

di Girolamo, P., 1968. Petrografia del somma-Vesuvio: le serie piroclastiche. Rend. Accad. Sci. Napoli, **35**: 1-68.

Heiken, G. H., 1971. Tuff rings: examples from the Fort Rock-Christmas Lake Valley basin, south central Oregon. J. Geophys. Res., **76**: 5615-5626.

Fisher, R. V. and Waters, A. C., 1970. Base surge bedforms in maar volcanoes. Am. J. Sci., **268**: 157-180.

LaCroix, A., 1904. *La Montagne Pelée et ses eruptions.* Masson et Cie., Paris.

Moore, J. G. and Peck, D. L., 1962. Accretionary lapilli in volcanic rocks of the western continental United States. J. Geol., **70**: 182-193.

Scrope, A. P. 1925. *Considerations on volcanoes.* W. Phillips and G. Yard Pub., London.

Self, S. and Sparks, R. S. J., 1978. Characteristics of widespread pyroclastic deposits formed by the interaction of silicic magma and water. Bull. Volcanol., **41**: 196-212.

Sheridan, M. F., Barberi, F., Rosi, M., and Santacroce, R., 1981. A model for Plinian eruptions of Vesuvius. Nature, **289**: 282-285.

Sheridan, M. F. and Wohletz, K. H., 1983. Origin of accretionary lapilli from the Pompeii and Avellino deposits of Vesuvius. *In*: R. Gooley (Ed.), *Microbeam analysis—1983.* San Francisco Press, pp. 35-38.

Stearns, H. T., 1925. The explosive phase of Kilauea volcano, Hawaii, in 1924. Bull. Volcanol., **5**: 1-16.

Wohletz, K. H., 1983. Mechanism of hydrovolcanic pyroclast formation: size, scanning electron microscopy, and experimental studies. *In*: M. F. Sheridan and F. Barberi (Eds.), *Explosive volcanism*, J. Volcanol. Geotherm. Res., **17**: 31-64.

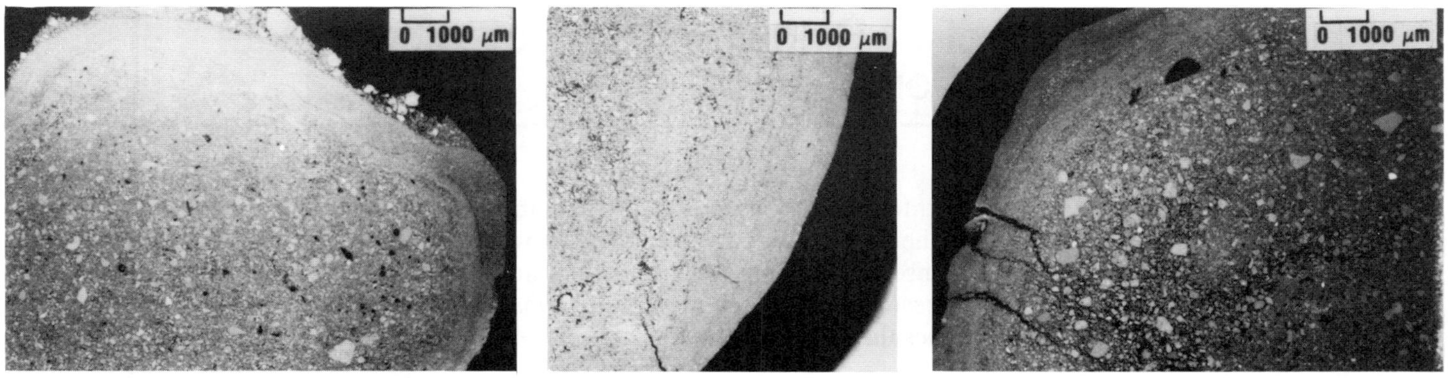

FIGURE 1. SEMs of representative accretionary lapilli that show the coarser grained centers and fine-grained rims.

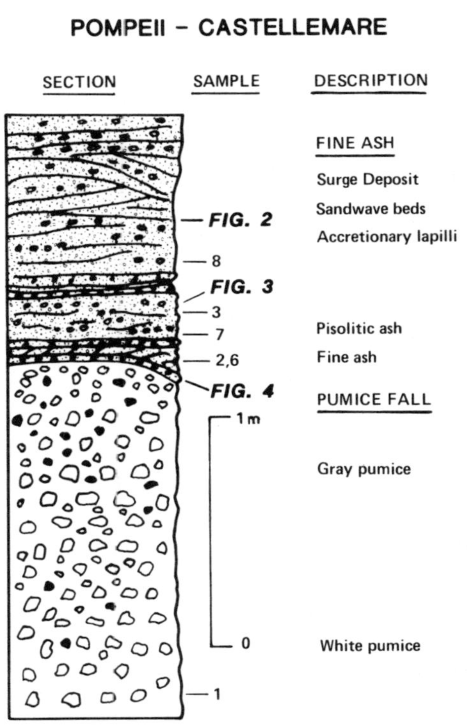

Columnar stratigraphic section of the A.D. 79 Vesuvius deposit at the Pompeii excavation near Castellemare. The lower half of the section is Plinian pumice fall products overlain by fine-ash deposits of pyroclastic surges in which accretionary lapilli occur. Sample numbers 1, 2, and 3 indicate levels from which di Girolamo (1968) took samples. Numbers 6, 7, and 8 indicate approximate levels of samples studied in this report.

FIGURE 2. SEMs of the core (*left*) and rim (*right*) of lapilli from sample 8.

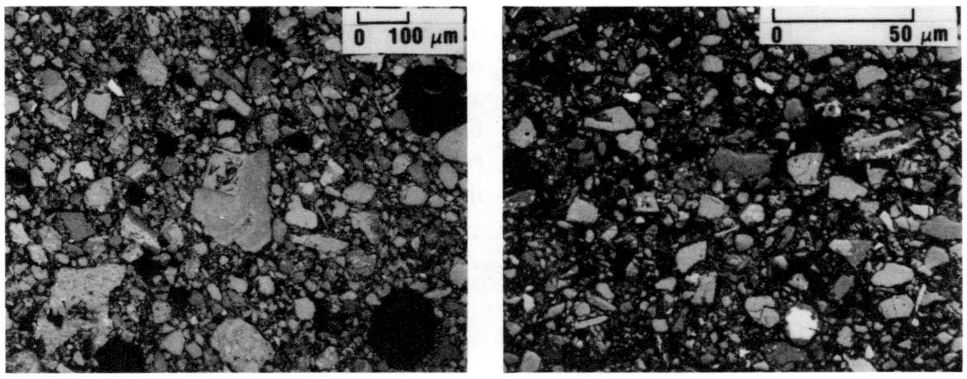

FIGURE 3. SEMs of the core (*left*) and rim (*right*) of lapilli from sample 7.

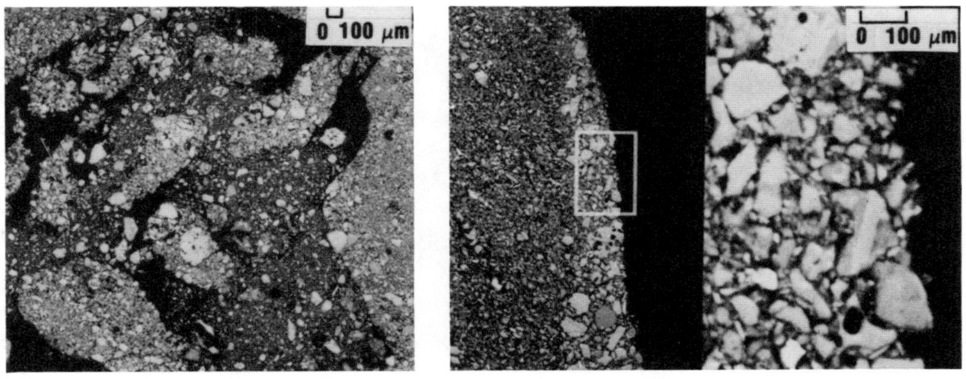

FIGURE 4. SEMs of the core (*left*) and rim (*right*) of lapilli from sample 6.

Tuff Rings (Magmas of Intermediate Composition)

Intermediate-composition (andesitic and dacitic) magmas generally have a higher viscosity than do basalts. The yield strength of such magmas is also higher. Consequently, brittle fracture dominates production of pyroclasts during phreatomagmatic eruption of these magmas. Tephra are typically angular, blocky ashes that are often coarse grained.

CLEAR LAKE VOLCANIC FIELD, CALIFORNIA

Introduction: The Clear Lake volcanic field covers a 400-km^2 area about 150 km north of San Francisco, California. This field has experienced eruptions from the late Pliocene to Holocene time (Hearn et al., 1972). The geology, geophysics, geochemistry, and geothermal potential has been well studied by Anderson (1936), Brice (1953), Hodges (1966), Chapman (1975), U.S. Geological Survey (1981), and Goff et al. (1977). Samples examined in this section were taken from pyroclastic deposits of small tuff rings that developed along the margins of the present-day lake during the last 0.03 m.y. (shown as Holocene pyroclastic deposits in Fig. 1). These pyroclastic deposits range in composition from basalt to dacite within individual tuff rings. This observation supports the hypothesis that mafic and silicic magmas were mixed during evolution of this field (Eichelberger, 1975).

Grain Size: Twenty-three samples taken from pyroclastic fall and surge beds were sieved, and results are shown in Table 1. The samples are moderately to well sorted, coarse ash-to-lapilli size with only a small contribution of fine ash. The small amount of fine ash is unusual for tuff ring deposits and may indicate that eruptions and tuff ring formation occurred beneath the level of Clear Lake. In this situation, fine ash, if produced, may have been segregated from coarse ash within the lake.

TABLE 1. GRAIN-SIZE DESCRIPTION OF CLEAR LAKE TEPHRA.

Sample	Bed Form*	Md_φ (mm)	σ_φ	Fine Ash (Wt. %)
1	M	0.98 (0.51)	1.37	1.51
2	S	−0.66 (1.58)	0.77	0.3
3	F	−2.67 (6.36)	0.54	0.10
4	S	−1.21 (2.31)	1.19	1.36
5	P	−1.00 (2.00)	1.67	2.17
6	S	0.25 (0.84)	1.47	1.46
7	P	−0.24 (1.18)	1.75	1.12
8	F	−1.68 (3.20)	1.88	0.54
9	P	−0.48 (1.39)	1.26	1.34
10	M	−0.81 (1.75)	1.41	2.42
11	F	−0.09 (1.06)	1.42	0.95
12	F	−1.96 (3.89)	1.28	0.34
13	P	0.56 (0.67)	1.30	1.99
14	F	−0.98 (1.97)	0.83	0.45
15	P	−1.78 (3.43)	1.53	0.99
16	S	−0.61 (1.53)	1.15	1.80
17	P	0.19 (0.88)	1.12	2.79
18	F	−1.59 (3.01)	1.22	2.65
19	S	0.72 (0.61)	1.29	4.96
20	F	−0.88 (1.84)	1.68	0.53
21	M	−0.57 (1.48)	1.14	0.45
22	S	−0.08 (1.06)	1.39	3.46
23	F	−0.07 (1.05)	1.36	3.90

*Bed forms are S = sandwave surge, M = massive surge, P = planar surge, and F = fall.

Tephra Description and Composition: Of the glass constituents, five color types were recognized—black, brown, orange, white, and gray (Table 2). Average chemical analyses of these types are shown in Table 3. The analyses partially reflect hydrothermal alteration (e.g., higher SiO_2 abundances). Black and brown glasses are basaltic andesite; the orange is basaltic; and the gray and white glasses are andesitic and dacitic, respectively.

Mixed magma eruption is shown in samples 1, 2, and 3. These samples were collected in ascending stratigraphic order from a tuff ring at Soda Bay. Early eruptions produced dominantly basaltic andesite (black and brown pyroclasts) with a strong lithic contribution (Table 2). These poorly vesicular pyroclasts show typical blocky shapes of phreatomagmatic tephra with fused surfaces. Later eruptions produced vesicular tephra that show planar breakage surfaces that are, in general, chemically altered. These tephra likely represent stronger magma/water interaction and post emplacement hydrothermal alteration of basalt. Final eruptions produced a dacitic tephra composed of white and gray pyroclasts.

Other variations in pyroclast shape are strongly correlated to chemical composition. Gray glass samples commonly show blocky texture with uneven surfaces and no vesicles, whereas orange glass is strongly vesicular with smooth surfaces. This variation is well illustrated by fine-fraction particles in which tiny blocky clasts of gray glass are covered with adhering dust whereas those of an adjacent sample of orange glass are curved splinters and chips of broken vesicle walls.

The Clear Lake fine ashes, where present, show many interesting features. Grain surfaces are commonly covered with adhering dust particles that are from 5 μm to less than 1 μm in diameter. Many of these tiny particles appear to be fused to surfaces, and some have the angular euhedral shapes of crystallites. Overgrowths of clay and zeolite alteration products on glass surfaces, especially in hollows, are very evident on a $<$1-μm scale. Yet other bulbous particles are stuck together as moss-like aggregates and are attached to grain surfaces; these particles have diameters of $<$0.05 μm.

TABLE 2. PETROGRAPHIC DESCRIPTION OF CLEAR LAKE TEPHRA.*

			Glass				
Sample	Crystal	Lithic	Gray	White	Black	Brown	Orange
1	---	14.9	---	---	37.3	22.3	---
2	---	---	---	14.9	1.9	---	83.1
3	17.4	9.9	12.6	60.0	---	---	---
4	3.9	---	21.2	---	---	---	74.8
5	4.5	---	27.5	---	---	---	67.9
6	9.9	22.3	15.0	36.6	---	---	16.0
7	22.7	---	48.0	---	---	---	29.2
8	23.3	6.5	45.4	---	---	---	24.6
9	---	---	9.1	79.2	---	---	11.6
10	10.9	11.9	77.1	---	---	---	---
11	---	9.0	20.0	70.0	---	---	---
12	4.5	6.4	---	32.5	56.4	---	---
13	11.0	14.2	59.0	15.5	---	---	---
14	---	---	54.5	---	32.4	---	12.9
15	3.9	---	31.4	---	---	---	64.6
16	6.4	---	21.4	72.1	---	---	---
17	---	4.5	5.6	89.8	---	---	---
18	---	---	80.1	---	5.6	---	14.1
19	8.9	7.5	19.9	---	63.6	---	---
20	---	---	39.6	16.9	---	---	43.4
21	---	---	9.4	33.9	---	---	56.7
22	7.8	--	18.8	64.6	---	8.6	---
23	---	37.1	31.4	31.4	---	---	---

*Sample grain count percentages.

TABLE 3. CHEMICAL COMPOSITION OF GLASSY TEPHRA FROM CLEAR LAKE.*

Element	Gray (65)	White (35)	Black (12)	Brown (7)	Orange (26)
SiO_2	63.30	63.94	60.99	61.38	59.55
TiO_2	1.10	1.18	0.97	1.02	1.06
Al_2O_3	19.03	18.94	20.50	22.24	27.21
FeO	6.69	7.80	4.20	3.93	4.91
MnO	0.80	0.23	0.61	0.15	0.34
MgO	1.92	1.88	3.53	2.57	2.08
CaO	3.05	3.62	3.71	3.90	0.84
Na_2O	0.97	0.69	1.28	1.11	0.61
K_2O	1.94	1.15	1.89	2.16	0.86
Total	98.80	99.42	97.68	98.16	97.46

*Normalized EDS analyses of grain surfaces. Results are the averages for the number of samples shown in parentheses.

Comments on Pyroclast Formation: Eruptions under Clear Lake during recent times have been mostly phreatomagmatic, forming tuff rings. The tuff deposits have mixed pyroclast assemblages consisting of basaltic andesite through dacite, which may be indicative of mixed magmas. Early eruptions produced nonvesicular ash of basaltic andesite composition and tended towards gas-rich eruptions of vesicular basaltic ash. Later products are from a more viscous, crystal-rich magma of dacitic affinity.

These materials generally were explosively erupted because they mixed with lake water. The variation in compositions also resulted in phreatomagmatic pyroclasts of different shapes, from blocky particles to vesicular splinters and broken bubble walls (basaltic andesite) to spongy (hydrated basalt) and angular grains (dacite). The stratigraphic sequence of erupted products suggests that a basaltic to basaltic-andesitic magma body intruded a crystal-rich dacite. The subsequent mixing may have stimulated vesicle growth in the basaltic magma.

REFERENCES: Anderson, C. A., 1936. Volcanic history of the Clear Lake area, California. Geol. Soc. Am. Bull., **47**: 629-664.

Brice, J. C., 1953. Geology of Lower Lake quadrangle, California. California Div. Mines Bull., **166**: 72 pp.

Chapman, R. H., 1975. Geophysical study of the Clear Lake region, California. California Div. Mines and Geol. Spec. Rept., **116**: 23 pp.

Eichelberger, J. C., 1975. Origin of andesite and dacite: evidence of mixing at Glass Mountain in California and at other circum-Pacific volcanoes. Geol. Soc. Am. Bull., **86**: 1381-1391.

Goff, F. E., Donnelly, J. M., Thompson, J. M., and Hearn, B. C., Jr., 1977. Geothermal prospecting in The Geysers-Clear Lake area, northern California. Geology, **5**: 509-515.

Hearn, B. C., Jr., Donnelly, J. M., and Goff, F. E., 1972. Geology and geochronology of the Clear Lake volcanics, California. Proc. 2nd U.N. Symp. on Develop. Use Geotherm. Res., San Francisco: 423-428.

Hodges, C. A., 1966. Geomorphic history of Clear Lake, California. Ph.D. Dissertation, Stanford University: 210 pp.

U.S. Geological Survey, 1981. Research in The Geysers-Clear Lake geothermal area, northern California. R. J. McLaughlin and J. M. Donnelly (Eds.), U.S. Geol. Surv. Prof. Paper **1141**: 191 pp.

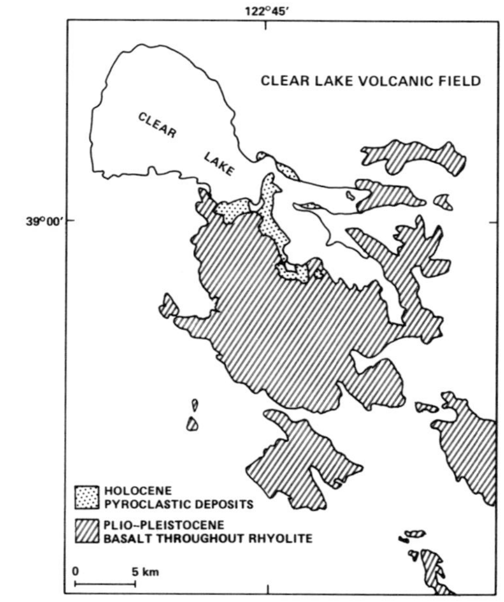

FIGURE 1. Generalized geologic map of Clear Lake showing distribution of the Clear Lake volcanic rocks, including Holocene pyroclastic deposits (adapted from Hearn et al., 1972). The pyroclastic rocks sampled for this section crop out along the shores of the lake as maar deposits.

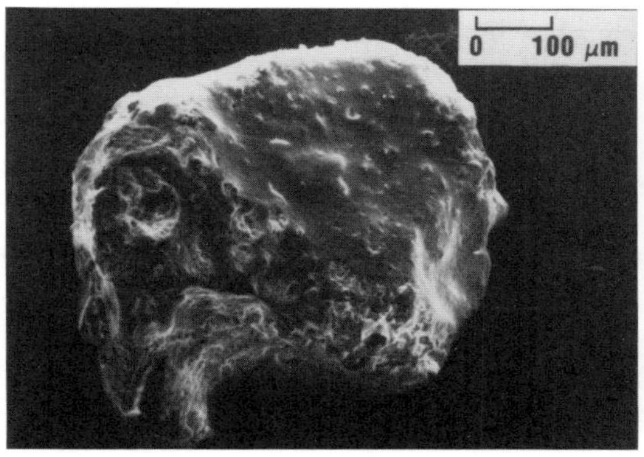

FIGURE 2a. SEMs of sample 1. Pyroclast of basaltic-andesitic black glass. Note the overall blocky shape with smooth, fused edges.

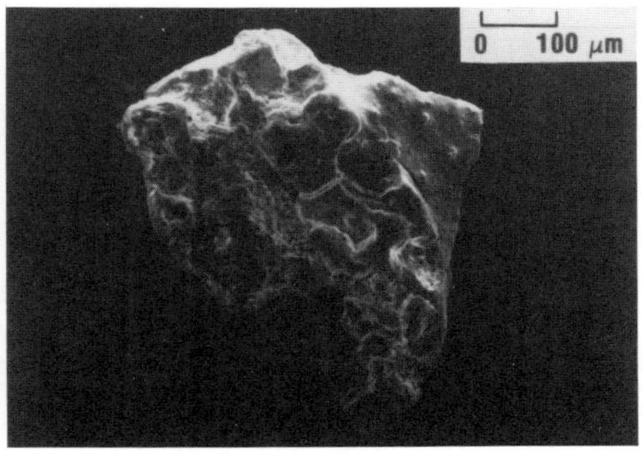

FIGURE 2b. Pyroclast of brown glass. Incipient alteration is shown by cracks on the surface. Smoothed surfaces result from glass quenching after fracture during water interaction.

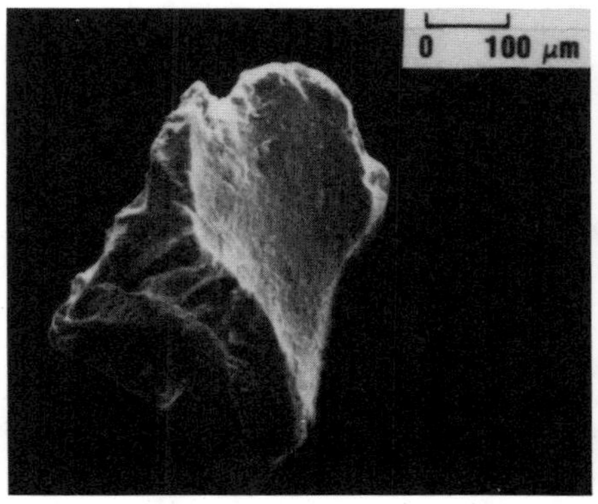

FIGURE 3a. SEMs of sample 2. Angular pyroclast of orange basaltic glass with a smooth vesicle surface that has been cut by a fracture.

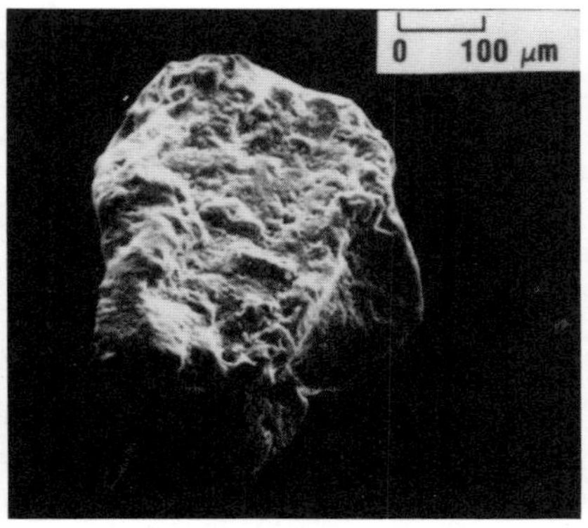

FIGURE 3b. Contrasting pyroclast of white dacitic glass with uneven surface formed by microcrystals.

FIGURE 3c. Fine fraction is composed of angular chips of nonvesicular glass that are splinters from larger vesicle walls.

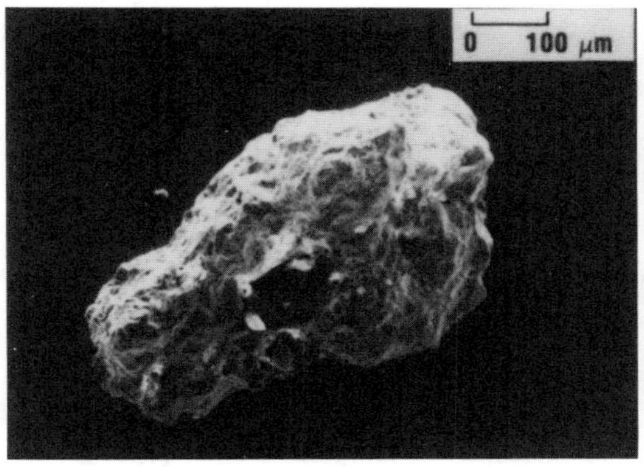

FIGURE 4a. SEMs of sample 3. Gray, dacitic glass with uneven, partly altered surface.

FIGURE 4b. White dacitic glass showing small vesicles and phenocrysts and altered intersertal glass; (*left* view is a secondary electron image; *right* is a backscattered view).

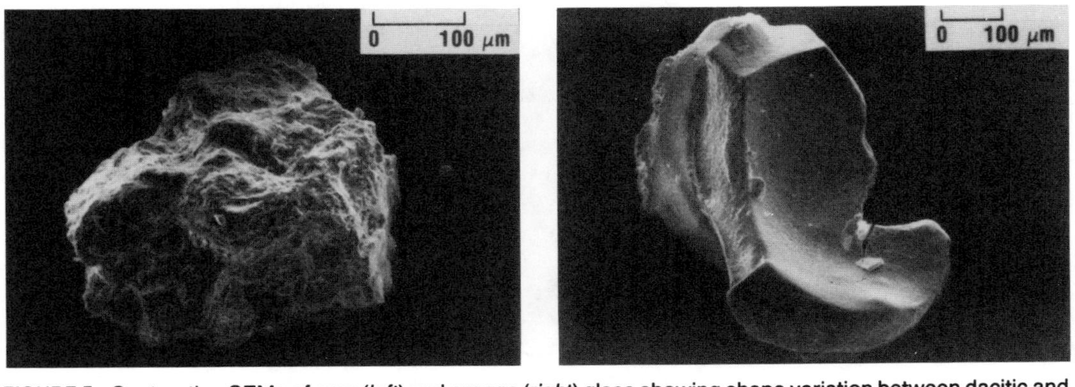

FIGURE 5. Contrasting SEMs of gray (*left*) and orange (*right*) glass showing shape variation between dacitic and basaltic glass. Both samples are from sample 2A and have broken surfaces from phreatomagmatic eruption; however, they show distinct differences in shape (blocky versus vesicular) and crystallinity.

FIGURE 6a. Sample 4 gray glass shows angular, blocky particles of dacitic glass with fine adhering dust.

FIGURE 6b. Sample 5, of dominantly orange glass, shows angular plate-like particles of orange glass that are somewhat finer grained than those of sample 4. Note smooth, curved surfaces with sharp points.

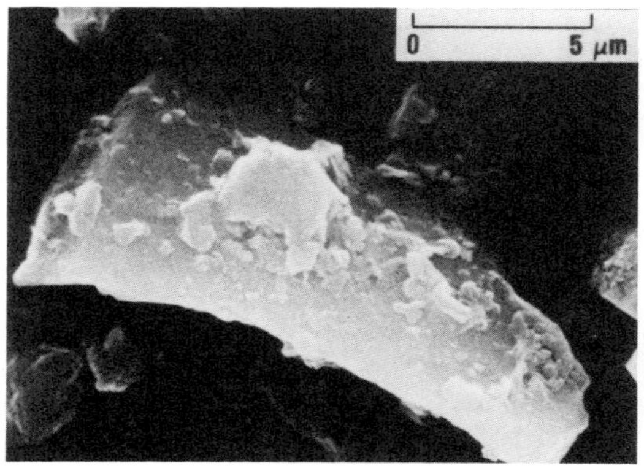

FIGURE 7a. High-resolution SEMs of the fine fraction of sample 16. Plate-like clast with adhering dust particles fused to the surface.

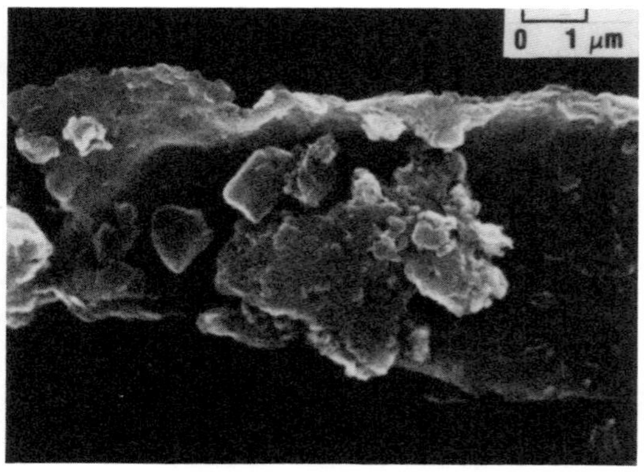

FIGURE 7b. Elongated vesicle hollow filled with clay-like debris and zeolite crystals (euhedral shapes). Vesicle edges are thin and irregular.

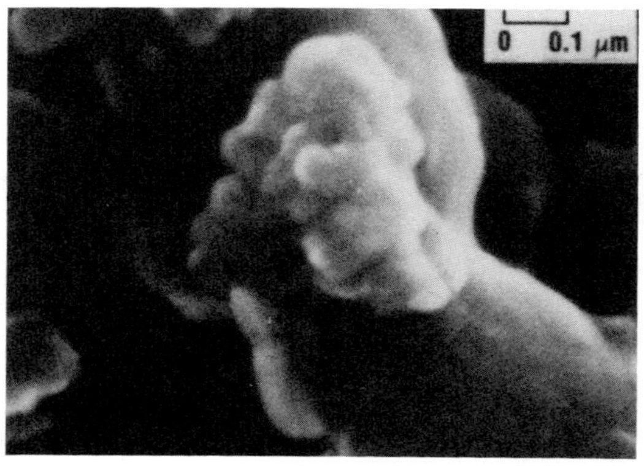

FIGURE 7c. Close-up image of 0.1-μm particle aggregates. The small cracks are likely formed by gold crazing of the conductive coat.

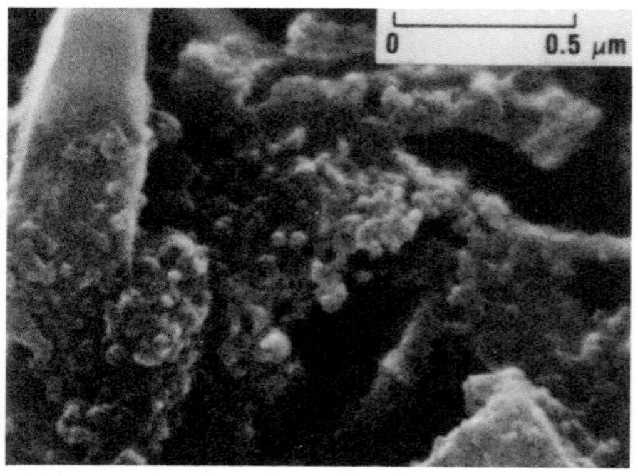

FIGURE 7d. Particle aggregate of 0.05-μm-diameter, sphere-like glass particles attached to flat surfaces of crystals.

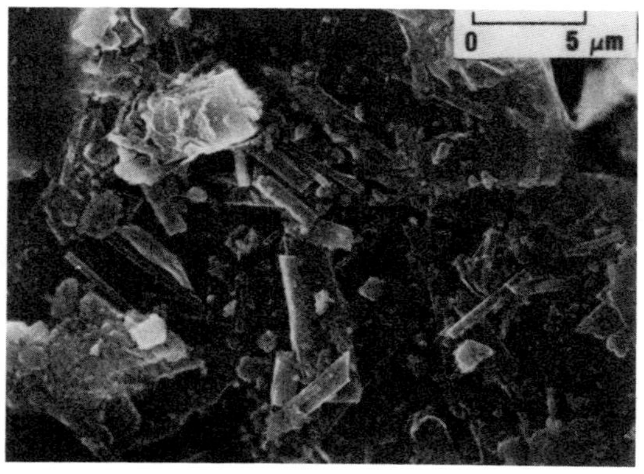

FIGURE 7e. Microphenocrysts of plagioclase attached to the surface of a glass particle.

Phreatoplinian Eruptions of Rhyolitic Magma

Phreatoplinian is a term first used by Self and Sparks (1978) for eruptions of extremely fine-grained rhyolitic ash deposited over large areas. This ash is formed by both vesiculation and interaction of magma with meteoric water. It is characterized by the presence of abundant vesicles and planar breakage features. Plinian eruptions at Taupo Volcano occurred within Lake Taupo, New Zealand; field evidence there lends further support to interpretations of the phreatoplinian eruption mechanism.

PANUM CRATER, CALIFORNIA

Introduction: Panum Crater is a Holocene rhyolite dome surrounded by a tuff ring 600 to 700 m in diameter. It is the northernmost volcano of the Mono Crater chain and is located in east-central California (Putnam, 1938; Kistler, 1966). Eruptions, dated at 1190 ± 80 y.b.p. (Wood, 1977), initially were explosive and were caused by extrusion of vesiculating rhyolite through the water-saturated alluvial deposits surrounding Mono Lake. Phreatomagmatic and magmatic explosions were interspersed during the eruptions that formed the tuff ring. Late-stage activity consisted of passive extrusion of a rhyolite dome in the tuff ring crater. Samples are from a pumice quarry on the southeast side of the tuff ring; several samples were provided by Wood (1977), whose tephra #2 was collected from meadow topsoils at several localities in the southern Sierra Nevada. Characteristics of phreatomagmatic activity and the sub-Plinian to Plinian dispersal of pumice and fine ash over more than 500 km^2 in southeastern California indicate that a phreatoplinian-type eruption occurred at Panum Crater.

Grain Size: Five samples taken from the measured stratigraphic sequence were sieved; the analysis results are given in Table 1. Opening activity produced the greatest abundance of fine material (samples 1 and 2). The final stages of pyroclastic activity (samples 3, 4, and 5) resulted in a gradual increase in median grain size, which is interpreted as a decrease in the explosive energy of the eruptions.

TABLE 1. GRAIN-SIZE CHARACTERISTICS OF PANUM CRATER TEPHRA.

Sample	Bed Form*	Md_φ (mm)	σ_φ	Fine Ash (Wt. %)
1	M	1.80 (0.29)	2.09	8.75
2	S	2.83 (0.14)	0.98	10.40
3	F	0.66 (0.63)	2.53	5.59
4	S	0.29 (0.82)	1.18	0.85
5	P	−0.18 (1.13)	2.06	1.69

*Bed forms are M = massive surge, S = sandwave surge, P = planar surge, and F = fall.

Tephra Description: The basal pyroclastic unit is an explosion breccia interstratified with massive surge beds. Grains are vesicular and bounded by nearly flat surfaces that cut bubble walls. Overall, these grains are blocky and have coatings of fine-grained plate-like material, especially in vesicular hollows. Some alteration products are apparent on grain surfaces and sharp vesicle edges have been smoothed by transport abrasion. The fine fraction is nonvesicular and consists of small, plate-like shards.

Above the basal unit is a 2-m-thick sandwave sequence. Grains sampled from this sequence are coarsely vesicular. They have blocky, equant shapes and many are tabular with elongated vesicles. Curved chips and dish-shaped fractures are abundant on edges of the grains, and surfaces are fresh and unaltered. The fine fraction is nonvesicular and consists dominantly of sharp-edged plates.

A sub-Plinian fall deposit above the lower surge sequence is reversely graded. Vesicular pyroclasts from this unit have blocky, equant surfaces with sharp vesicle edges, and they show no rounding from abrasion. The fine fraction consists of blocks and plate-like particles; as in Wood's (1977) tephra #2, vesicles are rare and the particles are bounded by the curved surfaces of vesicle walls.

Late-stage pyroclastic activity at Panum Crater produced blocky ash particles with highly elongated vesicles. Samples from coarse-grained pyroclastic surge deposits show strong grain-edge modification caused by transport. Both fresh and partly altered grain surfaces are evident.

Chemical Composition: The EDS analyses of sample grain surfaces are shown in Table 2. These analyses are similar to those reported by Loney (1968) and Lajoie (1968) and show variations in elemental abundances that are most likely related to variable chemical alteration on grain surfaces. Opening eruptions produced more lithic clasts and fewer crystals than the later eruptions did because of crater excavation and magma differentiation.

TABLE 2. CHEMICAL ANALYSES OF PANUM CRATER PYROCLAST SURFACES IN WT. %.*

	1	2	3	4	5
SiO_2	67.72	72.45	69.90	68.45	70.83
TiO_2	2.14	1.64	2.02	2.22	1.79
Al_2O_3	13.49	12.62	12.92	12.86	13.02
FeO	3.62	1.97	2.64	3.13	1.29
MnO	1.86	1.26	1.62	1.98	1.40
MgO	1.59	1.30	1.54	1.51	1.48
CaO	2.40	1.85	2.23	2.34	2.01
Na_2O	1.37	1.39	1.67	1.50	1.52
K_2O	5.80	5.55	5.94	6.01	5.77
Glass	85.9	74.0	83.0	87.0	83.0
Crystal	4.7	24.0	12.5	8.0	12.0
Lithic	9.4	2.0	4.5	4.8	4.0

*Normalized standardless EDS analyses.

Pyroclast Formation: Pyroclasts of Panum Crater are mostly of hydromagmatic origin, as is shown by their characteristic blocky morphology. High stands of Mono Lake (above the elevation of Panum Crater) occurred during Pleistocene times. Water from the lake and from aquifers in alluvial and glacial gravels underlying Panum Crater probably caused the hydrovolcanic eruptions. Explosive expansion of magmatic gases has also contributed to pyroclast formation, especially in early eruptions. The curviplanar fracture surfaces suggest that the magma fragmented in a brittle fashion. The foliation, caused by elongated vesicles, controlled much of the brittle fracture, and as a result, many of the grains are elongate slabs broken parallel to the foliation. The unaltered grain surfaces suggest that most of the hydrovolcanic eruptions producing the tuff ring were dry (there was little vapor condensation on pyroclasts). Vapor explosions that occurred at the surface of the magma column generated stress waves of sufficient energy to fracture the magma. Stress waves propagated down the vent and initially fragmented the lava. As the viscous particles of fragmented lava approached the surface, vaporized water ejected the particles in base surges.

REFERENCES:

Kistler, R. W., 1966. Geologic map of the Mono Craters quadrangle, Mono and Tuolomne Counties, California. U.S. Geol. Surv. Map GQ-462.

Lajoie, K. R., 1968. Late Quaternary stratigraphy and geologic history of Mono Basin, eastern California. Ph.D. Dissertation., Univ. California, Berkeley: 271 pp.

Loney, R. H., 1968. Structure and composition of the southern coulee, Mono Craters, California. Geol. Soc. Am. Mem., **116**: 415-440.

Putnam, W. L., 1938. The Mono Craters, California. Geog. Rev., **28**: 68-82.

Self, S. and Sparks, R. S. J., 1978. Characteristics of widespread pyroclastic deposits formed by the interaction of silicic magma and water. Bull. Volcanol., **41**: 196-212.

Wood, S. H., 1977. Distribution, correlation, and radiocarbon dating of late Holocene tephra, Mono and Inyo Craters, eastern California. Geol. Soc. Am. Bull., **88**: 89-95.

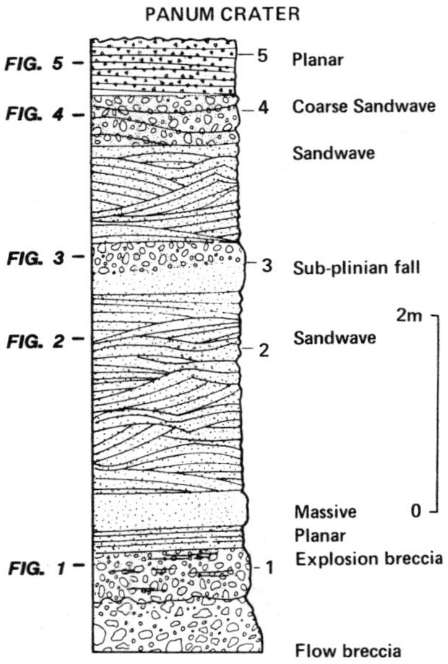

Stratigraphic section of the Panum Crater tuff ring measured in a pumice quarry southeast of the crater. The illustrated pyroclastic sequence overlies a lava breccia and continues nearly 20 m above the level at which sample 5 is shown. The upper section (not included) is covered but appears to be a continuation of the planar bedded deposits from which sample 5 was obtained.

FIGURE 1. SEM of a typical blocky, vesicular pyroclast of sample 1 (taken from a surge layer in the explosion breccia). The surface is slightly hydrated and covered with fine, adhering material, especially in vesicle hollows.

FIGURE 2. SEM of a sample 2 pyroclast. This vesicle-poor sample has broad, flat fracture surfaces and sharp edges. A few curved and dish-shaped spalled surfaces apparently were caused by transport abrasion. These spalled areas resemble conchoidal fractures.

FIGURE 3. SEM of sample 3 fine-fraction pyroclasts. Curviplanar surfaces form from broken vesicle bubble walls as well as from brittle fracture surfaces. Some tube-like vesicles are evident.

FIGURE 4. SEM of a sample 4 pyroclast showing numerous intersecting fracture surfaces that define the blocky morphology. Note the sharp, angular grain edge, unmodified by transport abrasion, that is typical of ash fall tephra.

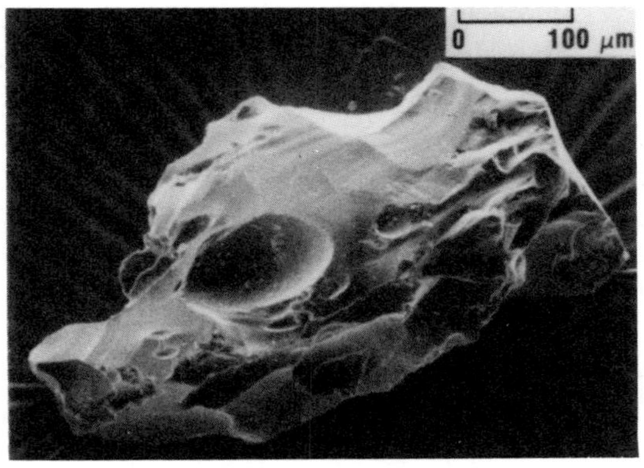

FIGURE 5. SEM of a sample 5 pyroclast. This poorly vesicular, blocky grain has edges that show numerous small chips and scratches resulting from abrasion.

FIGURE 6a. SEM of Wood's (1977) tephra #2 that was collected at several localities in the east-central Sierra Nevada range. The coarse fraction shows a few elongate vesicles that are absent in the fine fraction.

FIGURE 6b. Adhering dust is present on the surfaces of both examples; they strongly resemble grains from sample 3.

WAIRAKEI FORMATION, NEW ZEALAND

Introduction: The Wairakei Formation of North Island, New Zealand, is an excellent example of a phreatoplinian deposit. This deposit was erupted from the Taupo Volcanic center about 20,000 y.b.p. (Vucetich and Pullar, 1969; Self and Sparks, 1978; Self, 1983).

Of the six members of the Wairakei Formation that are exposed near Lake Taupo, four are widespread, fine-grained tephra fall deposits. Accretionary lapilli are common in two of these units. They are characterized by finely stratified, thin beds. The two other units are thick, fine-grained, nonwelded ignimbrites.

The samples here were collected near Lake Taupo by Stephen Self and from cores in peat bogs on the Chatham Islands, 800 km from the source, by New Zealand DSIR Soil Bureau scientists.

Grain Size: Median grain size for samples from the fine-grained tephra of the Wairakei Formation ranges from 1.4 mm to 31 μm (most are 31 to 62 μm). The tephra are mostly poorly sorted (Self and Sparks, 1978). Mean grain size does not change much downwind, which illustrates the uniform fine-grained nature of tephra generated during the eruption and suggests a deposition mechanism dominated by aggregate formation and rain-flushing.

Tephra Description: Most pyroclasts in these samples are fragments of vesicle walls and were produced by comminution of a vesicular rhyolitic melt during magma/water interaction. The vesicles are mostly ovoids; some have flattened to lense-like forms. The shattering of this melt produced curved plates, flat plates, and Y- or double-Y-shaped shards. Small pumice pyroclasts are also present, each with vesicles similar to those described above. Vesicle walls are fairly thick and are mostly greater than 10 μm.

Near-source samples are bimodal with 100- to 200-μm pumice pyroclasts and 12 to 14% phenocrysts (andesine, quartz, hypersthene, hornblende, and Fe-Ti oxides) in a matrix of fine-grained (<30-μm) shards.

Chemical Composition: Microprobe analyses of glass shards from near Lake Taupo and on the Chatham Islands are given in Table 1.

TABLE 1. CHEMICAL COMPOSITION OF GLASS SHARDS FROM THE WAIRAKEI FORMATION.		
	Lake Taupo (Wt. %)	Chatham Island (Wt. %)
SiO_2	72.40	72.25
Al_2O_3	11.49	11.40
FeO*	1.65	1.79
MgO	0.09	0.08
CaO	0.85	0.87
Na_2O	3.31	3.54
K_2O	3.00	2.84
TiO_2	0.04	0.07
Total	92.83	92.84
No. of Analyses	7	8

*FeO as total Fe.

Pyroclast Formation: Explosive interaction between a vesiculating rhyolitic magma and surface water has produced a finely comminuted mixture of shards (vesicle-wall fragments) and very small pumice pyroclasts. Self and Sparks (1978) and Self (1983) have provided an initial analysis of these eruption mechanisms.

REFERENCES:

Self, S., 1983. Large-scale phreatomagmatic volcanism: a case study from New Zealand. J. Volcanol. Geotherm. Res., **17**: 433-469.

Self, S. and Sparks, R. S. J., 1978. Characteristics of widespread pyroclastic deposits formed by the interaction of silicic magma and water. Bull. Volcanol., **41**: 196-212.

Vucetich, C. G. and Pullar, W. A., 1969. Stratigraphy and chronology of late Pleistocene volcanic ash beds in central North Island, New Zealand. N. Z. J. Geol. Geophys., **12**: 784-837.

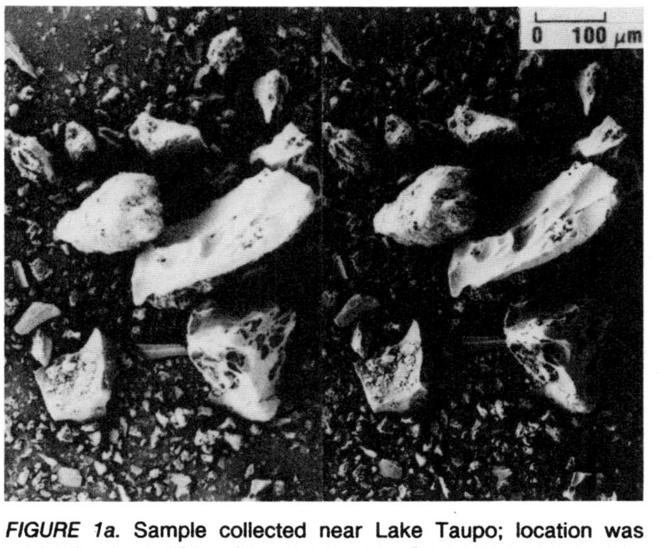

FIGURE 1a. Sample collected near Lake Taupo; location was probably about 15 km from the source. Stereomicrograph of an unsieved sample. The bimodal nature of this fine-grained ash is obvious; the vitric pyroclasts are in the 100- to 200-μm range and <30-μm range. A few rounded lithic fragments are visible. The larger pumice fragments are characterized by well-developed, coalesced vesicles; most are elongate ovoids. Vesicle walls are generally >10 μm thick. The fine-grained matrix consists of mostly finely comminuted fragments of vesicle walls; there are angular plates, wedges, and Y-shaped pyroclasts.

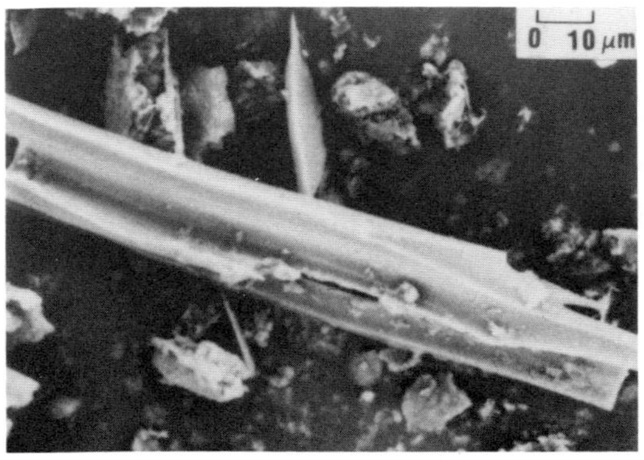

FIGURE 1b. The angular pumice fragment has elongate vesicles. The surrounding particles are angular, wedge-shaped, thin glass shards and clay(?) aggregates.

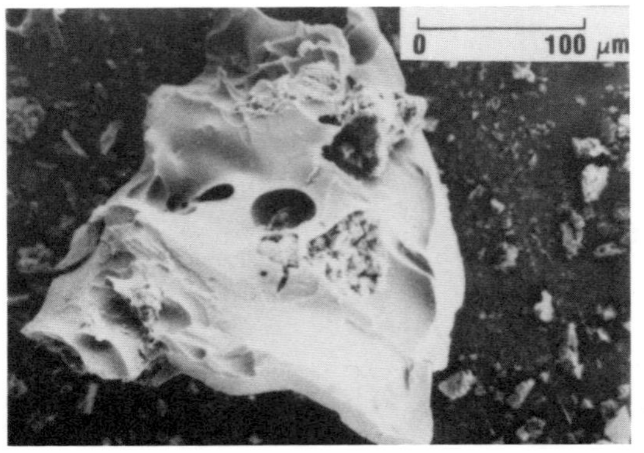

FIGURE 1c. Angular pumice pyroclast. Vesicles are irregular, flattened ovoids exhibiting some degree of coalescence. Most vesicle walls are thick (>10 μm). The pyroclast is surrounded by fine-ash size shards.

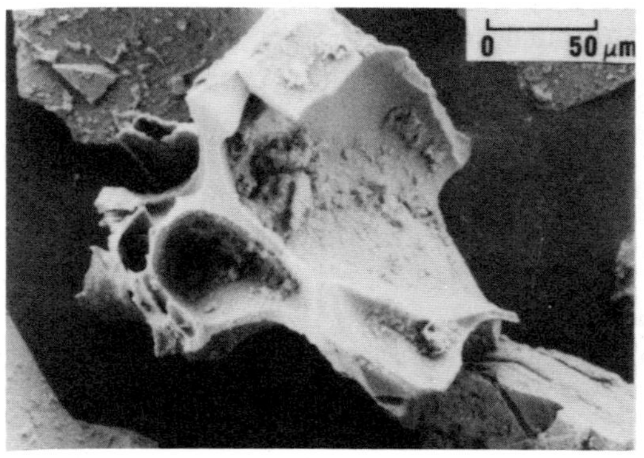

FIGURE 2a. Sample from a peat bog on the Chatham Islands 800 km southeast of the source. Note the better sorting of this sample (although it is similar in grain size) compared with the near-source sample illustrated in Fig. *1*. Angular pumice pyroclast with well-developed large vesicles. Fractures across vesicle walls are both planar and conchoidal. Particles adhering to grain surfaces are mostly smaller shards.

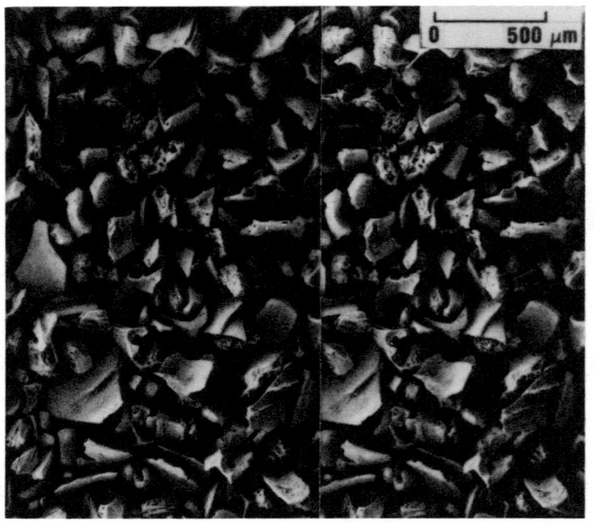

FIGURE 2b. Stereomicrograph of very angular, curved, plate-like, and Y-shaped shards broken from a vesiculated melt with large, well-developed vesicles. Some pyroclasts still retain a few flattened vesicles. Note the well-sorted nature of this sample. Most thin pyroclast edges are not abraded.

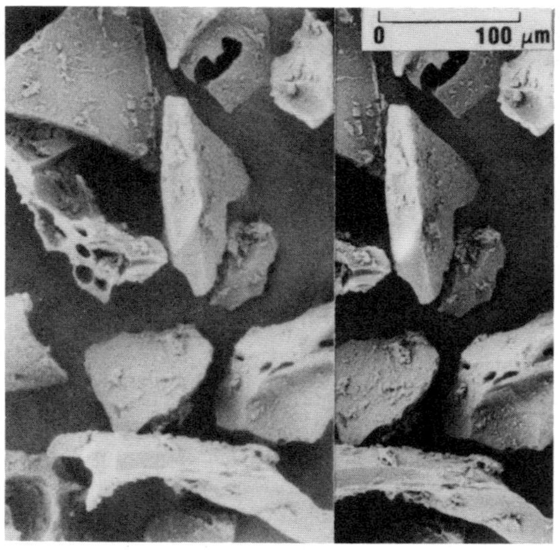

FIGURE 2c. Stereomicrographs of shards and small pumice pyroclasts that are typical of the distal Oruanui ash. These are mostly plate-like shards of 10 to 20 μm.

TAUPO VOLCANO, NEW ZEALAND

Introduction: The Taupo eruption sequence is the most recent explosive activity of the Lake Taupo volcanic area in the central North Island of New Zealand. The eruption, which produced a sequence of five tephra deposits, is dated between 130 and 186 A.D. by ^{14}C and historic atmospheric accounts (Wilson et al., 1980). The total volume of tephra erupted is in excess of 67 km^3. Various proportions of Lake Taupo waters mixed with erupting rhyolitic magma to produce both Plinian pumice falls and ignimbrite (at low lake levels) and phreatoplinian ash falls, as summarized in Table 1.

The late Quaternary ash deposits of this region were described by Healy et al. (1964). Self and Sparks (1978) recognized and discussed phreatoplinian eruptions from the Taupo Volcano. A recent series of papers by Walker and others is devoted to description of the deposits, including the fines-depleted nature of the Taupo Ignimbrite (Walker et al., 1980; Wilson and Walker, 1981, 1982), the exceptional dispersal of the Taupo Plinian deposit (Walker, 1980), the effect of phreatoplinian eruption in dispersing the Hatepe and Rotongaio Ashes (Walker, 1981a), and the Hatepe Plinian deposit (Walker, 1981b). This set of samples was kindly provided by Colin Wilson, University of Auckland.

TABLE 1. PHASES OF THE TAUPO ERUPTION AND THEIR VOLUMES.

Name	Sample No.*	Nature	Erupted Volume (km^3)
Taupo Ignimbrite	5	Nonwelded ignimbrite	$\simeq$30
Taupo Plinian	4	Plinian ash fall	$\simeq$24
Rotongaio Ash	3	Phreatoplinian ash fall	n.d.
Hatepe Ash	2	Phreatoplinian ash fall	n.d.
Hatepe Plinian	1	Plinian ash fall	6

*Samples were donated by C. J. N. Wilson, University of Auckland. Samples 1 and 4 were collected at the same location, approximately 55 km ENE of the vent; samples 2, 3, and 5 are from the same location 11 km SE of the vent.

Grain Size: Whole-deposit grain-size distributions (Fig. 1) display a fine-grained nature that is characteristic of the pre-Taupo ignimbrite deposits (Md $\simeq$ 60 to 120 μm; Walker, 1981a). The Taupo Ignimbrite varies in median diameter throughout the deposit from 100 to 150 μm (Walker et al., 1981). Like the Taupo Ignimbrite, the phreatoplinian deposits are moderately to poorly sorted (σ_φ = 1.6 to 2.7φ). The Plinian deposits are very poorly sorted ($\sigma_\varphi = 4.0\varphi$).

Tephra Descriptions: Ash particle shapes vary among samples from vesicular and pumiceous to nonvesicular and blocky; minor to abundant very fine ash (<10-μm diameter) adheres to grain surfaces.

The Hatepe Plinian deposit is characterized by highly vesicular, pumiceous pyroclasts with irregular, angular surfaces. Vesicles range from spherical to elongate and, in some places, are stretched and deformed. Vesicles are commonly 50 to 100 μm long and are separated by vesicle walls less than 10 μm thick.

In contrast to tephra of the Hatepe Plinian deposit, the Hatepe Ash is only poorly vesicular; most pyroclasts are blocky. Adhering dust and slight to moderate surface alteration are common on coarser ash particles. The fine ash consists of plate-like shapes that define parts of elongated vesicle walls. Breakage surfaces on the particles are curviplanar and form cuspate shapes. Little of the fine-grained fraction is finer than 10 μm; it is apparently limited by the smallest vesicle sizes.

The Rotongaio Ash is nonvesicular and is dominated by blocky shapes, hydrated surfaces on coarse ash particles, adhering fine dust, and abundant fine material of less than 1 μm. This abundance of aggregated material gives a mossy, irregular appearance to ash particles that are coarser than 10 μm.

Taupo Plinian pyroclasts are very vesicular. Spherical vesicles are 5 to 50 μm in diameter, and elongate, tube-shaped vesicles are 10 to 40 μm in long dimension. Finer grains are blocky and tabular but bounded on at least one face by curving vesicle walls. No adhering dust is present on these pyroclasts.

The Taupo Ignimbrite shows both partly vesicular and blocky pyroclast shapes. Adhering fine-grained ash is moderately abundant and covers either planar surfaces of larger blocky particles or vesicle hollows of poorly vesicular pyroclasts. Fine ash particles are dominated by curviplanar fracture surfaces. Some surface alteration is present on larger particles, and adhering fine-grained ash of less than 1 μm appears as tiny globules. Abrasion features include V-shaped pits and irregular chipped edges on sharp vesicle-wall projections.

Comments on Pyroclast Formation: Volatile supersaturation of a rhyolite magma and interaction of magma with the waters of Lake Taupo have contributed to the variety of ash textures described here. Walker (1981) proposed that the stratigraphic sequence of the Taupo eruption was related to the ratio of magma and water during each phase of activity. Self and Sparks (1978) described the phreatoplinian eruption and fragmentation mechanisms of the deposits of the Oruanui (Wairakei) Formation, which erupted from the Taupo Volcano. They suggested that vesiculation of the magma occurred earlier than the water interaction and thereby facilitated efficient fragmentation of the melt. Their conclusion is supported by SEM descriptions presented here that show fractures cutting across vesicles, apparently induced by magma/water interaction. Most of the particles in the phreatoplinian deposits of the Hatepe and Rotongaio Ashes show shapes indicative of brittle fracture, which may be attributed to the high viscosity of the rhyolitic magma. These same shapes, as well as adhering fine dust and some surface alteration of larger clasts (all indicative of magma/water interaction) are also apparent in the sample of Taupo Ignimbrite; this suggests that waters of Lake Taupo had access to the vent during the last eruption phase. This interpretation differs from that of Walker (1981), who concluded that the Taupo Ignimbrite was formed entirely by magmatic activity. However, limited magma/water interaction during this ignimbrite eruption may explain several features of this deposit: fines depletion caused by scavenging of small particles by accretion and adhesion to larger particles, the unusually wide distribution and formation of a "veneer deposit" (Walker et al., 1981) with cross-laminations typical of surging flow (Fisher et al., 1980), and perhaps, the "valley pond" deposits that could represent the portion of the ignimbrite from which steam had separated.

REFERENCES: Fisher, R., Smith, A., Wright, J., and Roobol, M.J., 1980. Ignimbrite veneer deposits or pyroclastic surge deposits? Nature, **286**: 912.

Healy, J., Vucetich, C., and Pullar, W., 1964. Stratigraphy and chronology of late Quaternary volcanic ash in Taupo, Rotorua and Gisborne Districts. N.Z. Geol. Surv. Bull., **73**: 1-88.

Self, S. and Sparks, R., 1978. Characteristics of widespread pyroclastic deposits formed by the interaction of silicic magma and water. Bull. Volcanol., **41**: 1-17.

Walker, G., 1980. The Taupo Pumice: product of the most powerful known (ultraplinian) eruption? J. Volcanol. Geotherm. Res., **8**: 62-94.

Walker, G., Wilson, C., and Froggatt, P., 1980. Fines-depleted ignimbrite in New Zealand - the product of a turbulent pyroclastic flow. Geol., **8**: 245-249.

Walker, G., 1981a. Characteristics of two phreatoplinian ashes and their water-flushed origin. J. Volcanol. Geotherm. Res., **9**: 395-407.

Walker, G., 1981b. The Maimihia and Hatepe Plinian deposits from the rhyolitic Taupo volcanic centre. N. Z. J. Geol. Geophys., **24**: 305-324.

Walker, G., Wilson, C., and Froggatt, P., 1981. An ignimbrite veneer deposit: the trail-marker of a pyroclastic flow. J. Volcanol. Geotherm. Res., **9**: 409-421.

Wilson, C., Ambraseys, N., Bradley, J., and Walker, G., 1980. A new date for the Taupo eruption, New Zealand. Nature, **288**: 252-253.

Wilson, C. and Walker, G., 1981. Violence in pyroclastic flow eruptions. *In*: S. Self and R. Sparks (Eds.), *Tephra Studies*, Reidel, Dordrecht: 441-448.

Wilson, C. and Walker, G., 1982. Ignimbrite depositional facies: The anatomy of a pyroclastic flow. J. Geol. Soc. London, **139**: 581-592.

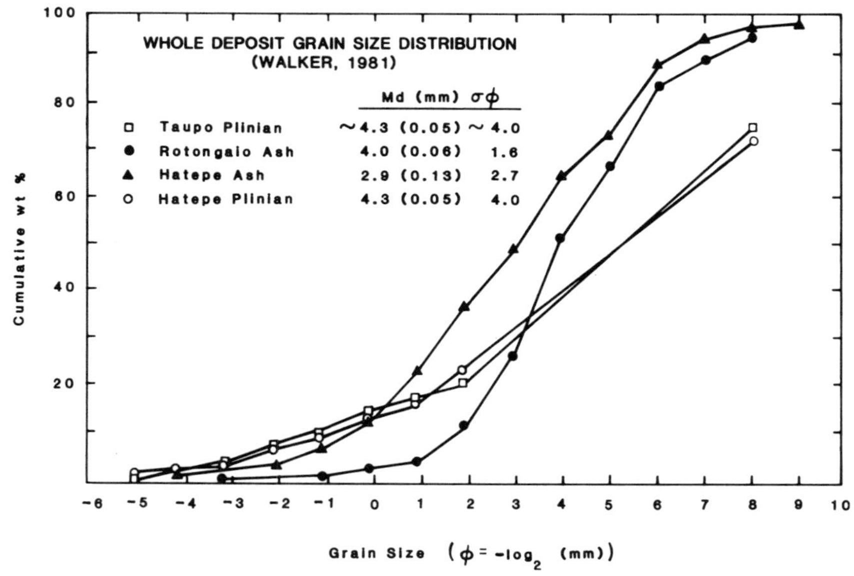

FIGURE 1. Plot of cumulative weight per cent versus grain size for pre-Taupo Ignimbrite deposits. These Plinian deposits are unusually fine grained, which attests to the extreme dispersive power of the eruptions.

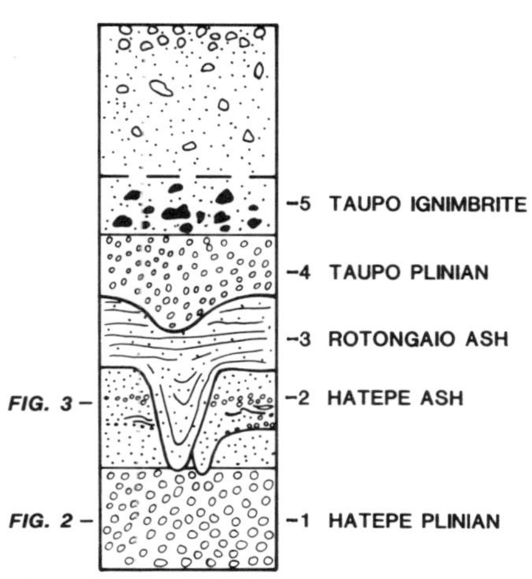

Schematic stratigraphic section of the Taupo eruption deposits; adapted from Walker (1981a).

FIGURE 2. SEM of pumiceous pyroclasts of the Hatepe Plinian deposit. Note the abundance of vesicles and irregular grain surfaces.

FIGURE 3a. SEMs of the Hatepe Ash: poorly vesicular, consisting of chips and splinters of fragmented material.

FIGURE 3b. Blocky grain showing moderate surface alteration.

FIGURE 3c. Finer grained pyroclasts consisting of angular, platy grains. Note the lack of particles less than about 5 μm in diameter.

FIGURE 4a. SEMs of the Rotongaio Ash: blocky, nonvesicular, altered ash with adhering dust.

FIGURE 4b. Fine fraction consisting of mossy and irregularly shaped particles of less than 10 μm—these shapes result from aggregation of <1-μm particles.

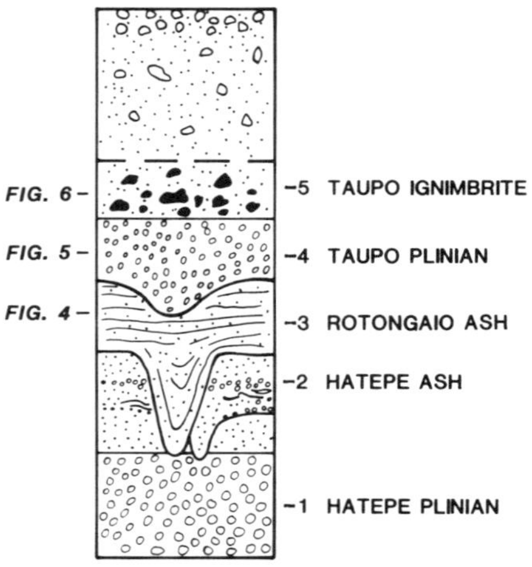

Schematic stratigraphic section of the Taupo eruption deposits; adapted from Walker (1981a).

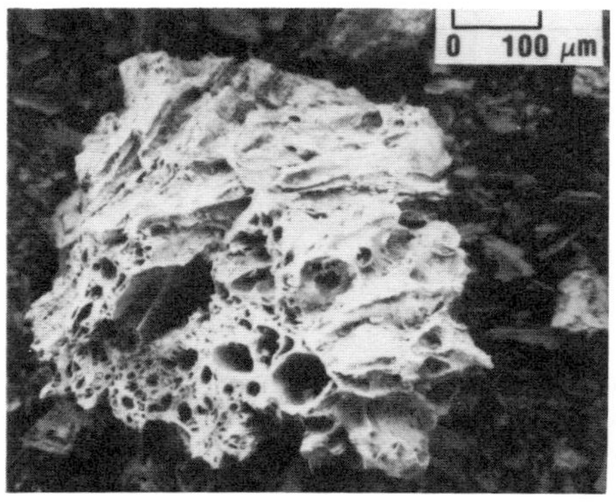

FIGURE 5a. SEMs of tephra from the Taupo Plinian deposit: highly vesicular pumice with elongate vesicles, delicate irregular surfaces, and little adhering dust.

FIGURE 5b. Fine fraction consisting of angular, plate-like, and cuspate fragments of larger particles, each showing at least one vesicle surface. Note the scarcity of <1-μm material.

FIGURE 6a. SEMs of the Taupo Ignimbrite: blocky particle with few visible vesicles, slight surface alteration, and some adhering dust.

FIGURE 6b. Angular, platy fine fraction with curviplanar fracture surfaces.

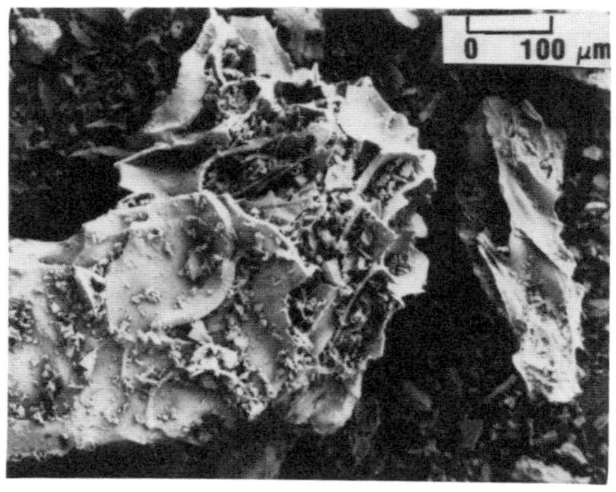

FIGURE 6c. Partly vesicular grain with adhering dust and vesicle-wall edges smoothed by transport abrasion.

Deep-Sea Hyaloclastites

Hyaloclastite is one of the most commonly sampled volcanic rock types from the ocean floor. This glassy, fragmental material is associated with pillow lavas and bedded deposits. The mode of hyaloclastite formation has not yet been fully explained. If hyaloclastite shards were erupted in relatively shallow portions of the ocean floor where there are low hydrostatic pressures, expansion of magmatic gases would be important in their genesis. If they are erupted at depths where vesicle growth is suppressed by hydrostatic pressure, other processes such as thermal spallation would be required to explain the formation of these deposits.

SEAMOUNT "F," EAST PACIFIC RISE

Introduction: Bedded hyaloclastites are associated with deep-sea basaltic lava flows and seamounts. Typically angular, nonvesicular, blocky shards probably formed during energetic mixing of magma and sea water (Carlisle, 1963; Honnorez, 1963; Honnorez and Kirst, 1975). The examples here are from a dredge haul of Research Vessel PLUTO, Leg 7; they were collected in the crater of Seamount "F" near the East Pacific Rise at 21°22′N, 108°37′W. Seamount "F" rises from ≃2900 m depth to a summit at 2100 m depth and is capped by a 2-km-diameter caldera. This caldera was investigated by a dive of the Submersible Vessel ALVIN during 1982 (Batiza et al., 1984); these investigators suggest that thin hyaloclastite pavements postdate the crater and overlie pelagic calcareous ooze, fine-grained crater-wall talus, and pillow basalts. The samples were generously provided by Rodey Batiza, Washington University.

Grain Size: There is a 5-μm to 2.5-mm range of shards within these samples. Within graded units described by Batiza et al. (1984), the median diameter is about 1 mm (0.0φ) for coarser, blocky shards and about 100 μm (3.3φ) for finer parts of the bed.

Tephra Descriptions: As described by Batiza et al. (1984), most of the hyaloclastites in these samples have vesicle-free, blocky, equant forms. These range from equant cubes to complex, angular polyhedrons. The second most common hyaloclastites are angular to slightly curved plates. Also present are rare irregular plates with conchoidal to hackly surfaces. Most of the shards in these samples exhibit some of the surface cracking and foliation that is characteristic of hydration and palagonitization. Some grains have pockmarked surfaces caused by solution.

Batiza et al. (1984) show a subcategory of shards that were broken from a chilled, vesiculated magma and exhibit some effects of surface tension (similar to those in Pele's hair). These are interbedded with the equant, blocky shards. The matrix contains small flakes of basaltic glass, which are mostly in a platy form.

As noted by Batiza et al. (1984), there are rare phenocrysts and hyalocrystalline shards in these samples; deep-sea hyaloclastites are composed of mostly homogeneous basaltic glass.

Chemical Composition: These samples of hyaloclastites are midocean ridge basalts (Table 1). Alteration by interaction with sea water has increased the silica concentrations at grain surfaces.

TABLE 1. CHEMICAL COMPOSITION OF SEAMOUNT "F" HYALOCLASTITE.*

	(Wt. %)
SiO_2	50.14
TiO_2	1.12
Al_2O_3	16.01
FeO**	9.16
MgO	8.45
CaO	12.59
Na_2O	2.42
K_2O	0.06
P_2O_5	0.10
Total	100.05

*R. Batiza, Washington University, analyst.
**Total Fe as FeO.

Comments on Pyroclast Formation: It is likely that most vesicle growth was suppressed and that rapid mixing of magma and water produced blocky and platy glass clasts when these deposits were erupted at a water depth of 2100 m. Batiza et al. (1984) proposed explosive mixing of basaltic magma and seawater for depths of less than 3100 m; boiling and large-volume changes are possible, although most vesicle growth is suppressed. Evidently, some vesicle growth was possible, perhaps within the short period of steam generation, because thin shards derived from vesicular melt have been described in deep-sea hyaloclastites, where they are mixed with nonvesicular, blocky shards.

Batiza et al. (1984) also suggest that hyaloclastites are formed by rapid quenching of magma and *not* by spallation of pillow rinds. Pillow rinds commonly have abundant microlites (quench crystals) and are not as glassy as hyaloclastites. Some of the thin glass slivers may be the result of thermal shock, as proposed by Carlisle (1963).

REFERENCES:

Batiza, R., Fornari, D., Vanko, D., and Lonsdale, P., 1984. Craters, calderas and hyaloclastites on young Pacific seamounts. J. Geophys. Res., **89**: 8371-8390.

Carlisle, D., 1963. Pillow breccias and their aquagene tuffs on Quadra Island, British Columbia. J. Geol., **71**: 48-71.

Honnorez, J., 1963. Sur l'origine des hyaloclastites. Bull. Volcanol., **25**: 253-258.

Honnorez, J. and Kirst, P., 1975. Submarine basaltic volcanism: morphomometric parameters for distinguishing hyaloclastites from hyalotuffs. Bull. Volcanol., **39**: 1-25.

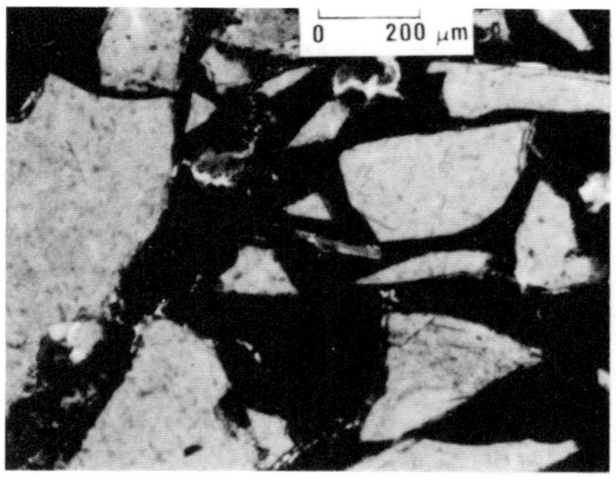

FIGURE 1. Transmitted-light photomicrograph of blocky, equant sideromelane shards.

FIGURE 2. Reflected-light photomicrograph of shards similar to those in Fig. *1*.

FIGURE 3. SEM of hyaloclastite sample 4-13. There are mostly hydrated, equant blocky glass shards of basaltic composition; some show surface pitting. A few irregular, platy shards are visible (*center-right*).

FIGURE 4. Close-up of Fig. 3. Blocky shards are broken by surface cracks and foliations that were caused by hydration and incipient palagonitization of the glass. Irregular clumps are remnants of calcareous cement that bonded the hyaloclastite before disaggregation and sample analysis.

FIGURE 5. SEM of flat, platy shards from sample 4-5 (best examples are near the center). All shards show some surface cracking caused by alteration.

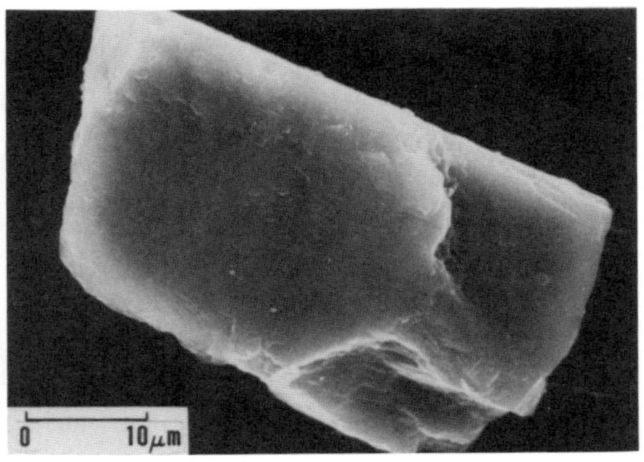

FIGURE 6. SEM of blocky glass shard from sample 4-5.

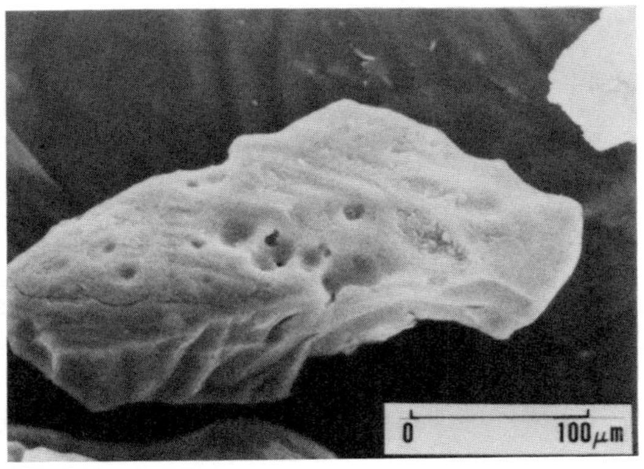

FIGURE 7. SEM of irregular glass shard from sample 4-5. This shard type is rare, with irregular, conchoidal surfaces rather than smooth, blocky surfaces. There is pitting caused by alteration, and irregular patches of carbonate cement are visible on grain surfaces.

Littoral Cones

One of the more curious observations about magma/water contact is that it is not always explosive. For example, when lava from Hawaiian volcanoes enters the sea, there is steam, but most of the activity is passive. A quenched surface allows lava to pour quietly downslope underwater to form pillow structures. However, passive behavior is not always the case, and explosive mixing of lava and water can form a tephra cone in the littoral zone.

PUU HOU AND SAND HILLS, HAWAII

Introduction: Ash similar to that ejected during phreatomagmatic eruptions can be produced when lava flows into a body of water (Moore and Ault, 1963; Fisher, 1968). If explosive mixing of lava and water occurs, small tephra cones (littoral cones) are formed at the water's edge, flanking the lava flow. Littoral cones are characteristically crescent-shaped ridges that, from the ground, appear to be cones without summit craters (Fisher, 1968). The examples described here are from the west coast of the Island of Hawaii, where littoral cones were constructed by basaltic lavas. Puu Hou was formed during the 1868 South Rift eruption of Mauna Loa, and Sand Hills resulted from an eruption of Kilauea in 1840.

Fisher (1968) suggests that the explosive, cone-building activity at Puu Hou occurred when steam was trapped beneath lava crusts. The explosions were in localized areas of the lava streams, and explosion centers migrated seaward as the flow built a lava delta into the ocean (Fisher, 1968). Flow surfaces between the cones were kept free of explosion debris as the flows moved seaward.

Some littoral cones are complete 50- to 100-m-high cones with summit craters and apparently were built above lava tubes (Fisher, personal communication). Peterson (1976) observed the development of a small littoral cone where waves had access to a lava tube through its breached top.

The samples studied here are from the collection of R. V. Fisher.

Grain Size: Mean grain size of the samples ranges from 0.7 to 1.3 mm (0.5 to -0.4φ); all are poorly sorted.

Tephra Description: Within the Puu Hou complex, hyalotuff layers and agglutinated zones (bomb-rich layers) are interbedded. Samples described here are from the hyalotuff layers. Fisher (1968) has distinguished three vitric pyroclast types within these deposits:

(1) angular scalloped fragments (intersecting large vesicles),
(2) brown glass droplets with curved, fluidal edges and a few small vesicles, and
(3) irregular to equant fragments with few or no vesicles.

All the sideromelane pyroclasts listed above have trace to abundant quench crystals (plagioclase, clinopyroxene, and olivine). The 1 to 9% olivine appears as phenocrysts in sideromelane and tachylite pyroclasts and as separate crystal pyroclasts. Tachylite pyroclasts make up 2 to 70% of the deposits. Most are subangular and nonvesicular.

Chemical Composition: All samples are of basaltic composition. Analyses are listed in Table 1.

TABLE 1. CHEMICAL ANALYSES OF PUU HOU AND SAND HILLS TEPHRA.*

	Puu Hou (Wt. %)	Sand Hills (Wt. %)
SiO_2	51.53	50.46
TiO_2	2.02	2.29
Al_2O_3	13.15	12.22
FeO_2O_3**	11.34	11.58
MnO	0.21	0.21
MgO	8.13	11.75
CaO	9.87	9.98
Na_2O	3.17	2.84
K_2O	0.42	0.41
P_2O_5	0.24	0.24
S	0.08	0.11
Total	100.16	102.09

*M. Rhodes and J. Evans, Johnson Space Center, analysts.
**Total Fe as Fe_2O_3.

Comments on Pyroclast Formation: Fisher (1968) determined that the littoral cones he described were formed by steam explosions that resulted from lava in contact with water. Explosion centers were confined to areas where lava was rapidly and continuously delivered. Pahoehoe and aa were the main lava types that reached the sea. Aa does mix with sea water, but its surface texture does not create the confinement necessary to produce explosions. At Puu Hou, pahoehoe and aa were comminuted where steam was trapped beneath rapidly delivered, broken lava crusts.

Comminution of the flow by water/lava interaction and subsequent steam explosions produced a variety of pyroclast types. Angular, chilled sideromelane fragments with large, dispersed vesicles came from the interior of the flow where vesicles were well developed during the lava flow's journey to the sea. Irregular, vesicular and nonvesicular droplets with smooth, fluidal surfaces were bits of melt ejected by the explosion but not chilled during water/lava interaction; they were chilled in flight by thermal radiation and flow of air over grain surfaces. Fragmentation of flow crusts may have produced vesicular and nonvesicular, angular fragments of tachylite.

REFERENCES:

Fisher, R. V., 1968. Puu Hou littoral cones, Hawaii. Geol. Rund. **57**: 837-864.

Moore, J. G. and Ault, W. V., 1965. Historic littoral cones in Hawaii. Pac. Sci., **19**: 3-11.

Peterson, D.W., 1976. Processes of island growth, Kilauea Volcano, Hawaii, 1969-1973. Proc. Symp. Andean and Antarctic Volcanology Problems, Santiago, Chile: 172-189.

FIGURE 1a. Pyroclasts from hyalotuff layers in the Puu Hou littoral cone, Hawaii. Blocky, equant pyroclasts with 10 to 20% spherical or ovoid vesicles are sideromelane with only a trace of quench crystals; these appear to have formed during the lava/water interaction. Broken sideromelane droplets with curved, fluidal edges are drops of melt from the flow interior that were thrown clear by a steam explosion. Ragged, irregular fragments with rough surfaces are bits of tachylite; the flow crust fragmented during the explosions.

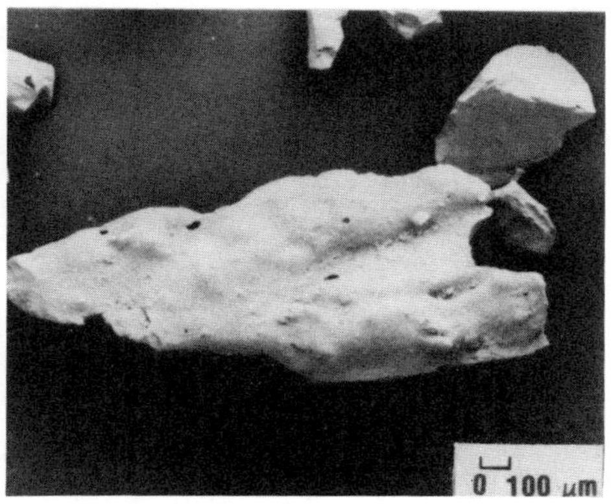

FIGURE 1b. Fragment of a sideromelane droplet characterized by curved, smooth surface (at least on one side). Pyroclasts such as these are vesicular.

FIGURE 1c. Close-up view of the finer grained fraction. It consists of mostly angular, blocky sideromelane pyroclasts with low vesicularity.

FIGURE 2a, b. A sample from Sand Hills littoral cones, Hawaii. General SEM views of hyaloclastites show the variety of pyroclasts, which is the same as for Puu Hou although these samples are predominantly blocky, angular sideromelane fragments with very low vesicularities. Some have curved surfaces where large, well-developed vesicles were intersected during comminution. Some curved sideromelane pyroclasts with smooth surfaces (droplets) are visible.

FIGURE 2c. Center—vesicular sideromelane droplet with elongate, coalesced vesicles. The pyroclast surface is smooth and curved where not broken. Most other pyroclasts are blocky sideromelane pyroclasts. Some of the thin grain edges are abraded, but generally, the grain surfaces show no signs of abrasion.

FIGURE 2d. Left—angular sideromelane fragment with no vesicles. *Center*—ragged, equant tachylite grain. *Right*—fragment of a sideromelane droplet.

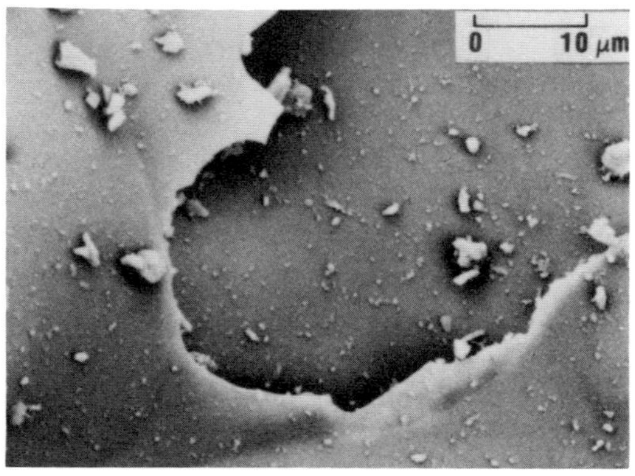

FIGURE 2e. Close-up of the smooth surface of the droplet fragment. A near-surface vesicle was a flattened ovoid before breaking. Note the <1-μm-thick vesicle wall (broken).

Experimental Production of Phreatomagmatic Ash

Experimental investigations provide a unique method for studying the formation of volcanic ash. For example, blast furnace slags that are quenched by water have textures similar to those of hyaloclastites and other phreatomagmatic ashes. This section describes particulates produced in some explosive melt/water mixing experiments.

THERMITE/WATER EXPLOSIONS

Introduction: Explosions caused by rapid vaporization of water during contact with molten material have been noted in industrial accidents for a number of years (Lipsett, 1966; Witte et al., 1970). This class of explosion is often termed *fuel-coolant interaction* (FCI) and has recently been the subject of considerable experimental and theoretical study in connection with nuclear reactor safety (Sandia Laboratories, 1975). The result of vapor explosions is fragmentation of the melt into micrometer and millimeter-size particles that are accelerated outward within a rapidly expanding steam cloud. The FCI mechanism is also applicable to phreatomagmatic (hydrovolcanic) eruptions (Peckover et al., 1973; Colgate and Sigurgeirsson, 1973; Sheridan and Wohletz, 1981).

This section provides a comparison of experimental debris with naturally occurring phreatomagmatic ash. Wohletz and McQueen (1984) described in detail experiments that use a thermite melt as a basalt analog in volcanic models. Additional information on FCI experiments using thermite and corium is given by Nelson et al. (1980). The explosive phenomena of volcano simulations are strikingly similar to a variety of eruptive activities, some of which are not normally considered phreatomagmatic. These activities include Hawaiian lava fountaining, Strombolian eruptions, Surtseyan and Vulcanian explosions, and littoral cone and submarine pillow lava formation.

Grain Size: Experimental phreatomagmatic ash closely resembles ash produced by Surtseyan eruptions. Figure 3 provides a comparison between Surtsey and Capelinhos ashes and those of experimental thermite and corium melts. Mean sizes are 0.13 to 0.14 mm, and particles are moderately to poorly sorted. Median grain size generally decreases with increasing explosive efficiency for thermite/water explosions. Nonexplosive interactions produce median sizes coarser than 2 to 4 mm; finer debris are recovered from explosive interactions. Median particle sizes reach the micrometer range in explosions of maximum energy.

Particle Description: Several dominant particle shapes are evident in experimentally produced ash. The most common are characterized as moss-like, convolute shapes with large surface areas. Less abundant particles have spheroidal or drop-like shapes and strongly resemble industrial fly-ash debris. Another common type of experimental ash shows angular, blocky, and equant shapes that are generally >50 µm in long dimension. Plate-like shapes are angular, flat chips and broken pieces of bubble walls. All the above particles form aggregates, and many irregularly shaped aggregates consist of droplets, blocky particles, and platelets bonded by twisted, deformed melt particles.

Chemical Composition: Thermite (Fe_3O_4 + Al), when reacted, forms iron and aluminum oxides. In volcano simulations, quartzo-feldspathic sand added to the thermite produced compositions more closely resembling a basaltic melt (Table 1). Although these compositions are still very different from natural silicate magmas, their similar density, liquidus temperature, and viscosity make them reasonable models.

TABLE 1. REPRESENTATIVE CHEMICAL ANALYSES OF THERMITE MELT DEBRIS.*

	1	2	3	4	5
SiO_2	10.07	10.73	14.34	36.55	18.25
TiO_2	---	---	2.25	1.44	1.90
Al_2O_3	31.24	25.09	11.36	34.41	42.46
FeO	57.94	64.18	57.01	17.32	23.28
MnO	---	---	1.58	0.96	1.33
MgO	---	---	6.44	3.70	5.84
CaO	0.45	---	2.05	1.65	1.90
Na_2O	---	---	3.07	1.98	3.19
K_2O	0.31	---	1.91	1.99	1.85

*Normalized standardless (EDS) analyses.
Sample descriptions:
1 - Iron-aluminum sphere, large,
2 - Iron-aluminum sphere, small,
3 - Blocky iron particle,
4 - Coating on iron particle,
5 - Iron-aluminum spindle.

Comments on Particle Formation: The blocky, equant pyroclasts from explosive water/magma eruptions have led previous investigators (Peacock, 1926; Carlisle, 1963; Fisher and Waters, 1969; Heiken, 1971; Honnorez and Kirst, 1975) to conclude that the magma is fragmented by thermal contraction and shattering of the glass. The strong similarity between experimentally produced debris and phreatomagmatic ash has provided insight into the mechanisms of ash formation. To create vapor explosions upon contact of magma with water, intimate mixing of the two is required. This mixing occurs when the melt fragments before explosion. Fragmentation increases the contact area of melt and water so that conductive heat exchange occurs at explosive rates. Experiment results indicate that the initial fragmentation of the melt occurs by several mechanisms, including (1) fluid instabilities between the water and melt, which promote turbulent mixing of the two; (2) oscillatory vapor film formation, expansion, and collapse, which produces stress waves at the melt surface and thereby fragments the melt; and (3) large compressive and tensile stress waves produced by vapor explosions—these stress waves propagate into the melt and fragment it. Each of the above mechanisms occurs on a millisecond time scale, increases the water/magma contact area, and leads to catastrophic heat exchange that is followed by vapor explosion (Wohletz, 1983).

REFERENCES:

Buchanan, D. J., 1974. A model for fuel-coolant interactions. J. Phys. D: Appl. Phys., **7**: 1441-1457.

Buxton, L. D. and Benedict, W. B., 1979. Steam explosion efficiency studies. Sandia National Laboratories, SAND79-1399, NUREG/CR-0947: 62 pp.

Colgate, S. A. and Sigurgeirsson, T., 1973. Dynamic mixing of water and lava. Nature, **244**: 552-555.

Carlisle, D., 1963. Pillow breccias and their aquagene tuffs, Quadra Island, British Columbia. J. Geol., **71**: 48-71.

Fisher, R. V. and Waters, A. C., 1969. Bed forms in base surge deposits: Lunar implications. Science, **165**: 1349-1352.

Heiken, G., 1971. Tuff rings: examples from the Fort Rock-Christmas Lake Valley, south-central Oregon. J. Geophys. Res., **76**: 5615-1626.

Honnorez, J. and Kirst, P., 1975. Submarine basaltic volcanism: Morphometric parameters for discriminating hyaloclastites from hyalotuffs. Bull. Volcanol., **32**: 441-465.

Lipsett, S. G., 1966. Explosions from molten materials and water. Fire Tech., May: 118-126.

Nelson, L. S., Buxton, L. D., and Planner, H. N., 1980. Steam explosion triggering phenomena, part 2: Corium-A and Corium-E and oxides of iron and cobalt studied with a floodable arc-melting apparatus. Sandia Laboratories, SAND77-0998, NUREG/CR-0122: 138 pp.

Peacock, M. A., 1926. The basic tuffs. *In*: G. W. Tyrell and M. A. Peacock (Eds.), *The petrology of Iceland*, Royal Soc. Edinburgh Trans., **45**: 51-76.

Peckover, R. S., Buchanan, D. J., and Ashby, D. E. T. F., 1973. Fuel-coolant interactions in submarine volcanism. Nature, **245**: 307-308.

Sandia Laboratories, 1975. Core-meltdown experimental review. SAND 74-0382: 472 pp.

Sheridan, M. F. and Wohletz, K. H., 1981. Hydrovolcanic explosions, the systematics of water-tephra equilibration. Science, **212**: 1387-1389.

Witte, L. C., Cox, J. E., and Bouvier, J. E., 1970. The vapor explosion. J. Metals, **22**: 39-44.

Wohletz, K. H., 1983. Mechanisms of hydrovolcanic pyroclast formation: size, scanning electron microscopy, and experimental studies. *In*: F. Barberi and M. Sheridan (Eds.), *Explosive Volcanism*. J. Volcanol. Geotherm. Res., **17**: 31-64.

Wohletz, K. H. and McQueen, R. G., 1984. Experimental studies of hydromagmatic volcanism. *In: Explosive volcanism: inception, evolution, and hazards*. Studies in Geophysics, National Academy Press, Washington, D.C.: 158-169.

FIGURE 1. Simulation of Surtseyan phreatomagmatic eruption. The vertical plume is 24 m high and composed of steam, sand, and thermite particles. Note the base-surge cloud spreading out horizontally.

FIGURE 2. Interaction of thermite melt with small amounts of water has produced explosions that resemble Strombolian activity. The centimeter-size fragments of melt are being ejected vertically in a billowing steam cloud.

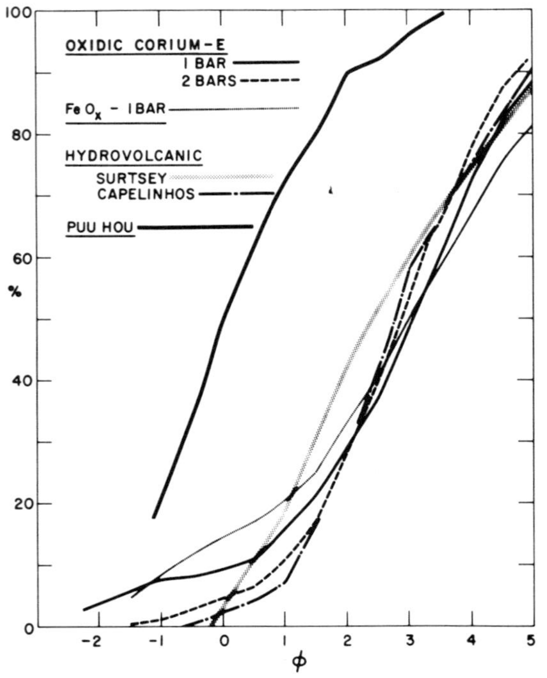

FIGURE 3. Comparison of grain-size analyses of experimentally produced ash with phreatomagmatic ash from Surtsey, Capelinhos, and Puu Hou. Plot shows cumulative weight per cent versus phi size ($\varphi = -\log_2$ *mm).*

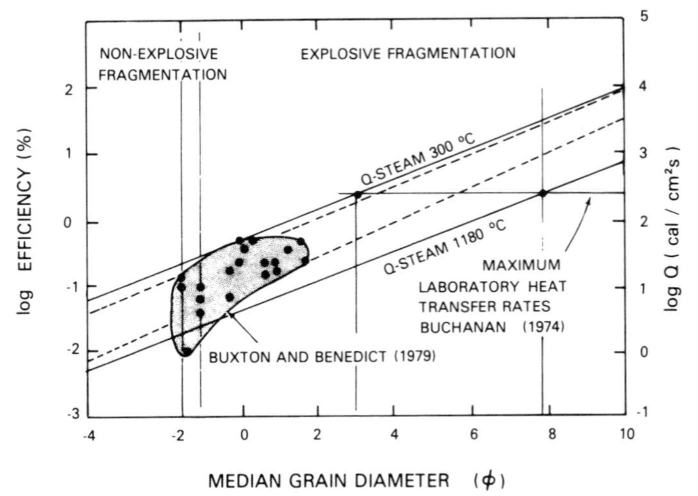

FIGURE 4. Plot of explosive efficiency (mechanical energy/thermal energy) versus median grain size for experimental ash. Points shown are from experiments by Buxton and Benedict (1979). Dashed lines extrapolate the plotted trend toward higher efficiencies and finer grain sizes. Lines plotted for Q, conductive heat transfer rates, are calculated from conduction of heat from a molten basalt sphere at 1200°C to a surrounding jacket of steam at both 300 and 1180°C. The maximum laboratory-measured heat transfer rates for fuel/coolant interactions (Buchanan, 1974) correspond to fragment sizes between 3φ (0.125 mm) and 4φ (0.063 mm).

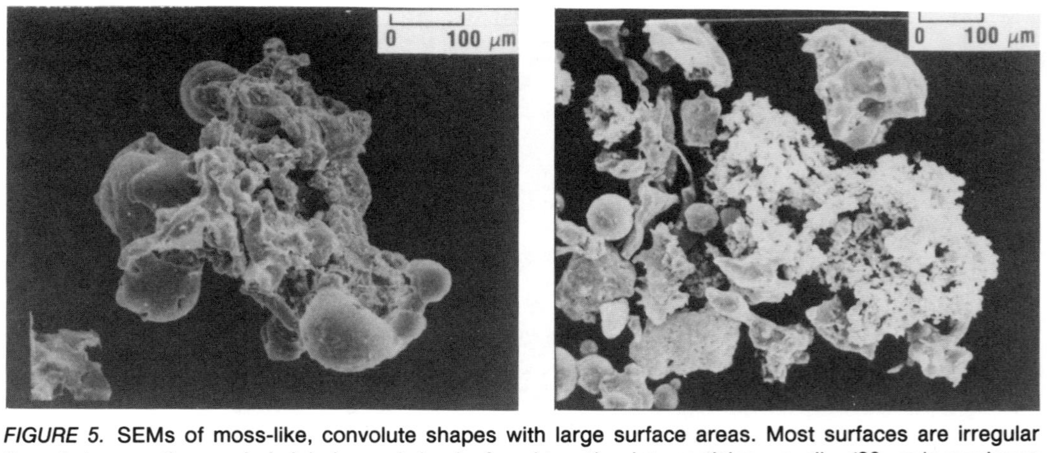

FIGURE 5. SEMs of moss-like, convolute shapes with large surface areas. Most surfaces are irregular threads to smooth, rounded globules and chunks fused together into particles, usually <20 μm in maximum dimension. Deep embayments separate globular lobes and tiny spheres that are attached to some particles. Jagged edges are rare except where ribbon-like or spongy material forms particle lobes.

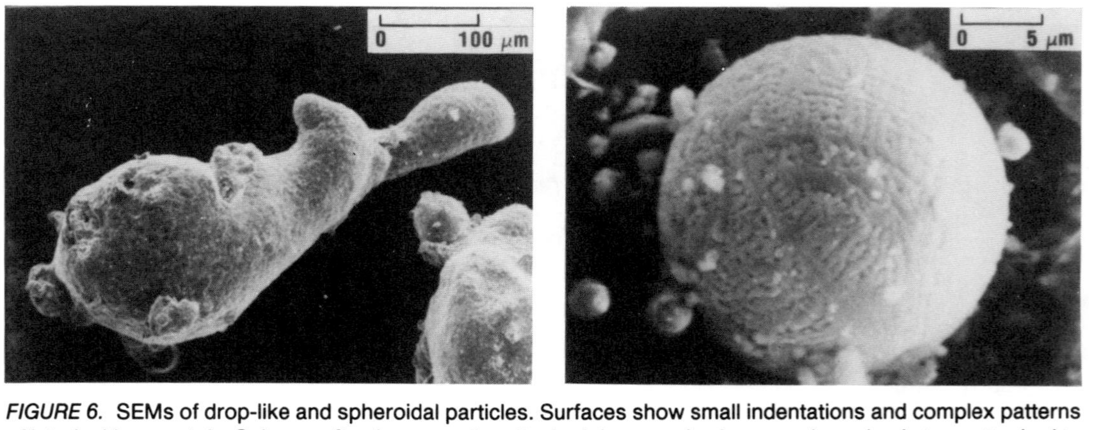

FIGURE 6. SEMs of drop-like and spheroidal particles. Surfaces show small indentations and complex patterns of interlocking crystals. Spheres often have smaller attached drops and spheres and may be, in turn, attached to a substrate of blocks and aggregate material. Many spheres have hollow centers and resemble industrial fly-ash debris.

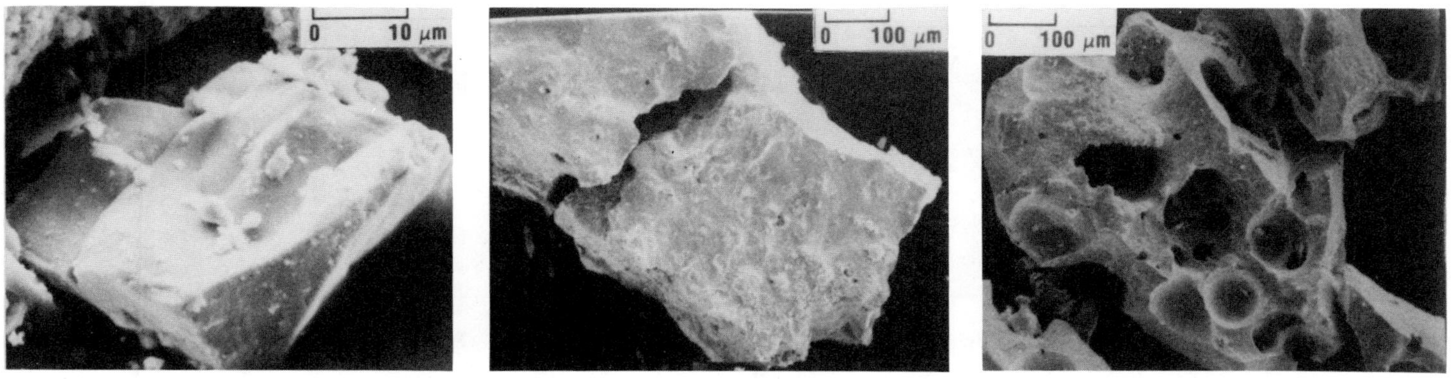

FIGURE 7. SEMs of angular, blocky particles that most resemble typical phreatomagmatic ash. Particles have one or more planar to curviplanar surfaces and may have vesicles, which are rare in experimental ash. Surfaces are often pitted but may show smooth, fluidal areas.

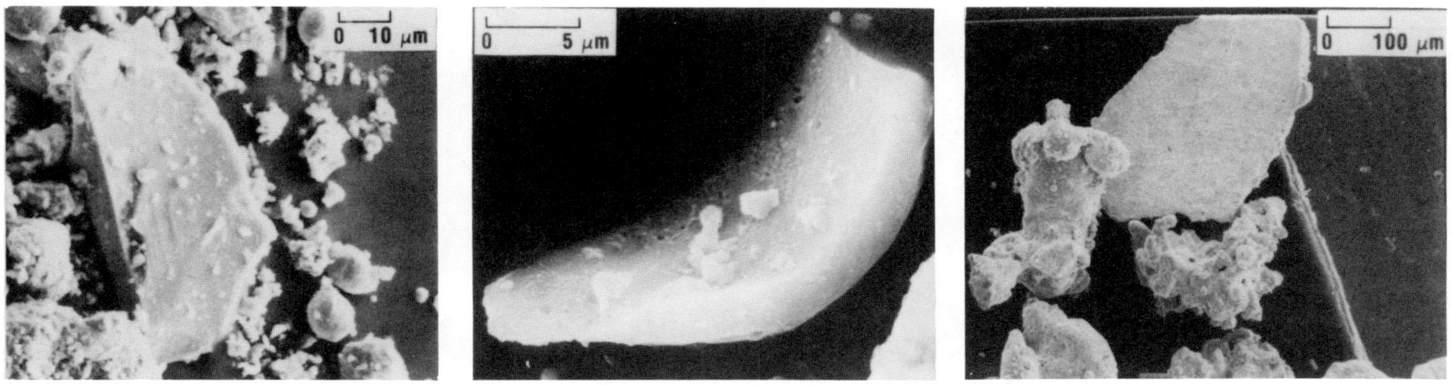

FIGURE 8. SEMs of plate-like particles that are flat, angular slabs. Where particles have curved surfaces, they are broken bubble-wall fragments.

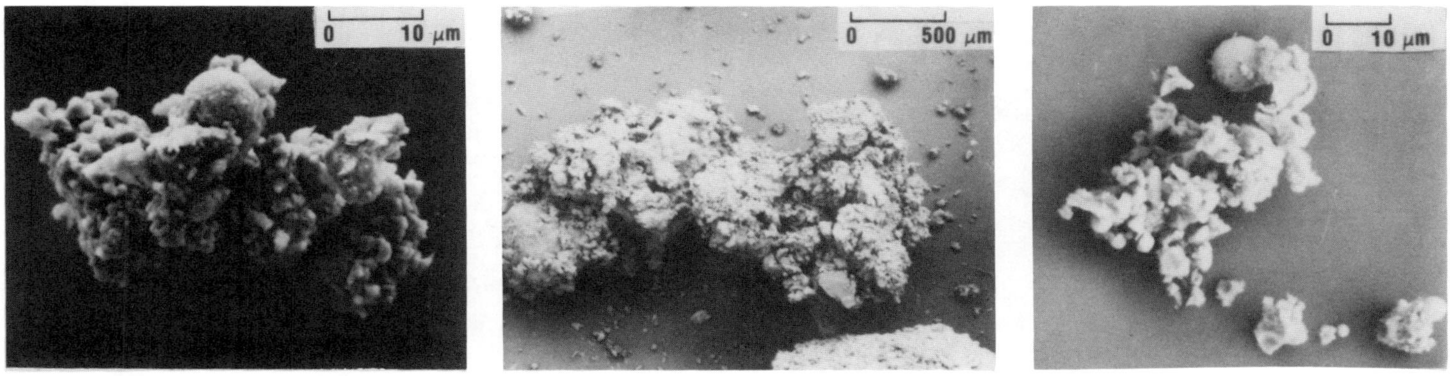

FIGURE 9. SEMs of aggregate particles common to both fine and coarse debris. These aggregates consists of droplets and spheres bonded to angular particles, often by ribbon-like, deformed melt particles. The coarse-grained particles (*middle*) strongly resemble volcanic aggregates found in wet phreatomagmatic deposits (see Cerro Colorado and Hopi Buttes sections).

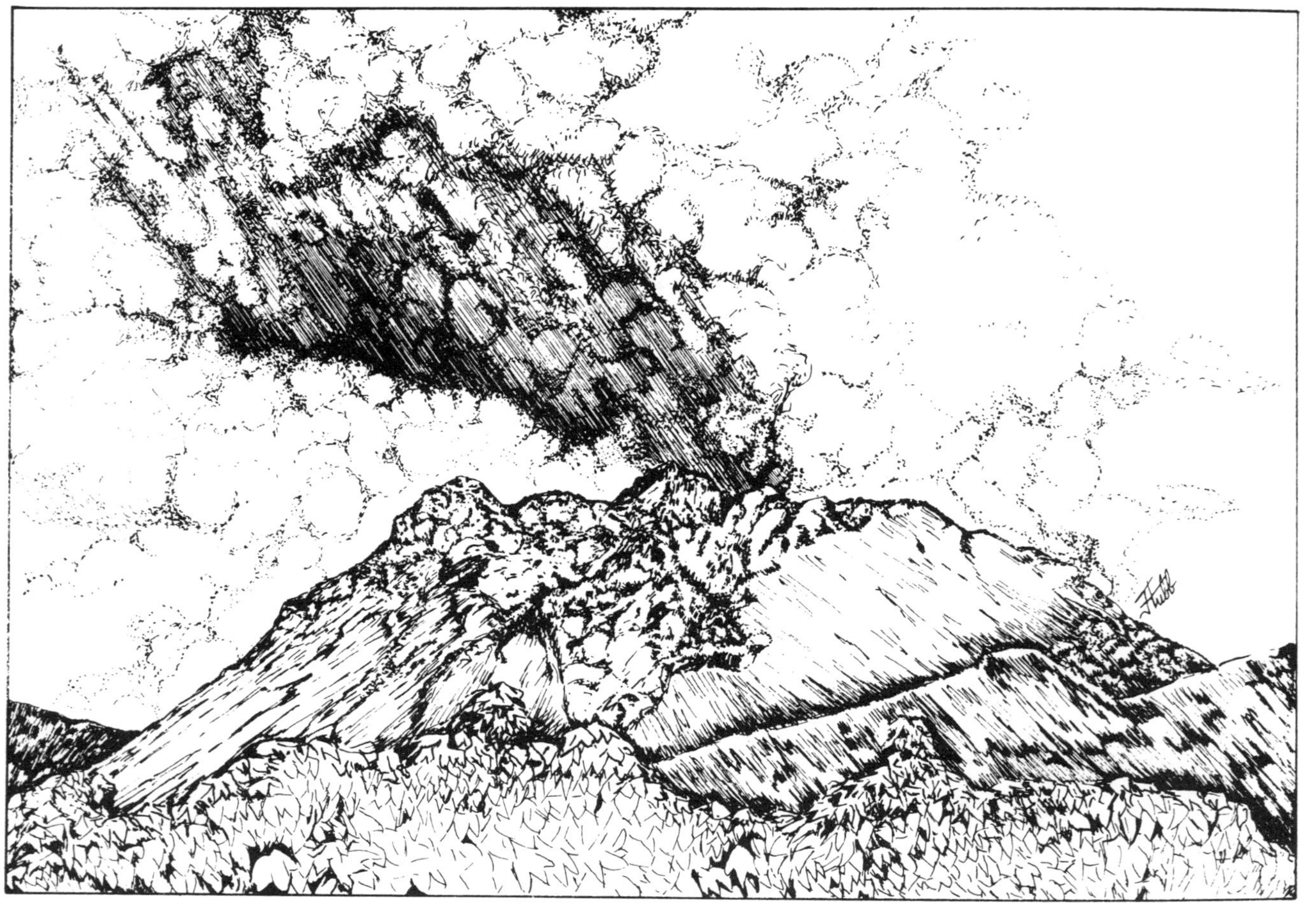

Phreatic explosions at Soufrière de Guadeloupe, 1976.

PHREATIC ERUPTIONS

Hydrothermal activity at fumeroles and hot springs can become energetic enough to erupt explosively. The volcanic ash from this activity consists of fragments from vent walls and crater fill, and it contains no juvenile ash. In some instances phreatic explosions have preceded magmatic or phreatomagmatic eruptions.

Historic Eruptions

LA SOUFRIÈRE DE GUADELOUPE, F.W.I.; 1976 ERUPTION

Introduction: La Soufrière de Guadeloupe is a steep-sided andesitic dome crossed by multiple fissures. There has been considerable explosive phreatic activity from these fissures during historic time. The last phreatic activity, during late summer and fall of 1976, caused the evacuation of thousands of residents from towns located on La Soufrière's slopes. Explosive steam eruptions from open fissures at the summit and along the southeastern flank of the dome entrained large amounts of fine-grained tephra. This material was deposited downwind, south and west of the volcano, as tephra fall and very dilute density currents (Heiken et al., 1980). These tephra formed a gray, thin, sticky deposit containing accretionary lapilli. During larger eruptions, thin streams of light gray steaming mud flowed down the flanks of the dome as lahars (Wohletz and Crowe, 1978).

During the phreatic activity, it was important to discover if fresh magmatic pyroclasts were present in the tephra because it was not possible to geophysically determine if magma had reached the surface at that time. Careful petrologic analysis of the tephra was, in this instance, of more than academic interest; it was needed, in addition to geophysical measurements, to assess the probability of a significant magmatic eruption following the period of phreatic activity.

Grain Size: Tephra from the eruption of October 2, 1976, is typical of tephra deposited throughout the fall of 1976. It is very poorly sorted and has a mean grain size of 38 μm (4.7φ).

Tephra Description: *Lithic Pyroclasts.* These are equant to subequant, subangular to subrounded dacitic and andesitic fragments that exhibit a wide range in alteration. The most common lithic tephra pyroclast consists of stubby plagioclase and orthopyroxene phenocrysts (up to 40 μm long) in a groundmass of microcrystalline (<1-μm) feldspar and pyroxene phases or colorless glass. The alteration in the lithic pyroclasts ranges from those that are nearly fresh with only a trace of hematite stain, through those laced with hematite or pyrite, to those that are mixtures of an authigenic silica phase (tridymite), pyrite, and anhydrite. This last type exhibits only relict textures that identify them as lava fragments.

Mineral pyroclasts. Plagioclase grains, with compositions of An_{50} to An_{88}, and in some cases exhibiting simple and oscillatory zoning, are present as broken, subhedral crystals. As was the case for lithic fragments, they range from fresh (slightly etched or weathered grain surfaces) to ragged skeletons. Some of the feldspars contain trains of 2- to 4-μm-long, square to circular glass inclusions.

Most of the pyroxene tephra grains are pale yellow to pale pink orthopyroxenes; some contain 20- to 200-μm-long brown glass inclusions. Small amounts of clinopyroxene are present in the tephra.

Amorphous masses of hematite and 30- to 50-μm pyrite grains are present in small amounts in many of the samples and in trace amounts in all the samples. These phases are, however, more abundant in the <38-μm fraction of the tephra. There are also traces of anhydrite and tridymite.

Glass. Glass shards are a minor component within the La Soufrière tephra. Most are angular, equant, colorless to pale brown, slightly vesicular to nonvesicular fragments ranging from andesitic to rhyolitic composition. The angular glass particles appear to be, in most cases, derived from the glassy groundmass of hyalocrystalline dome lavas.

Rare, colorless glass pyroclasts have 10 to 20% ovoid vesicles (long axes of 10 to 15 μm); some contain plagioclase phenocrysts. Nearly all are partly altered and therefore do not represent fresh juvenile pyroclasts.

Also present in the tephra are rare, orange-brown glass fragments (basaltic glass) that have palagonitic rims; these too are older, weathered, and hydrothermally altered pyroclasts.

Chemical Composition: No bulk analyses of the tephra are available. Electron microprobe analyses of the glass pyroclasts are included in Table 1.

TABLE 1. CHEMICAL ANALYSES OF GLASS PARTICLES FROM THE PHREATIC ERUPTION OF LA SOUFRIÈRE DE GUADELOUPE, 27 AUGUST 1976 (IN WT. %).*

	1	2	3
SiO_2	59.18	62.08	68.84
TiO_2	0.10	0.23	0.39
Al_2O_3	23.58	19.00	14.78
FeO	0.88	1.53	2.45
MnO	0.08	0.03	0.10
MgO	0.08	0.22	0.54
CaO	8.30	6.12	3.68
Na_2O	4.92	3.87	3.08
K_2O	0.75	1.10	1.76
Cr_2O_3	0.02	0.03	0.00
Total	97.80	94.21	95.62

*Sample 1: Small, rounded, slightly vesicular grain of colorless glass;
sample 2: slightly elongate, vesicular grain with plagioclase phenocrysts; and
sample 3: small, rounded grain; low vesicularity, but vesicles are large.

Comments on Pyroclast Formation: The tephra consist of lithic and mineral fragments that are similar in texture and composition to those of the dome lavas. Analyses of feldspars and pyroxenes from the tephra support this observation. The degree of comminution of dome lavas to fine-grained tephra ejected by the phreatic eruptions is most likely tied to the degree of hydrothermal alteration of those rocks. Explosive release of steam entrains the altered material as a relatively cool mixture of hydrothermally altered lithic and mineral fragments. Only a trace of the material was petrographically identified as older tephra that may have been derived from pyroclastic units underlying the dome lavas (the andesitic pumice mentioned by Reynal de Saint-Michel, 1966). Some of the tephra components, however, could have been derived from older, intermediate-composition lavas underlying the Soufrière dome, especially at the L'Echelle vent.

An abundance of slightly altered to nearly fresh plagioclase grains in the tephra may have resulted from the preferential alteration of a fine-grained or glassy groundmass of the lavas. This is the case, for example, in near-surface alteration of rock around hot springs at Yellowstone Park, U.S.A. (Allen and Day, 1935) and the superficial zone of acid-leached rocks at Wairakei, N.Z. (Ellis, 1967). None of the colorless pyroclastic fragments present in the Soufrière tephra are fresh enough to classify as new juvenile tephra. Incorrect identification of older, reworked vitric tephra deposited by phreatic eruptions as "fresh" tephra at La Soufrière in 1976 led some investigators to believe that the magma was, indeed, close to the surface. However, magma never did reach the surface.

Recent phreatic activity at Mount St. Helens, Washington, was followed by an explosive magmatic eruption (Sarna-Wojcicki et al., 1981). Phreatic tephra is often overlooked within the geologic record; if identified, it can be useful in determining if phreatic activity has preceded magmatic eruptions and if phreatic behavior is a typical precursor. This type of activity can occur at any volcano but appears to be most common at composite cones of intermediate composition.

REFERENCES:

Allen, E. T. and Day, Arthur L., 1935. Hot springs of the Yellowstone National Park. Carnegie Inst. Wash., Pub. 466: 515 pp.

Ellis, A. J., 1967. The chemistry of some explored geothermal systems. *In*: Barnes, H. L. (Ed.), *Geochemistry of Hydrothermal Ore Deposits.* Holt, Rinehart, and Winston, Inc., N.Y.: 465-514.

Heiken, G., Crowe, B., McGetchin, T., West, F., Eichelberger, J., Bartram, D., Peterson, R., and Wohletz, K., 1980. Phreatic eruption clouds: the activity of La Soufrière de Guadeloupe, F.W.I., August-October, 1976. Bull. Volcanol., **43**: 383-395.

de Reynal de Saint-Michel, Alain, 1966. Notice éxplicative des feuilles de Basse-Terre et des Saintes. Carte Géologique de la France, Départment de la Guadeloupe, Ministere de l'Industrie, Paris: 60 pp.

Sarna-Wojcicki, A.M., Waitt, R., Woodward, M., Shipley, S., and Rivera, J., 1981. Premagmatic ash erupted from March 27 through May 14, 1980—extent, mass, volume and composition. U.S. Geol. Surv. Prof. Paper 1250: 569-575.

Wohletz, K. H. and Crowe, B. M., 1978. Development of lahars during the 1976 activity of La Soufrière de Guadeloupe. Geol. Soc. Am. Abstracts with Programs, **10**: 154.

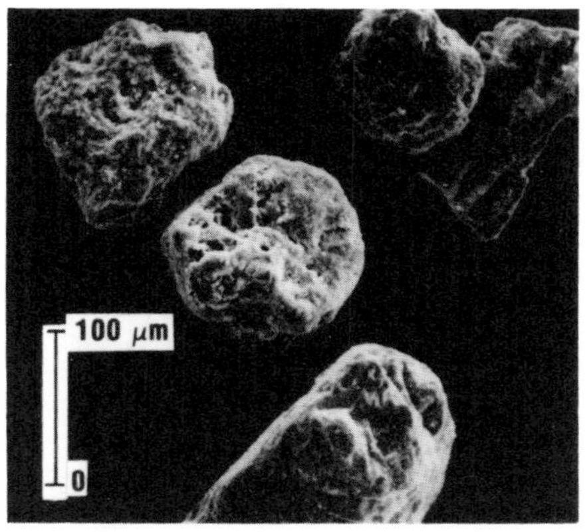

FIGURE 1. SEM of subrounded lithic fragments from the eruption of October 2, 1976. Rounding caused mainly by hydrothermal alteration of lava fragments derived from the dome.

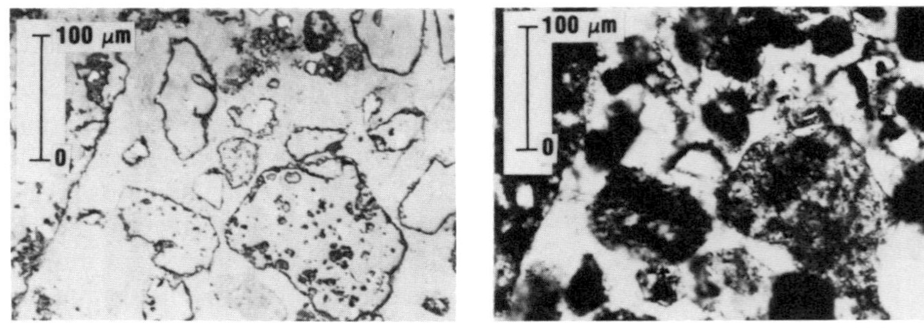

FIGURE 2. Photomicrographs (*left*—reflected light, *right*—transmitted light) of lithic and crystal pyroclasts from the eruption of August 22, 1976. Hydrothermally altered lithic pyroclasts (best examples are *lower right* and *left*) are similar in composition and texture to the dacitic lavas of the dome. Brightest phase in reflected light is pyrite.

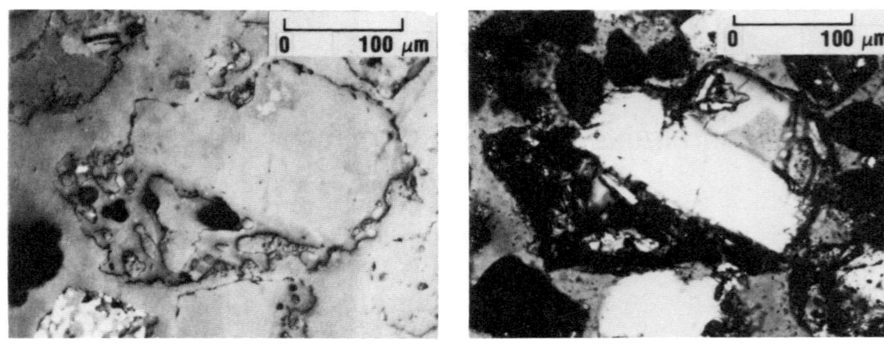

FIGURE 3. Photomicrographs (*left*—reflected light, *right*—transmitted light) of vitric and lithic pyroclasts from the eruption of September 22, 1976. *Center*—a vitric pyroclast; these are rare in the tephra of this eruption. The slightly pumiceous glass encloses a zoned plagioclase phenocryst. Traces of pyrite are indicative of some degree of hydrothermal alteration.

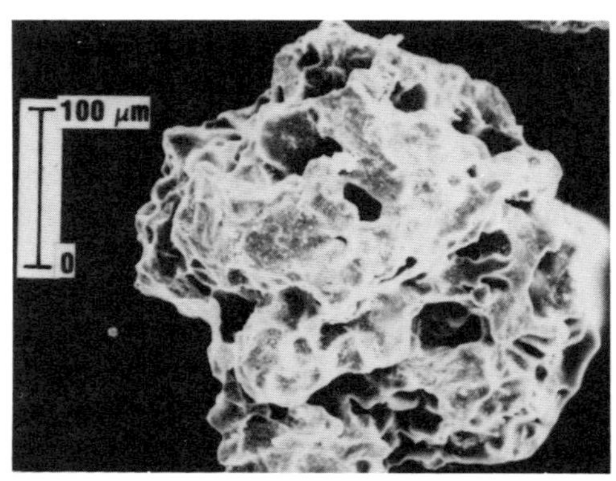

FIGURE 4a. Tephra from the eruption of August 12, 1976. SEM of a slightly altered glass pyroclast with low to medium vesicularity; vesicles are deformed ovoids. These are rare in tephra from this eruption and, at first glance, were thought to be juvenile pyroclasts; vesicular pyroclasts were described in older deposits.

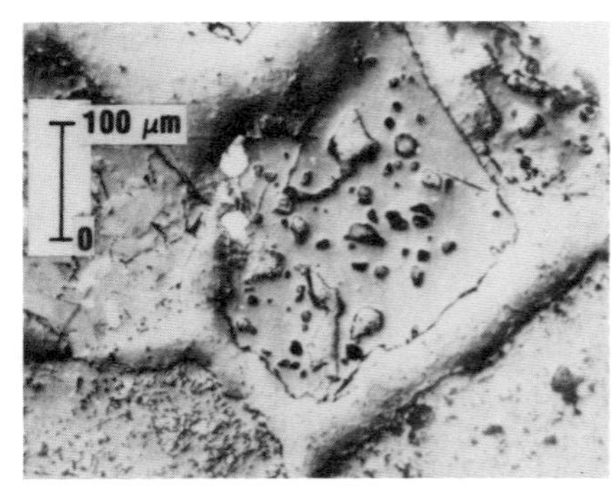

FIGURE 4b. Photomicrograph, in reflected light, of a vesicular, colorless glass pyroclast (*center*).

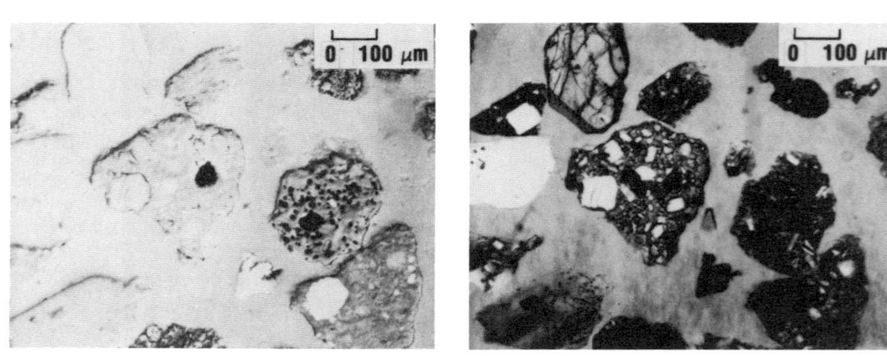

FIGURE 5. Tephra from the eruption of October 4, 1976. Photomicrographs in reflected light (*left*) and transmitted light (*right*). Most pyroclasts are microporphyritic dacite with equant, subangular to subrounded forms. Most are hydrothermally altered and contain pyrite grains. Lithic pyroclasts make up 75% of the sample. The remaining 23% consists of plagioclase, orthopyroxene, clinopyroxene, and a trace of glass.

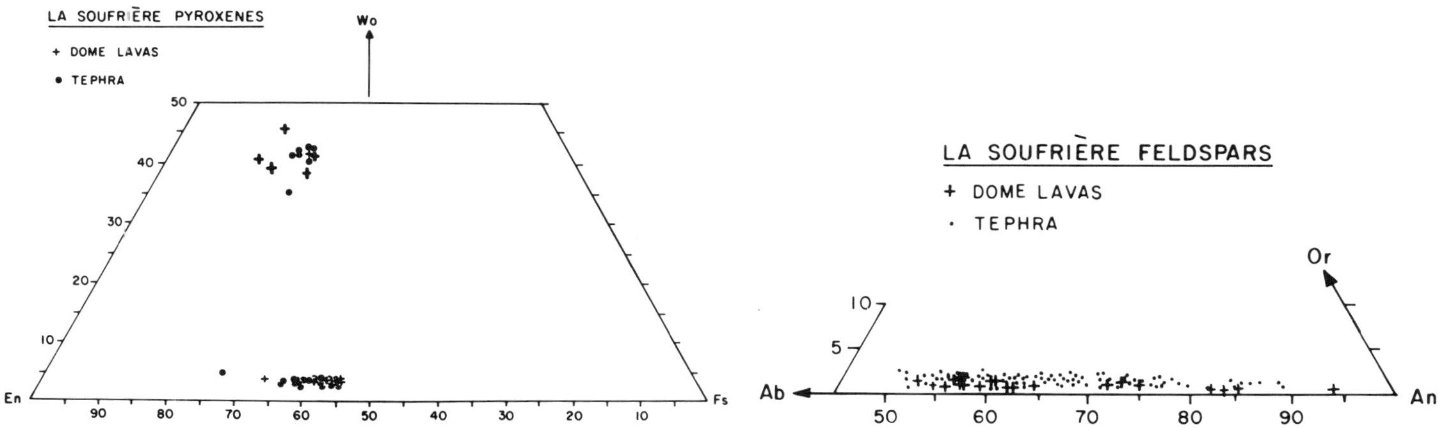

FIGURE 6. Plagioclase and pyroxene compositions from La Soufrière dome lavas and the phreatic tephra of the 1976 eruptions. This comparison supports the interpretation that all the pyroclasts were derived from hydrothermally altered dome lavas during steam explosions.

MT. BAKER, WASHINGTON; APRIL-JULY 1975

Introduction: Mount Baker, one of the northernmost volcanoes of the Cascade Range within the U.S., is a composite cone that rises above deeply dissected plutonic and metamorphic terranes. Most of its lavas and tephra are andesitic (Coombs, 1960). Over the last several hundred years, Mt. Baker has had periods of violent fumarolic activity, including what may have been a mild eruption in 1859 (Coombs, 1960).

The new fumarolic activity of 1975 was at vents located in Sherman Crater, which lies 300 m below the Summit Crater. Both craters are filled with ice. Particularly energetic fuming began during late March 1975, when steam plumes rose 200 to 300 m above the crater (Eichelberger et al., 1976). Steam issued from a 1- by 5-m fumarole in the eastern wall of Sherman Crater. Snow and ice melted by hot water ponded in the crater and formed a 100-m-diameter collapse feature (central ice plug).

Tephra from the fumarole was mostly confined to the crater and formed a continuous layer for about 100 m around the steam plume. Ash was concentrated by melting snow (Eichelberger et al., 1976). About 1-1/2 months after the activity began, the tephra layer was approximately 1 cm thick 100 m from the main fumarole.

Three and a half months after activity began, the total volume of material erupted was estimated at 500 m^3, which represents excavation within a 1- by 5- by 100-m fumarolic vent.

Tephra Description: *Lithic fragments.* The coarser tephra (>50 µm) consists of mostly equant to elongate, rounded to subrounded, granular aggregates of a 1- to 10-µm-diameter low birefringent mineral; there is 5 to 10% pyrite and a trace of hematite (Table 1). Many of the fragments are smeared with thin coats of white, very fine grained (probable) clay mineral. The granular aggregates were identified by XRD as mainly potassic analcite (J. R. Smyth, personal communication, 1975).

Many of the altered fragments retain relict textures. Fragments that appear to have been phenocrysts (feldspar?) of up to 150 µm long are set in a matrix of relict 10- to 30-µm-long tabular crystals (feldspars?). The same fragments have slightly elongate, 2- to 20-µm-long vesicles. These fragments may have been deposited as lithic or lithic-vitric ash before burial and subsequent alteration by fumarolic activity.

Rounded, granular aggregates of 1- to 10-µm analcite crystals are most common in the finer grain-size fractions (<50 µm) but still are abundant in coarser fractions. These aggregates are similar in composition to the altered lithic fragments but exhibit no relict textures. Individual, well-formed analcite crystals are present on the surface of the aggregates as octahedra of up to 10 µm in diameter. Less common spheroidal particles of similar size are composed of sulfur, and some have adhering pyrite cubes.

TABLE 1. COMPOSITION OF TEPHRA AND CLASTIC MATERIAL FROM SHERMAN CRATER, MOUNT BAKER.*

	Tephra (May 1, 1975) (%)	Clastic Material West Wall Fumaroles (%)
Granular aggregates (rock fragments and/or alteration products)	66	61
Clear glass (partially altered)	10	29
Brown glass (unaltered)	1	6
Opaque minerals	19	1
Quartz	1	2
Feldspar	3	1

*Percentages are based on grain counts of the >50-μm fraction.

Glass. Glass fragments make up a small part of the tephra; because of its fresh appearance, it is the only component of the tephra that could be interpreted as juvenile ash. Many glass fragments are colorless, equant, angular particles with low (<20%) vesicularity. In most cases, vesicles are filled with clusters of a very fine grained mineral with low birefringence (analcite). Colorless, elongate pumice grains are present in coarser grain size fractions; irregular, ovoid vesicles are up to 20 μm long. Some of the grains are stained with hematite. Altered and unaltered brown glass fragments also occur. Nearly equant fragments contain 5 to 20% ovoid to spherical vesicles that are up to 25 μm long. Some of these grains have smooth, fluidal grain surfaces.

Mineral fragments. The most common mineral type is pyrite; there are smaller amounts of alkali and plagioclase feldspars, hematite, quartz, chalcopyrite, sulfur, pyroxene, and gypsum(?).

Gas release study of the tephra: The tephra was analyzed by Everett K. Gibson, NASA Johnson Space Center, for gas content and weight loss during heating under a vacuum by the procedures outlined by Gibson and Johnson (1972). After drying at 60 to 70°C, the sample was heated at 6°C/minute to 1400°C. The 8.6% weight loss of H_20 at 350 to 500°C indicates the presence of zeolites. A 10.1% weight loss of SO_2 in the range of 650-750°C is most likely from breakdown of the pyrite observed in the sample.

Comments on Pyroclast Formation: Clastic material scraped from the west wall of Sherman Crater is similar in most aspects to the tephra. The coarser grain-size fraction (>50 μm) consists of mainly colorless, granular aggregates of analcite with thin coats of a white amorphous clay(?) mineral and equant to slightly elongate, highly altered lithic fragments. Brown, vesicular glass fragments are present in the sample and consist of broken droplets or angular, blocky fragments. Some varieties of the brown glass are pumiceous, exhibiting highly elongate "pipe" vesicles. Both unaltered and altered forms are present; vesicles in the unaltered form are filled with granular aggregates of analcite. Brown glass pyroclasts are similar to vitric ash fragments from Strombolian eruptions of basaltic or basaltic andesite lavas (Heiken, 1972). Colorless, spindle-shaped, pumiceous fragments are also present. Mineral fragments present include pyrite, plagioclase and alkali feldspars, hematite, quartz, sulfur, and possibly gypsum.

It is likely that clastic material present in the walls of the fumarole was originally deposited as lithic and lithic-vitric tuffs and was subsequently altered by fumarolic activity. More energetic fumarolic activity broke this material from the vent walls to form phreatic tephra. Glass pyroclasts within the 1975 tephra deposit represent reworked ash and not juvenile tephra.

REFERENCES:

Coombs, H. A., 1960. Active volcanoes and solfatara fields of the United States of America. *In*: *Catalogue of the active volcanoes of the world including solfatara fields.* Intern. Assoc. of Volcanol. and Chem. of the Earth's Interior, IX: 1-58.

Eichelberger, J. C., Heiken, G., Widdicombe, R., Wright, D., Keady, C., and Cobb, D., 1976. New fumarolic activity on Mt. Baker: observations during April through July, 1975. J. Volcanol. Geotherm. Res., **1**: 35-53.

Gibson, Jr., E. K. and Johnson, S. M., 1972. Thermogravimetric-quadrupole mass-spectrometric analysis of geochemical samples. Thermochim. Acta, **4**: 49-56.

Heiken, G. H., 1972. Morphology and petrography of volcanic ashes. Geol. Soc. Am. Bull., **83**: 1961-1988.

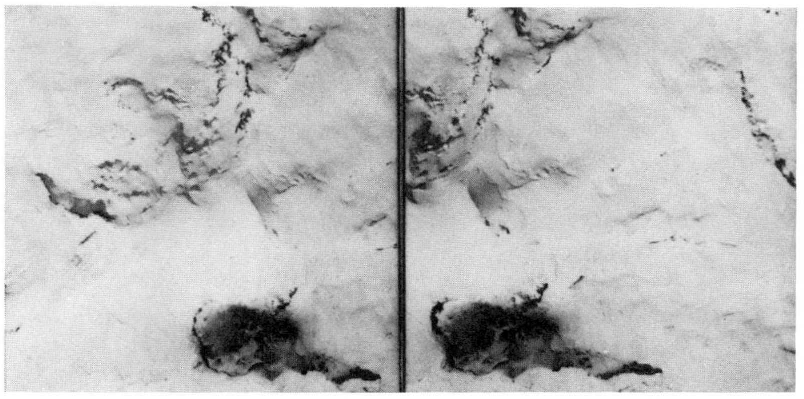

FIGURE 1. Aerial photographs (stereo pair) of Sherman Crater, Mt. Baker, taken April 11, 1975. Note the dark tephra cover within the crater and the steam plume rising from the NE crater wall.

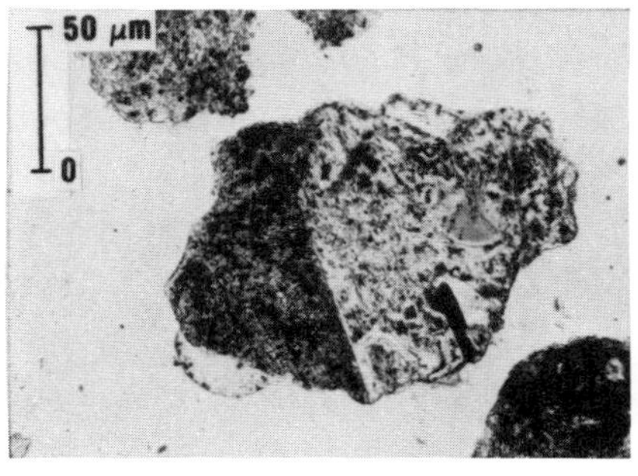

FIGURE 2a. Photomicrograph (transmitted light) of hydrothermally altered lithic pyroclasts. Most lithic fragments now consist of analcite but have faint relict textures.

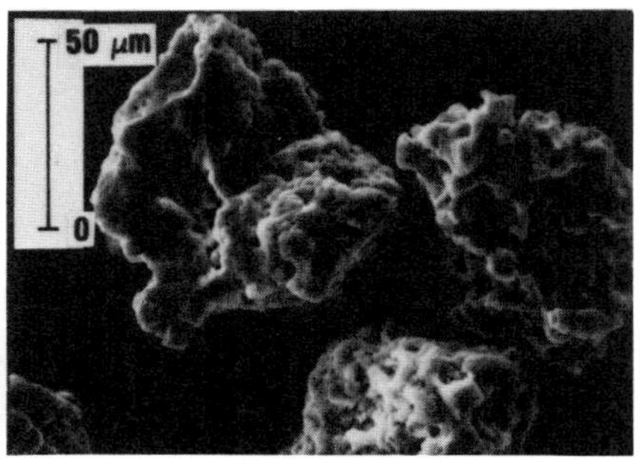

FIGURE 2b. SEM of hydrothermally altered lithic pyroclasts. Much of what may have been glassy matrix has been etched and removed.

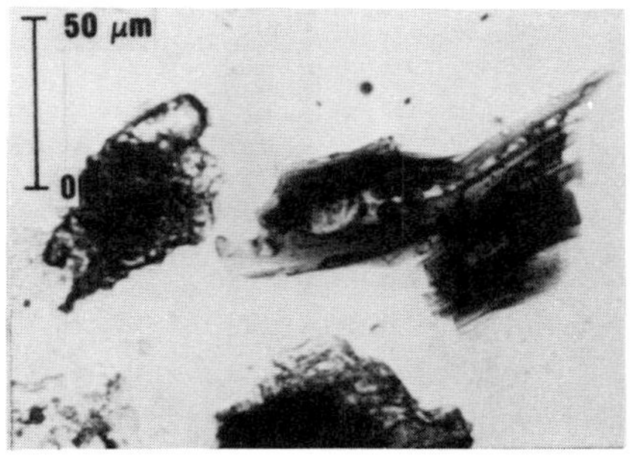

FIGURE 2c. Photomicrograph (transmitted light) of an angular vitric pyroclast that is typical of the brown-glass tephra fraction from fumarole walls and in the phreatic tephra.

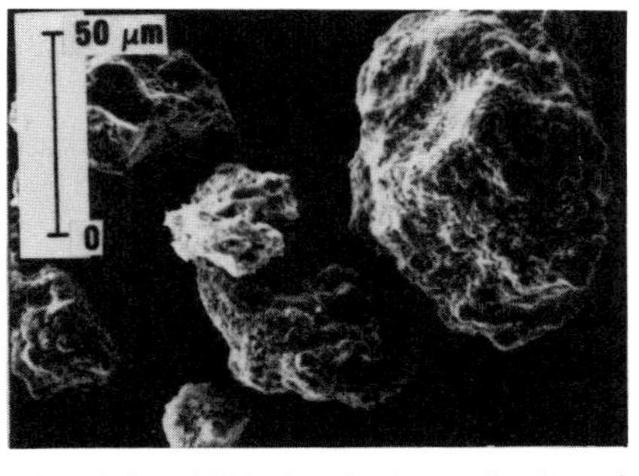

FIGURE 2d. SEM of hydrothermally altered lithic pyroclasts.

Prehistoric Eruptions

INYO CRATERS, CALIFORNIA

Introduction: The Inyo Craters consist of three explosion pits located at the southern end of the Inyo Domes chain, Long Valley caldera, eastern California (Mayo et al., 1936; Huber and Rinehart, 1967). Rinehart and Huber (1965) reported a radiocarbon age of 650 ± 200 years for the southern Inyo Crater. No juvenile ash or lava was ejected during the explosive excavation of the craters. Phreatic explosions likely resulted when rhyodacitic magma rose to a shallow depth and flashed ground water to steam (Bailey et al., 1976). Pyroclasts formed during these explosions are dominantly altered and broken pieces of the underlying basalt lavas and subordinant amounts of comminuted rhyolite.

The two largest Inyo Craters are bowl-shaped and have maximum diameters of nearly 200 m and depths of nearly 50 m. These craters are aligned along a fault trending N20W, which is the southern expression of the Inyo Dome chain (Mayo et al., 1936). The late basalts and moat rhyolites of Long Valley caldera (Bailey et al., 1976) are visible in crater walls. Overlying these lavas and exposed within crater rims is 15 to 25 m of tephra. The tephra consists of 2- to 3-m-thick, alternating layers of explosion breccia and finer grained ash fall that contains vesicles. Similar in composition and appearance to the Inyo Crater phreatic breccia and ash is the small (<100-m-diameter) phreatic tuff rampart on the south slope of Deer Mt., a rhyodacitic dome just north of the Inyo Craters.

The phreatic ash described here is from a thickly bedded ash layer in the explosion breccia at the rim of the southern Inyo Crater (sample 1) and from a massive lapilli layer at the rim of the north Inyo Crater (sample 2). These two samples are compared with a sample from the tuff rampart of Deer Mountain (sample 3). This last is an explosion breccia consisting of predominantly ash and lapilli-size clasts of lava with some blocks up to 20 cm in diameter. This phreatic crater probably has a origin similar to that of the Inyo Craters; it resulted from near-surface, explosive hydrothermal activity located along the N20W-trending fault.

Grain Size: Grain-size data for samples 1, 2, and 3 are shown in Table 1. The poor sorting (σ_φ = 237 to 2.53φ) of samples 2 and 3 indicates a pyroclastic surge emplacement, whereas the better sorting of sample 1 (σ_φ = 1.92φ) may be the result of ash fall. These phreatic tephra required some disaggregation before they were sieved. Disaggregation is often useful when analysing this material because phreatic tephra near their vents are typically lithified by the steam that condenses on particles during emplacement.

TABLE 1. GRAIN-SIZE ANALYSES.

Sample	Md_φ (mm)	σ_φ	Fine Ash (Wt. %)
1	−0.07 (1.04)	1.92	3.44
2	−1.05 (2.07)	2.37	1.78
3	0.25 (0.87)	2.53	6.90

Tephra Description: Probably the most notable characteristic of the phreatic pyroclasts is the surface aggregation of fine particles and their chemical composition, which shows the strong effects of hydrothermal alteration (Table 2). This surface aggregation and alteration gives individual grains irregular surfaces and a muddy or "dirty" appearance. Sample 1 contains pyroclasts that are rounded aggregates of smaller angular pieces. Fresh surfaces are rare, as are sharp points or edges, and vesicles are entirely absent. The fine fraction of sample 1 reveals individual blocky and angular clasts with abundant, weakly attached dust particles. Fresh surfaces are more common, as are flat or curviplanar surfaces. Clasts are commonly tabular or slab-like. Sample 2 demonstrates the surface alteration typical of phreatic ash.

Rare, vesicular basaltic clasts have a surface hydration rind that has contracted and cracked, especially within exposed vesicles. Hydration rinds tend to peel away from the substrate, like mud cracks. Sample 3 contains rhyodacitic lithic clasts that have somewhat less surface alteration than do samples 1 and 2, but these clasts still have considerable surface aggregation of fine material. Very fine grained ash is missing from sample 3, and its particles are much less slab-like than those of sample 1.

TABLE 2. CHEMICAL ANALYSES OF INYO CRATER PYROCLAST SURFACES.*

	1	3
SiO_2	67.50	55.41
TiO_2	1.59	5.38
Al_2O_3	20.08	15.02
FeO	9.71	6.45
MnO	---	4.64
MgO	0.18	3.71
CaO	0.42	1.72
Na_2O	---	1.58
K_2O	0.51	6.09

*Normalized EDS analyses of rhyolite and rhyodacitic lithic clasts.

Comments on Pyroclast Formation: Because juvenile tephra is absent in the pyroclastic deposits surrounding Inyo Craters, the explosions forming the crater are termed phreatic. Muffler et al. (1971) studied similar hydrothermal explosion craters in Yellowstone National Park. During phreatic explosions, rock overlying a magma body may be saturated with water, which fills tiny cracks and vesicle spaces. Shepherd and Sigurdsson (1982) studied the 1979 eruption of Soufrière Volcano, St. Vincent, in which the water-saturated upper part of the previously extruded lava dome was pulverized. The phreatic pulverization of lava probably occurs by hydraulic fracturing in which thermal expansion of water produces high-pressure gradients of about 1.5 MPa/C° during constant volume heating (Knapp and Knight, 1977). When the pressure of superheated pore water exceeds the strength of the rock, vapor explosion shatters the rock. The fracture density of the rock before eruption determines, in part, the size distribution of tephra. Hydraulic fracturing of the rock before eruption may increase fracture density and allow more water to infiltrate and contribute to the vapor explosion. Hydraulic fracturing followed by a large-scale vapor explosion produces fine-grained ejecta. Ejecta shapes are mainly determined by brittle fracture, and planar grain surfaces are produced. Aggregation of fine, cohesive dust and rounding during transport determines the typical shapes shown in these figures.

REFERENCES:

Bailey, R. A., Dalrymple, G. B., and Lanphere, M. A., 1976. Volcanism, structure, and geochronology of the Long Valley caldera, Mono County, California. J. Geophys. Res., **81**: 725-744.

Huber, N. K. and Rinehart, C. D., 1967. Cenozoic volcanic rocks of the Devil's Postpile quadrangle, eastern Sierra Nevada, California. U.S. Geol. Surv. Prof. Pap., 554-D: D1-D21.

Knapp, R. B. and Knight, J. E., 1977. Differential thermal expansion of pore fluids: fracture propagation and microearthquake production in a hot pluton. J. Geophys. Res., **82**: 2515-2522.

Mayo, E. B., Connant, L. C., and Chelikowsky, J. R., 1936. Southern extension of the Mono Craters, California. Am. J. Sci., **32**: 81-97.

Muffler, L. P. J., White, D. E., and Truesdale, A. H., 1971. Hydrothermal explosion craters in Yellowstone National Park. Geol. Soc. Am. Bull., **82**: 723-740.

Rinehart, C. D. and Huber, N. K., 1965. The Inyo Crater Lakes - a blast in the past. Calif. Div. Mines Geol. Mineral. Inform. Ser., **18**: 169-172.

Shepherd, J. B. and Sigurdsson, H., 1982. Mechanism of the 1979 explosive eruption of Soufrière Volcano, St. Vincent. J. Volcanol. Geotherm. Res., **13**: 119-130.

FIGURE 1. SEM of a pyroclast from sample 1 that shows rounded aggregates typical of phreatic tephra.

FIGURE 2. SEM of an aggregate from sample 1 with numerous angular particles cemented to the surface.

FIGURE 3. SEM of the fine fraction of sample 1 showing individual blocky and angular, slab-like particles with poorly attached, smaller dust grains.

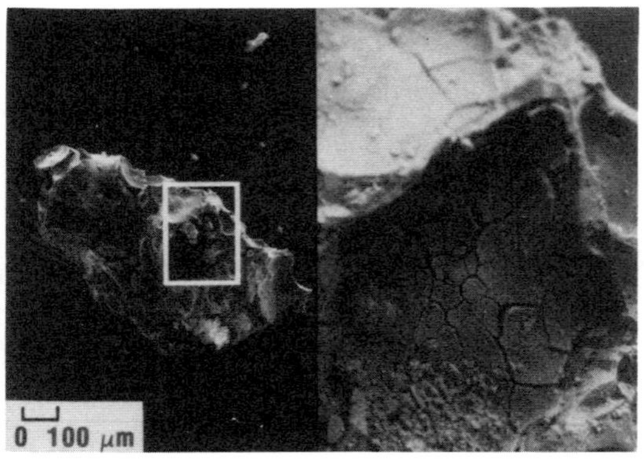

FIGURE 4. SEM of a vesicular basalt clast of sample 2 with 10X magnification in right view. Note the hydrated and cracked surface in vesicle embayment.

FIGURE 5. SEM of a sample 2 pyroclast covered with fine clay alteration products.

FIGURE 6. SEM of a rhyodacite clast of sample 3, phreatic tuff, which is a broken aggregate of finer material.

FIGURE 7. SEM of a sample 3 pyroclast showing a surface coating of fine, angular ash chips.

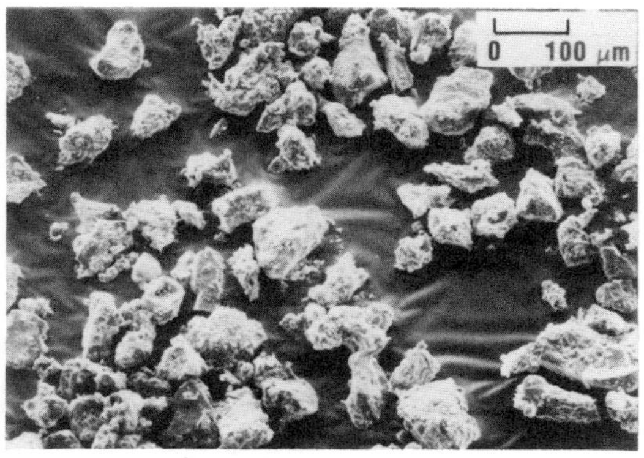

FIGURE 8. SEM of the fine fraction of sample 3. Note the equant shape and lack of fine dust, compared to particles in Fig. *3*.

Ash bed sequence, Oshima, Japan.

ERUPTION SEQUENCES

This section illustrates changes in eruption phenomena as they are reflected in ash properties from individual eruptions. Examples presented here include many of the eruption types presented in the previous section, but many of the samples exemplify the changes in eruption phenomena that can occur during an eruption, e.g., the progression from magmatic to phreatoplinian activity.

Historic Eruptions

A significant aspect of ash studies is the correlation of ash types with witnessed accounts of eruptions. Analysis of historic eruption sequences provides us with basic information on genetic aspects of tephra formation that can be used in interpretation of prehistoric eruption sequences.

LAKAGIGAR, ICELAND; 1783 ERUPTION

Introduction: The Lakagigar (Laki) eruption of 1783 may be the largest historic eruption of basaltic lava on Earth. It began on June 8, 1783, and lasted about 8 months. The eruption was from a 25-km-long line of fissures and craters and produced 12.3 km^3 of lava (Thorarinsson, 1970). The many craters include spatter cones, cinder cones, and maars; the maars are believed to have been constructed when the eruption fissure cut through lakes (Thorarinsson, 1958, 1970; S. Self and G. Larsen, verbal communication). Although the eruption is most famous for its voluminous lava flows, there was a considerable tephra component.

The tephra samples studied here were provided by G. Larsen (Nordic Volcanological Institute, Iceland) and S. Self (University of Texas, Arlington) from a site adjacent to the SW part of the Lakagigar fissure row near Laki Mountain; they are from a representative stratigraphic section in which the lowest and middle layers consist of basaltic phreatomagmatic ash and the top layer is basaltic scoria. The phreatomagmatic layers may be from only a few of the vents—perhaps from the larger maars along the line of craters. The uppermost scoria layer is from a nearby spatter and scoria rampart.

Tephra Description: The phreatomagmatic tephra layers are fine grained and consist of small, equant pyroclasts that are characterized by angular fracture surfaces and curved, smooth surfaces from vesicle walls.

The uppermost unit consists of mostly scoria lapilli. These are composed of highly vesicular basaltic glass. Many surfaces are characterized by a thin, lumpy skin with ridges and mounds developed over vesicles. Effects of surface tension are also evident in the strands of glass that formed when the outermost vesicle walls collapsed before chilling of the scoria pyroclasts. Most vesicles consist of large (30-µm to several millimeters in diameter) spheroidal to ovoid vesicles with thin vesicle walls. Large composite vesicles are present in many of the pyroclasts; they were formed when the walls of dozens of adjacent vesicles coalesced and collapsed.

Both tephra types are produced by the same magma; the marked difference is caused by interaction of magma with water.

Chemical Composition: The lava and tephra are tholeiitic basalts, which are poor in olivine but have traces of plagioclase (An_{60}-An_{64}) and clinopyroxene. For a chemical analysis, see Table 1.

TABLE 1. CHEMICAL ANALYSIS OF LAKAGIGAR TEPHRA.*

	(Wt. %)
SiO_2	50.30
TiO_2	2.11
Al_2O_3	14.16
Fe_2O_3	4.17
FeO	9.25
MnO	0.23
MgO	6.58
CaO	11.42
Na_2O	2.01
K_2O	0.53
Total	100.84

*From Thorarinsson, 1970.

Comments on Pyroclast Formation: The first two deposits (at least at this point in the line of craters) are phreatomagmatic. Quenching, mixing, and fragmentation occurred at the surface *after* complete or nearly complete vesicle development and growth. The result is a fine-grained tephra composed of vesicle-wall fragments.

During the last phase represented in this stratigraphic section, it appears that large lava fountains produced highly vesicular scoria and Pele's hair (described by Thorarinsson). The vesicles are large and equidimensional; vesicles continued to grow after eruption of molten droplets, and there was little vesicle elongation during ejection. Many of the delicate, thin-skinned lapilli broke upon hitting the ground. Lava fountains accompanied the effusion of lava flows that traveled toward the coast.

REFERENCES:

Thorarinsson, S., 1958. The Öraefajökull eruption of 1362. Acta Nat. Islandica, **II**: 1-99.

Thorarinsson, S., 1970. The Lakagigar eruption of 1783. Bull. Volcanol., **33**: 910-929.

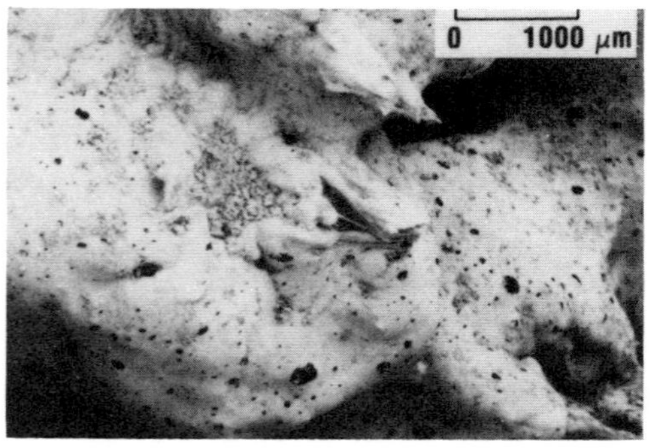

FIGURE 1a. Scoria fall unit at the surface, NW side of the Laki fissure. The surface of the scoria lapillus has a smooth, thin, glassy skin that was deformed after ejection of vesiculating melt. Many surface mounds are over vesicles; many have broken, leaving holes with ragged edges.

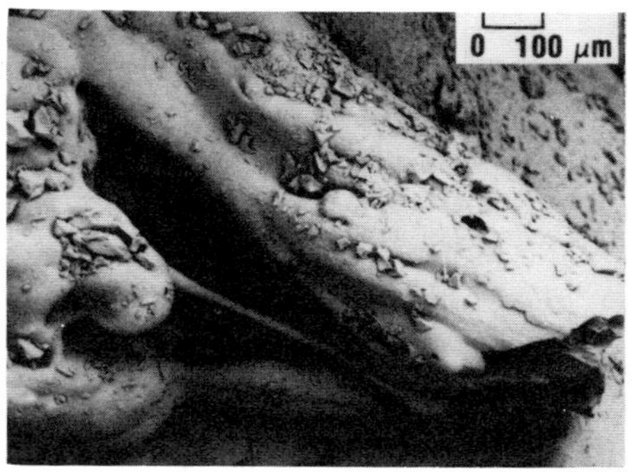

FIGURE 1b. Close-up of Fig. *1a.* The high vesicularity and thin vesicle walls are visible where the lapillus has been broken. Elongate ridges are present over elongate vesicles. Thin glass strands at the surface may have formed when still-molten broken vesicle walls were stretched.

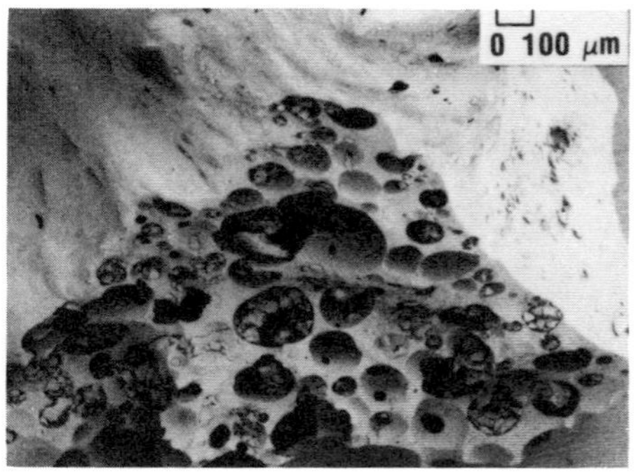

FIGURE 1c. Broken edge of a scoria lapillus shows development of spherical to ovoid vesicles within. There has been some degree of vesicle coalescence, as is noted in the vesicle walls with one or several holes.

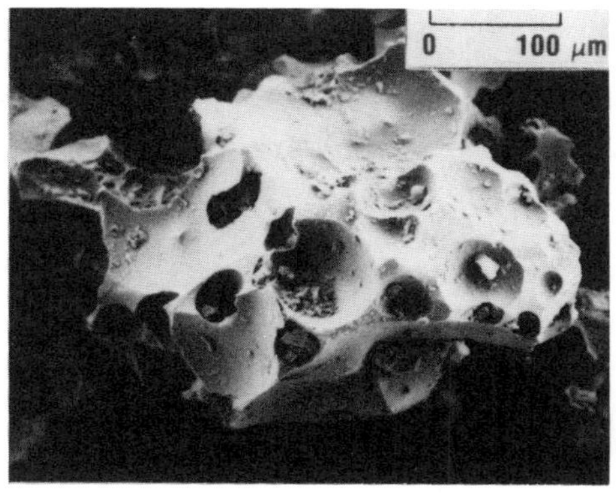

FIGURE 2a. Scoria layer at top of Laki tephra section. The broken surface of a scoria lapillus shows well-developed vesicularity and very thin vesicle walls. Larger cavities were formed during coalescence of dozens of vesicles.

FIGURE 2b. Close-up of Fig. *2a*. Basaltic glass can exist as very thin, delicate vesicle walls, such as those seen here. Thin flakes adhering to the scoria surface are only a few micrometers thick, as are the broken edges in the *center* and *top* of the photograph.

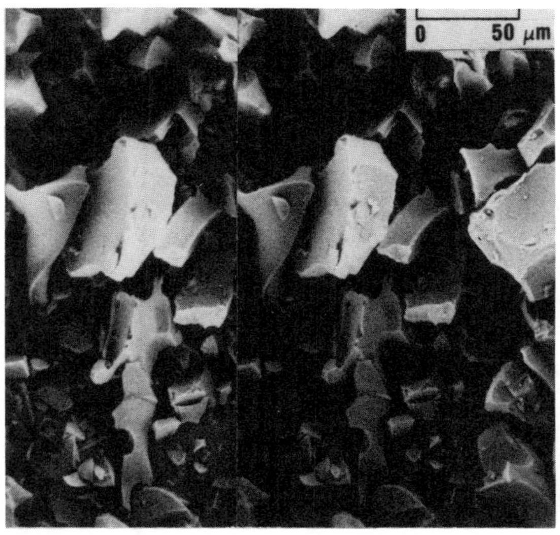

FIGURE 3a. Phreatomagmatic tephra layers at the base and middle of the Laki sequence, SW side of the fissure, near Middle (Laki) Hill. SEM of phreatomagmatic ash of basaltic composition. In contrast to the coarse scoria, this is a fine-grained tephra; most particles are less than 50 μm long. The small, angular particles have some smooth, curved surfaces that were vesicle walls before comminution. Interactions of melt with water and subsequent chilling and fragmentation occurred after vesiculation and may have been a near-surface phenomenon.

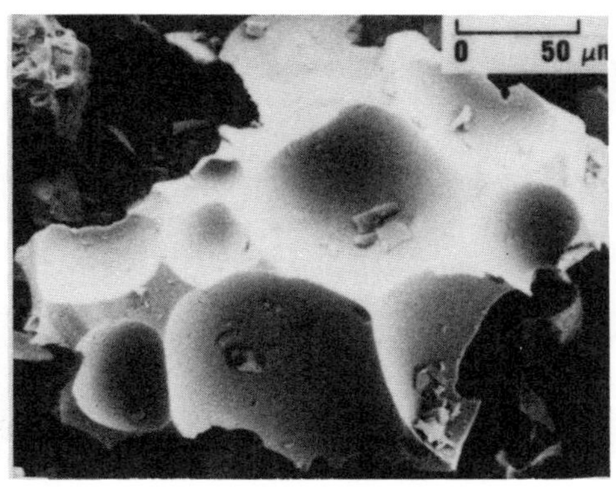

FIGURE 3b. One of the coarsest particles in this particular sample—a broken, angular scoria pyroclast. Vesicles in this pyroclast, as can be seen, were mostly spheroids and ovoids.

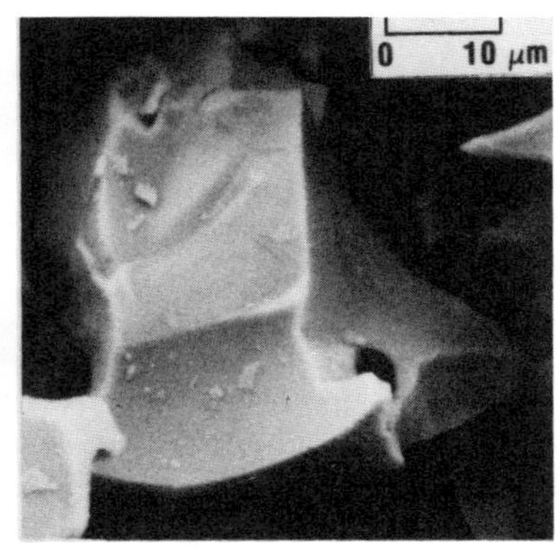

FIGURE 3c. Close-up of a typical glassy pyroclast from this tephra layer. It is a fragment from a wall between several vesicles; the former vesicle walls are curved and smooth, whereas fractures are conchoidal.

EL CHICHÓN, CHIAPAS, MEXICO; 1982 ERUPTION

Introduction: El Chichón, a small volcano located in Chiapas State, Mexico, and characterized by sub-alkaline eruption products, erupted in late March and early April 1982. The remarkably gas-rich eruption was characterized by high eruption clouds that penetrated the tropopause and were partly dispersed in the stratosphere. Volcanic ash and sulfate aerosols still remain in the stratosphere 1-1/2 years later, at the time of this writing, and have stimulated great interest within the community of atmospheric scientists. There were three main eruption phases: March 28-29, April 3, and April 4, 1982. The first phase dispersed crystal-rich ash and pumice over 37,000 km^2 (the area within the 1-mm isopach); the second enlarged the crater, producing lithic-fragment-rich ash and pumice; and the third phase spread ash to the ENE over 50,000 km^2 (1-mm isopach) (Varekamp et al., 1982; Luhr et al., 1982). Pyroclast surfaces are heavily laden with adsorbed S and Cl; Varekamp and Luhr (1982) estimate that about 2×10^6 metric tons of S and Cl were present on pyroclast surfaces. They also estimate that 0.4 to 0.43 km^3 (compacted) of tephra were erupted during the three main phases.

The samples described here were collected by J.C. Varekamp (Wesleyan University) and J.C. Luhr (University of California - Berkeley) and were given to the authors by Varekamp. SEMs of stratospheric samples were provided by D. S. McKay, NASA-Johnson Space Center, Houston, Texas.

Grain Size: Grain-size data for ash fall deposits (Table 1) were supplied by S. Carey (University of Rhode Island); detailed information on the grain-size variations may be found in a paper by Sigurdsson et al. (in press).

In most of the samples the tephra are bimodal with coarser pumice pyroclasts and very fine grained shards.

TABLE 1. GRAIN SIZE OF EL CHICHÓN TEPHRA.

	Grain Size at Two Downwind Locations	
Eruption	6.5 km (μm, φ)	22 km (μm, φ)
March 29	750, 0.5	200, 2.3
April 3	125, 3.0	20, 5.7
April 3-4(1)	4000, −2.0	450, 1.2
April 3-4(2)	500, 1.0	300, 1.7

Tephra Description: In general, tephra from the El Chichón eruption (in the samples provided) are fine grained. Pumice pyroclasts are subequant to slightly elongate, and they are partly coated with shards, anhydrite crystals, and small phenocrysts. Vesicles are ovoid with long axes of 1 to 18 μm and intermediate axes of 0.5 to 10 μm. Most vesicle walls are thick relative to the vesicle size (1 to 4 μm). Coalesced vesicles are rare.

Most shards are subequant and angular; the surfaces consist of fractured vesicle walls (smooth and conchoidal fractures). Some particles have sharp, pointed edges; they too are partly coated with sublimates, H_2SO_4 droplets, smaller shards, and small phenocryst fragments. Near-vent samples are bimodal and consist of coarse-grained, angular crystal and lithic pyroclasts and fine-grained vitric pyroclasts.

Chemical Composition: A chemical analysis of trachyandesitic pumice from El Chichón is shown in Table 2.

TABLE 2. CHEMICAL COMPOSITION OF EL CHICHÓN TRACHYANDESITIC PUMICE.*

	(Wt. %)		(ppm)
SiO_2	57.5	V	165
TiO_2	0.7	Cr	---
Al_2O_3	17.7	Ni	5
FeO	5.8	Cu	40
MnO	0.2	Zn	90
MgO	2.3	Th	8
CaO	8.2	Rb	90
Na_2O	4.4	Sr	1150
K_2O	3.0	Ba	800
P_2O_5	0.4	La	35
		Ce	60
Total	100.2		

*From Varekamp and Luhr (1982).

Comments on Pyroclast Formation: There are very few highly elongate vesicles, which implies little or no deformation and elongation of vesicles during their growth and ascent in magma within the vent. Either vesicle development was suppressed by high viscosity or it occurred very near the ground surface—at the magma/atmosphere interface.

The uniformly fine grained nature of the El Chichón tephra has led to some speculation that a phreatomagmatic component was involved, but there are not enough published data to verify this hypothesis. Extensive fragmentation may have been related to the fact that the magma was very rich in volatile phases (including abundant SO_2).

REFERENCES:

Luhr, J., Varekamp, J., and Prestegaard, K., 1982. The 1982 eruption of El Chichón Volcano, Chiapas, Mexico. Part II. Mineralogy and petrology of the ejecta. Geol. Soc. Am. Abs. with Programs, **14**: 551.

Sigurdsson, H., Carey, S., and Espindola, J., 1984. The 1982 eruptions of El Chichón Volcano, Mexico: Stratigraphy of pyroclastic deposits. J. Volcanol. Geotherm. Res. (in press).

Varekamp, J., Luhr, J., Prestegaard, K., and Carepa, J., 1982. The 1982 eruption of El Chichón Volcano, Chiapas, Mexico. Part I. Stratigraphy, volume, and volatile element characteristics of the ash fall deposits, Geol. Soc. Am. Abs. with Programs, **14**: 637.

Varekamp, J. and Luhr, J., 1982. The 1982 eruption of El Chichón Volcano: Nature of the eruption, ash and volatiles. Trans. Am. Geophys. Union, **63**: 1126.

ERUPTION OF MARCH 28-29, 1982:

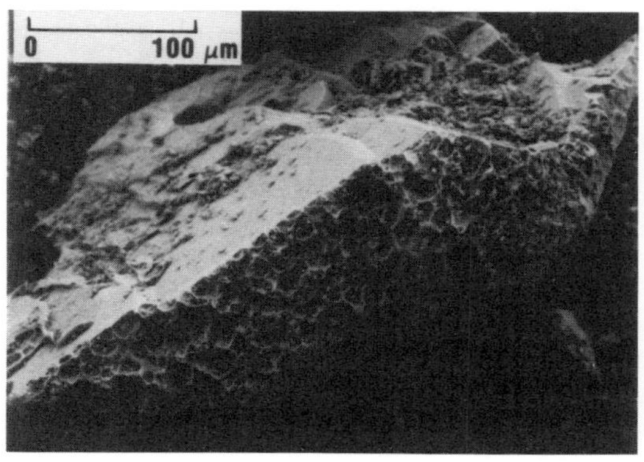

FIGURE 1a. SEMs of a sample collected 15 km E of El Chichón. Pyroxene pyroclast with a thin coating of vesicular glass (bubble-wall texture).

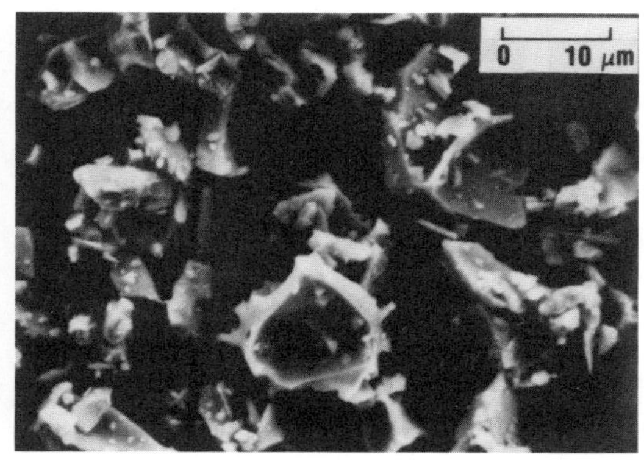

FIGURE 1b. Fine vitric ash. Most shards have been broken from a highly vesicular magma with small, slightly elongate vesicles. This particular ash is bimodal and consists of coarse-grained crystal and lithic pyroclasts (see Fig. *1a*) and very fine grained vitric pyroclasts. Some of the shards in this example are less than 1 µm long.

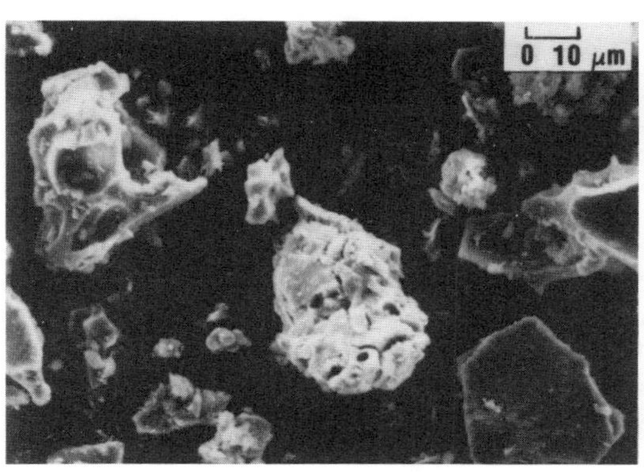

FIGURE 2a. SEMs of a sample collected 70 km SW of El Chichón. Vitric, lithic, and crystal components of fine ash. *Center* and *top right*—aggregates that appear to be hydrothermally altered lithic fragments. *Lower right*—plagioclase fragment. The remaining particles are mostly shards.

FIGURE 2b. *Center and lower right*—fractured pyroxene pyroclasts. Of the remaining particles, most of the larger (>10 µm) are shards.

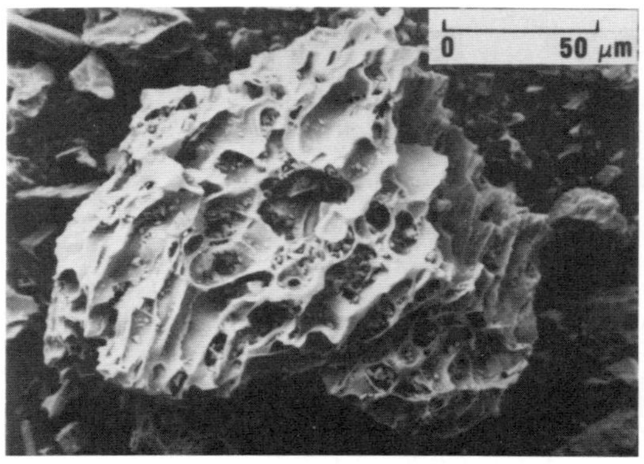

FIGURE 3a. SEMs of a sample collected 80 km NW of El Chichón. Subequant, angular pumice pyroclast. Most vesicles are ovoid and slightly elongate. Many are flattened. There is a range of vesicle sizes from 1 to about 40 μm long.

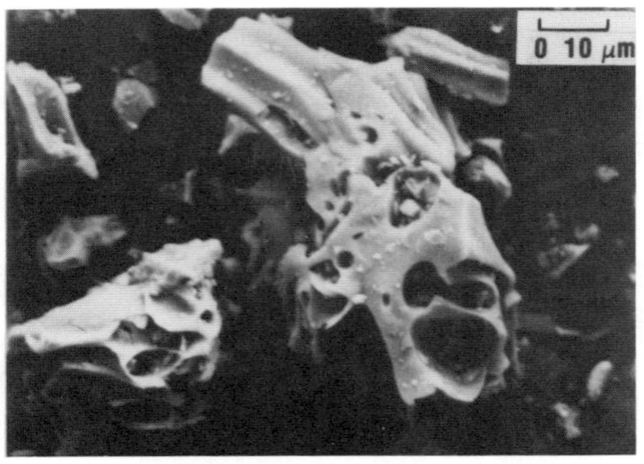

FIGURE 3b. Small pumice pyroclasts and shards from the fine ash matrix of this sample.

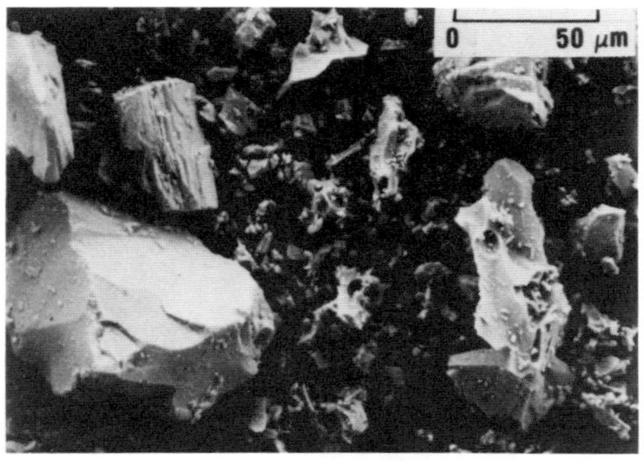

FIGURE 4. SEMs of a sample collected 125 km E of El Chichón. Even at this distance from the source, the ash is bimodal with crystal and pumice pyroclasts >40 μm long and shards <20 μm long. Most of the shards are <10 μm in size. *Bottom right*—plagioclase feldspar; *left* of the feldspar—pyroxene.

ERUPTION OF APRIL 3-4, 1982:

FIGURE 5a. SEMs of a sample collected 15 km E of El Chichón. A crystal-vitric tuff; like examples from the earlier eruption, it is bimodal with coarse crystal and lithic components and a fine-grained vitric fraction. Most of the irregular, coarse pyroclasts are finely vesicular pumices, and the blocky forms are phenocrysts. *Lower center and lower left*—plagioclase; *upper left*—Fe-Ti oxide; *upper right*—lithic fragment consisting of quartz and pyroxene phenocrysts.

FIGURE 5b. Subequant pumice pyroclast with small, slightly elongate vesicles and thick vesicle walls. It is surrounded by glass-free phenocrysts (crystal pyroclasts).

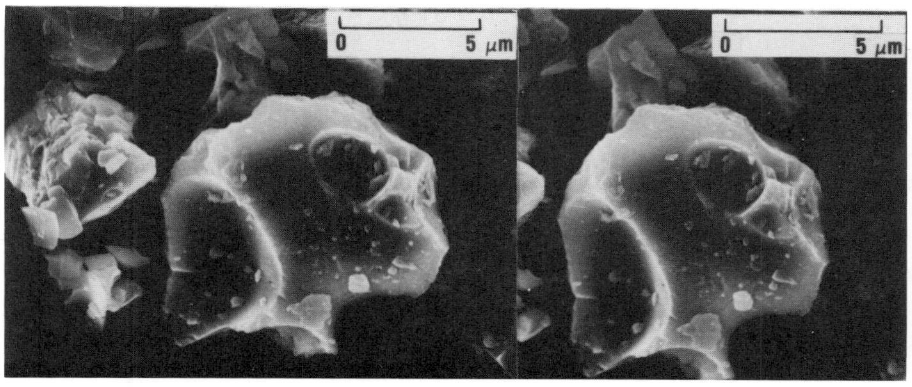

FIGURE 6a. SEMs of a sample collected 170 km E of El Chichón. Stereophotomicrograph of fine-grained vitric tephra.

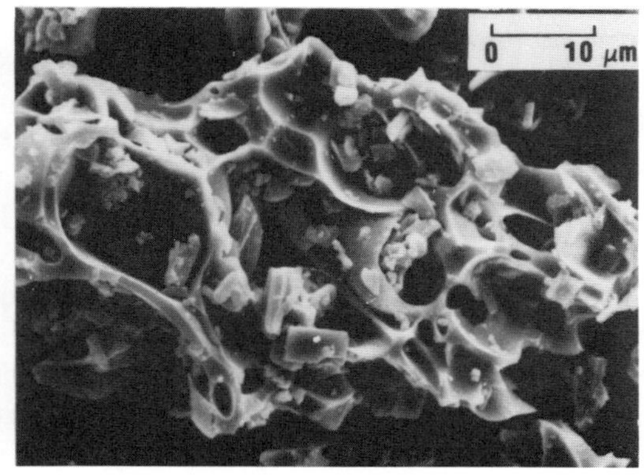

FIGURE 6b. Pumice pyroclast partly covered with fine-grained shards and tabular crystals. The tabular grains are anhydrite ($CaSO_4$) crystals.

STRATOSPHERIC SAMPLES:

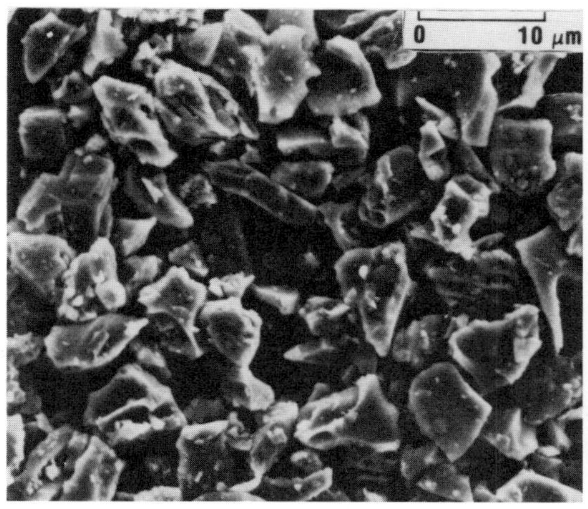

FIGURE 7a. These samples were collected by NASA aircraft in the eruption plume at an altitude of 18 km on May 7, 1982. SEMs were supplied by David S. McKay of NASA-Johnson Space Center. Equant to subequant shards, partly coated with sublimates and <1-μm shards. These are among the coarsest volcanic ash particles ever collected in an eruption plume at stratospheric heights.

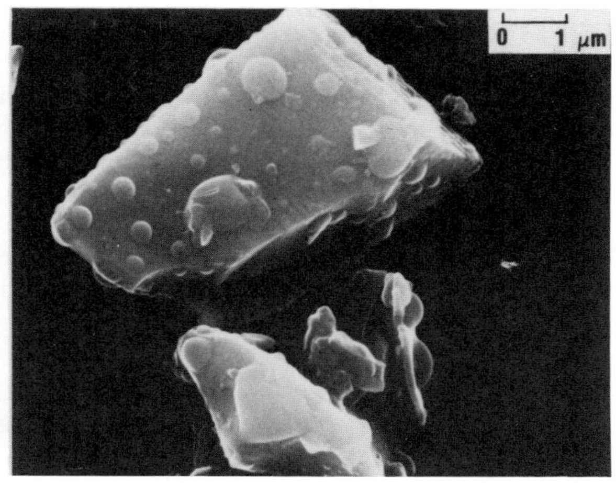

FIGURE 7b. Blocky phenocryst fragment (feldspar?) and shards. All pyroclasts are coated with droplets, which are most likely H_2SO_4 condensed on pyroclast surfaces in the eruption plume.

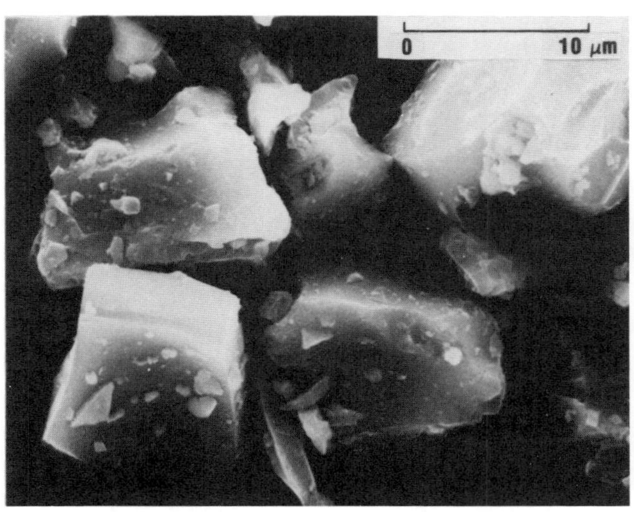

FIGURE 7c. Angular, blocky vitric pyroclasts. Curved surfaces are former vesicle walls.

FIGURE 7d. Fine-grained vitric ash coated with liquid droplets (most likely H_2SO_4).

MT. PELÉE, MARTINIQUE; MAY 8 AND 20, 1902 ERUPTIONS

Introduction: During the 1902 eruption of Mt. Pelée, Martinique, the town of St. Pierre was destroyed by nuées ardentes (LaCroix, 1904). Observations of this eruption led to some of the first studies of pyroclastic flows.

Recent stratigraphic and petrologic work on the deposits of the 1902 eruption have provided a clear picture of the eruption mechanism and characteristics of the Peléean sequence (Fisher et al., 1980; Fisher and Heiken, 1982). These eruptions followed periods of andesitic lava dome growth and mechanical disruption of the domes. Dense eruption columns collapsed and flowed outward through a cleft on the southwest side of the volcano. The pyroclastic flows separated into basal avalanches that left massive block and ash flow deposits and an overriding, expanding ash cloud that produced finer grained ash-cloud deposits (Fisher et al., 1980). The samples here are from the collection of R. V. Fisher.

Grain Size: Fisher and Heiken (1982) have sieved tephra samples from the Peléean deposits. Median grain sizes of the block and ash flows range from 0.43 to 0.36 mm (1.2 to 1.5 φ); ash cloud surges have median grain sizes of 0.38 to 0.1 mm (1.4 to 2.9 φ). Both types of deposits are poorly sorted.

Tephra Description: Most tephra in the devastating nuées ardentes consist of juvenile lithic fragments (43 to 81 vol.%). Most are blocky, tabular to irregular, hyalocrystalline andesite fragments consisting of 1-μm-wide, 2- to 3-μm-long feldspar and pyroxene microlites and a few feldspar phenocrysts (up to 30 μm long) in a glassy matrix ($\simeq$60%). Vesicularity is absent or very low (a few isolated, 1- to 10-μm irregular vesicles). A few pyroclasts have a diktytaxitic texture and about 10% vesicles.

There are traces of hydrothermally altered andesitic fragments in all samples; however, samples from gas pipes in block and ash flow deposits contain 18 to 35% of these fragments.

About 18 to 36 vol.% of the ash is composed of a crystal fraction; it is mostly plagioclase with smaller amounts of orthopyroxene, clinopyroxene, and Fe-Ti oxides. Most are cleanly fractured, and the forms range from splinters to euhedral crystals.

Vitric pyroclasts are present and range from splinter-shaped shards to small pumice fragments with 60 to 80% vesicularity. Pumice pyroclasts generally have thin-walled vesicles and ovoid vesicles 4 to 15 μm long. Vitric pyroclasts generally make up less than 7% of the deposits.

Chemical Composition: Table 1 shows the chemical analysis of a sample collected on deck of the ship *Roddam* in the harbor of St. Pierre on May 8, 1902.

TABLE 1. CHEMICAL COMPOSITION OF MT. PELÉE TEPHRA.*

	(Wt. %)
SiO_2	60.01
Al_2O_3	17.54
Fe_2O_3	2.88
FeO	4.30
MgO	2.26
CaO	6.80
Na_2O	3.41
K_2O	0.89
H_2O^+	0.30
H_2O^-	0.10
TiO_2	0.45
P_2O_3	0.13
MnO	0.23
Total	99.85

*Hillebrand, analyst (quoted in Diller, 1902).

Comments on Pyroclast Formation: Fisher and Heiken (1982) propose that nearly all pyroclasts in the Mt. Pelée pyroclastic flows are comminuted debris from the lava domes within the crater. Poorly vesiculated, juvenile (lithic) fragments were solidified, partly crystalline material from the dome. Vesicular glass shards and pumice are probably from molten or partly molten portions from the dome interior or from immediately below the dome. Explosive disruption and comminution of the dome could have been a combination of several processes: (1) overpressure within vesicles scattered throughout the dome during cooling and crystallization and (2) phreatomagmatic interaction of crater lake water and hot dome rocks. Fisher and Heiken (1982) note that there were heavy rains on May 7 and suggest that contact with ground water may have triggered the eruptions.

REFERENCES:

Diller, J. S., 1902. Volcanic rocks of Martinique and St. Vincent collected by Robert T. Hill and Israel C. Russell. Nat. Geogr. Mag., **13**: 285-296.

Fisher, R. V., Smith, A. L., and Roobol, M. J., 1980. Destruction of St. Pierre, Martinique by ash cloud surges, May 8 and 20, 1902. Geology, **8**: 472-476.

Fisher, R. V. and Heiken, G., 1982. Mt. Pelée, Martinique, May 8 and 20 pyroclastic flows and surges. J. Volcanol. Geotherm. Res., **13**, 339-371.

LaCroix, A., 1904. *La Montagne Pelée et ses eruptions.* Masson et Cie., Paris: 662 pp.

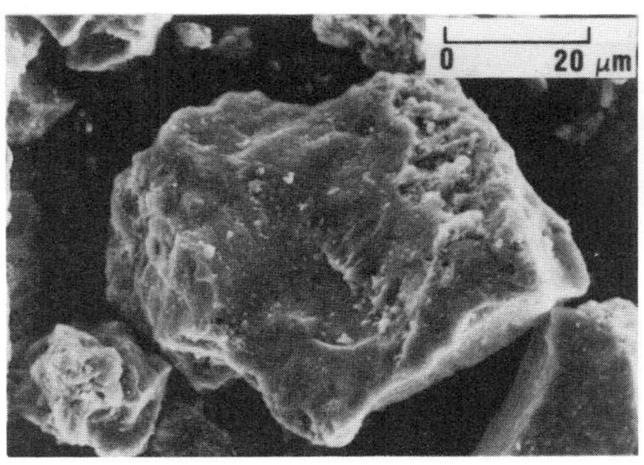

FIGURE 1a. Samples from near the base of block and ash flow deposits. Fine-grained fraction. SEM of hyalocrystalline, juvenile lithic fragments. Surfaces are irregular, with some conchoidal fractures. The equant, blocky clasts have no or rare vesicles.

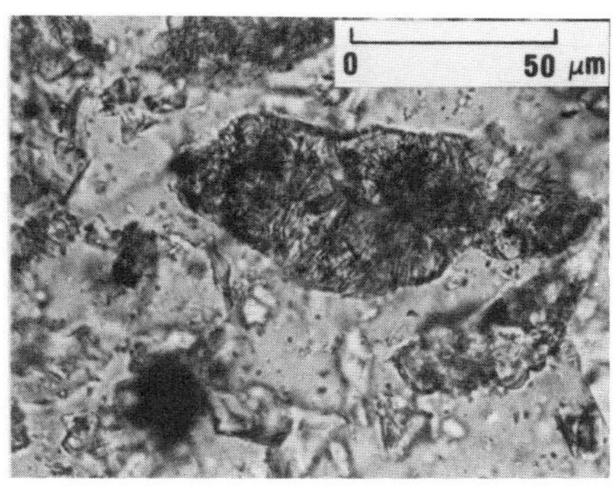

FIGURE 1b. Transmitted-light photomicrograph of mainly juvenile lithic fragments similar to those described in Fig. *2a*. Some 2- to 8-μm-wide, angular vitric shards are also visible. Microlites visible in glass groundmass are mainly feldspar.

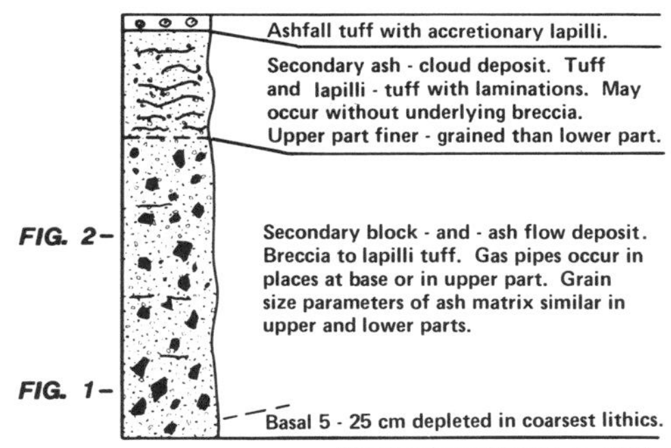

Depositional units from a Peléean ash cloud (from Fisher and Heiken, 1982).

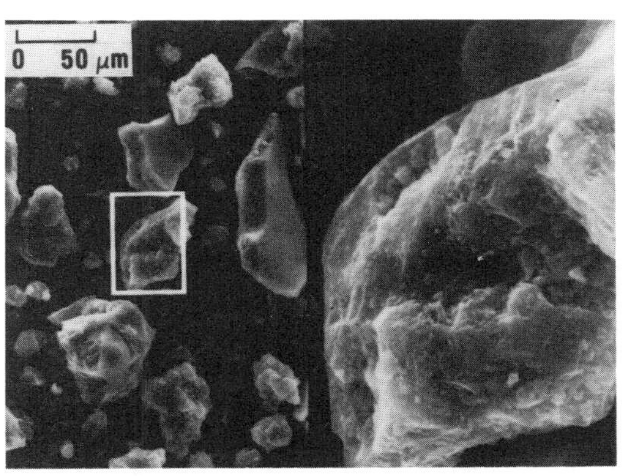

FIGURE 2a. Samples from the top of block and ash flow deposits. Fine-grained fraction. Mostly equant to slightly elongate lithic pyroclasts with rough, hackly grain surfaces. Some with a few vesicles have diktytaxitic textures. Smooth, flat pyroclasts (*right and top center*) are vitric shards. *Inset*—lithic pyroclast.

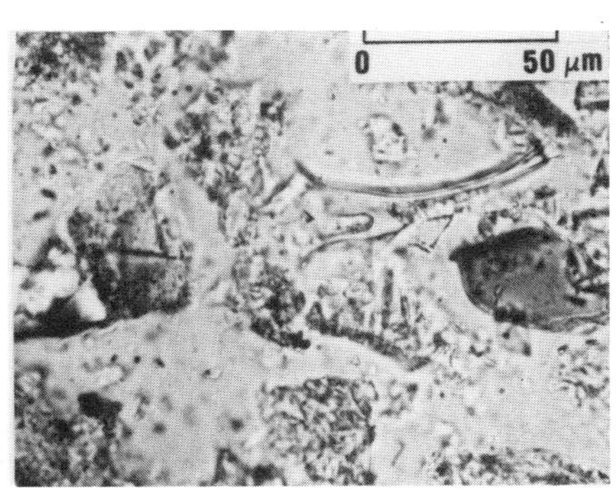

FIGURE 2b. Photomicrograph (transmitted light) of lithic ash. Mostly angular to subangular lithic and crystal pyroclasts. *Center right*—a delicate, elongate glass shard typical of the vitric fraction.

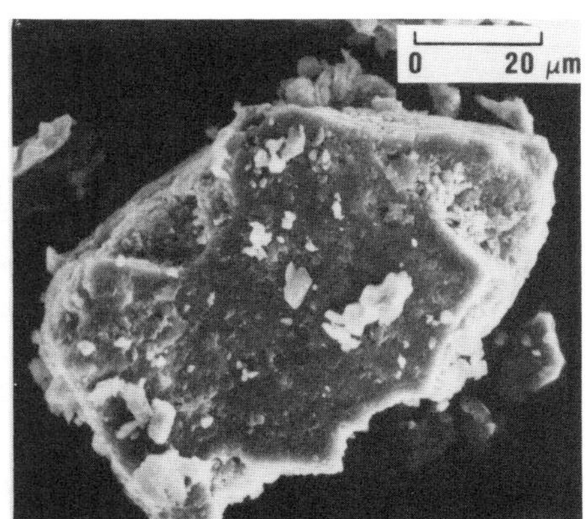

FIGURE 2c. Blocky, hyalocrystalline lithic pyroclast (juvenile). Smaller clasts on grain surface are small shards.

FIGURE 3. Sample from near the base of an ash-cloud deposit. Mostly equant, hyalocrystalline lithic pyroclasts; angular forms are bounded by rough fracture surfaces. Vesicles are rare and very small (<10 μm) when present. There are about 2% vitric shards that were broken from a pumiceous melt.

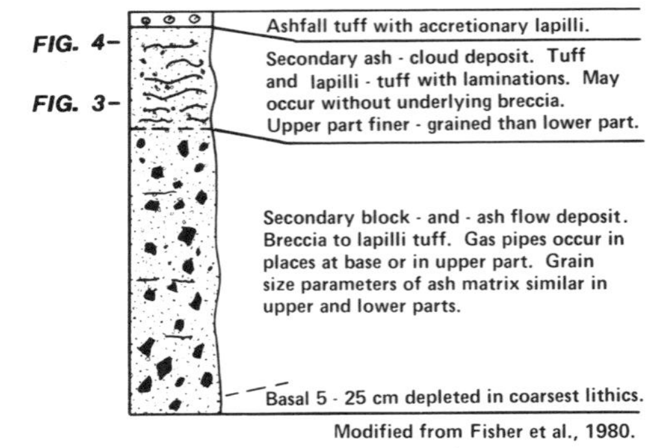

Depositional units from a Peléean ash cloud (from Fisher and Heiken, 1982).

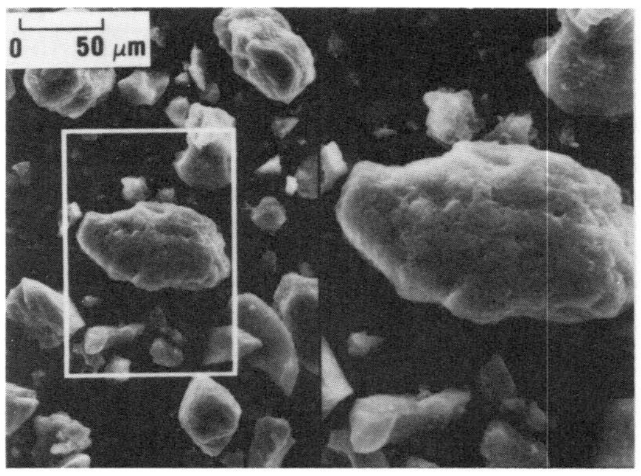

FIGURE 4a. Samples from the top of ash-cloud deposits; the upper part of this deposit is finer grained than the lower part (Fisher and Heiken, 1982). Equant to slightly elongate lithic fragments. *Inset*—clast with evidence of some rounding.

FIGURE 4b. Mostly equant lithic clasts; many of those with evidence of some rounding by abrasion. *Inset*—vesicular vitric shard. Ovoid vesicle remnants with thick vesicle walls.

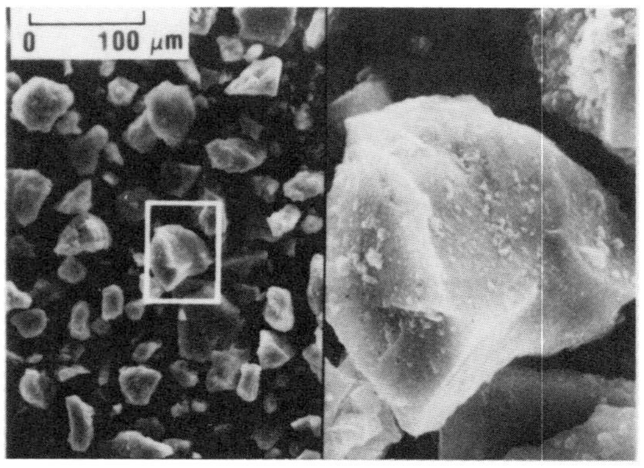

FIGURE 4c. Mostly 20- to 50-μm, equant, juvenile lithic and vitric pyroclasts. All are angular to subangular pyroclasts bounded by fracture surfaces. *Inset*—angular, juvenile lithic pyroclasts composed of hyalocrystalline andesite.

KRAKATAU, SUNDA STRAITS; 1883 ERUPTION

Introduction: Krakatau, an island volcano located in the Sunda Straits between Sumatra and Java, began erupting violently on May 20, 1883. The eruptions continued sporadically until the climactic eruption of August 27, 1883. Abundant tephra and volatiles were injected into the stratosphere and produced a worldwide haze and associated optical phenomena (Royal Society, 1888). Before 1883, the Krakatau group consisted of a 9- by 5-km island that included three overlapping volcanoes located along a fissure and two other islands that were remnants of the rim of a prehistoric caldera (Royal Society, 1883; Self and Rampino, 1981). Most of Krakatau Island was destroyed by caldera collapse during the eruption of about

20 km^3 of dacitic tephra (Self and Rampino, 1981). It has been proposed by Latter (1981) and Self and Rampino (1981) that tsunamis associated with this eruption were formed not by caldera collapse or submarine explosions, but by several cubic kilometers of pyroclastic flows entering the sea after each of several large explosions. The reader is referred to these two papers for information on the eruption sequence. Simkin and Fiske (1983) have prepared a comprehensive book on the eruption that includes eyewitness accounts from old manuscripts.

Grain Size: All the Plinian pumice fall deposits have mean grain sizes of 3 to 5 mm and are very poorly sorted (Self and Rampino, 1981). One of the pyroclastic surge deposits sampled has a 5-mm mean grain size and is also poorly sorted. The ignimbrite is finer grained, has a median of 0.2 mm, and is very poorly sorted.

Tephra Description: The Plinian pumice falls, associated with the intermittent explosive activity of May 20 to August 26, consist of mostly highly vesicular (65 to 70% vesicles) pumice pyroclasts. These are characterized by slightly elongate, flattened vesicles with thin (1- to 2-μm-thick) vesicle walls. There are many coalesced vesicles and some broken vesicle walls that have been rounded by viscous flow. Most of the finer grained pyroclasts are pumice.

Thin, possibly phreatomagmatic ash falls formed during the intermittent activity are interbedded with the pumice falls and pyroclastic surge deposits. The ash is bimodal, consisting of pumice pyroclasts in the 50- to 250-μm range and shards in the 5- to 100-μm range. Pumice vesicularities range from 40 to 70%. These are generally similar to pumices from the fall deposits but are finer grained.

An incipiently "welded" gray pumice fall that was deposited near-vent, just before eruption of the pyroclastic flows, is unique. It contains pumice pyroclasts with irregular, twisted forms whose vesicles are very well developed but curved, twisted, and flattened. Vesicle walls are 1 to 2 μm thick. Pyroclast surfaces are coated with small mounds and needle-shaped crystals (<1 μm wide); these may be sublimates. The coatings are largely responsible for cementing particles together.

Beginning at about 0530 on August 27, the first of several large explosions occurred. Self and Rampino (1981) suggest that each explosion produced a pyroclastic flow (ignimbrite). The ignimbrite sampled on Rakata Island consists of nonwelded, poorly sorted coarse ash and is a friable aggregate of pumice pyroclasts in a matrix of very fine grained (0.5- to 30-μm) shards. Many of the shards and pumice pyroclasts have fairly thick (15- to 30-μm) vesicle walls.

Chemical Composition: The 1883 ashes are dacitic; the phenocrysts, which are sparse, include plagioclase, augite, orthopyroxene and Fe-Ti oxides (Verbeek, 1886). For analyses of the pumices, see Table 1.

TABLE 1. MAJOR ELEMENT ANALYSES OF PUMICE CLASTS FROM KRAKATAU'S 1883 ERUPTION.*

	K2 (Wt. %)	K25 (Wt. %)
SiO_2	67.51	67.43
TiO_2	0.71	0.75
Al_2O_3	14.41	15.54
Fe_2O_3	4.02	3.91
MgO	1.14	0.76
MnO	0.14	0.15
CaO	2.65	2.92
Na_2O	5.26	5.16
K_2O	2.59	2.38
P_2O_5	0.10	0.09
Total	98.52	99.09

*S. Self, analyst.
K2 = gray pumice from the ignimbrite.
K25 = white pumice from the ignimbrite.

Comments on Pyroclast Formation: Nearly all the tephra were erupted during subaerial explosions—a fact noted in the field by Self and Rampino (1981). Nearly all eruption phases included pumice pyroclasts and shards that indicate complete vesicle growth; there are also high vesicularities and very thin vesicle walls. Variations in vesicle elongation and deformation indicate some variation in volatile content and/or viscosity throughout all eruption phases. The most unique pyroclasts are found in the near-vent "welded" pumice fall; twisted, deformed, very angular pumices indicate slightly lower viscosities and greater turbulence within the vesiculating magma.

The ignimbrite is strongly bimodal; it consists of pumice blocks in a poorly sorted ash matrix (Self and Rampino, 1981). The ash matrix is mostly very fine ash containing small, angular shards. It is possible that there was some magma/water interaction during this phase.

REFERENCES: Latter, J., 1981. Tsunamis of volcanic origin: summary of causes, with particular reference to Krakatau, 1883. Bull. Volcanol., **44**: 467-490.

Royal Society of London, Krakatau Committee, 1888. *The Eruption of Krakatau and Subsequent Phenomena.* Trübner and Co., London: 494 pp.

Self, S. and Rampino, M., 1981. The 1883 eruption of Krakatau. Nature, **294**: 699-704.

Simkin, T. and Fiske, R.S., 1983. *Krakatau 1883—The volcanic eruption and its effects.* Smithsonian Inst. Press, Washington: 464 pp.

Verbeek, R. D. M., 1886. *Krakatau.* Imprimerie de l'Etat Batavia.

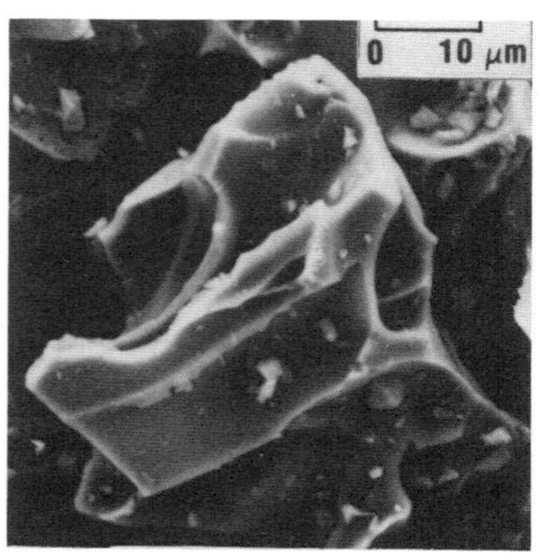

FIGURE 1a. SEM of pyroclasts from the coarsest Plinian phase. Collected on Rakata Kecil Island. Example of pyroclasts from the <62-μm fraction, which consists of mostly pumice pyroclasts in the 20- to 40-μm range and curved angular shards in the 5- to 10-μm range. Most are fragments of 1- to 6-μm-thick vesicle walls.

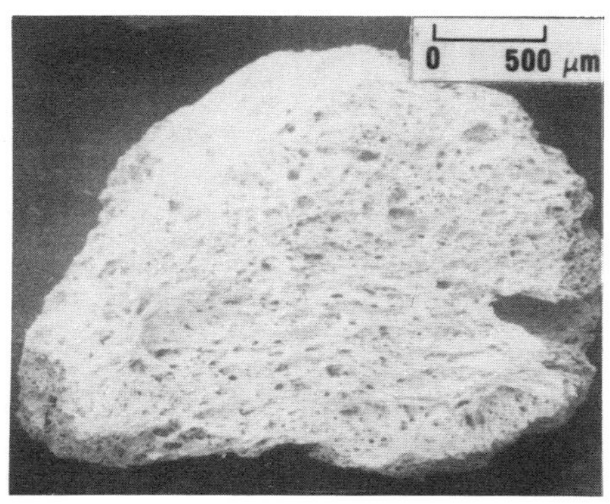

FIGURE 1b. Subrounded pumice pyroclast with about 65 to 70% vesicularity. Vesicles are well developed but were only slightly elongated by flow before eruption.

FIG. 1—

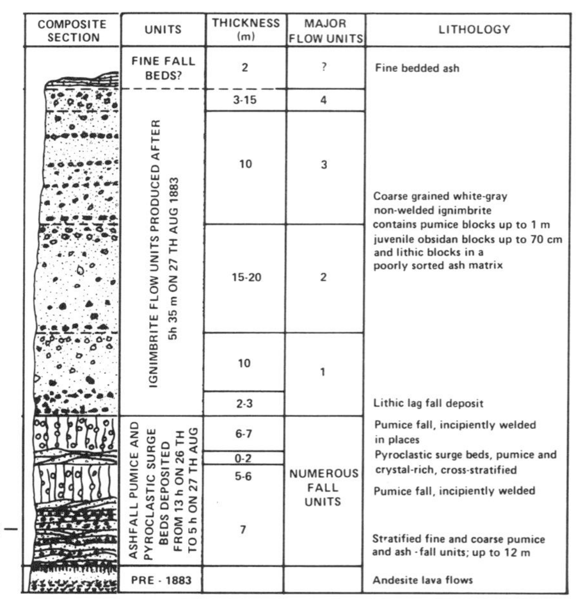

Composite section through the 1883 pyroclastic deposits of Krakatau (from Self and Rampino, 1981). All samples were collected by S. Self, University of Texas at Arlington.

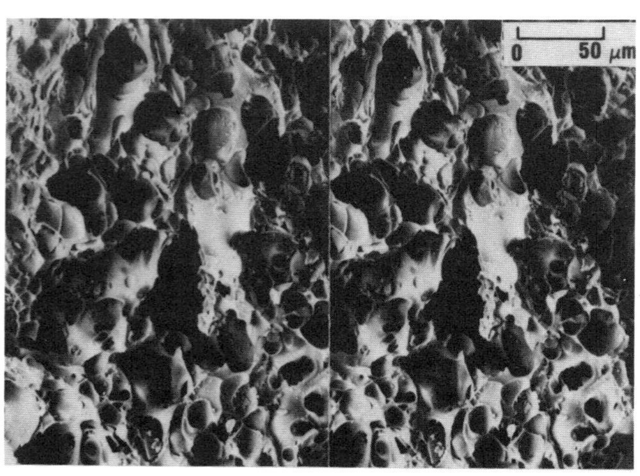

FIGURE 1c. Stereomicrographic detail of a pumice pyroclast. Most of the vesicles are slightly elongate and flattened, which produces parallel flattened ovoids. There are many coalesced vesicles with septae on vesicle walls (remnants of vesicle walls); some septae were rounded by viscous flow. Vesicle walls are mostly 1 to 2 μm thick.

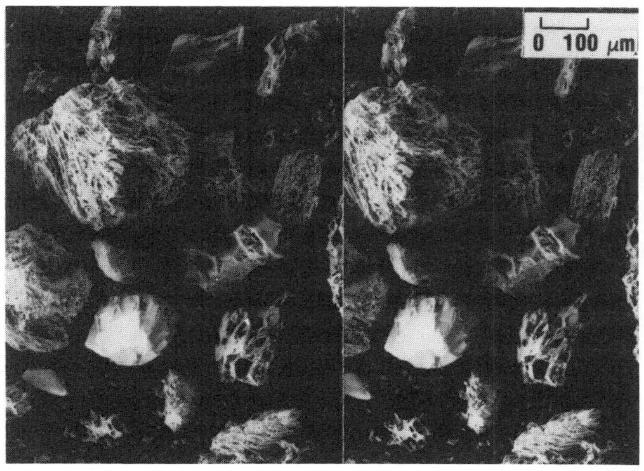

FIGURE 2a. Finest grained ash fall of phreatomagmatic origin. Collected on Rakata Kecil Island. Stereomicrograph of the tephra. The bimodality of this ash is evident; it consists of equant to slightly elongate pumice pyroclasts in the 50- to 250-μm range and glass shards in the 5- to 100-μm range. Most of the pumices are subangular to subrounded with vesicularities of less than 40 to 70%; some of this variation in vesicularity can occur within a single pyroclast. The equant pyroclast at center is a phenocryst covered with vesicular glass (bubble-wall texture).

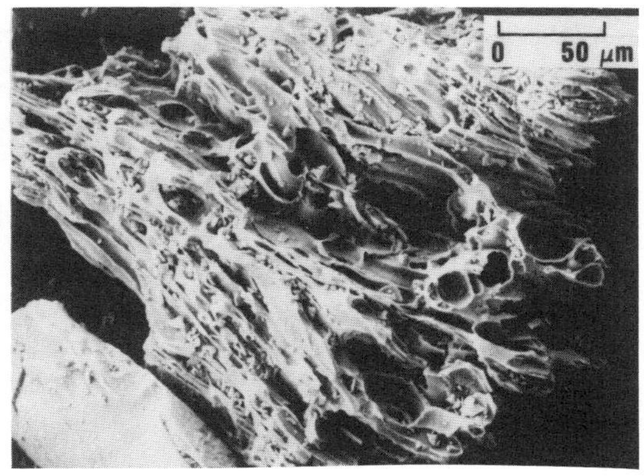

FIGURE 2b. Pumice pyroclast showing variation in vesiculation, coalescence of vesicles, and vesicle elongation. Vesicle-wall thicknesses vary from 1 to 8 μm.

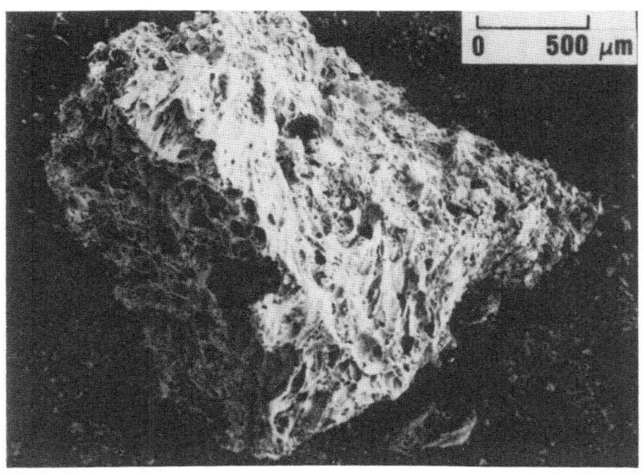

FIGURE 3a. "Welded" gray pumice fall collected on Rakata Kecil Island. Irregular, slightly elongate, ragged pumice pyroclast with a vesicularity of about 70%. Vesicles are curved, twisted, and flattened; vesicle walls are 1 to 2 μm thick. Many small particles are clinging to the pyroclast surfaces.

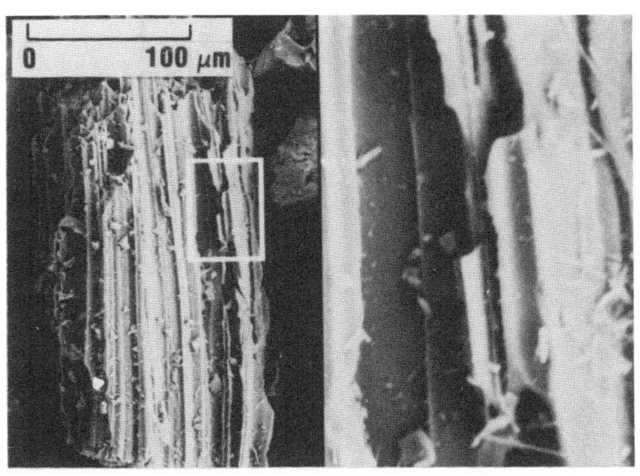

FIGURE 3b. A less common form of pumice pyroclast in this deposit has straight, parallel vesicles with thin walls. In the detailed inset, straight and slightly curved rods are visible. These are common on all pyroclasts in this deposit.

FIG. 3 —
FIG. 2 —

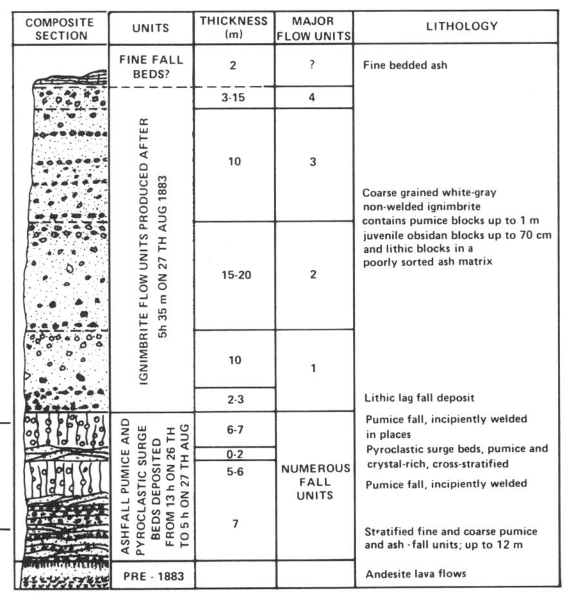

Composite section through the 1883 pyroclastic deposits of Krakatau (from Self and Rampino, 1981). All samples were collected by S. Self, University of Texas at Arlington.

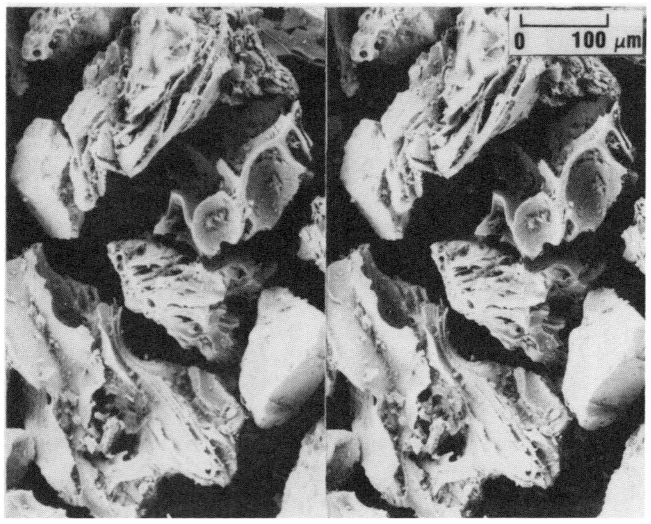

FIGURE 3c. Stereomicrograph of pyroclasts that are typical of this deposit. The most common pyroclasts are ragged, twisted, angular pumices. Vesicles are well developed (≃70% vesicularity); most are flattened ovoids separated by 2- to 20-μm-thick vesicle walls. Pyroclast surfaces are coated with 1-μm-wide, low mounds (sublimates?) and whiskers, 0.5 to 1 μm in diameter and 20 to 30 μm long. Clumps of these "whiskers" (zeolites?) bond some pyroclasts. It is probable that particle agglomeration was accomplished through cementation by secondary minerals and not through welding; perhaps it happened during release of volatiles from the cooling deposit.

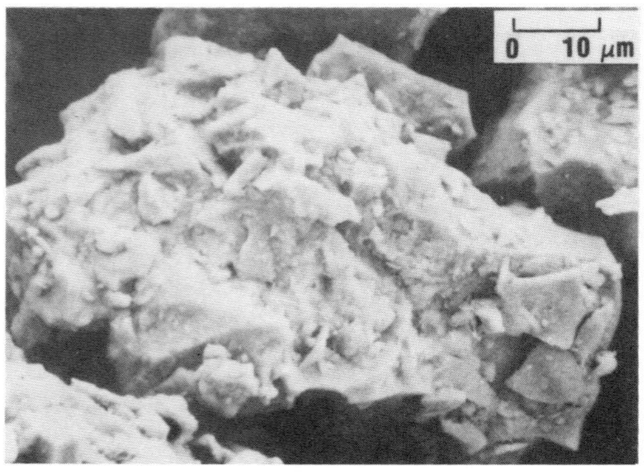

FIGURE 4a. Ignimbrite collected on Rakata Island. It is an aggregate composed of flat, curved, and irregular glass shards in a fine-grained matrix. Shards as small as 0.5 μm are present, some with visible broken walls (septae).

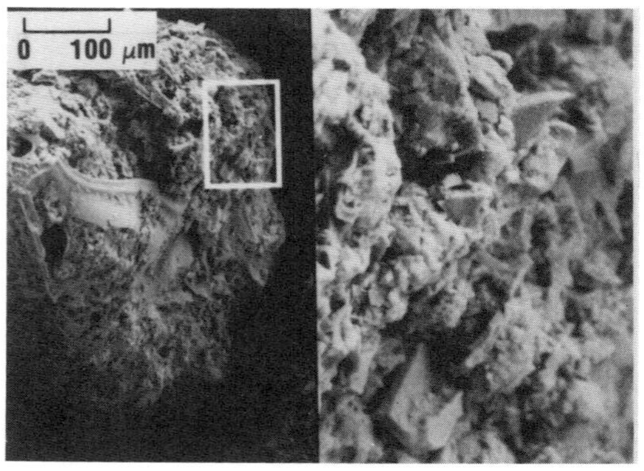

FIGURE 4b. Larger aggregate than the example in Fig. *4a* was broken and rounded by sieving of the tephra. A pumice pyroclast is visible in the aggregate and is characterized by well-developed, elongate vesicles and vesicle walls 15 to 30 μm thick. The inset is a closeup of the aggregate; there are small, curved and blocky equant shards in a very fine-grained matrix.

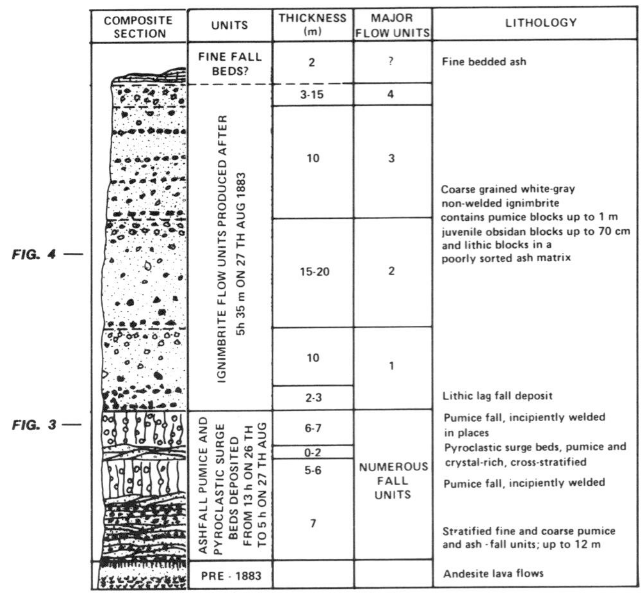

Composite section through the 1883 pyroclastic deposits of Krakatau (from Self and Rampino, 1981). All samples were collected by S. Self, University of Texas at Arlington.

FIGURE 4c. Stereomicrograph of the <70-μm fraction. The most common particle type is that of an aggregate of fine ash (probably a relict of sieving and incomplete disaggregation). Shards are similar to those described in Figs. *4a* and *b*. Small pumice pyroclasts, equant, blocky shards, and crystal fragments are also present.

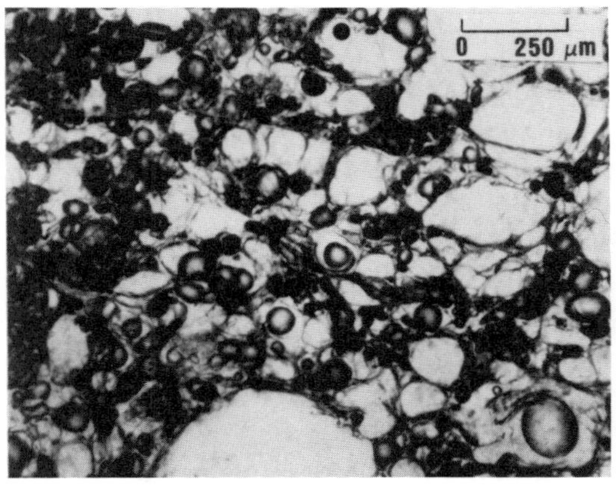

FIGURE 5a. Pumice pyroclasts from an ignimbrite collected on Rakata Island. Transmitted-light photomicrograph of a white pumice from the ignimbrite. There is a heterogeneous distribution of spherical to ovoid (deformed) vesicles in the 15- to 80-μm range. Some of the larger vesicles are in the 300- to 400-μm range. Vesicle walls, consisting of colorless glass, are 3 to 25 μm thick. There are 1% phenocrysts that consist of clinopyroxene, plagioclase, and hypersthene.

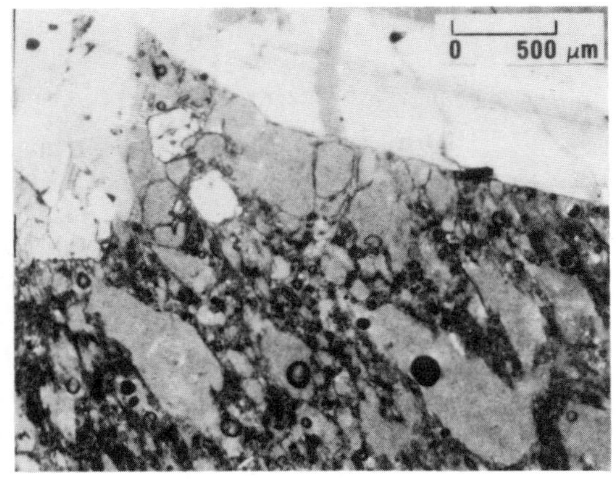

FIGURE 5b. Transmitted-light photomicrograph (partly crossed polarizers) of grey pumice from the ignimbrite. There are 5% phenocrysts and 95% colorless glass and vesicles. Phenocrysts are present as separate crystals and glomeroporphyritic clots of plagioclase, hypersthene, and Fe-Ti oxides. As can be seen here, vesicles have developed radially around the phenocrysts.

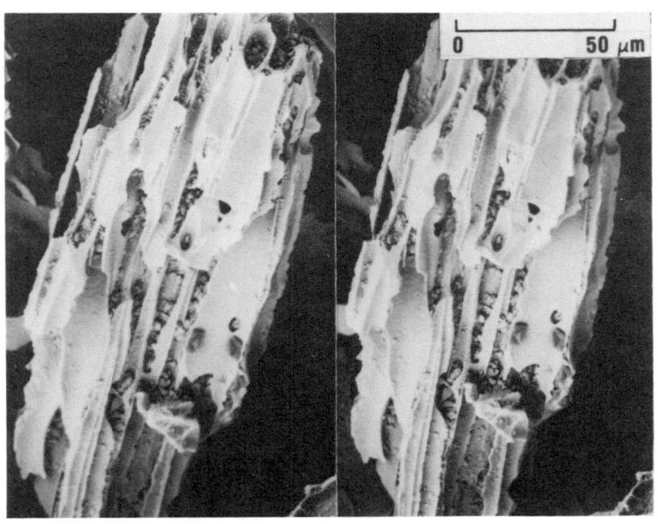

FIGURE 6a. Ash fall tephra collected on the vessel *Herschel*, in the Sunda Straits, just before the climactic phase of the eruption. These are similar to the Plinian deposits. Sample was provided by the U.S. Natural History Museum, Smithsonian Institution. Stereomicrograph of a pumice pyroclast. This pyroclast is characterized by high vesicularity and elongate vesicles. Vesicle aspect ratios range from 4:1 to 20:1.

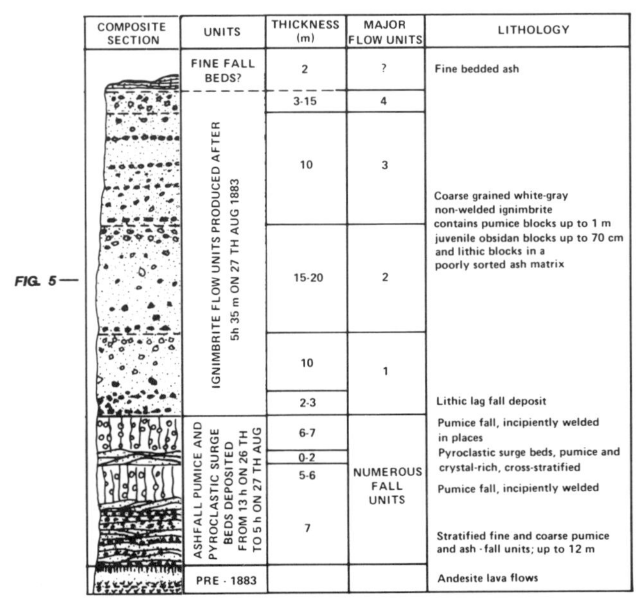

Composite section through the 1883 pyroclastic deposits of Krakatau (from Self and Rampino, 1981). All samples were collected by S. Self, University of Texas at Arlington.

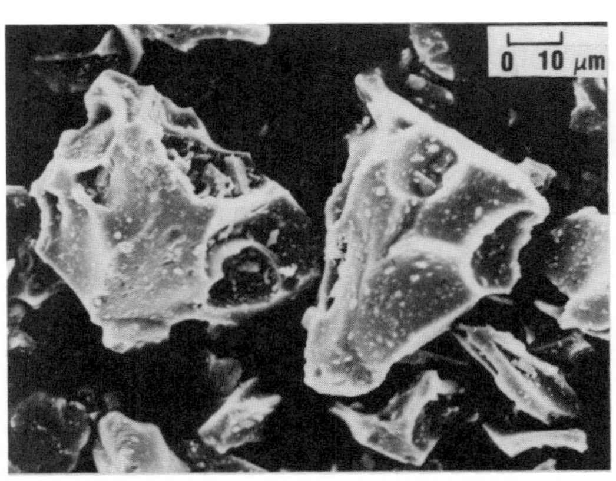

FIGURE 6b. An example of the fine-grained fraction of this tephra. It consists of mostly 10- to 70-μm-long shards and small pumice pyroclasts. Shards were broken from thick vesicle walls.

FIGURE 6c. Variety of pumice pyroclasts and thick-walled shards. Vesicles range from small ovoids to elongate, twisted tubes. Vesicle-wall thicknesses vary greatly within individual pumice pyroclasts.

MINOAN TUFF, THIRA, GREECE

Introduction: The Minoan eruption of about 1400 B.C. produced between 13 and 18 km^3 (dense rock equivalent) of rhyodacitic tephra (Watkins et al., 1978). The eruption sequence shows development from magmatic to phreatomagmatic eruption processes and back again to a mixture of both processes.

The eruption began at a vent located above sea level, where it produced a coarse-grained pumice deposit. This was followed by two later phreatomagmatic phases of the eruption from: (1) vents that propagated into an older, flooded caldera and (2) a collapsing Minoan caldera flooded by the sea. The final eruptive phase originated above sea level but may have been a mixture of magmatic and phreatomagmatic processes; during this phase, hot pyroclastic flows deposited a mixture of pumice pyroclasts and finely comminuted shards (Heiken and McCoy, 1984).

Grain Size: Samples from the Minoan tephra sequence were sieved by Bond and Sparks (1976). Plinian pumice deposits (unit 1) are poorly sorted and have a median grain size of 10 to 17 mm. Units 2 and 3 (phreatomagmatic deposits) are much finer grained and their median grain sizes range from 100 μm to 2 mm; all are poorly sorted. These deposits are all bimodal and consist of pumice bombs and lapilli in a very fine grained matrix. The ignimbrites are all very poorly sorted and have median grain sizes of 160 to 700 μm.

Tephra Description: Pumice pyroclasts (from unit 1) are mostly quite large (2- to 5-cm-long), angular to subangular forms. The pumices are characterized by bimodal vesicularity; pockets of large, coalesced ovoid vesicles are surrounded by highly elongate, tube-like vesicles. Vesicle walls are very thin. Within the upper part of unit 1 is a thin phreatomagmatic layer consisting of mostly fine-grained shards.

Deposits of units 2 and 3 are quite different. Unit 2 is characterized by plane beds and large base surge dunes, and unit 3 is massive and contains a high percentage of large lithic fragments. The tephra throughout these deposits are all similar, consisting of pumice lapilli and fragments in a matrix of fine white ash. Within the matrix, shards outnumber pumice pyroclasts 8:2. Shards are fine grained and flat to slightly curved with sharp, angular edges. There is little evidence of abrasion or edge modification on these particles. Most represent broken vesicle walls, 5 to 10 μm thick. Within the upper part of unit 3, small pumices and shards show decreasing vesicularity and thicker vesicle walls.

Within the ignimbrite there are 200- to 300-μm, subequant pumice pyroclasts; all are highly vesicular with well-developed, elongate vesicles. Some highly inflated pumices with polygonal vesicles and curved and flat shards with angular edges are also present. Most shards are in the 5- to 30-μm range.

Chemical Composition: Günther and Pichler (1973) published whole-rock chemical analyses of the Minoan tephra sequence. Analyses of the rhyodacitic tephra are shown in Table 1.

Comments on Pyroclast Formation: The most obvious change during this eruption sequence is from predominantly pumice pyroclasts in the coarse-grained Plinian deposits (shard-to-pumice ratio of 1:9) to predominantly shards in the phreatomagmatic deposits (shard-to-pumice ratio of 8:2).

All phases of the eruption demonstrate that vesicle growth was complete. The high vesicularities are characteristic of the Plinian pumices. Shards in later phases were broken from a vesiculated melt during magma/water interaction and produced thin, curved and flat, fine-grained, angular shards. Comminution was not complete, however, as a minor proportion of small pumice pyroclasts are present in all samples.

The last eruptive phase, which produced an ignimbrite, is a bimodal mixture of pumice pyroclasts and finely comminuted shards. Grain surfaces are coated with finer grained shards. It is possible that the very fine grained shards may have been produced by minor phreatomagmatic interaction during eruption of the ignimbrite.

TABLE 1. CHEMICAL ANALYSES OF PUMICE PYROCLASTS FROM THE MINOAN TUFF.*

	Unit 1 (pumice fall) (Wt. %)	Unit 3 (phreatomagmatic) (Wt. %)
SiO_2	68.5	69.12
Al_2O_3	14.2	13.93
Fe_2O_3	1.0	0.72
FeO	1.9	1.92
MnO	0.08	0.08
MgO	0.05	0.58
CaO	2.2	2.04
Na_2O	4.5	4.23
K_2O	2.8	3.01
TiO_2	0.4	0.41
H_2O	3.9	3.7
Total	99.61	99.36

*From Günther and Pichler (1973).

REFERENCES: Bond, A. and Sparks, R., 1976. The Minoan eruption of Santorini. J. Geol. Soc. London, **132**: 1-16.

Günther, D. and Pichler, H.. 1973. Die obere und untere bimsstein-folge auf Santorin. N. Jb. Geol. Palaönt. Mh. H., **7**: 394-415.

Heiken, G. and McCoy, F., Jr., 1984. Caldera development during the Minoan eruption, Thira, Cyclades, Greece. J. Geophys. Res., **89**: 8441-8462.

Watkins, N., Sparks, R., Sigurdsson, H., Huang, T. C., Federman, A., Carey, S., and Ninkovich, D., 1978. Volume and extent of the Minoan tephra from Santorini volcano: New evidence from deep-sea sediment cores. Nature, **271**: 122-126.

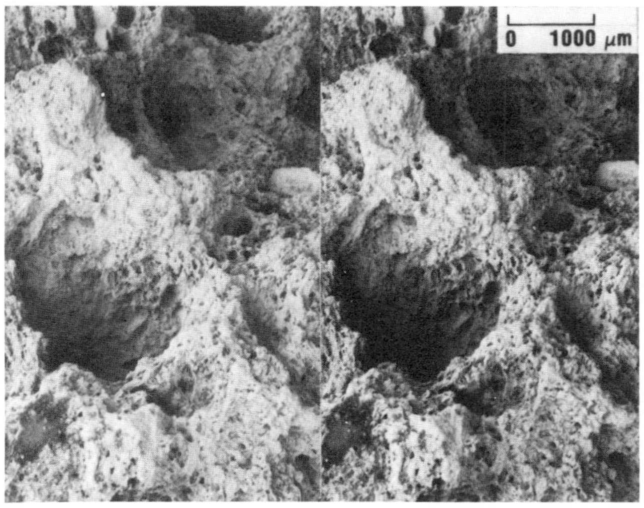

FIGURE 1a. Tephra from the Plinian pumice deposit (unit 1). SEM stereomicrograph of a typical pumice pyroclast. Lapilli-size and larger pumices are slightly elongate, blocky forms with surfaces broken by pockets of coalesced vesicles. The pockets were formed during heterogeneous vesicle growth throughout the vesiculated magma. Most of the vesicles are highly elongate and have thin vesicle walls. The tube-like vesicles are parallel to long axes of the pyroclasts; some are bent around larger cavities. The cavities, or pockets, are patches of highly inflated, coalesced vesicles. Most pockets are adjacent to pyroxene or feldspar phenocrysts.

FIG. 1

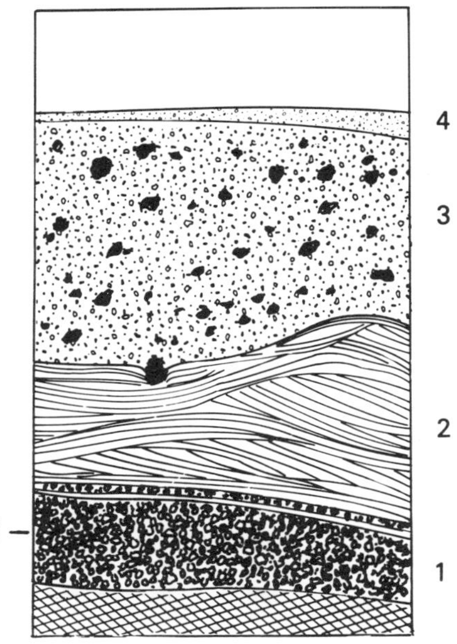

Composite section of the Minoan tephra sequence that was erupted in ≃1400 B.C. Unit 1 is a coarse-grained Plinian pumice fall; the thin bed near the top of this deposit is a fine-grained layer of phreatomagmatic origin. Unit 2 is a phreatomagmatic deposit characterized by large base surge dunes and plane beds. Unit 3 is a massive, lithic-fragment-rich phreatomagmatic deposit. Unit 4 is a nonwelded ignimbrite.

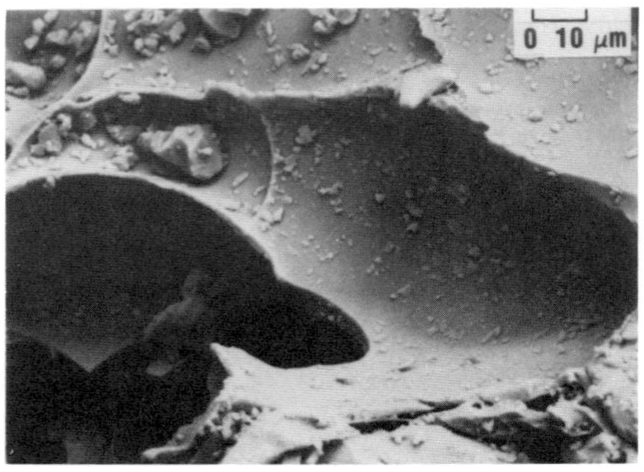

FIGURE 1b. Detail of coalesced vesicles in a pocket of the larger vesicles. Remnants of ovoid, coalesced vesicles are visible as septae or vesicle walls. Wall thicknesses vary from 1 to 5 μm.

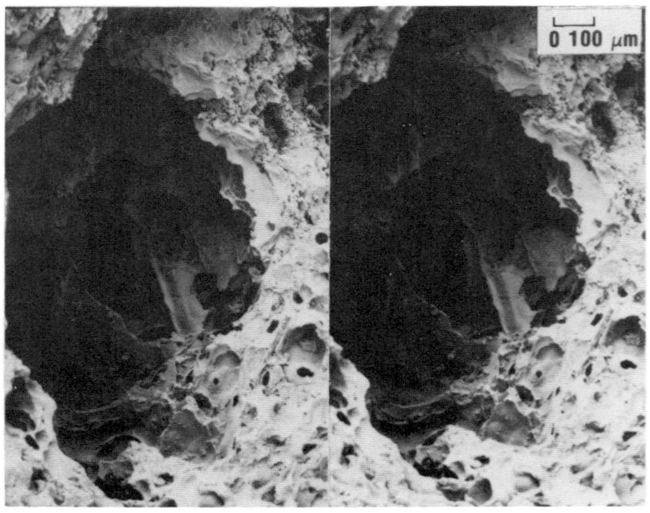

FIGURE 1c. Stereomicrograph of coalesced larger vesicles that illustrates the heterogeneity of these pumice pyroclasts.

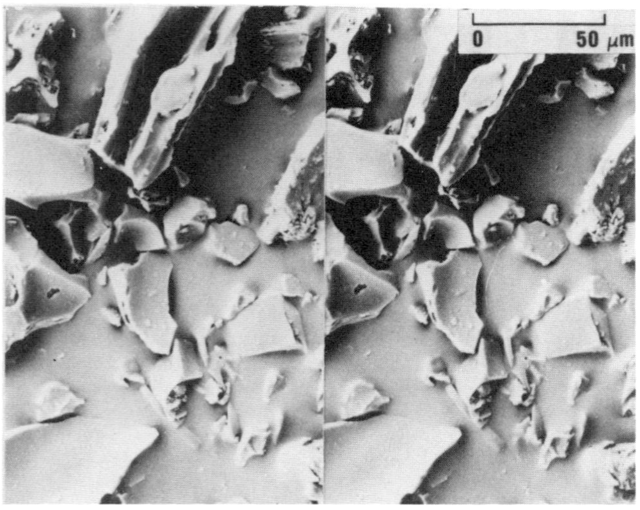

FIGURE 2a. Stereomicrograph of the tephra from a fine-grained phreatomagmatic layer within the upper part of the pumice fall. There are mostly broadly curved shards with very angular edges. Most were broken from a pumiceous magma with 50- to 70-μm-wide, flattened vesicles and 5- to 10-μm-thick vesicle walls. About 85% of the tephra consists of shards and 21% is pumice pyroclasts. The fine-grained nature (most shards are in the 20- to 40-μm range) and abundance of angular shards are typical of a highly vesicular melt that was fragmented during melt/water interactions.

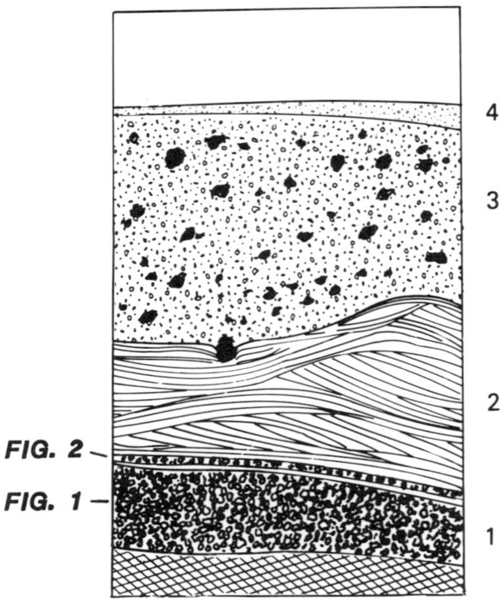

Composite section of the Minoan tephra sequence that was erupted in ≃1400 B.C. Unit 1 is a coarse-grained Plinian pumice fall; the thin bed near the top of this deposit is a fine-grained layer of phreatomagmatic origin. Unit 2 is a phreatomagmatic deposit characterized by large base surge dunes and plane beds. Unit 3 is a massive, lithic-fragment-rich phreatomagmatic deposit. Unit 4 is a nonwelded ignimbrite.

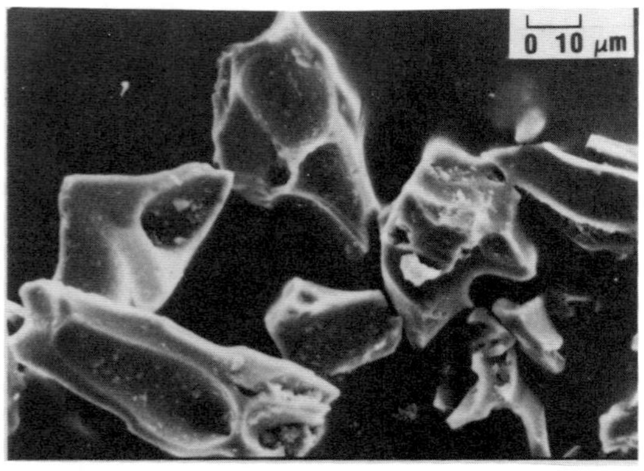

FIGURE 2b. Fine-grained shards typical of this layer.

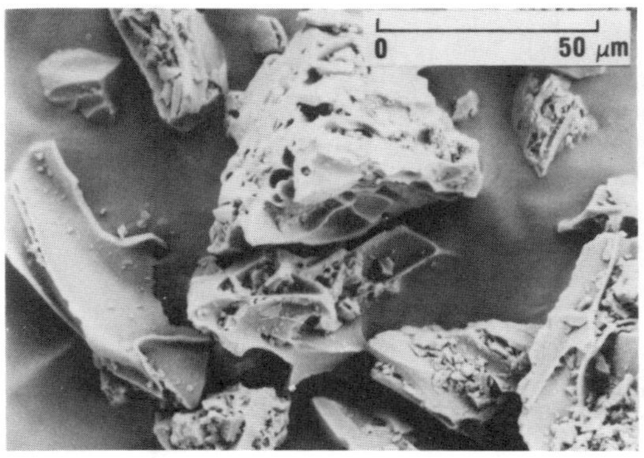

FIGURE 3a. Phreatomagmatic ash from unit 2 (not including the pumice lapilli and lithic fragments scattered through this ash matrix). The pyroclasts are very similar to those in Fig. *2*. Fine-grained ash that consists of mostly angular shards (75%) and small, angular pumice pyroclasts (25%). Most shards are broken from coalesced vesicles; thin, tapered ridges are remnants of vesicle walls broken during coalescence.

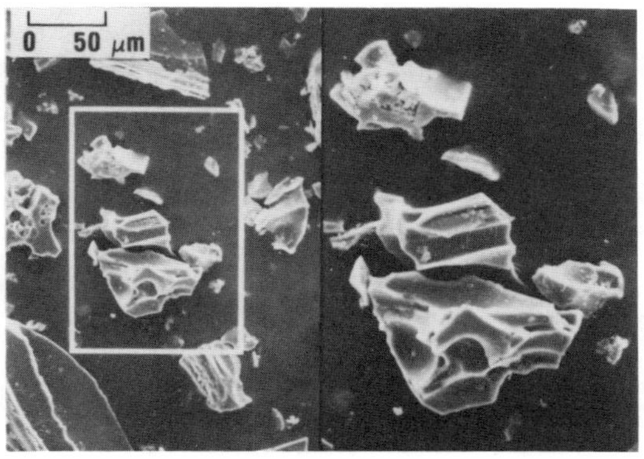

FIGURE 3b. Representative sample of shards and small pumice pyroclasts that are typical of this phreatomagmatic deposit.

FIG. 4 —
FIG. 3 —

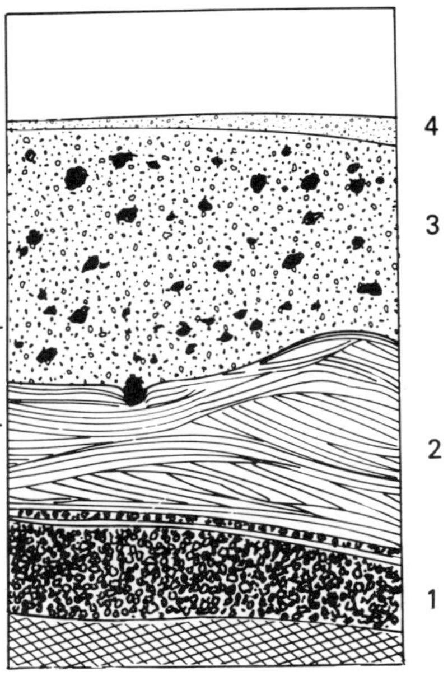

Composite section of the Minoan tephra sequence that was erupted in ≃1400 B.C. Unit 1 is a coarse-grained Plinian pumice fall; the thin bed near the top of this deposit is a fine-grained layer of phreatomagmatic origin. Unit 2 is a phreatomagmatic deposit characterized by large base surge dunes and plane beds. Unit 3 is a massive, lithic-fragment-rich phreatomagmatic deposit. Unit 4 is a nonwelded ignimbrite.

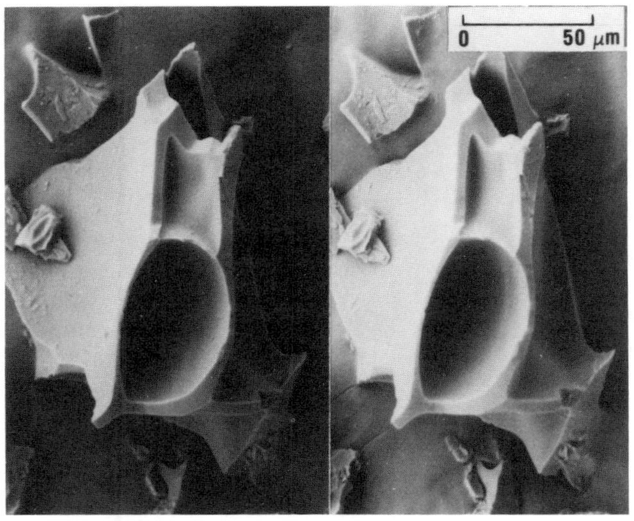

FIGURE 3c. Stereomicrograph of an angular, curved shard. Multiple vesicle remnants are visible. There is little or no abrasion of delicate edges on these pyroclasts.

FIGURE 4a. Matrix sample from the base of unit 3. Lithic fragments and pumice bombs, which make up 5 to 40% of this unit, are not included in this analysis. There are mostly (87%) slightly curved shards with angular edges. Small pumices make up about 13% of the matrix sample studied here. Vesicle walls are 5 to 20 μm thick.

FIGURE 4b. A closer look at some of the smaller shards in Fig. *4a*.

FIGURE 4c. A curved shard typical of products from finely comminuted vesiculated magma. Angular grain edges are not abraded.

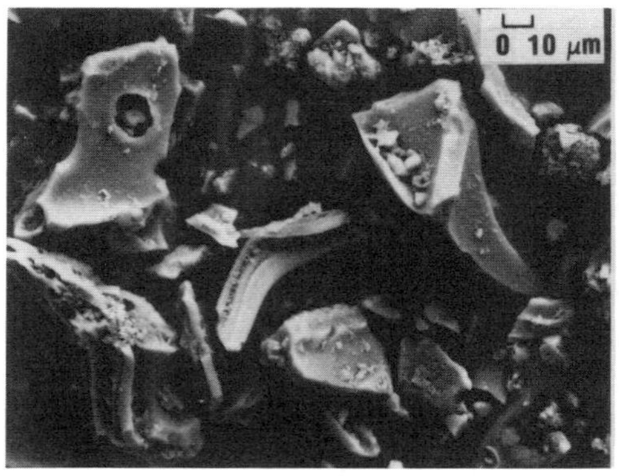

FIGURE 5a. Matrix sample from the uppermost part of unit 3. Small, slightly elongate to equant shards and pumice pyroclasts. The shard-to-pumice ratio (8:2) is nearly the same as that of a sample from the base of unit 3 (Fig. *4*), although the vesicularity of the pumice pyroclasts appears to be lower. Most vesicle walls are 5 to 10 μm thick. Smaller shards are mostly equant to slightly elongate.

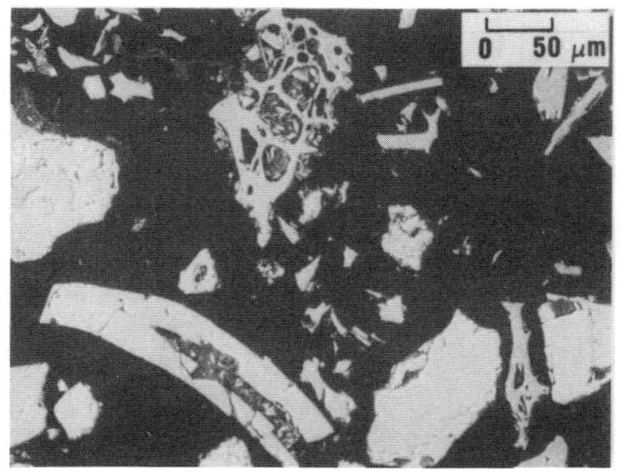

FIGURE 5b. Polished thin section of the coarser fraction from this sample. Nearly all the pyroclasts are glass; well-developed pumice (*top*) to coarse angular shards. Note the angularity of most of the smallest shards. A rounded lithic fragment (lava) is visible at the left of the photograph.

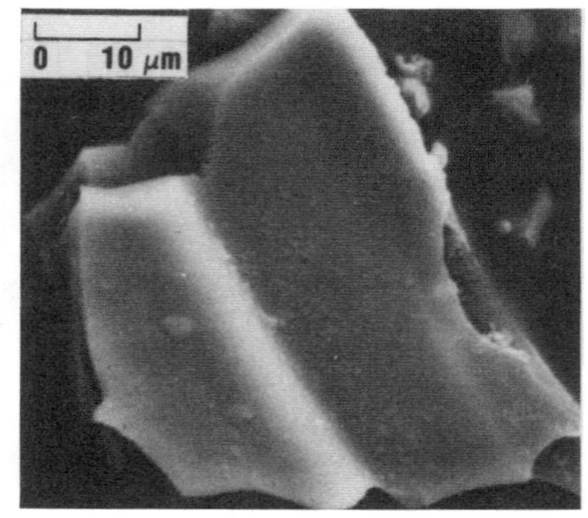

FIGURE 5c. Single angular pumice fragment. This finely comminuted pyroclast is typical of the phreatomagmatic phase of the eruption. Minor abrasion is visible only on the thinnest edges.

FIGURE 6a. Ignimbrite sample from unit 4. Bimodal distribution of pyroclasts is visible in this SEM. Subequant pumice pyroclasts are in the 200- to 300-μm range; most are highly vesicular with highly elongate vesicles and 3- to 15-μm-thick vesicle walls. There are also a few highly inflated pumices (*upper right*) with flat (polygonal) vesicle walls and some small, rounded lithic fragments (*center bottom*). Angular, curved and flat shards make up the 5- to 30-μm fraction (mostly about 20 μm long).

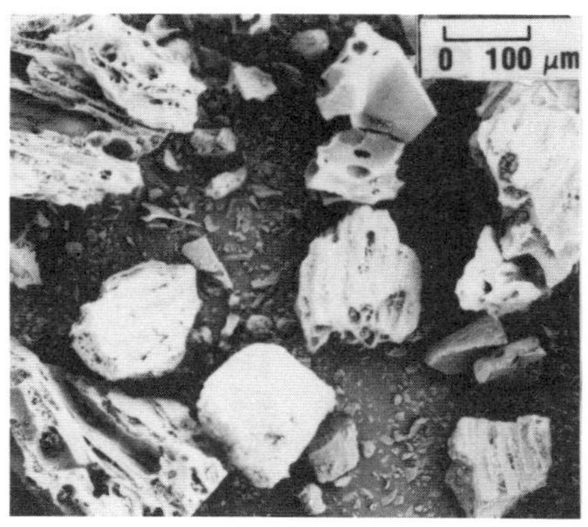

FIGURE 6b. View of the finer grained fraction, which is made up of equant, angular shards. Note that grain surfaces are coated with fine-grained particles (typical of many ignimbrites).

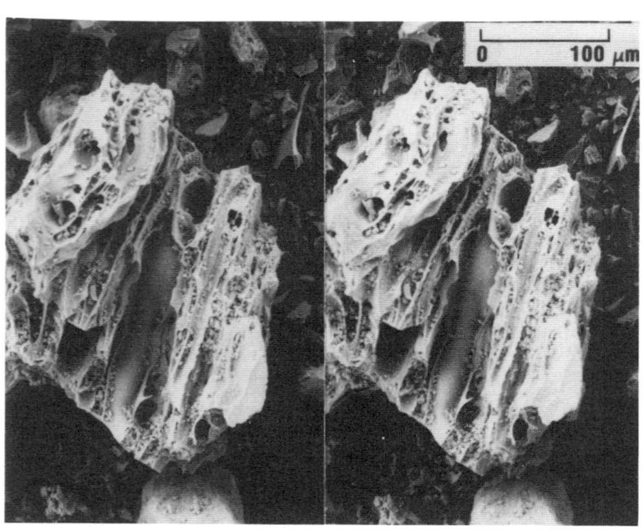

FIGURE 6c. Stereomicrograph of a pumice pyroclast that illustrates the well-developed elongate vesicles. There are some pockets (*center*) of coalesced vesicles. Note the thin vesicle walls.

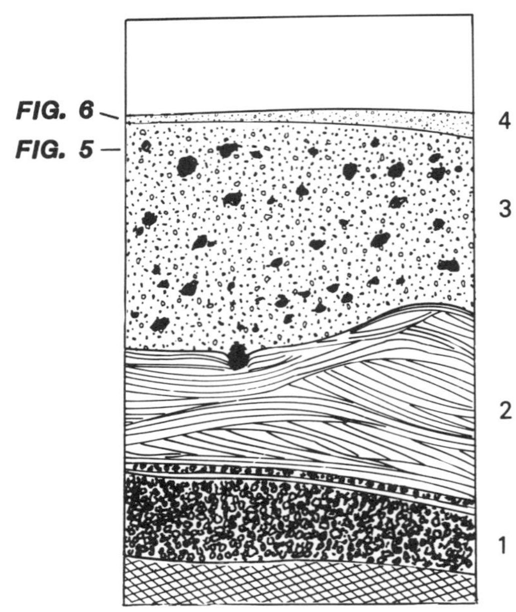

Composite section of the Minoan tephra sequence that was erupted in ≈1400 B.C. Unit 1 is a coarse-grained Plinian pumice fall; the thin bed near the top of this deposit is a fine-grained layer of phreatomagmatic origin. Unit 2 is a phreatomagmatic deposit characterized by large base surge dunes and plane beds. Unit 3 is a massive, lithic-fragment-rich phreatomagmatic deposit. Unit 4 is a nonwelded ignimbrite.

TAAL VOLCANO, PHILIPPINES; DRY PHREATOMAGMATIC ERUPTION

Introduction: Taal Volcano is located in Lake Taal, about 60 km S of Manila, Philippines. Continuous activity has been recorded there over the last several hundred years and up to recent activity in 1976 (Wolfe, 1980a). The activity of 1965 provided the first well-documented accounts of volcanic base surges, a type of pyroclastic surge (Moore et al., 1966). Surges that erupted from a vent on an island in the lake sandblasted objects as far as 6 km away. Approximately 9.0×10^4 kg of ash was ejected from the vent, of which only 10% was new basalt; the rest was older basaltic ash, cinders, and blocks from the new crater (Moore, 1967).

The stratigraphy of the 1965 deposits (Moore, 1967) supports witnessed accounts of the activity. Opening eruptions deposited ash falls of juvenile materials, including some spatter and lapilli. The second stage of eruption was marked by phreatomagmatic explosions and the excavation of a new crater when lake water gained access to the vent. Surge deposits resulting from this second stage of activity are breccias that are crudely stratified near the vent and well bedded away from the vent. The eruptions decreased in energy during the final stages and resulted in ash fall deposition and formation of a cinder cone within the earlier tuff ring.

Grain Size: Samples of Taal surge tephra from 1965 eruptions (collected by M. Sheridan) are dark gray, fine to medium ash. Sandwave beds (Table 1) are characterized by an abundant (over 35 wt.%) fine-ash fraction. Waters and Fisher (1971) give size data that show decreasing median grain size in sandwave beds farther from the vent, which possibly indicates a decrease in the surge's kinetic energy. The plotted grain-size groupings by bed form suggest different surge transport and depositional processes for each (Sheridan and Updike, 1975).

TABLE 1. GRAIN-SIZE CHARACTERISTICS OF TAAL TEPHRA.*

Sample	Bed Form*	Md_φ (mm)	σ_φ	Fine Ash (Wt. %)
1	M	0.65 (0.64)	1.42	2.53
2	S	3.10 (0.12)	2.14	36.07
3	S	2.93 (0.13)	1.99	28.42
4	M	0.51 (0.70)	1.98	4.02
5	S	2.37 (0.19)	1.57	15.39
6	S	2.08 (0.24)	1.20	8.19
7	P	0.33 (0.79)	1.55	3.01
8	M	1.12 (0.46)	2.35	12.07
9	P	−0.73 (1.66)	1.67	2.74
10	S	3.60 (0.08)	1.77	37.80
11	P	0.13 (0.91)	2.18	5.82
12	S	1.93 (0.26)	1.70	13.40
13	M	1.13 (0.46)	2.48	13.28
14	P	−0.57 (1.48)	1.28	0.33
15	M	1.35 (0.39)	1.60	6.40
16	M	1.40 (0.38)	1.40	4.71
17	M	0.53 (0.69)	1.62	2.32
18	S	1.48 (0.36)	1.36	4.88
19	M	1.06 (0.48)	1.47	3.31
20	S	2.98 (0.13)	1.83	28.80
21	S	2.67 (0.16)	1.93	23.26
22	S	2.33 (0.20)	1.96	18.33
23	M	0.96 (0.51)	2.17	9.68

*M. Maisanno and M. Sheridan, analysts. Bed forms: S = sandwave, M = massive surge, P = planar surge.

Tephra Description: The 1965 Taal ash is composed of: (1) fresh, dark gray sideromelane of basaltic andesite composition; (2) slightly altered to fresh, dark brown tachylite; and (3) plagioclase, olivine, and clinopyroxene crystals. The sideromelane is juvenile and composes up to 90% of fall deposits, whereas the tachylite is generally lithic and composes 90% of surge deposits. Crystals usually make up less than 10% of the ash.

Of the juvenile pyroclasts, three characteristic types dominate the 1965 products: (1) vesicular, (2) blocky, and (3) drop-like or vesicular ash with fluidal surfaces. The blocky ash samples are excellent examples of typical phreatomagmatic basaltic pyroclasts. The planar and curviplanar surfaces meet at nearly 90° angles and cut across vesicle walls. Vesicles, when present, are nearly spherical but may show some elongation; this shape contrasts with that of Strombolian-fall ash, which shows very irregularly shaped vesicles. Smooth, fluidal surfaces on ash are manifested as poorly vesicular, drop-like particles and, in vesicular ash, as gently curved areas that "flow" into vesicle hollows.

Lithic pyroclasts present are mostly tachylitic and exhibit diktytaxitic textures on vesicle walls. They are generally more altered than juvenile ash is, and alteration products fill the tube-like or elongated vesicles.

Alteration of juvenile pyroclasts shows two stages of development. Fine grained clays and zeolites form first in vesicle hollows because of interaction with steam that condensed there during emplacement. Further alteration is manifested by surface hydration, which forms a thin crust on the grain that often cracks after desiccation.

Fine-grained ash consists of both blocky and vesicular forms that are marked by abrasion chips and pits. Alteration products are commonly present in vesicle hollows. The late-stage growth of some small vesicles was arrested by chilling during their development. These tiny vesicles are 1 to 20 μm in diameter and represent a second stage of bubble growth; Vesicles from the first stage are 50 to 100 μm in diameter.

Comments on Pyroclast Formation: Taal magma was relatively volatile-rich, as is shown by the abundance of vesicles in both early- and late-stage ashes. These volatiles were abundant enough to tear apart the magma during their exsolution, producing Strombolian lava sprays and fountains early in the eruption, before they contacted lake water. In fact, Moore et al. (1966) described some pyroclasts formed during that stage as pumiceous. During phreatomagmatic stages, the magma was fragmented before much bubble-burst and bubble-coalescense occurred. The dominant blocky texture of the ash indicates that while it was in a nearly solid state and before it mixed with water, the magma was fragmented by propagating stress waves. However, some ash still must have been fluid when it contacted and mixed with water, as is indicated by the fluidal surfaces on vesicular and drop-like pyroclasts. Lithic ash expelled from the vent was abraded, resulting in smooth pyroclast edges. This older ash had formed mostly from a strongly crystallized magma with opaque oxide phases; therefore, it is tachylitic, unlike the juvenile material, which is glassy.

Although most vesicles attained diameters of 50 to 100 μm before bursting, a second stage of bubble nucleation and growth occurred, which is shown by the population of tiny vesicles. This second stage was probably triggered by stress waves propagating downward in the vent from a surface explosion source. This hypothesis is supported by the presence of small, linear discontinuities in glass shards, along which the tiny vesicles are aligned. Where the stress waves were tensile, they formed areas of renewed vesicle nucleation and growth in their wake. When the magma solidified, this second stage of growth was arrested before the bubbles grew to more than about 20 μm in diameter. Such secondary vesiculation areas may be preserved adjacent to a deformation plane.

A more detailed discussion of the formation of Taal ash is given by Wolfe (1980b), who described the explosive breccias of the 1976 eruptions.

REFERENCES:

Moore, J. G., 1967. Base surge in recent volcanic eruptions. Bull. Volcanol., **30**: 337-363.

Moore, J. G., Nakamura, K., and Alcaraz, A., 1966. The 1965 eruption of Taal Volcano. Science, **151**: 955-960.

Sheridan, M. F. and Updike, R. G., 1975. Sugarloaf Mountain tephra—a Pleistocene rhyolitic deposit of base-surge origin. Geol. Soc. Am. Bull., **56**: 571-581.

Waters, A. C. and Fisher, R. V., 1971. Base surges and their deposits: Capelinhos and Taal Volcanoes. J. Geophy. Res., **76**: 5596-5614.

Wolfe, J. A., 1980a. Phreatomagmatic eruption of Taal Volcano in 1976. Trans. Am. Geophys. Union., EOS, **61(7)**: cover photo.

Wolfe, J. A., 1980b. Fluidization versus phreatomagmatic explosions in breccia pipes. Econ. Geol., **75(7)**: 1105-1109.

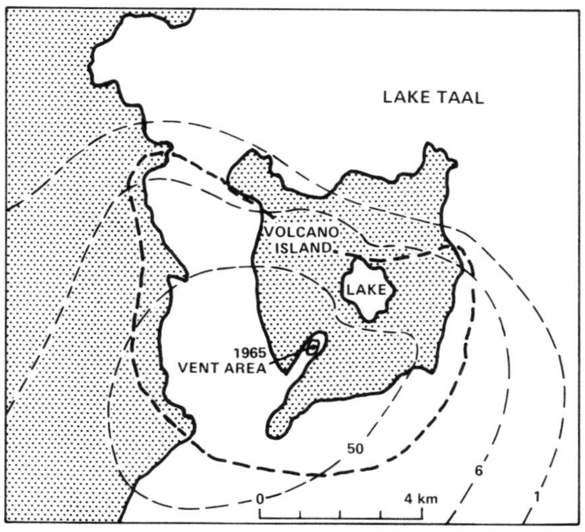

FIGURE 1. Map of Taal ash deposits adapted from Moore (1967). Dashed lines show 50-, 6-, and 1-cm isopachs of ash fall deposits from the 1965 eruption; the heavy dashed line shows the extent of base surge.

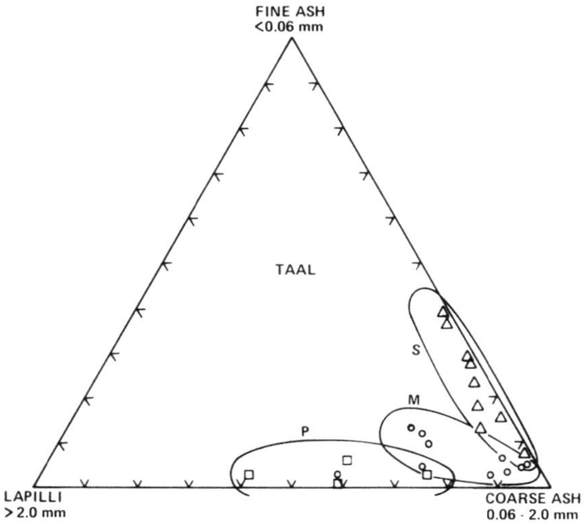

FIGURE 2. Ternary diagram of tephra sizes for 1965 Taal deposits. Size components are fine ash (<0.063 mm), ash (0.063 to 2.00 mm), and lapilli (>2.00 mm). Median sizes of samples are strongly correlated to bed form: S = sandwave, M = massive, and P = planar pyroclastic surge deposits.

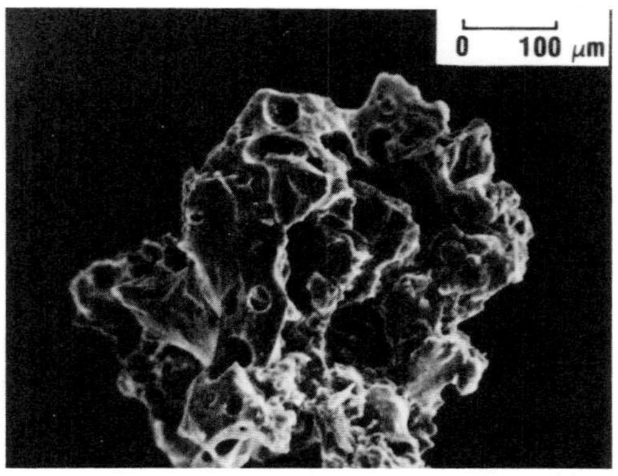

FIGURE 3. SEM of a vesicular pyroclast that is typical of Strombolian ash fall eruptions. Note the irregular shape of vesicles and their porous, spongy appearance.

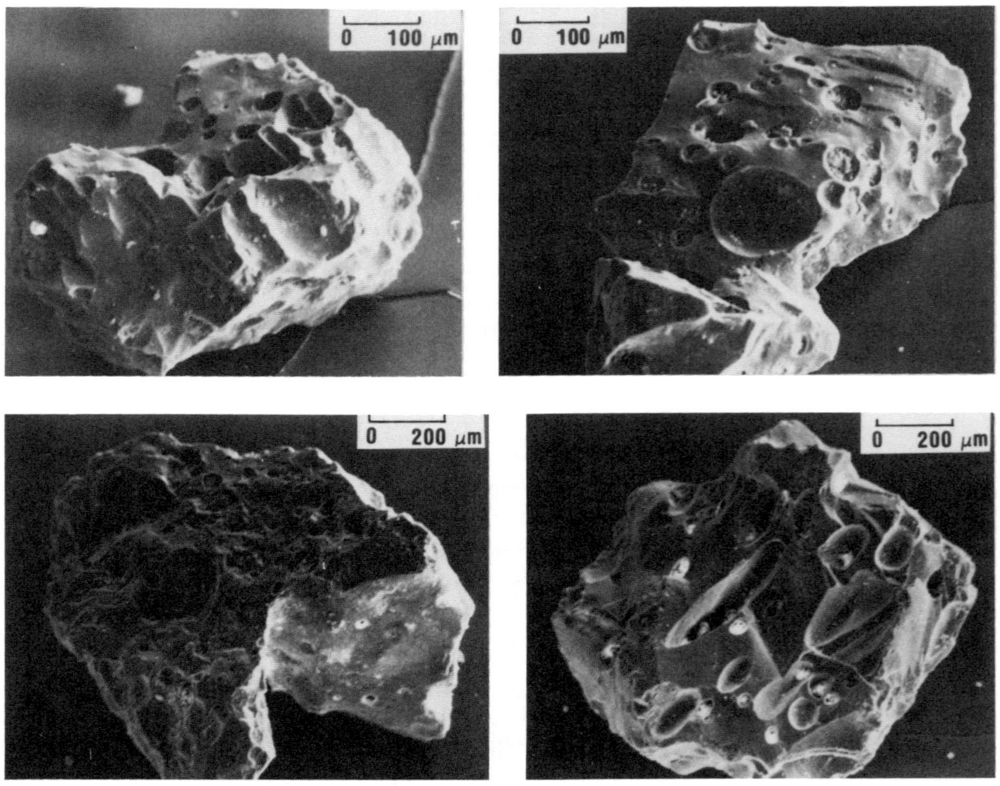

FIGURE 4. SEMs of blocky pyroclasts from surge deposits. Particles are poorly vesicular with vesicles ranging from spherical to ovoid shapes. Edges were chipped during transport, and surfaces are generally fresh.

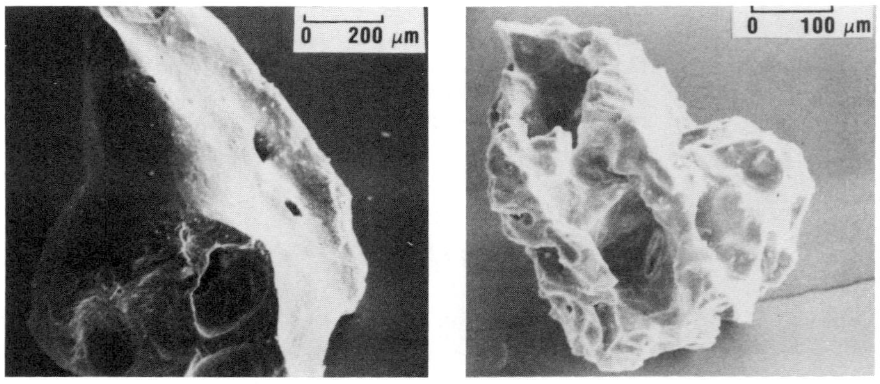

FIGURE 5. SEMs of drop-like and vesicular pyroclasts with fluid-formed surfaces. These shapes are found in deposits from the last stages of Surtseyan eruption. The shapes indicate that the magma was still partly fluid after it was fragmented by contact with water.

FIGURE 6. SEMs of lithic tachylite ash from early eruptions. These particles make up nearly 90% of the 1965 deposit. Phenocrysts produce the textured surface patterns that are apparent on the left and right pictures. *Center*—many older lithic ash pyroclasts contain elongated vesicles formed by plastic deformation during flow of lava within an older vent.

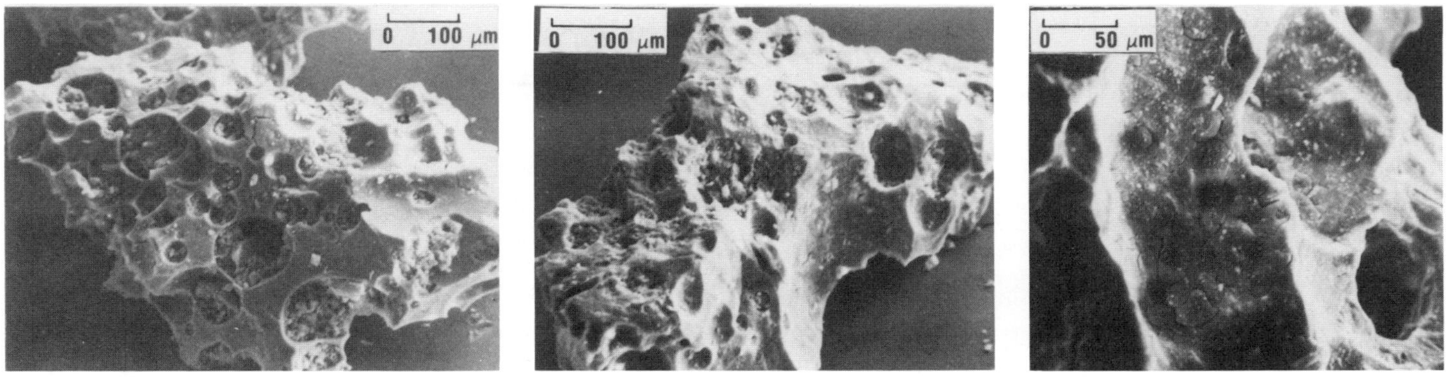

FIGURE 7. SEMs of altered ash from Taal. Clay and zeolite particles fill the vesicles of partly altered ash (*left picture*), and hydrated skin covers the particle surfaces in more strongly altered ash (*center and left pictures*). Note the cracks formed in hydrated skin.

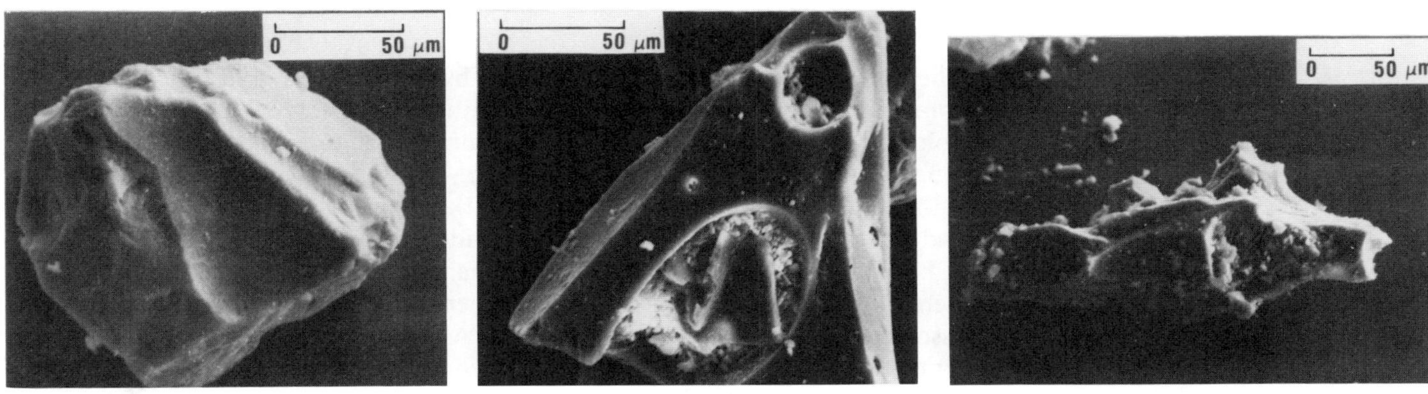

FIGURE 8. SEMs of fine-fraction sideromelane particles. Angular blocky shape is well developed (*left picture*), vesicles are filled with fine adhering dust (*center*), and slight alteration has developed within the adhering dust (*right*).

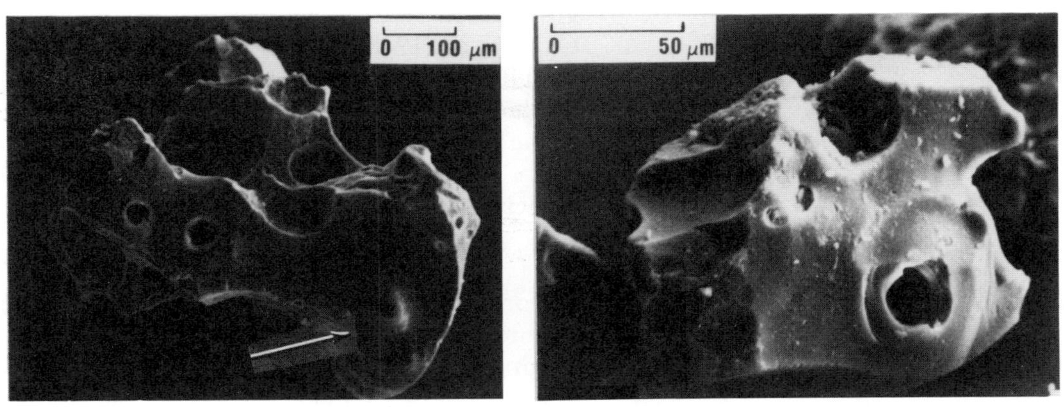

FIGURE 9. SEMs of pyroclasts that show arrested vesiculation. Left view (*arrow*) shows vesicle bulge over a bubble that was frozen as it expanded through the grain surface after fragmentation. Right view shows a vesicle that "popped" after fragmentation; the curved bubble walls are preserved.

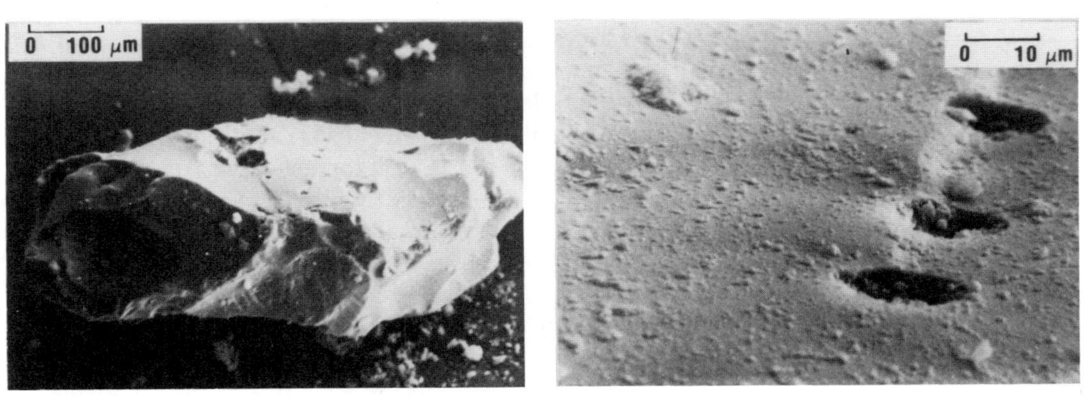

FIGURE 10. SEMs of a line of secondary vesicles formed in a pyroclast. Vesicle growth was arrested by rapid solidification after late-stage nucleation. A closer view (*right*) shows bubbles aligned along a surface irregularity. This texture may have been formed by a propagating stress wave that caused plastic deformation of the glass, and it was preserved as a result of quick solidification.

Prehistoric Eruptions

By studying old, well-dissected volcanoes, we can gain perspective on the many variations in eruption phenomena that may occur during a single period of activity. Deep erosion allows us to examine complete eruption sequences that sometimes are not visible at young vents. Using tephra characteristics that were defined on the basis of samples from observed eruptions, it is possible to reconstruct eruption histories of prehistoric volcanoes. Examples of this approach to tephra studies are presented in this section.

BANDELIER TUFF, JEMEZ MOUNTAINS VOLCANIC FIELD, NEW MEXICO

Introduction: The Bandelier Tuff, originally described by Smith (1938) and Griggs (1964), is composed of two major sequences of rhyolitic tuff that flank the Jemez Mountains volcanic field, New Mexico. The bulk volume of the tuff is about 600 km^3. It was erupted during two cycles, 1.4 and 1.1 m.y. ago and is associated with a major caldera complex. Much of the pioneer work on pyroclastic flow deposits by Ross and Smith (1961) was based upon their studies of the Bandelier Tuff.

Each of the two cycles began with a Plinian eruption of pumice and ash falls that were followed by pyroclastic flow sequences. Pyroclastic surge deposits are associated with the flows (Fisher, 1979). A facies model for ignimbrite-forming eruptions (Wright et al., 1981) was based chiefly on studies of the Bandelier Tuff.

Massive pyroclastic flow sequences, emplaced at temperatures of 600 to 800°C (Smith and Bailey, 1966), exhibit many secondary textures and mineralogical changes caused by welding, compaction, vapor-phase, and fumarolic alteration. For detailed descriptions of these petrologic changes, see Ross and Smith (1961) or Crowe et al. (1978).

Individual pyroclasts within this eruption sequence are often not preserved as a result of vapor-phase alteration. The examples used here are from a distal section of the tuff where there is little or no welding. These samples were collected for a study by Crowe et al. (1978) and are mainly from units where individual pyroclasts or relict textures are still preserved.

Grain Size: Unconsolidated samples from nonwelded zones within the upper member (Tshirege Member) are all poorly or very poorly sorted, with mean grain sizes ranging from 61 to 412 μm. All samples are within the pyroclastic flow field described by Walker (1971).

Tephra Description: The variety of pyroclast types is similar for both upper and lower members of the Bandelier Tuff. Pumice pyroclasts are most common, and there are fewer shards, crystals, and lithic clasts. Sanidine and quartz phenocrysts are "clean" or are encased by a thin, vesicular glass coating. Vesicles are flattened ovoids that are parallel to phenocryst surfaces. These are similar to the bubble-wall texture described by Fisher (1963) and Meyer (1972).

There are at least three varieties of pumice pyroclasts:

(1) Those with thin, elongate vesicles whose aspect ratios are >50:1. These are mostly small, elongate pyroclasts with a "silky" appearance. The vesicles have been flattened (they are ovoid to lens-like in cross-section) and have thin vesicle walls.

(2) Pyroclasts with large ovoid vesicles that range from 100 to 500 μm in length and have low aspect ratios (30:1 to 1:1). The vesicle walls are very thin. The pyroclasts are equant to slightly elongate and are mostly angular.

(3) Dense pumice pyroclasts, characterized by ovoid to spherical vesicles that are 5 to 80 μm long (or in diameter if spherical). These vesicle walls are thick relative to those of the two other pumice types.

Most shards are fragments of comminuted pumice pyroclasts that range from thin, curved plates to Y-shaped forms derived from the mutual intersection of three vesicles. Ash-size lithic fragments include a variety of types but are mostly equant, subrounded lava clasts of intermediate composition; some have been hydrothermally altered.

Chemical Composition: The tephra are rhyolitic with only minor variations in major elements. For variations in trace elements, see Smith (1979) and Smith and Bailey (1966). Analyses of glassy pyroclasts are shown in Table 1.

TABLE 1. ELECTRON MICROPROBE ANALYSES OF GLASS COMPOSITIONS, LOWER (OTOWI) MEMBER, BANDELIER TUFF.*

Sample	Basal Pumice Fall	Lower, Pyroclastic Flow	Center, Pyroclastic Flow	Top, Pyroclastic Flow	Top, Pyroclastic Flow
SiO_2	73.78	74.03	73.59	73.92	74.94
TiO_2	0.10	0.08	0.10	0.11	0.07
Al_2O_3	11.74	11.92	12.05	11.73	12.05
FeO**	1.15	1.06	1.11	1.09	1.14
MgO	0.05	0.04	0.04	0.05	0.03
CaO	0.30	0.29	0.29	0.29	0.25
Na_2O	3.38	2.82	3.23	2.95	3.29
K_2O	4.42	4.68	4.65	4.89	4.70
F	0.1	0.09	0.06	0.04	0.05
Total	95.02	95.01	95.12	95.07	96.52
No. of analyses	6	6	6	5	6

*From Crowe et al., 1978.
**Total iron as FeO.

Comments on Pyroclast Formation: Meyer (1972) used the amount of glass remaining on phenocrysts as an indicator of eruption energy; the presence of fewer phenocrysts with glass "crusts" indicates increased explosivity. Within the lower member of the Bandelier Tuff, this general observation holds. For example, within the base of the Plinian pumice fall, only a few per cent of the crystal pyroclasts have glassy coatings, whereas in the top and throughout the pyroclastic flow, 25 to 30% of all pyroclasts are phenocrysts with glass coatings. With the exception of the lowest tephra fallout unit, the same variation exists in the upper member.

Pumice pyroclasts with thin, elongate vesicles, which indicate considerable laminar flow in the vent of a vesiculating melt before eruption, are abundant (25%) in the lowest tephra fall units of each member and decrease to 10 to 13% throughout overlying deposits. Dense, finely vesicular pyroclasts dominate (60%) only the lowest tephra fall in the lower member and decrease to 5 to 25% in overlying pyroclastic flows. Pumice pyroclasts with large, ovoid vesicles, indicative of lower viscosity and vesiculation and fragmentation near the magma/atmosphere interface, are present in varying proportions in all samples.

REFERENCES:

Crowe, B. M., Linn, G. W., Heiken, G., and Bevier, M. L., 1978. Stratigraphy of the Bandelier Tuff in the Pajarito Plateau. Los Alamos Scientific Laboratory Report LA-7225-MS: 57 pp.

Fisher, R. V., 1963. Bubble-wall texture and its significance. J. Sed. Petrol., **33**: 224-235.

Fisher, R.V., 1979. Model for pyroclastic surges and pyroclastic flows. J. Volcanol. Geotherm Res., **6**: 305-318.

Griggs, R. L., 1964. Geology and ground water resources of the Los Alamos area, New Mexico. U.S. Geol. Surv. Water Supply Paper 1753: 107 pp.

Meyer, J. D., 1972. Glass crust on intratelluric phenocrysts in volcanic ash as a measure of eruptive violence. Bull. Volcanol., **35**: 358-368.

Ross, C. S. and Smith, R. L., 1961. Ash-flow tuffs: their origin, geologic relations and identification. U.S. Geol. Surv. Prof. Paper 366: 81 pp.

Smith, H. T. U., 1938. Tertiary geology of the Abiquiu Quadrangle, New Mexico. J. Geol., **46**: 933-965.

Smith, R. L., 1979. Ash-flow magmatism. Geol. Soc. Am. Spec. Paper 180: 5-27.

Smith, R.L. and Bailey, R.A., 1966. The Bandelier Tuff: a study of ash-flow eruption cycles from zoned magma chambers. Bull. Volcanol., **29**: 83-104.

Walker, G. P. L., 1971. Grain-size characteristics of pyroclastic deposits. J. Geol. **79**: 696-714.

Wright, J. V., Self, S., and Fisher, R. V., 1981. Towards a facies model for ignimbrite-forming eruptions. *In*: S. Self and S. Sparks (Eds.), *Tephra Studies*, Reidel Pub. Co., Dordrecht: 433-439.

FIGURE 1. Pumice pyroclasts from the lower member. Most are angular, relatively dense pumices with small vesicles and thick vesicle walls.

FIGURE 2. Tephra sample from the upper half of the tephra fall bed, lower member. *Center*—a quartz(?) phenocryst coated with vesicular glass; vesicles are broad ovoids flattened parallel to crystal faces. The remaining pyroclasts are composed of a mixture of curved vesicles and fractured mineral fragments.

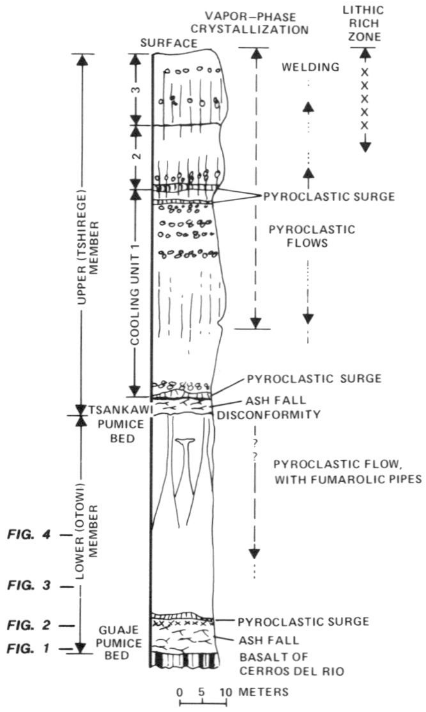

Stratigraphic section of the Bandelier Tuff.

FIGURE 3. Sample from near the base of the pyroclastic flow, lower member. Pumice pyroclasts of all three types described in the text are present. *Center*—a pyroclast with large, flattened ovoid vesicles. *Bottom*—a pumice with straight, elongate, narrow vesicles. All pumice pyroclasts are coated with fine-grained shard fragments.

FIGURE 4. Disaggregated tephra from the middle of the pyroclastic flow, lower member. Equant to slightly elongate phenocrysts (with and without glass coatings) and a variety of pumice pyroclasts are visible. Pumices include those with thin, elongate vesicles (elongate, angular pyroclasts) and dense pumices with small vesicles (*upper right*).

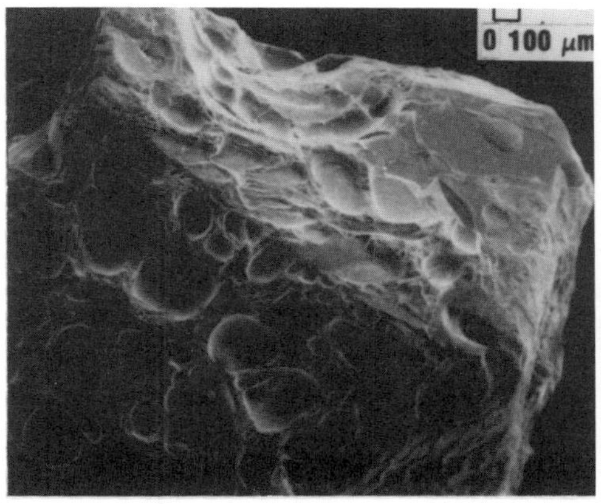

FIGURE 5a. Samples from thin but widespread tephra fall beds at the base of the upper member of the Bandelier Tuff. Large phenocryst coated with a thin layer of vesicular glass. Note that vesicles are flattened parallel to crystal faces.

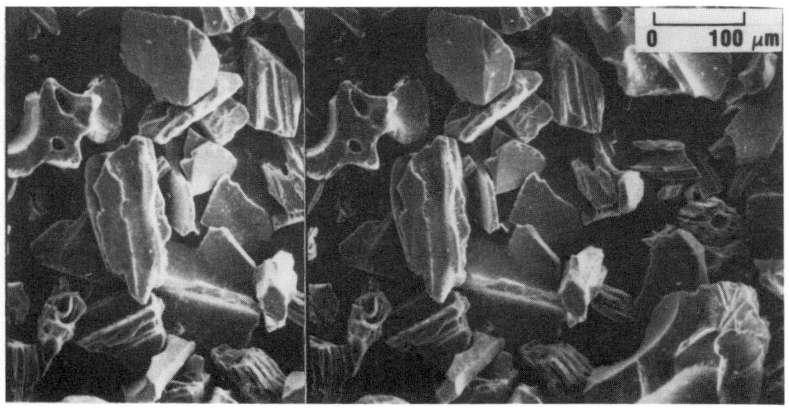

FIGURE 5b. Stereomicrograph of ash from a relatively fine grained tephra fall bed. It is composed chiefly of shards derived from a vesicular melt with well-developed, elongate flattened vesicles. Note the large, curved vesicle walls (now shard surfaces). There are a few small pumice pyroclasts with narrow, tube-shaped vesicles.

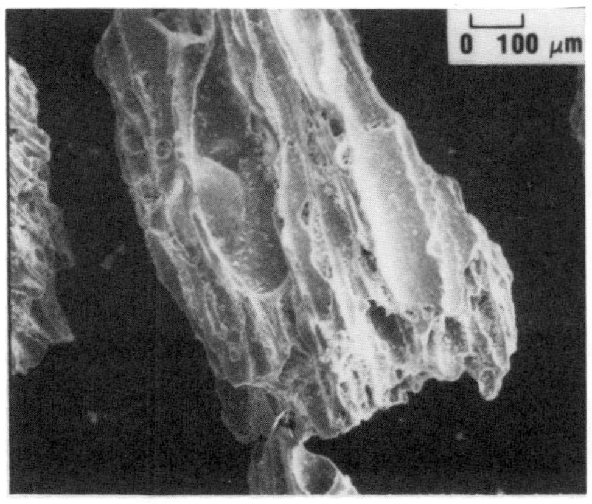

FIGURE 5c. Pumice pyroclast with well-developed, elongate ovoid vesicles; most have been flattened.

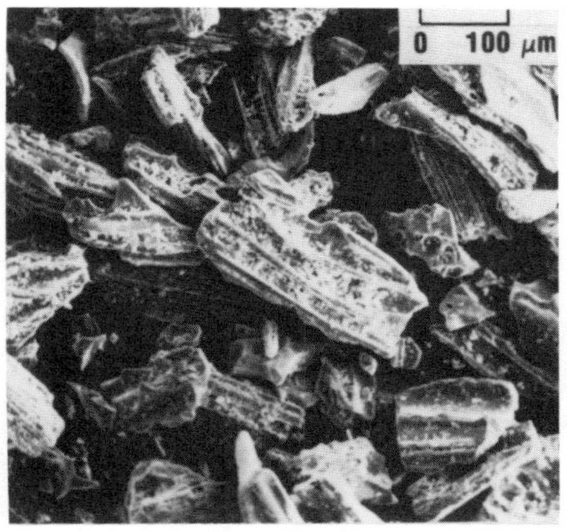

FIGURE 5d. These pyroclasts are mostly small pumices with narrow, elongate parallel vesicles; the shapes of these pumices reflect the nature of the vesicles.

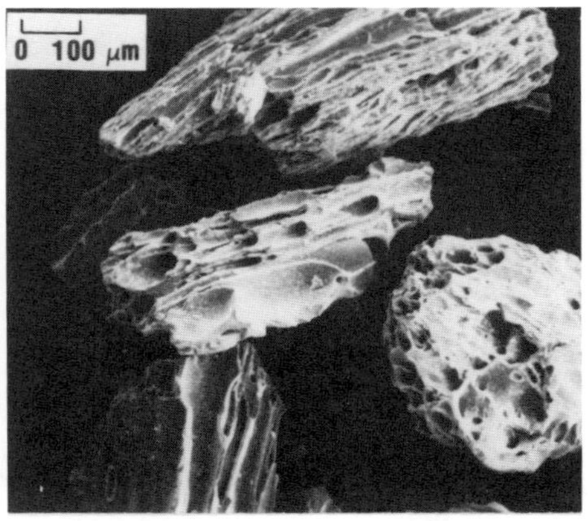

FIGURE 5e. Angular to subangular pumice pyroclasts with mostly large, well-developed ovoid vesicles (*lowest 4 particles*) and thin vesicle walls. The uppermost pumice contains both elongate ovoid vesicles and thin, tube-like vesicles with high aspect ratios.

FIGURE 6. Tephra from a ground surge bed at the base of the lowest pyroclastic flow, upper member. There are mostly equant to slightly elongate shards, each with curved or parallel ridges remaining from adjacent vesicles. There are also a few needle- or wedge-shaped shards.

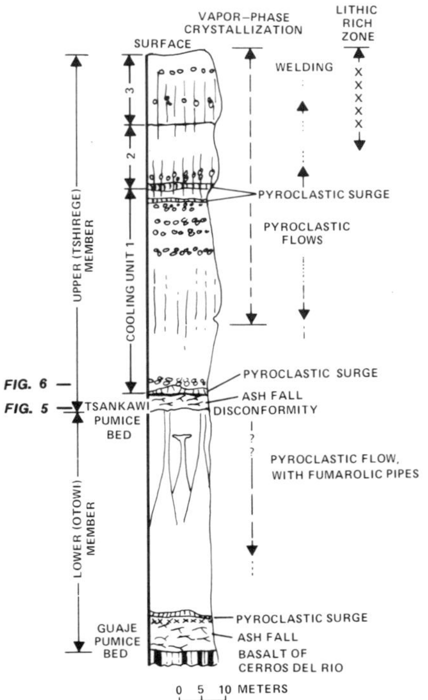

Stratigraphic section of the Bandelier Tuff.

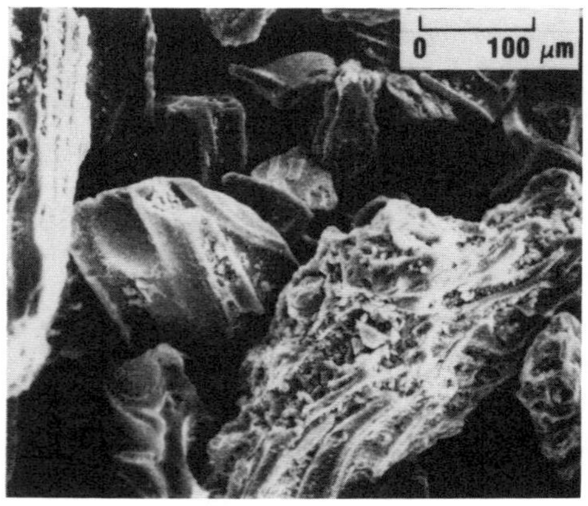

FIGURE 7a. Sample from near the base of the lowest pyroclastic flow, upper member. Mixture of elongate pumice pyroclasts and nearly equant shards. Note the fine-grained shards clinging to pyroclast surfaces—a feature that is typical of pyroclastic flow deposits.

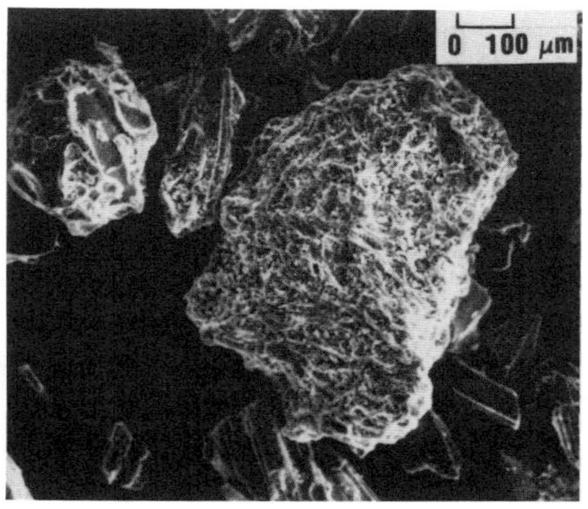

FIGURE 7b. Large, subequant pumice pyroclast (*center*) with finely vesicular, thick-walled vesicles. *Lower left*—quartz phenocrysts coated with vesiculated glass (bubble-wall texture described by Fisher, 1963).

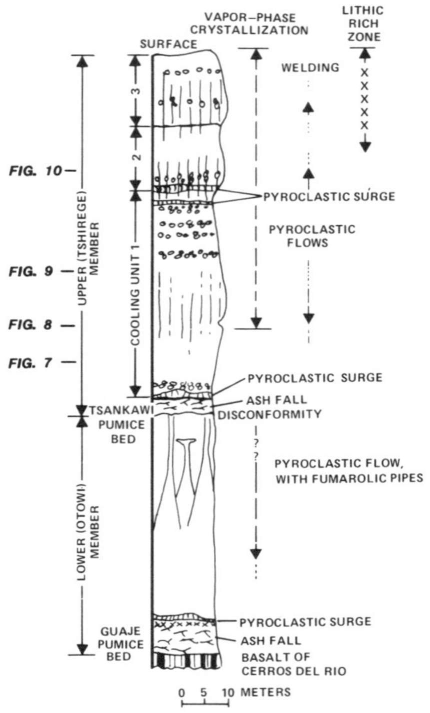

Stratigraphic section of the Bandelier Tuff.

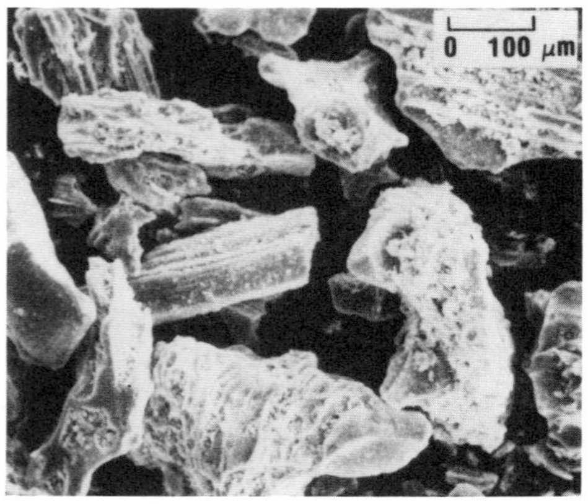

FIGURE 8a. From the lower half of the lowest pyroclastic flow, upper member. Small pumice pyroclasts and shards partly coated with fine-grained shards.

FIGURE 8b. Mostly elongate pumice pyroclasts with thin, tube-like vesicles. All pyroclasts are coated with fine-grained shards and alteration products.

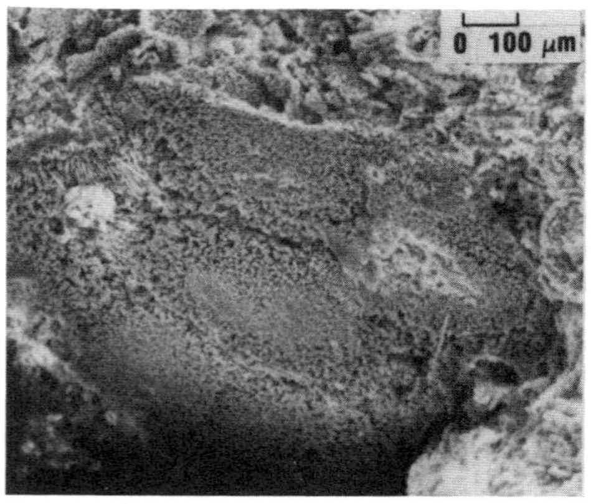

FIGURE 9a. Consolidated, cemented sample from cooling unit 1, upper member. This partially welded tuff is in the vapor-phase zone, where all glass pyroclasts have been replaced by authigenic alkali feldspar and silica. The flat surface is the "mold" of a shard—a relict form or "fossil" pyroclast. Relicts of smaller glass shards are visible to the left of the large relict pyroclast.

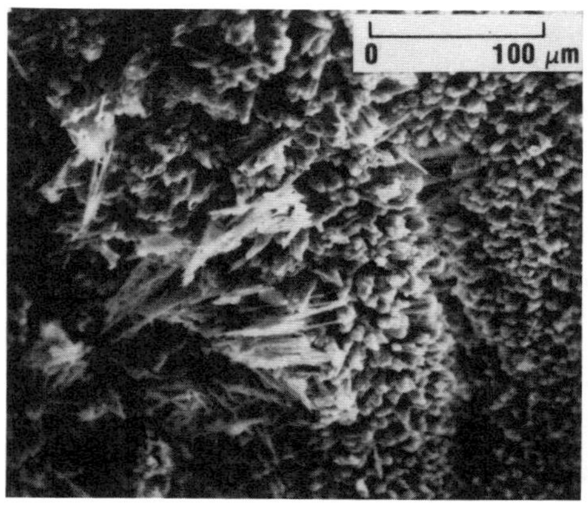

FIGURE 9b. Close-up of the relict shard surface in Fig. *9a*, which shows parallel intergrowths of alkali feldspar and cristobalite.

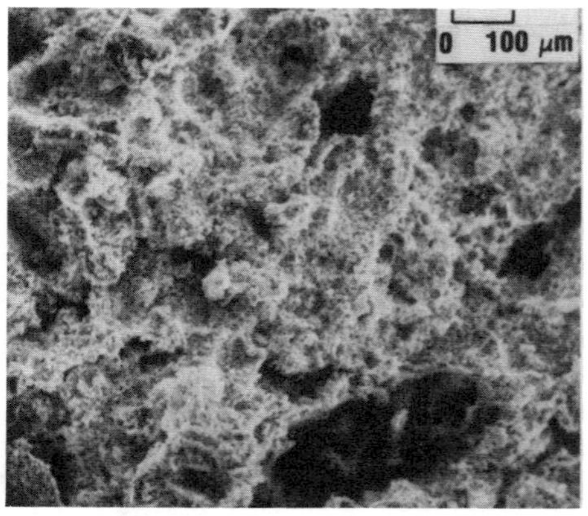

FIGURE 10. Sample from cooling unit 2, upper member. Vapor-phase alteration of the tuff has replaced vitric pyroclasts and filled void space with alkali feldspar, cristobalite, and tridymite. Grain boundaries and relict textures have been completely destroyed in this unit; there is nothing left to indicate the types of pyroclasts involved in this phase of the eruption.

HOPI BUTTES, ARIZONA; VULCANIAN ERUPTION OF DIATREMES

Introduction: The geology of the Hopi Buttes has been described by Williams (1936), Hack (1942), Shoemaker et al. (1962), and Sutton (1974). This area in northeastern Arizona is primarily underlain by the Pliocene Bidahochi Formation, which includes six members that are composed mostly of lake sediments. The fifth member (White Cone Member) is mostly volcanic flows and tuffs up to 100 m thick; it was derived from 200 to 300 vents that erupted within the 600-km^2 area during Pliocene time. Most of the vents are diatremes exposed at various depths by erosion. Vents that show little erosion are maar craters, whereas those that have been eroded are breccia dikes and necks. Some of the maars are capped by lava. Lavas of the Hopi area are of monchiquitic affinity (sodium-amphibole-bearing lamprophyre), and the tuffs have been termed limburgites (extrusive lamprophyre that contains glass). The structure and composition of these diatremes is very similar to those of the Schwabian Alb, Germany (Cloos, 1941).

Most of the Hopi vents erupted along a lake shore—a setting that promoted steam explosions during initial phases of eruption. Williams (1936) noted that the angular and subangular chips of black limburgite and monchiquite in the pyroclastic deposits indicated that lavas were solid or nearly so before their eruption. Hence, the eruptions were a Vulcanian type, erupting both pyroclasts and large volumes of pulverized sandstone and shale. Shoemaker et al. (1962) explained the diatreme formation by magmatic explosions that followed exsolution of magmatic volatiles from a volatile-rich magma. Evidence discussed in this section strongly supports the idea that phreatomagmatic eruptions occurred, during which meteoric water caused fragmentation of the magma.

The pyroclast samples are from two vents, Bidahochi Butte and Coliseum Diatreme.

Grain Size: Both tuff rings studied are composed of well-indurated, bedded pyroclasts that are lapilli-bearing tuffs to fine-grained tuffs. They are moderately sorted except for lens-shaped concentrations of pebbles. Table 1 lists median sizes, sorting, and fine-ash abundances for the studied samples. The samples from Bidahochi Butte are somewhat finer grained and have more abundant fine ash than those of Coliseum Diatreme; however, they are slightly less sorted. This distinction is manifested by the smaller degree of lithification at Bidahochi Butte.

TABLE 1. GRAIN-SIZE CHARACTERISTICS OF HOPI BUTTES TEPHRA.*

Sample	Bed Form*	Md_φ (mm)	σ_φ	Fine Ash (Wt. %)
1	F	1.45 (0.37)	1.72	6.39
2	M	2.03 (0.24)	1.54	11.65
3	S	0.16 (0.89)	1.28	0.34
4	M	0.68 (0.62)	1.56	4.45

*Bed forms: F = fall, M = massive surge, S = sandwave surge.

Tephra Description: Williams (1936) noted that Hopi Butte pyroclasts are mostly angular. There is a marked paucity of vesicular fragments in these samples. Bidahochi Butte ash fall sample 1 consists of abundant aggregated particles. Most of the small particles are prismatic richterite crystals. The fine fraction of this sample contains curved splinters and chips of relatively fresh glass with little fine, adhering dust. In contrast to sample 1, sample 2 is from a massive surge bed and shows abundant flat surfaces of blocky particles. Where present, vesicles are cut by these flat surfaces. The alteration of grain surfaces to clay and zeolite is minimal, and a few aggregates are evident in the sample. The finer grained particles of sample 2 have a mossy appearance, high surface area, and abundant small dust particles clinging to surfaces. No abrasion or edge modification features are observed in this sample.

Sample 3, from Coliseum Diatreme, is from a coarse sandwave surge bed. It consists of very angular grains with broken, flat surfaces and nearly spherical, drop-like particles. Some of these spherical particles have adhered to the surfaces of larger grains. Hydration cracks and crusty surface overgrowths caused by alteration are abundant, and fine particles show abundant tiny crystals adhering to glass surfaces. Sample 4 is from a massive surge bed and is made up of subangular to subrounded grains with smooth surfaces that are partly covered by crusty alteration. Botryoidal surfaces formed from the aggregation of small, drop-like particles. Abrasion features such as v-shaped scratches and chipped edges are common and contribute to grain rounding. The fine fraction contains many angular, blocky particles of glass and crystals.

Chemical Composition: Analyses listed in Table 2 show the differentiation trends of the monchiquite to trachybasalt and syenite-like compositions, as well as the effects of surface alteration after emplacement in wet pyroclastic surges. The limburgite tuffs were derived from a silica-undersaturated melt that differentiated to a more silica-rich composition during extrusion. Abundant crystals of richterite and analcime occur in the tuffs and lavas.

TABLE 2. GRAIN SURFACE CHEMISTRY OF HOPI BUTTES TEPHRA.*

	A	1	2	3	4
SiO_2	47.62	61.03	55.6	66.98	73.57
TiO_2	8.45	6.51	7.21	5.66	4.32
Al_2O_3	8.38	9.78	9.40	7.76	7.58
FeO	11.96	8.15	9.36	7.31	4.33
MnO	1.67	1.50	1.53	2.09	1.53
MgO	6.02	4.05	4.88	2.46	2.39
CaO	13.14	6.02	8.66	4.43	3.00
Na_2O	1.03	0.95	1.03	0.75	0.75
K_2O	1.73	2.02	2.27	2.57	2.52
Glass		79.73	87.01	88.06	
Crystal		15.86	9.45	7.46	
Lithic		4.41	3.54	4.48	

*Normalized standardless EDS analyses. A = analysis of freshest grain from fall sample 2, a trachybasalt similar to that provided by Williams (1936).

Comments on Pyroclast Formation: The absence of vesicles and the ubiquity of blocky pyroclasts with planar surfaces is strong evidence for phreatomagmatic eruptions. Drop-like and spherical particles indicate that some magma erupted while partly fluid. The occurrence of surge deposits (indicative of wet and cohesive tephra and particle aggregates) supports the contention that phreatomagmatic eruptions occurred when magma interacted with ground or lake water. Vulcanian eruptions consist of expulsions of steam and particles derived from previously solidified lava; they are often interspersed with phreatomagmatic bursts of fresh lava (Frazzetta et al., 1983). Furthermore, drop-like particles produced during phreatomagmatic eruptions indicate very fluid lava and strong, turbulent mixing of water with magma before eruption. At Vesuvius, this mixing occurs at depths as great as 5 km (Sheridan et al., 1981) and thus may be responsible for the admixture of lithic fragments from underlying Mesozoic strata.

REFERENCES:

Frazzetta, G., La Volpe, L., and Sheridan, M. F., 1983. Evolution of the Fossa Cone, Vulcano. *In*: M. F. Sheridan and F. Barberi (Eds.), *Explosive Volcanism*, J. Volcanol. Geotherm. Res., **17**: 329-361.

Cloos, H., 1941. Bau and Tätigkeit von Tuffschloten—Untersuchungen an dem schwäbischen Vulkan. Geol. Rundschau, **32(H6-8)**: 709-800.

Hack, J. I., 1942. Sedimentation and volcanism in the Hopi Buttes, Arizona. Geol. Soc. Am. Bull., **53**: 335-372.

Sheridan, M. F., Barberi, F., Rosi, M., and Santacroce, R., 1981. A model for the Plinian eruptions of Vesuvius. Nature, **289**: 282-285.

Shoemaker, E. M., Reach, E. H., and Byers, F. M., Jr., 1962. Diatremes and uranium deposits in the Hopi Buttes, Arizona. *In: Petrologic Studies—A Volume to Honor A. F. Buddington*, Geol. Soc. Am. Mem., Buddington Vol.: 317-355.

Sutton, R. L., 1974. The geology of Hopi Buttes, Arizona. *In:* Guidebook to northern Arizona. Geol. Soc. Amer. Rocky Mt. Section Meeting, 1974.

Williams, H., 1936. Pliocene volcanoes of the Navajo-Hopi country. Geol. Soc. Am. Bull., **47**: 111-172.

FIGURE 1a. SEMs of sample 1, an irregularly shaped aggregate particle of elongate richterite crystals. These crystals were winnowed from erupted ash falls because of their contrasting density with glass particles. They adhered to each other while tumbling around in a steam-rich eruption cloud.

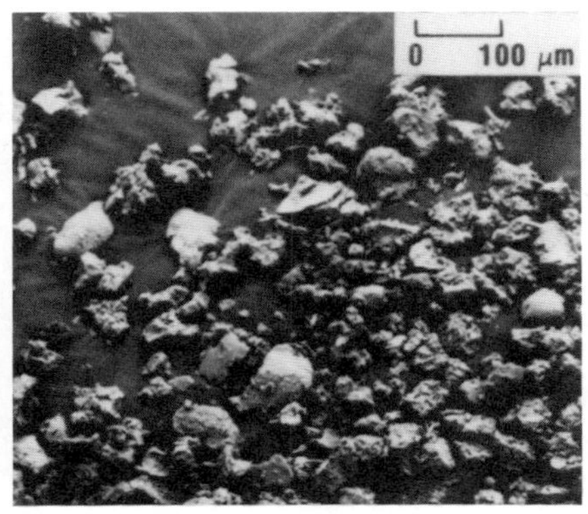

FIGURE 1b. Fine-fraction particles are crescent-shaped and curved splinters of glass with fresh surfaces.

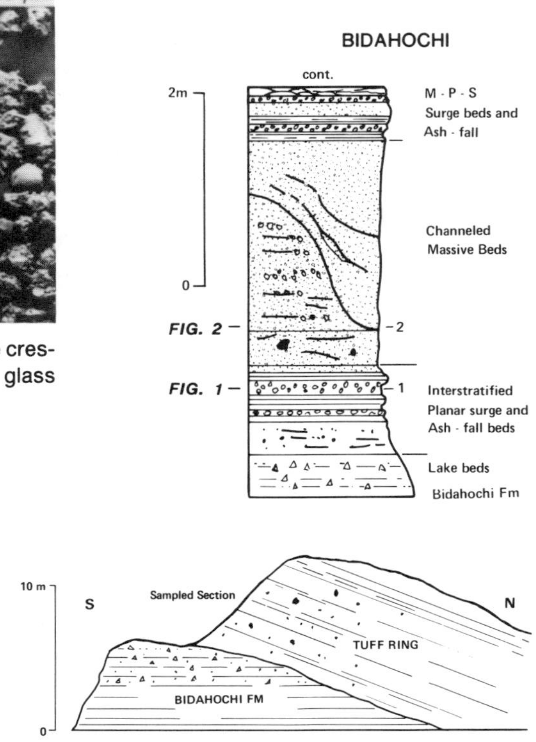

Stratigraphic columnar section and cross section of the S tuff rampart of Bidahochi Butte. The Bidahochi Formation is dominantly flat-lying lake sediments and is overlain by massive, consolidated tuff beds dipping 12 to 15° towards the vent. The tuff consists of ash and lapilli fall beds interlayered with dominantly massive, channeled pyroclastic surge tephra.

FIGURE 2a. SEMs of sample 2, a blocky pyroclast with flat surfaces, patchy altered overgrowths, and angular, chipped corners.

FIGURE 2b. Fine-fraction particles are mossy aggregates of tiny crystals, dust, and angular glass fragments.

FIGURE 3a. SEMs of sample 3, an analcime crystal with surface accumulation of glass droplets that have been altered to clay. Note the crystal faces, abraded chips, and hydrated glass cracks.

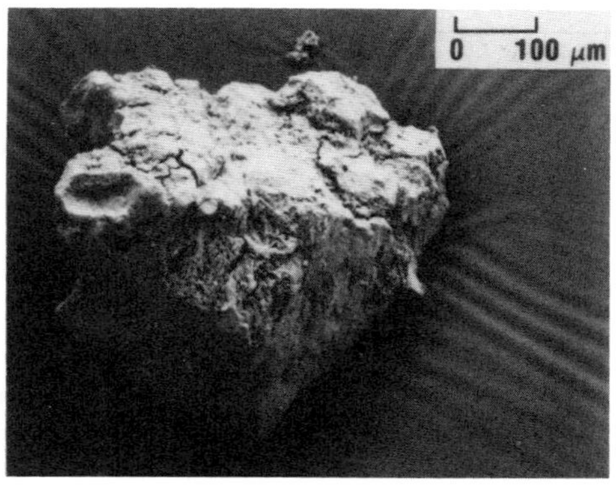

FIGURE 3b. Blocky, altered pyroclast with cracked, altered skin and clay minerals attached to embayed areas.

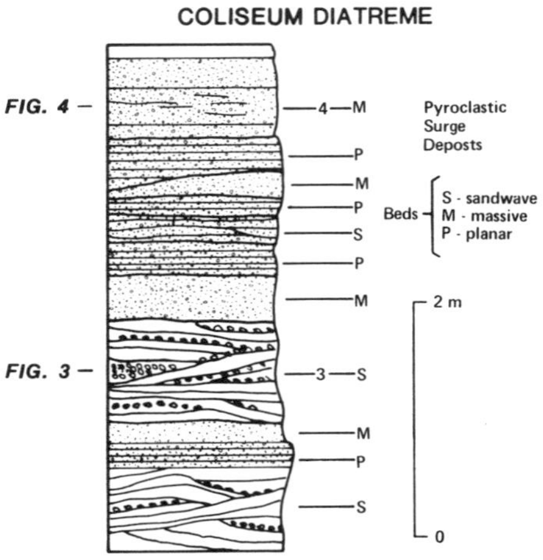

Stratigraphic columnar section of a part of the Coliseum Diatreme tuff rampart. This well-lithified tuff is well bedded and primarily consists of surge beds with interspersed ash fall layers.

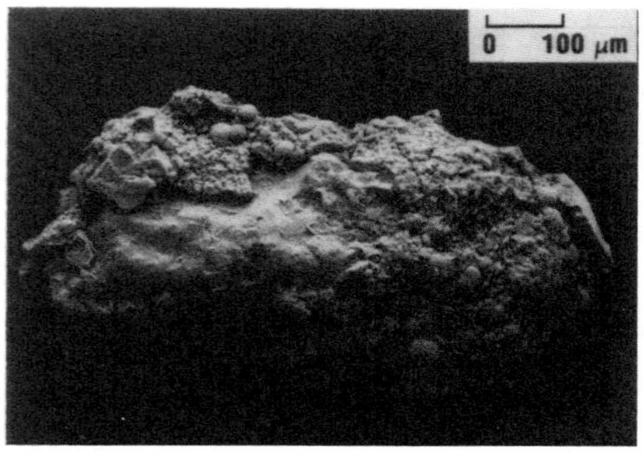

FIGURE 4a. SEMs of sample 4. This glassy, rounded pyroclast shows surface agglutination of drop-like particles and crystalites. The sample is highly altered.

FIGURE 4b. Angular chips and blocks of the fine fraction include numerous, elongate richterite crystals.

BISHOP TUFF, CALIFORNIA; PLINIAN AND PHREATOPLINIAN ERUPTIONS

Introduction: The Bishop Tuff, which crops out over 900 km^2 in eastern California, was first described by Gilbert (1938). This calc-alkaline rhyolitic tuff was erupted from the Long Valley Caldera 0.71 m.y. ago (Bailey et al., 1976), and Bishop ash beds have been found as far away as Nebraska 1800 km to the E (Izett, 1981). The total volume of the Bishop tuff is calculated to be at least 500 km^3, 100 km^3 of which is a pumiceous tephra fall overlain by 150 km^3 of welded ash flow tuff that is exposed as outflow units around the caldera (Bailey et al., 1976). Descriptions of the Bishop Tuff include a paper by Bateman (1965), maps by Crowder and Sheridan (1972) and Huber and Rinehart (1967), and mineralogic and petrologic studies by Sheridan (1965) and Hildreth (1979). The samples and their SEM photos described in this report were donated by M. Sheridan, who collected them from the basal ash fall and lower ash flow sections exposed in the Owen's River Gorge near Tom's Place, California.

Tephra Description: The sample of the Plinian lapilli fall and the two samples of unconsolidated ash flow are representative of early eruptions of the Bishop Tuff. The chemical and petrologic zonation throughout the total thickness of the Bishop Tuff are well documented in articles by Sheridan (1965), Ragan and Sheridan (1972), and Hildreth (1979). These zonation features contribute to ash textures that vary upward in the sequence; however, some textures are difficult to identify because of welding and crystallization of the tuff.

Glass. Pyroclasts from the pyroclastic flow deposit are only poorly vesicular and consist of dominantly blocky and equant grains. Plates and curved splinters from broken pieces of larger bubble walls are subordinate. Hydration cracks are evident on a few grains, and fine dust fills vesicles of pumiceous particles.

Fine-ash samples of the fall deposit are delicate, prong-like shards of vesicle walls with thicknesses of 30 to 50 μm. Vesicle surfaces are spherical to ellipsoidal and have dimensions >50 μm. The fine ash shows clean surfaces with no hydration or adhering fine dust. Coarser grained ash shows elongate and curved vesicles of from 0.1 to >1.0 mm long. Fine dust and alteration products stick to vesicle surfaces. The overall grain shape is blocky to subrounded.

Crystals. Crystalline pyroclasts of the Bishop lapilli fall and ash flow have been described by de Rita et al. (in press). The edges of quartz and feldspar grains from the pyroclastic flow deposit have been rounded by transport abrasion. Numerous small chips, v-shaped pits, and a frosted appearance distinguish these grain edges from those in fall deposits. The dish-shaped concavities and smooth, planar faces of cleavages are well developed.

Ash and lapilli fall crystals consist of nearly equal amounts of quartz and feldspar; they are irregularly shaped grains with delicate edges and have clean surfaces. No abrasion features are apparent, although the original crystal faces are rare. The grain surfaces have small, dish-shaped fractures and are coated with solution and precipitation surfaces. The feldspars show small cleavage surfaces, whereas quartz grains have large, conchoidal fractures.

Comments on Pyroclast Formation: The initial eruptions of the Bishop Tuff produced ash and lapilli fall, surge, and pyroclastic flow deposits. The delicate, prong-like, vesicular shapes of glass and pumice in fall products indicate magmatic eruptions where decompression of a volatile-rich magma caused violent vesiculation and fragmentation. The initial Plinian eruption probably widened a vent to the point that the eruption column collapsed (Wilson et al., 1980), first in pulses that formed pyroclastic surge deposits (Fisher, 1979), and then in a continuous fashion that produced pyroclastic flows. The dominantly blocky, angular glass and pumice fragments of the flow deposit, which are less vesicular than those of fall deposits, most likely attained their shape by large-scale interaction of meteoric water with vesiculating magma in the vent (Self and Sparks, 1978). This phreatoplinian eruption of early Bishop products may have been initiated as the vent widened and allowed water access to the magma. This process is probable because a Pleistocene lake may have occupied the Long Valley Caldera area before eruption of the Bishop Tuff. Later Bishop eruption products dominantly consist of welded tuff, and they make up most of the total erupted volume of the Bishop Tuff. Because these tuffs are compacted and welded, samples were not investigated and their origin by phreatoplinian or Plinian mechanisms is not known.

REFERENCES:

Bailey, R. A., Dalrymple, G. B., and Lanphere, M. A., 1976. Volcanism, structure, and geochronology of Long Valley Caldera, Mono County, California. J. Geophys. Res., **81**: 725-744.

Bateman, P. C., 1965. Geology and tungsten mineralization of the Bishop district, California. U.S. Geol. Sur. Prof. Paper 470: 465-482.

Crowder, D. F. and Sheridan, M. F., 1972. Geologic map of the White Mountain Peak quadrangle, Mono County, California. U.S. Geol. Sur. Map GQ-1012.

Fisher, R. V., 1979. Models for pyroclastic surges. J. Volcanol. Geotherm. Res., **6**: 305-318.

Gilbert, C. M., 1938. Welded tuff in eastern California, Geol. Soc. Am. Bull., **49**: 1829-1862.

Hildreth, W., 1979. The Bishop Tuff: Evidence for the origin of compositional zonation in silicic magma chambers. Geol. Soc. Am. Spec. Paper 180: 43-75.

Huber, N. K. and Rinehart, C. D., 1965. Geologic map of the Devil's Postpile quadrangle, California. U.S. Geol. Surv. Prof. Paper 554-D: D1-D21.

Izett, G. A., 1981. Volcanic ash beds: Recorders of upper Cenozoic silicic pyroclastic volcanism in the western United States. J. Geophys. Res., **86**: 10200-10222.

Ragan, D. M. and Sheridan, M. F., 1972. Compaction of the Bishop Tuff, California. Geol. Soc. Am. Bull., **83**: 95-106.

de Rita, D., Sheridan, M. F., and Marshall, J. R., in press. SEM surface textural analysis of phenocrysts from pyroclastic deposits in the Sabatini volcanic field, Latium, Italy. *In*: W. B. Whalley and D. H. Krinsley (Eds.), *Scanning Electron Microscopy in Geology*, Geoabstracts, Norwich, England.

Self, S. and Sparks, R. S. J., 1978. Characteristics of widespread pyroclastic deposits formed by the interaction of silicic magma and water. Bull. Volcanol., **41**: 196-212.

Sheridan, M. F., 1965. The mineralogy and petrology of the Bishop Tuff, Ph.D. dissertation, Stanford University, Palo Alto, California: 165 pp.

Wilson, L., Sparks, R. S. J., and Walker, G. P. L., 1980. Explosive volcanic eruptions-IV, the control of magma properties and conduit geometry on eruption column behaviour. Geophys. J. Roy. Astron. Soc., **63**: 117-148.

FIGURE 1. SEMs of vitric pyroclasts from the Bishop Tuff pyroclastic flows. Several size fractions are shown (left to right): 250 to 500 μm and 125 to 250 μm (*top*), 63 to 125 μm and <63 μm (*bottom*). Note the mixture of blocky, nonvesiculated glass and pumiceous particles in the coarser size fractions. The fine fraction shows very few vesicle surfaces. Plate-like and angular shards are characteristic of phreatoplinian fine ash.

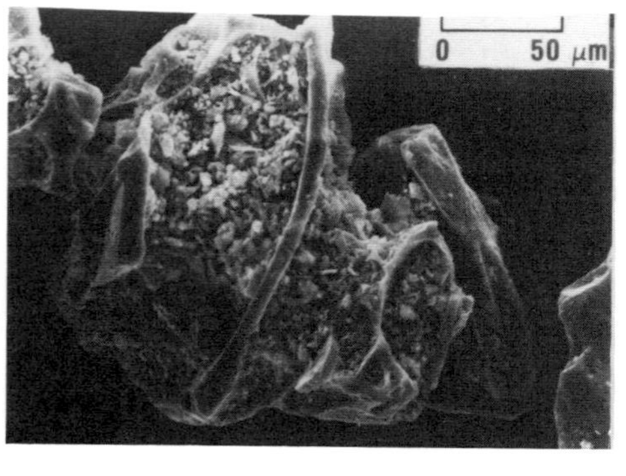

FIGURE 2. SEM of a vesicular ash particle from the flow deposits shows vesicles completely filled with the fine, adhering dust that is characteristic of hydrovolcanism.

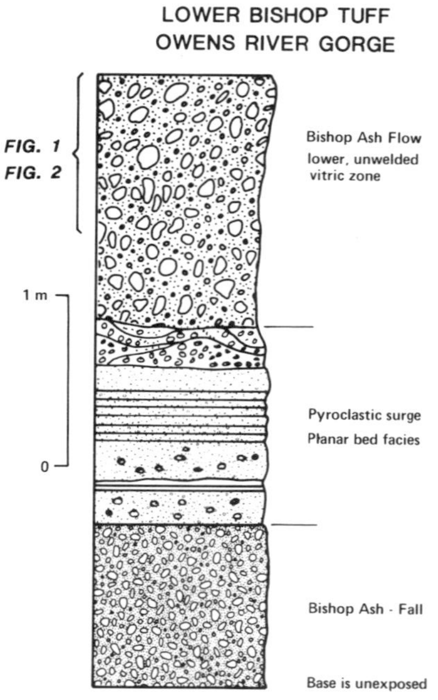

Stratigraphic columnar section of the lower part of the Bishop Tuff near Tom's Place, California. Samples studied were taken from the pyroclastic fall and pyroclastic flow units.

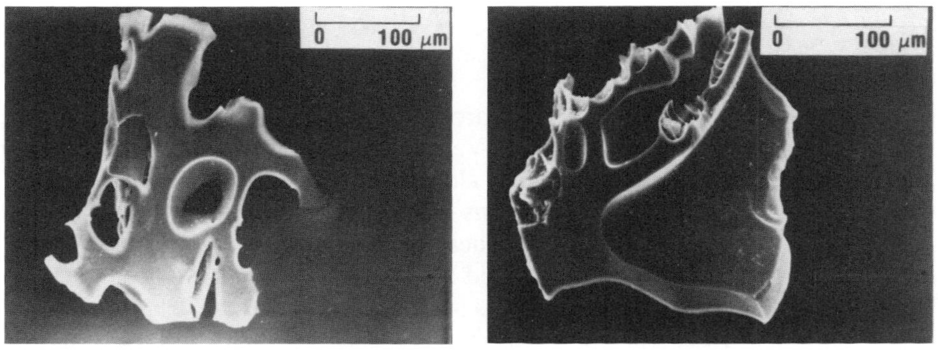

FIGURE 3. SEMs of vesicular, prong-like shards of fine ash in Bishop fall deposits. Note the delicate shapes and sharp edges determined by bubble walls; compare with fine ash in Fig. *1* (*lower right picture*).

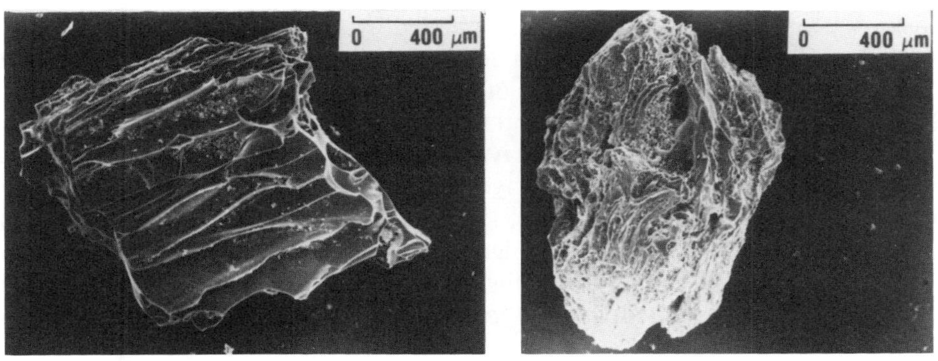

FIGURE 4. SEMs of pumiceous ash particles that dominate the fall deposit materials and are typified by high vesicularity, elongated vesicles, and some fine dust adhering to vesicle surfaces. Note that these pumices are more vesicular than those of the flow products shown in Fig. *1*.

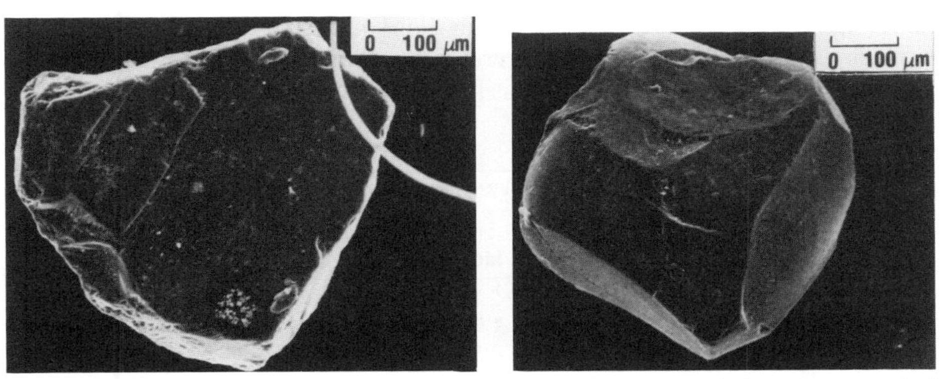

FIGURE 5. SEMs of crystals from the flow deposit show that they are rounded and have abundant abrasion chips and indentations along edges. A feldspar is shown on the *left* and quartz on the *right*. Note that large conchoidal fractures on the quartz have smoothed edges and that the cleavage of the feldspar is well developed.

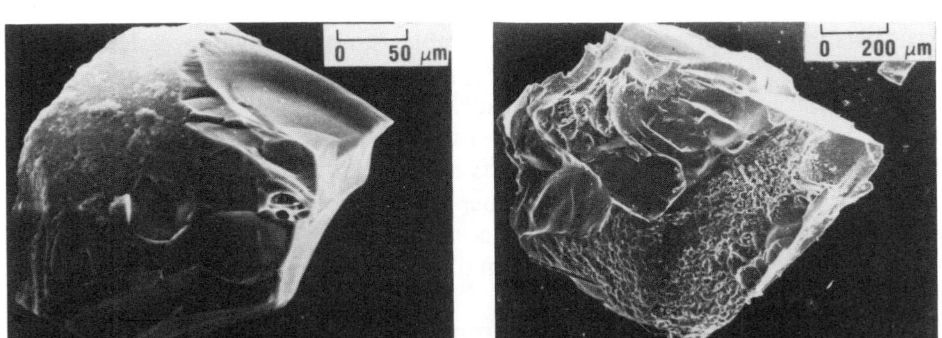

FIGURE 6. SEM of a quartz (*left*) and a feldspar (*right*) crystal from the fall deposit. Note the sharp edges of conchoidal fractures and dish-shaped fractures. Cleavage is only partly developed on the feldspar. Edges of both grains show little modification from percussion.

LOWER BISHOP TUFF
OWENS RIVER GORGE

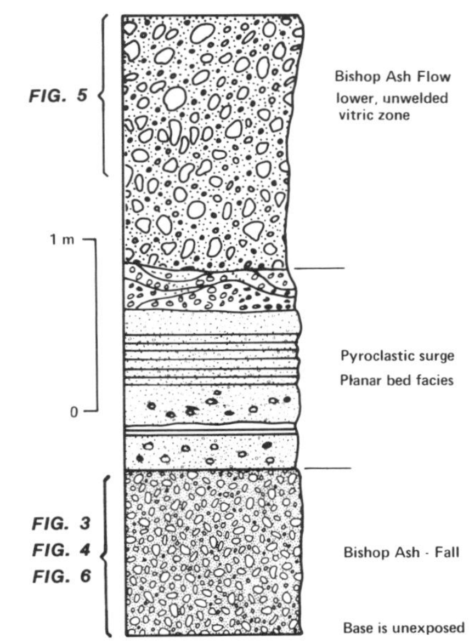

Stratigraphic columnar section of the lower part of the Bishop Tuff near Tom's Place, California. Samples studied were taken from the pyroclastic fall and pyroclastic flow units.

OBSIDIAN DOME, INYO DOMES, CALIFORNIA; SUB-PLINIAN AND PHREATOMAGMATIC ERUPTION

Introduction: Obsidian Dome is one of the five rhyolitic to rhyodacitic lava domes that are the youngest volcanic features of the Long Valley caldera, eastern California (Bailey et al., 1976). These domes are aligned along a N-trending fracture on the NW margin of the caldera, and their eruptions have been bracketed between 720 ± 60 and 1040 ± 250 y.b.p. (Wood, 1977). These vents have been studied by Mayo et al. (1936), and eruptions of the associated Mono Crater chain have been dated by Dalrymple (1967) and Friedman (1968).

The domes show evidence of a polygenetic history. The initial sub-Plinian to phreatomagmatic eruptions were caused by venting of a volatile-rich magma with an increasing hydromagmatic component as ground water gained access to the vent. These eruption phases were followed by passive lava extrusion. Our diagrammatic cross section illustrates a typical crater-rim profile. Initial explosions created the shallow crater surrounded by a tuff ring rampart, and final extrusions filled the crater with a lava dome. Samples described below were taken from the crater rim on the SW side of Obsidian Dome (the north Glass Creek Dome mentioned by Wood, 1977) and the tuff rampart of north Deadman Creek Dome (Wood, 1977). The erupted lavas are composed of a fluidal mixture of pumiceous, coarsely porphyritic rhyodacite and dark, aphyric to sparsely porphyritic rhyolitic obsidian (Lajoie, 1968; Jack and Carmichael, 1968). There is evidence that suggests mixing of Long Valley magmas with those of the Mono Craters. An increasing proportion of the rhyolitic obsidian in the lavas of the Inyo Domes from S to N is noted, and the rhyolite resembles that of rhyolite of Mono Craters.

Grain Size: Grain-size analyses for three samples taken from the Obsidian Dome tuff ring are shown in Table 1. Note the decrease in median diameter from the sample of early sub-Plinian fall deposit (sample 1) through those of the later phreatoplinian surge eruptions (samples 2 and 3). This decrease in grain size is accompanied by an increase in fine ash, which is evident as fine, adhering dust on larger pyroclasts.

TABLE 1. GRAIN-SIZE ANALYSIS OF OBSIDIAN DOME TEPHRA.*

Sample	Bed Form*	Md_φ (mm)	σ_φ	Fine Ash (Wt. %)
1	F	−2.11 (4.32)	1.74	1.77
2	P	1.68 (0.31)	1.52	7.80
3	M	2.10 (0.23)	1.75	5.68

*Bed forms: F = fall, P = planar surge, and M = massive surge.

Tephra Description: Sample 1 is a pumiceous fall deposit of highly vesicular rhyodacitic glass fragments. Pyroclasts have a jagged outline that consists of delicate, sharp points. Pyroclasts from a planar surge deposit (sample 2) have elongated vesicles. Grain shapes are not determined by vesicle surfaces but are blocky with curviplanar fracture surfaces. Alteration of vesicle surfaces is moderate and vesicle walls are broken. Sample 3 is from a rhyolitic vesiculated tuff that was deposited as a massive surge bed. Individual pyroclasts are weakly vesicular and are blocky in shape. Surfaces were smoothed by transport abrasion and are strongly altered; each has a hydration rind. The fine fraction of sample 3 is composed of broken bubble walls, chips, and flakes. These small pieces are covered with adhering dust particles 1 to 10 μm in diameter.

Dominantly rhyodacitic pyroclasts from north Deadman Creek Dome show both blocky and vesicular shapes. Vesicles are elongated in the mixed sample (as described for tephra 1; Wood, 1977) and are characteristic of rhyodacitic clasts formed during the sub-Plinian eruptions. In contrast to the vesicular rhyodacite, obsidian pyroclasts are nonvesicular (for example, those from Devil's Punch Bowl crater on the S end of the Mono Crater chain near its intersection with the Inyo chain).

Chemical Composition: Representative EDS chemical analyses are shown for Obsidian Dome samples in Table 2. The change in composition going up section from sample 1 to sample 3 is reflected as a increase in SiO_2 from 65 to 70 wt.%. There are two explanations for this change: (1) a progressive increase in the rhyolitic component of the glass during the eruption (glass analyses of rhyodacite from south Inyo Dome show SiO_2 at an average 62 wt.%, and analyses of Devil's Punch Bowl rhyolitic obsidian show SiO_2 at 76 wt.%); or (2) progressive pyroclast alteration caused by phreatomagmatic eruptions in which vaporized water condensed on pyroclasts and hydrated their surfaces. Both explanations are supported by field and microscopic evidence. Strong surface alteration is evident in sample 3, as observed by SEM, and its origin as a wet surge deposit suggests that surface hydration contributed to the observed chemical composition. However, the decrease in crystal content in sample 3 suggests an increasing rhyolitic component in the tephra during eruption.

TABLE 2. SURFACE CHEMICAL COMPOSITION, OBSIDIAN DOME.*

	1 (Wt. %)	2 (Wt. %)	3 (Wt. %)
SiO_2	65.56	68.58	70.02
TiO_2	2.33	0.57	0.66
Al_2O_3	14.44	17.68	16.34
FeO	3.15	2.32	1.54
MnO	1.85	0.02	---
MgO	1.60	0.37	0.39
CaO	3.03	2.25	2.30
Na_2O	1.37	0.91	0.94
K_2O	6.68	7.29	7.72
Glass	74.0	83.0	88.0
Crystal	19.0	15.0	9.0
Lithic	7.0	2.0	3.0

*Normalized standardless EDS analyses of grain surfaces.

Comments on Pyroclast Formation: Phreatomagmatic and phreatoplinian eruptions formed tuff rings around the Inyo Domes. Magmatic volatiles were a strong contributing factor in earlier eruptions. The volatile-rich top of the magma column decompressed as it vented, and that action initiated strong vesiculation and fragmentation of the melt. These early eruptions deposited mostly pumice falls. During this phase of sub-Plinian activity, rapid bubble growth was likely aided by propagation of expansion waves from the surface downward into the vent. These stress waves could have both shattered the viscous melt and promoted the formation of vesicular pyroclasts. As volatiles in the magma were depleted, there was fragmentation during interaction with ground water. Abundant water gained access to the vent when the local aquifer was disrupted by the initial magmatic bursts. Stress waves resulting from rapid water vaporization exceeded the strength of the melt as they propagated through vent. The blocky, nonvesicular pyroclasts formed were ejected from the vent as pyroclastic surges. In view of these ash formation mechanisms (Wohletz, 1983), sample 1 is probably magmatic fragmentation, sample 2 indicates a transitional phase of phreatomagmatic activity, and sample 3 is wholly phreatomagmatic. Variations in volatile content caused by mixing of porphyritic, volatile-rich, rhyodacitic and aphyric, dry rhyolite caused fluctuations in the degree of magmatic fragmentation. Magma mixing has been suggested as a trigger for silicic explosive eruptions (Sparks et al., 1977). The lack of crystals in the rhyolitic magma suggests that it was most likely hotter than the rhyodacite. If the rhyodacite was saturated with volatiles, an increase in temperature resulting from mixing with the hotter rhyolite would have caused supersaturation. Rapid dissolution of water from supersaturation then initiated vesiculation and magmatic eruption.

REFERENCES:

Bailey, R. A., Dalrymple, G. B., and Lanphere, M. A., 1976. Volcanism, structure, and geochronology of the Long Valley caldera, Mono County, California. J. Geophys. Res., **81**: 725-744.

Dalrymple, G. B., 1967. Potassium-argon ages of recent rhyolites of the Mono and Inyo Craters, California. Earth Planet. Sci. Letters, **3**: 289-298.

Friedman, I., 1968. Hydration rind dates rhyolite flows. Science, **159**: 878-880.

Jack, R. N. and Carmichael, I. S. E., 1968. The chemical 'fingerprinting' of acid volcanic rocks. Calif. Div. Mines Geol. Spec. Report 100: 17-32.

Lajoie, K. R., 1968. Late Quaternary stratigraphy and geologic history of Mono Basin. Ph.D. dissertation, Univ. of Calif., Berkeley: 271 pp.

Mayo, E. B., Conant, L. C., and Chelikowsky, J. R., 1936. Southern extension of the Mono Craters, California. Am. J. Sci., **32**: 81-97.

Sparks, R. S. J., Sigurdsson, H., and Wilson, L., 1977. Magma mixing: a mechanism for triggering acid explosive eruptions. Nature, **267**: 315-318.

Wohletz, K. H., 1983. Mechanisms of hydrovolcanic pyroclast formation: size, scanning electron microscopy, and experiments. J. Volcanol. Geotherm. Res., **17**: 31-64.

Wood, S. H., 1977. Distribution, correlation, and radiocarbon dating of late Holocene tephra, Mono and Inyo Craters, eastern California. Geol. Soc. Am. Bull., **88**: 89-95.

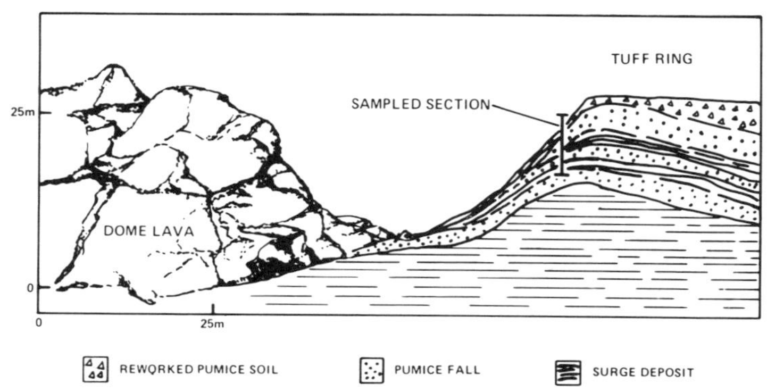

FIGURE 1. Schematic cross section of the rim section of a lava dome surrounded by a tuff ring. The tuff ring forms a collar of pyroclastic material that is often draped over an earlier crater rim. The section sampled at Obsidian Dome is located on the inner slope of a tuff ring that forms a moat around the dome.

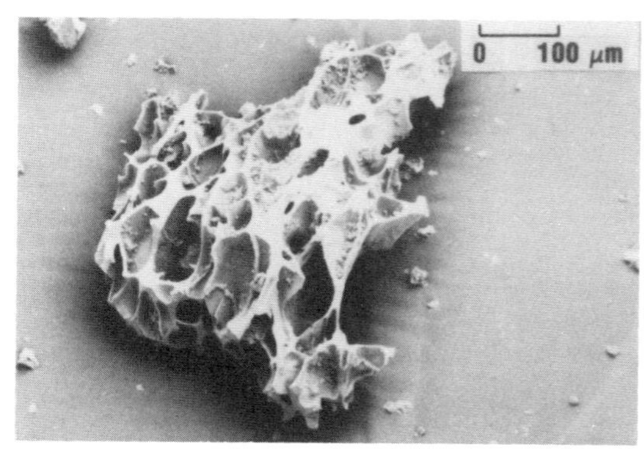

FIGURE 2. SEM of a rhyolite pyroclast from sample 1 that shows the strongly vesicular, delicate shape.

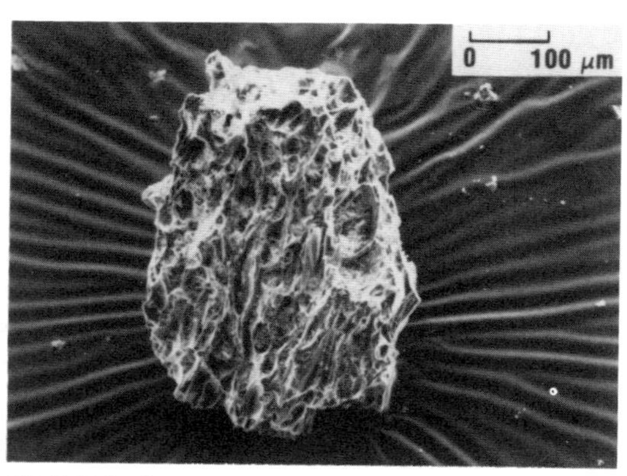

FIGURE 3. SEM of a blocky, vesicular pyroclast from sample 2. Note the curviplanar outline and vesicle fillings.

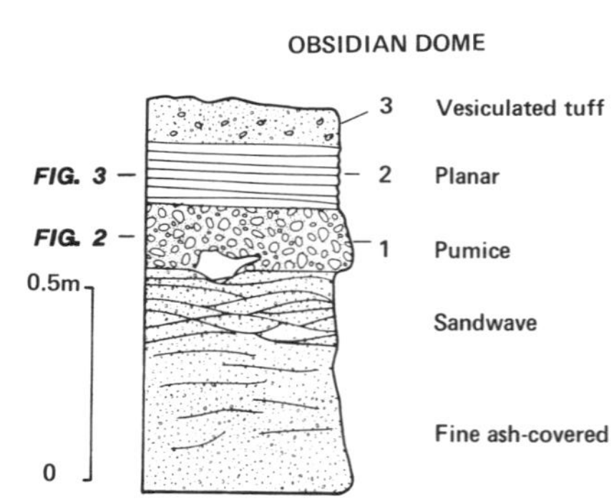

Stratigraphic section of the tuff ring at Obsidian Dome. Sample 1 is from a sub-Plinian pumice fall deposit, and samples 2 and 3 are from phreatoplinian deposits composed of planar and vesiculated massive pyroclastic surge beds.

FIGURE 4. SEM of a blocky pyroclast from sample 3 that shows rounded abraded edges and altered surface.

FIGURE 5. SEM of the fine fraction from sample 3. Note the vesicle chips and blocky particles covered with adhering dust.

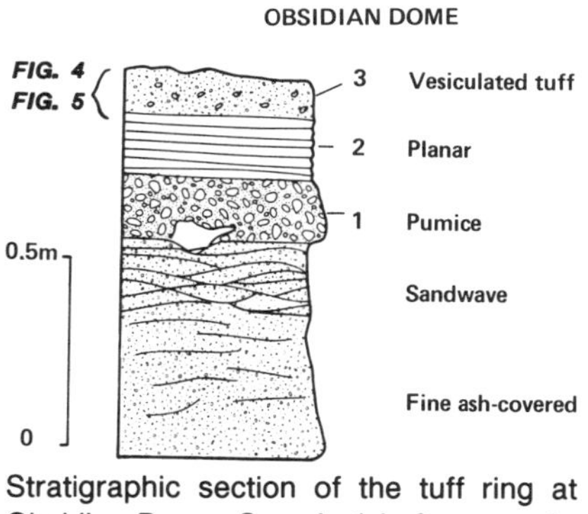

Stratigraphic section of the tuff ring at Obsidian Dome. Sample 1 is from a sub-Plinian pumice fall deposit, and samples 2 and 3 are from phreatoplinian deposits composed of planar and vesiculated massive pyroclastic surge beds.

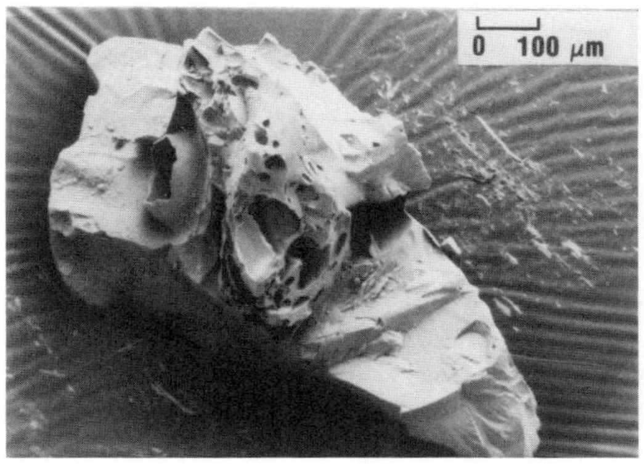

FIGURE 6. SEM of a pyroclast from north Deadman Creek Dome that shows both blocky and vesicular surfaces.

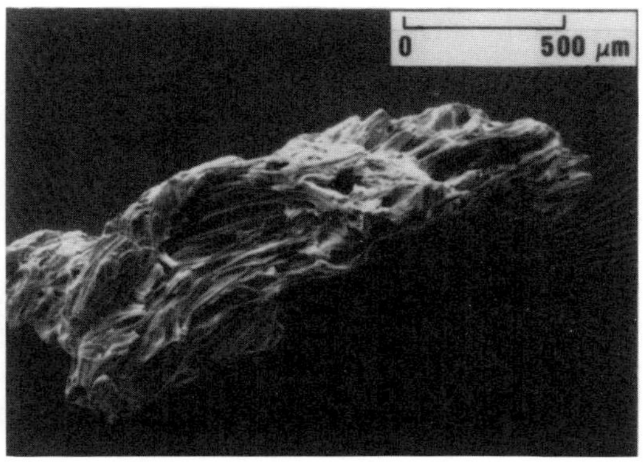

FIGURE 7. SEM of a blocky pyroclast with elongated vesicles that formed during fluid mixing of vesicular and nonvesicular lavas at north Deadman Creek Dome.

FIGURE 8. SEM of the fine fraction from north Deadman Creek Dome; it shows dominantly broken bubble-wall chips and splinters.

FIGURE 9. SEM of the fine fraction of Devil's Punch Bowl ash. Note the blocky, nonvesicular nature of this rhyolitic ash.

CERRO COLORADO, SONORA, MEXICO; WET PHREATOMAGMATIC ERUPTION

Introduction: Cerro Colorado is a tuff cone located in the Pinacate volcanic field near Crater Elegante, Sonora, Mexico. It has been studied by Jahns (1959); its stratigraphy is discussed by Gutmann and Sheridan (1978) and Wohletz and Sheridan (1983); and preliminary SEM analyses of tephra are shown by de Rita et al. (in press). Eruptions that built the steep-sided cone have been tentatively dated at 10,000 years (Shakel and Harris, 1972). The tuff cone deposits consist of highly palagonitized and indurated pyroclastic surge and fall beds that contain abundant accretionary lapilli; these features are indicative of wet eruptions in which steam condensed on particle surfaces and resulted in wet deposits. The cone encircles a crater nearly 1 km in diameter, and it rises over 100 m above the crater floor. The rim stratigraphy consists of a 10-m-thick basal unit of thinly bedded, slightly palagonitized tephra that overlies lake beds. This basal unit is, in turn, overlain by massive, thickly bedded, highly palagonitized tuff that makes up the bulk of the cone.

Grain Size: Tephra grain-size characteristics are given in Table 1 and show a gradual fining of median grain diameter upward in the section. The tephra are moderately sorted, and accretionary lapilli account for the lack of abundant fine ash, which is noted especially in samples of fall and planar bedded ash.

TABLE 1. GRAIN-SIZE CHARACTERISTICS OF CERRO COLORADO TEPHRA.

Sample	Bed Form*	Md_φ (mm)	σ_φ	Fine Ash (Wt. %)
1	S	1.40 (0.38)	1.64	7.38
2	M	1.96 (0.26)	1.80	10.53
3	F	0.38 (0.77)	1.64	1.07
4	M	1.39 (0.38)	1.74	8.91
5	P	0.44 (0.74)	1.52	0.91

*Bed forms: S = sandwave, M = massive, P = planar, and F = fall.

Tephra Description: Table 2 lists textural features of the samples described below. This tephra sequence is characterized by ash that has been extensively hydrated as a result of palagonitization during and after emplacement. Subangular to subrounded grains in the surge deposits show edge modification that resulted from abrasion of the original curviplanar, blocky pyroclast shapes. The sample of lapilli fall tuff shows few chipped and abraded edges. All samples except sample 1 show surface alteration coatings and desiccation cracks on hydrated skins and palagonitic surface material. Sample 1 from the basal, thinly bedded, poorly consolidated, dry surge deposit shows pyroclasts that are relatively fresh.

TABLE 2. DESCRIPTIONS OF TEPHRA FROM CERRO COLORADO.

Sample	Bed Form*	Grain	Edges	Alteration	Fine Fraction (Wt. %)	Distinguishing Features
1	S	blocky, drop-like	subrounded, smoothed, chipped	crusty hydration rind	blocky with abundant adhering dust (7%)	adhering dust on coarse grains
2	M	accretionary, blocky	rounded, dish-shaped fractures	high; granular material covering surface	blocky cores with accreted fine dust (11%)	glass enclosing crystals, accretionary lapilli
3	F	vesicular	subangular, few chipped surfaces	a few vesicle fillings, fresh surfaces	angular plates and blocks (1%)	solution pits
4	M	blocky vesicular	subrounded, grooves and chips	moderate to high; granular encrustation	angular, blocky, and drop-like (9%)	crystal molds in glass
5	P	blocky	subrounded, grooves, dish-shaped fractures	moderate; hydration cracks and skin, fresh areas	plate-like (1%)	glass enclosing crystals

*Bed forms: S = sandwave, M = massive, P = planar, and F = fall.

Chemical Composition: The EDS analyses of pyroclast surfaces are shown in Table 3. Sample A is an analysis of a fresh, basaltic fall deposit from nearby Crater Elegante for comparison. Note that Colorado samples all show greater SiO_2 values and smaller values for most of the other oxides. These differences arise from extensive palagonitization that affects most pyroclast surfaces—even those in the relatively fresh example, sample 1. Ash constituent analyses of glass, crystal, and lithic abundances show little variation throughout the eruption sequence, with the exception of the fall deposit, sample 3, in which crystals are separated from glass and lithics (possibly during emplacement).

TABLE 3. GRAIN SURFACE CHEMISTRY OF CERRO COLORADO TEPHRA.*

	A	1	2	3	4	5
SiO_2	50.45	69.67	62.55	69.55	69.88	71.03
TiO_2	4.26	2.97	3.56	2.63	2.68	2.43
Al_2O_3	16.85	10.78	13.54	11.58	12.48	11.58
FeO	6.70	4.33	6.57	8.17	5.21	5.94
MnO	2.21	1.96	1.66	---	0.83	---
MgO	3.84	2.49	2.99	1.67	1.63	1.62
CaO	8.31	3.64	4.92	5.45	5.24	5.22
Na_2O	2.36	1.14	1.28	---	0.88	0.62
K_2O	2.84	3.02	2.78	1.95	1.68	1.58
Glass	86.00	87.00	86.00	92.50	88.00	82.00
Crystal	11.00	12.00	14.00	4.50	10.00	17.00
Lithic	3.50	1.00	1.00	2.00	1.50	1.00

*Normalized standardless EDS analyses. Sample A is a fresh pyroclast from nearby Crater Elegante.

Comments on Pyroclast Formation: The following discussion illustrates probable eruption and emplacement variations that occurred throughout the course of activity and that resulted in the observed sequence of textural features. A more detailed version of this interpretation is available in Wohletz and Sheridan (1983).

Opening eruptions were high-energy explosions that produced fine-grained tephra and dry steam. These eruptions produced pyroclastic surges in which steam separated from pyroclasts before their emplacement. Hence, particles were deposited with relatively dry surfaces. The generally slight surface alteration of these pyroclasts contrasts with that of pyroclasts emplaced in wet deposits and is evidence that hydration of pyroclast surfaces occurs during and shortly after emplacement because of the wetness of the deposit. Blocky pyroclasts were smoothed by abrasion during emplacement. These pyroclasts are covered with fine adhering dust, which also attests to high-energy explosions that were capable of producing abundant fine-grained particles.

Later eruptions were wetter; they produced grains that stuck together because of the cohesion of wet surfaces. The strong alteration of these particles is possibly related to the presence of steam droplets that condensed on pyroclast surfaces before and during emplacement. Glass that clings to the crystals likely indicates a less energetic disruption of the magma than that which produced the glass-free crystal surfaces common in earlier eruptions.

Phreatomagmatic activity was punctuated by explosive pulses in which there was little magma/water interaction. Sample 3 is from an eruption caused by vesiculation. Grains in the fine fraction of this sample show little hydration, which suggests a relatively dry environment after emplacement. Those of the coarse fraction, however, have cracked, hydrated skins, which indicates that some water was probably present in the deposit after emplacement.

Samples 4 and 5 are from planar and massive bedded deposits emplaced by surges during wet eruptions; the massive surge deposits are characteristic of tuff cones. Sample 4 originated from a more energetic eruption than that which produced sample 5; it is finer grained, and phenocryst surfaces are completely separated from glass coatings. Other evidence for a more energetic eruption is seen in the blocky and drop-like fine fraction of sample 4, which indicates explosive mixing on a fine scale (see section in this book on the experimental production of ash). Sample 5 fine fraction pyroclasts formed more as a result of larger pyroclasts breaking than by the energetic mixing with water. Larger clasts in these samples often form the nuclei of accretionary lapilli—another feature that typically indicates wet emplacement of surges at tuff cones.

REFERENCES:

Gutmann, J. T. and Sheridan, M. F., 1978. Geology of the Pinacate volcanic field: State of Arizona Bureau of Geology and Mineral Technology, Special Paper No. 2: 47-59.

Jahns, R. H., 1959. Collapse depressions of the Pinacate volcanic field, Sonora, Mexico. Arizona Geol. Soc. Southern Arizona Guidebook II: 165-184.

de Rita, D., Sheridan, M. F., and Marshall, J., in press. SEM surface textural analysis of phenocrysts from pyroclastic deposits at Baccano and Sacrofano volcanoes, Latium, Italy. *In*: W. B. Whalley and D. H. Krinsley (Eds.), *Scanning Electron Microscopy in Geology*, Geo Abstracts, Inc., Norwich, England.

Shakel, D. W. and Harris, K. M., 1972. Potsherd evidence for minimum age of Cerro Colorado Crater, Pinacate volcanic field, northwestern Sonora, Mexico, Geol. Soc. Am. Abs. with Programs **4**: 408.

Wohletz, K. H. and Sheridan, M. F., 1983. Hydrovolcanic explosions II: evolution of basaltic tuff rings and tuff cones. Am. J. Sci., **283**: 385-413.

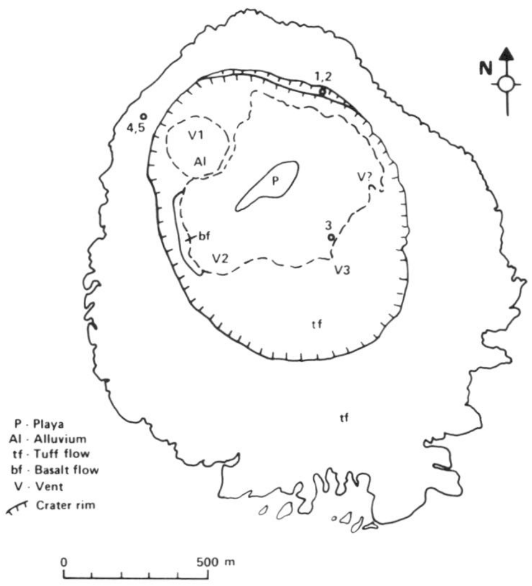

FIGURE 1. Geologic sketch map of Cerro Colorado with map units shown by key (adapted from Jahns, 1959). Samples were taken from locations 2 and 3.

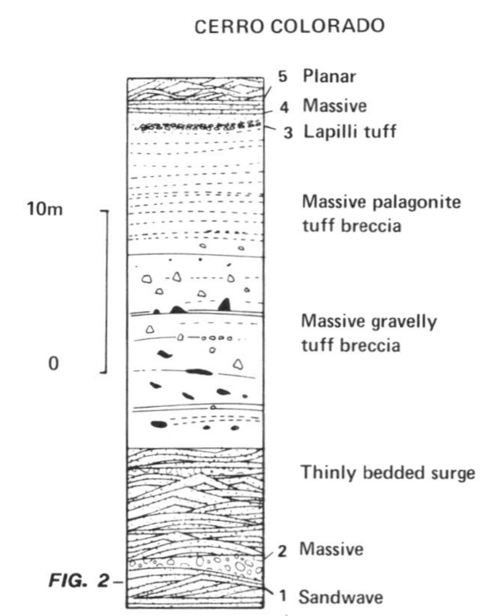

FIG. 2– Composite stratigraphic section of the rim of Cerro Colorado that shows a thinly bedded basal unit, thickly bedded upper unit, sample locations, and deposit characteristics.

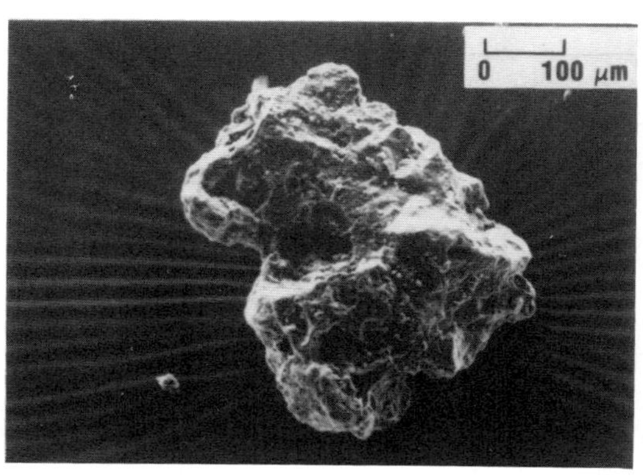

FIGURE 2a. SEMs of sample 1. Coarse-fraction pyroclast with blocky shape and several vesicle surfaces. It has relatively fresh-appearing surfaces and fine, adhering dust particles.

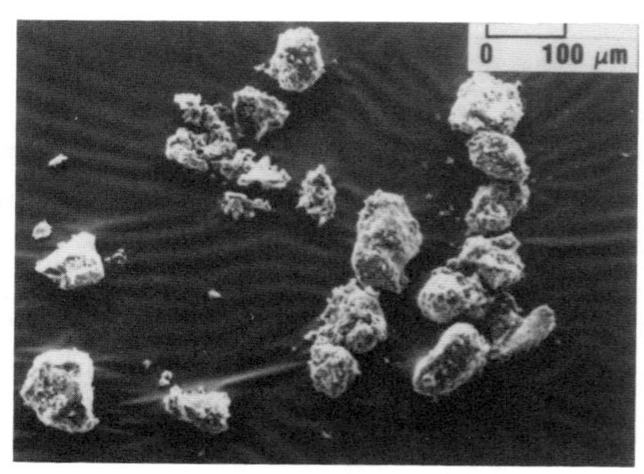

FIGURE 2b. Fine fraction consisting of blocky pyroclasts with abundant adhering dust.

FIGURE 3a. SEMs of sample 2, a glassy pyroclast with crystal molds. Note the smooth phenocryst surfaces and crusty glass coating with an uneven surface.

FIGURE 3b. Accretionary ash grain of hydrated particles bonded tightly by fine particles and altered glass. Note the rounded grain shape.

FIGURE 4a. SEMs of sample 3, a blocky, vesicular pyroclast with hydrated glass skin showing desiccation cracks. The right hand view is a 5X enlargement of the cracked surface that shows vestiges of a crystal mold.

FIGURE 4b. Fine-fraction particles are angular plates and blocks with little adhering dust.

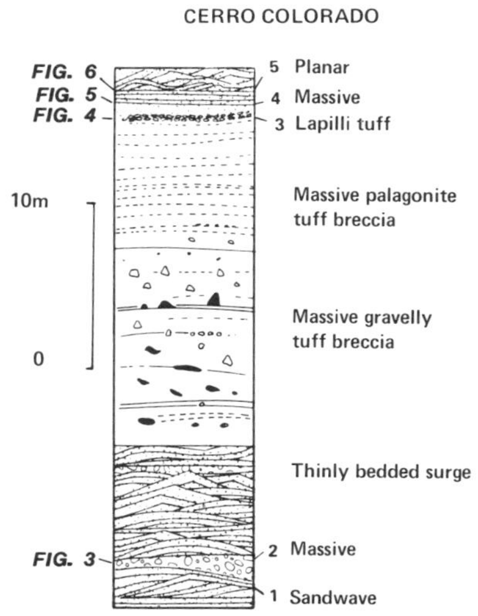

Composite stratigraphic section of the rim of Cerro Colorado that shows a thinly bedded basal unit, thickly bedded upper unit, sample locations, and deposit characteristics.

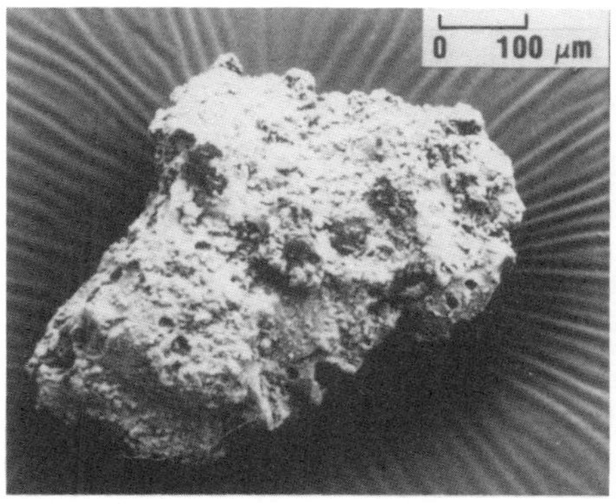

FIGURE 5a. SEMs of sample 4, a coarse-fraction pyroclast with highly altered surface and surface encrustation of clay materials. This grain is highly rounded and shows few vesicles.

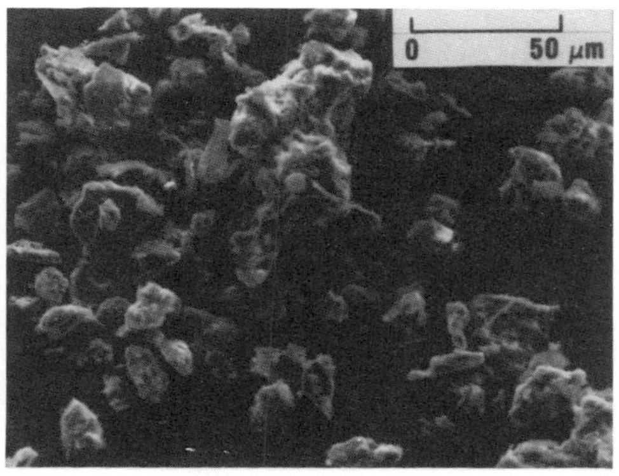

FIGURE 5b. Fine-fraction particles are irregularly shaped and angular. Grains show highly irregular shape.

FIGURE 6a. SEMs of sample 5, a crystal enclosed in hydrated glass coating. Note the crystal faces and desiccation cracks.

FIGURE 6b. The fine fraction is composed of plate-like fragments with sharp, angular edges.

CRATER ELEGANTE, SONORA, MEXICO; DRY PHREATOMAGMATIC ERUPTION

Introduction: Crater Elegante is in the Pinacate volcanic field, northern Sonora, Mexico (Jahns, 1959). This maar crater, 1.6 km in diameter and 0.24 km in depth, was studied in detail by Gutmann (1976). The crater is surrounded by a 50-m-thick tuff ring. The tuff ring is composed mostly of pyroclastic surge deposits that were mapped and described by Wohletz and Sheridan (1979; 1983). The earliest, vent-opening pyroclastic eruptions at Crater Elegante produced Strombolian fall deposits and were followed by phreatomagmatic eruptions; this sequence is visible in crater-rim stratigraphic sections. Samples from a typical stratigraphic section are discussed below.

Grain Size: Samples are moderately sorted and are fine-ash to ash size with grain-size distributions that are typical of surge deposits (Table 1). An ash fall sample (1) has a median size of 7.88 mm, whereas surge samples range from 0.12 to 0.88 mm. Sandwave beds are the finest grained, and planar beds are the coarsest grained of the surge deposits. The abundance of fine ash (Md < 0.06 mm) is a measure of the fine fragmentation caused where the basalt lava mixed explosively with water during eruption.

TABLE 1. GRAIN-SIZE CHARACTERISTICS, CRATER ELEGANTE.

Sample	Bed Form*	Md_φ (mm)	σ_φ	Fine Ash (Wt. %)
1	AF	−2.98 (7.88)	2.07	1.97
2	S	2.56 (0.17)	1.81	14.54
3	S	1.81 (0.29)	2.06	11.83
4	M	3.02 (0.12)	1.77	25.12
5	P	0.17 (0.88)	1.55	0.78
6	P	0.72 (0.61)	1.54	1.86
7	M	2.71 (0.15)	1.85	13.36
8	S	1.76 (0.29)	1.35	4.53

*Bed forms: AF = ash fall, S = sandwave, M = massive, P = planar.

Tephra Description: The tephra samples of Crater Elegante illustrate a spectrum of SEM textures that are typical of basaltic phreatomagmatic ash. These textural features are summarized in Table 2 and are classified under overall grain shape, edge modification features produced by transport abrasion, alteration caused by postdepositional diagenetic processes, fine-fraction characteristics, and distinguishing textural features. The typical basaltic phreatomagmatic tephra (samples 2 through 8) consist of angular and blocky pyroclasts with curviplanar surfaces; they contrast with tephra from Strombolian deposits (sample 1) that are characterized by vesicular ash, surfaces bounded by bubble walls, and a lack of very fine ash particles.

Above the basal Strombolian fall deposit (sample 1), tephra were emplaced during phreatomagmatic eruptions; ash particles are weakly vesicular and show broken surfaces, surface alteration, rounding during transport abrasion, and abundant dust-size material in the fine fraction. Features distinctive of ash from this deposit include jagged edges and moderately fresh surfaces for pyroclasts from sandwave beds, rounded particles with highly altered surfaces from massive beds, and blocky, angular shapes with moderately fresh surfaces from planar beds.

Chemical Composition: The EDS analyses of grain surfaces are given in Table 3. Analysis of sample 1 is an average of several grain surfaces and represents typical unaltered hawaiite basalt compositions of the Crater Elegante area (Gutmann and Sheridan, 1978). Pyroclast surfaces remained mostly unaltered during Strombolian eruptions in which little water interacted with the melt. Samples 2 through 8 have EDS analyses showing strong chemical changes. These changes reflect alteration of grain surfaces by the abundant water in phreatomagmatic eruptions that was emplaced as steam with the tephra.

TABLE 2. DESCRIPTIONS OF CRATER ELEGANTE TEPHRA.

Sample	Bed Form*	Grain Shape	Modification	Alteration	Fine Fraction (Wt. %)	Distinguishing Features
1	F	vesicular	none	none	broken, clean surfaces (2%)	delicate shape preserved
2	SW	blocky, planar surfaces; few vesicles	rounded, large chips	slight to moderate	moss-like (15%)	abundant adhering micron-size surface material
3	SW	blocky, planar surfaces; few vesicles	rounded, large chips	moderate, hydration cracks	moss-like (12%)	adhering material, but less than sample 2
4	M	blocky accretionary masses	rounded, smooth	high, totally altered to palagonite	blocky, little adhering material (25%)	uneven surfaces of many small blocks and plates stuck together
5	P	blocky and vesicular	subangular, many small chips on edges	moderate; some fresh surfaces	broken, angular chips with fine adhering grains (1%)	abrasion features well preserved, many fracture surfaces
6	P	blocky with few vesicle surfaces	subrounded, chipped surface	moderate to high, vesicles filled with alteration material	broken, angular chips with fine adhering grains (2%)	abrasion features well preserved, many fracture surfaces
7	M	blocky and vesicular	subangular, small chips on edges	moderate vesicle fillings	blocky, abundant adhering material (13%)	alteration masking abrasion features
8	SW	blocky with few vesicles	subangular, smoothed edges	slight to moderate with some fresh surfaces	clean, blocky, nonvesicular (5%)	hydration rind in vesicles

*Bed forms: F = fall, SW = sandwave, M = massive, and P = planar.

TABLE 3. GRAIN SURFACE CHEMISTRY OF CRATER ELEGANTE TEPHRA.*

	1	2	3	4	5	6	7	8
SiO_2	50.45	62.69	63.78	58.59	67.07	60.81	52.35	62.61
TiO_2	4.26	3.29	3.67	3.46	3.28	3.50	3.58	2.36
Al_2O_3	16.85	14.41	16.60	19.61	14.36	17.18	21.66	15.89
FeO	6.70	9.37	2.93	3.67	4.21	5.50	5.62	6.01
MnO	2.21	0.06	1.11	1.60	0.44	0.89	1.37	0.62
MgO	3.80	1.70	2.72	3.28	2.50	3.74	4.94	2.68
CaO	8.31	6.83	5.05	5.06	4.68	3.70	5.12	6.68
Na_2O	2.36	0.11	1.50	1.37	0.93	1.36	1.88	1.23
K_2O	2.84	2.03	2.64	3.36	2.55	3.24	3.50	1.93
Glass	86.0	22.0	88.0	89.0	80.0	87.0	89.0	67.0
Crystal	11.0	12.0	7.0	9.0	13.0	9.0	9.0	26.0
Lithic	3.5	67.0	5.0	2.0	7.0	4.0	2.0	7.0

*Normalized standardless EDS analyses.

Comments on Pyroclast Formation: Phreatomagmatic ashes owe their overall shapes to fragmentation of the magma by contact with water. The thermal shock and vaporization pressure waves break up the melt into blocky, irregular, and sometimes drop-like and moss-like shapes. These shapes are strongly dependent upon the amount of water interacting with melt, the chemical composition, and physical characteristics of the melt, such as viscosity and yield strength. Subsequent dispersal of the ash in base surges promotes mechanical abrasion of the surfaces and particle rounding. Crater Elegante tephra were deposited as dry surge deposits in which much of the steam had segregated from the ash before condensation. In this fashion, ash particles were emplaced with relatively dry surfaces; subsequent alteration of the glass is minor, except in areas where there is abundant ground water. These dry deposits can be contrasted to wet surge deposits discussed in the section on Cerro Colorado.

REFERENCES: Gutmann, J. T., 1976. Geology of Crater Elegante, Sonora, Mexico, Geol. Soc. Am. Bull. **87**: 1718-1729.

Gutmann, J. T. and Sheridan, M. F., 1978. Geology of the Pinacate volcanic field. State of Arizona Bureau of Geology and Mineral Technology Spec. Paper No. 2: 47-59.

Jahns, R. H. 1959. Collapse depressions of the Pinacate volcanic field, Sonora, Mexico. Arizona Geol. Soc. Southern Arizona Guidebook II: 165-184.

Wohletz, K. H. and Sheridan, M. F., 1979. A model of pyroclastic surge. Geol. Soc. Am. Spec. Paper 180: 177-193.

Wohletz, K. H. and Sheridan, M. F., 1983. Hydrovolcanic explosions II: evolution of basaltic tuff rings and tuff cones. Am. J. Sci., **283**: 385-413.

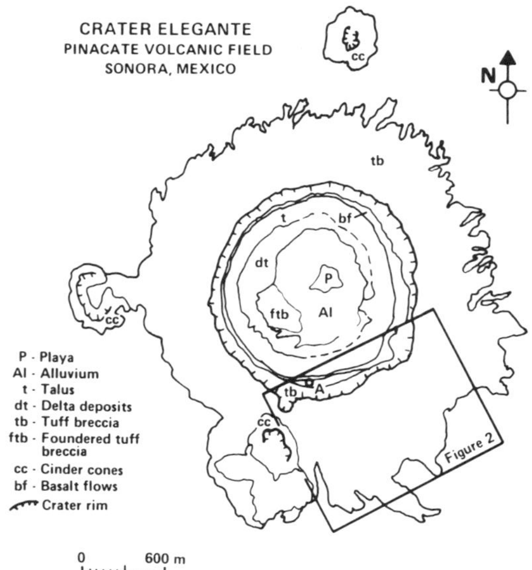

FIGURE 1. Geologic map of Crater Elegante showing the relationship of the surge deposit (tb) to the crater. Stratigraphic section and samples described in this chapter are from the S rim at location A.

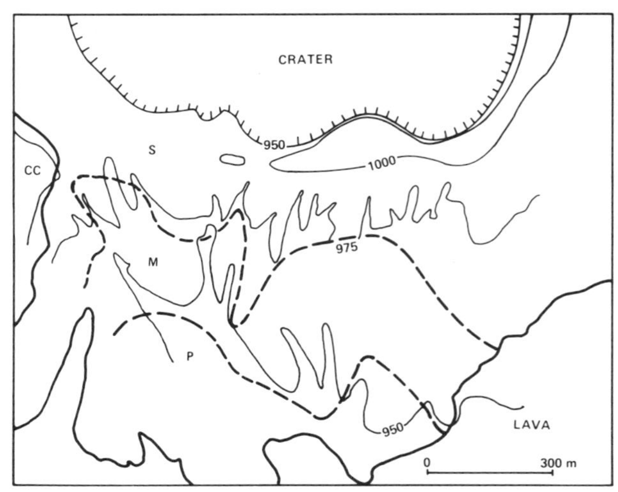

FIGURE 2. Pyroclastic surge facies map of the S rim of Crater Elegante (from Wohletz and Sheridan, 1979), which shows the extent of depositional facies characterized by dominantly sandwave (S), massive (M), and planar (P) bed forms.

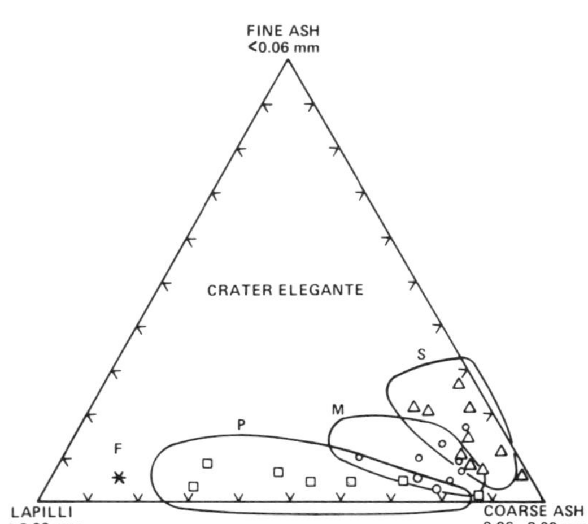

FIGURE 3. Triangular plot of sample median diameters by bed form: S = sandwave, M = massive, and P = planar surge beds.

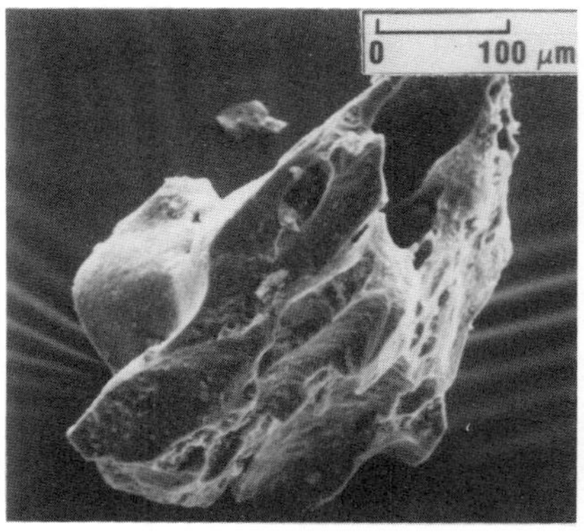

FIGURE 4. SEM of a characteristic pyroclast from sample 1 of the Strombolian fall deposit. The typical vesicular shape with fresh surfaces was formed by fragmentation during vesicle bursts. This texture indicates little interaction with meteoric water during eruption and marks a dominantly magmatic pyroclast.

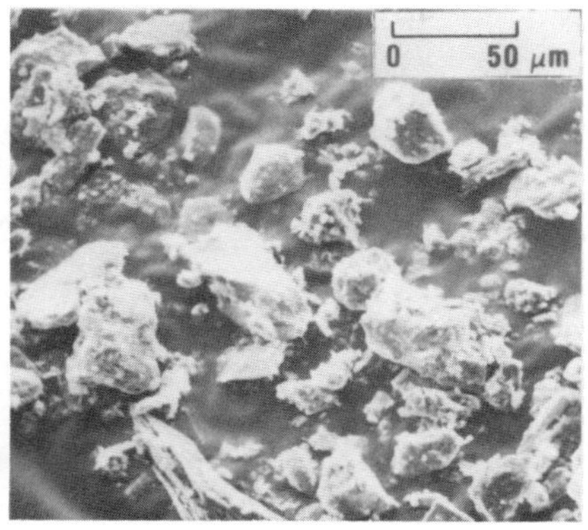

FIGURE 5. SEM of the fine fraction of sample 1, which accounts for less than 2 Wt.% of the total sample. Note the absence of very fine material adhering to grain surfaces. Blocky shapes with few vesicle surfaces are characteristic of particle sizes that are smaller than average vesicle diameters.

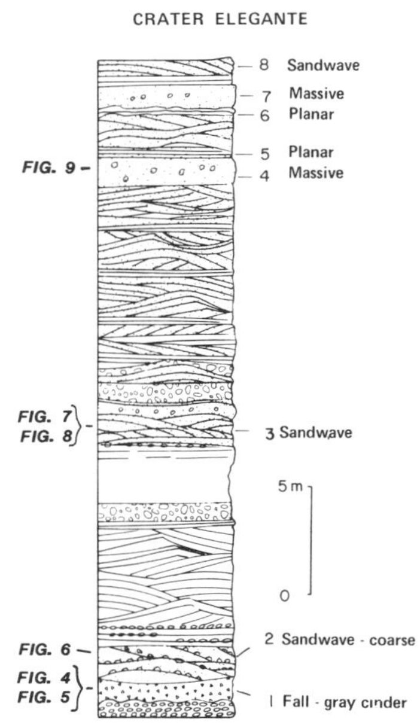

Rim stratigraphic section measured at location A (Fig. 1) that shows the sampled beds, their bed forms, and their grain-size characteristics. Sample 1 is at the base of the tuff ring deposit and is a Strombolian fall deposit. Sample 2 marks the beginning of phreatomagmatic activity that persisted throughout the emplacement of the rest of this stratigraphic section.

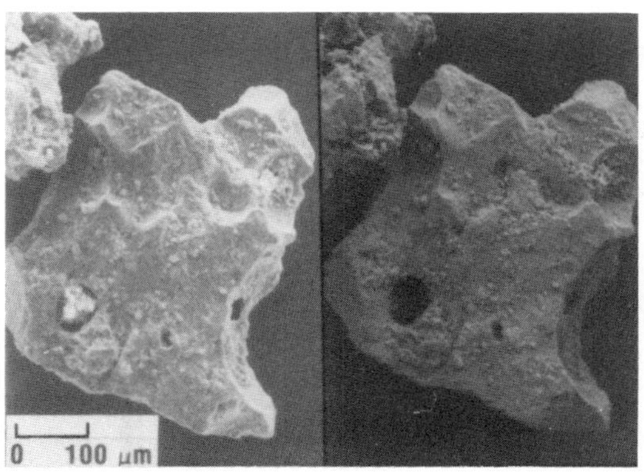

FIGURE 6. SEM of basaltic pyroclast from sample 2 that shows a blocky shape with few vesicle surfaces (typical of phreatomagmatic "Surtseyan" ash). The left view is a secondary electron image and the right view is a back-scattered image. Jagged edges result from abrasion during emplacement by pyroclastic surges. Fine, adhering dust on this grain is evidence of strong explosions, which tend to form abundant fine ash. Note that this deposit contains dominantly lithic material that is characteristic of deposition during the opening stages of phreatomagmatic eruptions (Table 2).

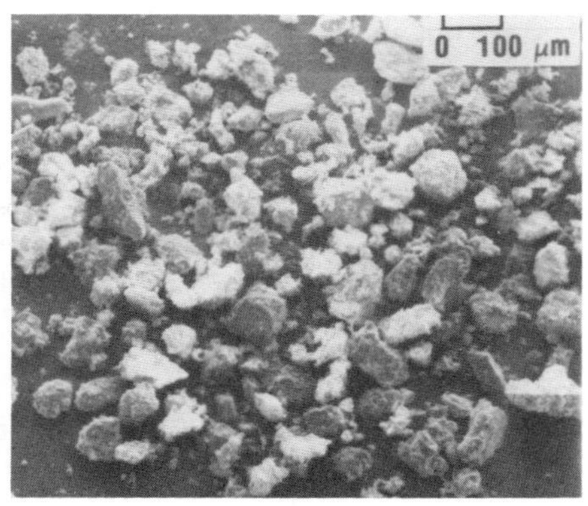

FIGURE 7. SEM of a pyroclast from sample 3; note the rounded and frosted surfaces.

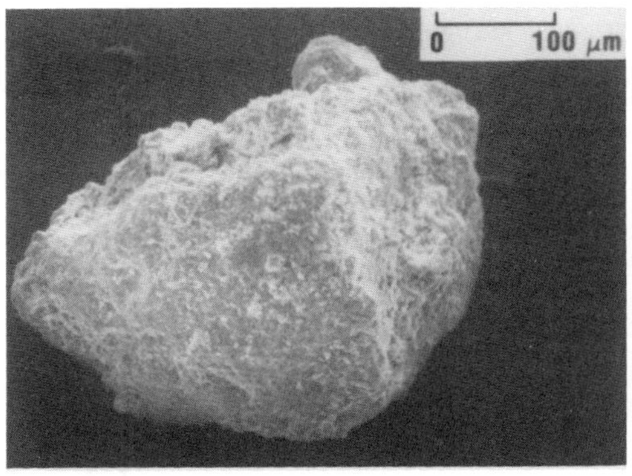

FIGURE 8. SEM of the fine fraction of sample 3, which has irregular, moss-like surfaces and abundant, fine, adhering dust.

FIGURE 9. SEM of a pyroclast from sample 4, a massive-bedded, pyroclastic surge deposit. The aggregation of small particles and strong surface alteration are evidence of emplacement in a wet surge cloud.

FIGURE 10. SEM of a pyroclast from a planar bedded surge deposit (sample 5). Both a secondary electron image (*left view*) and a back-scattered image (*right view*) are shown. Note the blocky shape with curviplanar surfaces showing dish-shaped fractures.

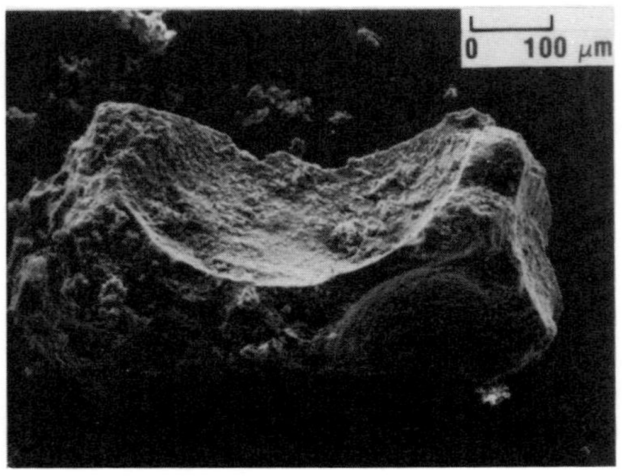

FIGURE 11. SEM of of blocky, vesicular pyroclast from a planar surge bed. Note the angular surfaces and slight incrustation of alteration products on vesicle surfaces.

FIGURE 12. SEM of a pyroclast from sample 7, a massive bedded surge deposit. The grain is well rounded and covered with palagonite alteration.

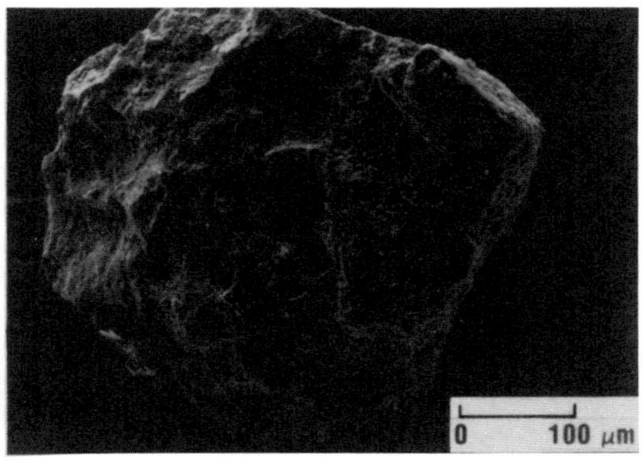

FIGURE 13. SEM of a blocky pyroclast from sample 8. Numerous pits and angular fractures that cover the grain surface have resulted from transport abrasion. Note the relatively fresh surface.

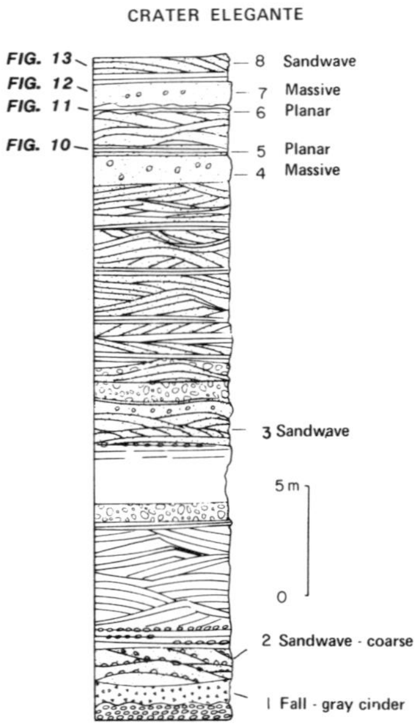

Rim stratigraphic section measured at location A (Fig. 1) that shows the sampled beds, their bed forms, and their grain-size characteristics. Sample 1 is at the base of the tuff ring deposit and is a Strombolian fall deposit. Sample 2 marks the beginning of phreatomagmatic activity that persisted throughout the emplacement of the rest of this stratigraphic section.

Eruption plume, Ngauruhoe, New Zealand.

ERUPTION PLUMES AND DISTAL ASH FALLS

Because gravity segregates the coarser grained and denser pyroclasts near the volcano, atmospheric scientists and tephrochronologists usually see only the finest grained volcanic ash. The ash examples used here are from observed historic eruptions and well-documented prehistoric eruptions. The samples were collected in eruption plumes or from ash beds located over 500 km from their sources.

Eruption Plumes

The examples here were collected by aircraft within dispersing eruption plumes in the troposphere and stratosphere.

VOLCÁN PACAYA AND VOLCÁN FUEGO, GUATEMALA; FEBRUARY 1978 ERUPTION

Introduction: In 1978, a U.S. research expedition to Guatemala used aircraft to sample volcanic eruption plumes from Volcán Pacaya and Volcán Fuego. Samplers were mounted on the aircraft, which flew through dispersing eruption plumes.

Volcán Fuego, characterized by violent, short-lived eruptions, was in a mild Vulcanian phase during the sampling period. It produced plumes of vapor and pyroclasts that rose to elevations of 4000 to 5000 m before being swept downwind. Fuego was erupting olivine-bearing, high-alumina basalts at that time.

During the sampling expedition, Volcán Pacaya was in a state of intense vapor emission that began in November 1977 and continued through May 1978. Lava was visible in the crater; recent lavas there are olivine-bearing, high-alumina basalts.

Both examples used here were provided by W. I. Rose, Jr.; the plates are from a paper by Rose et al. (1980).

Grain Size: The Fuego samples have a trimodal distribution with peaks at 0.4 to 0.8, 3, and 12 μm. Because the Pacaya plume consisted of liquid droplets only, no size data are available.

Tephra Descriptions: Glassy pyroclasts in the Fuego eruption plume have irregular, lumpy forms similar to the much larger pyroclasts that are typical of Vulcanian and some Strombolian eruptions. Angular cleavage fragments of plagioclase are also present. Most of the plumes from Fuego and Pacaya consist of impure acid droplets (mostly H_2SO_4).

Chemical Composition: Table 1 presents a whole-rock chemical analysis of Fuego basalt.

TABLE 1. WHOLE-ROCK CHEMICAL ANALYSIS OF 1974 FUEGO BASALT.*

	(Wt. %)
SiO_2	50.3
Al_2O_3	19.4
FeO_3	10.4
MgO	5.0
CaO	9.0
Na_2O	3.4
K_2O	0.8
TiO_2	0.9
Total	99.4

*From A. Anderson, University of Chicago, unpublished data.

Comments on Pyroclast Formation: Mild but gas-rich eruptions of high-alumina basalt similar to those observed by the sampling team produce glassy pyroclasts with few vesicles. Grain surfaces are usually relatively smooth areas formed by surface tension effects or rough, hackly surfaces formed by brittle fracture. Heterogeneities in the erupting magma are evident in the mixture of glassy and tachylitic pyroclasts. There may be some degree of interaction between magma and ground water during these explosive eruptions, but this is very difficult to verify. The abundance of sulfuric acid droplets in the plume is evidence for the $S0_2$-rich magma at Fuego.

REFERENCES:

Cadle, R. D., Lazrus, A. L., Huebert, B. J., Heidt, L. E., Rose, W. I., Jr., Woods, D. C., Chuan, R. L., Stoiber, R. E., Smith, D. B., and Zielinski, R. A., 1979. Atmospheric implications of studies of Central American volcanic eruption clouds. J. Geophys. Res., **84**: 6961-6968.

Rose, W. I., Jr., Chuan, R. L., Cadle, R. D., and Woods, D. C., 1980. Small particles in eruption clouds. Am. J. Sci., **280**: 671-696.

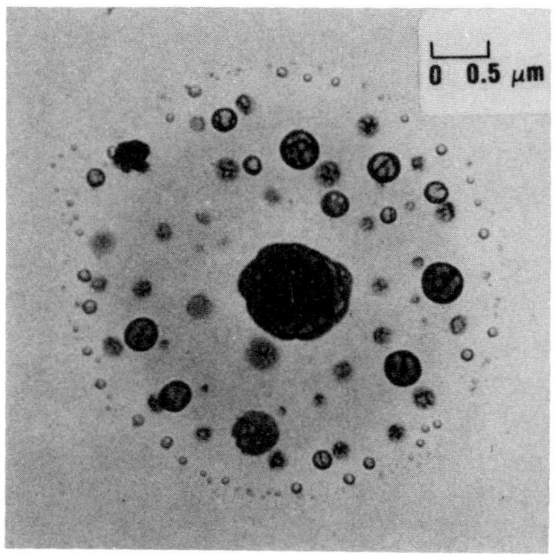

FIGURE 1a. Transmission electron micrographs of conventional impactor samples from the Fuego eruption plume of February 11, 1978 (from Rose et al., 1980). The impactor is a sampling tool used on an aircraft to sample particulates in the atmosphere. Probably an H_2SO_4 droplet with impurities, which is surrounded by spatter formed on impact with the collector.

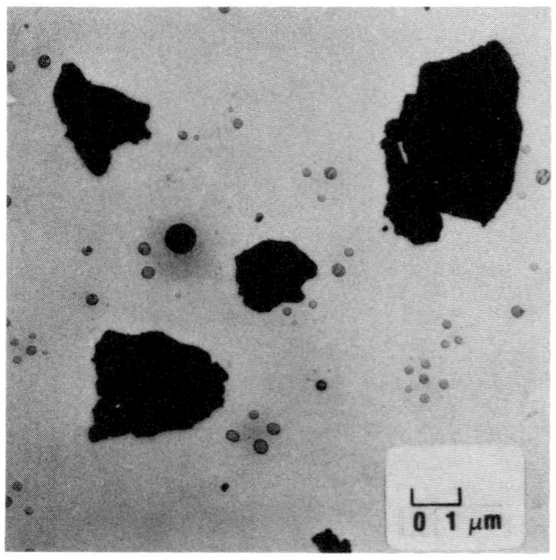

FIGURE 1b. Angular silicate pyroclasts and acid droplets. In another sample (see Cadle et al., 1979), similar particles were a mixture of plagioclase and glass.

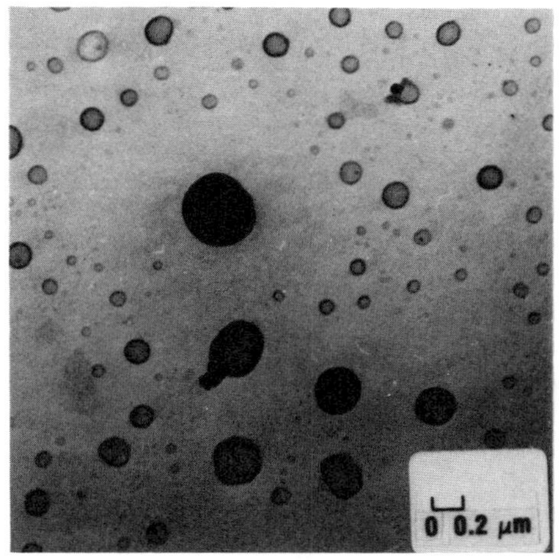

FIGURE 1c. Acid droplets, the most common component of dilute volcanic eruption plumes.

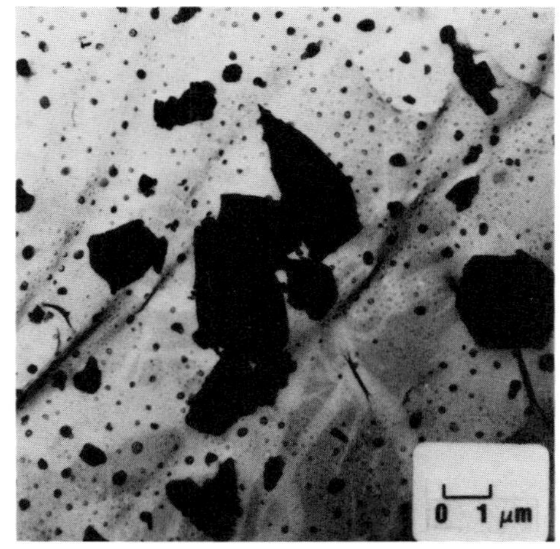

FIGURE 1d. Angular silicate particles and much smaller acid droplets. Those pyroclasts with well-developed cleavage are probably plagioclase and those with lumpy, botryoidal surfaces are most likely glass pyroclasts.

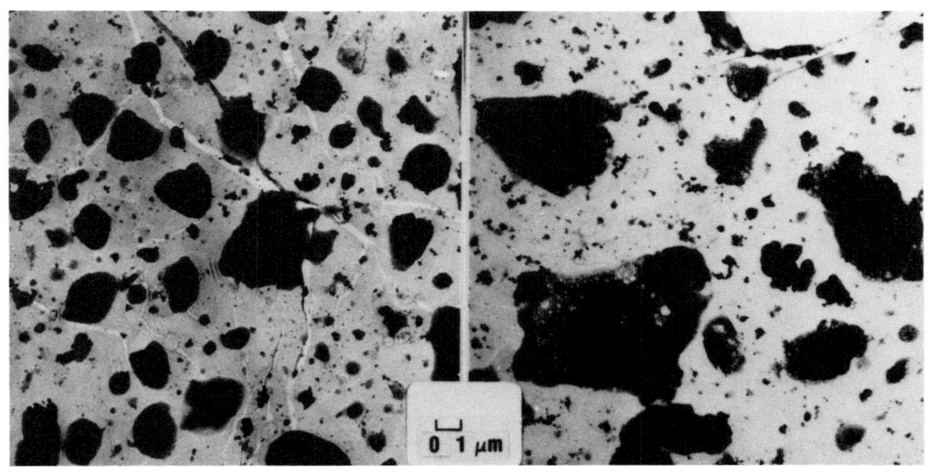

FIGURE 2. Transmission electron micrographs of impactor samples from the eruption plume of Pacaya Volcano, February 1978. Both samples illustrated here consist of impure acid droplets. Rose et al. (1980) note that silicate pyroclasts were rare in this eruption plume.

SANTIAGUITO, GUATEMALA; RECENT ERUPTIONS

Introduction: Santiaguito, located in western Guatemela, was the third volcano sampled by the 1978 U.S. research expedition. Santiaguito volcano consists of a group of dacitic lava domes that have been growing continuously since 1922; since April 1975, there have been daily explosive eruptions from the Caliente vent on the east side of the volcano. During the period of investigation, several large, ash-laden clouds rose to more than 5 km above sea level (2.5 km above the summit of the volcano) (Rose et al., 1980). The information for this example and the photographs were generously supplied by W. I. Rose, Jr., Michigan Technological University. Samples were collected with a 10-stage, quartz crystal, microbalance cascade impactor; these 10 stages sample 2 g/cm^3-density-equivalent particles of 25 to 0.05 μm in size.

Grain Size: Grain-size distributions within the cloud of February 25, 1978, are bimodal, with peaks at 0.12 to 0.4 and 3 to 12 μm. The sample was collected between 5.5 and 11 km downwind from the vent. These samples represent ~2% by weight of the ash erupted; most of the ash falls to the ground near the volcano and is not carried along in the plume.

Tephra Description: Tephra sampled with conventional and cascade impactors consist of glass pyroclasts of rhyolitic composition, plagioclase, magnetite, anhydrite, and sylvite. These pyroclasts are partly or completely coated with sulfuric acid. There are also numerous sulfuric acid droplets. There are aggregates present that consist of all the types of pyroclasts described above.

Glass pyroclasts are blocky, angular, equant to slightly elongate forms. Pyroclasts are indistinct due to the sulfuric acid coatings. Some vesicular pyroclasts exhibit irregular, equant to elongate vesicles and vesicle walls that range from <1 to 5 μm thick.

On lower stages of the impactor, there are mostly <1-μm angular fragments of plagioclase and magnetite, which are encrusted with liquid coatings (acids and Fe, K, Na, and Al salts).

Chemical Composition: Table 1 presents chemical composition data for Santiaguito dacite.

TABLE 1. CHEMICAL COMPOSITION OF SANTIAGUITO DACITE.*

	Glass (Wt. %)	Whole Rock (Wt. %)
SiO_2	72.8	63.45
Al_2O_3	13.5	16.94
FeO	1.3	4.57
MgO	0.4	1.87
CaO	1.1	5.03
Na_2O	4.2	4.91
K_2O	2.8	1.59
TiO_2	0.3	0.47
P_2O_5	0.3	0.22
MnO	0.0	0.14
H_2O	---	0.29
Total	96.7	99.53

*From Rose et al. (1980).

Comments on Pyroclast Formation: Rose et al. (1980) note that the vesiculated melt was fragmented by some other mechanism than bursting of vesicles; several possibilities are outlined, including particle collisions in the vent, explosion of overpressured glass inclusions in crystals, pressure-wave mechanisms, and melt/ground water interaction. They note that the >1-μm pyroclast population is dominantly glass, whereas the <1-μm pyroclasts are mostly phenocryst fragments. It is possible that phenocrysts with perfect cleavages are more easily fragmented. Liquid has accumulated on nearly all the silicate particles, which demonstrates the condensation and scavenging of dilute acid droplets by pyroclasts in the eruption plumes.

REFERENCE:
Rose, W. I., Jr., Chuan, R. L., Cadle, R. D., and Woods, D. C., 1980. Small particles in volcanic eruption clouds. Am. J. Sci., **280**: 671-696.

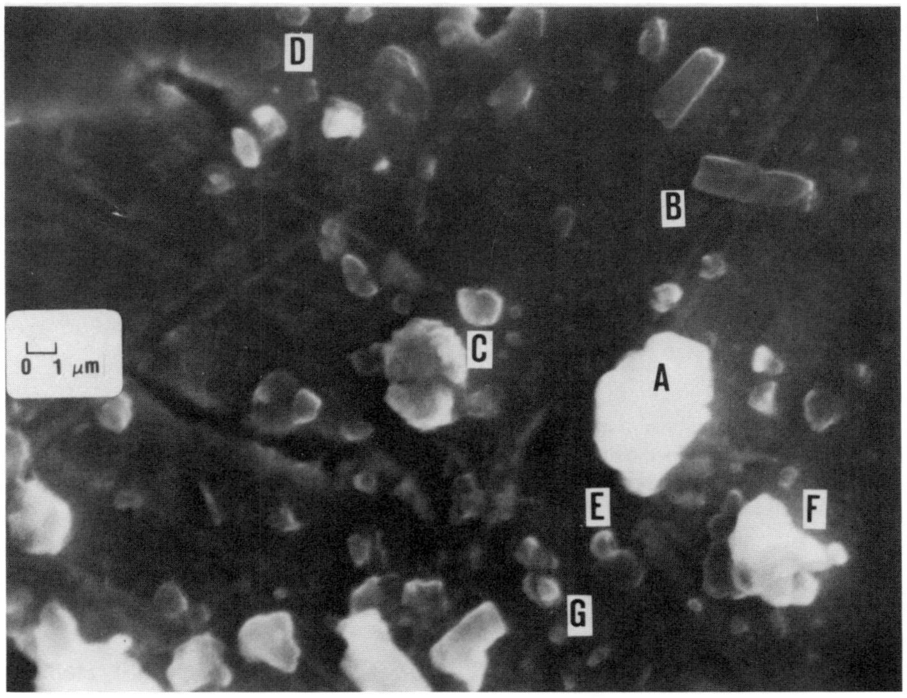

FIGURE 1. SEM of eruption cloud particles of February 22, 1978. A and F are rhyolitic glass pyroclasts; B is a tabular $CaSO_4$ particle; C a plagioclase fragment; and D, E, and G are acid droplets (probably H_2SO_4).

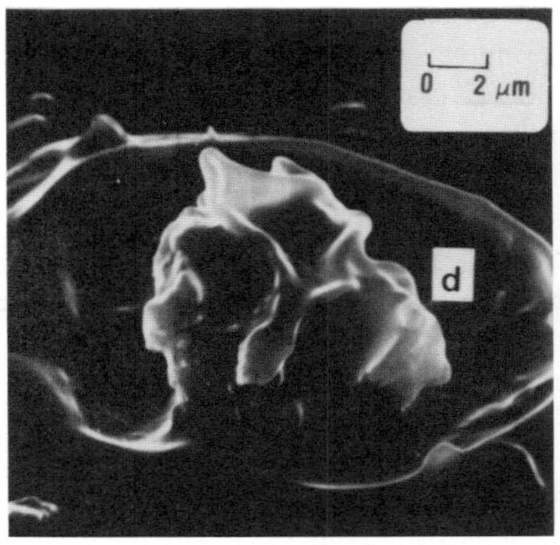

FIGURE 2a. Particles from the February 28, 1978 eruption. SEM of a vesicular shard of rhyolitic glass that is coated and surrounded by a ring of dilute sulfuric acid (spatter formed when droplet or coated pyroclast hit the impactor surface).

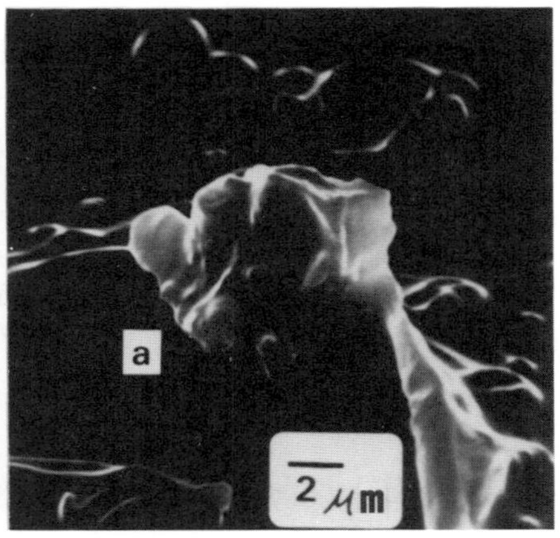

FIGURE 2b. Angular shard, which may be glass of rhyolitic composition, coated with $AlCl_3$.

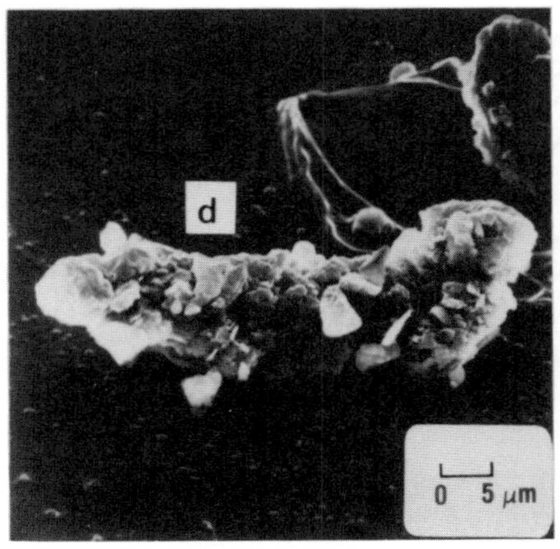

FIGURE 3a. Particles from the eruption of February 28, 1978, collected on stage 2 of the cascade impactor. These "clusters" (agglomerates) of silicate particles, including magnetite and $CaSO_4$, are bonded by acids.

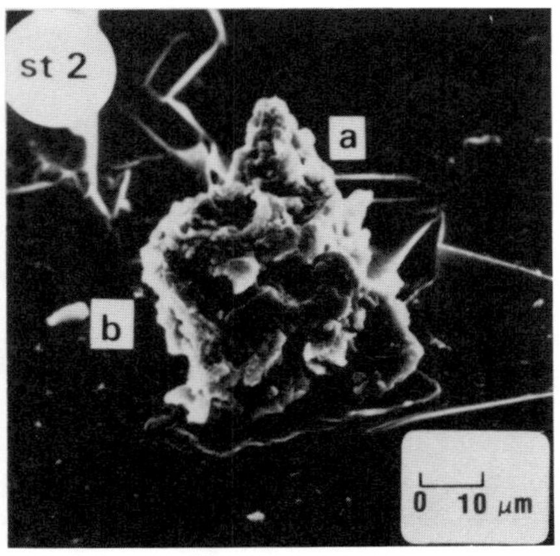

FIGURE 3b. Another "cluster." Analysis of particle *A* indicates a mixture of plagioclase and groundmass; *B* is a nonsilicate crystalline particle.

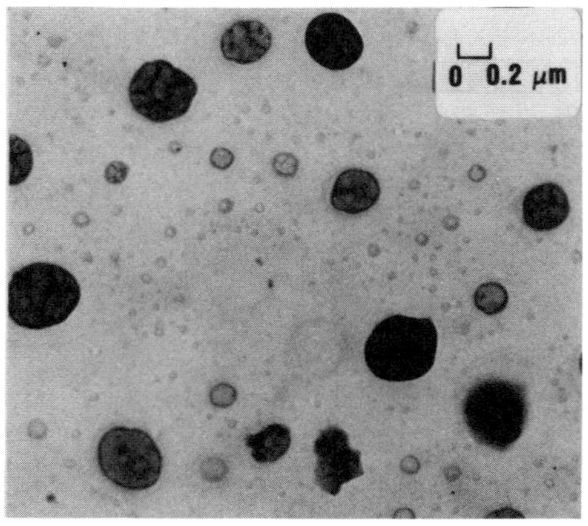

FIGURE 4a. Finer grained (0.4- to 0.2-μm) particles from the eruption cloud of February 28, 1978. Transmission electron micrograph of acid droplets collected on a conventional impactor.

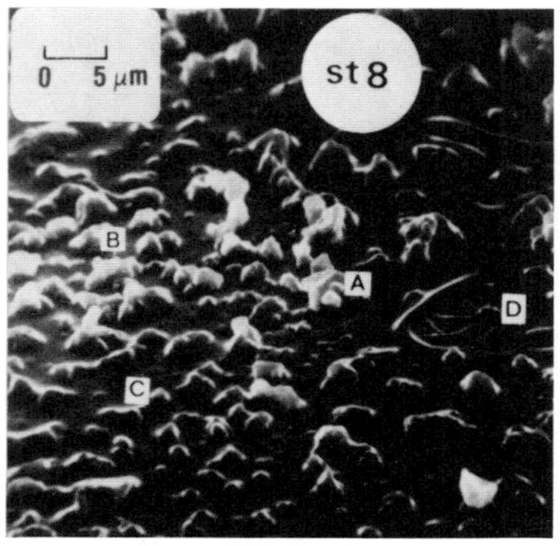

FIGURE 4b. Liquid-coated silicates. Fragments *A* and *C* are plagioclase; *B* and *D* are rhyolitic glass. The ring around *D* may be liquid spatter.

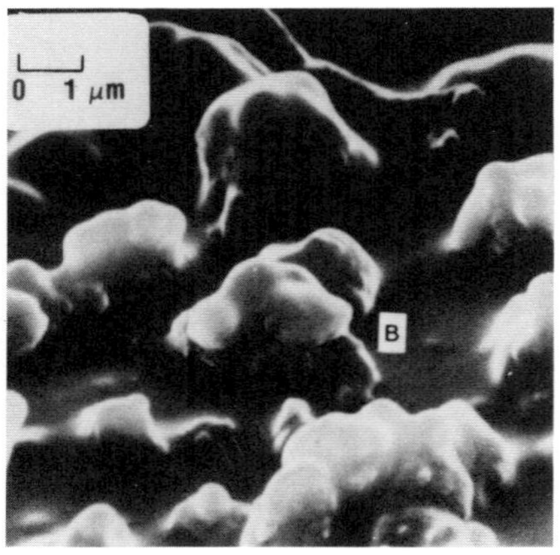

FIGURE 4c. Close-up of particle *B*, showing smooth surfaces due to liquid coatings.

ST. HELENS, WASHINGTON; MAY 18, 1980 ERUPTION

Introduction: The 1980 eruption of Mount St. Helens was the first whose eruption clouds were studied at the volcano and from aircraft, balloons, and spacecraft. The eruption brought together workers from many disciplines, including volcanologists, atmospheric physicists and chemists, meteorologists, and life scientists. Eruption plumes from the May 18 eruption were followed around the earth through satellite observations, lidar (laser radar), and twilight observations. *In situ* sampling was done at lower elevations by aircraft and in the stratosphere by high-altitude aircraft and balloons. For comprehensive syntheses of this information, see publications edited by Lipman and Mullineaux (1981), Newell and Deepak (1982), and Deepak (1982). As the plume moved E and SE across the U.S. only days after that eruption, aircraft were being prepared for sampling missions. Samples described here were collected in the stratosphere on May 20 (altitude of 16.8 km over SW Kansas) and May 21 (altitude of 18.3 km over S-central Wyoming and W-central Colorado). Detailed information on these flights is available in a paper by Sedlacek et al. (1982).

Particle Size: Silicate particles adhering to filters were measured by SEM. Samples from the May 20 flight (altitudes of 15.2 to 16.8 km) range from 0.5 to 10 μm and have a median size of 1.5 μm. Those collected on May 21 (altitudes of 15.2 to 18.3 km) range from 0.5 to 4 μm and have medians of 1.0 and 1.5 μm. Estimated concentrations in the central part of the plume (with an assumed median grain size of 2.0 μm) were 15 particles per cm^3 (May 20) and 1.5 particles per cm^3 (May 21).

Tephra Description: Silicate particles in the stratospheric plume from the May 18 eruption exhibit very little variety. The most common pyroclast types are slightly elongate, angular glass shards. Most are blocky, but some are irregular, scalloped forms; curved surfaces are vesicle-wall fragments. Most are partly coated with smaller glass fragments or liquid droplets.

Farlow et al. (1982) observed, in addition to glass shards, smaller amounts of fine-grained plagioclase, hornblende, pyroxenes, biotite, and chromitite fragments in the stratospheric plume a few days after the eruption. However, these particular clasts were not observed in the samples used here.

Chemical Composition: Chemical compositions of St. Helens ash are shown in Table 1. The variations in silicate and surface chemistry of Mount St. Helens ash are discussed in books edited by Newell and Deepak (1982) and Lipman and Mullineaux (1981).

TABLE 1. CHEMICAL COMPOSITION OF MOUNT ST. HELENS MAY 18, 1980 ASH.*

	1 (Wt. %)	2 (Wt. %)
SiO_2	67.00	71.40
Al_2O_3	16.10	14.60
TiO_2	0.64	0.37
Fe_2O_3	4.50	2.40
CaO	4.10	2.60
MgO	1.50	0.53
K_2O	1.70	2.00
Na_2O	4.60	4.30
P_2O_5	0.13	0.99
Total	101.07	99.19

*Samples are: (1) Bulk ash collected in Missoula, Montana, and analyzed by Rose and Hoffman (1980) by XRF. (2) Glass shard analyzed by Nelson et al. (1980) with an electron microprobe.

Comments on Pyroclast Formation: Interaction of ground and surface water after vesiculation of the dacitic magma is probably responsible for the very fine grained ash produced during the St. Helens eruption. Other lines of evidence for phreatomagmatic interaction are outlined in Rose et al. (1982). Sulfuric acid droplets, although relatively low in this eruption plume, were still much more voluminous than silicate particles.

REFERENCES:

Deepak, A. (Ed.), 1982. Atmospheric effects and potential climatic impact of the 1980 eruptions of Mount St. Helens. NASA Conf. Pub. 2240: 314 pp.

Farlow, N. H., Snetsinger, K. G., Oberbeck, V. R., Ferry, G. V., Polkowski, G., and Hayes, D. M., 1982. Time variations of aerosols in the stratosphere following Mount St. Helens eruptions. *In*: A. Deepak (Ed.), Atmospheric effects and potential climatic impact of the 1980 eruptions of Mount St. Helens, NASA Conf. Pub. 2240: 55-82.

Lipman, P. W. and Mullineaux, D. R. (Eds.), 1981. The 1980 eruptions of Mount St. Helens, Washington. U.S. Geol. Surv. Prof. Paper 1250: 844 pp.

Newell, R. E. and Deepak, A. (Eds.), 1982. Mount St. Helens eruptions of 1980. NASA Spec. Pub. SP-458: 119 pp.

Rose, W. I., Jr., Harris, D. M., Heiken, G., Sarna-Wojcicki, A., and Self, S., 1982. Volcanological description of the 18 May 1980 eruption of Mount St. Helens. *In*: R. Newell and A. Deepak (Eds.), Mt. St. Helens eruptions of 1980, NASA Spec. Pub. SP-458: 1-36.

Sedlacek, W. A., Heiken, G. H., Mroz, E. J., Gladney, E. S., Perrin, D. R., Leifer, R., Fisenne, I., Hinchliffe, L., and Chuan, R., 1982. Physical and chemical characteristics of Mount St. Helens airborne debris. *In*: A. Deepak (Ed.), Atmospheric effects and potential climatic impact of the 1980 eruptions of Mt. St. Helens, NASA Conf. Pub. 2240: 83-107.

FIGURE 1a. These silicate pyroclasts were collected on a filter at an altitude of 16.8 km over SW Kansas, on May 20, 1980. View of filter fibers that are coated with glassy pyroclasts and sulfuric acid droplets.

FIGURE 1b. The angular glass shard in this image is one of the largest from this sampling mission. Curved surfaces are vesicle-wall fragments. Vesicle walls visible in this pyroclast are 0.2 to 2 μm thick. Small, angular, and subrounded particles adhering to the shard are also glass fragments.

FIGURE 1c. Irregular glass shard with smaller shards adhering to the pyroclast surface. Only one remnant is visible on this blocky pyroclast. The substrate is a filter fiber.

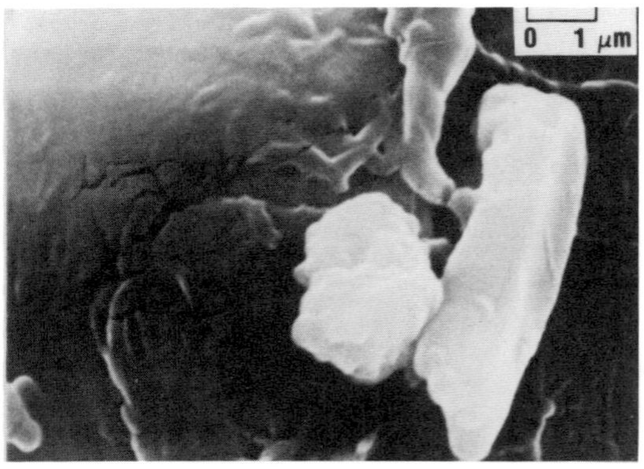

FIGURE 1d. Glass shards; because of liquid droplets adhering to grain surfaces, both have a rounded, lumpy appearance.

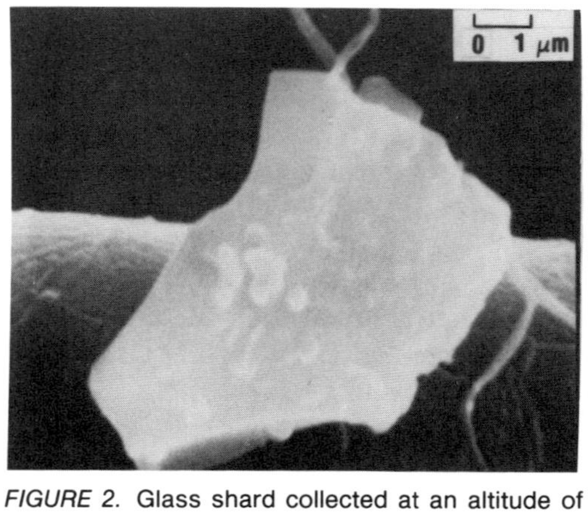

FIGURE 2. Glass shard collected at an altitude of 18.3 km over S-central and W-central Colorado on May 21, 1980. The angular, subequant pyroclast is partly coated with smaller shard fragments.

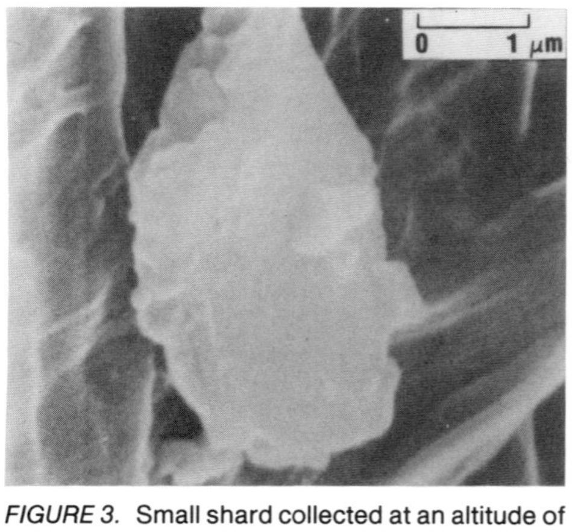

FIGURE 3. Small shard collected at an altitude of 9.5 km over central Kansas on May 20, 1980.

Distal Ash Falls

Plumes from larger eruptions can deposit ash fall layers over thousands to millions of square kilometers and these layers are convenient stratigraphic markers. Older ash beds provide information on the magnitude of large prehistoric eruptions. The "fingerprinting" and stratigraphic utility of ash beds has grown into the discipline of *tephrochronology*. For reviews of this subject, see Wilcox (1965), a bibliography by Westgate and Gold (1974), and papers in a volume edited by Self and Sparks (1981).

BANDELIER TUFF; BLANCO, TEXAS

Introduction: This example of a distal ash fall is a tephra layer correlated with the lower member of the Bandelier Tuff. It was collected near Mt. Blanco, Texas, 528 km SE of the Jemez volcanic field, New Mexico, which is the source of the Bandelier Tuff. Izett et al. (1972) made a correlation between this ash bed and the lower member of the Bandelier Tuff, proposing that it was related to the Plinian phase pumice fall of the lower member, whereas Wright et al. (1981) suggest that at least part is a co-ignimbrite ash fall deposit associated with pyroclastic flows of the lower member. Although it is not a definitive means of correlation, the shard shapes suggest that there is a closer affinity between this ash bed and the fine-grained fraction of the Plinian pumice fall. For a description of the Bandelier Tuff, see the example in Eruption Sequences.

Sample was provided by J. V. Wright of the British Petroleum Co.

Grain Size: Pyroclasts in this sample range from 1 to 100 μm.

Tephra Descriptions: The sample consists of nearly all glass shards and small pumice pyroclasts. They appear to have been derived from a physically heterogeneous magma and are of two pyroclast types.

(1) Slightly curved, plate-shaped shards represent wall fragments from large, flattened vesicles and clusters of coalesced vesicles. Reconstruction of these vesicles indicates that they were flattened, elongate ovoids with maximum intermediate axes of 50 to 130 μm. Thin ridges preserved on smooth shard surfaces represent remnants of vesicle walls (septae) that burst while still fluid. Surviving vesicle walls are 3 to 20 μm thick.

(2) Small pumice pyroclasts are characterized by highly elongate vesicles; in cross section, the vesicles are flattened ovoids with intermediate axes of 3 to 20 μm. Fracture surfaces across vesicle walls are both smooth and conchoidal. There is very little abrasion of these delicate shards.

Chemical Composition: Table 1 presents an electron microprobe analysis of the glass pyroclasts from a near-source Plinian pumice fall deposit of the Bandelier Tuff (from Crowe et al., 1978).

TABLE 1. COMPOSITION OF THE LOWER MEMBER OF THE BANDELIER TUFF.*

	(Wt. %)
SiO_2	73.78
TiO_2	0.10
Al_2O_3	11.74
FeO	1.15
MgO	0.05
CaO	0.30
Na_2O	3.38
K_2O	4.42
F	0.10
Total	95.02

*From Crowe et al., 1978.

Comments on Pyroclast Formation: See the comments on pyroclast formation within the section on the Bandelier Tuff sequence. Considering the age (1.4 m.y.) of this ash deposit, it is interesting to note that there is little alteration or chemical etching of glass shards and small pumices.

REFERENCES:

Crowe, B. M., Linn, G. W., Heiken, G., and Bevier, M. L., 1978. Stratigraphy of the Bandelier Tuff in the Pajarito Plateau. Los Alamos Scientific Laboratory Report LA-7225-MS: 57 pp.

Izett, G. A., Wilcox, R. E., and Borchardt, G. A., 1972. Correlation of a volcanic ash bed in Pleistocene deposits near Mount Blanco, Texas, with the Guaje pumice bed of the Jemez Mountains, New Mexico. Quat. Res., **2**: 554-578.

Self, S. and Sparks, R. (Eds.), 1981. *Tephra Studies.* Reidel Pub. Co., Dordrecht: 481 pp.

Westgate, J. and Gold, C., 1974. *World Bibliography and Index of Quaternary Tephrochronology.* U. of Alberta, Edmonton: 528 pp.

Wilcox, R., 1965. Volcanic ash chronology, *In*: H. E. Wright, Jr., and D. C. Frey (Eds.), *The Quaternary of the United States.* Princeton U. Press, Princeton: 807-816.

Wright, J., Self, S., and Fisher, R. V., 1981. Towards a facies model for ignimbrite-forming eruptions. *In*: S. Self and R. Sparks, (Eds.), *Tephra Studies.* Reidel Pub. Co., Dordrecht: 433-439.

FIGURE 1a. SEMs of an unsieved ash sample collected near Mount Blanco, Texas, 528 km SE of its source in the Jemez Mountains, New Mexico. Slightly curved, smooth, plate-like shards dominate this view; some are crossed by thin septae that are remnants of vesicle walls that burst while still fluid.

FIGURE 1b. Similar to Fig. *1a.* At the base of the image is a vitric pyroclast with a conchoidal fracture surface. All the smaller fragments are glass shards as well.

FIGURE 1c. Blocky, angular shards and small pumice pyroclasts. The pyroclasts were produced by fragmentation of a brittle, solid vesicular magma. Most fracture surfaces are normal to elongate vesicles. Fracture surfaces are both smooth and conchoidal. Smooth, slightly curved to semicircular surfaces were vesicle walls.

FIGURE 1d. This image illustrates the two types of pyroclasts in this sample: slightly curved, plate-like shards broken from large vesicles and small pumice pyroclasts with thin, elongate vesicles.

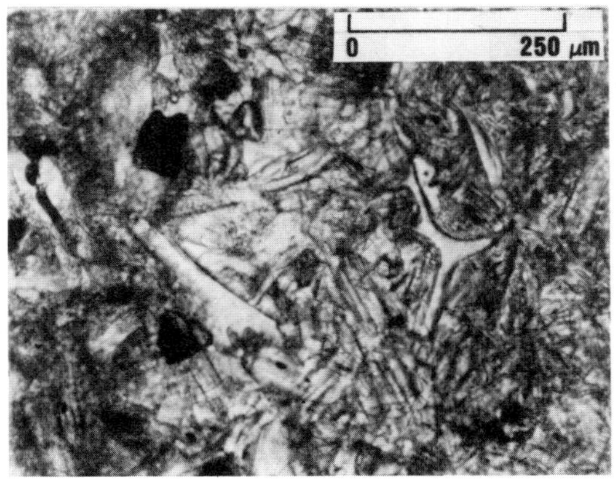

FIGURE 2a. Plane-light photomicrographs of the sample shown in Fig. *1*. Nearly all the ash consists of small, colorless shards. Note slightly curved, plate-like, and Y-shaped shards.

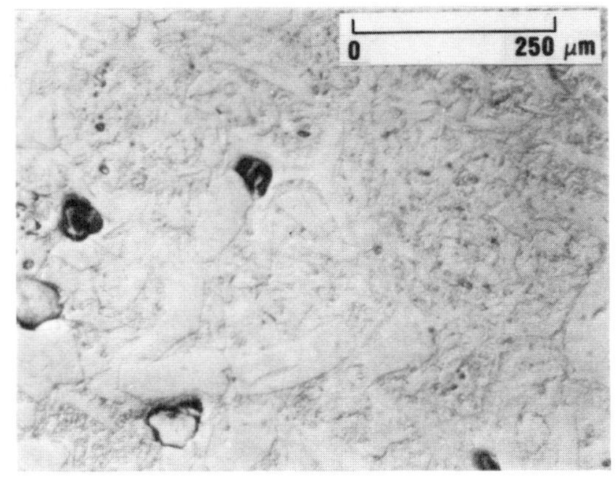

FIGURE 2b. Reflected-light photomicrograph of the sample seen in Fig. *2a*.

MAZAMA TEPHRA; STEPTOE CANYON, WASHINGTON

Introduction: Crater Lake, located in S-central Oregon, is one of the better known calderas of the Cascade Range. A large eruption that occurred about 7,000 years ago (^{14}C dates, as cited by Kittleman, 1973; Bacon, 1983) erupted 51 to 59 km^3 of rhyodacite to mafic andesite as Plinian pumice falls, pyroclastic flows, and preclimactic lava flows (Williams and Goles, 1968; Kittleman, 1973; Bacon, 1983). The extensive ash fall associated with the climactic phase covered much of the Pacific Northwest. Caldera collapse associated with the climactic phase formed the crater now occupied by Crater Lake.

Measurable ash fall deposits cover at least 900,000 km^2 (Fryxell, 1965), making this deposit an important stratigraphic marker that can be used by stratigraphers and archeologists. This sample of Mazama tephra was collected by R. Fryxell in Steptoe Canyon, Whitman County, Washington (600 km NE of Crater Lake).

Grain Size: Thorough analyses of grain-size variations within the Mazama tephra are given by Fisher (1964) and Kittleman (1973). Median grain size decreases from 23 mm (-4.5φ), at a distance of 5 km, to 350 μm (1.5φ), at a distance of 190 km. This deposit is moderately sorted near the source (5 km) and well sorted at a distance of 100 km.

Tephra Descriptions: An "average" sample, as described by Kittleman (1973), is a dacitic crystal ash (up to 60 Wt.% crystals). The distal ash fall sample described here was collected 600 km from its source; it is a vitric ash that consists of colorless flat and curved shards and small pumice pyroclasts (up to 100 μm long). Shards range in size from 1 to 100 μm long and have a variety of shapes, including Y-shapes, curved plates, and flat plates.

Chemical Composition: Table 1 gives an average composition of Mazama ash.

TABLE 1. AVERAGE COMPOSITION OF MAZAMA ASH.*

	(Wt. %)	
SiO_2	72.59	±0.27
Al_2O_3	14.42	±0.16
FeO	2.08	±0.06
CaO	1.71	±0.09
MgO	0.54	±0.08
Na_2O	5.15	±0.16
K_2O	2.70	±0.06
TiO_2	0.48	±0.02
Total	99.67	

*Average of 58 microprobe analyses from Westgate et al., 1969.

Comments on Pyroclast Formation: This vitric ash appears to have been derived from a heterogeneous, highly vesicular magma. Shards and pumice pyroclasts broken from a pumiceous melt have highly elongate, narrow vesicles and large, flattened, irregular vesicles. Much of the fine-grained ash may have been derived partly through a phreatomagmatic interaction of magma and surface water, but there are no data available at this time to confirm this interpretation.

REFERENCES: Bacon, C., 1983. Eruptive history of Mount Mazama and Crater Lake caldera, Cascade Range, U.S.A.. J. Volcanol. Geotherm. Res., **18**: 57-115.

Fisher, R., 1964. Maximum size, median diameter, and sorting of tephra. J. Geophys. Res., **69**: 341-355.

Fryxell, R., 1965. Mazama and Glacier Peak volcanic ash layers; relative ages. Science, **147**: 1288-1290.

Kittleman, L., 1973. Mineralogy, correlation and grain-size distributions of Mazama tephra and other postglacial pyroclastic layers, Pacific Northwest. Geol. Soc. Am. Bull., **84**: 2957-2980.

Westgate, J., Smith, D., and Nichols, H., 1969. Late Quaternary pyroclastic layers in the Edmonton area, Alberta. *In*: Pedology and Quaternary Research, Symp. Proc. University of Alberta: 179-186.

Williams, H. and Goles, G., 1968. Volume of the Mazama ash-fall and the origin of Crater Lake Caldera. *In*: Dole, H. (Ed.), Andesite Conference Guidebook. Ore. Dept. Geol. and Min. Indus., Bull. **62**: 37-41.

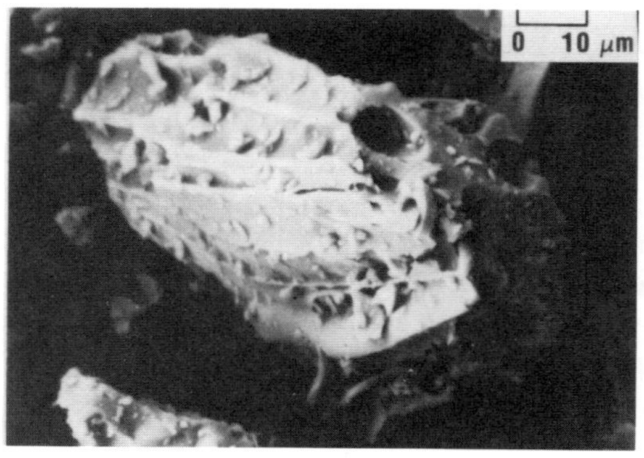

FIGURE 1a. Pumice pyroclasts from the Mazama tephra. Pumice pyroclast with highly elongate vesicles. They are slightly flattened ovals in cross section. Vesicle walls range in thickness from <1 to 10 μm. Small fragments clinging to the pumice pyroclasts are shards.

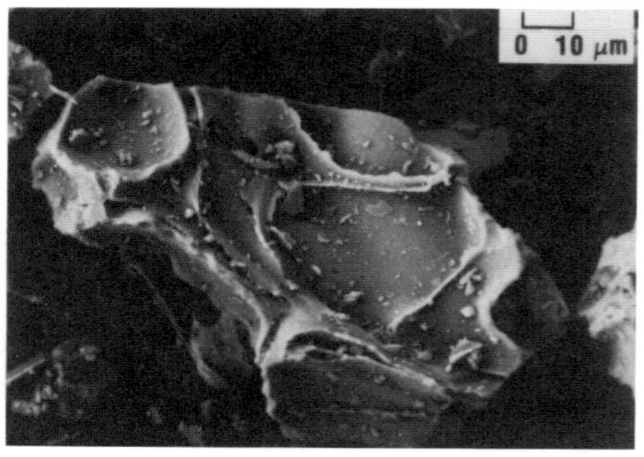

FIGURE 1b. A pumice pyroclast characterized by well-developed, flattened vesicles; many flat or slightly curved shards may have been derived from pumices similar to this one.

FIGURE 2a. Shards from the Mazama tephra. There are a variety of shards visible in this SEM photo. They range from 1 to 70 μm and vary from broad, flat plates to complex Y-shaped forms. The variety of shapes reflects heterogeneities in the size, shape, and degree of vesicle coalescence in the magma. The "lump" visible on a large plate near the center of the photo is most likely over a small phenocryst.

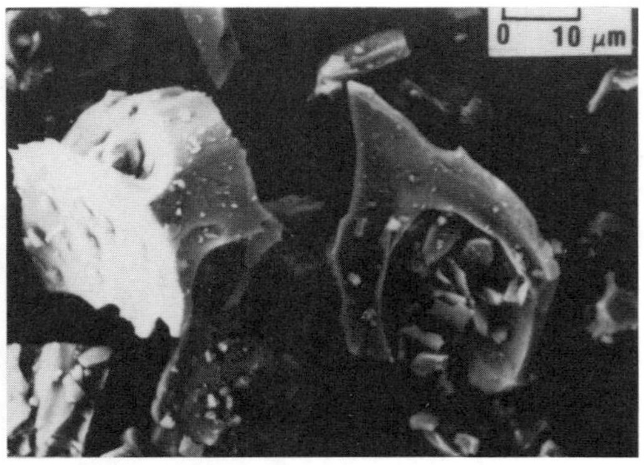

FIGURE 2b. Angular, blocky shards that are partly coated with aggregates of very fine-grained, thin shards.

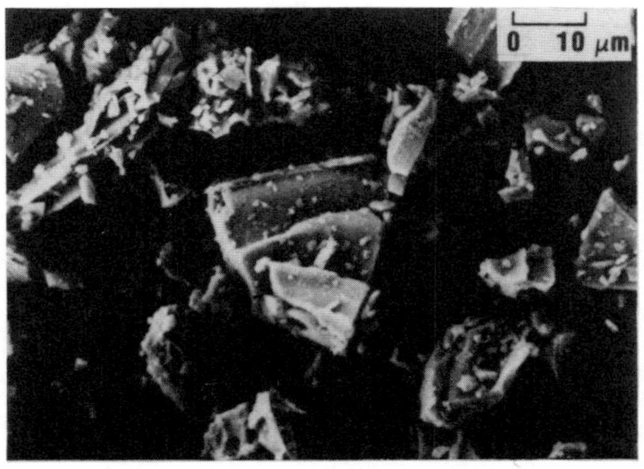

FIGURE 3. Aggregates or clusters of very fine grained vitric ash.

This weathered tuff from St. Vincent, West Indies, shows a 600-μm-long weathered pyroxene in a clay matrix.

WEATHERED AND METAMORPHOSED TUFFS AND RELICT PYROCLASTIC TEXTURES

Volcanic ash deposits form an important part of the geologic record. Little is done to identify weathered ash deposits and ashy soils. Most pre-Tertiary ash deposits are not recognized and consequently are classified as sedimentary rocks or lava flows. Only recently, investigators in Canada have begun to recognize the diversity of pyroclastic deposits that are preserved in Archean-age volcanic terranes. It is possible to identify relict pyroclastic textures in metamorphosed and deeply weathered tuffs, and several examples are provided here.

Introduction

The following section discusses the features of ash that result from secondary processes of hydration and palagonitization. The examples are drawn from other sections in this book. Instructive literature concerning alteration of volcanic ash is directed towards distinguishing a volcanic origin of clays and specific examples of ash diagenesis (Wise and Weaver, 1979; Wise et al., 1973). In this section we have attempted to present a general treatment of the topic.

ALTERATION OF VOLCANIC ASH: AN OVERVIEW OF HYDRATION AND PALAGONITIZATION STUDIES DONE BY SEM

Introduction: The processes of hydration and palagonitization (see the discussion by Fisher and Schmincke, 1984) involve physical and chemical changes that are intermediate between those of weathering (diagenesis) and metamorphism. Hydration of glass occurs when silicic pyroclasts or lavas are altered by water or steam (Friedman et al., 1966) and may be compared with the gradual alteration of basaltic glass during the process of palagonitization. Palagonite is formed by the alteration of basaltic glass to smectite clays and zeolites (Peacock, 1926; Peacock and Fuller, 1928; Honnorez, 1962; Hay and Iijima, 1968). Some investigators, including Nayudu (1964) and Bonatti (1965), suggest that palagonitization involves hydration of glass. The difference between the alteration of basaltic glass and that of silicic glass is not well understood, but experimental studies (Marshall, 1961; Hawkins and Rustim, 1963) show that basaltic glasses are more reactive to aqueous solutions than silicic glasses are. This reactivity may be a reflection of the greater concentration of silicon/aluminum polymer networks in silicic glasses.

Hydrated obsidian (perlite) is valuable for dating silicic rocks because it can be easily distinguished from fresh glass with an optical microscope. Because hydration "fronts" move into glass from an initially fresh, exposed surface at a constant rate, knowledge of local climatic conditions makes it possible to calibrate hydration rind thickness to age (Friedman and Smith, 1960). The alteration of silicic ash deposits of tuff cones and tuff rings that are commonly associated with lava domes is poorly understood; these deposits have not been studied as thoroughly as their basaltic counterparts.

Palagonite tuffs were first recognized over a century ago (von Walterhausen, 1845); they are most commonly associated with pillow lavas (Moore, 1966) and basaltic ash deposits of tuff rings and tuff cones such as Surtsey (Jakobsson, 1978). Tuff cone and tuff ring deposits are generally intensely altered around the craters by posteruptive hydrothermal activity. The widely dispersed tuffs from caldera-forming eruptions and Strombolian ash fall blankets generally show only minor diagenetic and weathering effects unless they were deposited in a tropical environment.

The descriptions below are based upon SEM observations of hydration and palagonitization textures and upon EDS analyses of the surface chemical compositions of many altered pyroclastic samples discussed in other sections of this book. For most cases, sample disaggregation was necessary to separate particles cemented by alteration products. The samples were prepared for SEM examination by a thorough cleansing with acetone and ultrasound, which removed loosely adhering material from grains and exposed individual grain surfaces.

Grain Size: Strongly palagonitized or hydrated ash that still contains individual particles is generally too consolidated to be accurately sieved. Less altered ash, however, can be disaggregated and sieved for a rough estimate of size distribution. Regardless of the degree of hydration or palagonitization, the grain size of altered pyroclastic deposits appears to deviate very little from that of unaltered equivalents. It appears that hydration and palagonitization do not affect grain size significantly. In general, within deposits from an individual eruption, the degree of palagonitization increases slightly with increasing grain size, which indicates that porosity and permeability are important controls of alteration. This relationship does not hold true, however, for massive and planar-bedded surge deposits of tuff rings, where the degree of palagonitization increases slightly with decreasing grain size and probably is a function of total grain-surface area. These trends of alteration versus grain size are shown in Fig. 1. The trends are not strictly conclusive for all pyroclastic deposits because the data base is limited to 50 sample analyses of basaltic tuff cones and tuff rings, including Zuni Salt Lake, Koko Crater, Kilbourne Hole, Taal Volcano, and Surtsey.

Alteration Description: Peacock (1926) described two main varieties of palagonite: "gel-palagonite," a banded isotropic material that encloses cores of unaltered glass and "fibrous palagonite," which is birefringent and more strongly colored, contains opaque material, and sometimes shows microtubes penetrating into the glass. Although these terms are difficult to apply to SEM images of altered grain surfaces, two distinct palagonite textures are evident on grains and exhibit various combinations of phases that correspond to several stages of alteration. These stages of alteration noted below are associated with hydration and palagonitization but not with solution and precipitation of secondary phases.

Cracks. The first stage of basaltic glass alteration is the formation of a skin on grain surfaces. This skin, or rind, is generally lighter or more reflective than fresh glass surfaces and first develops in vesicle hollows. With progressive alteration, the rind will cover the entire grain surface. Palagonite rinds are usually broken by straight or slightly curved cracks (Noble, 1968). These cracks strongly resemble desiccation cracks in mud but are distinguishable from those formed by eruptive fragmentation and grain collisions during transport; both of the latter crack types tend to project through the grain.

In silicic glasses, the cracks tend to be curved and circular, forming what has been called the "onion skin" texture of perlite (Ross and Smith, 1955). Perlitic cracks are formed by the secondary hydration of glass (Noble, 1968) and develop more rapidly in phenocryst-bearing glasses than in aphyric glasses.

Close observation of cracks in ashes of both basaltic and silicic compositions shows that they are one to several micrometers wide and they separate thin polygonal blocks, which curl at the edges. The curled-back skin is only several micrometers thick and may overlie more altered skin layers beneath the surface. Cracks appear to grow in stages and they intersect other cracks in a polygonal fashion. The first cracks appear in vesicles and then are visible over the entire grain. Finally, they form pull-aparts of the grain skin and skin-plates with curled edges. Crack orientation may be controlled by the presence of microlites such as feldspar laths. The cracked glass skin probably represents a hydration front that progressively moves into the grain (Moore, 1966).

Cracks probably result from volume changes that take place during hydration. It is not clear why they resemble desiccation features, but this resemblance suggests that hydrated skin gradually dehydrates after it separates from fresh, underlying glass. Perhaps gradual crystallization of the hydrated skin accounts for a second volume change.

Microcrystalline Material. In the second stage of glass alteration, microcrystalline material forms, initially in vesicle hollows and then over the entire grain surface. The material may be superimposed over a hydrated skin or occur alone.

Because microcrystalline material is so typical of samples taken from the most highly altered basaltic deposits, it seems likely that it represents an advanced stage of alteration. In addition, formation of microcrystalline material corresponds to the greatest degree of chemical change documented in glasses, as we discuss in the next section. In the most advanced stages of alteration, both cracked skin and microcrystalline material cover the entire grain surface, and solution of the altered glass takes place, which results in a wormy appearance and a variety of relict textures.

Alteration stages that are concurrent with the hydration of silicic pyroclasts can be seen in the microcrystalline material on grain surfaces. Secondary phases appear first in vesicle hollows and then cover progressively more of the grain surface.

At high magnification (>20,000X), the euhedral forms of microcrystalline material may be observed. Rhomb-shaped chabazite and prismatic and spherulitic phillipsite are two common zeolites associated with palagonite. Fibrous to fan-like crystals of the natrolite-thompsonite series are less common. Clay minerals (mostly smectites) are usually too small (<0.01 μm) to distinguish easily, even at very high magnifications (greater than 100,000X).

Solution and Precipitation. In addition to the more common aspects of alteration (hydration-palagonitization-crystallization), solution and precipitation often affect glassy pyroclasts. These processes are common where circulating ground water encounters lithic materials such as sandstone or limestone and tephra. Commonly, solution forms tiny pits and frosted coatings on grain surfaces, whereas opal coatings or carbonate fillings are precipitated between grains.

Chemical Composition: Bunsen (1851) first recognized the chemical differences between fresh basalt and palagonite. In general, alteration (especially palagonitization) involves addition of H_2O—as much as 30 Wt.%. Hydration of glass has an effect on the total elemental abundances within individual pyroclasts. A number of studies document the chemical changes that occur during alteration; they all find that the relative mobility of each major element and trace element is different (for example, Hay and Iijima, 1968; Furnes, 1975; Allen et al., 1981). Relative gain or loss of elements during alteration varies with the degree of hydration.

To illustrate the effects of alteration on the chemical composition of tephra, averaged, normalized EDS analyses of fresh and altered grain surfaces were compared. Samples were taken from individual beds in stratigraphic sequences where the relative degree of alteration in bed forms could be assessed in the field. Two basaltic volcanoes (Crater Elegante tuff ring and Cerro Colorado tuff cone) and two rhyolitic tuff rings (Panum Crater and Obsidian Dome) were studied. The tabulated analyses of these samples are given in their respective sections of this book. Figure 14 illustrates the gain and loss of major elements in samples from these stratigraphic sections. Note that the analyses were not recalculated for an assumed "immobile" element; however, they were calculated on a water-free basis. For each of the two basaltic examples, a series of analyses is shown in which the first is from a lower fall deposit that represents a fresh composition and subsequent analyses represent altered compositions. The effect of chemical alteration increases upwards in individual deposits and is thought to be related to water involved in hydrovolcanic explosions; higher in the sequence, steam has condensed on ash particles during emplacement. Wet eruption deposits (massive beds) show the greatest alteration, and dry eruption deposits (sandwave beds) the least. The two rhyolitic examples demonstrate the reduced chemical effect of alteration by hydration upon silicic glasses. Again, the degree of alteration is related to bed form.

Comments on Alteration Mechanisms: Silicic glass clearly shows the effects of hydration. It is not well defined on SEM images but is clearly evident by optical inspection of thin sections. Although basaltic glass becomes hydrated, it simultaneously changes to palagonite, which is unstable and gradually crystallizes to form smectites and zeolites.

There is still some debate as to whether palagonitization occurs at ambient temperatures or high temperatures during and shortly after eruptions (Peacock, 1926; Jonsson, 1961; Fuller, 1931; Bonatti, 1965). Both options are likely. Field evidence supports the hypothesis that the rate of palagonitization depends on temperature and bulk water content of the deposit. If water is present during eruption and emplacement, glass becomes hydrated at accelerated rates. Where magmatic fall deposits (dry eruptions) are interstratified with phreatomagmatic surge deposits (wet eruptions), the phreatomagmatic deposits are generally altered, whereas the magmatic deposits are fresh. On the other hand, alteration zones may be independent of vent location and bedding. These zones appear to be caused by water-table fluctuations after emplacement. The rate and type of alteration is affected by Eh-pH conditions, vent locations, deposition (subaerial or subaqueous), and temperature.

In summary, three mechanisms of ash alteration are recognizable. (1) Alteration that is caused by glass reaction with high-temperature water during eruption and deposition shows a bed-form dependence. Successive layers of ash show varying amounts of alteration that may be correlated to mode of emplacement (such as fall versus surge) and/or moisture in the eruption plume. (2) Postemplacement hydrothermal activity near the vent may progressively alter all near-vent deposits. This alteration is not controlled by either bedding planes or water-table horizons, so individual bedding sets can be traced from their fresh outcrops to the highly altered outcrops at near-vent locations. (3) Slow alteration occurs when ground water percolates through porous and permeable deposits. This alteration shows a horizontal and topographic control, and it may cut across bedding planes regardless of grain size.

Of the above mechanisms, bed-form control of alteration may be the most ambiguous, albeit the most instructive about eruption conditions. This ambiguity results from the nonlinear effects that porosity, grain size, and permeability have upon aqueous movement. For example, consider the following complications: smaller grain sizes promote alteration as a result of increased surface area—however, permeability usually decreases with decreasing grain size and sorting—but fine-grained deposits may retain longer the water that is emplaced with them.

Fall deposits from a magmatic burst often alter by mechanisms 2 and 3, above. Therefore, coarse-grained, well-sorted fall deposits (emplaced during dry eruptions) often show more alteration than do fine-grained deposits. In contrast, steam-rich eruptions produce relatively fine-grained tephra, which may be emplaced with wet steam that condenses on ash surfaces or dry steam that separates from the ash before condensation. Ash grains are subjected to hot fluids from the time of their formation until they finally cool to ambient temperatures within the deposit (mechanism 1). Under these circumstances, alteration may progress quickly and it is most rapid for the finest grain sizes. This relationship is especially true for vesicular, massive surge beds, which often contain accretionary lapilli and are composed of very fine ash. However, many sandwave surge beds are emplaced when steam separates from the ash before condensation, and the resulting bed remains relatively unaltered.

REFERENCES: Allen, C. C., Gooding, J. L., Jercinovic, M., and Keil, K., 1981. Altered basaltic glass: A terrestrial analog of the soil of Mars. Icarus, **45**: 347-369.

Bonatti, E., 1965. Palagonite, hyaloclastite and alteration of volcanic glass in the ocean. Bull Volcanol., **28**: 1-15.

Bunsen, R., 1851. Ueber die prozesse der vulkanischen gesteinbildungen islands. Prog. Ann., **83**: 197-272.

Fisher, R. V. and Schmicke, H.-U., 1984. *Pyroclastic rocks*. Springer-Verlag, New York: 472 pp.

Friedman, I. and Smith, R. L., 1960. A new dating method using obsidian - Pt. 1, the development of the method. Am. Antiq., **25(4)**: 476-522.

Friedman, I. I., Smith, R. L., and Long, W. D., 1966. Hydration of natural glass and formation of perlite. Geol. Soc. Am. Bull., **77(3)**: 323-327.

Furnes, H., 1975. Experimental palagonitization of basaltic glasses of varied composition. Contrib. Min. Petrol., **50**: 105-113.

Fuller, R. E., 1931. The aqueous chilling of basaltic lava on the Columbia River Plateau. Am. J. Sci., **21**: 281-300.

Hawkins, D. B. and Rustim, R., 1963. Experimental hydrothermal studies on rock alteration and clay mineral formation. Geochim. Cosmochim. Acta, **27**: 1047-1054.

Hay, R. L. and Iijima, A., 1968. Nature and origin of palagonite tuff of the Honolulu group on Oahu. Geol. Soc. Am., Mem., **116**: 331-376.

Honnorez, J., 1962. Sur l'origine des hyaloclastites. Bull. Soc. Belge Geol., **70**: 407-412.

Jakobsson, S. P., 1978. Environmental factors controlling the palagonitization of Surtsey tephra, Iceland. Geol. Soc. Denmark Bull., **27**: 91-105.

Jonsson, J., 1961. Some observations on the occurrence of sideromelane and palagonite. Bull. Geol. Inst. Uppsala, **40**: 81-86.

Marshall, R. R., 1961. Devitrification of natural glass. Geol. Soc. Am. Bull., **21**: 273-291.

Moore, J. G., 1966. Rate of palagonitization of submarine basalt adjacent to Hawaii. U.S. Geol. Sur. Prof. Paper, 550-D: D163-D171.

Nayudu, Y. R., 1964. Palagonite tuffs (hyaloclastites) and the products of post-eruptive processes. Bull. Volcanol., **27**: 392-410.

Noble, D. P., 1968. Stress-corrosion failure and the hydration of glassy silicic rocks. Am. Mineral., **53**: 1756-1759.

Peacock, M. A., 1926. The volcano glacial palagonite formation of Iceland. Geol. Mag., **63**: 385-399.

Peacock, M. A. and Fuller, R. D., 1928. Chlorophaeite, sideromelane and palagonite from the Columbia River Plateau. Am. Mineral, **13**: 360-382.

Ross, C. S. and Smith, R. L., 1955. Water and other volatiles in volcanic glasses. Am. Mineral., **40**: 1071-1089.

von Waltershausen, W. S., 1845. Über die submarinen Ausbrüch in der tertiären Formation des Val di Noto im verwandten Erscheinungen am Ätna. Gött. Stud., **1**: 371-431.

Wise, S. W. and Weaver, F. M.,1979. Volcanic ash: Examples of devitrification and early diagenesis. *In*: *Scanning Electron Microscopy/1979/I*. SEM Inc., AMF O'Hare, Illinois: 511-518.

Wise, S. W., Weaver, F. M., and Gûven, N., 1973. Early silica diagenesis in volcanic and sedimentary rocks: Devitrification and replacement phenomena. *In*: C. J. Arceneaux (Ed.), *Proceedings, 31st Ann. Meet. of the Electron Microscopy Soc. Am.*. Claitor's Publishing Division, Baton Rouge, Louisiana: 206-207.

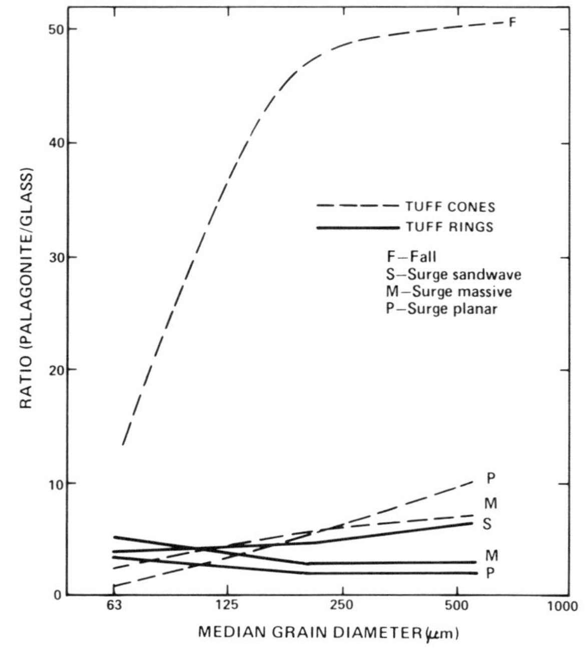

FIGURE 1. Plot of palagonite-to-glass-volume ratio versus median grain diameter for samples of basaltic tephra. Results of 50 sample analyses from tuff rings and tuff cones are grouped by bed form to show dependency of palagonitization on eruption and emplacement mechanisms. Overall palagonitization increases with grain size as a probable result of increased porosity and permeability. Samples from tuff rings, however, show a slight increase in palagonitization with decreasing grain sizes, probably because of increased grain-surface area for fine deposits emplaced in wet conditions.

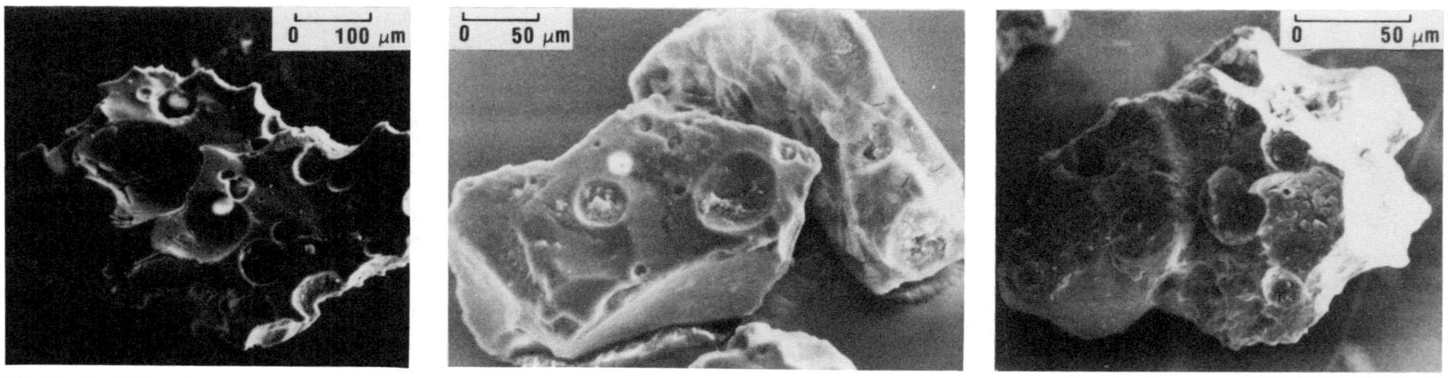

FIGURE 2. SEMs show the initial stages of alteration of basaltic pyroclasts. Alteration is indicated by a cracked skin forming first in vesicle hollows and then progressively covering the entire grain (as in Figs. *3* and *4*).

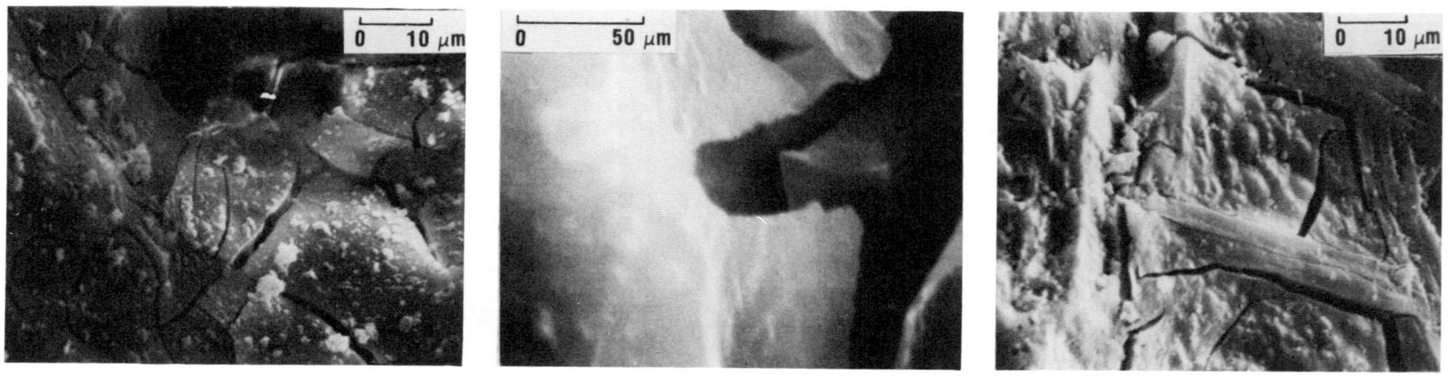

FIGURE 3. SEMs of hydration-crack detail. Note the polygonal structure in left view, the formation of two skin layers in the middle view, and microlitic control of crack orientation (feldspar laths) in right view.

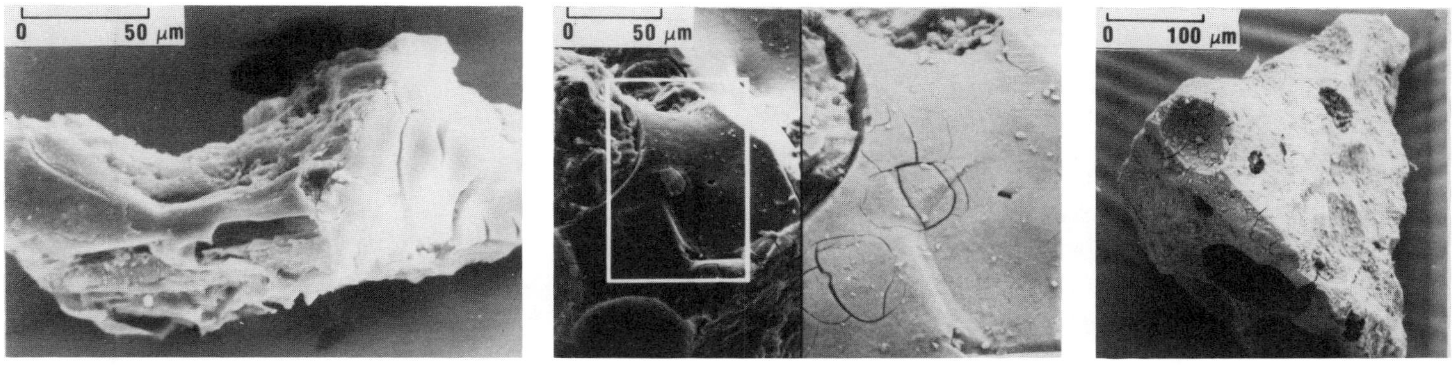

FIGURE 4. SEMs showing completed development of hydrated or palagonitized skin; it gives the grain a bright, lustrous appearance in secondary-electron images (left picture). In the middle picture, the left half (secondary-electron image) shows a bright, hydrated area surrounded by cracks, which, when viewed under 2X magnification (right half), shows a detailed crack structure with a back-scattered electron image. The right picture of this set is a back-scattered image of a hydrated grain. Note the dull, earthy appearance of the altered glass in this view.

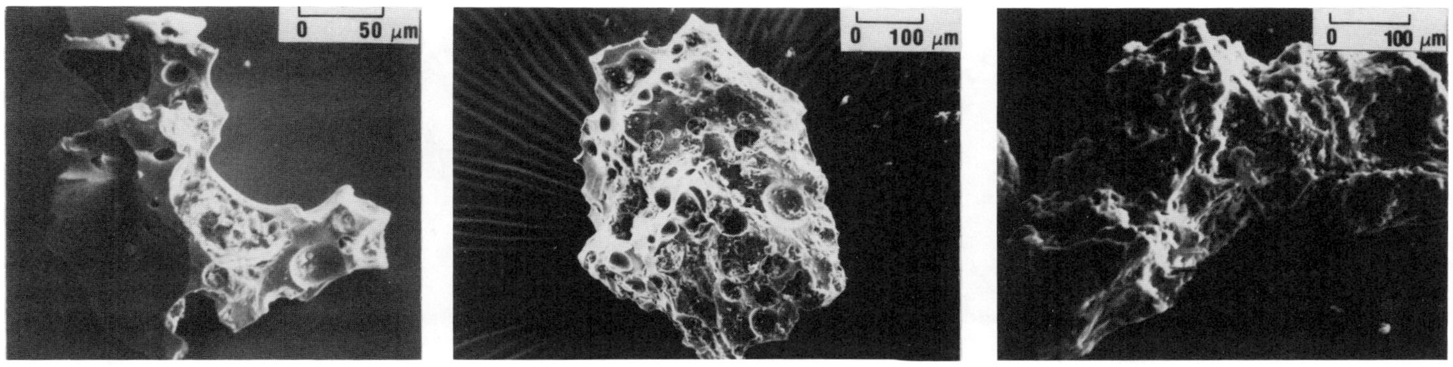

FIGURE 5. SEMs show stages of partial to extensive (left to right) alteration of pyroclasts by microcrystalline secondary minerals filling vesicles and partly covering the grain surface.

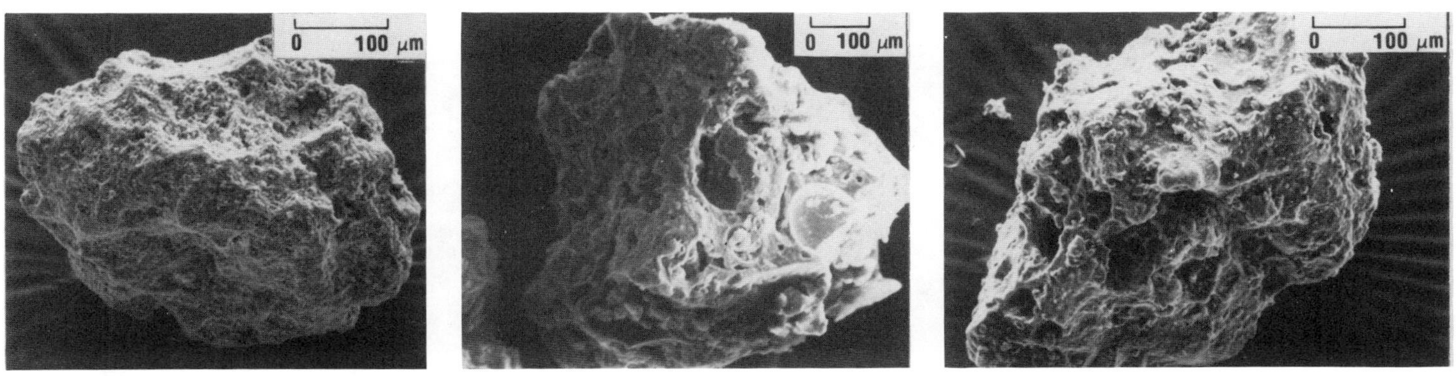

FIGURE 6. SEMs of strongly altered glassy pyroclasts; surface is rough, with smooth, glassy surfaces completely altered to microcrystalline minerals and palagonite.

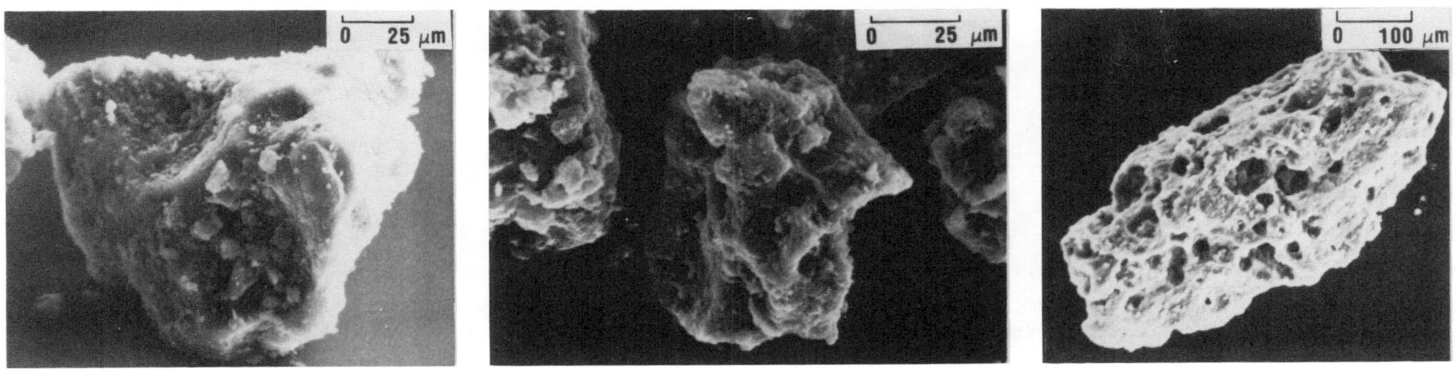

FIGURE 7. SEMs of grains that are strongly altered; they appear light gray to white in both SEM (shown) and light microscope (not shown) views. Strong hydration of the glass gives shards a sugary appearance.

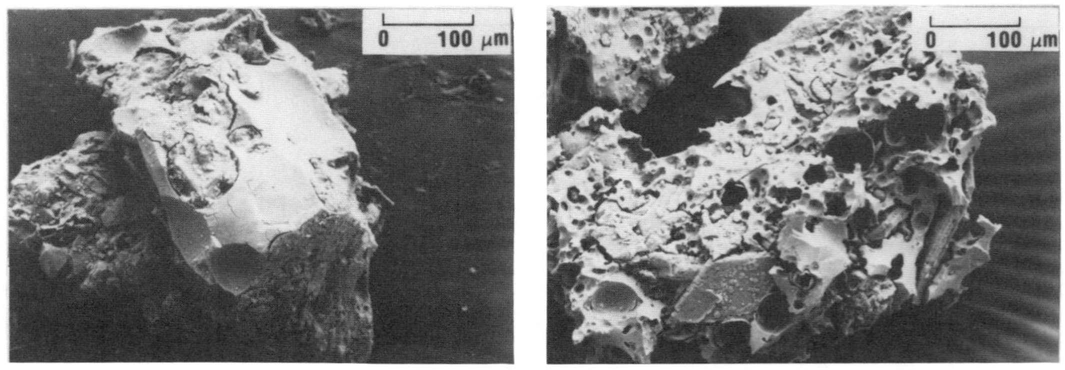

FIGURE 8. SEMs show that complete palagonitization of ash particles is often followed by solution of altered material; solid grains (left view) have a wormy appearance (compare with descriptions of the weathered Pahala ash).

FIGURE 9. SEMs of basaltic ash from strongly lithified, tuff cone, palagonite breccias of the Hopi Buttes. These consist of aggregates of clay-like and microcrystalline material. Vesicles are no longer visible, the matrix is cracked, and tiny crystals form a network on grain surfaces where hydrated glass has been dissolved.

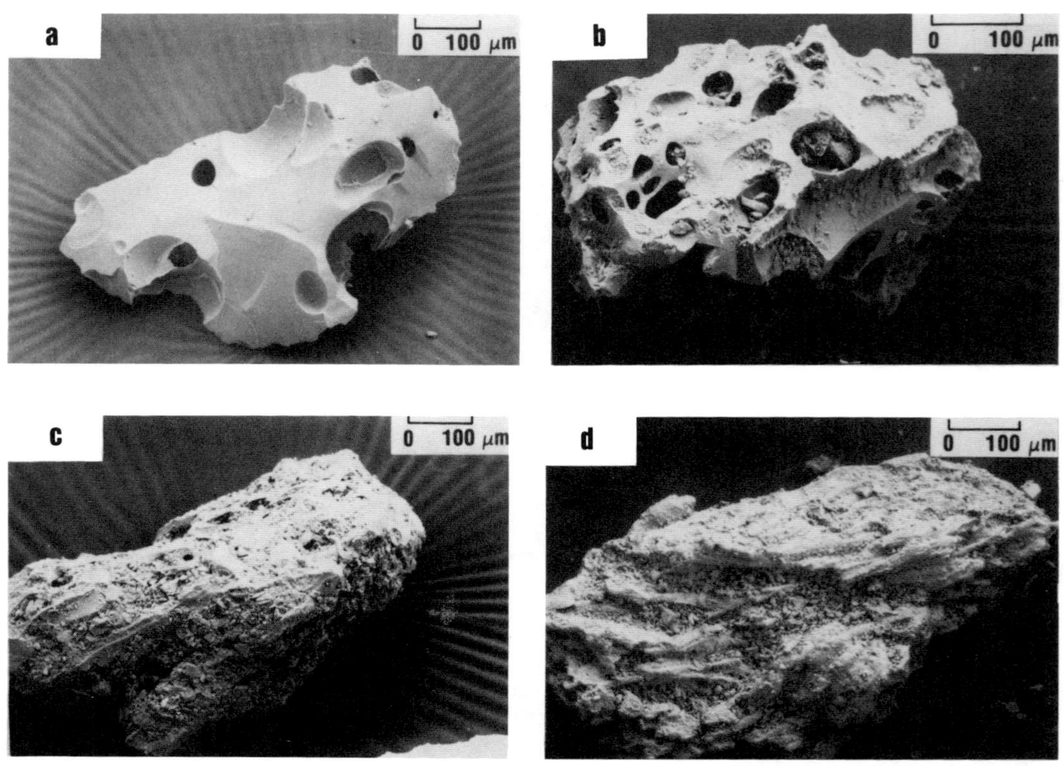

FIGURE 10. Photos a through d are SEMs of the alteration stages of silicic glass, from a fresh pyroclast on the upper left to altered ones on the lower right. The pyroclast on the lower right is from the green tuff, Pantelleria, and shows hydration typical of these pyroclasts.

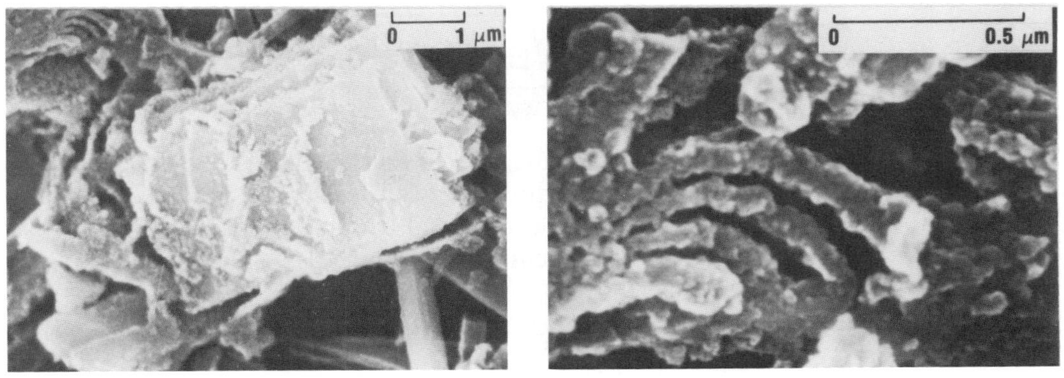

FIGURE 11. SEM high-resolution images of secondary minerals, showing a rhomb-like zeolite covered by tiny spheres that are probably smectite clays.

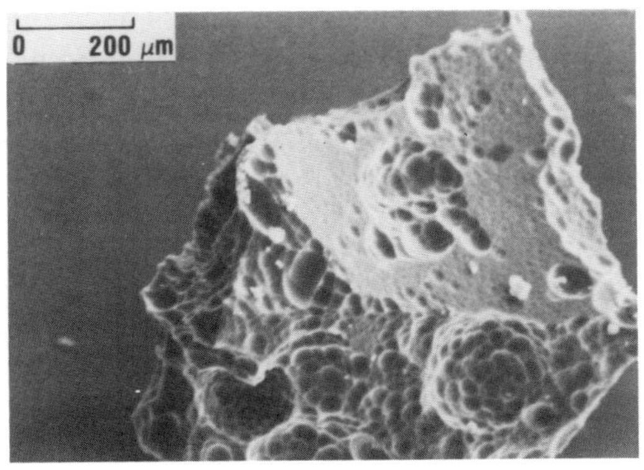

FIGURE 12. SEMs of numerous pits and rough, frosted surfaces formed during solution and precipitation. Note the original vesicle hollows that have been pitted by solution. Surfaces between vesicles have a crusty coat of precipitated material that is probably opal.

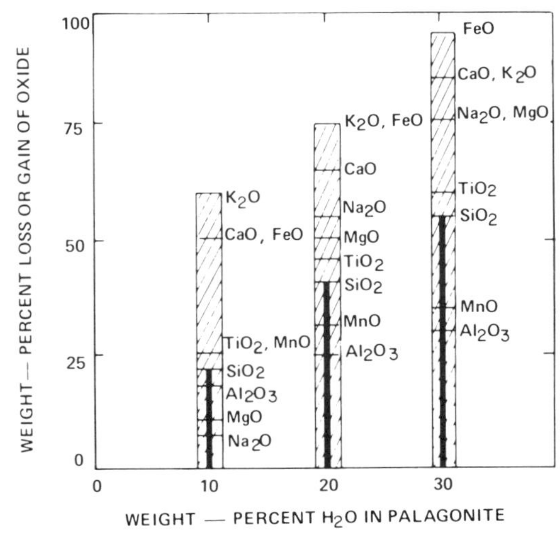

FIGURE 13. Plot of weight percentage loss or gain of major elements in palagonites with various water contents (adapted from Furnes, 1975).

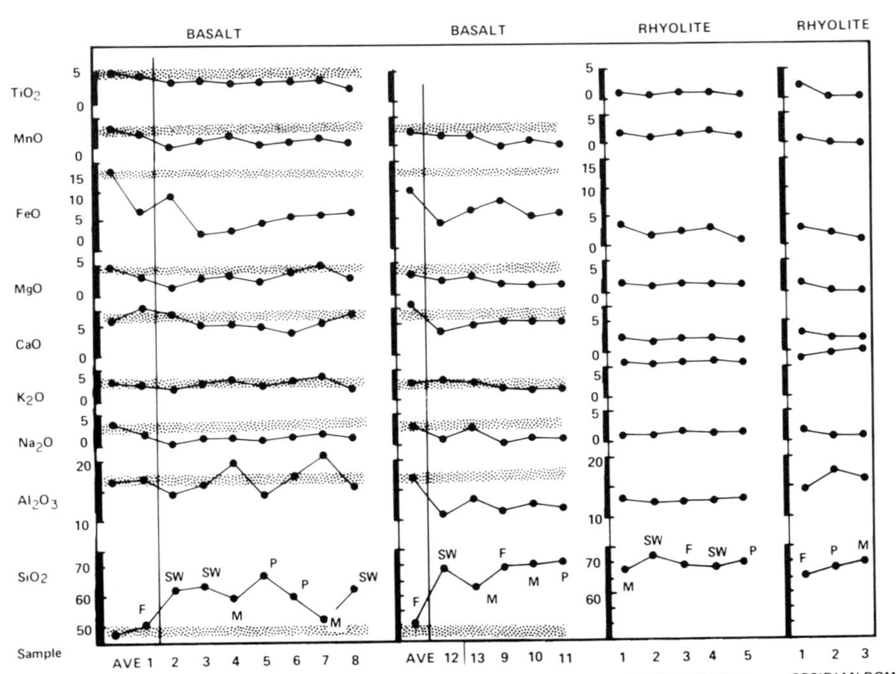

FIGURE 14. Plot of chemical analyses of pyroclast surfaces; these samples are from four stratigraphic sections where alteration varies from fresh ash to highly altered ash. The samples are numbered in ascending stratigraphic order; samples 1 and AVE (average Crater Elegante and Cerro Colorado) are fresh basalts. Note the variation in surface chemistry as a function of alteration, which varies from dry, magmatic fall deposits (F) to wet, hydromagmatic surge deposits: SW = sandwave, M = massive, and P = planar bed forms.

Weathered Tuffs

Weathered volcanic ash is often considered to be an important component in fertile soils. Many of the rich agricultural areas are located on volcanic islands and in continental arcs such as Central America and other regions of the Circum-pacific "ring of fire."

PAHALA ASH, HAWAII—AN EXAMPLE OF WEATHERED BASALTIC TEPHRA

Introduction: The Pahala ash is a widespread, weathered basaltic tephra unit on the island of Hawaii (Hay and Jones, 1972). Present rainfall over the Pahala ash ranges from 25 to 635 cm/year. The Pahala ash consists of deposits from many eruptions between 10,000 and 17,000 years ago (Hay and Jones, 1972); all are of basaltic composition. Most pyroclasts in the Pahala ash are (or were) vitric pumices, droplets, and shards. Most examples used here were probably erupted from Kilauea Volcano, although examples from South Point may have been erupted from the south rift of Mauna Loa Volcano (Macdonald and Abbott, 1970). For this study, deposits ranging in thickness from 1.5 to 3 m were sampled during a traverse across southern Hawaii.

Weathering of sideromelane produces palagonite and clay. Palagonite is a general term for a very fine grained, intergrown mixture of smectite clays, iron oxides, carbonate, and zeolites; it replaces sideromelane and forms a cement between grains. In the examples used here, the increased alteration of glass to palagonite can be correlated with increasing rainfall. All samples were provided by R. L. Hay, University of Illinois.

Grain Size: Because the deposits are weathered, only a subjective evaluation of grain size is available. Field descriptions indicate that particles range from clay size (<4 μm) to lapilli size (2 to 64 mm).

Tephra Description: Samples from sites 1 and 2 (Figs. 1 and 2) consist of fine-grained, flat to blocky pyroclasts 1 to 20 μm long that are bonded into clusters by montmorillonitic clay. These clusters, plus shards and phenocrysts coated with fine-grained particles and clay, form rounded aggregates that make up yellowish-orange, clayey deposits (Hay and Jones, 1972).

Weathered tephra from sampling sites 3 through 5 appear to have originally consisted of lapilli-size basaltic pumice (present as droplets and broken droplets) and shards. (Refer to the section on Hawaiian eruptions for examples of fresh pyroclasts.) These pyroclasts range in vesicularity from 50 to 70% by volume; most vesicles are 10- to 160-μm ovoids or spheres. There are some curved, plate-like shards that may have been derived from much larger vesicles.

Palagonitic alteration products increase from site 3 to site 5. At site 3, the outer surfaces of pumice pyroclasts have been altered to single layers of palagonite that are 1 to 2 μm thick; this surface is rough and lumpy. Where broken away, the glass at the palagonite/glass contact is smooth. At site 4, multiple layers of palagonite have formed on the outer surfaces and vesicle walls of the pyroclasts and have pseudomorphed the original grain shape. Some palagonite surfaces are lumpy, others are botryoidal, and still others are covered with a network of desiccation cracks. Some retain their cores of basaltic glass. Palagonite rinds are 20 to 50 μm thick on the outer surfaces of pyroclasts and 5 to 10 μm thick on interior vesicles. At site 5, little if any glass is present; most has been dissolved, leaving only the palagonite. If the original vesicles are regarded as molds, only the clayey palagonite casts remain, the "mold" having been dissolved. These spherical and ovoid linings remain; they resemble cracked fish eggs. Pumice pyroclasts with elongate vesicles have undergone a similar process and only stacks of tubular palagonite linings remain.

Chemical Composition: Fresh basaltic glass and palagonite rinds were separated by Hay and Jones (1972) for chemical analyses. This example of the Pahala ash is from near site 4, Kilauea Caldera (Table 1).

TABLE 1. CHEMICAL COMPOSITION OF PAHALA ASH.*

	Glass from Palagonitic Pumice Lapilli	Palagonite Rind
SiO_2	49.31	29.43
TiO_2	2.57	3.68
Al_2O_3	12.58	17.12
Fe_2O_3	1.65	16.29
FeO	10.51	1.82
MnO	0.17	0.08
MgO	10.81	1.38
CaO	10.29	1.70
Na_2O	1.81	0.22
K_2O	0.36	0.11
P_2O_5	0.20	0.23
H_2O	0.36	14.46
H_2O^-	0.21	13.59
Total	100.20	100.13

*K. Aoki, analyst (from Hay and Jones, 1972).

Pyroclast Formation: Little comment is required here, as pyroclast formation in basaltic eruptions is described in other sections. The Pahala ash, however, is a composite unit that is made up of tephra from many eruptions. Among the samples described here, it appears that those from sites 3 to 5 were the products of lava fountains. Samples from sites 1 and 2 are very poorly preserved but may have been from littoral or phreatomagmatic activity; the tephra deposits there are very fine grained, and the samples were collected from near the coast. Palagonitic rinds are preserved and pseudomorph the original pyroclasts. Where no glass remains, only the castings are visible as thin palagonitic linings.

REFERENCES: Hay, R. L. and Jones, B. F., 1972. Weathering of basaltic tephra on the Island of Hawaii. Geol. Soc. Am. Bull., **83**: 317-332.

Macdonald, G. A. and Abbott, A. T., 1970. *Volcanoes in the Sea.* Univ. of Hawaii Press, Honolulu: 441 pp.

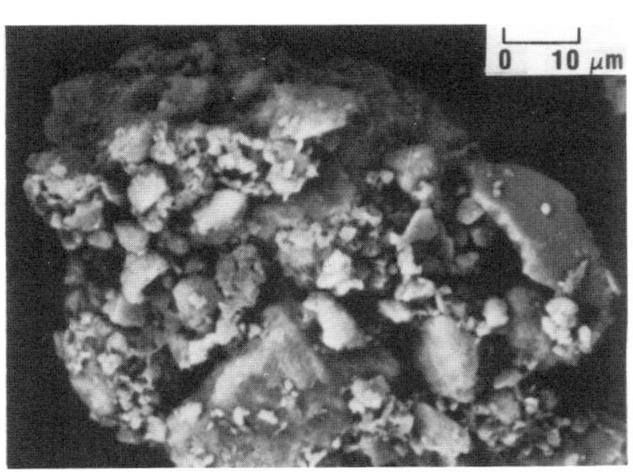

FIGURE 1. Clayey, yellowish-brown ash from South Point. These are mostly grainy pellets that consist of aggregates of fine ash and montmorillonite with and without cores consisting of coarse ash.

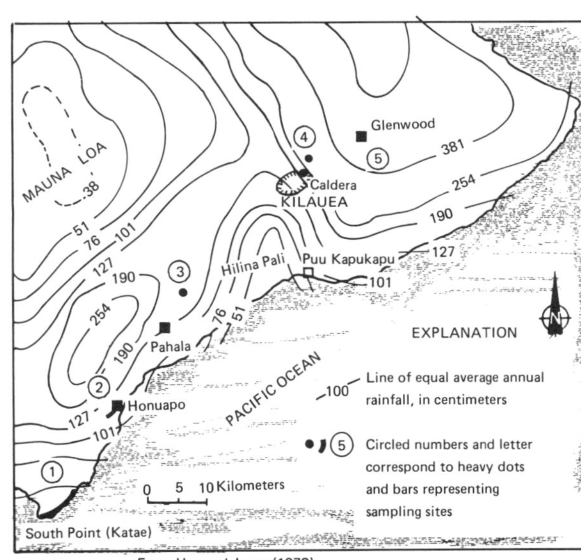

From Hay and Jones (1972)

Map of Southeastern Hawaii shows sampling sites, Mauna Loa and Kilauea volcanoes, and lines of equal average rainfall. Circled numbers 1 through 5 correlate to samples shown in Figs. 1 through 5.

FIGURE 2. Clayey, yellowish-brown ash from Honuapo. Close-up of a rounded pellet (similar to those in Fig. *1*) that consists of flat and blocky 1- to 20-µm-long vitric pyroclasts. The aggregate is probably bonded by montmorillonitic clay. Although it is difficult to see, some shards have pitted surfaces.

FIGURE 3a. Ash and lapilli from 7.5 km NE of Pahala are mostly glass, but some palagonitic rims are visible on vitric pyroclasts. Stereophotomicrograph of coarse basaltic ash. Smooth, angular surfaces are basaltic glass, and rough, lumpy surfaces are palagonitic rinds on glassy pyroclasts.

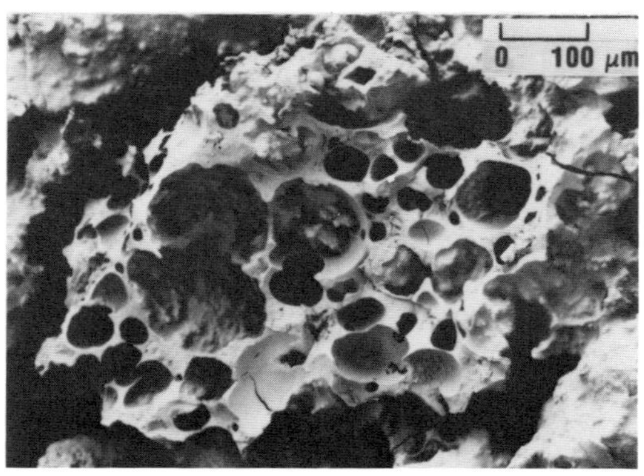

FIGURE 3b. Basaltic pumice with a thin palagonitic rind. Rough surfaces are palagonite and smooth surfaces are glass. The pyroclast is bonded to adjacent pyroclasts by palagonite.

FIGURE 3c. Basaltic pyroclasts with thin (1- to 2-µm) rinds of palagonite. The palagonite has spalled off a few grain surfaces, exposing the smooth underlying glass surface.

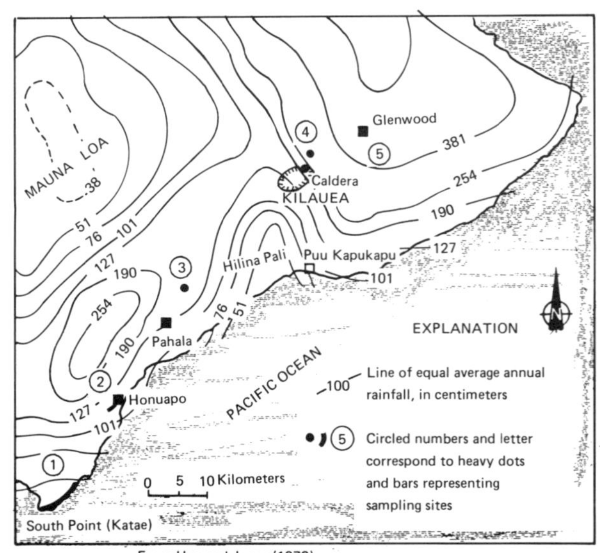

From Hay and Jones (1972)

Map of Southeastern Hawaii shows sampling sites, Mauna Loa and Kilauea volcanoes, and lines of equal average rainfall. Circled numbers 1 through 5 correlate to samples shown in Figs. 1 through 5.

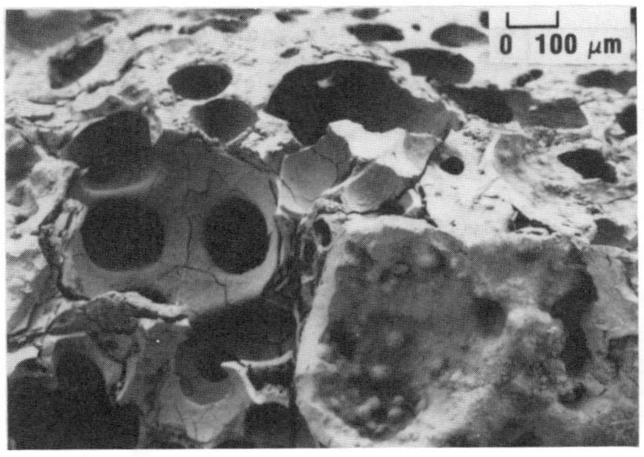

FIGURE 4a. Reddish-brown, clayey ash and lapilli that were collected 2.4 km NNE of Kilauea Caldera. Surface of a basaltic pumice lapillus. Outer grain surfaces have been altered; a 20- to 30-μm palagonite rind pseudomorphs the original pyroclast. Inner surfaces (vesicle walls) have a thinner (5- to 10-μm) palagonite rind; it is relatively smooth and crossed by a network of desiccation cracks.

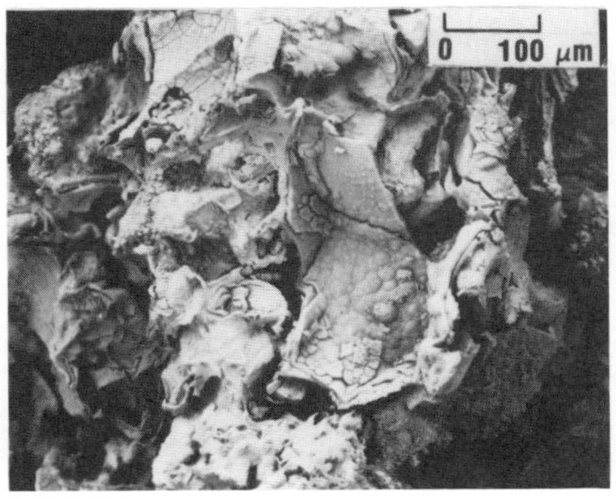

FIGURE 4b. Box-like, multi-layered palagonite rinds are all that is left of most of these pyroclasts. The agglomeration of pyroclasts is bonded by palagonite cement. Palagonite surfaces vary from those crossed with desiccation cracks to those with a botryoidal appearance.

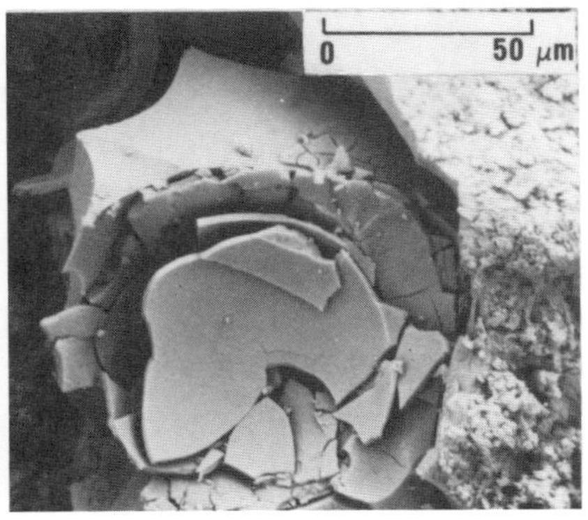

FIGURE 4c. Vitric pyroclast remnant with a 5- to 10-μm-thick void space separating it from a well-developed palagonite rind. Some alteration of the pyroclast has produced exfoliation features and desiccation cracks on the surface.

FIGURE 5. Reddish-brown clay and palagonitic lapilli tuff from a location 1 km SE of Glenwood. The round forms that look like fish eggs are palagonitic shells from the interior vesicles of a scoria fragment; the glass surrounding these vesicles has been dissolved. These vesicle "casts" are only 1 to 3 μm thick. The tubular objects (*left*) are palagonite casts of elongate vesicles from another scoria pyroclast. Holes that connected adjacent vesicles are visible as circular apertures in these delicate shells.

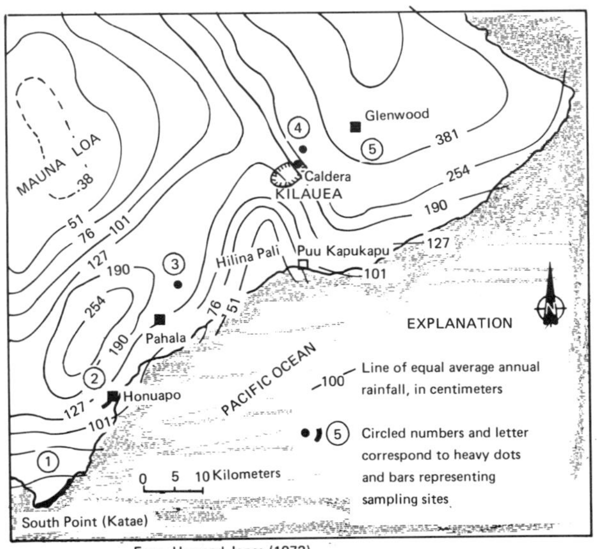

From Hay and Jones (1972)

Map of Southeastern Hawaii shows sampling sites, Mauna Loa and Kilauea volcanoes, and lines of equal average rainfall. Circled numbers 1 through 5 correlate to samples shown in Figs. 1 through 5.

ST. VINCENT, WEST INDIES; WEATHERED VOLCANIC ASH OF ANDESITIC COMPOSITION

Introduction: Tephra deposits of late Pleistocene age cover much of the island of St. Vincent, Lesser Antilles; there are many lateral facies variations that are based upon depositional mode and grain size as well as facies based upon rainfall and degree of weathering. Most of these ash deposits were erupted from the Soufriere, a large composite cone located at the northern end of the island. Ash fall deposits drape the rugged countryside, and pyroclastic flow deposits are present in some valleys (Hay, 1959). Two papers have been published on the weathering of volcanic ash in the Lesser Antilles (Hay, 1959; 1960). The first discussed a sequence of late Pleistocene ash deposits of andesitic composition, and the second concerned soil formation on a 4000-year-old pyroclastic flow sequence. R. L. Hay of the University of Illinois provided the samples used for this example of ash weathering and relict textures.

One sequence of pyroclastic deposits, dated at 4090 ± 50 y.b.p., forms a fan on the east side of the Soufriere (Hay, 1960). These deposits, which underlie deposits of the 1902 eruption, are as much as 23 m thick at the coast. Soil has developed on the 4000-year-old ash deposit to depths of 1.3 to 2.6 m. Samples from unaltered ash at a depth of 3.4 m and samples in the soil at depths of 1.3 and 0.3 m are used here to demonstrate the weathering of andesitic ash. This type of ash provides the base for widespread, fertile soils in many tropical subduction zones.

Grain Size: The late Pleistocene ash falls of St. Vincent have median grain sizes of ~80 to 200 μm (Hay, 1959); these sizes are approximate, however, as they have been altered to clay and are very difficult to sieve.

Tephra Description: *Late Pleistocene Ash.* These examples from the eastern coastal facies were described in the field by Hay (1959) as a moderately indurated, yellowish-brown, poorly sorted medium- and coarse-grained ash. The original ash appears to have been mostly porphyritic andesite pumice with phenocrysts of plagioclase, augite, hypersthene, and olivine (Table 1). Some lithic-rich ashes are also present (Fig. 3c). All glass phases and parts of the plagioclase phenocrysts have been altered to halloysite, but relict textures are very clear; the original vesicularity of pumices and the relation of vesicles to phenocrysts are visible in all samples. Ovoid vesicles of all sizes are visible (Figs. 3a, b, and d).

TABLE 1. CHEMICAL COMPOSITION OF LATE PLEISTOCENE ASH: CHIBARABU PT., ST. VINCENT.*

	(Wt. %)
SiO_2	53.03
Al_2O_3	18.78
FeO	5.49
Fe_2O_3	3.76
TiO_2	1.07
MnO	0.12
CaO	8.72
MgO	4.02
K_2O	0.44
Na_2O	3.14
P_2O_5	0.05
H_2O^+	1.45
Total	100.07

*W. H. Herdsman, analyst.

Pyroxene phenocrysts are deeply etched and are characterized by multiple needle-like terminations. Olivine is etched but retains its original shape. Plagioclase cores are, in some examples, altered to clay. Within hyalocrystalline lithic fragments, matrix minerals are altered to clay, phenocrysts are etched, and glass phases are dissolved (Fig. 3c). Pore space is filled with halloysite (Hay, 1959).

4000-Year-Old Ash. This is a crystal-rich ash (crystals form 60 to 80%) consisting of plagioclase, hypersthene, augite, olivine, magnetite, andesitic pumice, and andesitic lithic fragments (Hay, 1960). Most phenocrysts are not weathered (near the top of the soil the feldspars are only slightly etched), but vitric pyroclasts from the top of the soil horizon are totally altered to halloysite. Relict forms of vitric pyroclasts are difficult to identify in the most thoroughly weathered zone.

Comments on the Weathering of Ash: These examples from St. Vincent illustrate the physical changes that occur in pyroclasts during weathering. As shown, vitric pyroclasts are replaced by clay, but the weathering process does not always destroy pumice or shard shapes. In most of the samples studied here, relict shards and pumices are easily recognized; size, shape, and distribution of vesicles can be measured. The features of pyroclasts from ashy soils and deeply weathered tephra deposits can be described: this is especially useful when unraveling the history of volcanoes located in the tropics.

REFERENCES:

Hay, R. L., 1959. Origin and weathering of late Pleistocene ash deposits on St. Vincent, B.W.I. J. Geol., **67**: 65-87.

Hay, R. L., 1960. Rate of formation and mineral alteration in a 4000-year-old volcanic ash soil on St. Vincent, B.W.I. Am. J. Sci., **258**: 354-368.

FIGURES 1 AND 2. Andesitic crystal-lithic ash from a 4000-year-old pyroclastic flow deposit along the Rabaka River, on the slopes of Soufriere of St. Vincent. Variations in the degree of weathering of a massive pyroclastic flow sequence are demonstrated here. Samples were collected 3.4, 1.3, and 0.3 m below the surface of a 4000-year-old soil (Figs. 1 and 2 *a*, *b*, and *c*, respectively).

FIGURE 1a. Mostly lithic fragments and phenocrysts. Slightly weathered pyroclasts are very well preserved and have only a thin coating of fine-grained particles. *Center*—vesicular porphyritic lithic fragment with no alteration of pyroxene and plagioclase phenocrysts. *Top*—Pumice fragment. Other pyroclasts are phenocrysts and lithic fragments.

FIGURE 1b. Center—well-preserved, unaltered plagioclase phenocryst. Subrounded, porphyritic andesite fragments are also visible.

FIGURE 1c. Center—unaltered pyroxene fragment; irregular surfaces are remnants of glass that are attached to the phenocrysts but are now altered to halloysite. *Top*—partly etched plagioclase with irregular, wormy holes. *Lower left*—porphyritic scoria pyroclast that is altered to halloysite. There are also andesitic lithic fragments that have been altered, in part, to halloysite.

FIGURE 2a. Pumice pyroclasts (juvenile). A fresh, unaltered vitric pyroclast with low vesicularity but well-developed vesicles (deformed ovoids). Finer grained clasts clinging to the surfaces are typical of pyroclastic flow deposits.

FIGURE 2b. Vitric pumice pyroclast. The glass has been altered to halloysite but relict textures remain, including vesicles that surround a phenocryst (*center*).

FIGURE 2c. Aggregate of plagioclase and pyroxene phenocrysts and halloysite. Individual aggregates might be relict pumice pyroclasts, but there are few relict textures remaining to confirm this hypothesis. It is possible that ovoid holes in the aggregate were vesicles. Pyroxene phenocrysts exhibit little or no alteration.

FIGURE 3a. Late Pleistocene andesitic ash from the east coast of St. Vincent. Stereophotomicrograph of a relict pumice pyroclast. The vitric portions of the pyroclast have been totally replaced by clay, but the original form remains. Most vesicles are 5- to 120-μm-long, irregular ovoids except where adjacent to phenocrysts (*left*), where they are as much as 200 μm long. Center of the plagioclase phenocryst has been partly replaced by halloysite.

FIGURE 3b. Surface of a pumice pyroclast where much of the original vesicularity remains. Glass has been replaced by halloysite. Notice the desiccation cracks in the clay.

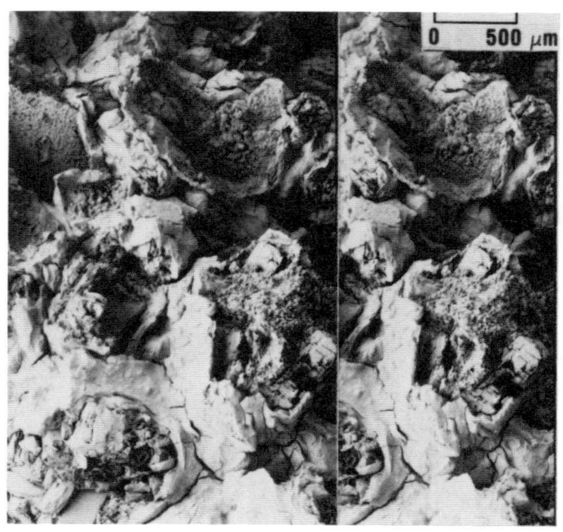

FIGURE 3c. Stereophotomicrograph of a lithic-fragment-rich andesitic tuff. Relict lithic fragments are visible as subrounded "aggregates;" these consist of etched phenocrysts and void space (perhaps originally glass between groundmass minerals, now dissolved away), surrounded by a network of groundmass minerals that is replaced by clay. Each relict grain is coated with irregular layers of clay that filled the void spaces between pyroclasts.

FIGURE 3d. Pyroxene phenocryst in a pumice pyroclast. Although the glass has been replaced by clay, vesicles are very well preserved in this example. The pyroxene has been etched both along a twin plane and at random (note the honeycomb weathering in lower left corner of the phenocryst).

FIGURE 3e. Etched pyroxene with needle-like terminations that are typical of weathered pyroxenes.

FIGURES 1 AND 2. Andesitic crystal-lithic ash from a 4000-year-old pyroclastic flow deposit along the Rabaka River, on the slopes of Soufriere of St. Vincent. Variations in the degree of weathering of a massive pyroclastic flow sequence are demonstrated here. Samples were collected 3.4, 1.3, and 0.3 m below the surface of a 4000-year-old soil (Figs. 1 and 2 *a*, *b*, and *c*, respectively).

FIGURE 1a. Mostly lithic fragments and phenocrysts. Slightly weathered pyroclasts are very well preserved and have only a thin coating of fine-grained particles. *Center*—vesicular porphyritic lithic fragment with no alteration of pyroxene and plagioclase phenocrysts. *Top*—Pumice fragment. Other pyroclasts are phenocrysts and lithic fragments.

FIGURE 1b. *Center*—well-preserved, unaltered plagioclase phenocryst. Subrounded, porphyritic andesite fragments are also visible.

FIGURE 1c. *Center*—unaltered pyroxene fragment; irregular surfaces are remnants of glass that are attached to the phenocrysts but are now altered to halloysite. *Top*—partly etched plagioclase with irregular, wormy holes. *Lower left*—porphyritic scoria pyroclast that is altered to halloysite. There are also andesitic lithic fragments that have been altered, in part, to halloysite.

FIGURE 2a. Pumice pyroclasts (juvenile). A fresh, unaltered vitric pyroclast with low vesicularity but well-developed vesicles (deformed ovoids). Finer grained clasts clinging to the surfaces are typical of pyroclastic flow deposits.

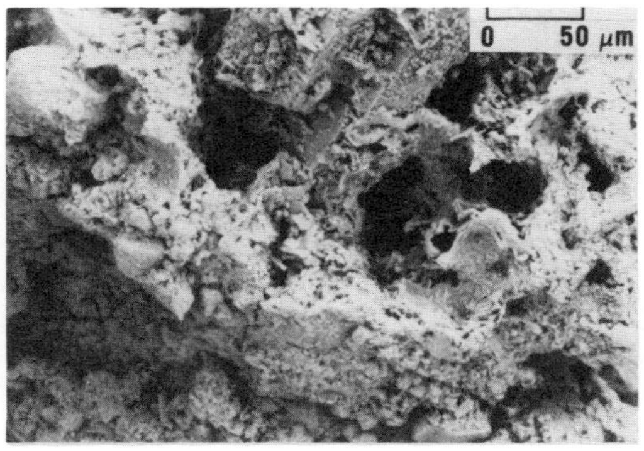

FIGURE 2b. Vitric pumice pyroclast. The glass has been altered to halloysite but relict textures remain, including vesicles that surround a phenocryst (*center*).

FIGURE 2c. Aggregate of plagioclase and pyroxene phenocrysts and halloysite. Individual aggregates might be relict pumice pyroclasts, but there are few relict textures remaining to confirm this hypothesis. It is possible that ovoid holes in the aggregate were vesicles. Pyroxene phenocrysts exhibit little or no alteration.

FIGURE 3a. Late Pleistocene andesitic ash from the east coast of St. Vincent. Stereophotomicrograph of a relict pumice pyroclast. The vitric portions of the pyroclast have been totally replaced by clay, but the original form remains. Most vesicles are 5- to 120-μm-long, irregular ovoids except where adjacent to phenocrysts (*left*), where they are as much as 200 μm long. Center of the plagioclase phenocryst has been partly replaced by halloysite.

FIGURE 3b. Surface of a pumice pyroclast where much of the original vesicularity remains. Glass has been replaced by halloysite. Notice the desiccation cracks in the clay.

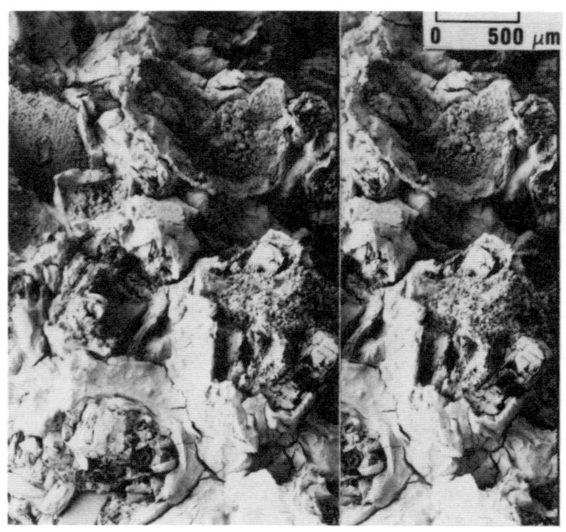

FIGURE 3c. Stereophotomicrograph of a lithic-fragment-rich andesitic tuff. Relict lithic fragments are visible as subrounded "aggregates;" these consist of etched phenocrysts and void space (perhaps originally glass between groundmass minerals, now dissolved away), surrounded by a network of groundmass minerals that is replaced by clay. Each relict grain is coated with irregular layers of clay that filled the void spaces between pyroclasts.

FIGURE 3d. Pyroxene phenocryst in a pumice pyroclast. Although the glass has been replaced by clay, vesicles are very well preserved in this example. The pyroxene has been etched both along a twin plane and at random (note the honeycomb weathering in lower left corner of the phenocryst).

FIGURE 3e. Etched pyroxene with needle-like terminations that are typical of weathered pyroxenes.

RELICT TEXTURES IN BENTONITE

Introduction: *Bentonite* is a term first used for clay-like material from the Benton Formation (Cretaceous) of Wyoming. Hewitt (1917) and Wherry (1917) established that these clays are the alteration products of volcanic ash. There have been dozens of bentonite studies over the past 60 years, including a comprehensive account of the Mowry bentonites by Slaughter and Early (1965). Bentonite beds have been useful as a source of clay for drilling muds and as marker beds for stratigraphic correlation over large areas. Slaughter and Early (1965) wrote the following field description: "a typical bentonite is a relatively thin and distinctly light-colored unindurated rock composed dominantly of clay." The major component is montmorillonite with minor relict phenocryst phases, zeolites, and authigenic silica phases.

Several SEM studies of bentonite have revealed well-preserved relict textures of pumice pyroclasts and shards (Khoury and Eberl, 1979; Wise and Weaver, 1979). The example used here to illustrate relict pyroclastic textures in bentonites is from the upper Tesuque Formation (Pliocene) of the Santa Fe Group and was collected by J. Stix near Ojo Caliente, New Mexico.

Tephra Description: The mold-cast relationship (described in the section on Pahala Ash) is well demonstrated in this example. The most apparent pyroclasts are small pumices in which long, thin parallel rods are casts of elongate vesicles. The curved, plate-like surfaces are casts of large shards. Alteration of vitric pyroclasts leaves a "rind" that is preserved, whereas the glassy core is dissolved, leaving voids.

Chemical Composition: No chemical analyses are available for the sample shown in Figs. 1 through 3. However, analyses of bentonite deposits near Santa Rita, New Mexico, and Otay, California, were published by Slaughter and Early (1965) (Table 1). Variations from bentonite to bentonite reflect the original magma composition and alteration environment.

TABLE 1. CHEMICAL ANALYSES OF TWO BENTONITES.

	Santa Rita, New Mexico (Wt. %)	Otay, California* (Wt. %)
SiO_2	58.10	57.80
Al_2O_3	17.15	16.25
Fe_2O_3	1.02	1.10
TiO_2	0.14	0.16
CaO	0.15	0.06
MgO	4.33	6.56
Na_2O	2.80	3.87
K_2O	0.14	0.14
Ignition loss	16.11	14.10
Total	100.04	100.04

*From Osthaus (1956).

Comments on Weathering Alteration of Ash: Most relict pyroclasts in this and other described examples appear to have been rhyolitic or rhyodacitic pumice pyroclasts and shards. The widespread distribution of bentonitic deposits is consistent with derivation from silicic tephra layers deposited during large-volume eruptions.

The alteration of silicic glass to bentonite (mixture of clays, silica phases, and zeolites) is described by Slaughter and Early (1965) and Grim and Güven (1978).

REFERENCES:

Grim, R. and Güven, N., 1978. *Bentonites; Geology, Mineralogy, Properties and Uses*. Elsevier, Amsterdam: 256 pp.

Hewitt, D., 1917. The origin of bentonite. J. Wash. Acad. Sci., **7**: 196-198.

Khoary, H. and Eberl, D., 1979. Bubble-wall shards altered to montmorillonite. Clays Clay Miner., **27**: 291-292.

Osthaus, B., 1956. Kinetic studies on montmorillonite and nontronite by the acid dissolution technique. 4th Nat'l. Conf. on Clays and Clay Min., Proc.: 301-321.

Slaughter, M. and Early, J., 1965. Mineralogy and geological significance of the Mowry Bentonites, Wyoming. Geol. Soc. Am. Spec. Paper 83: 116 pp.

Wherry, E., 1917. Clay derived from volcanic dust in the Pierre of South Dakota. J. Wash. Acad. Sci., **7**: 576-583.

Wise, Jr., S. and Weaver, F., 1979. Volcanic ash: examples of devitrification and early diagenesis. Scanning Electron Microsc.: 511-518.

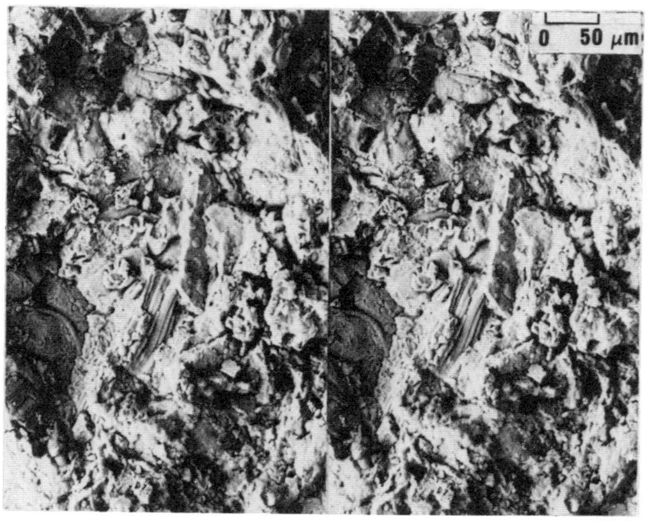

FIGURE 1. Stereophotomicrograph of bentonite from the Santa Fe Formation, Ojo Caliente, New Mexico. Casts of relict pyroclasts in clay (the pyroclasts themselves were the molds). The most apparent pyroclast is a 160-μm-long pumice located in the photo center; the long, thin, parallel rods are casts of elongate vesicles. Curved, plate-like surfaces are casts of large shards (these are less obvious).

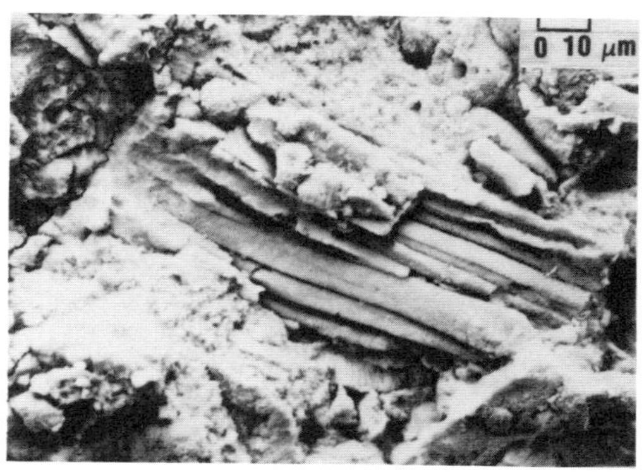

FIGURE 2a. Relict pumice pyroclasts. This is a closer view of the relict pumice pyroclast of Fig. 1. Clays have filled and pseudomorphed elongated vesicles and the original glass of the pyroclast was dissolved. The pseudomorph is that of a 90- by 35-μm, angular pumice fragment.

FIGURE 2b. Relict pumice pyroclast similar to the one in Fig. *2a.*, but broken, perhaps in sample preparation.

FIGURE 3a. Relict shards. The curved form visible here (*center*) is the casting of a shard (the glass has dissolved). Fuzzy, irregular clumps on the right wall are most likely montmorillonitic clays.

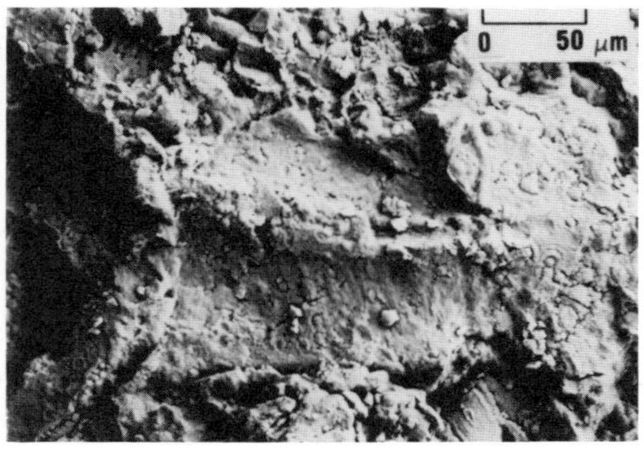

FIGURE 3b. Relict texture consisting of a "casting" of one or several shards (one multi-walled shard or several closely spaced shards).

FIGURE 3c. A more three-dimensional view of a shard casting. The original shard may have been tri-cornered and about 60 µm long. The 4-µm-thick "casting" may represent alteration products on the shard surface; the glass core probably dissolved after a clay "rind" formed.

Metamorphosed Tuffs

Metamorphosed volcanic ash deposits may be difficult to recognize. However, examples of "metatuffs" have been found in rocks from zeolite facies to amphibolite facies in Precambrian terranes. For example, rocks bearing a strong textural resemblance to pyroclastic flows have been found in Precambrian greenschists of central Arizona (Thorpe and Burt, 1978) and in the Precambrian Grenville metamorphic sequence of northern New York (Carl and van Diver, 1975). Knowledge of original pyroclast compositions and morphologies permits detailed analysis of mineral paragenesis and mechanical textural variations that accompany regional metamorphism. The two examples below illustrate common features associated with low-grade metatuffs of ultramafic and silicic compositions.

ZEOLITE FACIES IN SILICIC TUFFS; TIMBER MOUNTAIN—OASIS VALLEY VOLCANIC FIELD, NEVADA

Introduction: Pyroclasts in pyroclastic flow deposits (ash flow tuffs) may undergo many physical and chemical changes after deposition. In the well-insulated interior of a pyroclastic flow, individual pyroclasts may be flattened by the weight of the overlying deposit and, subsequently, weld together. Blocky pumice pyroclasts are flattened into pancake-shape disks. The slow release of volatiles allows hot gases to interact with glass pyroclasts in the upper parts of the flow; the glass is devitrified and often leaves only friable pyroclast relicts. After the flow has cooled, interaction of ground water with remaining vitric pyroclasts will form authigenic feldspars, silica phases, smectite clays, and zeolites. Within these rocks, the search for relict pyroclastic textures is often difficult, but identifying those textures may help in characterizing the tuffs.

For a comprehensive review of pyroclastic flow deposits, see Ross and Smith (1961) and Lipman et al. (1966). A mineralogical approach to the study of alteration products in tuffs is provided by a volume on zeolite minerals edited by Mumpton (1978).

The examples used here are from the Timber Mountain-Oasis Valley volcanic field of Nevada; these 9.5- to 16-million-year-old tuff deposits cover 11,000 km^2 and have a total volume of over 5000 km^3 (Byers et al., 1976). Detailed analyses of authigenic minerals in these deposits have been made in support of the U.S. Nuclear Waste Isolation Program (e.g., Sykes et al., 1979).

Chemical Composition: Most of the examples here are tuffs that have quartz latitic to rhyolitic compositions.

Tephra Description: In pyroclastic flow deposits affected by zeolite-facies metamorphism (diagenesis), relict textures are best seen in thin sections by transmitted light (plane) at low magnification. Outlines of pyroclasts and shards are often visible because of <1-μm hematite grains located on former pyroclast surfaces and on vesicle walls. In many cases, cores of large pyroclasts have been dissolved, leaving open vugs, and authigenic phases (silica, K-feldspar, and zeolites) may line or fill these voids. Finer grained phases are frequently replaced by finely crystalline, intergrown silica phases, authigenic feldspars, smectite clays, and zeolites. Relict textures are often completely destroyed by spherulite growth.

Phenocrysts and lithic fragments in these tuffs are rarely affected by postdepositional vapor-phase alteration and interaction with ground water.

REFERENCES:

Byers, Jr., F., Carr, W., Orkild, P., Quinlivan, W., and Sargent, K., 1976. Volcanic suites and related cauldrons of Timber Mountain-Oasis Valley Caldera complex, Southern Nevada. U.S. Geol. Surv. Prof. Paper 919: 70 pp.

Carl, J. D. and van Diver, B. B., 1975. Precambrian Grenville alaskite bodies as ash-flow tuffs, northwest Adirondacks, New York. Geol. Soc. Am. Bull., **86**: 1691-1707.

Lipman, P., Christiansen, R., and O'Conner, J., 1966. A compositionally zoned ash-flow sheet in Southern Nevada. U.S. Geol. Surv. Prof. Paper 524-F: 47 pp.

Mumpton, F. (Ed.), 1978. *Natural Zeolites.* Pergamon, Oxford: 546 pp.

Ross, C. and Smith, R., 1961. Ash-flow tuffs: their origin, geologic relations and identification. U.S. Geol. Surv. Prof. Paper 366: 81 pp.

Sykes, M., Heiken, G., and Smyth, J., 1979. Mineralogy and petrology of tuffs from the UE25a1 drill site, Yucca Mountain, Nevada. Los Alamos Scientific Laboratory report LA-8139-MS: 76 pp.

Thorpe, D. G. and Burt, D. M., 1978. Precambrian metavolcanic rocks of the Squaw Peak area, Maricopa County, Arizona. *In*: Burt, D. M. and Péwé, T. L. (Eds.), *Guidebook to the geology of Central Arizona*, Ariz. Bur. Geol. Mineral. Tech. Spec. Pap. No. 2: 101-106.

APPENDIX: CLASSIFICATION OF COMMON IGNEOUS ROCKS

The creation of igneous rock categories, complete with names, is artificial; in reality, igneous rocks grade into one another chemically, texturally, and mineralogically.

There is a great variety of igneous rocks, but only the most common are included in this simplified classification. Rock names used for both the rapidly cooled, finely crystalline or glassy volcanic rocks and the slowly cooled, coarsely crystalline plutonic rocks are included in Table 1.

For a complete review of igneous petrology, the reader is referred to the works of Carmichael, Turner, and Verhoogen (1974) and MacDonald (1972).

TABLE 1. SIMPLE CLASSIFICATION OF COMMON VOLCANIC ROCKS AND SOME PHYSICAL AND CHEMICAL PROPERTIES.

	Silicic	Intermediate		Mafic	
Main mineral phases	K-feldspar, quartz oligoclase	Plagioclase biotite hornblende and pyroxenes, quartz	Plagioclase augite, orthopyroxene, hornblende	Plagioclase augite olivine	Feldspathoid, K-feldspar plagioclase, hornblende, biotite
Rock name	**Rhyolite**	**Dacite**	**Andesite**	**Basalt**	**Phonolite trachyte**
		Average Chemical Composition* (Wt.%)			
SiO_2	73.66	63.58	57.7	50.83	56.90
TiO_2	0.22	0.64	0.6	2.03	0.59
Al_2O_3	13.45	16.67	17.0	14.07	20.17
Fe_2O_3	1.25	2.24	2.4	2.88	2.26
FeO	0.75	3.00	3.6	9.05	1.85
MnO	0.03	0.11	0.15	0.18	0.19
MgO	0.32	2.12	5.1	6.34	0.58
CaO	1.13	5.53	7.5	10.42	1.88
Na_2O	2.99	3.98	3.4	2.23	8.72
K_2O	5.35	1.40	1.6	0.82	5.42
Estimated extrusion temperatures (°C)**	800-900	900-1000	900-1000	1000-1200	1000
Viscosities at extrusion temperatures (Poise)†	10^{10}-10^{12}	10^{10}	10^{5}-10^{7}	10^{2}-10^{4}	10^{5}

*From Nuckolds (1954) and Williams (1932).
**From Carmichael et al. (1974).
†From MacDonald (1972).

REFERENCES FOR APPENDIX:

Carmichael, I. S. E., Turner, F. J., and Verhoogen, J., 1974. *Igneous Petrology.* McGraw-Hill, New York: 739 pp.

MacDonald, G. A., 1972. *Volcanoes.* Prentice-Hall, Inc., Englewood Cliffs: 510 pp.

Nuckolds, S. R., 1954. Average chemical compositions of some igneous rocks. Bull. Geol. Soc. Am., **65**: 1007-1032.

Williams, H., 1932. Geology of the Lassen Volcanic National Park. Calif. Univ. Pubs. in Geol. Sci., **21**: 195-385.

INDEX

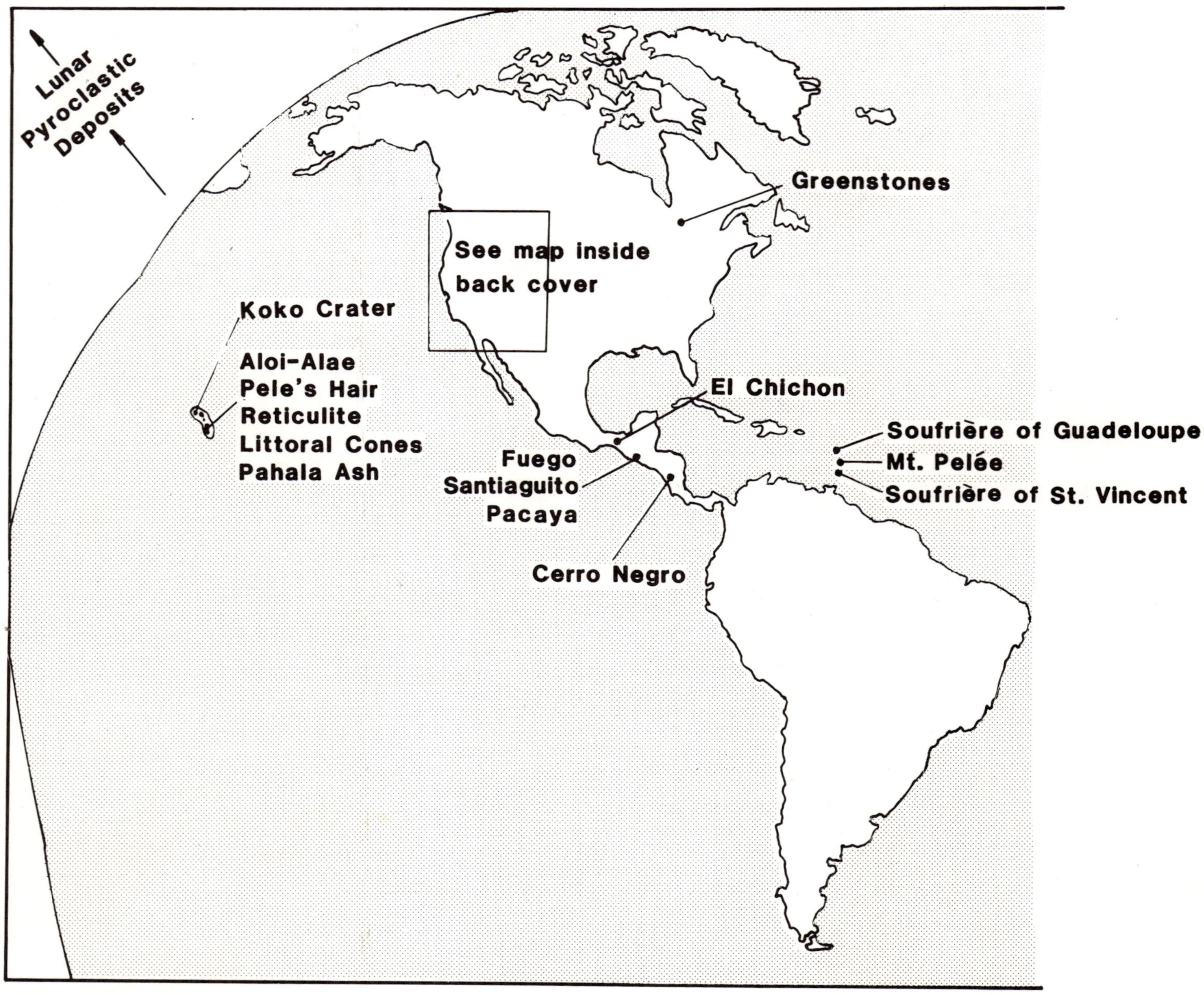

Lunar Pyroclastic Deposits
Greenstones
See map inside back cover
Koko Crater
Aloi-Alae
Pele's Hair
Reticulite
Littoral Cones
Pahala Ash
El Chichon
Soufrière of Guadeloupe
Mt. Pelée
Soufrière of St. Vincent
Fuego
Santiaguito
Pacaya
Cerro Negro